Nanocarriers in Plant Science and Agriculture

Jen-Tsung Chen
National University of Kaohsiung, Taiwan

IGI Global
Scientific Publishing
Publishing Tomorrow's Research Today

Vice President of Editorial	Melissa Wagner
Director of Acquisitions	Mikaela Felty
Director of Book Development	Jocelynn Hessler
Production Manager	Mike Brehm
Cover Design	Jose Rosado

Published in the United States of America by
IGI Global Scientific Publishing
701 East Chocolate Avenue
Hershey, PA, 17033, USA
Tel: 717-533-8845
Fax: 717-533-7115
Website: https://www.igi-global.com E-mail: cust@igi-global.com

Library of Congress Cataloging-in-Publication Data

Names: Chen, Jen-Tsung editor
Title: Nanocarriers in plant science and agriculture / edited by Jen-Tsung Chen.
Description: Hershey, PA : IGI Global Scientific Publishing, [2025] | Includes bibliographical references and index. | Summary: "This book attempts to fill the knowledge gap by providing a series of comprehensive chapters including fundamentals and molecular mechanisms of nanocarriers and highlighting the subtopics of their applications on genetic engineering, genome editing such as CRISPR-edited crops, delivering chemicals, biomolecules, biostimulants, pesticides, fertilizers"-- Provided by publisher.
Identifiers: LCCN 2025005859 (print) | LCCN 2025005860 (ebook) | ISBN 9798369398692 hardcover | ISBN 9798369398708 paperback | ISBN 9798369398715 ebook
Subjects: LCSH: Agricultural biotechnology | Nanobiotechnology | Plant genetic engineering
Classification: LCC S494.5.B563 N335 2026 (print) | LCC S494.5.B563 (ebook) | DDC 338.1/6--dc23/eng/20250519
LC record available at https://lccn.loc.gov/2025005859
LC ebook record available at https://lccn.loc.gov/2025005860

British Cataloguing in Publication Data
A Cataloguing in Publication record for this book is available from the British Library.

Table of Contents

Detailed Table of Contents

diverse cargoes, including pesticides, herbicides, nutrients, PGRs, genetic material, and biostimulants. Through passive (root absorption, stomatal penetration, cuticular diffusion) and active (endocytosis, plasmodesmata transport) uptake mechanisms, nanocarriers deliver cargo directly to plant tissues. Surface functionalization optimizes interaction with plant cells, enhancing adhesion, permeability, and controlled release. This technology improves nutrient uptake, pest control, and stress tolerance, while minimizing environmental impact. However, safety and environmental concerns necessitate rigorous research and regulatory frameworks. Future directions include bioengineered nanocarriers, AI integration, and scalable production for sustainable agriculture.

Delivering functional biomolecules, such as RNA, DNA, and proteins, into plants offers a valuable opportunity to manipulate plant traits and address environmental sustainability challenges. Existing methods for biomolecule delivery face significant hurdles. Techniques like gene guns and Agrobacterium-mediated transformation are often inefficient. Nanocarriers have emerged as a promising alternative, offering precise, efficient, and versatile delivery systems. They can bypass the cell wall of plants without causing damage and enable broader applications in plant genetic engineering and biotechnology. This chapter examines the types of nanocarriers used for DNA, RNA, and protein delivery, exploring their uptake and transport mechanisms in plants. Nanocarrier-based delivery, system has improved efficiency, targeted delivery, and controlled release, alongside potential limitations like toxicity. The chapter summarizes future directions for optimizing nanocarriers, focusing on advancing plant genetic engineering and promoting sustainable agricultural practices.

The integration of plant biostimulants and nanocarriers in the agri-food industry is gaining attention, yet most research has focused on their independent applications. This chapter explores how combining biostimulant technology with nanotechnology can enhance agricultural sustainability and food quality. By leveraging nanotechnology's capabilities, plant biostimulants—compounds, microbes, or complex mixtures that improve plant growth and nutrition—can be more effectively delivered and optimized for food security, safety, and quality. This chapter provides a tutorial-style overview of nanotechnology's fundamental principles, highlighting nanoencapsulation and other advanced techniques to create nano-biostimulants. It discusses sustainability, including biostimulant production from waste materials, supporting a circular economy. Ultimately, this integration offers a long-term, eco-friendly solution to food insecurity and enhances the value of biostimulants for sustainable agriculture.Keywords: Biostimulants; Plant growth; Nanotechnology; Nanocarriers

Nucleic acid nanocarriers represent a transformative approach in plant biotechnology, facilitating the efficient delivery of genetic material into plant cells. This review explores the various types of nanocarriers, including gold nanoparticles, silica nanoparticles, and carbon nanotubes, highlighting their unique advantages, such as enhanced cellular uptake, targeted delivery, and controlled release. By overcoming the physical barriers inherent in plant cell walls, these nanocarriers enable precise genetic modifications and the introduction of traits that enhance growth, stress tolerance, and metabolic profiles. Furthermore, the application of nucleic acid therapeutics, including RNA interference and genome editing technologies, is significantly advanced through the use of nanocarriers. The future of plant biotechnology is poised for innovative breakthroughs as researchers continue to refine these delivery systems, unlocking new possibilities for crop improvement and sustainable agricultural practices.

The increasing world population and development have led to a heightened demand for food production, often met through the extensive application of chemical fertilizers and pesticides. To mitigate these adverse effects, sustainable agricultural practices are essential, emphasizing the need for alternatives to chemical inputs. Plant growth-promoting microorganisms (PGPM), particularly plant growth-promoting bacteria (PGPBs), have emerged as sustainable solutions to enhance plant growth. Despite their potential, the practical application of PGPMs is often compromised by issues such as instability, reduced efficacy, and environmental sensitivity. Nanocarriers such as biochar, clay, alginate, starch, chitosan, and gelatin have shown a potential to increase the viability and beneficial properties of PGPMs. This chapter reviews the potential of nanocarriers in delivering PGPMs, highlighting the advancements and benefits of integrating nanotechnology in sustainable agriculture.

Nanocarriers have emerged as a promising solution for pesticide delivery, aiming to enhance efficiency, reduce environmental contamination, and improve targeted pest control. Traditional pesticide applications often result in excessive use, rapid degradation, and non-specific distribution, leading to environmental and health concerns. Nanocarriers, such as polymeric nanoparticles, lipid-based carriers, inorganic nanoparticles, and carbon-based materials, offer controlled release, enhanced bioavailability, and improved pesticide stability. These nanoformulations protect active ingredients from premature degradation, facilitate targeted delivery, and minimize off-target effects. Recent advancements in stimuli-responsive and biodegradable nanocarriers offer promising solutions for eco-friendly pest management. This review explores the types, synthesis, mechanisms, applications, challenges, and future perspectives of nanocarriers in pesticide delivery, emphasizing their potential in advancing precision agriculture.

Nanocarrier-enabled agricultural biotechnology indicates an innovative approach to solving fundamental difficulties in modern agriculture, ranging from crop yield improvement to insect resistance and environmental stress reduction. Nanocarriers are nanosized delivery devices capable of precisely delivering agrochemicals, nutrients, or genetic material to plant cells. This method brings up new possibilities for crop breeding, genetic engineering, and pest management by providing unparalleled molecular control over biological processes. Nanocarriers are getting highly used in agriculture for gene delivery and pesticide applications. They are capable of transporting fertilizers, insecticides, and herbicides efficiently, decreasing environmental contamination and the need for repeated treatments. They have the potential to improve agricultural traits such as drought resistance and pest tolerance through gene transfer.

The current approaches to controlling plant diseases rely primarily on agrochemicals that may harm both people and the environment. Nanotechnology can assist in minimizing the adverse effects of pesticides in a sustainable and environmentally friendly way by improving the solubility of pesticides, extending their shelf life, and lowering their toxicity. Similarly, biopesticides are great options for ecologically friendly agriculture. In agriculture, the application of NCs (nanocarriers), or nanoparticles that may encapsulate medications and release them selectively, is a developing field. The active components of biopesticides can be loaded into nanocarriers via encapsulation, entrapment, attachment, or absorption to create nano-biopesticides. Different types of nanocarriers, such as metal oxides, metals, non-metal oxides, carbon, polymers, lipids, etc., are used to make nano-biopesticides and RNAi pesticides. This chapter explores the use of nanocarriers to manage plant pests and deals with existing obstacles and future opportunities.

Almost 30-40% of the harvest is lost due to plant diseases worldwide. When traditional disease management strategies become less efficient, the advancement in management and incorporation of modern domains like nanotechnology can be the key factor to increase the efficiency of disease management. Nanocarriers are the particles in the nano range that are greatly diverse in makeup and transporting agents that target the specific site to induce biochemical interactions by releasing active compounds and defensive responses. The mechanism, despite being completely immunological, is a combination of chemical, physical, and biochemical interactions. A diverse range of nanocarriers is used to deliver bioactive molecules to control fungi, bacteria, nematodes, and other biotic limiting factors in agriculture. The usual makeup of nanocarriers varies from liposomes, chitosan, to inorganic metal carriers or carbon, silica-based materials. This chapter is a comprehensive analysis of morphology, loading capacity, biocompatibility of nanocarriers, and the host plant for effective disease control.

The application of nanocarriers in agriculture has become a groundbreaking method to increase nutrient delivery, boost plant health, and successfully fight fungal infections. These nanoscale structures increase agrochemical stability, solubility, and bioavailability, guaranteeing long-term efficacy while minimizing negative environmental effects. The synthesis and characterization of nanocarriers involve sophisticated techniques such as sol-gel processes, self-assembly, and in situ polymerization, ensuring optimal performance. Despite their vast potential, challenges such as production scalability, regulatory compliance, and long-term environmental safety must be addressed to facilitate widespread adoption. This chapter provides an in-depth analysis of nanocarrier technology in agriculture, highlighting its role in improving plant nutrient uptake, disease resistance, and environmental sustainability. With continued research and development, nanocarriers are poised to revolutionize modern agriculture, paving the way for sustainable and high-yield farming practices.

Nanotechnology is a multidisciplinary field that manipulates materials at the nanoscale, typically between 1 and 100 nanometers. It bridges various disciplines, including physics, chemistry, biology, engineering, and materials science, creating materials with unique properties that classical theories cannot predict. Nanotechnology is central to advancements in energy, medical therapies, environmental solutions, and electronics. Nanotechnology's ability to create more efficient energy storage systems, such as high-capacity batteries, and its potential to revolutionize healthcare by enabling personalized medicine further demonstrate its vast impact. These innovations offer solutions to pressing issues like clean energy, health care, and ecological preservation. This chapter focuses on green synthesis methods in plant science, highlighting environmentally friendly techniques for producing nanomaterials. These sustainable methods not only contribute to eco-friendly agricultural practices but also promote environmental conservation and sustainability in the field of nanotechnology.

Preface

Over the past few decades, nanotechnology has emerged as a transformative force across a range of scientific disciplines, and its integration into the fields of plant science and agriculture holds immense promise. Among the various innovations, nanocarriers—engineered nanoscale vehicles designed to deliver active agents to specific targets—have become especially significant. Their capacity to transport agrochemicals, biomolecules, and genetic materials with precision has opened new avenues for improving crop performance, enhancing stress tolerance, and enabling sophisticated genetic and epigenetic modifications. Despite the growing body of literature on nanotechnology in agriculture, there remains a critical need for a dedicated and comprehensive reference that systematically explores the fundamentals, mechanisms, and practical applications of nanocarriers in plant systems.

This edited volume, *Nanocarriers in Plant Science and Agriculture*, is developed to fill that gap. It brings together a collection of chapters authored by experts from diverse disciplines, aiming to provide an in-depth exploration of nanocarriers from their basic principles to advanced applications. The book covers a wide range of topics including the use of nanocarriers for the targeted delivery of fertilizers, pesticides, biostimulants, and growth regulators, as well as their role in modern plant biotechnology—particularly in the delivery of nucleic acids for genome editing technologies such as CRISPR. Moreover, this volume highlights critical categories of nanocarriers such as biogenic, metallic, polymeric, and carbon-based nanomaterials, examining their unique properties and suitability for various agricultural and biotechnological purposes.

CHAPTER OVERVIEWS

This volume begins with **Chapter 1**, *Nanocarriers in Plant Science and Agriculture: An Introductory Overview*, which lays the foundational understanding of nanocarrier systems in agriculture. It introduces key types such as polymeric, metallic, and liposomal carriers, and discusses their mechanisms of delivery, role in improving nutrient uptake, genetic engineering, and potential to reduce environmental waste. This chapter serves as a vital primer for readers seeking a broad yet insightful entry into plant nanotechnology.

Chapter 2, *Metallic Nanocarriers in Plant Sciences: Types and Applications*, delves into metal-based nanoparticles and their unique properties, including high reactivity and controlled delivery potential. It underscores their utility in precise agrochemical application, stress mitigation, and plant growth enhancement while promoting reduced chemical input and improved sustainability outcomes in agriculture.

In **Chapter 3**, *Carbon Nanotubes as Carriers in Plant Science*, the focus shifts to carbon nanotubes (CNTs) and their diverse applications, including gene delivery, nano-pesticide formulation, and biosensing. The chapter presents both the promise and the challenges—particularly toxicity and environmental impact—and explores future integration with AI and omics approaches for enhanced precision agriculture.

Chapter 4, *Nanocarriers in Plant Genetic Engineering*, explores the crucial role of nanocarriers in facilitating gene transfer and genome editing. The chapter contrasts nanoparticle-mediated delivery with traditional techniques, emphasizing their efficiency, versatility, and compatibility with multiple plant species in the pursuit of high-yield, stress-resilient crops.

In **Chapter 5**, *Nanocarriers for Targeted Delivering and Translocation of Chemical Cargoes in Plant Cells*, readers will find a comprehensive treatment of passive and active uptake pathways, surface functionalization strategies, and cargo specificity. This chapter highlights how nanocarriers can enhance agrochemical precision, uptake efficiency, and reduce unintended environmental consequences.

Chapter 6, *Nanocarriers for Delivering Biomolecules in Plants: Mechanisms and Applications*, addresses the delivery of RNA, DNA, and proteins. It evaluates nanocarrier systems that bypass the plant cell wall without damage, offering innovative pathways for manipulating plant traits and enhancing genetic engineering.

Chapter 7, *Nanocarriers for Delivering Biostimulants in Plants: Mechanisms and Applications*, explores the synergistic integration of nanotechnology with biostimulants to improve crop quality and food security. The chapter highlights nanoencapsulation techniques and sustainable practices, including the use of biostimulants derived from waste materials, thus supporting a circular economy.

In **Chapter 8**, *Nanocarriers for Delivering Nucleic Acids in Plants: Mechanisms and Applications*, the focus is on overcoming cellular barriers to genetic delivery. The chapter provides insights into the use of nanocarriers such as gold and silica nanoparticles to enhance uptake, enable precise genome editing, and support emerging nucleic acid-based therapeutics in plant systems.

Chapter 9, *Nanocarriers for Delivering Plant Growth Promoting Microorganisms*, investigates how nanocarriers improve the viability and stability of beneficial microbes like PGPBs. With carriers such as chitosan, alginate, and biochar, the chapter illustrates the promise of biologically enhanced soil fertility and plant growth as an alternative to chemical fertilizers.

Chapter 10, *Nanocarriers for Delivering Pesticides*, details innovative nanoformulations designed for improved stability, controlled release, and targeted pest control. By reducing off-target effects and environmental risks, these systems contribute to a more responsible and sustainable pesticide strategy.

Chapter 11, *Nanocarrier-Enabled Agricultural Biotechnology and Crop Breeding: Advances and Prospects*, expands on the integration of nanocarriers in genetic modification and crop breeding. This chapter demonstrates how nanotechnology advances the precision, efficacy, and sustainability of agricultural biotechnology, particularly in enhancing traits like drought tolerance and pest resistance.

In **Chapter 12**, *The Role of Nanocarriers in Phytopathogen Management*, readers are introduced to the use of nanocarriers in developing nano-biopesticides and RNAi-based tools for disease control. The chapter discusses various nanocarrier types and their application methods, presenting a forward-looking strategy for environmentally safe pest management.

Chapter 13, *Nanocarriers in Plant Disease Management*, presents a comprehensive review of nanocarrier systems designed to combat a wide range of plant pathogens. The chapter offers detailed insights into their morphology, host interaction, and mechanisms of disease suppression through a fusion of biochemical and physical means.

Chapter 14, *Potential Impact of Nanocarriers in Prevention and Management of Plant Fungal Diseases*, further narrows the focus to fungal pathogens. This chapter covers synthesis techniques, environmental impact considerations, and the effectiveness of nanocarriers in delivering antifungal agents, ensuring crop protection with reduced ecological disruption.

Finally, **Chapter 15**, *Nanotechnological Green Synthesis and Its Impact on Plant Science*, concludes the volume by exploring eco-friendly synthesis methods for nanomaterials. It highlights how green nanotechnology can support sustainable agriculture, reduce toxic byproducts, and align with broader goals of environmental conservation and responsible innovation.

In addition to these applications, the book also discusses strategic uses of nanocarriers in crop breeding programs and in mitigating abiotic stresses and plant diseases, showcasing the versatility and potential of nanotechnology in ensuring sustainable agricultural practices. This reference aims to support researchers, educators, and professionals working in plant nanotechnology, biotechnology, agriculture, and materials science by offering both foundational knowledge and inspiration for future innovation.

As the editor, I sincerely hope that this volume will serve as a valuable resource for advancing research and fostering new ideas in the evolving field of nanocarriers in plant science. I would like to thank all the contributing authors for their insightful work and commitment, and the readers for their interest in this growing area of scientific inquiry.

Jen-Tsung Chen
National University of Kaohsiung, Taiwan

Chapter 1
Nanocarriers in Plant Science and Agriculture: An Introductory Overview

Sarfraz Ahmad
https://orcid.org/0000-0002-6197-2116
Swami Keshwanand Rajasthan Agriculture University, Bikaner, India

Rubina Khan
https://orcid.org/0009-0007-8087-606X
ICAR-National Institute for Plant Biotechnology, India

Saniya
National Research Centre for Grapes, Pune, India

Vikas Belwal
VCS Garhwali Uttarakhand University of Horticulture and Forestry, Ranichauri, India

Mohammad Ghani Raghib
Nangarhar University, Afghanistan

Pankaj Kumar
VCS Garhwali Uttarakhand University of Horticulture and Forestry, Ranichauri, India

ABSTRACT

Nanocarriers, engineered nanoparticles designed to deliver fertilizers, pesticides, and genetic materials, are emerging as essential tools in modern agriculture. This chapter explores nanocarrier types, mechanisms, and applications in plant science. Key systems such as liposomes, polymeric, metallic, and silica nanoparticles enable targeted delivery, enhancing nutrient uptake, pest control, and stress resilience. By ensuring precise agrochemical release, nanocarriers reduce wastage, contamination, and application frequency, supporting sustainable farming. They also facilitate genetic transformation and gene editing for stress-tolerant, non-transgenic crops. However, challenges in cost, regulation, and environmental impact remain. With ongoing research, nanocarriers show promise in advancing food security and sustainable agriculture.

1. INTRODUCTION

Agriculture, as one of the most vital sectors of human civilization, plays a vital role in nourishing the global population by providing food, fiber, and raw materials for several industries. However, modern agriculture faces numerous challenges that threaten its ability to meet the growing demand for food

DOI: 10.4018/979-8-3693-9869-2.ch001

in a sustainable and environmentally responsible manner. Issues such as declining soil fertility, water scarcity, climate change, pest infestations, and nutrient inefficiencies are some of the significant hurdles that need to be overcome to ensure agricultural productivity and food security (Khan et al., 2018). Due to these challenges, the agricultural sector has increasingly turned towards innovative technologies that can enhance crop yields, improve resource use efficiency, and reduce the environmental impact of farming practices.

Nanotechnology, a cutting-edge field that deals with the manipulation and application of atoms and molecules at the nanoscale (ranging between 1 to 100 nanometers), has emerged as a promising solution to address the complex challenges of modern agriculture (Rai et al., 2016). The uniqueness of nanomaterials, such as their small size, high surface area-to-volume ratio, and ability to be engineered for specific functions, enable them to interact with biological systems in novel ways, offering unprecedented opportunities for enhancing agricultural productivity (Nair et al., 2010). Nanotechnology's entry into agriculture initially focused on the development of nanopesticides and nanofertilizers designed to improve the efficiency and precision of traditional chemical inputs (Kah et al., 2013). Over time, the scope of nanotechnology in agriculture has expanded significantly, encompassing a wide array of nanoscale tools and devices that enhance plant growth, protect against biotic and abiotic stresses, and improve overall crop resilience (El-Saadony et al., 2022).

Nanocarrier development is one of the most promising applications of nanotechnology in agriculture, which are engineered nanoparticles that act as delivery systems for active substances such as fertilizers, pesticides, herbicides, and genetic materials. Nanocarriers are designed to encapsulate, protect, and transport these active ingredients, releasing them in a controlled manner that enhances their effectiveness and reduces their environmental impact (Chariou et al., 2020). Unlike conventional delivery methods, which often involve the blanket application of agrochemicals with significant losses to the environment, nanocarriers offer precise and targeted delivery directly to the site of action within plants. This precision minimizes the overuse of agrochemicals, enhances their bioavailability, and reduces the potential for off-target effects, making agriculture more sustainable and efficient (Jiang et al., 2021).

The advantages of nanocarriers in agriculture extend beyond their ability to improve the delivery of agrochemicals. These nanoscale delivery systems also offer innovative solutions to persistent agricultural problems that have been difficult to address with traditional methods. For instance, nutrient loss and runoff are major issues allied with conventional fertilizers, leading to water pollution and reduced nutrient availability to crops (Rico et al., 2022). Nanocarriers, specifically designed for slow and controlled nutrient release, can mitigate these issues by ensuring that nutrients are released in sync with plant uptake rates, significantly enhancing nutrient use efficiency. This targeted release not only maximizes the effectiveness of fertilizers but also reduces the environmental footprint of agricultural practices, aligning with the goals of sustainable agriculture (Usman et al., 2020).

Similarly, in pest control, nanocarriers have showed considerable potential in improving the effectiveness of pesticides. Traditional pesticide formulations often suffer from low stability, rapid degradation, and significant losses caused by environmental factors such as UV radiation, rain, and microbial activity (Slomberg & Schoenfisch, 2012). Nanocarriers protect the active ingredients from these adverse conditions, extending their shelf life and ensuring that they remain active until they reach their intended targets. Moreover, nanocarriers may be designed to release pesticides in reaction to certain environmental causes, for example changes in pH or the presence of particular enzymes, this allows for a more controlled and effective pest control strategy (Mittal et al., 2020). This regulated release reduces need for frequent pesticide applications, lowers the total chemical burden on the environment, and minimizes the risk of developing resistance in insect populations (Singh et al., 2020).

In addition to their role in nutrient and pest management, nanocarriers are revolutionizing the field of plant biotechnology, particularly in the delivery of genetic material for plant transformation and crop improvement. Carbon-based nanocarriers, such as carbon nanotubes and graphene oxide, have demonstrated the ability to penetrate plant cells and deliver genetic material, including DNA and RNA, directly into plant tissues (Demirer et al., 2019). This capability opens up new possibilities for precision breeding and genetic engineering, allowing scientists to produce crops with improved characters, such as drought tolerance, disease resistance, and increased nutritional value. The use of nanocarriers for genetic delivery bypasses some of the restrictions of conventional transformation methods, offering a faster, more efficient, and less labor-intensive approach to crop improvement (Nair et al., 2010).

The use of nanocarriers in agricultural techniques is also aligns with the wider aims of sustainable agriculture, which tries to strike a balance between the demand for increasing food production and the need to maintain natural resources and prevent environmental destruction. By improving the efficiency of input use, reducing waste, and minimizing chemical exposure to non-target organisms, nanocarriers contribute to more sustainable farming systems (Usman et al., 2020). For example, the enhanced nutrient uptake provided by nanofertilizers reduces nutrient runoff into waterways, mitigating issues such as eutrophication and algal blooms that are commonly associated with conventional fertilizer use (Subramanian et al., 2015). Similarly, the use of nanocarriers in biopesticide formulations can enhance the stability and efficacy of these biological agents, making them a more viable alternative to synthetic pesticides and supporting the transition to eco-friendly pest management practices (Patra et al., 2018).

Despite their numerous advantages, the extensive application of nanocarriers in agriculture is not without challenges. Concerns related the potential environmental and health impacts of nanoparticles, the scalability and cost-effectiveness of nanoparticle production, and the need for clear regulatory frameworks are some of the key hurdles that need to be addressed to facilitate their use in agriculture (Rico et al., 2011). As research in this subject progresses, it is critical to develop standards for the safe and responsible use of nanotechnology in agriculture, combining the benefits of increased agricultural output with the need to preserve human health and the environment.

To summarize, nanocarriers are a transformational technology with the potential to alter agriculture by improving agrochemicals delivery and efficacy, increasing crop resilience, and contributing to the creation of more sustainable agricultural techniques. As the agricultural industry continues to face the problems of feeding a growing global population in the face of climate change and resource scarcity, nanocarriers provide a potent tool for increasing agricultural output while reducing environmental effect. This chapter provides an introductory overview of the types of nanocarriers used in agriculture, their mechanisms of action, and their applications in plant science, along with a discussion of the challenges and future perspectives of this promising technology. By exploring the current state of nanocarrier technology and its potential to shape the future of agriculture, this chapter aims to highlight the critical role that nanocarriers will play in the ongoing evolution of sustainable agricultural practices.

2. OVERVIEW OF NANOCARRIERS IN AGRICULTURE

Nanocarriers are specialized nanoscale systems designed to deliver active substances such as fertilizers, pesticides, herbicides, and genetic material to plants in a controlled and targeted manner. They have developed as powerful tools in agriculture, offering a range of benefits, including improved delivery efficiency, enhanced bioavailability of active ingredients, reduced environmental impact, and minimized risks of non-target effects. The versatility and adaptability of nanocarriers stem from their ability to encapsulate and protect active compounds, control the release rate, and deliver substances precisely to their intended targets within the plant or soil environment. Further section provides an overview of the types of nanocarriers used in agriculture, their mechanisms of action, and their specific applications, highlighting their potential to revolutionize modern farming practices.

3. TYPES OF NANOCARRIERS

Nanocarriers are a diverse group of nanoscale materials engineered to deliver active ingredients like agrochemicals, nutrients, or genetic material to crops in a regulated and targeted manner. The type of nanocarrier selected for a particular application depends on its structural and functional properties, including stability, biocompatibility, and ability to encapsulate and release the active ingredient. The main types of nanocarriers used in agriculture include polymeric nanoparticles, lipid-based nanoparticles, inorganic nanoparticles, carbon-based nanocarriers, and hybrid nanocarriers (Figure 1). Each type offers unique benefits tailored to specific agricultural needs.

Figure 1. The figure illustrating different types of nanocarriers

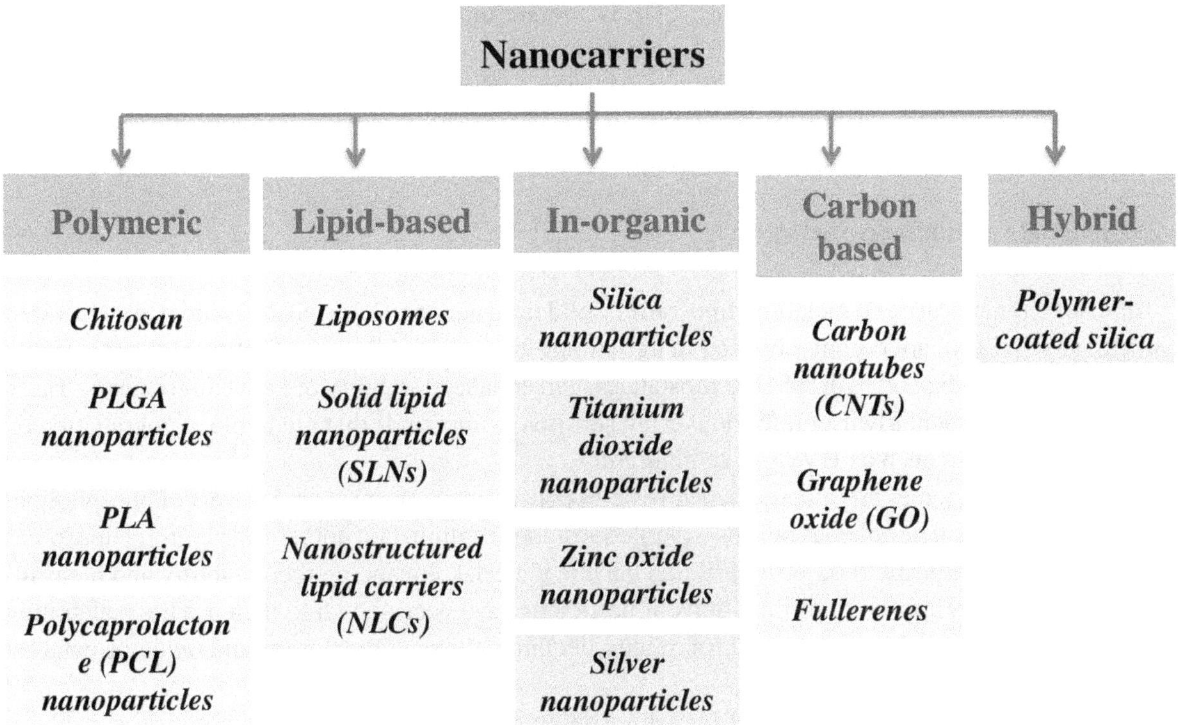

3.1. Polymeric Nanoparticles

Polymeric nanoparticles are the most commonly used nanocarriers in agriculture because of their versatility, biodegradability, and capability to encapsulate a broad range of active substances. These are made up of natural or synthetic polymers viz, chitosan, poly lactic-co-glycolic acid (PLGA), polylactic acid (PLA), and polycaprolactone (PCL). Polymeric nanocarriers are particularly useful for controlled release applications, where they gradually release the encapsulated agrochemicals over time, improving the efficiency and reducing the environmental effect of these substances.

Chitosan nanoparticles: Chitosan, a natural polymer derived from chitin found in the exoskeletons of crustaceans, is widely used due to its biocompatibility, biodegradability, and inherent antimicrobial properties. In agriculture, chitosan nanoparticles are used to deliver fertilizers, pesticides, and biostimulants, enhancing nutrient uptake, boosting plant immunity, and protecting against pathogens (Ingle et al., 2022). The positive charge of chitosan nanoparticles allows them to interact effectively with negatively charged plant surfaces, facilitating better adhesion and uptake.

PLGA and PLA nanoparticles: PLGA and PLA are synthetic, biodegradable polymers approved by regulatory bodies such as the FDA for use in medical and agricultural applications. These nanoparticles offer controlled and continued discharge of active substances, reducing frequency of agrochemical applications and minimizing losses due to environmental factors. In agriculture, PLGA have been used to supply pesticides, hormones, and growth hormones, providing a steady release that synchronizes with plant needs, thereby enhancing overall productivity (Shakiba et al., 2020).

Polycaprolactone (PCL) nanoparticles: PCL is another biodegradable polymer commonly used in agricultural nanocarriers. PCL nanoparticles are particularly suitable for delivering hydrophobic compounds, such as certain pesticides and herbicides, enhancing their solubility and bioavailability (Buluş et al., 2020). Their slow degradation rate makes them ideal for applications that require prolonged release, such as the sustained delivery of nutrients and pest control agents.

3.2. Lipid-Based Nanoparticles

Lipid-based nanocarriers, including liposomes, solid lipid nanoparticles (SLNs), and nanostructured lipid carriers (NLCs), are commonly used in agriculture because of biocompatibility, ability to encapsulate both hydrophilic and hydrophobic substances, and enhanced stability of active ingredients. These nanocarriers are particularly useful for delivering sensitive compounds that are liable to degradation by environmental causes such as UV light and humidity.

Liposomes: Liposomes are sphere-shaped vesicles consisting of one or more bilayers of phospholipid that is used to encapsulate aqueous or lipid-soluble substances within their core. In agriculture, liposomes are used to introduce pesticides, fertilizers, and genetic material, enhancing their stability and bioavailability. For example, liposomal formulations of pesticides have been shown to increase the penetration and retention of active substances on plant surfaces, declining the required dosage and reducing external contamination (De Leo et al., 2022).

Solid lipid nanoparticles (SLNs): SLNs are made up of solid lipids that form a matrix capable of encapsulating active compounds. SLNs are known for their ability to protect encapsulated substances from environmental degradation and provide controlled release profiles. In agriculture, SLNs have been used to deliver fungicides, herbicides, and growth regulators, offering improved stability and prolonged activity compared to conventional formulations (Campos et al., 2015).

Nanostructured lipid carriers (NLCs): NLCs are advanced lipid-based nanocarriers that combine solid and liquid lipids, resulting in a more flexible and less crystalline structure compared to SLNs. This hybrid structure allows for higher drug loading capacity and more controlled release properties (Tortorici et al., 2022). NLCs have been used in agriculture for delivering various agrochemicals, enhancing their solubility, bioavailability, and effectiveness while reducing their environmental impact.

3.3. Inorganic Nanoparticles

Silica, titanium dioxide, zinc oxide, and silver nanoparticles are inorganic nanocarriers, characterized by their stability, high surface area, and unique physicochemical properties. These nanocarriers are especially useful in applications that require robust and stable delivery systems that can withstand harsh environmental conditions.

Silica nanoparticles: Silica nanoparticles are extensively used in agriculture based on their biocompatibility, high surface area, and tunable pore sizes, which make them ideal for encapsulating and delivering a variety of agrochemicals. They are commonly employed as carriers for fertilizers, pesticides, and growth hormoness, increasing their stability and targeted delivery. Mesoporous silica nanoparticles, in particular, are valued for their quality to give regulated delivery, improving the efficiency of agrochemical inputs and minimizing environmental contamination (Rastogi et al., 2019).

Titanium dioxide nanoparticles: Titanium dioxide (TiO_2) nanoparticles are known for their photocatalytic properties, which can be harnessed to degrade harmful substances in the environment and protect plants from pathogens. TiO_2 nanoparticles are also used as carriers for fertilizers and pesticides, enhancing their stability and bioavailability (Javed et al., 2022). Their photocatalytic activity can improve the decomposition of organic pollutants on plant surfaces, providing additional protective benefits.

Zinc oxide nanoparticles: Zinc oxide (ZnO) nanoparticles serve as both micronutrient fertilizers and carriers for other agrochemicals. As a source of zinc, an essential micronutrient for plants, ZnO nanoparticles improve seed germination, root expansion, and total plant growth. Their antimicrobial properties also help protect plants from bacterial and fungal pathogens, making them a dual-purpose nanocarrier in agricultural applications (Zhou et al., 2023).

Silver nanoparticles: Silver nanoparticles are renowned for their potent antimicrobial properties, which are utilized in agriculture to protect crops from pathogenic microorganisms. They can be used as carriers for fungicides and bactericides, providing enhanced protection from various crop diseases. Additionally, silver nanoparticles have been shown to induce systemic resistance in plants, further bolstering their defense against biotic stresses (Kale et al., 2021).

3.4. Carbon-Based Nanocarriers

Carbon-based nanocarriers, including carbon nanotubes (CNTs), graphene oxide (GO), have distinctive structural, electrical, and mechanical qualities that allow them particularly effective in agricultural applications. These nanocarriers are often used for delivering genetic material, agrochemicals, and growth regulators directly into plant cells, enhancing their absorption and bioactivity.

Carbon nanotubes (CNTs): CNTs are tubular nanostructures consisting carbon atoms in a hexagonal matrix. Due to high surface area, mechanical strength, and penetration into cell walls, CNTs are used as carriers for genetic material, such as DNA and RNA, facilitating gene transfer in crop biotechnology (Rudakiya et al., 2019). This property makes CNTs invaluable for genetic engineering and improvement, allowing the crops development with improved characters viz, higher yield, disease resistance, and stress tolerance.

Graphene oxide (GO): GO is a single layer of carbon atoms arranged in a honeycomb manner with various functional groups having oxygen. GO nanoparticles have been explored for their ability to deliver agrochemicals and genetic material to plants based on their high surface area, biocompatibility, and tunable surface chemistry. GO has been used to improve water retention in soils, enhance nutrient uptake, and deliver pesticides with controlled release, offering multiple benefits for sustainable agriculture (Bhattacharya et al., 2023).

Fullerenes: Fullerenes are spherical carbon nanostructures that can encapsulate various molecules within their hollow cores. In agriculture, fullerenes have been studied for their ability to deliver nutrients, antioxidants, and plant growth regulators (Husen & Siddiqi 2014). Their unique structure allows them to act as radical scavengers, protecting plants from oxidative stress and improving overall plant health and resilience.

3.5. Hybrid Nanocarriers

Hybrid nanocarriers combine materials from different classes, such as polymers and inorganic nanoparticles, to create multifunctional delivery systems with enhanced properties. These nanocarriers are designed to maximize the strengths of each component, offering improved stability, targeted delivery, and controlled release of agrochemicals. For example, polymer-coated silica nanoparticles can provide the biocompatibility of polymers (Machado et al., 2022) and the high loading capacity of silica, which is ideal in delivering a various agrochemicals in a sustainable manner.

4. MECHANISMS OF ACTION OF NANOCARRIERS IN PLANTS

Nanocarriers enhance the delivery of agrochemicals, nutrients, and genetic material to plants through various mechanisms that improve their bioavailability, stability, and targeted delivery. The effectiveness of nanocarriers in agriculture lies in their ability to encapsulate active ingredients, control the release rate, target specific plant tissues, and protect the active compounds from environmental degradation. These mechanisms enhance the efficacy of agrochemical inputs and also reduces the environmental effect and improve plant health and productivity. A conceptual flow diagram outlining their mechanisms of action is presented in the Figure 2. The key mechanisms by which nanocarriers function in plants, including encapsulation and protection, controlled release, targeted delivery, enhanced uptake, and environmental interaction are as follows.

Figure 2. A conceptual flow diagram outlining mechanisms of action of nanocarriers

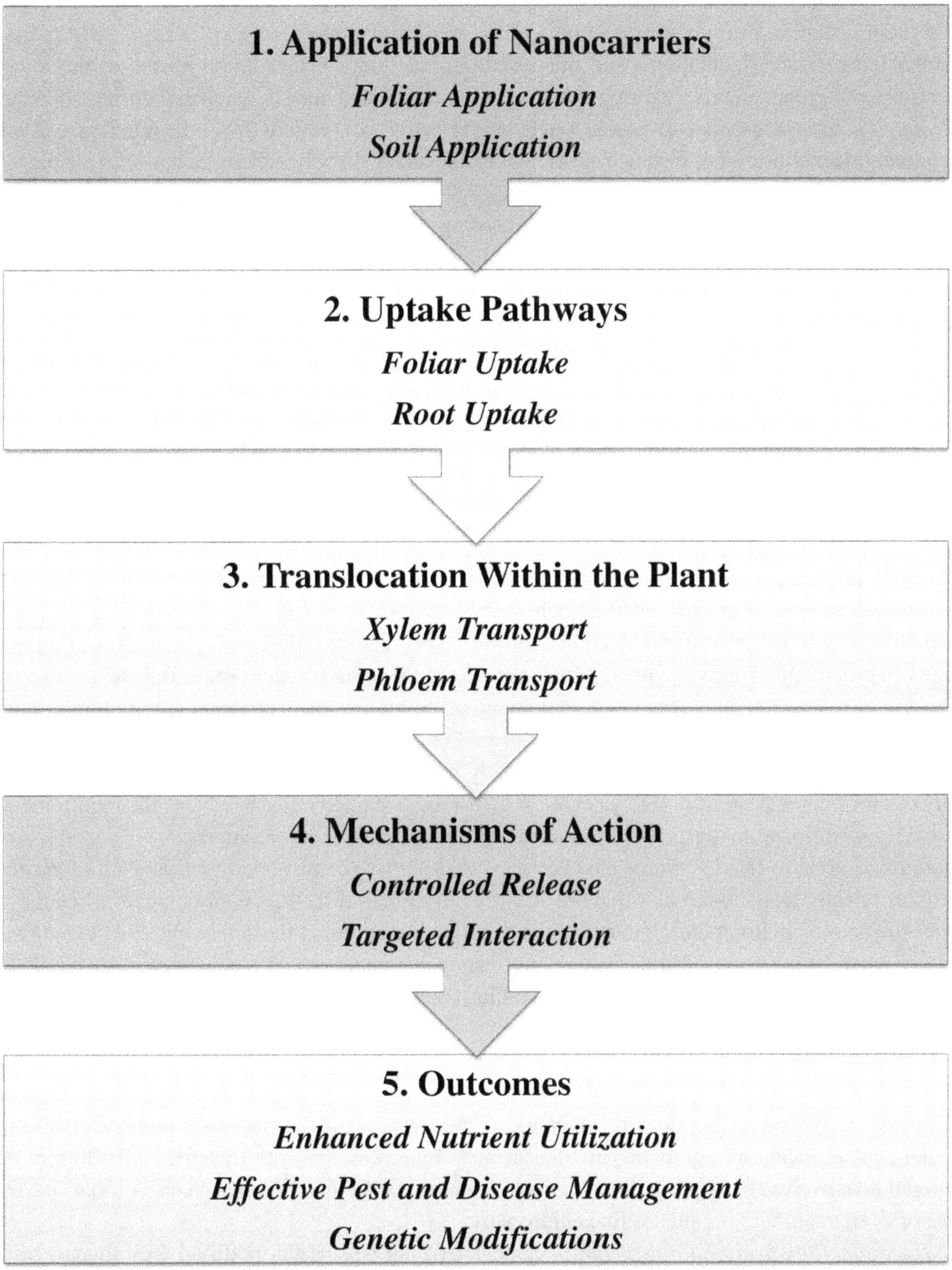

4.1. Encapsulation and Protection

One of the fundamental mechanisms of nanocarriers is their ability to encapsulate active ingredients within their structure, providing protection against environmental aspects for example UV radiation, moisture, temperature fluctuations, and microbial degradation. Encapsulation preserves the integrity and activity of agrochemicals, ensuring that they remain effective until they reach their intended target.

Protection from degradation: Nanocarriers shield active compounds from degradation caused by environmental conditions. For instance, pesticides encapsulated in polymeric nanoparticles are protected from UV light, which often leads to the rapid breakdown of conventional pesticide formulations (Perlatti et al., 2013). This protection extends the shelf life of agrochemicals and maintains their potency during application.

Enhanced solubility and stability: Many active ingredients, particularly hydrophobic pesticides and herbicides, suffer from poor solubility and stability in aqueous environments. Nanocarriers improve the solubility of these compounds by encapsulating them in lipid or polymeric matrices, enhancing their dispersion in water and increasing their bioavailability to plants. Lipid-based nanoparticles, such as liposomes and solid lipid nanoparticles (SLNs) are particularly effective in solubilizing hydrophobic agrochemicals, facilitating their uptake by plant tissues (Rehman et al., 2022).

4.2 Controlled Release

Controlled release is one of the most valuable mechanisms of nanocarriers, allowing for the gradual and sustained release of encapsulated agrochemicals as a result of specific elicits such as changes in pH, enzymatic activity and environmental conditions. This regulated delivery optimizes the timing and dosage of agrochemical delivery, decreasing waste and enhancing resource use efficiency.

Sustained release: Nanocarriers can be engineered to release their contents for the longer period, corresponding to release rate with the crop's needs. For example, polymeric nanoparticles made from materials such as PLGA or chitosan degrade slowly in the plant environment, providing a continuous supply of nutrients or pesticides (Divya et al., 2018). This sustained release reduces the requirement for recurrent applications, saving costs and minimizing environmental contamination.

Stimuli-responsive release: Some nanocarriers are designed to deliver cargo as a result of particular stimuli. For example, nanoparticles that are sensitive to pH can discharge their contents when they encounter the acidic environment of plant tissues or soils, ensuring that the active ingredient is delivered precisely where it is needed. Similarly, enzyme-responsive nanocarriers can release agrochemicals in response to specific plant enzymes, allowing for highly targeted and efficient delivery (Shishir et al., 2021).

4.3 Targeted Delivery

Targeted delivery is a key feature of nanocarriers, enhancing the precision and effectiveness of agrochemical applications. By modifying the surface of nanocarriers with ligands, antibodies, or other functional groups, researchers can direct nanocarriers to specific plant tissues, such as roots, leaves, or stems, or even to cellular organelles like chloroplasts.

Surface functionalization for targeting: Nanocarriers can be surface-modified with ligands or peptides that recognize specific receptors on plant cell surfaces, facilitating targeted delivery. For instance, nanocarriers modified with carbohydrate ligands can selectively bind to lectins on the surface of plant cells, enhancing the uptake of encapsulated nutrients or pesticides (Smith & Gilbertson, 2018). This directed approach decreases the dose of agrochemical and reduces off-target effects, such as damage to beneficial organisms.

Chloroplast targeting: Targeting nanocarriers to chloroplasts, the photosynthetic machinery of plant cells, is an emerging area of interest. Lipid-based nanocarriers, such as liposomes, can be engineered to deliver nutrients or herbicides directly to chloroplasts, improving photosynthesis and enhancing the plant's growth and productivity (Alghuthaymi et al., 2021). This approched delivery enhances the efficacy of agrochemical inputs and reduces their environmental footprint.

4.4 Enhanced Uptake and Translocation

Nanocarriers enhance the uptake and translocation of agrochemicals within plants, facilitating their movement from the site of application (e.g., leaves or roots) to other parts of the plant. The nanoscale size, surface charge, and surface properties of nanocarriers play crucial roles in defining how effectively they enter plant tissues and move within the plant.

Penetration of plant tissues: The minor size of nanocarriers makes them to easily enter into plant tissues, including the cuticle and cell walls, which are natural barriers to conventional agrochemicals. Carbon-based nanocarriers like carbon nanotubes (CNTs) and graphene oxide (GO), have been shown to penetrate plant cell walls and deliver DNA or RNA directly into plant cells, making them valuable tools for genetic transformation and gene editing in plants (Mostafavi & Zare, 2022).

Facilitated translocation: Once inside the plant, nanocarriers can facilitate the movement of agrochemicals to other parts of the plant through the vascular system. For example, nanoparticles can be taken up by the roots and transported via the xylem to the leaves, ensuring uniform distribution of nutrients or pesticides throughout the plant (Wang et al., 2023). This enhanced translocation improves the overall effectiveness of agrochemicals and ensures that all parts of the plant receive the required treatment.

4.5 Interaction With Plant Metabolism and Physiology

Nanocarriers can interact with plant metabolism and physiology in ways that enhance the absorption, utilization, and effectiveness of agrochemicals. These contacts can lead to improved nutrient absorption, enhanced stress tolerance, and better overall plant health.

Enhanced nutrient uptake: Nanocarriers may enhance the bioavailability of essential elements, such as iron, zinc, and phosphorus, which are often limited in soils. For example, zinc oxide nanoparticles are used to release zinc to plant roots, enhancing root uptake and improving plant growth. Similarly, mesoporous silica nanoparticles can encapsulate phosphorus and release it in a form that is readily available to plants, overcoming the limitations of conventional phosphorus fertilizers (Iqbal et al., 2019).

Induction of stress tolerance: Some nanocarriers can induce plant stress responses that enhance biotic and abiotic stress tolerance. For instance, silver nanoparticles may trigger the production of reactive oxygen species (ROS) in plants, it can activate defense pathways against pathogens (Abasi et al., 2022). This dual role of nanocarriers, as delivery vehicles and inducers of stress responses, can enhance plant resilience and reduce the reliance on chemical pesticides.

Modulation of plant hormones: Nanocarriers can also be used to deliver plant growth regulators, such as gibberellins, auxins, and cytokinins, which modulate various physiological processes. By providing controlled and targeted delivery of these hormones, nanocarriers can influence seed germination, root development, and flowering, optimizing plant growth and increasing crop yields (Tripathi et al., 2022).

4.6 Environmental Interaction and Reduced Contamination

Nanocarriers help mitigate the environmental impact of agrochemicals by reducing volatilization, leaching, and runoff, these are problems associated with traditional formulations. By encapsulating active ingredients, nanocarriers minimize their release into the environment, protecting non-target organisms and reducing contamination of soil and water resources.

Reduction of leaching and runoff: Encapsulation of fertilizers in nanocarriers, such as silica nanoparticles, reduces nutrient leaching into groundwater, protecting water quality and minimizing environmental pollution (Iqbal et al., 2019). This mechanism is mainly important for nitrogen and phosphorus fertilizers, which are prone to leaching and can contribute to eutrophication in aquatic ecosystems.

Minimization of volatilization: Many pesticides and herbicides are lost to the atmosphere through volatilization, reducing their effectiveness and posing risks to human and environmental health. Nanocarriers can reduce volatilization by encapsulating volatile compounds and controlling their release, ensuring that the active ingredients remain on the plant surface where they are needed (Singh et al., 2022).

Enhanced biodegradability and reduced persistence: The use of biodegradable polymers in nanocarriers ensures that they break down into non-toxic byproducts after delivering their cargo, reducing the persistence of agrochemicals in the environment (Pascoli et al., 2018). This biodegradability minimizes the accumulation of residues in soil and water, supporting the sustainable agriculture and environment.

5. APPLICATIONS OF NANOCARRIERS IN PLANT SCIENCE

Nanocarriers emerging as potential tools in plant science and agriculture, contributing advanced solutions to improve crop productivity, enhance nutrient use efficiency, and develop sustainable pest management strategies. By facilitating the regulated release of agrochemicals, nutrients, and genetic material, nanocarriers address some of the critical challenges faced in food security, environmental sustainability, and the need for precision farming. This section explores the key applications of nanocarriers in plant science, highlighting their roles in pest control, nutrient delivery, stress management, and crop improvement through genetic engineering (Table 1).

Table 1. Various applications of nanocarriers in plant science and agriculture

Applications	Benefits	Examples	References
Pesticide Delivery Systems	- Effective pest and disease control - Reduced pesticide usage - Minimized environmental contamination	Nanocarriers can coat pesticides on their surfaces and then release them onto target plants, providing more effective pest and disease control.	Bueno et al., 2022 Tao et al., 2023
Smart Fertilizer Release	- Increased nutrient uptake - Reduced fertilizer usage - Minimized environmental effect	Nano-zinc fertilizers improve zinc uptake in crops, addressing micronutrient deficiencies while enhancing yield.	Yuvaraj et al., 2023 Singh, 2025
Nanocarriers for Gene Delivery	- Efficient genetic modification - Improved crop traits - Development of stress-resistant crops	Gold nanoparticles are used to deliver genetic material into plants for developing stress-resistant crops.	Rani et al., 2025 Munir et al., 2023
Plant Growth Promotion	- Improved germination rates - Enhanced stress resistance - Increased crop yield and quality	Nanosilica particles can enhance the stress resistance of plants and improve their growth environment.	Ismail et al., 2022 Wang et al., 2022
Crop Preservation	- Extended shelf life - Reduced food loss - Improved food safety	Nanomaterials can be used to prepare crop preservation packaging to prolong the storage life of food by reducing moisture loss and bacterial growth.	Sridhar et al., 2021 Manikandan & Min, 2023
Nanosensors for Precision Agriculture	- Real-time monitoring - Resource optimization - Early disease detection	Carbon nanotube-based sensors detect ethylene gas emissions, indicating fruit ripening or plant stress.	Korah et al., 2022 Shaw & Honeychurch, 2022
Nano-Coated Seeds	- Improved germination rates - Disease resistance - Enhanced early-stage plant development	Silver nanoparticles in seed coatings offer antimicrobial protection while enhancing seed viability.	Zaim et al., 2023 Shelar et al., 2023
Soil Pollution Remediation	- Remediation of contaminated soils - Restoration of soil health - Promotion of sustainable agriculture	Nanomaterials such as nano-iron and nano-oxides can be used for soil pollution remediation by adsorbing or breaking down harmful pollutants.	Solomon et al., 2024 Marcon et al., 2021
Water Purification and Irrigation	- Efficient water management - Removal of contaminants - Sustainable agriculture practices	Nanofiltration membranes remove contaminants like heavy metals, pesticides, and microbes from irrigation water.	Buonomenna et al., 2022 Joseph et al., 2023

5.1 Pest Management and Control

The effective management of pests and diseases is crucial for sustainable agriculture, as pests can significantly reduce crop yields and quality. Nanocarriers offer innovative approaches to enhance the efficiency and reduce the environmental effect of pesticides through enhanced delivery systems and targeted action.

Nanopesticides: Nanocarriers enable the formulation of nanopesticides, which are pesticides encapsulated within nanoscale carriers. These nanopesticides exhibit improved solubility, stability, and controlled release properties compared to traditional formulations. For example, liposomal formulations of insecticides can enhance their penetration and retention on plant surfaces, leading to improved efficacy while minimizing the required dosage (Saini et al., 2020). By targeting specific pest species and reducing non-target effects, nanopesticides contribute to more sustainable pest management practices.

Targeted delivery to pests: Functionalized nanocarriers can be engineered to release pesticides to specific pest species. For instance, nanoparticles coated with plant-derived molecules or pheromones can attract and bind to pests, ensuring that the active ingredient is delivered precisely where it is needed (An et al., 2022). This controlled approach increases the efficiency of pest control measures and also decreases the overall use of agrochemicals, lowering the risk of environmental contamination.

Biological control agents: Nanocarriers can also be used to deliver biological control agents, such as beneficial bacteria and fungi, to plants. Encapsulating these agents in nanocarriers protects them from environmental stressors and enhances their survival and efficacy. For example, chitosan nanoparticles can be used to deliver beneficial microorganisms that promote plant growth and enhance resistance to pathogens, making environmentally suitable alternative to chemical pesticides (Riseh et al., 2022).

5.2 Nutrient Delivery and Fertilization

Effective nutrient management is necessary for maximizing yield and quality while reducing environmental effect. Nanocarriers are crucial in enhancing the supply of essential nutrients to plants, improving nutrient uptake, and reducing losses associated with conventional fertilization methods.

Nano-fertilizers: Nano-fertilizers are fertilizers formulated with nanocarriers that facilitate the controlled release of nutrients. By encapsulating nitrogen, phosphorus, and potassium in nanocarriers, these fertilizers can deliver a steady supply of essential elements over an extended period (Solanki et al., 2015). This slow-release mechanism enhances nutrient use efficiency, decreases the frequency of applications, and nutrient leaching into the environment.

Micronutrient delivery: Microelements, such as zinc, iron, and manganese, are critical for growth but are often deficient in many soils. Nanocarriers can effectively deliver these micronutrients to plants, improving their bioavailability and uptake. For example, zinc oxide have been used to enhance zinc accessibility in soil, important to plant growth and development (Dimkpa et al., 2019). This targeted delivery of micronutrients helps address nutritional deficiencies in crops, backing improved yield and quality.

Enhanced root uptake: Nanocarriers improve nutrient uptake by promoting root colonization and facilitating nutrient transport within the plant. For instance, polymeric nanoparticles can enhance root absorption of nutrients by increasing the contact area between the roots and the nutrient source (Zhao et al., 2021). This enhanced uptake leads to improved plant growth, higher nutrient efficiency, and reduced fertilizer application rates.

5.3 Stress Management and Resilience

Plants face abiotic and biotic stresses, such as drought, salinity, extreme temperatures, and pathogens, which can adversely affect growth and yield. Nanocarriers can enhance plant resilience to these stresses through targeted delivery of stress-inducing compounds and growth regulators.

Stress-responsive nanocarriers: Plants adapt to drought stress through secondary metabolites and phytohormones, which changes physiological, biochemical, and molecular behaviour. PHs, like abscisic acid, important in drought stress signaling pathways, reducing transpiration losses, promoting leaf senescence, root growth, and activating genes for water stress avoidance (Ahmad et al., 2022). Nanocarriers can be designed to release growth regulators or stress hormones in response to environmental stimuli. For example, pH-sensitive nanocarriers can release auxins or gibberellins when plants are subjected to stress conditions, promoting adaptive responses and improving stress tolerance (Ijaz et al., 2023). This mechanism helps plants cope with adverse conditions, enhancing their resilience and productivity.

Antioxidant delivery: Oxidative stress caused by environmental factors can lead to cellular damage in plants. Nanocarriers used to release antioxidants straightly to plant tissues, declining oxidative stress and enhancing plant health. For instance, fullerenes and carbon nanotubes have been explored for their ability to scavenge reactive oxygen species (ROS) in plants, protecting them from oxidative damage and improving overall resilience (Husen & Siddiqi, 2014).

Enhanced water retention: Some nanocarriers can enhance soil water retention and improve plant access to water during periods of drought. Nanoparticles, such as superabsorbent polymers or silica nanoparticles, can retain water in the soil, providing a reservoir for plants during dry spells (Oladosu et al., 2022). This enhanced water availability can lead to improved growth and yield under drought conditions, contributing to agricultural sustainability.

5.4 Crop Improvement and Genetic Engineering

Nanocarriers have opened new ways for crop breeding through genetic engineering and biotechnology. Their unique properties enable the effective release of genetic material into plant cells, facilitating the improvement of crops with superior traits. The identification of quantitative trait loci (QTLs) and genes related to drought and heat tolerance has contributed to the design of stress-tolerant cultivars. Transgenic and 'omics' technologies provide new insights into the metabolic pathways associated with QTLs and candidate genes, paving the way for further utilization (Azameti et al., 2023; Jain et al., 2023).

Gene delivery: Carbon-based nanocarriers like carbon nanotubes (CNTs) and graphene oxide (GO), used to effectively deliver nucleic acids, such as DNA and RNA, into plant cells (Mostafavi & Zare, 2022). This capability is particularly valuable for genetic transformation, for the transfer of desirable characters, such as disease resistance or improved nutritional content, into crop species. The use of nanocarriers in gene delivery enhances transformation efficiency and reduces the reliance on conventional methods, such as Agrobacterium-mediated transformation.

RNA interference (RNAi) technology: RNA interference is a potential tool for regulating gene expression in crops. Nanocarriers can deliver small interfering RNA (siRNA) molecules to silence particular genes associated with undesirable character, such as susceptibility to insects and diseases (Mujtaba et al., 2021). This targeted approach to gene regulation offers a precise method for crop improvement without altering the overall genetic composition of the plant.

CRISPR/Cas9 delivery: The CRISPR/Cas9 system has modernized genetic engineering, allowing for precise gene editing in plants. Nanocarriers can facilitate the supply of CRISPR/Cas9 components, including guide RNA and Cas9 protein, into plant cells, enabling targeted modifications to the plant genome (Duan et al., 2021). This method holds great promise for improving crops with enhanced attributes, such as higher yield, stress resistance, and nutritional quality.

5.5 Sustainable Agriculture Practices

The incorporation of nanocarriers into agriculture contributes the growth of sustainable and environmental friendly farming systems. By improving the efficiency of input use, reducing chemical loads, and reducing environmental effects, nanocarriers play a critical role in advancing sustainable agriculture.

Precision agriculture: Nanocarriers enable precision agriculture practices by providing targeted and controlled delivery of agrochemicals and nutrients. This approach makes farmers to apply resources only where and when they are needed, minimizing waste and environmental pollution. By optimizing resource use, nanocarriers favours sustainable farming practices and enhance the economic viability of agricultural systems (Yadav et al., 2023).

Reduced chemical use: By enhancing the efficacy of agrochemicals through improved delivery systems, nanocarriers can reduce the overall quantity of pesticides and fertilizers required for effective crop management (Wani et al., 2019). This reduction in chemical use minimizes the pesticide tolerance, decreases the impact on off target organisms, and protects water quality.

Environmental stewardship: The biodegradable nature of many nanocarriers ensures that they break down into non-toxic byproducts after use, reducing their persistence in the environment. This attributes stands with the principles of environmental stewardship and supports the goal of sustainable agriculture by minimizing the accumulation of agrochemical residues in soil and water (Pandit et al., 2023).

6. CHALLENGES AND FUTURE PERSPECTIVES

The incorporation of nanocarriers in plant science presents significant opportunities for enhancing crop productivity, improving nutrient delivery, and advancing sustainable pest management. However, despite their potential benefits, several challenges must be addressed before nanocarriers can be widely adopted in agricultural practices. This section explores these challenges and offers insights into future perspectives and directions for research and development in the field of nanocarriers in agriculture.

6.1 Challenges in the Application of Nanocarriers

Regulatory framework and safety concerns: One of the principal constraints in the application of nanocarriers in agriculture is the lack of comprehensive regulatory frameworks governing their use. The exclusive properties of nanomaterials rise concerns about their possible environmental cause and effects on human health. Regulatory agencies must establish guidelines for the evaluation of the safety, efficacy, and environmental behavior of nanocarriers before they can be approved for commercial use.

Ensuring that these nanomaterials do not accumulate in the environment or enter the food chain is crucial for addressing public concerns about their safety (Kumari et al., 2023).

Cost of production: The production of nanocarriers can be costly and complex, which may limit their accessibility for smallholder farmers. Many nanocarriers are synthesized using specialized techniques that require advanced equipment and materials, increasing production costs (Vega-Vásquez et al., 2020). Developing cost-effective, scalable methods for the synthesis of nanocarriers will be essential to make these technologies more accessible and affordable for farmers, particularly in developing regions.

Environmental fate and persistence: The environmental fate of nanocarriers after application is not completely known. Things regarding their persistence, potential bioaccumulation, and toxicity to non-target organisms pose significant challenges. Research is needed to assess the degradation pathways of nanocarriers in soil and water, as well as their potential interactions with soil microbiota and ecosystems (Kah et al., 2013). Understanding how these materials interact in the environment will be critical to confirming their safe and sustainable usage in agriculture.

Limited knowledge of plant-nanocarrier interactions: The interactions between plants and nanocarriers are complex and not fully understood. Aspects like plant species, developmental stage, and environmental situations can influence the uptake, translocation, and physiological responses of plants to nanocarriers (Ali et al., 2021). Additional research is required to elucidate these interactions and optimize the design of nanocarriers for specific plant applications. Understanding the mechanisms governing plant-nanocarrier interactions will enhance the effectiveness of these delivery systems and enable tailored solutions for different crops and environments.

Public perception and acceptance: Public perception of nanotechnology plays important role in its acceptance in agriculture. Misconceptions and lack of knowledge regarding nanotechnology may lead to resistance from consumers and farmers alike. Educating stakeholders about the benefits and safety of nanocarriers, as well as addressing concerns related to environmental and health impacts, will be essential for fostering acceptance and promoting the incorporation of these innovative technologies in agriculture (Parisi et al., 2015).

Societal implications of using nanotechnology in agriculture: The integration of nanotechnology in agriculture particularly through nanocarriers for targeted delivery of agrochemicals, nutrients, and genetic materials presents significant opportunities for sustainable crop production. However, societal acceptance remains a critical barrier to its widespread adoption. Consumer perception is shaped not only by the scientific evidence of safety and efficacy but also by the transparency and credibility of the institutions promoting these technologies. Studies have revealed that a lack of awareness and limited understanding of nanoscale processes among the general public often leads to skepticism and resistance, especially when applications involve food systems (Handford et al., 2015). Public concerns tend to revolve around potential toxicity, long-term environmental impacts, and ethical considerations related to manipulating natural systems. Therefore, transparent communication is essential in bridging the gap between scientific innovation and public acceptance. Transparent communication must involve clear, accessible information about the development, application, and regulation of nanotechnologies in agriculture. This includes disclosing safety assessments, engaging with public concerns, and providing evidence-based explanations of benefits and risks. Proactive engagement through public forums, participatory policy-making, and transparent labeling can enhance consumer trust and social license to operate. Moreover, interdisciplinary collaboration among scientists, regulators, ethicists, and communication specialists is vital to address societal concerns holistically (Tawiah et al., 2024). Incorporating social science perspectives into the innovation process ensures that ethical, cultural, and psychological dimensions are considered alongside technical performance. Ultimately, fostering public trust in nanotechnology requires not only scientific rigor but also a commitment to openness, inclusivity, and accountability. As nanocarriers become increasingly prevalent in plant science and agriculture, aligning technological advancement with societal expectations will be crucial for ensuring long-term acceptance and responsible innovation.

6.2 Future Perspectives and Research Directions

Development of biodegradable nanocarriers: Advance research should focus on the production of biodegradable and environmentally friendly nanocarriers. Using natural polymers, such as chitosan, alginate, and starch, for nanocarrier synthesis can reduce environmental impact and improve biodegradability (Kucuk et al., 2023). These biopolymers can be engineered to provide targeted delivery and controlled release of agrochemicals while minimizing potential risks associated with synthetic materials.

Smart nanocarrier systems: The advancement of smart nanocarrier systems that respond to specific environmental stimuli offers exciting opportunities for improving the precision and effectiveness of agrochemical delivery. By integrating stimuli-responsive elements, such as pH-sensitive or temperature-sensitive materials, researchers can design nanocarriers that release their payloads only under specific conditions, optimizing resource use and enhancing crop performance (Kah et al., 2013).

Interdisciplinary research approaches: Addressing the challenges associated with nanocarriers in agriculture will require interdisciplinary research approaches that integrate knowledge from materials science, plant biology, environmental science, and agronomy. Collective efforts of researchers, industry stakeholders, and policymakers will be vital for developing innovative solutions that address the complex challenges of modern agriculture while ensuring sustainability and safety (Demirer et al., 2019).

Field trials and real-world applications: While laboratory readings have demonstrated the potential of nanocarriers in plant science, real-world applications and field trials are essential to assess their efficacy and practicality in agricultural settings (Nair et al., 2010). Conducting field trials will offer important insights into the performance of nanocarriers under varying environmental circumstances and will help optimize their formulations and application methods for different crops and regions.

Integration with precision agriculture technologies: The addition of nanocarriers with precision agriculture, viz, remote sensing, soil monitoring, and data analytics, offers opportunities to enhance the efficiency and effectiveness of agrochemical applications. By combining nanocarrier technology with real-time data on soil and crop conditions, farmers can optimize their inputs and reduce waste, contributing to sustainable and productive farming practices (Rico et al., 2011).

Public engagement and education: Growing public engagement and learning about the benefits and safety of nanocarriers in agriculture is essential for fostering acceptance and promoting their use. Outreach efforts aimed at educating farmers, consumers, and policymakers about the possible advantages of nanotechnology in sustainable agriculture can help in constructing trust and support for these innovative approaches (Ghormade et al., 2011).

7. CONCLUSION

The integration of nanocarriers in plant science holds transformative potential for modern agriculture. These innovative delivery systems can significantly enhance crop productivity, improve nutrient efficiency, and facilitate sustainable pest management practices. By enabling targeted and controlled delivery of agrochemicals, nutrients, and genetic material, nanocarriers address pressing challenges faced by the agricultural sector, including food security, environmental sustainability, and the need for precision farming. However, the successful implementation of nanocarrier technology requires overcoming various challenges, such as regulatory hurdles, cost considerations, and public perception. Ensuring the safety and environmental impact of nanocarriers is paramount, necessitating robust research and development efforts. Moreover, fostering interdisciplinary collaboration among researchers, policymakers, and industry stakeholders is necessary to advance this field.

The prospect of nanocarriers in agriculture looks promising. Continued innovation, coupled with increased awareness and education, can facilitate the extensive approval of these technologies. By embracing the potential of nanocarriers, the agricultural community can add to sustainable and efficient food production, ultimately profiting farmers, consumers, and the environment. The route ahead is one of cooperation and exploration, paving the way for a robust agricultural future.

REFERENCES

Abasi, F., Raja, N. I., Mashwani, Z. U. R., Amjad, M. S., Ehsan, M., Mustafa, N., Haroon, M., & Proćków, J. (2022). Biogenic silver nanoparticles as a stress alleviator in plants: A mechanistic overview. *Molecules (Basel, Switzerland)*, *27*(11), 3378. DOI: 10.3390/molecules27113378 PMID: 35684312

Ahmad, S., Belwal, V., Punia, S. S., Ram, M., Dalip, , Rajput, S. S., Kunwar, R., Meena, M. K., Gupta, D., Kumawat, G. L., Hussain, T., & Mohamed, H. I. (2022). Role of plant secondary metabolites and phytohormones in drought tolerance: A review. *Gesunde Pflanzen*, *75*(4), 729–746. DOI: 10.1007/s10343-022-00795-z

Alghuthaymi, M. A., Ahmad, A., Khan, Z., Khan, S. H., Ahmed, F. K., Faiz, S., Nepovimova, E., Kuča, K., & Abd-Elsalam, K. A. (2021). Exosome/liposome-like nanoparticles: New carriers for CRISPR genome editing in plants. *International Journal of Molecular Sciences*, *22*(14), 7456. DOI: 10.3390/ijms22147456 PMID: 34299081

Ali, S., Mehmood, A., & Khan, N. (2021). Uptake, translocation, and consequences of nanomaterials on plant growth and stress adaptation. *Journal of Nanomaterials*, *2021*(1), 6677616. DOI: 10.1155/2021/6677616

An, C., Sun, C., Li, N., Huang, B., Jiang, J., Shen, Y., Wang, Y., Zhao, X., Cui, B., Wang, C., Li, X., Zhan, S., Gao, F., Zeng, Z., Cui, H., & Wang, Y. (2022). Nanomaterials and nanotechnology for the delivery of agrochemicals: Strategies towards sustainable agriculture. *Journal of Nanobiotechnology*, *20*(1), 11. DOI: 10.1186/s12951-021-01214-7 PMID: 34983545

Azameti, M. K., & Imoro, A. W. M. (2023). Nanotechnology: A promising field in enhancing abiotic stress tolerance in plants. *Crop Design*, 100037.

Bhattacharya, N., Cahill, D. M., Yang, W., & Kochar, M. (2023). Graphene as a nano-delivery vehicle in agriculture–current knowledge and future prospects. *Critical Reviews in Biotechnology*, *43*(6), 851–869. DOI: 10.1080/07388551.2022.2090315 PMID: 35815813

Bueno, V., Gao, X., Abdul Rahim, A., Wang, P., Bayen, S., & Ghoshal, S. (2022). Uptake and translocation of a silica nanocarrier and an encapsulated organic pesticide following foliar application in tomato plants. *Environmental Science & Technology*, *56*(10), 6722–6732. DOI: 10.1021/acs.est.1c08185 PMID: 35467849

Buluş, E., Bulus, G. S., & Yakuphanoglu, F. (2020). Production and characterization of novel nature-friendly organic fertilizer covers based on nanotechnology for the agricultural sector. *Journal of Materials and Electronic Devices*, *5*(1), 12–16.

Buonomenna, M. G., Mousavi, S. M., Hashemi, S. A., & Lai, C. W. (2022). Water cleaning adsorptive membranes for efficient removal of heavy metals and metalloids. *Water (Basel)*, *14*(17), 2718. DOI: 10.3390/w14172718

Campos, E. V. R., Oliveira, J. L. D., da Silva, C. M. G., Pascoli, M., Pasquoto, T., Lima, R., Abhilash, P. C., & Fernandes Fraceto, L. (2015). Polymeric and solid lipid nanoparticles for sustained release of carbendazim and tebuconazole in agricultural applications. *Scientific Reports*, *5*(1), 13809. DOI: 10.1038/srep13809 PMID: 26346969

Chariou, P. L., Ortega-Rivera, O. A., & Steinmetz, N. F. (2020). Nanocarriers for the delivery of medical, veterinary, and agricultural active ingredients. *ACS Nano*, *14*(3), 2678–2701. DOI: 10.1021/acsnano.0c00173 PMID: 32125825

De Leo, V., Maurelli, A. M., Giotta, L., & Catucci, L. (2022). Liposomes containing nanoparticles: Preparation and applications. *Colloids and Surfaces. B, Biointerfaces*, *218*, 112737. DOI: 10.1016/j.colsurfb.2022.112737 PMID: 35933888

Demirer, G. S., Zhang, H., Matos, J. L., Goh, N. S., Cunningham, F. J., Sung, Y., Chang, R., Aditham, A. J., Chio, L., Cho, M.-J., Staskawicz, B., & Landry, M. P. (2019). High aspect ratio nanomaterials enable delivery of functional genetic material without DNA integration in mature plants. *Nature Nanotechnology*, *14*(5), 456–464. DOI: 10.1038/s41565-019-0382-5 PMID: 30804481

Dimkpa, C. O., Singh, U., Bindraban, P. S., Elmer, W. H., Gardea-Torresdey, J. L., & White, J. C. (2019). Zinc oxide nanoparticles alleviate drought-induced alterations in sorghum performance, nutrient acquisition, and grain fortification. *The Science of the Total Environment*, *688*, 926–934. DOI: 10.1016/j.scitotenv.2019.06.392 PMID: 31726574

Divya, K., & Jisha, M. S. (2018). Chitosan nanoparticles preparation and applications. *Environmental Chemistry Letters*, *16*(1), 101–112. DOI: 10.1007/s10311-017-0670-y

Duan, L., Ouyang, K., Xu, X., Xu, L., Wen, C., Zhou, X., Qin, Z., Xu, Z., Sun, W., & Liang, Y. (2021). Nanoparticle delivery of CRISPR/Cas9 for genome editing. *Frontiers in Genetics*, *12*, 673286. DOI: 10.3389/fgene.2021.673286 PMID: 34054927

El-Saadony, M. T., Saad, A. M., Soliman, S. M., Salem, H. M., Desoky, E. S. M., Babalghith, A. O., El-Tahan, A. M., Ibrahim, O. M., Ebrahim, A. A. M., Abd El-Mageed, T. A., Elrys, A. S., Elbadawi, A. A., El-Tarabily, K. A., & AbuQamar, S. F. (2022). Role of nanoparticles in enhancing crop tolerance to abiotic stress: A comprehensive review. *Frontiers in Plant Science*, *13*, 946717. DOI: 10.3389/fpls.2022.946717 PMID: 36407622

Ghormade, V., Deshpande, M. V., & Paknikar, K. M. (2011). Perspectives for nano-biotechnology enabled protection and nutrition of plants. *Biotechnology Advances*, *29*(6), 792–803. DOI: 10.1016/j.biotechadv.2011.06.007 PMID: 21729746

Handford, C. E., Dean, M., Spence, M., Henchion, M., Elliott, C. T., & Campbell, K. (2015). Awareness and attitudes towards the emerging use of nanotechnology in the agri-food sector. *Food Control*, *57*, 24–34. DOI: 10.1016/j.foodcont.2015.03.033

Husen, A., & Siddiqi, K. S. (2014). Carbon and fullerene nanomaterials in plant system. *Journal of Nanobiotechnology*, *12*(1), 1–10. DOI: 10.1186/1477-3155-12-16 PMID: 24766786

Ijaz, M., Khan, F., Ahmed, T., Noman, M., Zulfiqar, F., & Rizwan, M.. (2023). Nanobiotechnology to advance stress resilience in plants: Current opportunities and challenges. *Materials Today. Bio*, •••, 100759. DOI: 10.1016/j.mtbio.2023.100759 PMID: 37600356

Ingle, P. U., Shende, S. S., Shingote, P. R., Mishra, S. S., Sarda, V., Wasule, D. L., Rajput, V. D., Minkina, T., Rai, M., Sushkova, S., Mandzhieva, S., & Gade, A. (2022). Chitosan nanoparticles (ChNPs): A versatile growth promoter in modern agricultural production. *Heliyon*, *8*(11), e11893. DOI: 10.1016/j.heliyon.2022.e11893 PMID: 36468119

Iqbal, M., Umar, S., & Mahmooduzzafar, F. (2019). Nano-fertilization to enhance nutrient use efficiency and productivity of crop plants. *Nanomaterials and plant potential*, 473-505.

Ismail, L. M., Soliman, M. I., Abd El-Aziz, M. H., & Abdel-Aziz, H. M. (2022). Impact of silica ions and nano silica on growth and productivity of pea plants under salinity stress. *Plants*, *11*(4), 494. DOI: 10.3390/plants11040494 PMID: 35214827

Jain, S. K., Wettberg, E. J. V., Punia, S. S., Parihar, A. K., Lamichaney, A., Kumar, J., Gupta, D. S., Ahmad, S., Pant, N. C., Dixit, G. P., Sari, H., Sari, D., Ma'ruf, A., Toker, P., & Toker, C. (2023). Genomic-mediated breeding strategies for global warming in chickpeas (Cicer arietinum L.). *Agriculture*, *13*(9), 1721. DOI: 10.3390/agriculture13091721

Javed, R., Ain, N. U., Gul, A., Arslan Ahmad, M., Guo, W., Ao, Q., & Tian, S. (2022). Diverse biotechnological applications of multifunctional titanium dioxide nanoparticles: An up-to-date review. *IET Nanobiotechnology / IET*, *16*(5), 171–189. DOI: 10.1049/nbt2.12085 PMID: 35411585

Jiang, M., Song, Y., Kanwar, M. K., Ahammed, G. J., Shao, S., & Zhou, J. (2021). Phytonanotechnology applications in modern agriculture. *Journal of Nanobiotechnology*, *19*(1), 1–20. DOI: 10.1186/s12951-021-01176-w PMID: 34930275

Joseph, T. M., Al-Hazmi, H. E., Śniatała, B., Esmaeili, A., & Habibzadeh, S. (2023). Nanoparticles and nanofiltration for wastewater treatment: From polluted to fresh water. *Environmental Research*, *238*, 117114. DOI: 10.1016/j.envres.2023.117114 PMID: 37716387

Kah, M., Beulke, S., Tiede, K., & Hofmann, T. (2013). Nanopesticides: State of knowledge, environmental fate, and exposure modeling. *Critical Reviews in Environmental Science and Technology*, *43*(16), 1823–1867. DOI: 10.1080/10643389.2012.671750

Kale, S. K., Parishwad, G. V., & Patil, A. S. H. A. S. (2021). Emerging agriculture applications of silver nanoparticles. *ES Food & Agroforestry*, *3*, 17–22.

Khan, M. N., Mobin, M., Abbas, Z. K., & Alamri, S. A. (2018). Fertilizers and their contaminants in soils, surface and groundwater. *Encyclopedia of the Anthropocene*, *5*, 225-240.

Korah, B. K., Chacko, A. R., Mathew, S., & Mathew, B. (2022). Carbon Nanotubes as Sensors in Food and Agricultural Applications. In *Carbon Nanotubes for Energy and Environmental Applications* (pp. 115–159). Apple Academic Press. DOI: 10.1201/9781003189619-6

Kucuk, N., Primožič, M., Knez, Z., & Leitgeb, M. (2023). Sustainable biodegradable biopolymer-based nanoparticles for healthcare applications. *International Journal of Molecular Sciences*, *24*(4), 3188. DOI: 10.3390/ijms24043188 PMID: 36834596

Kumari, R., Suman, K., Karmakar, S., Mishra, V., Lakra, S. G., Saurav, G. K., & Mahto, B. K. (2023). Regulation and safety measures for nanotechnology-based agri-products. *Frontiers in Genome Editing*, *5*, 1200987. DOI: 10.3389/fgeed.2023.1200987 PMID: 37415849

Machado, T. O., Grabow, J., Sayer, C., de Araújo, P. H., Ehrenhard, M. L., & Wurm, F. R. (2022). Biopolymer-based nanocarriers for sustained release of agrochemicals: A review on materials and social science perspectives for a sustainable future of agri-and horticulture. *Advances in Colloid and Interface Science*, *303*, 102645. DOI: 10.1016/j.cis.2022.102645 PMID: 35358807

Manikandan, V., & Min, S. C. (2023). Roles of polysaccharides-based nanomaterials in food preservation and extension of shelf-life of food products: A review. *International Journal of Biological Macromolecules*, *252*, 126381. DOI: 10.1016/j.ijbiomac.2023.126381 PMID: 37595723

Marcon, L., Oliveras, J., & Puntes, V. F. (2021). In situ nanoremediation of soils and groundwaters from the nanoparticle's standpoint: A review. *The Science of the Total Environment*, *791*, 148324. DOI: 10.1016/j.scitotenv.2021.148324 PMID: 34412401

Mittal, D., Kaur, G., Singh, P., Yadav, K., & Ali, S. A. (2020). Nanoparticle-based sustainable agriculture and food science: Recent advances and future outlook. *Frontiers in Nanotechnology*, *2*, 579954. DOI: 10.3389/fnano.2020.579954

Mostafavi, E., & Zare, H. (2022). Carbon-based nanomaterials in gene therapy. *OpenNano*, *7*, 100062. DOI: 10.1016/j.onano.2022.100062

Mujtaba, M., Wang, D., Carvalho, L. B., Oliveira, J. L., Espirito Santo Pereira, A. D., Sharif, R., Jogaiah, S., Paidi, M. K., Wang, L., Ali, Q., & Fraceto, L. F. (2021). Nanocarrier-mediated delivery of miRNA, RNAi, and CRISPR-Cas for plant protection: Current trends and future directions. *ACS Agricultural Science & Technology*, *1*(5), 417–435. DOI: 10.1021/acsagscitech.1c00146

Munir, N., Gulzar, W., Abideen, Z., Hancock, J. T., El-Keblawy, A., & Radicetti, E. (2023). Nanotechnology improves disease resistance in plants for food security: Applications and challenges. *Biocatalysis and Agricultural Biotechnology*, *51*, 102781. DOI: 10.1016/j.bcab.2023.102781

Nair, R., Varghese, S. H., Nair, B. G., Maekawa, T., Yoshida, Y., & Kumar, D. S. (2010). Nanoparticulate material delivery to plants. *Plant Science*, *179*(3), 154–163. DOI: 10.1016/j.plantsci.2010.04.012

Oladosu, Y., Rafii, M. Y., Arolu, F., Chukwu, S. C., Salisu, M. A., Fagbohun, I. K., Muftaudeen, T. K., Swaray, S., & Haliru, B. S. (2022). Superabsorbent polymer hydrogels for sustainable agriculture: A review. *Horticulturae*, *8*(7), 605. DOI: 10.3390/horticulturae8070605

Pandit, M. A., Bhardwaj, K., & Kaur, J. (2023). Microbial nanotechnology: a potential tool for a sustainable environment. In *Environmental applications of microbial nanotechnology* (pp. 217–230). Elsevier.

Parisi, C., Vigani, M., & Rodríguez-Cerezo, E. (2015). Agricultural nanotechnologies: What are the current possibilities? *Nano Today*, *10*(2), 124–127. DOI: 10.1016/j.nantod.2014.09.009

Pascoli, M., Lopes-Oliveira, P. J., Fraceto, L. F., Seabra, A. B., & Oliveira, H. C. (2018). State of the art of polymeric nanoparticles as carrier systems with agricultural applications: A minireview. *Energy, Ecology & Environment*, *3*(3), 137–148. DOI: 10.1007/s40974-018-0090-2

Patra, J. K., Das, G., Fraceto, L. F., Campos, E. V. R., Rodriguez-Torres, M. D. P., Acosta-Torres, L. S., Diaz-Torres, L. A., Grillo, R., Swamy, M. K., Sharma, S., Habtemariam, S., & Shin, H.-S. (2018). Nano based drug delivery systems: Recent developments and future prospects. *Journal of Nanobiotechnology*, *16*(1), 1–33. DOI: 10.1186/s12951-018-0392-8 PMID: 30231877

Perlatti, B., de Souza Bergo, P. L., Fernandes, J. B., & Forim, M. R. (2013). Polymeric nanoparticle-based insecticides: a controlled release purpose for agrochemicals. In *Insecticides-Development of safer and more effective technologies*. IntechOpen. DOI: 10.5772/53355

Rai, M., Ingle, A. P., Birla, S., Yadav, A., & Santos, C. A. D. (2016). Strategic role of selected noble metal nanoparticles in medicine. *Critical Reviews in Microbiology*, *42*(5), 696–719. PMID: 26089024

Rani, N., Kumari, K., & Hooda, V. (2025). The role of nanoparticles in transforming plant genetic engineering: Advancements, challenges and future prospects. *Functional & Integrative Genomics*, *25*(1), 1–22. DOI: 10.1007/s10142-025-01528-x PMID: 39841261

Rastogi, A., Tripathi, D. K., Yadav, S., Chauhan, D. K., Zivcak, M., Ghorbanpour, M., et al. (2019). Application of silicon nanoparticles in agriculture. *3 Biotech, 9*, 1-11.

Rehman, A., Qunyi, T., Usman, M., Manzoor, M. F., Zhao, L., & Feng, J.. (2022). Nano-enabled agrochemicals for sustainable agriculture. In *Nano-enabled agrochemicals in agriculture* (pp. 291–306). Academic Press. DOI: 10.1016/B978-0-323-91009-5.00020-3

Rico, C. M., Majumdar, S., Duarte-Gardea, M., Peralta-Videa, J. R., & Gardea-Torresdey, J. L. (2011). Interaction of nanoparticles with edible plants and their possible implications in the food chain. *Journal of Agricultural and Food Chemistry*, *59*(8), 3485–3498. DOI: 10.1021/jf104517j PMID: 21405020

Rico, C. M., Ofoegbu, P. C., Kirwa, N. J., Wagner, D. C., Abolade, O. M., & Jia, S.. (2022). Changes in metabolite profile of plants exposed to engineered nanomaterials. In *Plant Exposure to Engineered Nanoparticles* (pp. 119–149). Academic Press. DOI: 10.1016/B978-0-323-85032-2.00003-8

Riseh, R. S., Tamanadar, E., Hajabdollahi, N., Vatankhah, M., Thakur, V. K., & Skorik, Y. A. (2022). Chitosan microencapsulation of rhizobacteria for biological control of plant pests and diseases: Recent advances and applications. *Rhizosphere*, *23*, 100565. DOI: 10.1016/j.rhisph.2022.100565

Rudakiya, D., Patel, Y., Chhaya, U., & Gupte, A. (2019). Carbon nanotubes in agriculture: production, potential, and prospects. *Nanotechnology for Agriculture: Advances for Sustainable Agriculture*, 121-130.

Saini, R. K., Patel, S., Bajpai, J., & Bajpai, A. K. (2020). Advanced controlled nanopesticide delivery systems for managing insect pests. *Controlled Release of Pesticides for Sustainable Agriculture*, 155-184.

Shakiba, S., Astete, C. E., Paudel, S., Sabliov, C. M., Rodrigues, D. F., & Louie, S. M. (2020). Emerging investigator series: polymeric nanocarriers for agricultural applications: synthesis, characterization, and environmental and biological interactions. *Environmental Science. Nano*, *7*(1), 37–67. DOI: 10.1039/C9EN01127G

Shaw, D. S., & Honeychurch, K. C. (2022). Nanosensor applications in plant science. *Biosensors (Basel)*, *12*(9), 675. DOI: 10.3390/bios12090675 PMID: 36140060

Shelar, A., Nile, S. H., Singh, A. V., Rothenstein, D., Bill, J., Xiao, J., Chaskar, M., Kai, G., & Patil, R. (2023). Recent advances in nano-enabled seed treatment strategies for sustainable agriculture: Challenges, risk assessment, and future perspectives. *Nano-Micro Letters*, *15*(1), 54. DOI: 10.1007/s40820-023-01025-5 PMID: 36795339

Shishir, M. R. I., Gowd, V., Suo, H., Wang, M., Wang, Q., Chen, F., & Cheng, K.-W. (2021). Advances in smart delivery of food bioactive compounds using stimuli-responsive carriers: Responsive mechanism, contemporary challenges, and prospects. *Comprehensive Reviews in Food Science and Food Safety*, *20*(6), 5449–5488. DOI: 10.1111/1541-4337.12851 PMID: 34668321

Singh, A., Dhiman, N., Kar, A. K., Singh, D., Purohit, M. P., Ghosh, D., & Patnaik, S. (2020). Advances in controlled release pesticide formulations: Prospects to safer integrated pest management and sustainable agriculture. *Journal of Hazardous Materials*, *385*, 121525. DOI: 10.1016/j.jhazmat.2019.121525 PMID: 31740313

Singh, G., Ramadass, K., Sooriyakumar, P., Hettithanthri, O., Vithange, M., Bolan, N., Tavakkoli, E., Van Zwieten, L., & Vinu, A. (2022). Nanoporous materials for pesticide formulation and delivery in the agricultural sector. *Journal of Controlled Release*, *343*, 187–206. DOI: 10.1016/j.jconrel.2022.01.036 PMID: 35090962

Singh, S. (2025). Optimizing Nutrient Uptake: Nano-Fertilizers and Soil Amendments. In *Food and Industry 5.0: Transforming the Food System for a Sustainable Future* (pp. 229–248). Springer Nature Switzerland. DOI: 10.1007/978-3-031-76758-6_15

Slomberg, D. L., & Schoenfisch, M. H. (2012). Silica nanoparticle phytotoxicity to Arabidopsis thaliana. *Environmental Science & Technology*, *46*(18), 10247–10254. DOI: 10.1021/es300949f PMID: 22889047

Smith, A. M., & Gilbertson, L. M. (2018). Rational ligand design to improve agrochemical delivery efficiency and advance agriculture sustainability. *ACS Sustainable Chemistry & Engineering*, *6*(11), 13599–13610. DOI: 10.1021/acssuschemeng.8b03457

Solanki, P., Bhargava, A., Chhipa, H., Jain, N., & Panwar, J. (2015). Nano-fertilizers and their smart delivery system. *Nanotechnologies in food and agriculture*, 81-101.

Solomon, N. O., Kanchan, S., & Kesheri, M. (2024). Nanoparticles as detoxifiers for industrial wastewater. *Water, Air, and Soil Pollution*, *235*(3), 214. DOI: 10.1007/s11270-024-07016-5

Sridhar, A., Ponnuchamy, M., Kumar, P. S., & Kapoor, A. (2021). Food preservation techniques and nanotechnology for increased shelf life of fruits, vegetables, beverages and spices: A review. *Environmental Chemistry Letters*, *19*(2), 1715–1735. DOI: 10.1007/s10311-020-01126-2 PMID: 33192209

Subramanian, K. S., Manikandan, A., Thirunavukkarasu, M., & Rahale, C. S. (2015). Nano-fertilizers for balanced crop nutrition. *Nanotechnologies in food and agriculture*, 69-80.

Tao, R., You, C., Qu, Q., Zhang, X., Deng, Y., Ma, W., & Huang, C. (2023). Recent advances in the design of controlled-and sustained-release micro/nanocarriers of pesticide. *Environmental Science. Nano*, *10*(2), 351–371. DOI: 10.1039/D2EN00446A

Tawiah, B., Ofori, E. A., & George, S. C. (2024). Nanotechnology in societal development. In *Nanotechnology in Societal Development* (pp. 1–64). Springer Nature Singapore. DOI: 10.1007/978-981-97-6184-5_1

Tortorici, S., Cimino, C., Ricupero, M., Musumeci, T., Biondi, A., Siscaro, G., Carbone, C., & Zappalà, L. (2022). Nanostructured lipid carriers of essential oils as potential tools for the sustainable control of insect pests. *Industrial Crops and Products*, *181*, 114766. DOI: 10.1016/j.indcrop.2022.114766

Tripathi, D., Singh, M., & Pandey-Rai, S. (2022). Crosstalk of nanoparticles and phytohormones regulate plant growth and metabolism under abiotic and biotic stress. *Plant Stress*, *6*, 100107. DOI: 10.1016/j.stress.2022.100107

Usman, M., Farooq, M., Wakeel, A., Nawaz, A., Cheema, S. A., Rehman, H., Ashraf, I., & Sanaullah, M. (2020). Nanotechnology in agriculture: Current status, challenges and future opportunities. *The Science of the Total Environment*, *721*, 137778. DOI: 10.1016/j.scitotenv.2020.137778 PMID: 32179352

Vega-Vásquez, P., Mosier, N. S., & Irudayaraj, J. (2020). Nanoscale drug delivery systems: From medicine to agriculture. *Frontiers in Bioengineering and Biotechnology*, *8*, 79. DOI: 10.3389/fbioe.2020.00079 PMID: 32133353

Wang, L., Ning, C., Pan, T., & Cai, K. (2022). Role of silica nanoparticles in abiotic and biotic stress tolerance in plants: A review. *International Journal of Molecular Sciences*, *23*(4), 1947. DOI: 10.3390/ijms23041947 PMID: 35216062

Wang, X., Xie, H., Wang, P., & Yin, H. (2023). Nanoparticles in plants: Uptake, transport and physiological activity in leaf and root. *Materials (Basel)*, *16*(8), 3097. DOI: 10.3390/ma16083097 PMID: 37109933

Wani, T. A., Masoodi, F. A., Baba, W. N., Ahmad, M., Rahmanian, N., & Jafari, S. M. (2019). Nanoencapsulation of agrochemicals, fertilizers, and pesticides for improved plant production. In *Advances in Phytonanotechnology* (pp. 279–298). Academic Press. DOI: 10.1016/B978-0-12-815322-2.00012-2

Yadav, A., Yadav, K., Ahmad, R., & Abd-Elsalam, K. A. (2023). Emerging frontiers in nanotechnology for precision agriculture: Advancements, hurdles and prospects. *Agrochemicals*, *2*(2), 220–256. DOI: 10.3390/agrochemicals2020016

Yuvaraj, M., Subramanian, K. S., & Cyriac, J. (2023). Efficiency of zinc oxide nanoparticles as controlled release nanofertilizer for rice (*Oryza sativa* L). *Journal of Plant Nutrition*, *46*(18), 4477–4493. DOI: 10.1080/01904167.2023.2233561

Zaim, N. S. H. B. H., Tan, H. L., Rahman, S. M. A., Abu Bakar, N. F., Osman, M. S., Thakur, V. K., & Radacsi, N. (2023). Recent advances in seed coating treatment using nanoparticles and nanofibers for enhanced seed germination and protection. *Journal of Plant Growth Regulation*, *42*(12), 7374–7402. DOI: 10.1007/s00344-023-11038-4

Zhao, F., Xin, X., Cao, Y., Su, D., Ji, P., Zhu, Z., & He, Z. (2021). Use of carbon nanoparticles to improve soil fertility, crop growth and nutrient uptake by corn (*Zea mays* L.). *Nanomaterials (Basel, Switzerland)*, *11*(10), 2717. DOI: 10.3390/nano11102717 PMID: 34685156

Zhou, X. Q., Hayat, Z., Zhang, D. D., Li, M. Y., Hu, S., Wu, Q., Cao, Y.-F., & Yuan, Y. (2023). Zinc oxide nanoparticles: Synthesis, characterization, modification, and applications in food and agriculture. *Processes (Basel, Switzerland)*, *11*(4), 1193. DOI: 10.3390/pr11041193

Chapter 2
Metallic Nanocarriers in Plant Sciences:
Types and Applications

Sana Fatima
The Government Sadiq College Women University, Bahawalpur, Pakistan

Rabia Kousar
The Government Sadiq College Women University, Bahawalpur, Pakistan

Misbah Naeem
The Government Sadiq College Women University, Bahawalpur, Pakistan

Waiza Kalsoom
The Government Sadiq College Women University, Bahawalpur, Pakistan

Majid Anwar
The Government Sadiq College Women University, Bahawalpur, Pakistan

Ansar Mehmood
University of Poonch, Rawalakot, Pakistan

Khawaja Shafique Ahmad
https://orcid.org/0000-0001-8814-9878
University of Poonch, Rawalakot, Pakistan

ABSTRACT

The rapidly evolving field of metallic nanocarriers explores the use of metal-based nanoparticles (NPs) in plant science, providing innovative strategies for sustainable agriculture. These nanocarriers, typically 1 to 100 nm in size, exhibit high surface area, enhanced reactivity, and quantum effects, making them highly efficient in agricultural and plant protection applications. Metals are widely utilized in formulating metallic nanocarriers to enable the precise and controlled delivery of fertilizers, agrochemicals, and bioactive compounds. Their incorporation into controlled release systems enhances plant growth and strengthens resistance to pests and environmental stressors, reducing excessive chemical use. Additionally, metallic nanocarriers improve targeted delivery mechanisms and minimize environmental contamination. This chapter highlights their role in advancing precision agriculture, promoting sustainability, and reducing the ecological footprint of agrochemicals.

DOI: 10.4018/979-8-3693-9869-2.ch002

METALLIC NANO-CARRIERS IN PLANT SCIENCE

Nanotechnology is the field of science that focuses on the study of systems at the nanoscale. The word "nano" comes from the Latin word "nanus," meaning "dwarf" or "tiny." It represents a scale reduction factor of 109, according to the International System of Units (SI), which means that 1 nanometer (nm) is equal to 10^{-9} meters (Kumari and Sarkar, 2021).

"Nano" refers to a scale that extends down to the nanometer (nm) range. The normal size of nanoparticles (NPs), whether they be solid, liquid, or gas, ranges from 1 nm to 100 nm. NPs demonstrate unique properties due to effects such as surface effect, small size effect, quantum effect, and macroscopic quantum tunneling effect. (Hu and Xianyu, 2021).

Although "nanotechnology" has been used for a long time in many other industries, developments in nanoparticles now have a lot of potential for agriculture, which is a relatively new use. Environmental research, agriculture, and medicine have all benefited from recent advancements that provide exact control over particle size and form and, consequently, material qualities (Haris et al., 2023).

HISTORICAL BACKGROUND

American physicist Richard Feynman is known as the father of nanotechnology. Although he did not use the term "nanotechnology" itself, he outlined fundamental concepts of the field in his 1959 speech "There's Plenty of Room at the Bottom," which focused on the possibility that scientists may control and manage individual atoms and molecules. This idea set the stage for later developments in the discipline (Omran, 2020).

The invention of the scanning tunneling microscope in 1981 allowed scientists and engineers to view and work with individual atoms, laying the groundwork for contemporary nanotechnology. Gerd Binnig and Heinrich Rohrer, two IBM researchers who won the Nobel Prize in Physics in 1986 for their innovation, made this accomplishment. At the Binnig and Rohrer Nanotechnology Center in Zurich, Switzerland, which is committed to expanding research and creating cutting-edge applications in the field of nanotechnology, their contributions are still felt today (Moriarty, 2022).

SYNTHESIS OF NANOPARTICLES

According to Mazhar et al., (2017) numerous techniques have been developed for creating nanomaterials, which can be broadly divided into two categories (Fig. 1).

1.Top-Down Method

By using techniques like cutting, milling, or etching to reduce bigger, bulk materials into nanoscale particles, top-down synthesis produces nanoparticles. This method successfully reduces material size to achieve nanoscale dimensions.

2. Bottom-Up Method

Bottom-up synthesis builds up to the required structure from the atomic level by assembling nanoparticles atom by atom or molecule by molecule. By precisely controlling particle generation, this technique makes it possible to create distinct nanoscale structures.

Figure 1. Techniques of synthesis of nanoparticles

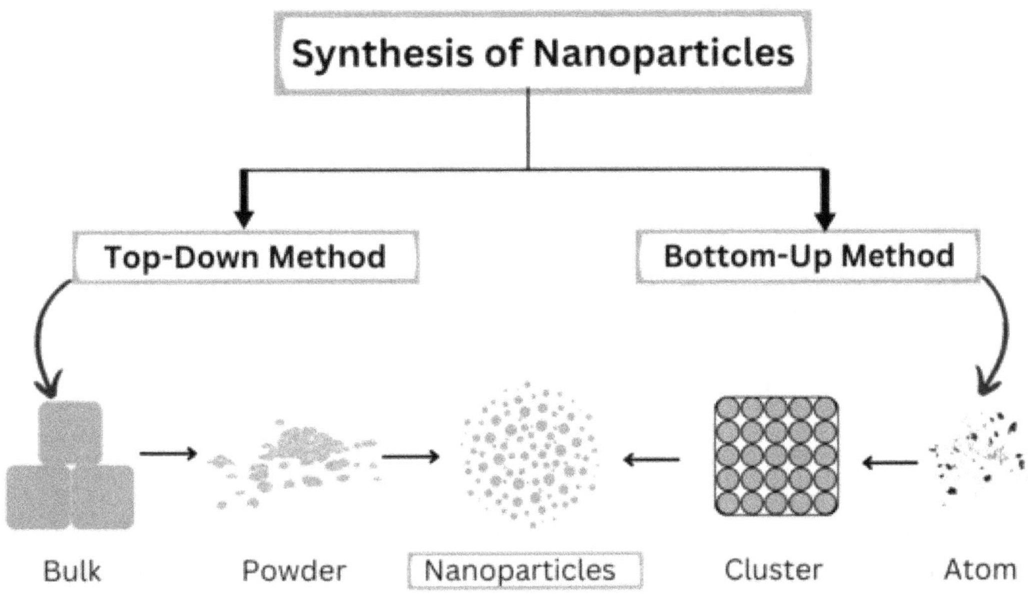

STRUCTURE OF NANOPARTICLES

According to Shin et al., (2016) two or three separate layers make up the complicated structure of nanoparticles (NPs). These consist of:

1. **Surface Layer**: A variety of tiny molecules, metal ions, surfactants, or polymers functionalize this outermost layer.
2. **Shell Layer**: Chemically distinct from the core, this layer can be purposefully added.
3. **Core Material**: The core of the nanoparticle, which is essential to its distinctive characteristics.

Since the core material largely determines the features of NPs, this core component is frequently the only basis for naming them.

NANOTECHNOLOGY IN SUSTAINABLE AGRICULTURE

Nanoparticles (NPs) have transformed the field of medicine and have significant applications in various other areas, including agriculture. According to a report from the Economic Times in 2019, food grain production in an agricultural economy like India is expected to reach 291.95 million tonnes (Pandey and Kumari, 2023). One of the United Nations' Sustainable Development Goals is to eradicate hunger and poverty. Nanomaterials, also known as engineered nanomaterials (ENMs), are utilized in fertilizers and pesticides (Ahmad et al., 2022).

In order to produce a range of crops for food and feed, agriculture is essential. The necessity for the agriculture industry to create effective methods to satisfy the nutritional needs of the growing human population is growing. Although crops can grow and produce at their best with traditional or chemical fertilizers reliance on these methods has raised serious issues (Singh et al., 2020).

Nanoencapsulation for Agrochemicals

Agrochemicals' rate of dissolution is slowed by nanoencapsulation, allowing for a progressive release that improves plant absorption. Compared to traditional bulk formulations, this approach lowers the overall quantity of agrochemicals required and helps to mitigate biotic stress. Furthermore, encapsulation shields medications, fertilizers, and other active ingredients from deterioration brought on by light, oxidation, and other adverse circumstances (Sinha et al., 2019).

Nanoformulations of Agrochemicals

Traditional plant protection techniques could be revolutionized by the application of nanotechnology in agriculture, especially when it comes to the targeted delivery of biomolecules like proteins, activators, and nucleotides and the regulated release of agrochemicals (Jiang et al., 2019).

Furthermore, there are several ways that nanotechnology can be used in agriculture, such as through the development of plant nanobionics and the use of nanopesticides, nanoherbicides, and nanofertilizers (Yadav et al., 2023).

UPTAKE, TRANSPORT AND DISTRIBUTION OF NANOPARTICLES IN PLANTS

Uptake of Nanoparticles in Plants

Pathways for Foliar Uptake

Nanoparticles are typically sprayed onto leaf surfaces in agricultural applications, where they settle and are then taken up by plants via their stomata or cuticle. As the initial natural barrier against the entry of nanoparticles, the waxy cuticle of the leaf epidermis, which is mainly made up of wax, cutin, and pectin, protects leaves by halting water loss during growth (Yang et al., 2015).

The surface of the waxy cuticle has two different kinds of channels: lipophilic and hydrophilic. These channels have widths between 0.6 and 4.8 nm. While lipophilic channels allow lipophilic nanoparticles to be absorbed into leaves by infiltration and diffusion, hydrophilic channels allow hydrophilic nanoparticles smaller than 4.8 nm to diffuse (Avellan et al., 2019). Using high-resolution confocal fluorescence microscopy, Hu et al., 2021 recently showed that carbon dots smaller than 2 nm could enter cotton leaves through the cuticular channel.

Due to the restricted size of pore channels in the cuticle, plant absorption of nanoparticles through the epidermis is limited. Research has shown that nanoparticles can accumulate in the leaf epidermis and vascular tissue after being applied to the leaf surface (Lv et al., 2019). Additionally, numerous studies have observed that nanoparticles can translocate to other plant tissues. Researchers suggest that nanoparticles may also be absorbed through an alternative route: the stomatal pathway.

Pathways for Root Uptake

Adsorption on the root surface is how nanoparticles initially contact plant roots. The root surface is negatively charged due to the discharge of mucus and organic acids by the root hairs, which increases the likelihood that positively charged nanoparticles will collect and stick to the root surface (Rajput et al., 2020). Furthermore, additional adsorption sites are produced by the growth of lateral roots, which raises the possibility that nanoparticles will enter the root cortex.

The root epidermis functions and is composed similarly to the leaf surface. Nevertheless, the primary and secondary roots' root hairs and the epidermis on their tips are still developing. Nanoparticles can immediately interact with and penetrate the root epidermis when they come into touch with these regions. Larger particles cannot pass through the semi-permeable cell walls of root epidermal cells because of their tiny pores (Ali et al., 2021).

Nanoparticles can enter the central column or xylem of roots without an exodermis. According to some research, some nanoparticles may be able to break down the plasma membrane, opening up new holes in the cell wall of the epidermis so that bigger nanoparticles can enter. Ion channels, endocytosis, attachment to cell membrane proteins, and physical disruption are some of the ways that nanoparticles can be absorbed by plant cells once they are inside plant tissue (Rani et al., 2022)

Studies have shown that nanoparticles are primarily absorbed by plant roots and enter cells through the hydrophilic pathway. However, due to the small pore size, this route restricts the entry of nanoparticles into cells (Li et al., 2020). Another key absorption method for nanoparticles in plant cells is endocytosis, where the plasma membrane invaginates to internalize nanoparticles into the cell (Sun et al., 2021).

According to research, plant protoplasts have the ability to endocytose particles smaller than 1 μm, which enables the absorption of nanoparticles without regard to particle size selectivity. Liu et al., (2009) proposed that endocytosis is the mechanism by which *Catharanthus roseus* root cells absorb carbon-based nanoparticles and carbon nanotubes. By attaching themselves to transport proteins on the outer epidermis, nanoparticles can also penetrate plants (Grillo et al., 2021).

Transport and Distribution of Nanoparticles in Leaves

Transport Pathway of Nanoparticles in Leaves

Nanoparticles can travel vast distances within the plant by either the extracellular or plastid pathways after entering mesophyll cells through the cuticle and stomata. Depending on particle size and surface charge, the extracellular pathway enables nanoparticles to migrate through the extracellular space and along structures such as cell walls, intercellular channels, the middle lamella, and the xylem (Parkinson et al., 2022). The protoplast route, on the other hand, transports materials mostly through plasmodesmata, which are intercellular channels with a diameter of 2–20 nm. Following their passage through plasmodesmata, nanoparticles gather in the cytoplasm and proceed in the direction of the Casparian strip and endodermal layer (Miralles et al., 2012).

According to research, the plastid pathway is usually responsible for the movement of nanoparticles smaller than 50 nm within plants, whereas the apoplast pathway is mostly responsible for the movement of nanoparticles between 50 and 200 nm (Rallia et al., 2018). The vascular system of the plant, more especially the phloem transport channels, then transports the nanoparticles from leaf to root. Particles up to around 0.405 μm in size can be effectively carried via the phloem because of the comparatively wide diameter of phloem sieve tubes (Bussières, 2014).

Distribution of Nanoparticles Absorbed via Leaves

Nanoparticles applied to leaves have been shown in numerous experiments to be carried to the stems, roots, flowers, fruits, and even the rhizosphere soil (Sun et al., 2021). For example, deposits on the leaf epidermis were discovered when CuO nanoparticles were sprayed on maize leaves. CuO nanoparticles were transported between cells by plasmodesmata after entering the cells (Adhikari et al., 2015).

In another study, watermelon leaves were sprayed with gold nanoparticles, and ICP-MS verified their absorption and buildup in the leaf cells 48 hours later. Additionally, gold was found in the watermelon's stems and roots, suggesting that the phloem carried the nanoparticles from the leaves to the roots (Raliya et al., 2016). Furthermore, Hong et al. (2014) revealed that CeO_2 nanoparticles were detected in cucumber root slices using transmission electron microscopy following a 72-hour exposure of cucumber leaves to the nanoparticles. This offered concrete proof that CeO_2 nanoparticles were transported from the cucumber's leaves to its stems and roots (Hong et al., 2014).

Porous hollow silica nanoparticles loaded with azoline oxalate, were created by Bueno et al., 2022 and applied to tomato plants. On the second and fourth days after treatment, the amount of silicon in different plant sections was identified and measured. According to the findings, the order of silicon levels in the tomato tissues was stem root, young leaf to mature leaf, suggesting that the transport efficiency of PHSNs differed among tissues. Furthermore, the study discovered that nanoparticles moved more efficiently from the stem to the root than they did across different leaves (Bueno et al., 2022).

METALLIC NANOPARTICLES IN AGRICULTURE

Numerous metals, such as aluminum (Al), gold (Au), silver (Ag), cadmium (Cd), cobalt (Co), copper (Cu), iron (Fe), lead (Pb), and zinc (Zn), can be used to create metal-based nanoparticles. The most widely used of these are silver, gold, copper, iron, and zinc. Because of their partially filled d-orbitals, which increase their redox activity, transition metals are especially well-suited for creating metal-based nanoparticles (Elena et al., 2020). This characteristic encourages the agglomeration of nanoparticles. The average size range of metal-based nanoparticles is 10–100 nm, and they can be spherical or cylindrical in shape. A high surface area-to-volume ratio, a variety of pore diameters, surface charge, surface charge density, and crystalline or amorphous morphologies are some of its distinctive features. They also exhibit increased sensitivity and responsiveness (Kumari and Sarkar, 2021).

Since ancient times, metals have been used to treat a variety of illnesses in humans, animals, and plants. Metallic nanoparticles are used as antimicrobial agents and are efficient substitutes for synthetic fungicides due to their remarkable physicochemical characteristics, which include nanoscale dimensions, a high surface-to-volume ratio, structural stability, and specific affinity for targets (Aziz et al., 2016).

Plant research has made extensive use of metallic nanoparticles (NPs), including gold (Au) and silver (Ag) NPs. Conventional chemical synthesis techniques frequently use dangerous chemicals and can be costly (Rastogi et al., 2019). Nevertheless, new ecologically friendly techniques have surfaced, including ionizing radiation in aqueous solutions and plant extracts. Furthermore, because of their superior electrical, catalytic, and light-absorbing qualities, oxidized nanomaterials such as MgO, CaO, ZnO, and TiO_2 have been suggested more and more. Because of their biocompatibility, affordability, and responsiveness to external stimuli, polymeric nanoparticles have garnered increasing attention in recent years (Baskar et al., 2018). Additionally, core/shell NPs are accessible and producible.

Properties and Characterization of Metallic Nanoparticles

In contrast to their bulk counterparts, metallic nanoparticles are among the most studied because of their ease of synthesis, high surface-to-volume ratio, varied optical properties, distinct surface chemistry, and ease of functionalization (Mondal et al., 2021). The optical characteristics of NMNPs are greatly improved by localized surface plasmon resonance, which is the oscillation of free electrons brought on by light.

Behavior MNPs in Agricultural Soil

Direct exposure and indirect exposure are the two main ways that MNPs usually reach the soil ecosystem. When direct exposure occurs, active nano-agrochemicals are sprayed on the soil as antimicrobial solutions, nanopesticides, nanoherbicides, or nanofertilizers (Lv *et al.,* 2019). By shielding crops against insects, weeds, pests, and plant diseases, these nano-agrochemicals eventually increase crop productivity and health (Chen, 2018).

Types of Metallic Nanoparticles

Metallic nanoparticles are nanoscale metals that fall between 1 to 100 nm in length, dimension, and thickness. Faraday initially looked into the presence of metallic nanoparticles in solution in 1857. Mie provided a quantitative explanation of their colour in 1908. These nanomaterials can now be made and altered with different chemical functional groups, enabling them to attach to ligands, antibodies, and medications. Metallic nanoparticles have several uses in biotechnology, medicine, and as medication and gene delivery vehicles.

Metallic nanoparticles are typically solid colloidal metal particles between 10 and 1000 nm in size. They contain a therapeutic molecule that is either entrapped in a polymer shell, dispersed within the polymer carrier matrix, or adsorbed or attached to the particle surface by a covalent bond. Recently, metallic nanoparticle probes have become innovative contrast and tracking agents for a range of cancer therapies. Researchers have shown that metallic nanoparticles have several benefits for the diagnosis and treatment of a wide variety of illnesses. They offer a high surface area-to-volume ratio that aids in chemical modification, may improve their penetration into cellular compartments, shields the medications from the hostile biological environment, and enhances the therapeutic moieties' stability and bioavailability.

Metal Nanoparticles Formation

To investigate the distinct chemical and physical characteristics of nanoparticles, regulated synthesis of nanoparticle size, shape, and monodispersity is crucial in the study of nanotechnology. Several physical and chemical synthesis techniques are now being used to regulate the size and dispersion of NPs (Huang and Yang, 2005). However, the majority of them use costly and hazardous chemicals, and issues with nanoparticle stability, particle aggregation, and the inability to regulate crystal development occur frequently.

Exploring the distinct chemical and physical properties of nanoparticles in the subject of nanotechnology requires the controlled synthesis of nanoparticle size, shape, and monodispersity (Govender et al., 2010). The size and variation of NPs can currently be controlled using a variety of chemical and physical synthesis approaches.

Problems with nanoparticle stability, particle aggregation, and the inability to regulate crystal development are common, though, and the majority of them use complex and hazardous chemicals. As a specific type of nanoparticle, MNPs can be synthesized practically using chemical, physical, and biological processes. An innovative technique for creating different nanoparticles in an effort to regulate their size, shape, composition, and monodispersity is the use of biological systems. The previously mentioned "green biosynthesis," which uses vascular plants (mostly leaf broth) as well as algae, bacteria, yeasts, fungi, and actinomycetes, is one of the most significant pathways for the manufacture of MNPs.

The regulated synthesis of nanoparticle size, shape, and monodispersity is crucial in the study of nanotechnology (including metal nanoparticles) in order to investigate their particular chemical and physical properties. The size and dispersion of nanoparticles can be controlled using a variety of physical and chemical techniques, as was previously noted. For instance, Fe nanoparticles with diameters of 10–30 nm can be created by mechanical techniques including inert gas condensation, extreme plastic deformation, high-energy ball milling, and ultrasonic shot peeling (Li and Zhang, 2006).

Chemical techniques include pulse electrodeposition, chemical wet reduction, chemical coprecipitation, chemical vapour condensation, and microemulsion (Li and Zhang, 2006). Leaf reductants found in leaf extracts are the primary ingredient employed in biological approaches for the synthesis of metal nanoparticles Nonetheless, MNPs can also be directly produced in living plants through the reduction of ingested metal ions as a soluble salt, suggesting that plants are an appropriate medium for MNP formation (Luangpipat et al., 2011).

Figure 2. Metal nanoparticles formation methods

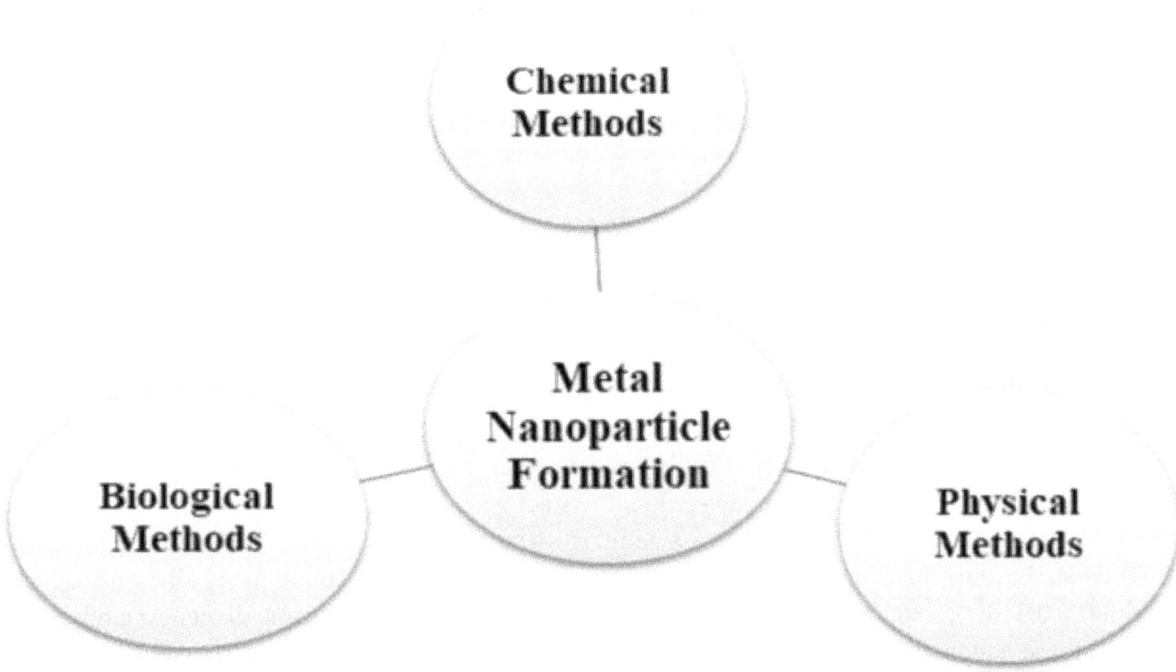

CHEMICAL AND PHYSICAL METHODS OF NANOPARTICLE SYNTHESIS

There are numerous approaches to synthesize metallic nanoparticles (Fig. 2). It is necessary to have a thorough understanding of the existing chemical and physical procedures in order to compare them and identify areas for improvement while studying biological ways of synthesizing nanoparticles (NPs).

Cube and tetrahedron-shaped silver nanoparticles (AgNPs) were successfully created. To do this, they heated ethylene glycol and $AgNO_3$ to 148°C. The resultant single-crystal nanoparticles ranged in diameter from 20 to 80 nm. Only ten minutes passed during the reaction, and the AgNPs were comparatively monodispersed (Wiley *et al.,* 2004). However, a considerable amount of energy would be required to produce commercial amounts due to the relatively high temperature required for the reaction to occur. AgNPs were successfully created utilizing a technique that used polyamide textiles (Montazer et al., 2013). The NPs had an average size of 90 nm and ranged in diameter from 20 to 150 nm, according to SEM observations. The boiling point temperature during the process was responsible for the few aggregates that were observed.

The produced NPs were discovered to be stabilized by the polyamide fabric itself, which was employed as the Ag+ reducing agent. This significant outcome rendered the use of further stabilising medications unnecessary. After that, the fabric's antibacterial activity was examined, and even after 30 washing cycles, it had strong bactericidal effects (Montazer et al., 2013). A physical technique has been used to create stable platinum nanoparticles (PtNPs), which eliminates the need for reducing chemicals that contaminate the final product (Mafuné et al., 2003).

The Different Types of Bio-Synthesized Metallic Nanoparticles

Silver

Because of its strong antibacterial activity against bacteria, viruses, and other eukaryotic microorganisms, silver nanoparticles have shown themselves to be the most effective (Gong *et al.,* 2007, Mahendra et al., 2009). Since they are without a doubt the most frequently utilised nanomaterials, they are utilised in the textile industry, as antimicrobial agents, for water treatment, sunscreen lotions, and other applications (Rai et al., 2009; Sharma et al., 2009). Research has already documented that plants including *Capsicum annuum*, *Carica papay*, and *Azadirachta indica* can successfully biosynthesis silver nanoparticles.

When geranium (*Pelargonium graveolens*) leaf extract was used to create silver nanoparticles, the particles formed quickly and stabilized at a size of 16–40 nm (Shankar et al., 2003). Safaepour et al., (2009) created evenly formed silver nanoparticles in the size range of 1–10 nm using silver nitrate and geraniol (C10H18O), a naturally occurring monoterpene alcohol present in some plants. At a dosage of 5 μg/mL, these nanoparticles were reported to inhibit the growth of the fibrosarcoma Wehi 164 cancer cell line in vitro by up to 60% (Safaepour *et al.,* 2009). According to Kaviya et al., (2011), *Crossandra infundibuliformis* leaf extract was used to create silver nanoflakes.

There are different types of nanoparticles which are illustrated in Fig. 3.

Gold

A leaf extract of *Coriandrum sativum* (coriander) was used by Narayanan and Sakthivel (2008) to create gold nanoparticles. The particles were spherical, triangular, and decahedral, and their sizes varied from roughly 7 to 58 nm. Edison and Sethuraman (2012) created gold nanoparticles varying in size from 6 to 60 nm using an aqueous extract of *Terminalia chebula.* Both Gram-positive *S. aureus* and Gram-negative *E. coli* were susceptible to the effects of these nanoparticles. Chrysanthemum and tea extracts used to create gold nanoparticles (Liu et al., 2012). To measure the antioxidant qualities of teas, an assay based on nanoparticles was created (Liu et al., 2012). Daisy and Saipriya (2012) used an aqueous extract of *Cassia fistula* to create gold nanoparticles (55–98 nm).

Palladium and Platinium Nanoparticles

In 2009, synthesized palladium nanoparticles using an extract from the bark of *C. zeylanicum*, or cinnamon (Sathishkumar et al., 2009). It was discovered that altering the temperature, reaction pH, and concentration of bark extract during synthesis had no effect on the morphology or particle size (15–20 nm). While soybean (*Glycine max*) leaf extract has been able to synthesize nanoparticles with a mean size of 15 nm (Kumar et al., 2012), *Annona squamosa* (Custard apple) peel extract has also been used to create palladium nanoparticles with a size range of 75 to 85 nm (Roopan et al., 2011). Additionally, palladium nanoparticles with faced-centered cubic crystal symmetry, ranging in size from 20 to 60 nm, have been synthesised using extracts from common commercial items such as coffee (*Coffea arabica*) and tea (*Camellia sinensis*) (Nadagouda et al., 2008).

Furthermore, it was shown that antioxidants such geniposide, chlorogenic acid, crocins, and crocetin functioned as both stabilising and reducing agents when an extract from *Gardenia jasminoides* (Cape jasmine) was utilised to synthesise palladium nanoparticles (Jia et al., 2009). Particle sizes ranged from 3 to 5 nm, according to further investigation, and the study also discovered that particle size was influenced by reaction temperature. Using a leaf extract from *Diospyros kaki* (Persimmon).

The first synthesis of platinium nanoparticles in 2010. With a 10% concentration of leaf biomass at 95 ˝C, 90% of the platinum ions in solution were transformed by the resulting nanoparticles, which ranged in size from 2 to 12 nm (Song et al., 2010). Platinum nanoparticles with a mean particle size of 23 nm have also been synthesised from aqueous chloroplatinic acid at a reaction temperature of 100 ˝C using a leaf extract from *Ocimum sanctum* (basil) (Soundarrajan et al., 2012). Additionally, it has recently been discovered that plant wood nanometer-scale materials can be used to biologically synthesise platinum nanoparticles with controllable particle size and shape (Lin *et al.*, 2011). For instance, recently described a one-pot synthesis method that uses lignin extracted from red pine (*Pinus resinosa*) to produce platinum and palladium nanoparticles (Coccia et al., 2012).

Copper Nanoparticles

Plant extracts of various kinds have been used to create copper (Cu) and copper oxide (CuO) nanoparticles. Cu nanoparticles with stable sizes between 40 and 100 nm have been produced naturally by employing extract from magnolia leaves. According to antimicrobial research, Cu nanoparticles may exhibit antibacterial properties against the common pathogen *Escherichia coli* cells (Lee et al., 2011). Extracts from *Syzygium aromaticum* (Clove) can yield Cu nanoparticles with a spherical to granular shape and a mean particle size of 40 nm (Subhankari et al., 2013).

Iron (Magnetic) Nanoparticles

It is well known that magnetic nanoparticles such as Fe_3O4 (magnetite) and Fe_2O3 (maghemite) are biocompatible. They have been actively studied for guided drug administration, gene therapy, magnetic resonance imaging (MRI), DNA analysis, stem cell sorting and manipulation, targeted cancer treatment (magnetic hyperthermia), and more.

Titanium Nanoparticles

Using certain plant extracts, titanium dioxide hydrate can be reduced to produce TiO_2 NP (Peralta-Videa et al., 2016). Velayutham et al., (2011) created titanium dioxide nanoparticles using a *Catharanthus roseus* leaf extract. The particles ranged in size from 25 to 110 nm and had an uneven form.

Figure 3. Different types of bio-synthesized metallic nanoparticles

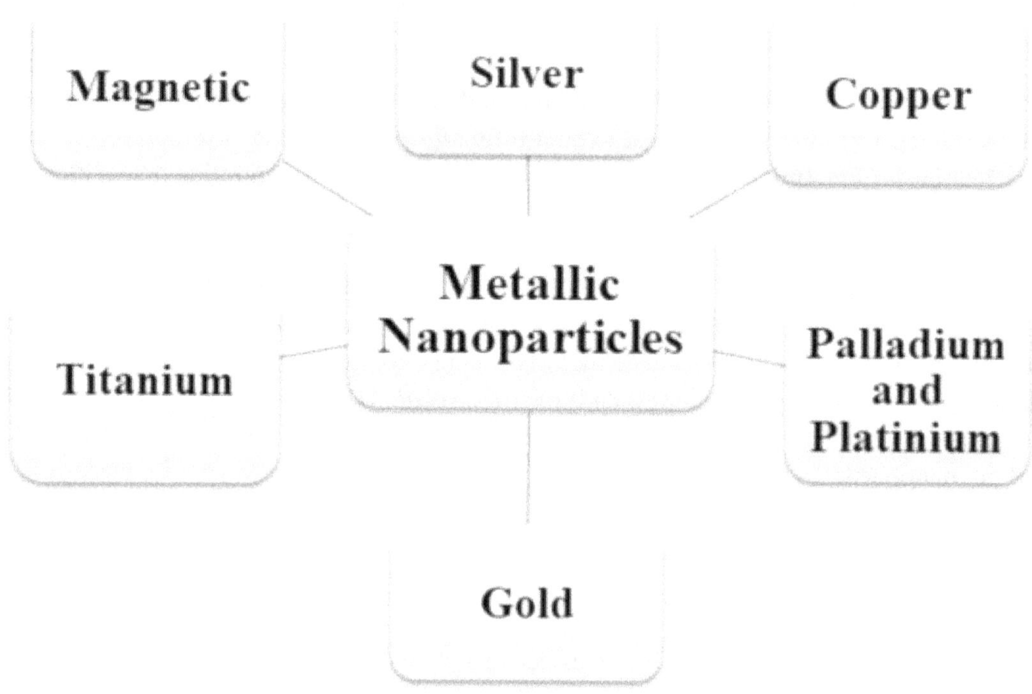

METALLIC NANOCARRIERS APPLICATIONS IN PLANT SCIENCES

Metallic nanoparticles provide a versatile platform for a range of uses. This section provides a summary of some of the most significant uses of these metallic nanoparticles in plant sciences.

Agriculture

In order to improve food production and security, nanotechnology has demonstrated its ability to help the agriculture industry by solving environmental and agricultural issues (Usman et al., 2020).

Nanomaterials' Effect on Plants

According to multiple research studies, seed germination is insensitive to NPs, and the U.S. Environmental Protection Agency has proposed root elongation and seed germination as two common hazardous markers (Stampoulis et al., 2009). Nano-growth stimulants help in seed germination and subsequent growth stages (Aslani et al., 2014). Nanoparticles' small size and huge surface area allow them to enter seed pores and eventually activate the phytohormones needed for seed germination and growth (Kasote et al., 2019). When applied to soybean seedlings, for instance, nano-TiO_2 and nano-SiO_2 increased nitrate reductase activity, which in turn boosted seed germination. Nevertheless, it was more beneficial to combine the two nanomaterials (NMs) (Lu et al., 2002).

In spinach (*Spinacia oleracea* L.), seed treatment with 0.25 percent TiO_2 increased the rate of photosynthesis and nitrogen uptake, leading to improved growth (Yang et al., 2006). Fe_2O_3 NPs enhanced germination and triggered plant growth and fruiting behavior in watermelon (*Citrullus lanatus*) seeds (Li et al., 2013). Tomato seed germination was enhanced by applying low-concentration SiO_2 NPs (Siddiqui et al., 2014). Suriyaprabha et al., (2012) claim that nano-SiO_2 improved seed germination by giving maize seeds better access to nutrients and improving the growth medium's pH and conductivity. Nano-SiO_2 improved seed germination and activated the antioxidant system under NaCl stress, according to studies by Haghighi et al., (2012) in tomatoes and Siddiqui et al., (2014) in squash.

In tobacco (*Nicotiana tabacum* L.), zinc as a ZnO nanomaterial led to an increase in calli growth and physiological markers. NM application may be harmful to plants (Mazaheri et al., 2020), despite the fact that nanoparticles have been routinely used to promote plant growth (Raliya & Tarafdar, 2013). However, the NM's dosage, size, duration, exposure, and kind (biological or synthetic) all affect its effects, both beneficial and detrimental (Noori et al., 2020).

Application of NM that has been chemically or physically produced has been shown to stimulate plant development in recent years; nevertheless, because it seeps into the soil, it poses a higher hazard to ecology and the environment (Anjum et al., 2016). Therefore, the safety and viability of green/biological NM have made them useful in agriculture (Irshad et al., 2021). Biologically produced Au, Ag, Ti, N, and Fe nanoparticles have been used more frequently, either as nanopesticides or as nanofertilizers (Hasan et al., 2020).

Application of Metallic Nanocarriers in the Field of Agriculture

Nanomaterials have been applied to agriculture in two main ways in recent years: either as nanopesticides to get rid of weeds, pests, and diseases that are inhibiting crop plant growth or as nanofertilizers to increase agricultural output. The importance of nanopesticides and fertilizers to modern farming methods will be discussed in this chapter.

1) Nanofertilizers

Numerous studies have documented the beneficial effects of nanomaterials (NMs) on a range of crops, and they have significant implications for sustainable agricultural crop production. One of the most practical ways to greatly boost agricultural productivity and support the world's continuously expanding population is to use cutting-edge nanotechnology in agriculture, including the production of fertilizer (Lal, 2008). Because of their enhanced surface and adhesion effects, as well as their enhanced characterization, absorption, and responsiveness, nanoparticles are used as fertilizers in agriculture (Qureshi et al., 2018). Nanofertilizers are fertilizers with a particle size of less than 100 nm that include macro- or micronutrients and are used to boost agricultural production.

According to Chipa (2017), nanofertilizers are nanomaterials that give developing plants one or more types of nutrients to boost their growth and increase yield. Two varieties of them are offered. On the one hand, the nanomaterials help plants grow and produce more by providing them with nutrients;

on the other hand, they are the carriers of nutrients and only help move and release nutrients; they are not utilized as a source of nutrients themselves (Liu and Lal 2015). The method that nanofertilizers are given to plants and how they are absorbed and stored by them both affect how efficient they are at increasing crop productivity.

Nanofertilizers can be applied above or below ground via irrigation or foliar spray to increase plant growth and output. Furthermore, seeds can be primed or supplemented with biosynthesized nanoparticles (Mittal et al., 2020). The type of plant and the nanoparticles' size, concentration, chemical makeup, stability, and rate of change following biological contact all affect how easily the nanoparticles are absorbed and accumulate to improve crop growth (Chen, 2018). Through the root epidermis and endodermis, nanofertilizers enter the xylem vessels and reach the plant's aerial parts. Furthermore, the phloem and leaf stomata allow these nanoparticle nutrients to be transported to various parts of the plant (Prasad et al., 2017).

a) **Silicon Nanoparticles**

According to Hussain et al., (2020), Si NPs (10 and 20 mg/L) applied manually successfully reduced the buildup of Cd in grains and enhanced yield. Nano-SiO_2 enhances proline accumulation, chlorophyll content, and fresh and dry weight of leaves under salinity stress. Increased proline accumulation, free amino acid content, nutritional content, and antioxidant enzyme activity as a result of nano-SiO_2, which enhances plants' ability to withstand abiotic stress (Kalteh et al., 2014). It has been demonstrated that NPs increase the synthesis of structural protection agents. For example, exogenous Si NPs increased the synthesis of organic acids and decreased the damage caused by Cd to the plant. Si NPs treatments reduced Cd and drought stress and enhanced wheat development (Khan et al., 2020).

According to Xie et al., (2011), nano-SiO_2 improves plant growth and development by increasing gas exchange and chlorophyll fluorescence parameters, including net photosynthetic rate, transpiration rate, stomatal conductance, PSII potential activity, effective photochemical efficiency, actual photochemical efficiency, electron transport rate, and photochemical quench.

Changbai larch (*Larix olgensis*) seedlings were subjected to an exogenous application of nano-SiO_2 by Bao-shan et al., (2004). They discovered that nano-SiO_2 enhanced the growth and quality of the seedlings, as evidenced by improvements in mean height, root collar diameter, main root length, and the number of lateral roots, as well as the induction of chlorophyll synthesis.

b) **Zinc Nanoparticles**

Zinc oxide nanoparticles (ZnONPs) have been shown in numerous studies to promote plant growth and development. According to studies by Prasad et al., (2012) in peanuts, Sedghi et al., (2013) in soybeans, Ramesh et al., (2014) in wheat, and Raskar and Laware (2014) in onions, seed germination was positively impacted by reduced ZnONP concentrations. Higher ZnONP dosages, however, hindered seed germination. The impact of NPs on germination varies from plant to plant and is dependent on NP concentrations. When de la Rosa et al., (2013) applied varying concentrations of ZnONPs to tomato, alfalfa, and cucumber, they discovered that only the germination of cucumber seeds was improved.

According to Raliya and Tarafdar (2013), ZnONPs significantly increased the biomass of *Cyamopsis tetragonoloba* plants, as well as their shoot and root growth, root area, chlorophyll and protein synthesis, rhizospheric microbial population, and phytase, acid phosphatase, and alkaline phosphatase activity in the rhizosphere of cluster beans. According to the Umavathi et al., study (2021), using ZnO nanoparticles in *Sesamum indicum* improved the plant's vegetative growth and seed germination. By improving the absorption of potassium and nitrogen, ZnO NPs can alleviate the drought stress in sorghum (Dimkpa et al., 2019).

c) Silver Nanoparticles

Boswellia ovaliofoliolata trees' seed germination and seedling growth were improved by biologically produced AgNPs (Savithramma et al., 2012). AgNPs enhanced the biochemical characteristics (chlorophyll, carbohydrate and protein contents, antioxidant enzymes) and growth profile (shoot and root length, leaf area) of *Brassica juncea*, common bean, and maize plants (Salama 2012). However, according to Gruyer et al., (2013), depending on the species of plant, AgNPs can have both positive and negative effects on root elongation. They found that while root length was suppressed in lettuce, it was boosted in barley.

When examined, plants' morphology and physiology are improved by the Ag ions of silver nanoparticles, increasing their resistance to worm, bacterial, and fungal infections (Gruyer et al., 2013). Numerous studies have shown that encouraging seed germination and plant growth requires the right concentrations of silver nanoparticles (Shelar et al., 2015). Furthermore, the efficiency of water and fertilizer use (Lu et al., 2002) as well as the concentration of chlorophyll and photosynthetic quantum efficiency have been improved (Hatami et al., 2013).

However, it was shown that high concentrations of the 25 nm silver nanoparticles had a harmful effect on *Oryza sativa* root cells, rupturing the cell wall and damaging the vacuoles (Mazumdar et al., 2011). Mirzajani et al., (2013) found that at low concentrations of up to 30 g/mL, the silver was unable to penetrate the root cells of *Oryza sativa*; but, at higher concentrations, the silver was successful in destroying the cell structure and having a detrimental effect. According to a number of studies, there is a direct correlation between the size of silver nanoparticles and their toxicity to plants; smaller nanoparticles are regularly proven to be more dangerous to plants than larger ones (Cvjetko et al., 2017).

d) Gold Nanoparticles

The relationship between gold nanoparticles (AuNPs) and plants has not been well investigated. According to Shah and Belozerova (2009), some researchers discovered that AuNPs cause toxicity in plants by impairing the function of aquaporins, a class of proteins that aid in the movement of a variety of molecules, including water. Nonetheless, AuNPs have been shown to enhance seed germination in *Boswellia ovalifoliolata* by Savithramma et al., (2012), Arora et al., (2012), and Barrena et al., (2009) in lettuce and cucumber. AuNPs increase plant height, leaf area, number of leaves, sugar content, and chlorophyll content, all of which increase crop output (Gopinath et al., 2014). Therefore, more research is necessary to examine the physiological and molecular mechanisms behind AuNPs' beneficial effects. According to Kumar et al., (2013), AuNPs significantly affect the antioxidant system and seed germination in *Arabidopsis thaliana*. They also change the expression of microRNAs, which control a number of physiological, metabolic, and morphological processes in plants.

e) Iron Nanoparticles

The use of iron oxide nanoparticles in agriculture is a new technology that has shown promise, although it still needs some refinement. By controlling the phytohormone levels and antioxidant enzyme activity in peanuts, for example, Fe_2O_3 nanoparticles enhanced development by increasing the amount of iron in the soil and allowing it to accumulate in plant cells (Rui et al., 2016). The two most common ways to apply iron oxide nanoparticles to plants are foliar application and soil saturation, typically as a source of Fe nutrition (Zia-ur-Rehman et al., 2018). In response to various environmental stressors, nanoparticles can also boost crop yields (Askary et al., 2017). By interacting with chemicals within plant cells, iron oxide nanoparticles can improve nutritional intake (Rahmatizadeh et al., 2019).

Numerous studies have documented improvements in plant growth metrics and dry matter material when iron oxide nanoparticles are applied to various crops. Iron oxide nanoparticles improved tomato plant development metrics (Shankramma et al., 2016). The comparable results were noted by Tawfik et al., (2021), who found that spraying iron oxide nanoparticles on moringa plants greatly improved the plants' growth performance, photosynthetic pigments, indole acetic acid, proline content, free amino acids, and total soluble sugars. According to Sebastian et al., (2019), Fe_3O_4 NPs reduced the mobility of Cd and other HMs in the soil. Numerous investigations have shown that nano Fe has a great deal of promise for use in fertilizers. Furthermore, it has been demonstrated that the use of low dose Fe_2O_3 NPs pre-soaking increased the growth of legume roots by 88–366% (Palchoudhury et al., 2018).

f) Titanium Nanoparticles

TiO_2 nanoparticles, for example, have been utilized in agriculture as fertilizers, growth-regulating agents, and antimicrobials. They have a lot of promise for boosting plant growth and preventing food poisoning in humans. TiO_2 nanoparticles kill a variety of plant and fruit diseases. Furthermore, in hydroponic cultures and under controlled conditions, TiO_2 nanoparticles can mineralize leftover contaminants, herbicides, and organic substances (Rodriguez-Gonzalez et al., 2019). Research has indicated that TiO_2 positively affects plant output and growth. Seed treatment with TiO_2 before planting, the study conducted by Zheng et al., (2005) demonstrated an increase in the dry weight, chlorophyll content, and photosynthetic rate of spinach plants.

Moreover, applying TiO_2 to *Zea mays* boosted their intake of micro and macronutrients, while greater concentrations reduced their dry biomass (Daghan et al., 2018). According to Mahmoodzadeh et al., (2013), titanium dioxide nanoparticle-treated canola plants exhibited improved radicle and plumule development as well as higher germination rates. According to Jaberzadeh et al., (2013), under conditions of water deficiency stress, TiO_2 NPs increased the growth and yield of wheat plants. According to Wang et al., (2010), the addition of TiO_2 NPs to the nutritional solution enhanced the activity of antioxidant enzymes in the tissues of maize. According to Singh et al., (2016), adding TiO_2 NPs to soil improved the physiological parameters and photosynthetic rate of soybeans, hence reducing the toxicity of Cd to soybean plants.

2) Nanopesticides

Sessile plants are vulnerable to a range of biotic and abiotic stressors, including diseases, pests, heat, drought, pollution, and more. These pressures have an impact on a plant's overall yield, either directly or indirectly. The detrimental impact of infections on a whole crop, which can result in famine like conditions, has been well investigated. To combat these biotic stresses, the use of pesticides has been recommended.

The effectiveness of insecticides in eliminating dangerous pests and diseases from agricultural fields has been demonstrated, protecting crops (Jampilek & Kralova, 2017). Pesticide use, however, has been shown to have negative impacts on both human health and the environment.

Because of this, several pesticides have been outlawed by national or international governments. Therefore, it is imperative that safe and effective insecticides be developed. Despite being a breakthrough, the extremely expensive cost of production has restricted the application of biopesticides. Through the development of nanopesticides, nanotechnology provides an improved and novel strategy (Sasson et al., 2007). A common example of a nanopesticide is nanostructured alumina (NSA). NSA functions as a negatively charged pesticide that interacts with insects' positively charged bodies to cause dehydration. In insects, dehydration causes the cuticle to separate, which ultimately ends in death (Mittal et al., 2020).

The antibacterial activity of TiO_2 and ZnO NPs against *X. perforans*, an incidental organism of tomato spot disease, was investigated by Paret et al., (2013). It has been demonstrated that Imidacloprid (IMI) is an effective systemic insecticide against a variety of sucking insects, including *Martianus dermestoides*; additionally, the use of nano-IMI, which is more photodegradable than the conventional formulation, enhances its efficacy and environmental safety (Guan et al., 2008). In a recent study, Zhao et al., (2020) examined the insecticidal efficacy of Cu NPs and found that, even at low doses, Cu NPs may enhance resistance against bollworms by upregulating exogenous microbial proteins in plant tissue. These results show that nanopesticides may effectively combat a variety of crop pests, making them a promising tool for future agricultural pest management strategies (Usman et al., 2020).

3) **Biosensors**

NMs have been utilized as "sensing materials" or to create biosensors in the agricultural, food, and crop biotechnology sectors (Chaudhry et al., 2018). Plants have been used to test a variety of nanosensor types, including plasmonic, carbon-based electrochemical, fluorescence resonance energy transfer (FRET)-based, and antibody-based nanosensors. Despite the fact that plant nanosensors are still in their infancy (Rai et al., 2012), intriguing studies have suggested using NMs as instruments for the detection and measurement of plant metabolic flux, pesticide residue in food, and bacterial, viral, and fungal diseases. The ability to quickly identify and precisely quantify bacteria, viruses, and fungi in plants makes NMs-based biosensors extremely promising (Duhan et al., 2017). Fluorescent silica nanoparticles and antibodies, for instance, were developed to diagnose bacterial spot disease in Solanaceae plants caused by *Xanthomonas axonopodis pv. vesicatoria* (Yao et al., 2009). In order to detect mycotoxin residues in corn, wheat, oats, and barley, smart nanosensors are also available. For example, the 4mycosensor is a competitive antibody-based assay that was successfully introduced to the market (Lattanzio and Nivarlet, 2017).

4) **Nanomaterials for Plant Genetic Engineering**

In plant cells, the cell walls act as a barrier to the entry of foreign biomolecules. Various techniques based on *Agrobacterium* transformation or biolistic procedures are employed globally for DNA delivery in plant cells in order to get over this obstacle and accomplish plant genetic transformation. These methods' limitations include their limited host range and significant plant damage, which frequently impede plant growth. Plant cell cultures have been the site of the majority of the groundbreaking research on nanomaterial-based plant genetic engineering.

For instance, it has been shown that Silicon Carbide-Mediated Transformation is an effective method for delivering DNA in several callus, including cotton, soybean, rice, maize, and tobacco (Lau et al., 2017). In addition to providing systemic protection to freshly emerged, unsprayed leaves on viral challenge 20 days following a single spray treatment in tobacco, the dsRNAs and/or their RNA breakdown products offer protection against the *Cauliflower Mosaic Virus* (CMV) in sprayed tobacco leaves (Mitter et al., 2017).

Cotton plants have successfully undergone a durable genetic alteration thanks to magnetic nanoparticles (MNPs). Without affecting pollen viability, the β-glucuronidase (GUS) reporter gene-MNP complex was magnetically inserted into cotton pollen grains. Exogenous DNA was successfully incorporated into the genome, efficiently expressed, and persistently transmitted in the offspring acquired by selfing in cotton transgenic plants created by pollination with magnetofected pollen (Zhao et al., 2017). Recently, the first and most promising method of genome editing using mesoporous silica nanoparticles (MSNs) was put forth. MSNs, which have *loxP* sites incorporated into chromosomal DNA, have been utilized as carriers to deliver Cre recombinase in *Zea mays* immature embryos. A successful genome editing was established when the *loxP* was appropriately recombined following the biolistic introduction of modified MSNs in plant tissues (Valenstein et al., 2013).

5) **Environmental Remediation**

Nanoparticles are crucial in the environmental sector (Fig. 4), because of their surface area to mass ratio, which allows them to bond with precipitates, debris, and heavy metals to purify water (Khan et al., 2019). The nanoparticles' absorbency, shape, and composition all affect this binding. Three distinct approaches are used to apply nanoparticles in the environmental field. First, avoiding pollution by using green chemistry to create environmentally friendly products (Wilson et al., 2009). The bioremediation of environmental pollutants comes in second (Benjamin et al., 2019). In order to detect changes in environmental phases, nanoparticles are employed as sensors (Abel et al., 2021).

An efficient photocatalytic agent for treating water is TiO_2 nanoparticles. It has been investigated to employ these nanoparticles to remove organic pollutants from a number of water reservoirs (Wu et al., 2011). FeNPs' ability to bioremediate heavy metals from water, including lead, mercury, arsenic, cadmium, and thallium, has drawn attention (Thomas et al., 2020). Photo-degradation by NiO and ZnO nanoparticles has also been achieved in addition to bioremediation (Bashir et al., 2019). Nanoparticles (10–50 nm) were responsible for the effective photo-degradation (Rogozea et al., 2017).

Figure 4. Application of metallic nanocarriers in plant sciences

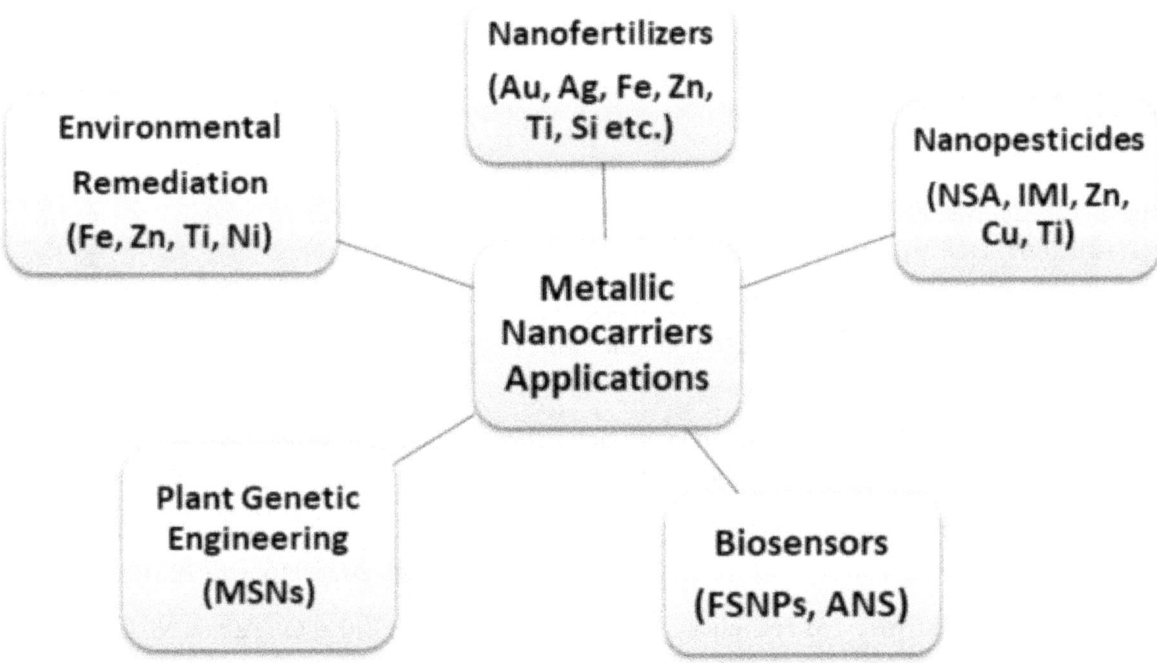

REFERENCES

Abel, S., Jule, L. T., Belay, F., Shanmugam, R., Dwarampudi, L. P., Nagaprasad, N., & Krishnaraj, R. (2021). Application of Titanium Dioxide Nanoparticles Synthesized by Sol-Gel Methods in Wastewater Treatment. *Journal of Nanomaterials*, *2021*(1), 3039761. DOI: 10.1155/2021/3039761

Adhikari, T., Kundu, S., Biswas, A. K., Tarafdar, J. C., & Rao, A. S. (2015). Characterization of Zinc Oxide Nano Particles and Their Effect on Growth of Maize (*Zea mays* L.) Plant. *Journal of Plant Nutrition*, *38*(10), 1505–1515. DOI: 10.1080/01904167.2014.992536

Ahmad, Z., Tahseen, S., Wasi, A., Ganie, I. B., Shahzad, A., Emamverdian, A., Ramakrishnan, M., & Ding, Y. (2022). Nanotechnological Interventions in Agriculture. *Nanomaterials (Basel, Switzerland)*, *2022*(12), 2667. DOI: 10.3390/nano12152667 PMID: 35957097

Ali, S., Mehmood, A., & Khan, N. (2021). Uptake, Translocation, and Consequences of Nanomaterials on Plant Growth and Stress Adaptation. *Journal of Nanomaterials*, *2021*, 6677616. DOI: 10.1155/2021/6677616

Alphandéry, E. (2020). Natural metallic nanoparticles for application in nano oncology. *International Journal of Molecular Sciences*, *21*(12), 4412. DOI: 10.3390/ijms21124412 PMID: 32575884

Altammar, K. A. (2023). A review on nanoparticles: Characteristics, synthesis, applications, and challenges. *Frontiers in Microbiology*, *14*, 1155622. DOI: 10.3389/fmicb.2023.1155622 PMID: 37180257

Anjum, N. A., Rodrigo, M. A. M., Moulick, A., Heger, Z., Kopel, P., Zítka, O., Adam, V., Lukatkin, A. S., Duarte, A. C., Pereira, E., & Kizek, R. (2016). Transport phenomena of nanoparticles in plants and animals/humans. *Environmental Research*, *151*, 233–243. DOI: 10.1016/j.envres.2016.07.018 PMID: 27504871

Arora, S., Sharma, P., Kumar, S., Nayan, R., Khanna, P. K., & Zaidi, M. G. H. (2012). Gold-nanoparticle induced enhancement in growth and seed yield of *Brassica juncea.Plant Growth Regulation*, *66*(3), 303–310. DOI: 10.1007/s10725-011-9649-z

Askary, M., Talebi, S. M., Amini, F., & Bangan, A. D. B. (2017). Effects of iron nanoparticles on *Mentha piperita* L. under salinity stress. *Biologija (Vilnius, Lithuania)*, *63*(1). Advance online publication. DOI: 10.6001/biologija.v63i1.3476

Aslani, F., Bagheri, S., Muhd Julkapli, N., Juraimi, A. S., Hashemi, F. S. G., & Baghdadi, A. (2014). Effects of engineered nanomaterials on plants growth: An overview. *TheScientificWorldJournal*, *2014*(1), 641759. DOI: 10.1155/2014/641759 PMID: 25202734

Avellan, A., Yun, J., Zhang, Y., Spielman-Sun, E., Unrine, J. M., Thieme, J., Li, J., Lombi, E., Bland, G., & Lowry, G. V. (2019). Nanoparticle Size and Coating Chemistry Control Foliar Uptake Pathways, Translocation, and Leaf-to-Rhizosphere Transport in Wheat. *ACS Nano*, *13*(5), 5291–5305. DOI: 10.1021/acsnano.8b09781 PMID: 31074967

Balla, A., Silini, A., Cherif-Silini, H., Chenari Bouket, A., Alenezi, F. N., & Belbahri, L. (2022). Recent advances in encapsulation techniques of plant growth-promoting microorganisms and their prospects in the sustainable agriculture. *Applied Sciences (Basel, Switzerland)*, *12*(18), 9020. DOI: 10.3390/app12189020

Bao-Shan, L., shao-qi, D., Chun-hui, L., Li-jun, F., Shu-chun, Q., & Min, Y. (2004). Effect of TMS (nanostructured silicon dioxide) on growth of *Changbai larch* seedlings. *Journal of Forestry Research, 15*(2), 138–140. DOI: 10.1007/BF02856749

Barrena, R., Casals, E., Colón, J., Font, X., Sánchez, A., & Puntes, V. (2009). Evaluation of the ecotoxicity of model nanoparticles. *Chemosphere, 75*(7), 850–857. DOI: 10.1016/j.chemosphere.2009.01.078 PMID: 19264345

Bashir, A. K. H., Razanamahandry, L. C., Nwanya, A. C., Kaviyarasu, K., Saban, W., Mohamed, H. E. A., Ntwampe, S. K. O., Ezema, F. I., & Maaza, M.H. E. A. (2019). Biosynthesis of NiO nanoparticles for photodegradation of free cyanide solutions under ultraviolet light. *Journal of Physics and Chemistry of Solids, 134*, 133–140. DOI: 10.1016/j.jpcs.2019.05.048

Benjamin, S. R., Lima, F. D., Florean, E. O. P. T., & Guedes, M. I. F. (2019). Current trends in nanotechnology for bioremediation. *International Journal of Environment and Pollution, 66*(1-3), 19–40. DOI: 10.1504/IJEP.2019.104526

Bueno, V., Gao, X., Rahim, A. A., Wang, P., Bayen, S., & Ghoshal, S. (2022). Uptake and Translocation of a Silica Nanocarrier and an Encapsulated Organic Pesticide Following Foliar Application in Tomato Plants. *Environmental Science & Technology, 56*(10), 6722–6732. DOI: 10.1021/acs.est.1c08185 PMID: 35467849

Bussières, P. (2014). Estimating the number and size of phloem sieve plate pores using longitudinal views and geometric reconstruction. *Scientific Reports, 4*(1), 4929. DOI: 10.1038/srep04929 PMID: 24816291

Chaudhry, N., Dwivedi, S., Chaudhry, V., Singh, A., Saquib, Q., Azam, A., & Musarrat, J. (2018). Bio-inspired nanomaterials in agriculture and food: Current status, foreseen applications and challenges. *Microbial Pathogenesis, 123*, 196–200. DOI: 10.1016/j.micpath.2018.07.013 PMID: 30009970

Chen, C. J. (2021). *Introduction to Scanning Tunneling Microscopy* (3rd ed., Vol. 69). Oxford university press. DOI: 10.1093/oso/9780198856559.001.0001

Chen, H. (2018). Metal based nanoparticles in agricultural system: Behavior, transport, and interaction with plants. *Chemical Speciation and Bioavailability, 30*(1), 123–134. DOI: 10.1080/09542299.2018.1520050

Chhipa, H. (2017). Nanofertilizers and nanopesticides for agriculture. *Environmental Chemistry Letters, 15*(1), 15–22. DOI: 10.1007/s10311-016-0600-4

Coccia, F., Tonucci, L., Bosco, D., Bressan, M., & d'Alessandro, N. (2012). One pot synthesis of lignin-stabilized platinum and palladium nanoparticles and their catalytic behaviours in oxidation and reduction reactions. *Green Chemistry, 14*(4), 1073–1078. DOI: 10.1039/c2gc16524d

Cvjetko, P., Milošić, A., Domijan, A. M., Vrček, I. V., Tolić, S., Štefanić, P. P., & Balen, B. (2017). Toxicity of silver ions and differently coated silver nanoparticles in *Allium cepa* roots. *Ecotoxicology and Environmental Safety, 137*, 18–28. DOI: 10.1016/j.ecoenv.2016.11.009 PMID: 27894021

Daghan, H. A. T. İ. C. E. (2018). Effects of TiO_2 nanoparticles on maize (*Zea mays* L.) growth, chlorophyll content and nutrient uptake. *Applied Ecology and Environmental Research, 16*, 6873–6883.

Daisy, P., & Saipriya, K. (2012). Biochemical analysis of *Cassia fistula* aqueous extract and phytochemically synthesized gold nanoparticles as hypoglycemic treatment for diabetes mellitus. *International Journal of Nanomedicine*, *7*, 1189–1202. DOI: 10.2147/IJN.S26650 PMID: 22419867

de la Rosa, G., López-Moreno, M. L., de Haro, D., Botez, C. E., Peralta-Videa, J. R., & Gardea-Torresdey, J. L. (2013). Effects of ZnO nanoparticles in alfalfa, tomato, and cucumber at the germination stage: root development and X-ray absorption spectroscopy studies. *Pure and Applied Chemistry 4th*, 85(12), 2161-2174.

Desai, S., Singh, M., Chavan, A., Wagh, N. S., & Lakkakula, J. (2022). Micro-and nanoencapsulation techniques in agriculture. In *Agricultural nanobiotechnology* (pp. 297–323). Woodhead Publishing. DOI: 10.1016/B978-0-323-91908-1.00009-2

Dheyab, M. A., Aziz, A. A., Moradi Khaniabadi, P., Jameel, M. S., Oladzadabbasabadi, N., Mohammed, S. A., Abdullah, R. S., & Mehrdel, B. (2022). Monodisperse gold nanoparticles: A review on synthesis and their application in modern medicine. *International Journal of Molecular Sciences*, *23*(13), 7400. DOI: 10.3390/ijms23137400 PMID: 35806405

Dimkpa, C. O., Singh, U., Bindraban, P. S., Elmer, W. H., Gardea-Torresdey, J. L., & White, J. C. (2019). Zinc oxide nanoparticles alleviate drought-induced alterations in sorghum performance, nutrient acquisition, and grain fortification. *The Science of the Total Environment*, *688*, 926–934. DOI: 10.1016/j.scitotenv.2019.06.392 PMID: 31726574

Duhan, J. S., Kumar, R., Kumar, N., Kaur, P., Nehra, K., & Duhan, S. (2017). Nanotechnology: The new perspective in precision agriculture. *Biotechnology Reports (Amsterdam, Netherlands)*, *15*, 11–23. DOI: 10.1016/j.btre.2017.03.002 PMID: 28603692

Edison, T., & Sethuraman, M. (2012). Instant green synthesis of silver nanoparticles using *Terminalia chebula* fruit extract and evaluation of their catalytic activity on reduction of Methylene Blue. *Process Biochemistry (Barking, London, England)*, *47*(9), 1351–1357. DOI: 10.1016/j.procbio.2012.04.025

Elena, S. L., Daniela, G., Gerard, E., Lorena, B., Ana, L. L. M., Ruth, G., Amanda, C., Marta, E., Miren, E., Antoni, C., Amélia, M. S., Alessandra, D., Antonello, S., Maria, L. G., & Eliana, B. S. (2020). Metal-Based Nanoparticles as Antimicrobial Agents: An Overview. *Nanomaterials (Basel, Switzerland)*, *10*(2), 292. DOI: 10.3390/nano10020292 PMID: 32050443

Gong, P., Li, H., He, X., Wang, K., Hu, J., Tan, W., Tan, S., & Zhang, X. Y. (2007). Preparation and antibacterial activity of Fe3O4@Ag nanoparticles. *Nanotechnology*, *18*(28), 604–611. DOI: 10.1088/0957-4484/18/28/285604

Gopinath, K., Gowri, S., Karthika, V., & Arumugam, A. (2014). Green synthesis of gold nanoparticles from fruit extract of *Terminalia arjuna*, for the enhanced seed germination activity of *Gloriosa superba*. *Journal of Nanostructure in Chemistry*, *4*(3), 1–11. DOI: 10.1007/s40097-014-0115-0

Govender, Y., Riddin, T. L., Gericke, M., & Whiteley, C. G. (2010). On the enzymatic formation of platinum nanoparticles. *Journal of Nanoparticle Research*, *12*(1), 261–271. DOI: 10.1007/s11051-009-9604-3

Grillo, R., Mattos, B. D., Antunes, D. R., Forini, M. M. L., Monikh, F. A., & Rojas, O. J. (2021). Foliage adhesion and interactions with particulate delivery systems for plant nanobionics and intelligent agriculture. *Nano Today*, *37*, 101078. DOI: 10.1016/j.nantod.2021.101078

Gruyer, N., Dorais, M., Bastien, C., Dassylva, N., & Triffault-Bouchet, G. (2013, October). Interaction between silver nanoparticles and plant growth. *InInternational Symposium on New Technologies for Environment Control, Energy-Saving and Crop Production in Greenhouse and Plant* 1037 (pp. 795-800).

Guan, Y. F., Pearce, R. C., Melechko, A. V., Hensley, D. K., Simpson, M. L., & Rack, P. D. (2008). Pulsed laser dewetting of nickel catalyst for carbon nanofiber growth. *Nanotechnology*, *19*(23), 235604. DOI: 10.1088/0957-4484/19/23/235604 PMID: 21825799

Haghighi, M., Afifipour, Z., & Mozafarian, M. (2012). The effect of N-Si on tomato seed germination under salinity levels. *Journal of Biological and Environmental Sciences*, *6*(16), 87–90.

Haris, M., Hussain, T., Mohamed, H. I., Khan, A., Ansari, M. S., Tauseef, A., Khan, A. A., & Akhtar, N. (2023). Nanotechnology–A new frontier of nano-farming in agricultural and food production and its development. *The Science of the Total Environment*, *857*, 159639. DOI: 10.1016/j.scitotenv.2022.159639 PMID: 36283520

Hasan, M., Rafique, S., Zafar, A., Loomba, S., Khan, R., Hassan, S. G., Khan, M. W., Zahra, S., Zia, M., Mustafa, G., Shu, X., Ihsan, Z., & Mahmood, N. (2020). Physiological and anti-oxidative response of biologically and chemically synthesized iron oxide: *Zea mays* a case study. *Heliyon*, *6*(8), 1–7. DOI: 10.1016/j.heliyon.2020.e04595 PMID: 32923707

Hatami, M., & Ghorbanpour, M. (2013). Effect of nanosilver on physiological performance of pelargonium plants exposed to dark storage. *Journal of Horticultural Research*, *21*(1), 15–20. DOI: 10.2478/johr-2013-0003

Hong, J., Peralta-Videa, J. R., Rico, C., Sahi, S., Viveros, M. N., Bartonjo, J., Zhao, L., & Gardea-Torresdey, J. L. (2014). Evidence of Trans location and Physiological Impacts of Foliar Applied CeO_2 Nanoparticles on Cucumber (*Cucumis sativus*) Plants. *Environmental Science & Technology*, *48*(8), 4376–4385. DOI: 10.1021/es404931g PMID: 24625209

Hu, J., & Xianyu, Y. (2021). When nano meets plants: A review on the interplay between nanoparticles and plants. *Nano Today*, *38*, 101143. DOI: 10.1016/j.nantod.2021.101143

Huang, H., & Yang, X. (2005). One-step, shape control synthesis of gold nanoparticles stabilized by 3-thiopheneacetic acid. *Colloids and Surfaces. A, Physicochemical and Engineering Aspects*, *255*(1-3), 11–17. DOI: 10.1016/j.colsurfa.2004.12.020

Hussain, B., Lin, Q., Hamid, Y., Sanaullah, M., Di, L., Khan, M. B., & Yang, X. (2020). Foliage application of selenium and silicon nanoparticles alleviates Cd and Pb toxicity in rice (*Oryza sativa* L.). *The Science of the Total Environment*, *712*, 136497. DOI: 10.1016/j.scitotenv.2020.136497 PMID: 31945526

Irshad, M. A., Nawaz, R., Ur Rehman, M. Z., Adrees, M., Rizwan, M., Ali, S., & Tasleem, S. (2021). Synthesis, characterization and advanced sustainable applications of titanium dioxide nanoparticles: A review. *Ecotoxicology and Environmental Safety*, *212*, 111978. DOI: 10.1016/j.ecoenv.2021.111978 PMID: 33561774

Jaberzadeh, A., Moaveni, P., Tohidi Moghadam, H. R., & Zahedi, H. (2013). Influence of bulk and nanoparticles titanium foliar application on some agronomic traits, seed gluten and starch contents of wheat subjected to water deficit stress. *Notulae Botanicae Horti Agrobotanici Cluj-Napoca*, *41*(1), 201–207. DOI: 10.15835/nbha4119093

Jampílek, J., & Kráľová, K. (2017). *Nanopesticides: preparation, targeting, and controlled release.* In New pesticides and soil sensors *(pp. 81-127).* Academic Press.

Jia, L., Zhang, Q., Li, Q., & Song, H. (2009). The biosynthesis of palladium nanoparticles by antioxidants in garden *Jasminoids ellis*: Long life time catalyst for p-nitrotoluene hydrogenation. *Nanotechnology*, *20*(38), 385601. DOI: 10.1088/0957-4484/20/38/385601 PMID: 19713585

Jiang, M., Song, Y., Kanwar, M. K., Ahammed, G. J., Shao, S., & Zhou, J. (2021). Phytonanotechnology applications in modern agriculture. *Journal of Nanobiotechnology*, *19*(1), 1–20. DOI: 10.1186/s12951-021-01176-w PMID: 34930275

Kalteh, M., Alipour, Z. T., Ashraf, S., Aliabadi, M. M., & Nosratabadi, A. F. (2014). Effect of silica nanoparticles on basil (*Ocimum basilicum*) under salinity stress. *Journal of Chemical Health Risks*, *4*, 49–55.

Kasote, D. M., Lee, J. H., Jayaprakasha, G. K., & Patil, B. S. (2019). Seed priming with iron oxide nanoparticles modulate antioxidant potential and defense-linked hormones in watermelon seedlings. *ACS Sustainable Chemistry & Engineering*, *7*(5), 5142–5151. DOI: 10.1021/acssuschemeng.8b06013

Kaviya, S., Santhanalakshmi, J., & Viswanathan, B. (2011). Biosynthesis of silver nano-flakes by *Crossandra infundibuliformis* leaf extract. *Materials Letters*, *67*(1), 64–66. DOI: 10.1016/j.matlet.2011.09.023

Khan, S. T., & Malik, A. (2019). Engineered nanomaterials for water decontamination and purification: From lab to products. *Journal of Hazardous Materials*, *363*, 295–308. DOI: 10.1016/j.jhazmat.2018.09.091 PMID: 30312926

Khan, Z. S., Rizwan, M., Hafeez, M., Ali, S., Adrees, M., Qayyum, M. F., Khalid, S., ur Rehman, M. Z., & Sarwar, M. A. (2020). Effects of silicon nanoparticles on growth and physiology of wheat in cadmium contaminated soil under different soil moisture levels. *Environmental Science and Pollution Research International*, *27*(5), 4958–4968. DOI: 10.1007/s11356-019-06673-y PMID: 31840219

Kumar, H., Venkatesh, N., Bhowmik, H., & Kuila, A. (2018). Metallic nanoparticle: A review. *Biomedical Journal of Scientific & Technical Research*, *4*(2), 3765–3775.

Kumar, P. R., Vivekanandhan, S., Misra, M., Mohanty, A., & Satyanarayana, N. (2012). Soybean (*Glycine max*) leaf extract based green synthesis of palladium nanoparticles. *Journal of Biomaterials and Nanobiotechnology*, *3*(1), 14–19. DOI: 10.4236/jbnb.2012.31003

Kumar, V., Guleria, P., Kumar, V., & Yadav, S. K. (2013). Gold nanoparticle exposure induces growth and yield enhancement in *Arabidopsis thaliana.The Science of the Total Environment*, *461*, 462–468. DOI: 10.1016/j.scitotenv.2013.05.018 PMID: 23747561

Kumari, S., & Sarkar, L. (2021). A review on nanoparticles: Structure, classification, synthesis & applications. *Journal of Scientific Research*, *65*(8), 42–46. DOI: 10.37398/JSR.2021.650809

Lal, R. (2008). Soils and sustainable agriculture. A review. *Agronomy for Sustainable Development*, *28*(1), 57–64. DOI: 10.1051/agro:2007025

Lattanzio, V. M. T., & Nivarlet, N. (2017). Multiplex dipstick immunoassay for semiquantitative determination of fusarium mycotoxins in oat in *Methods. Molecular Biology*, *1536*, 137–142. DOI: 10.1007/978-1-4939-6682-0_10 PMID: 28132148

Lau, H. Y., Wu, H., Wee, E. J. H., Trau, M., Wang, Y., & Botella, J. R. (2017). Specific and sensitive isothermal electrochemical biosensor for plant pathogen DNA detection with colloidal gold nanoparticles as probes. *Scientific Reports*, *7*(1), 38896. DOI: 10.1038/srep38896 PMID: 28094255

Lee, H. J., Lee, G., Jang, N. R., Yun, J. H., Song, J. Y., & Kim, B. S. (2011). Biological synthesis of copper nanoparticles using plants extract. *Nanotechnology*, *1*, 371–374.

Li, J., Chang, P. R., Huang, J., Wang, Y., Yuan, H., & Ren, H. (2013). Physiological effects of magnetic iron oxide nanoparticles towards watermelon. *Journal of Nanoscience and Nanotechnology*, *13*(8), 5561–5567. DOI: 10.1166/jnn.2013.7533 PMID: 23882795

Li, L., Luo, Y., Li, R., Zhou, Q., Peijnenburg, W. J. G. M., Yin, N., Yang, J., Tu, C., & Zhang, Y. (2020). Effective uptake of submicrometre plastics by crop plants via a crack-entry mode. *Nature Sustainability*, *3*(11), 929–937. DOI: 10.1038/s41893-020-0567-9

Li, X. Q., & Zhang, W. X. (2006). Iron nanoparticles: The core– shell structure and unique properties for Ni (II) sequestration. *Langmuir*, *22*(10), 4638–4642. DOI: 10.1021/la060057k PMID: 16649775

Lin, X., Wu, M., Wu, B., Kuga, S., Endo, T., & Huang, Y. (2011). Platinum nanoparticles using wood nanomaterials: Eco-friendly synthesis, shape control and catalytic activity for p-nitrophenol reduction. *Green Chemistry*, *13*(2), 283–287. DOI: 10.1039/C0GC00513D

Liu, Q., Chen, B., Wang, Q., Shi, X., Xiao, Z., Lin, J., & Fang, X. (2009). Carbon Nanotubes as Molecular Transporters for Walled Plant Cells. *Nano Letters*, *9*(3), 1007–1010. DOI: 10.1021/nl803083u PMID: 19191500

Liu, R., & Lal, R. (2015). Potentials of engineered nanoparticles as fertilizers for increasing agronomic productions. *The Science of the Total Environment*, *514*, 131–139. DOI: 10.1016/j.scitotenv.2015.01.104 PMID: 25659311

Lu ChangMei, L. C., Zhang ChaoYing, Z. C., Wen JunQiang, W. J., Wu GuoRong, W. G., & Tao MingXuan, T. M. (2002). Research of the effect of nanometer materials on germination and growth enhancement of Glycine max and its mechanism.

Lu, L., Wang, H., Zhou, Y., Xi, S., Zhang, H., Hu, J., & Zhao, B. (2002). Seed-mediated growth of large, monodisperse core–shell gold–silver nanoparticles with Ag-like optical properties. *Chemical Communications*, (2), 144–145. DOI: 10.1039/b108473a PMID: 12120344

Luangpipat, T., Beattie, I. R., Chisti, Y., & Haverkamp, R. G. (2011). Gold nanoparticles produced in a microalga. *Journal of Nanoparticle Research*, *13*(12), 6439–6445. DOI: 10.1007/s11051-011-0397-9

Lv, J., Christie, P., & Zhang, S. (2019). Uptake, translocation, and transformation of metal-based nanoparticles in plants: Recent advances and methodological challenges. *Environmental Science. Nano*, *6*(1), 41–59. DOI: 10.1039/C8EN00645H

Machado, T. O., Grabow, J., Sayer, C., de Araújo, P. H., Ehrenhard, M. L., & Wurm, F. R. (2022). Biopolymer-based nanocarriers for sustained release of agrochemicals: A review on materials and social science perspectives for a sustainable future of agri-and horticulture. *Advances in Colloid and Interface Science*, *303*, 102645. DOI: 10.1016/j.cis.2022.102645 PMID: 35358807

Mafuné, F., Kohno, J. Y., Takeda, Y., & Kondow, T. (2003). Formation of Stable Platinum Nanoparticles by Laser Ablation in Water. *The Journal of Physical Chemistry B*, *107*(18), 4218–4223. DOI: 10.1021/jp021580k

Mahmoodzadeh, H., Nabavi, M., & Kashefi, H. (2013). Effect of nanoscale titanium dioxide particles on the germination and growth of canola (*Brassica napus*). *Ornamental Plants*, *3*, 25–32.

Mazaheri-Tirani, M., & Dayani, S. (2020). In vitro effect of zinc oxide nanoparticles on *Nicotiana tabacum* callus compared to ZnO micro particles and zinc sulfate ($ZnSO_4$). [PCTOC]. *Plant Cell, Tissue and Organ Culture*, *140*(2), 279–289. DOI: 10.1007/s11240-019-01725-0

Mazhar, T., Shrivastava, V., & Tomar, R. S. (2017). Green synthesis of bimetallic nanoparticles and its applications: A review. *Journal of Pharmaceutical Sciences and Research*, *9*(2), 102.

Mazumdar, H., & Ahmed, G. U. (2011). Phytotoxicity effect of silver nanoparticles on *Oryza sativa*. *International Journal of Chemtech Research*, *3*(3), 1494–1500.

Miralles, P., Church, T. L., & Harris, A. T. (2012). Toxicity, Uptake, and Translocation of Engineered Nanomaterials in Vascular plants. *Environmental Science & Technology*, *46*(17), 9224–9239. DOI: 10.1021/es202995d PMID: 22892035

Mirzajani, F., Askari, H., Hamzelou, S., Farzaneh, M., & Ghassempour, A. (2013). Effect of silver nanoparticles on *Oryza sativa* L. and its rhizosphere bacteria. *Ecotoxicology and Environmental Safety*, *88*, 48–54. DOI: 10.1016/j.ecoenv.2012.10.018 PMID: 23174269

Mittal, A. K., Chisti, Y., & Banerjee, U. C. (2013). Synthesis of metallic nanoparticles using plant extracts. *Biotechnology Advances*, *31*(2), 346–356. DOI: 10.1016/j.biotechadv.2013.01.003 PMID: 23318667

Mittal, D., Kaur, G., Singh, P., Yadav, K., & Ali, S. A. (2020). Nanoparticle-based sustainable agriculture and food science: Recent advances and future outlook. *Frontiers in Nanotechnology*, *2*, 579954. DOI: 10.3389/fnano.2020.579954

Mittal, D., Kaur, G., Singh, P., Yadav, K., & Ali, S. A. (2020). Nanoparticle-based sustainable agriculture and food science: Recent advances and future outlook. *Frontiers in Nanotechnology*, *2*, 579954. DOI: 10.3389/fnano.2020.579954

Mitter, N., Worrall, E. A., Robinson, K. E., Li, P., Jain, R. G., Taochy, C., Fletcher, S. J., Carroll, B. J., Lu, G. Q., & Xu, Z. P. (2017). Clay nanosheets for topical delivery of RNAi for sustained protection against plant viruses. *Nature Plants*, *3*(2), 16207. DOI: 10.1038/nplants.2016.207 PMID: 28067898

Mohammed, N., Nawar, S. H., Etawy, M. S., Nassar, G. E., & Hassabo, A. G. (2024). Nanotechnology and its applications in industry and product design. *Journal of Textiles. Coloration and Polymer Science*, *21*(2), 273–284.

Molahalli, V., Sharma, A., Bijapur, K., Soman, G., Shetty, A., Sirichandana, B., & Hegde, G. (2024). Properties, synthesis, and characterization of cu-based nanomaterials. In *Copper-Based Nanomaterials in Organic Transformations* (pp. 1–33). American Chemical Society. DOI: 10.1021/bk-2024-1466.ch001

Montazer, M., Shamei, A., & Alimohammadi, F. (2014). Synthesis of nanosilver on polyamide fabric using silver/ammonia complex. *Materials Science and Engineering C*, *38*, 170–176. DOI: 10.1016/j.msec.2014.01.044 PMID: 24656365

Moriarty, P. (2022). *Nanotechnology: A Very Short Introduction*. Oxford University Press. DOI: 10.1093/actrade/9780198841104.001.0001

Nadagouda, M. N., & Varma, R. S. (2008). Green synthesis of silver and palladium nanoparticles at room temperature using coffee and tea extract. *Green Chemistry*, *10*(8), 859–862. DOI: 10.1039/b804703k

Narayanan, K. B., & Sakthivel, N. (2008). Coriander leaf mediated biosynthesis of gold nanoparticles. *Materials Letters*, *62*(30), 4588–4590. DOI: 10.1016/j.matlet.2008.08.044

Noori, A., Donnelly, T., Colbert, J., Cai, W., Newman, L. A., & White, J. C. (2020). Exposure of tomato (*Lycopersicon esculentum*) to silver nanoparticles and silver nitrate: Physiological and molecular response. *International Journal of Phytoremediation*, *22*(1), 40–51. DOI: 10.1080/15226514.2019.1634000 PMID: 31282192

Omran, B. A., & Omran, B. A. (2020). Fundamentals of nanotechnology and nanobiotechnology. *Nanobiotechnology: A Multidisciplinary Field of Science*, 1-36.

Osman, Y. A., Eltanahy, E., Hegazy, A. S., & Razak, A. A. (2023). Nanoformulations of agrochemicals used to apply pesticides and fertilizers for crop improvement. *Nanoformulations for sustainable agriculture and environmental risk mitigation*, 103-117.

Palchoudhury, S., Jungjohann, K. L., Weerasena, L., Arabshahi, A., Gharge, U., Albattah, A., Miller, J., Patel, K., & Holler, R. A. (2018). Enhanced legume root growth with pre-soaking in α-Fe_2O_3 nanoparticle fertilizer. *RSC Advances*, *8*(43), 24075–24083. DOI: 10.1039/C8RA04680H PMID: 35539206

Pandey, S., & Kumari, N. (2023). Valorisation of Agricultural and Food Waste Biomass for Production of Bioenergy. In *Recent Technologies for Waste to Clean Energy and its Utilization* (pp. 197–218). Springer Nature Singapore. DOI: 10.1007/978-981-19-3784-2_10

Pantidos, N., & Horsfall, L. E. (2014). Biological synthesis of metallic nanoparticles by bacteria, fungi and plants. *Journal of Nanomedicine & Nanotechnology*, *5*(5), 1. DOI: 10.4172/2157-7439.1000233

Paret, M. L., Vallad, G. E., Averett, D. R., Jones, J. B., & Olson, S. M. (2013). Photocatalysis: Effect of light-activated nanoscale formulations of TiO_2 on *Xanthomonas perforans* and control of bacterial spot of tomato. *Phytopathology*, *103*(3), 228–236. DOI: 10.1094/PHYTO-08-12-0183-R PMID: 23190116

Parkinson, S. J., Tungsirisurp, S., Joshi, C., Richmond, B. L., Gifford, M. L., Sikder, A., Lynch, I., O'Reilly, R. K., & Napier, R. M. (2022). Polymer nanoparticles pass the plant interface. *Nature Communications*, *13*(1), 7385. DOI: 10.1038/s41467-022-35066-y PMID: 36450796

Peralta-Videa, J. R., Huang, Y., Parsons, J. G., Zhao, L., Lopez-Moreno, L., Hernandez-Viezcas, J. A., & Gardea-Torresdey, J. L. (2016). Plant-based green synthesis of metallic nanoparticles: Scientific curiosity or a realistic alternative to chemical synthesis? *Nanotechnology for Environmental Engineering*, *1*(1), 1–29. DOI: 10.1007/s41204-016-0004-5

Prasad, R., Bhattacharyya, A., & Nguyen, Q. D. (2017). Nanotechnology in sustainable agriculture: Recent developments, challenges, and perspectives. *Frontiers in Microbiology*, *8*, 1014. DOI: 10.3389/fmicb.2017.01014 PMID: 28676790

Prasad, T. N. V. K. V., Sudhakar, P., Sreenivasulu, Y., Latha, P., Munaswamy, V., Reddy, K. R., Sreeprasad, T. S., Sajanlal, P. R., & Pradeep, T. (2012). Effect of nanoscale zinc oxide particles on the germination, growth and yield of peanut. *Journal of Plant Nutrition*, *35*(6), 905–927. DOI: 10.1080/01904167.2012.663443

Qureshi, A., Singh, D. K., & Dwivedi, S. (2018). Nano-fertilizers: A novel way for enhancing nutrient use efficiency and crop productivity. *International Journal of Current Microbiology and Applied Sciences*, *7*(2), 3325–3335. DOI: 10.20546/ijcmas.2018.702.398

Rahmatizadeh, R., Arvin, S. M. J., Jamei, R., Mozaffari, H., & Reza Nejhad, F. (2019). Response of tomato plants to interaction effects of magnetic (Fe_3O_4) nanoparticles and cadmium stress. *Journal of Plant Interactions*, *14*(1), 474–481. DOI: 10.1080/17429145.2019.1626922

Rai, M., Yadav, A., & Gade, A. (2009). Silver nanoparticles as a new generation of antimicrobials. *Biotechnology Advances*, *27*(1), 76–83. DOI: 10.1016/j.biotechadv.2008.09.002 PMID: 18854209

Rai, V., Acharya, S., & Dey, N. (2012). Implications of nanobiosensors in agriculture. *Journal of Biomaterials and Nanobiotechnology*, *03*(2), 315–324. DOI: 10.4236/jbnb.2012.322039

Rajput, V., Minkina, T., Mazarji, M., Shende, S., Sushkova, S., Mandzhieva, S., Burachevskaya, M., Chaplygin, V., Singh, A., & Jatav, H. (2020). Accumulation of nanoparticles in the soil-plant systems and their effects on human health. *Annals of Agricultural Science*, *65*(2), 137–143. DOI: 10.1016/j.aoas.2020.08.001

Raliya, R., Franke, C., Chavalmane, S., Nair, R., Reed, N., & Biswas, P. (2016). Quantitative Understanding of Nanoparticle Uptake in Watermelon Plants. *Frontiers in Plant Science*, *7*, 1288. DOI: 10.3389/fpls.2016.01288 PMID: 27617020

Raliya, R., Saharan, V., Dimkpa, C., & Biswas, P. (2018). Nanofertilizer for Precision and Sustainable Agriculture: Current State and Future Perspectives. *Journal of Agricultural and Food Chemistry*, *66*(26), 6487–6503. DOI: 10.1021/acs.jafc.7b02178 PMID: 28835103

Raliya, R., & Tarafdar, J. C. (2013). ZnO nanoparticle biosynthesis and its effect on phosphorous-mobilizing enzyme secretion and gum contents in Clusterbean (*Cyamopsis tetragonoloba* L.). *Agricultural Research*, *2*(1), 48–57. DOI: 10.1007/s40003-012-0049-z

Ramesh, M., Palanisamy, K., Babu, K., & Sharma, N. K. (2014). Effects of bulk & nano-titanium dioxide and zinc oxide on physio-morphological changes in *Triticum aestivum* Linn. *Jouranl of Global Biosciences*, *3*(2), 415–422.

Rani, S., Kumari, N., & Sharma, V. (2022). Uptake, translocation, transformation and physiological effects of nanoparticles in plants. *Archives of Agronomy and Soil Science*, •••, 1–21.

Raskar, S. V., & Laware, S. L. (2014). Effect of zinc oxide nanoparticles on cytology and seed germination in onion. *International Journal of Current Microbiology and Applied Sciences*, *3*(2), 467–473.

Rastogi, A., Tripathi, D. K., Yadav, S., Chauhan, D. K., Živčák, M., Ghorbanpour, M., et al., (2019). Application of silicon nanoparticles in agriculture. 3 Biotech 9:90. DOI: 10.1007/s13205-019-1626-7

Rodríguez-González, V., Terashima, C., & Fujishima, A. (2019). Applications of photocatalytic titanium dioxide-based nanomaterials in sustainable agriculture. *Journal of Photochemistry and Photobiology C, Photochemistry Reviews*, *40*, 49–67. DOI: 10.1016/j.jphotochemrev.2019.06.001

Rogozea, E. A., Petcu, A. R., Olteanu, N. L., Lazar, C. A., Cadar, D., & Mihaly, M. (2017). Tandem adsorption-photodegradation activity induced by light on NiO-ZnO p–n couple modified silica nanomaterials. *Materials Science in Semiconductor Processing*, *57*, 1–11. DOI: 10.1016/j.mssp.2016.10.006

Roopan, S. M., Bharathi, A., Kumar, R., Khanna, V. G., & Prabhakarn, A. (2011). Acaricidal, insecticidal, and larvicidal efficacy of aqueous extract of *Annona squamosa* L peel as biomaterial for the reduction of palladium salts into nanoparticles. *Colloids and Surfaces. B, Biointerfaces*, *92*, 209–212. DOI: 10.1016/j.colsurfb.2011.11.044 PMID: 22205064

Rui, M., Ma, C., Hao, Y., Guo, J., Rui, Y., Tang, X., Zhao, Q., Fan, X., Zhang, Z., Hou, T., & Zhu, S. (2016). Iron oxide nanoparticles as a potential iron fertilizer for peanut (*Arachis hypogaea*). *Frontiers in Plant Science*, *7*, 815. DOI: 10.3389/fpls.2016.00815 PMID: 27375665

Safaepour, M., Shahverdi, A. R., Shahverdi, H. R., Khorramizadeh, M. R., & Gohari, A. R. (2009). Green synthesis of small silver nanoparticles using geraniol and its cytotoxicity against Fibrosarcoma–Wehi 164. *Avicenna Journal of Medical Biotechnology*, *1*, 111–115. PMID: 23407598

Salama, H. M. (2012). Effects of silver nanoparticles in some crop plants, common bean (*Phaseolus vulgaris* L.) and corn (*Zea mays* L.). *International Research Journal of Biotechnology*, *3*(10), 190–197.

Sasson, Y., Levy-Ruso, G., Toledano, O., & Ishaaya, I. (2007). Nanosuspensions: emerging novel agrochemical formulations. In Insecticides design using advanced technologies (pp. 1-39). Springer Berlin Heidelberg.

Sathishkumar, M., Sneha, K., Kwak, I. S., Mao, J., Tripathy, S. J., & Yun, Y. S. (2009). Phyto-crystallization of palladium through reduction process using *Cinnamon zeylanicum* bark extract. *Journal of Hazardous Materials*, *171*(1-3), 400–404. DOI: 10.1016/j.jhazmat.2009.06.014 PMID: 19576689

Savithramma, N., Ankanna, S., & Bhumi, G. (2012). Effect of nanoparticles on seed germination and seedling growth of *Boswellia ovalifoliolata* an endemic and endangered medicinal tree taxon. *Nano Vision*, *2*, 61–68.

Sebastian, A., Nangia, A., & Prasad, M. N. V. (2019). Cadmium and sodium adsorption properties of magnetite nanoparticles synthesized from *Hevea brasiliensis* Muell. Arg. bark: Relevance in amelioration of metal stress in rice. *Journal of Hazardous Materials*, *371*, 261–272. DOI: 10.1016/j.jhazmat.2019.03.021 PMID: 30856436

Sedghi, M., Hadi, M., & Toluie, S. G. (2013). Effect of nano zinc oxide on the germination parameters of soybean seeds under drought stress. *Annals of West University of Timisoara. Series of Biology*, *16*(2), 73.

Shah, M., Fawcett, D., Sharma, S., Tripathy, S. K., & Poinern, G. E. J. (2015). Green synthesis of metallic nanoparticles via biological entities. *Materials (Basel)*, *8*(11), 7278–7308. DOI: 10.3390/ma8115377 PMID: 28793638

Shah, V., & Belozerova, I. (2009). Influence of metal nanoparticles on the soil microbial community and germination of lettuce seeds. *Water, Air, and Soil Pollution*, *197*(1-4), 143–148. DOI: 10.1007/s11270-008-9797-6

Shankar, S. S., Ahmad, A., & Sastry, M. (2003). Geranium leaf assisted biosynthesis of silver nanoparticles. *Biotechnology Progress*, *19*(6), 1627–1631. DOI: 10.1021/bp034070w PMID: 14656132

Shankramma, K., Yallappa, S., Shivanna, M. B., & Manjanna, J. (2016). Fe_2O_3 magnetic nanoparticles to enhance *S. lycopersicum* (tomato) plant growth and their biomineralization. *Applied Nanoscience*, *6*(7), 983–990. DOI: 10.1007/s13204-015-0510-y

Sharma, V. K., Ria, A. Y., & Lin, Y. (2009). Silver nanoparticles: Green synthesis and their antimicrobial activities. *Advances in Colloid and Interface Science*, *145*(1-2), 83–96. DOI: 10.1016/j.cis.2008.09.002 PMID: 18945421

Shelar, G. B., & Chavan, A. M. (2015). Myco-synthesis of silver nanoparticles from *Trichoderma harzianum* and its impact on germination status of oil seed. *Biolife : An International Quarterly Journal of Biology and Life Sciences*, *3*(1), 109–113.

Shin, W. K., Cho, J., Kannan, A. G., Lee, Y. S., & Kim, W. (2016). Cross- linked composite gel polymer electrolyte using mesoporous methacrylate functionalized SiO2 nanoparticles for lithium- ion polymer batteries. *Scientific Reports*, *6*(1), 26332. DOI: 10.1038/srep26332 PMID: 27189842

Siddiqui, M. H., & Al-Whaibi, M. H. (2014). Role of nano-SiO_2 in germination of tomato (*Lycopersicum esculentum* seeds Mill.). *Saudi Journal of Biological Sciences*, *21*(1), 13–17. DOI: 10.1016/j.sjbs.2013.04.005 PMID: 24596495

Singh, J., & Lee, B. K. (2016). Influence of nano-TiO_2 particles on the bioaccumulation of Cd in soybean plants (*Glycine max*): A possible mechanism for the removal of Cd from the contaminated soil. *Journal of Environmental Management*, *170*, 88–96. DOI: 10.1016/j.jenvman.2016.01.015 PMID: 26803259

Singh, M., Malhotra, N., & Sharma, K. (2020). Buckwheat (Fagopyrum sp.) genetic resources: What can they contribute towards nutritional security of changing world? *Genetic Resources and Crop Evolution*, *67*(7), 1639–1658. DOI: 10.1007/s10722-020-00961-0

Sinha, T., Bhagwatwar, P., Krishnamoorthy, C., & Chidambaram, R. (2019). Polymer based micro-and nanoencapsulation of agrochemicals. *Polymers for agri-food applications*, 5-28.

Song, J. Y., Kwon, E. Y., & Kim, B. S. (2010). Biological synthesis of platinum nanoparticles using *Diopyros kaki* leaf extract. *Bioprocess and Biosystems Engineering*, *33*(1), 159–164. DOI: 10.1007/s00449-009-0373-2 PMID: 19701776

Soundarrajan, C., Sankari, A., Dhandapani, P., Maruthamuthu, S., Ravichandran, S., Sozhan, G., & Palaniswamy, N. (2012). Rapid biological synthesis of platinum nanoparticles using *Ocimum sanctum* for water electrolysis applications. *Bioprocess and Biosystems Engineering*, *35*(5), 827–833. DOI: 10.1007/s00449-011-0666-0 PMID: 22167464

Stampoulis, D., Sinha, S. K., & White, J. C. (2009). Assay-dependent phytotoxicity of nanoparticles to plants. *Environmental Science & Technology*, *43*(24), 9473–9479. DOI: 10.1021/es901695c PMID: 19924897

Subhankari, I., & Nayak, P. L. (2013). Synthesis of copper nanoparticles using *Syzygium aromaticum* (Cloves) aqueous extract by using green chemistry. *World Journal of Nano Science and Technology*, *2*, 14–17.

Sun, H., Lei, C., Xu, J., & Li, R. (2021). Foliar uptake and leaf-to-root translocation of nanoplastics with different coating charge in maize plants. *Journal of Hazardous Materials*, *416*, 125854. DOI: 10.1016/j.jhazmat.2021.125854 PMID: 33892383

Suriyaprabha, R., Karunakaran, G., Yuvakkumar, R., Rajendran, V., & Kannan, N. (2012). Silica nanoparticles for increased silica availability in maize (*Zea mays*. L) seeds under hydroponic conditions. *Current Nanoscience*, *8*(6), 902–908. DOI: 10.2174/157341312803989033

Takemura, K. (2021). Surface plasmon resonance (SPR)-and localized SPR (LSPR)-based virus sensing systems: Optical vibration of nano-and micro-metallic materials for the development of next-generation virus detection technology. *Biosensors (Basel)*, *11*(8), 250. DOI: 10.3390/bios11080250 PMID: 34436053

Tawfik, M. M., Mohamed, M. H., Sadak, M. S., & Thalooth, A. T. (2021). Iron oxide nanoparticles effect on growth, physiological traits and nutritional contents of *Moringa oleifera* grown in saline environment. *Bulletin of the National Research Center*, *45*(1), 1–9. DOI: 10.1186/s42269-021-00624-9

Thakur, A., & Kumar, A. (2022). Recent advances on rapid detection and remediation of environmental pollutants utilizing nanomaterials-based (bio) sensors. *The Science of the Total Environment*, *834*, 155219. DOI: 10.1016/j.scitotenv.2022.155219 PMID: 35421493

Thomas, P., Rumjit, N. P., George, P. J., Lai, C. W., Tyagi, P., Johan, M. R. B., & Saravanakumar, M. P. (2020). Remediation of heavy metal Ions using nanomaterials sourced from wastewaters. *Nanotechnology for Food. Agriculture, and Environment*, •••, 255–296.

Umavathi, S., Mahboob, S., Govindarajan, M., Al-Ghanim, K. A., Ahmed, Z., Virik, P., Al-Mulhm, N., Subash, M., Gopinath, K., & Kavitha, C. (2021). Green synthesis of ZnO nanoparticles for antimicrobial and vegetative growth applications: A novel approach for advancing efficient high quality health care to human wellbeing. *Saudi Journal of Biological Sciences*, *28*(3), 1808–1815. DOI: 10.1016/j.sjbs.2020.12.025 PMID: 33732066

Usman, M., Farooq, M., Wakeel, A., Nawaz, A., Cheema, S. A., Rehman, H., Ashraf, I., & Sanaullah, M. (2020). Nanotechnology in agriculture: Current status, challenges and future opportunities. *The Science of the Total Environment*, *721*, 137778. DOI: 10.1016/j.scitotenv.2020.137778 PMID: 32179352

Valenstein, J. S., Lin, V. S.-Y., Lyznik, L. A., Martin-Ortigosa, S., Wang, K., & Peterson, D. J.. (2013). Mesoporous silica nanoparticle-mediated intracellular cre protein delivery for maize genome editing via loxP site excision. *Plant Physiology*, *164*(2), 537–547. DOI: 10.1104/pp.113.233650 PMID: 24376280

Wang, Z. Y., Yu, X. L., Gao, D. M., Feng, W. Q., Xing, B. S., & Li, F. M. (2010). Effect of nano-rutile TiO_2 and multiwalled carbon nanotubes on the growth of maize (*Zea mays* L.) seedlings and the relevant antioxidant response. *Huan Jing ke Xue=. Huan Jing Ke Xue*, *31*(2), 480–487. PMID: 20391721

Wilson, M. P., & Schwarzman, M. R. (2009). Toward a new US chemicals policy: Rebuilding the foundation to advance new science, green chemistry, and environmental health. *Environmental Health Perspectives*, *117*(8), 1202–1209. DOI: 10.1289/ehp.0800404 PMID: 19672398

Wu, W., Xiao, X., Zhang, S., Ren, F., & Jiang, C. (2011). Facile method to synthesize magnetic iron oxides/TiO_2 hybrid nanoparticles and their photodegradation application of methylene blue. *Nanoscale Research Letters*, *6*(1), 1–15. DOI: 10.1186/1556-276X-6-533 PMID: 21961891

Xie, Y., Li, B., Zhang, Q., Zhang, C., Lu, K., & Tao, G. (2011). Effects of nano-TiO_2 on photosynthetic characteristics of *Indocalamus barbatus.Dongbei Linye Daxue Xuebao*, *39*, 22–25.

Yadav, N., Garg, V. K., Chhillar, A. K., & Rana, J. S. (2023). Recent advances in nanotechnology for the improvement of conventional agricultural systems: A review. *Plant Nano Biology*, *4*, 100032. DOI: 10.1016/j.plana.2023.100032

Yang, C., Powell, C. A., Duan, Y., Shatters, R., & Zhang, M. (2015). Antimicrobial Nanoemulsion Formulation with Improved Penetration of Foliar Spray through Citrus Leaf Cuticles to Control Citrus Huanglongbing. *PLoS One*, *10*(7), e0133826. DOI: 10.1371/journal.pone.0133826 PMID: 26207823

Yang, F., Hong, F., You, W., Liu, C., Gao, F., Wu, C., & Yang, P. (2006). Influence of nano-anatase TiO_2 on the nitrogen metabolism of growing spinach. *Biological Trace Element Research*, *110*(2), 179–190. DOI: 10.1385/BTER:110:2:179 PMID: 16757845

Yao, K. S., Li, S. J., Tzeng, K. C., Cheng, T. C., Chang, C. Y., Chiu, C. Y., Liao, C. Y., Hsu, J. J., & Lin, Z. P. (2009). Fluorescence silica nanoprobe as a biomarker for rapid detection of plant pathogens. *Advanced Materials Research*, *79–82*, 513–516. . DOI: 10.4028/www.scientific.net/AMR.79-82.513

Zhang, Y., Long, H., Wang, M. Y., Li, Y., Ma, L., Chen, K., Zheng, Y., & Jiang, T. (2020). The hidden mechanism of chemical fertiliser overuse in rural China. *Habitat International*, *102*, 102210. DOI: 10.1016/j.habitatint.2020.102210

Zhao, L., Lu, L., Wang, A., Zhang, H., Huang, M., Wu, H., Xing, B., Wang, Z., & Ji, R. (2020). Nanobiotechnology in agriculture: Use of nanomaterials to promote plant growth and stress tolerance. *Journal of Agricultural and Food Chemistry*, *68*(7), 1935–1947. DOI: 10.1021/acs.jafc.9b06615 PMID: 32003987

Zhao, X., Meng, Z., Wang, Y., Chen, W., Sun, C., Cui, B., Cui, J., Yu, M., Zeng, Z., Guo, S., Luo, D., Cheng, J. Q., Zhang, R., & Cui, H. (2017). Pollen magnetofection for genetic modification with magnetic nanoparticles as gene carriers. *Nature Plants*, *3*(12), 956–964. DOI: 10.1038/s41477-017-0063-z PMID: 29180813

Zheng, L., Hong, F., Lu, S., & Liu, C. (2005). Effect of nano-TiO_2 on strength of naturally aged seeds and growth of spinach. *Biological Trace Element Research*, *104*(1), 83–91. DOI: 10.1385/BTER:104:1:083 PMID: 15851835

Zia-ur-Rehman, M., Naeem, A., Khalid, H., Rizwan, M., Ali, S., & Azhar, M. (2018). Responses of plants to iron oxide nanoparticles. In *Nanomaterials in plants, algae, and microorganisms* (pp. 221–238). Academic Press.

KEY TERMS AND DEFINITIONS

Nanotechnology: The study and use of materials and systems at the nanoscale (1-100 nm) is known as nanotechnology (Mohammed et al., 2024).

Nanoparticles (NPs): Due to their small size and quantum effects, these tiny particles have special characteristics (Altammar, 2023).

Scanning Tunneling Microscope (STM): Gerd Binnig and Heinrich Rohrer created the scanning tunneling microscope (STM), a tool that allows researchers to view and work with individual atoms (Chen, 2021).

Surface Effect: The phenomenon whereby nanoparticles' high surface-to-volume ratio causes them to display distinct features (Molahalli et al., 2024).

Nanoencapsulation: A method to improve plant absorption and efficiency by delaying the release of agrochemicals (Desai et al., 2022).

Nanoformulations: The application of nanoparticles in agriculture to distribute biomolecules, fertilizers, and insecticides in a controlled manner (Osman et al., 2023).

Localized Surface Plasmon Resonance (LSPR): This phenomenon increases the optical properties of metal nanoparticles by causing free electrons to vibrate when exposed to light (Takemura, 2021).

Monodispersed Nanoparticles: These particles have a consistent size distribution, which is essential for uses like medication administration and catalysis (Dheyab et al., 2022).

Biosensors: Tools that use nanomaterials to identify pollutants, pathogens, or changes in plant metabolism for better agricultural monitoring and disease control (Thakur and Kumar 2022).

Chapter 3
Carbon Nanotubes as Carriers in Plant Science

Sahithi Venkata Naga Baru
Christ University, India

Gicy Mathew
https://orcid.org/0009-0007-4282-6117
Christ University, India

Dhevi B. Maanasa
https://orcid.org/0009-0008-6701-768X
Christ University, India

Vasantha Veerappa Lakshmaiah
Christ University, India

Praveen Nagella
https://orcid.org/0000-0001-6971-0527
Christ University, India

ABSTRACT

Carbon nanotubes have emerged as promising nanomaterials in plant sciences, offering innovative solutions for enhancing sustainable agricultural production. This chapter provides a comprehensive analysis of the structural variations of CNTs, including single-walled, stacked-cup, and multi-walled nanotubes, and their applications in gene delivery, biosensing, and development of nano-pesticides and nano-insecticides. CNTs facilitate improved nutrient uptake, promote plant growth, and enhance stress tolerance through their unique physicochemical properties. Additionally, their interactions with plant-associated microbes and soil microbial communities, plant health and soil fertility are discussed. However, challenges related to CNT toxicity, environmental persistence, and their effects on plant physiology necessitate further research. The chapter also explores the integration of machine learning, artificial intelligence, and multi omics approaches in optimizing CNT applications for plant breeding, precision agriculture, and crop improvement.

DOI: 10.4018/979-8-3693-9869-2.ch003

INTRODUCTION

The integration of nanotechnology with plant sciences has helped in addressing a variety of issues which include pest and disease control, stress tolerance, and improved soil fertility. However it had significantly reduced the use of agrochemical application owing to its effective delivery of nutrients and pesticides to plants. Thus nanotechnology ensures a sustainable agricultural practice which tackles both environmental and agricultural concerns. Nanotechnology has paved the way for advancements in plant sciences by introducing nanoformulations, nano devices and nano sensors improving agricultural practices and productivity while using minimum agricultural inputs. Nanoformulations involve integration of nanoparticles to produce products which maximize the efficiency of plant functions. Nano devices, nanoformulations and nano sensors back up agriculture for better yield, quick and early detection and eradication of disease and engineering plants for desired traits. Carbon molecules have the property of catenation which allows them to form bonds with other carbon atoms and also form multiple bonds with other atoms. Carbons exist as different allotropes having two or more crystalline structures. The structure of carbon nanotubes was given by Lijima (Ajayan & Lijima, 1993). Carbon nanotubes are allotropes of carbon whose diameter is made up of nanoparticle size graphene with hexagonal lattices coiled as hollow cylinders. Equivalent lattice points align together, forming bonds and hence providing a stable structure. The coiling of carbon into nanotube is dependent on chiral vector represented by $C = na1+ ma2$, where n and m determine how the edges of graphene would align and a1, a2 describe the arrangement of carbon atoms arranged in graphene's hexagonal structure which help in determining the shape and properties of the nanotube. If $n = m$, then the nanotube would have an armchair configuration with metallic properties. Whereas if $m = 0$, then the nanotube would have a zig zag pattern with metallic or semiconducting properties and a chiral pattern for other values. Carbon nanotubes have greater thermal conductivity compared to their allotropes due to the presence of sp2 covalent bonding allowing them to transfer heat efficiently. The presence of sp2 covalent bonds also contributes to the mechanical strength of structure preserving them even after bending into various conformations (Patel et al., 2019).

Attractive properties such as small size, high surface area, greater thermal conductivity and mechanical strength allow carbon nanotubes to significantly contribute to agriculture reducing water, pesticide use and contribute to seed germination, plant growth, resistance against biotic and abiotic stresses, development of biosensors for improved crop productivity and disease management (Safdar et al., 2022a). Precise farming using CNTs can significantly reduce environmental damage by minimizing the use of agrochemicals in field, delivery the drugs to specific sites of the plants and production of improved quality and quantity of food grains to meet the demands of growing population (Shojaei et al., 2019).

TYPES OF CARBON NANOTUBES

Carbon nanotubes are classified into two types based on their structural arrangement and the direction of coiling.

Based on the number of layers, carbon nanotubes are classified into:

1. Single Walled Carbon Nanotubes (SWCNTs)
2. Multi Walled Carbon Nanotubes (MWCNTs)
    a. Conical Stacked cup Carbon Nanotubes (CSCNTs)

Single Walled Carbon Nanotubes (SWCNTs)

Single Walled Carbon Nanotubes have a distinct one dimensional tubule structure with strong covalent bonding between the carbon atoms. They have a single cylindrical carbon layer whose diameter is approximately 0.4 to 3 nm which depends on temperature at which they are synthesized. These nanotubes are synthesised with a catalyst and are prone to defects during functionalization. Single walled carbon nanotubes find its potential use in agricultural technology by promotion of various qualities such as root elongation and gene delivery in several species of plants including *Daucus carota, Cucumis sativus, Allium cepa and Lactuca sativa.* SWCNTs are also used as potential carriers for agrochemicals that can act as nano-delivery systems providing controlled release of fertilizers and pesticides necessary for the plants. This results in targeted chemical waste reductions and pollution as well as improved effectiveness in agricultural treatment applications. Their unique property of adsorption at root surfaces with little or no absorption into plant tissues further ensures that active compounds they carry would be delivered much more efficiently than without the compounds by restricting risk associated with nanoparticle accumulation (Cañas et al., 2008).

Apart from these, SWCNTs will also help improve the resistance of plants to the effect of different environmental stress agents. They have shown a lot of promise in terms of oxidative stress mitigation in plants, which could augment resistance to drought, salinity, or pathogen infliction. These features, combined with great surface area and electrical conductivity, allow them to modulate cellular stresses in plants, which may help have a better survival rate in adverse conditions. However, despite these promising applications, the effects of SWCNTs on different plant species are quite different. Some plants might show growth enhancement, while others, including lettuce and tomato plants, exhibit inhibited root elongation upon treatment with certain concentrations (Samadi, Saharkhiz, et al., 2021).

Multi Walled Carbon Nanotubes (MWCNTs)

These carbon nanotubes are stacked over one another and are present as clusters which are uniformly packed with size 20 to 30 nm. Multiple sheets of graphene are concentrically rolled which forms the structure. These nanotubes' diameter can vary with graphene layers which can be synthesized without a catalyst. These nanoparticles are functionalized with minimal defects (Mathew et al., 2021; Arunkumar et al., 2020). There is emerging evidence that the capabilities of nanotechnology can be harnessed for advanced agricultural strategies by applying it in ways that improve growth and productivity of plants. Among the available nanomaterials, multi-walled carbon nanotubes (MWCNTs) have currently attracted a great deal of interest because they promise enhancement of root elongation and total plant biomass. Recently, it has been claimed that oxidized MWCNTs (o-MWCNTs) induce root elongation and increased vegetative biomass along with increased dehydrogenase activity in seedlings, all of which are developmental for wheat (*Triticum aestivum*). Such results indeed indicate that MWCNTs could be transformed into tools of modern agriculture to enhance the vigor of plants (Wang et al., 2012).

The o-MWCNTs, when applied to wheat seedlings for root stimulation, depicted its cell proliferative effects necessary for germination. The o-MWCNTs, between 40-160 µg/mL, showed significant signs for root elongation. Such longer roots make possible optimum nutrient and water uptake for a healthier, bigger-growing plant. MWCNTs have stimulated root elongation possibly because of their penetrating ability into root cells causing an alteration in the mechanisms of the cells favoring elongation. The notable increment was attributed to improved water uptake aided by MWCNTs. Penetration of root cells by these nanotubes effectively interfered with physiological mechanisms, wherein dehydrogenase activity had an important contribution in plant metabolism. Dehydrogenase activity increased in a concentration-dependent manner, implying that o-MWCNTs probably aid in root function, which in turn makes the plant healthier and stronger (Zhao et al., 2017).

At the cellular level, o-MWCNTs promoted cell elongation in wheat roots. Microscopic analysis revealed that cells in the root elongation zone of treated seedlings were significantly longer than those in untreated plants. This evokes an MWCNT influence on hormone distribution and cellular organization leading to accelerated root development. In a study, it was confirmed that at least o-MWCNTs could enter root cells, but were not moved to the machine; therefore, translocation trouble of these in the plant system is minimal. Thus, the findings indicate the great potential of MWCNTs for agricultural application, especially in plant growth and resilience. However, further research is required to appreciate their long-term effects better and develop protocols for sustainable utilization. Harnessing their merits promises a huge boost in agricultural productivity, providing an avenue for modern strategies to enhance plant growth (Hu et al., 2021).

Conical Stacked Cup Carbon Nanotubes (CSCNTs)

CSCNTs have a conical shape along with a hollow core. These are a component of multiwalled carbon nanotubes composed of numerous graphite layers. The thickness of their walls increase minutely along the length. The presence of multiple layers provides them with the property of semi conductance (Thakur et al., 2022; Liu et al., 2019). Carbon nanotubes (CNTs) have aroused maximum attention in agricultural research because of their promises toward better enhancing and promoting growth, nutrient uptake, and stress tolerance. The nanomaterials include single-walled (SWCNTs) or multi-walled (MWCNTs) variants that could interact with plant cells at the molecular level, thanks to their property of unique physicochemical features. Because of studies on CNTs that will lead to properties suggesting penetration into plant cell walls, protective barriers made from cellulose microfibrils, hemicellulose, and pectin are probably breached. SWCNTs have a smaller diameter which, through passive diffusion, enhances their ability to penetrate cell wall pores; however, for MWCNTs, other penetration strategies have been described, such as enzymatic hydrolysis. Inside the cell, nanotubes then localize into different compartments such as cytoplasm, vacuoles, and the nucleus, influencing metabolic pathways and physiological functions (Serag et al., 2013).

Evidence from research indicates that direct exposure to oxidized MWCNTs (o-MWCNTs) could render root elongation and biomass accumulation improvements in wheat *Triticum aestivum*. In controlled conditions, o-MWCNT-treated wheat seedlings at concentrations from 10 to 160 μg/ml show their roots were significantly longer and had increased biomass compared to untreated plants. Root cell elongation and increased dehydrogenase activity, which makes possible water uptake and metabolic efficiency, are believed to encourage this growth. Apart from enhancing root development, CNTs also improve nutrient absorption by acting as carriers for the movement into plant roots of essential minerals such as nitrogen, phosphorus, and potassium (Abdulhussein et al., 2016). This favors nutrient-deficient soils where CNTs enhance the bioavailability of vital elements with better plant health overall and improved nutrient uptake (Karthik et al., 2014).

Their possible use in genetic engineering also provides a new avenue for the delivery of DNA and RNA into plant cells, circumventing conventional transformation techniques while allowing for more exact genetic modifications. Furthermore, CNTs have been found to influence plant immune responses associated with their activation due to signal transduction regulation, thereby mediating defense responses against biotic stresses. The retention of water through capillary action mediated by CNTs typically affects drought tolerance in plants treated with CNT because such plants have been noted to show water retention. Another potential implication for CNTs is to act as the nanocarrier for agrochemicals. Functionalizing the CNTs can ensure precise control over the timing of delivery of these agrochemical agents while ensuring minimum waste and averting environmental pollution, thus enhancing the overall efficiency of agrochemical treatments. Several studies have proved their antimicrobial activity, which can help in enhancing resistance to bacterial and fungal pathogens. Nonetheless, even with numerous benefits, CNTs are becoming controversial regarding their long-term environmental consequences and social impacts (Pourhashem et al., 2020).

EFFECTS OF CNT ON PLANT PHYSIOLOGY

Carbon nanotubes have the ability to change plant morphology and physiology which includes breaking seed dormancy, growth promotion and passing through cell walls for smart delivery of agrochemicals (Liu et al., 2009). Studies had reported both positive and negative effect of CNTs on plants wherein some species had improved biomass with enhanced shoot and root growth on other hand on few it had retarded the growth of root and shoot system (Lin & Xing, 2007; Majeed et al., 2020a). When tomato seeds were grown in media containing carbon nanotubes, it was shown to affect both germination and germination rate (Haghighi & Teixeira da Silva, 2014a). In another study CNTs were added while watering the tomato plants that improved reproductive efficiency with more flowers and fruit production. This showed that carbon nanotubes have the ability to penetrate the cell wall and the cell membrane. When dry seeds were placed in a medium containing carbon nanotubes, it was found that there was increased moisture content in the former compared to control leading to better yield and faster germination (Torney et al., 2007). Many other plant species including *Raphanus sativus, Zea mays, Cucumis sativus* were shown to break seed dormancy when exposed to carbon nanotubes (Lin & Xing, 2007). Unlike the traditional methods, which involves damaging the seed coat to break dormancy, carbon nanotubes can break dormancy without doing so and could enhance germination (Torney et al., 2007). SCWNT have the ability to increase seed germination by increasing the water uptake and gas diffusion. SCWNTs have the ability to induce increased expression of aquaporins, which further facilitates the uptake of water. When seeds were in contact with carbon nanotubes i.e, when soaked in jellies containing SCWNT or SCWNT solutions it was shown to improve germination (Vithanage et al., 2017).

Interaction of Carbon Nanotubes with Plants

Mechanism and Transport Within Tissues

The uptake and interaction of carbon nanotubes by the plants depends on their shape, size and their functional properties (Kim et al., 2005). If the size of the carbon nanotube is very small, they would diffuse out of the cell whereas if they are too large, they would remain unabsorbed by the cell wall (Li et al., 2001). The main sources for carbon nanotubes are from the environment, industrial by-products such as soot and fertilizers used during agriculture (Baruah & Dutta, 2009). CNTs are tubular in nature and are effectively taken up while watering the plants. They enter the lumen of tracheal elements to form capillaries that improve the water uptake through xylem. When released in the soil, these carbon nanotubes interact with soil which further enhances their uptake through the plant cell walls (Etxeberria et al., 2009). The carbon nanotubes are taken up by the root through capillary action from where they are transported to the leaves. The carbon nanotubes are transported from one cell to another through plasmodesmata (Rico et al., 2011). The external application of these nanotubes can be through the stomatal route or through cuticles (Eichert & Goldbach, 2008). When these applied, it was shown to reduce variability in germination and increased reproducibility of agricultural plants (Patel et al., 2020). Under drought conditions, SWCNTS are known to activate defense mechanisms to protect the plant by reducing oxidative stress by preventing the formation of free radicals like H_2O_2 (Vithanage et al., 2017). Carbon nanotubes have shown to influence growth rate and metabolism. It is shown that when carbon nanotubes are used in lower concentrations (10-30 mg/L), they are known to promote growth, whereas higher concentrations (40 mg/L) inhibit the growth of plants (Samadi, Asgari Lajayer, et al., 2021). The photosynthetic efficiency of plants is known to increase by increasing the rate of light absorption contributing to increased electron transport which further allows for efficient energy conversion in plants (Ali et al., 2021). MCWNT are known to promote shoot and root development in *Glycine max, Triticum aestivum, Brassica juncea.* These carbon nanotubes have the ability to promote reproduction (Majeed et al., 2020b). CNTs have the ability to alter the concentrations of essential nutrients like Ca, Mg, K, P and S which can directly impact the growth rate and stress responses (Martinez-Ballesta et al., 2020).

Modulation of Plant Signalling Pathways by CNTs

Carbon nanotubes, particularly MWCNTs, have the ability to modulate plant signalling pathways, significantly influencing the growth, development process and stress responses. Studies conducted by Zhao et al on *Arabidopsis thaliana* and *Brassica napus* have shown that MWCNTs enhanced root hair formation through activation of nitric oxide (NO) and ethylene signalling pathways (Zhao et al., 2020). Similarly, in tomato (*Solanum lycopersicum*), MWCNTs promote lateral root formation by including NO accumulation (Cao et al., 2020). Beyond growth, CNTs also influence stress-related pathways such as salicylic acid (SA) and jasmonic acid (JA), enhancing stress tolerance and resistance to pathogens (Sharma et al., 2025). Additionally, CNTs affect nutrient assimilation and transport (Hu et al., 2021).

EFFECT OF CNTs ON PLANT ASSOCIATED SOIL MICROBES

CNTs have the ability to interfere with the soil microorganisms. Microbial growth is inhibited through mechanisms such as disrupting cell walls, cell membranes and generating reactive oxygen species and in some cases, there is active expression of degradation genes which negatively affect the microbial flora. Exposure to CNTs can alter the microbial community and its biomass. However, these soil microorganisms also have the ability to counteract these effects (Zhang et al., 2020). Bacterial colonies could be revived, but fungal growth was inhibited for a longer period of time. MWCNT has the ability to inhibit microbial growth at a higher concentration (Vithanage et al., 2017). CNTs have the ability to attack plant pathogens, reducing the dependency on pesticides. Also they are known to reduce the reproductive ability of viruses like Tobacco Mosaic Virus. The efficiency of these CNTs depends on its concentration and contact time. Its mode of action is disruption of cellular structure by rupturing its cell wall and cell membrane. But persistence of CNT in soil, allows indigenous microorganisms like

Pseudomonas aeruginosa to form aggregates and develop antibiotic resistance (Maksimova, 2019). Soil is typically considered to be one of the main recipients for nanomaterials, as they are known to directly interact with the ecosystem. Studies have shown that the soil composition and the microflora present in soil are significantly affected when exposed to nanoparticles. The effect of well-defined CNTs when applied to soil and its effect on extracellular enzyme activity was studied indicating significant effects like suppression of activity and biomass of microbes in soil when MWCNTs are applied in higher concentrations of 50, 500 and 5000μg (Chung et al., 2011). Studies done on the toxicity of Carbon Nano Materials in plant based effects have shown that exposure of plants to high concentration of MWCNTs and C_{60} have significant effects on the enzyme activity of Peroxidase (POD) and Superoxide Dismutase (SOD) in the roots. The study also highlights the result that application of Carbon Nano Materials (CNM) may lead to increased toxicity in the environment specifically to rice species disrupting the rhizosphere soil microbe community which could pose potential risk to food safety. The findings were mainly based on paired -end sequencing strategies which are utilized to analyze whether the exposure to CNTs at variable concentration will affect the soil bacterial community (Hao et al., 2018).

CARBON NANOTUBES IN PRE-HARVEST TECHNOLOGY

CNTs as Carriers for Nano Fertilizers

CNTS are preferred as carriers due to their ability to control and accurately deliver nutrients to the plant and improve their immunity against pathogens mechanisms (Aacharya & Chhipa, 2020). CNTs can be applied as thin coats of polymer films or can be delivered directly as emulsions to plants (Thavaseelan & Priyadarshana, 2021). CNTs are able to penetrate the plant cell wall and ensure enhanced nutrient absorption and precise delivery of agrochemicals (Aacharya & Chhipa, 2020). CNTs when integrated in soil can readily absorb nitrogen from ammonia releasing hydrogen which induces the plant to absorb more water and other nutrients from soil (Xu et al., 2024). Application of CNTs will improve the uptake potential of plants and hence discourage the excess application of fertilizers to soil for a sustainable agricultural practice. When CNTs are used as carriers in precise farming their uptake by plants varies from species to species. However, the key factor influencing its uptake is the ratio of its size to the cell wall porosity in plants (Thavaseelan & Priyadarshana, 2021). Functionalized nanoparticles with carboxylic groups had reported better solubility, stability and uptake by plants. When CNTs are applied in soil at low concentration they are effective in improving the germination and plant growth which could be attributed to ROS and RNS produced by plants (Safdar et al., 2022a). Another theory explains their growth promoting action directly due to their interaction with DNA and other transcription factors (Patel et al., 2020).

Carbon Nanotubes in Biosensors

Nanobiosensors are preferred over traditional biosensors like metal oxide or silica owing to its unique features like greater surface area, fast detection, high stabilities and enhanced luminous efficiency. More specifically, carbon nanobiosensors are used in agriculture to detect soil humidity, hazardous materials, pesticides, plant pathogens or any other substances which can significantly contribute to improved agricultural practices (Safdar et al., 2022b). These carbon nanotubes, especially SWCNT which are flexible, can be directly introduced into the plant through needleless syringe transferring them into the parenchyma tissues after where it directly interacts with the plant. CNT based biosensors once inside the plant system can detect chemical changes in plants under stress conditions (Safdar et al., 2022a). CNT can be combined with other nanoparticles of gold, platinum to increase the electrical conductivity in electrodes which improves the sensitivity in detection. CNTs can easily detect analytes based on changes in atoms resulting in variation in electrochemical, and optical effects (Magar et al., 2017). Electrochemical biosensors are mostly preferred due to simplicity and low cost. When a molecule interacts with electrochemical CNT, due to change in conductance, a signal is generated which helps in monitoring the plant environment. Whereas, optical biosensors use light to detect changes (Hammond et al., 2016). Bioluminescent sensors present in the rhizosphere have the ability to detect varying nutrient conditions which monitor nutrient uptake by the plant. When there is an increased load of bacteria, there is decreased intensity in bioluminescence indicating less nutrient availability for the plant (Ferrier & Honeychurch, 2021). Near infrared fluorescence sensors constructed with SWCNTs for the analysis of hydrogen peroxide in plants as the ROS accumulates under stressed conditions. In A*rabidopsis thaliana* leaves where H_2O_2 has a quenching effect on the fluorescence of the nanosensor that was monitored to report the concentration of H_2O_2 in the plant system (Wu et al., 2020). PdNPs-SWCNTs@Cu-MOF-74 nanocomposite based nanosensor was constructed for detection of ethylene which is the indicator plant hormone for monitoring fruit ripening (Yan et al., 2023). Plant metabolism can be studied when CNT are used to detect glucose concentrations which directly contributes to efficiency of photosynthesis and overall plant growth (Majeed et al., 2020b).

Carbon Nanotubes in Post Harvest Technology

The quality of crop produced is often affected by external parameters such as oxidative stress, high moisture content, temperature, availability of nutrients etc which could significantly contribute to post harvest losses. Treating the crop plants during the pre- harvest period with Carbon Nano tubing and Nanographene oxide have shown to promote better tolerance in water deficit conditions as well as in upregulating antioxidant defense (Ahmadi-Majd, Rezaei Nejad, et al., 2022). Studies have shown that grapes are highly susceptible to oxidative stress and damage especially during transportation and storage, affecting their commercial value. This bring into effect the utilization of Multi-Wall Carbon NanoTubes (MWCNT) at low concentrations of 25 and 50 mg/l led to the increased production of peroxidase, catalase and superoxide mutase which lowered the oxidative damage (Sha et al., 2022). To address the adverse effects of environmental stress on cut flower longevity, 0-80 mg/l concentration of Single-Wall Carbon NanoTubes (SWNT) and Multi-Wall Carbon NanoTubes were found to be effective in prolonging the shelf life as well as flower blooming (Ahmadi-Majd, Mousavi-Fard, et al., 2022). Some of the applications of carbon nanotubes in the pre harvest and post harvest technology has been depicted in figure 1.

Carbon Nanotubes as Effective Agents in Preservation

Cryopreservation also referred to as Cryobanking is one of the most practiced techniques in long term preservation of seeds. Despite its effective preservation, studies have shown that Reactive Oxygen Species (ROS) induced stress and membrane damage could evidently cause cell death in samples preserved. In order to improve upon this drawback, addition of Single-Wall Carbon NanoTubes (SWCNT) in Cryopreservation of *Agapanthus praecox* showed potential in interfering with the ROS activity thereby limiting its adverse effects (Ren et al., 2020). Studies conducted in *Allium cepa* demonstrate that along with seed preservation, carbon nanotubes also find use in promoting seed germination on treatment with ZnO/MWCNT at concentrations of 15μg/ml (Kumar et al., 2018). Carbon NanoTube aids in increasing the penetration of cell walls, facilitating faster uptake of air and water through the testa of seed enabling better germination and growth. This was observed in studies conducted in *Triticum aestivum* where plants were subjected to treatment with MWCNTs of 80 μg/ml (Haghighi & Teixeira da Silva, 2014b).

Carbon Nanotubes as Carriers in Seed Priming

Seed priming techniques are mainly utilized to enhance the germination process in plants facilitating all necessary metabolic processes required for germination without the emergence of radicals. Studies conducted on utilizing the effect of Multiwall Carbon Nanotubes in *Zea mays* demonstrated modulated germination of seeds and early germination carried out under cadmium toxicity. Different concentrations of MWCNTs ranging from 0 mg/l, 100 mg/l and 200 mg/l were studied along with supplementation of cadmium which improves the antioxidant defense mechanism in plants. The presence of MWCNTs were observed to improve the efficiency of peroxidase (POD), superoxide dismutase (SOD) as well as several antioxidant enzymes (Chen et al., 2021). Treatment of carbon nanomaterials (1000 mg/l) for seed priming in trials conducted on tomatoes had a significant effect on decreasing height of plant (29%) and shoot biomass (63%) (Pandey et al., 2018). Seed priming utilizing graphene has led to increment in vitamin C levels up to 78% and β-carotene levels to 11 fold (López-Vargas et al., 2020). In plants which are highly susceptible to environmental stress like salinity and arid conditions, seed priming with carbon nanotubes has shown to mitigate the negative effects (Liu et al., 2009). On treatment of tomato seeds with Carbon nanotubes (CNTs) and graphene in varying concentrations of 50 mg/l, 250 mg/l and 500 mg/l have reported a protective action by altering the bioactive compounds (González-García et al., 2022). Seed priming involving the combination of multi-walled carbon nanotubes (MWCNTs) along with cold plasma in *Momordica charantia* conferred an increase in growth of shoots, roots and increase in dry mass, leaf length and vigor index. The study also highlights the non toxic nature of MWCNTs at high dosage and long term exposure (Seddighinia et al., 2020). The effects of OH-MWCNTS and SiO_2 after application on seeds of *Brassica juncea* were shown to produce significant agronomical traits when applied in concentration of 25 μg/ml, 50 μg/ml, 75 μg/ml, 100 μg/ml and 125μg/ml. The screening of the test plants showed an increase in 3 traits of the plants including leaf petiole length, number of seeds per siliqua and siliquae length. In contrast, five traits were observed to show a significant decrease including overall leaf length, harvest index, main length of inflorescence, biological yield and number of siliquae per plant (Dhingra et al., 2022). Studies conducted on *Triticum aestivum* utilizing Oxidized Multi-Walled Carbon NanoTubes (O-MWCNT) at concentration ranging from 10 μg/ml to 160 μg/ml resulted in increase of dehydrogenase activity. This suggested that the application of O-MWCNT to wheat plants promotes cell elongation as well as increase the concentration of dehydrogenase resulting in better biomass production and faster root growth (Wang et al., 2012). Seed priming also presents high germination rates in plants exposed to drought stress, along with increase in seed vigour index and overall elongation in cells of root and stem. The effects of priming technique involving germination of *Alnus subcordata* under drought stress conditions of (0,-2,-4,-6,-8 and -10 bar) were done after priming the seeds with MWCNT at varying concentrations. The seed showed rapid germination rates at concentration of 100 mg/l along with elevated resistance against drought stress (Rahimi et al., 2016).

Figure 1. Application of carbon nanotubes in pre harvest and post harvest technology in plants

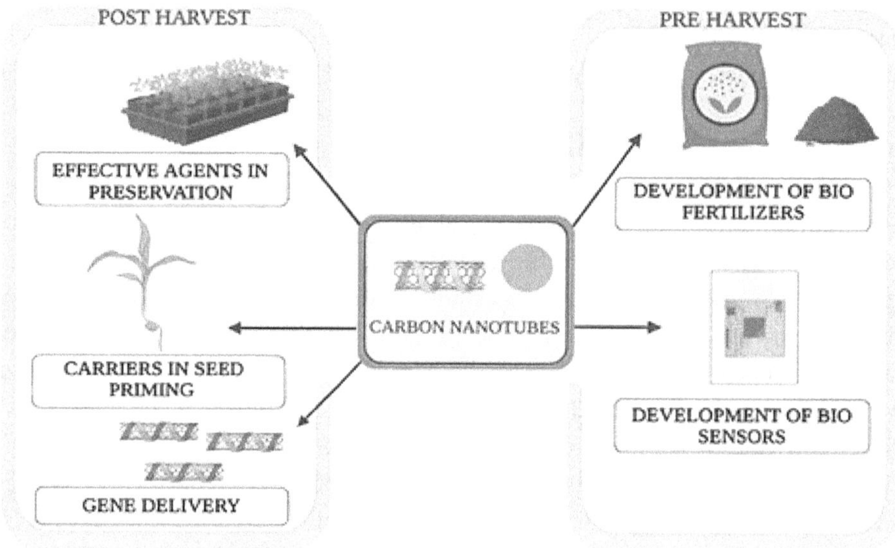

Carbon Nanotubes in Gene Delivery

In plants, traditional methods of gene delivery methods like electroporation, biolistics always had certain limitations such as tissue damage, low transformation efficiency and uncontrolled integration of DNA. In order to overcome these drawbacks which could potentially affect the efficacy of gene transfer, CNTs are utilized to facilitate productive DNA delivery into plant species. This approach of gene delivery with the aid of Carbon NanoTubes as mentioned in figure 2 provided higher biocompatibility and tensile strength along with high aspect ratio in plant cells (Demirer et al., 2019). Studies have shown transformation of Chloroplast with chitosan- complexed SWCNTs when utilized as nanocarriers in delivery of pDNA has effective tracking efficiency in cells. Significant increase in the efficiency of pDNA delivery was observed when CS-SWCNT were prepared from deacetylated chitosan (Kwak et al., 2019). Functionalized single or multi walled carbon nanotubes with amino groups are promising candidates to deliver DNA, siRNA, plasmid DNA. CNT functionalized with amino and carboxylated CNTs are used for effective transfection *in vivo* and *in vitro* conditions. As carboxylated CNTs possess negative charge inorder to prevent their electrostatic repulsion from DNA. Some of the studies have been mentioned in table 1. The nucleic acids are amino modified and coupled with CNTs (Bates & Kostarelos, 2013).

Figure 2. Gene delivery using carbon nanotubes in plants

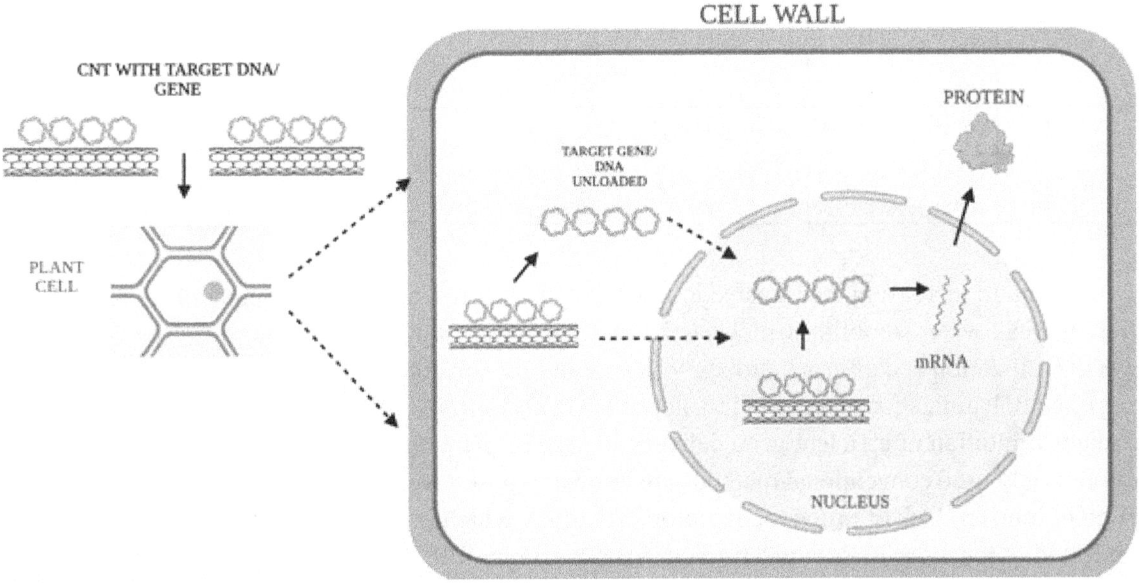

Table 1. Gene delivery efficiency in various plants using carbon nanotubes

Carbon Nanotube Type	Target Plant	Concentration	Gene	Genetic Fate	Reference
Arg- SWCNT	*Nicotiana tabacum*	10µg, 20µg, 40µg, 80µg, 250µg, 400 µg	GFP	↑ Root cell wall porosity ↑ Transfection rate	(Golestanipour et al., 2018).
SWCNT MWCNT	*Nicotiana tabacum*	20 µg 15 µg	yfp	Transient expression in protoplast Preserved functional activity ↑ dissociation from nanocarriers	(Burlaka et al., 2015) .
SWCNT MWCNT	*Nicotiana tabacum*	40 µg 30 µg	nptII	Transformation of callus and explants	(Burlaka et al., 2015) .
PEI-SWCNT	*Nicotiana benthamiana* *Triticum aestivum* *Eruca sativa* *Gossypium hirsutum*	500 ng	GFP	↑ Transfection rate Transient expression	(Demirer et al., 2019).

continued on following page

Table 1. Continued

Carbon Nanotube Type	Target Plant	Concentration	Gene	Genetic Fate	Reference
pDNA-CNT	*Oryza sativa*	1.5ng pDNA/ µl	GFP	↑ Transcription and translation of gene ↑ Fluorescence	(Dunbar et al., 2022).
Arg- SWCNT	*Nicotiana tabacum*	10µg, 80µg, 200µg	GFP	↑ Transfection rate ↓Transfection at very high concentration of SWNT ↑ Root cell wall porosity	(Golestanipour et al., 2018).

In addition to increasing the transfection efficiency, the presence of carbon nanotubes results in formation of weak water solubility which can be modified later on to increase the solubility. Studies have shown the utilization of carbon nanotubes with ammonium polyethylenimine and amphiphilic molecules like pyrenyl polyamine for successful transfer of DNA and siRNA into cells (Golestanipour et al., 2018). The major limitation of efficient gene delivery in case of plant cells is the presence of rigid cell walls. The commonly used conventional methods are known to pose several demerits such as damage to tissue, inefficient transfer, lack of proper penetration efficiency which is effectively addressed by using CNTs. Several studies have also highlighted the benefits of arginine-SWCNTs in transfer of bioactive molecules like GFP coded plasmid into plants like tobacco root cells (Singh et al., 2020).

MACHINE LEARNING AND ARTIFICIAL INTELLIGENCE IN PLANT SCIENCES

Machine learning typically focussed on relying on a set of computational approaches to determine the predictive patterns in data. This development ensures the progress in plant breeding programmes aimed at validating the connection of genotypes to phenotypes and at the biochemical, macroscopic levels (Van Dijk et al., 2021). Studies have shown that over the course of time there has been a significant increase in the number of genome sequencing projects which demands the need for better analysis of available data and utilizing them in co-integration with other fields such as crop improvement, nutrigenomics and analysis of biochemical data. This brings the need for utilization of Artificial Intelligence involving the usage of an algorithm for Machine learning to integrate and analyse biological data (Osama et al., 2015).

APPLICATION OF AI AND MACHINE LEARNING IN PLANT BREEDING

Over the years, advancement in the technology has enabled implementation of Machine Learning algorithms in accurate identification of Genotype and accuracy in estimating the yield of crops using XGBoost and Random Forests (RF) (Eftekhari et al., 2024). Studies conducted on the utilization of ML models like eXtreme Gradient Boosting (XGBoost) were done to estimate the overall yield in peanut yields enabling the determination of better strategies for improving efficiency of plant breeding programs. The utilization of Random Forest model is shown to effectively minimize the occurrence of Type I and Type II errors (Pugh et al. 2024). Several other studies focus on utilizing techniques which incorporate various integrated feature selection methods such as PCRF-RFE (Pearson Correlation Coefficient (PC) and Random Forest) to essentially evaluate the vegetation index (Li et al., 2023). Application of ML algorithms (RF and Artificial Neural Networks (ANN)) along with FAO-Agro-Climate method were studied to understand its effectiveness in identification of environmental factors and soil properties for predicting the best suitable parameters of increased yield of wheat (Mousavi et al., 2024). The technique of somatic embryogenesis (SE) is a valuable tool for developmental analysis where the somatic cells could be utilized for regeneration of bipolar structures. The analysis of this is efficaciously carried out by instituting the usage of Deep Learning methodologies like Convolutional Neural Networks (CNN) to obtain structural and spatial information from images generated (Davidson et al., 2024). The technique of Convolutional Neural Networks mainly aims to develop a natural visual perception of plants facilitating easier detection, tracking, visual saliency aiding in better regression analysis of biological data (Niazian & Niedbała, 2020). The utilization of ML techniques like CFNet-VoV-GCSP-LSKNet-YOLOv8s which is shown to be effective in identification as well as control of pests and diseases (Li et al., 2024). Studies have shown that advancement in development of new technologies for Machine Learning and computational vision technology have significantly improved several aspects of automation, disease monitoring and real time processing (Wang et al., 2023).

INTEGRATION OF CNT- BASED TECHNOLOGY WITH MACHINE LEARNING

The utilization of carbon nanotubes (CNTs) as a tool for efficient gene delivery in plant breeding programs presents significant opportunities for enhancing genetic transformation and crop improvement. However, challenges such as the complexity of CNT synthesis, variability in functionalization methods, and concerns regarding cytotoxicity and biocompatibility limit their practical applications. To overcome these challenges, researchers are exploring the integration of CNT-based gene delivery systems with Machine Learning (ML) algorithms, which can optimize experimental conditions, predict outcomes, and enhance overall efficiency in plant genetic engineering. CNT can also assist in synthesis of the carbon nanotube and to capture their high resolution images to predict their mechanical, thermal, electronic and optical properties (Chen and Tang 2024).

Machine Learning (ML) and Artificial Intelligence (AI)-driven models have demonstrated significant potential in integrating complex biological datasets, improving data interpretation, and increasing confidence levels in plant breeding research. By leveraging AI-based models, researchers can analyze large-scale genomic, transcriptomic, and phenotypic data, facilitating more precise selection of desirable traits and improving breeding efficiency (Zhang et al. 2021). Among the various ML techniques, methods such as Random Forest (RF), Extreme Gradient Boosting (XGBoost), and Artificial Neural Networks (ANNs) have been extensively validated for their ability to manage complex and heterogeneous biological data. These algorithms effectively handle high-dimensional datasets, reduce noise, and improve predictive accuracy by balancing feature weights and minimizing dimensional effects. RF, an ensemble learning method, enhances classification reliability, while XGBoost optimizes decision trees through gradient boosting. ANNs, inspired by neural networks, can detect intricate patterns within datasets, making them highly useful for plant breeding applications. By employing these ML techniques, researchers can streamline data processing, refine trait selection, and enhance the efficiency of modern breeding programs (Deng et al. 2023).

AI IN DRUG DEVELOPMENT FOR PLANT HEALTH

One of the major obstacles faced during efficient plant breeding techniques is the proper storage and pathogen susceptibility of the crops. To overcome this several traditional techniques have been practiced over the years which are relatively time consuming and resource intensive. To combat this need, advancement in technology has enabled the implementation of Artificial Intelligence (AI) and Machine Learning (ML) algorithms for pathogen detection and management through drug screening (James & Owen, 2022). Studies conducted assessing the feasibility of this technology have tested the utilization of microfluidic platforms which are integrated with AI based algorithms for pathogen detection and screening (Negi & Anand, 2024).

The platform of microfluidics enables the drug screening of plants for thousands of plants simultaneously which is beneficial for speeding up the process of drug development. The combination of AI based tools and ML algorithms helps in making prediction and analysis of the complex data provided. This enables high throughput of expression patterns and increased efficacy of potential drugs (Joe & Keith, 2022). Over the recent years, utilization of Nanoarchitectonics which involves experimental and statistical analysis to optimize nanomaterials have been widely used. The application of Artificial Intelligence along with the branch of Nanoarchitectonics to produce an analysis of mechanical, chemical and physical aspects of material used (Oviedo et al., 2022). The concept of Nano Based Drug Delivery Systems (DDSs) have been known to elevate the efficacy of drug potential as well as reducing the side effects. The implications of AI are merged with DDSs to develop personalized drug delivery catering to the personal needs of an individual, by optimization of the drug formula and its properties (Heydari et al., 2024).

MULTI OMICS DATA AND CNT APPLICATION

The exponential growth in the global population has led to an increased demand for food security, necessitating the development of high-yielding and climate-resilient crop varieties. Addressing these challenges requires the integration of cutting-edge technologies to enhance agricultural productivity while ensuring sustainability. Over the years, significant advancements in Multi-Omics approaches, including proteomics and transcriptomics, have played a crucial role in revolutionizing plant breeding strategies. These approaches have facilitated the identification of key genes, proteins, and regulatory elements that contribute to stress tolerance, growth, and yield enhancement in crops, thereby aiding in the development of improved cultivars tailored to withstand environmental stressors (Mahmood et al. 2022). Among the various Omic disciplines, metabolomics has emerged as a powerful tool for understanding plant metabolic networks, particularly those activated in response to biotic (pathogen-induced) and abiotic (drought, salinity, temperature fluctuations) stresses. By analyzing small-molecule metabolites, metabolomics provides crucial insights into plant physiological and biochemical responses under varying environmental conditions. The ability to characterize metabolic pathways has greatly enhanced the screening and selection of stress-tolerant crop genotypes at a molecular level, enabling the identification of key biomarkers associated with resilience and productivity (Razzaq et al. 2019). In recent years, advancements in Next-Generation Sequencing (NGS) technology have significantly enhanced the precision and efficiency of Omics-based studies. NGS has revolutionized high-throughput data generation, allowing for comprehensive genomic, transcriptomic, and epigenomic analyses with unprecedented accuracy and sensitivity. This cutting-edge technology has facilitated large-scale screening of plant genetic diversity, enabling researchers to decode complex regulatory networks and identify novel targets for genetic improvement. Additionally, NGS-based platforms have minimized the likelihood of errors in Omics experiments, ensuring robust and reliable data for predictive modeling and functional annotation of stress-responsive genes (Schmidt et al. 2020).

INTEGRATION OF OMIC DATA FOR BETTER UNDERSTANDING OF CNT- PLANT INTERACTION

With the increasing global demand for agricultural produce, the application of Engineered Nanomaterials (ENMs) is being explored for their potential to enhance crop yield while mitigating risks associated with environmental stressors. ENMs have shown promising effects in improving plant resilience, growth, and productivity by modulating physiological and biochemical pathways. However, to comprehensively understand the metabolic interactions and underlying mechanisms governing ENM-plant interactions, advanced discovery-driven Omic technologies are extensively employed (Majumdar and Keller 2021). These high-throughput approaches facilitate the analysis of global molecular changes, enabling researchers to decipher complex networks of gene expression, protein dynamics, and metabolite fluxes. Among the various Omic approaches, transcriptomics has emerged as a powerful tool for predictive analysis, particularly in identifying transcription factors that could serve as biomarkers for assessing ENM-induced toxicity. By profiling gene expression patterns, transcriptomics enables researchers to predict potential molecular disruptions and regulatory shifts in plant systems following ENM exposure (Ruotolo et al. 2018). Additionally, metabolomics—a comprehensive analysis of plant metabolites—provides valuable insights into the biochemical alterations associated with ENM interactions. This approach allows for the real-time monitoring of metabolic changes in plant cells, thereby serving as a crucial indicator of plant physiological responses (Pérez-Labrada et al. 2019). A specific case study involving the multi-walled carbon nanotubes (MWCNTs) in *Solanum nigrum* under induced stress conditions demonstrated the sensitivity of plants to MWCNT exposure. Through transcriptomic and metabolomic analyses, researchers observed significant metabolic reprogramming, indicating activation of plant defense mechanisms in response to ENM-induced stress. The Omic-based approach provided a detailed understanding of stress-responsive pathways, shedding light on the adaptive strategies employed by plants to cope with ENM-related challenges (Chen et al. 2022).

CONCLUSION

The integration of carbon nanotubes (CNTs) into plant sciences has demonstrated significant potential in revolutionizing agricultural practices through enhanced nutrient delivery, stress tolerance, and genetic advancements. Their physicochemical properties make them suitable candidates for improving plant growth, facilitating efficient gene transfer, and developing nanoscale agricultural inputs such as nano-pesticides and nano-insecticides. Additionally, their interactions with plant-associated microbes and soil ecosystems highlight their role in modulating soil fertility and plant health. However, despite these advantages, concerns regarding CNT toxicity, environmental persistence, and unintended physiological effects necessitate further investigation to ensure their safe and sustainable application. Future research should focus on establishing standardized protocols for CNT synthesis, functionalization, and application to minimize risks associated with their use. Comprehensive toxicological assessments at molecular, cellular, and ecological levels are essential to determine long-term implications. Additionally, the integration of machine learning (ML), artificial intelligence (AI), and multi omics approaches offers promising avenues for optimizing CNT utilization in precision agriculture and plant breeding. These computational tools can facilitate predictive modeling, data-driven decision-making, and improved efficiency in agricultural applications. Moving forward, interdisciplinary collaborations between nanotechnology, plant sciences, and computational biology will be crucial in developing targeted, eco-friendly CNT-based solutions. Sustainable nanotechnology framework should be prioritized to harness the full potential of CNTs while ensuring environmental safety and agricultural resilience.

NOTE

Sahithi, Gicy, and Manasa have contributed equally for this chapter.

REFERENCES

Aacharya, R., & Chhipa, H. (2020). Nanocarbon fertilizers: Implications of carbon nanomaterials in sustainable agriculture production. In Abd-Elsalam, K. A. (Ed.), *Carbon Nanomaterials for Agri-Food and Environmental Applications* (pp. 297–321). Elsevier., DOI: 10.1016/B978-0-12-819786-8.00015-3

Abdulhussein, A. T., Kannarpady, G. K., Wright, A. B., Ghosh, A., & Biris, A. S. (2016). Current trend in fabrication of complex morphologically tunable superhydrophobic nano scale surfaces. *Applied Surface Science*, *384*, 311–332. DOI: 10.1016/j.apsusc.2016.04.186

Ahmadi-Majd, M., Mousavi-Fard, S., Rezaei Nejad, A., & Fanourakis, D. (2022). Carbon nanotubes in the holding solution stimulate flower opening and prolong vase life in carnation. *Chemical and Biological Technologies in Agriculture*, *9*(1), 15. DOI: 10.1186/s40538-021-00264-1

Ahmadi-Majd, M., Rezaei Nejad, A., Mousavi-Fard, S., & Fanourakis, D. (2022). Postharvest application of single, multi-walled carbon nanotubes and nanographene oxide improves rose keeping quality. *The Journal of Horticultural Science & Biotechnology*, *97*(3), 346–360. DOI: 10.1080/14620316.2021.1993755

Ajayan, P. M., & Lijima, S. (1993). Capillarity-induced filling of carbon nanotubes. *Nature*, *361*(6410), 333–334. DOI: 10.1038/361333a0

Ali, S., Mehmood, A., & Khan, N. (2021). Uptake, translocation, and consequences of nanomaterials on plant growth and stress adaptation. *Journal of Nanomaterials*, *2021*(1), 6677616. DOI: 10.1155/2021/6677616

Arunkumar, T., Karthikeyan, R., Ram Subramani, R., Viswanathan, K., & Anish, M. (2020). Synthesis and characterisation of multi-walled carbon nanotubes (MWCNTs). *International Journal of Ambient Energy*, *41*(4), 452–456. DOI: 10.1080/01430750.2018.1472657

Baruah, S., & Dutta, J. (2009). Nanotechnology applications in pollution sensing and degradation in agriculture: A review. *Environmental Chemistry Letters*, *7*(3), 191–204. DOI: 10.1007/s10311-009-0228-8

Bates, K., & Kostarelos, K. (2013). Carbon nanotubes as vectors for gene therapy: Past achievements, present challenges and future goals. *Advanced Drug Delivery Reviews*, *65*(15), 2023–2033. DOI: 10.1016/j.addr.2013.10.003 PMID: 24184373

Burlaka, O. M., Pirko, Ya. V., Yemets, A. I., & Blume, Ya. B. (2015). Plant genetic transformation using carbon nanotubes for DNA delivery. *Cytology and Genetics*, *49*(6), 349–357. DOI: 10.3103/S009545271506002X PMID: 26841488

Cañas, J. E., Long, M., Nations, S., Vadan, R., Dai, L., Luo, M., Ambikapathi, R., Lee, E. H., & Olszyk, D. (2008). Effects of functionalized and nonfunctionalized single-walled carbon nanotubes on root elongation of select crop species. *Environmental Toxicology and Chemistry*, *27*(9), 1922–1931. DOI: 10.1897/08-117.1 PMID: 19086209

Cao, Z., Zhou, H., Kong, L., Li, L., Wang, R., & Shen, W. (2020). A novel mechanism underlying multi-walled carbon nanotube-triggered tomato lateral root formation: The involvement of nitric oxide. *Nanoscale Research Letters*, *15*(1), 49. DOI: 10.1186/s11671-020-3276-4 PMID: 32103348

Chen, J., Zeng, X., Yang, W., Xie, H., Ashraf, U., Mo, Z., Liu, J., Li, G., & Li, W. (2021). Seed priming with multiwall carbon nanotubes (MWCNTs) modulates seed germination and early growth of maize under cadmium (Cd) toxicity. *Journal of Soil Science and Plant Nutrition*, *21*(3), 1793–1805. DOI: 10.1007/s42729-021-00480-6

Chung, H., Son, Y., Yoon, T. K., Kim, S., & Kim, W. (2011). The effect of multi-walled carbon nanotubes on soil microbial activity. *Ecotoxicology and Environmental Safety*, *74*(4), 569–575. DOI: 10.1016/j.ecoenv.2011.01.004 PMID: 21310485

Davidson, S. J., Saggese, T., & Krajňáková, J. (2024). Deep learning for automated segmentation and counting of hypocotyl and cotyledon regions in mature *Pinus radiata* D. Don. Somatic embryo images. *Frontiers in Plant Science*, *15*, 1322920. Advance online publication. DOI: 10.3389/fpls.2024.1322920 PMID: 38495377

Demirer, G. S., Zhang, H., Goh, N. S., González-Grandío, E., & Landry, M. P. (2019). Carbon nanotube–mediated DNA delivery without transgene integration in intact plants. *Nature Protocols*, *14*(10), 2954–2971. DOI: 10.1038/s41596-019-0208-9 PMID: 31534231

Dhingra, P., Sharma, S., Singh, K. H., Kushwaha, H. S., Barupal, J. K., Haq, S., Kothari, S. L., & Kachhwaha, S. (2022). Seed priming with carbon nanotubes and silicon dioxide nanoparticles influence agronomic traits of Indian mustard (*Brassica juncea*) in field experiments. *Journal of King Saud University. Science*, *34*(4), 102067. DOI: 10.1016/j.jksus.2022.102067

Dunbar, T., Tsakirpaloglou, N., Septiningsih, E. M., & Thomson, M. J. (2022). Carbon nanotube-mediated plasmid DNA delivery in rice leaves and seeds. *International Journal of Molecular Sciences*, *23*(8), 8. Advance online publication. DOI: 10.3390/ijms23084081 PMID: 35456898

Eftekhari, M., Ma, C., & Orlov, Y. L. (2024). Editorial: Applications of artificial intelligence, machine learning, and deep learning in plant breeding. *Frontiers in Plant Science*, *15*, 1420938. Advance online publication. DOI: 10.3389/fpls.2024.1420938 PMID: 38841285

Eichert, T., & Goldbach, H. E. (2008). Equivalent pore radii of hydrophilic foliar uptake routes in stomatous and astomatous leaf surfaces – further evidence for a stomatal pathway. *Physiologia Plantarum*, *132*(4), 491–502. DOI: 10.1111/j.1399-3054.2007.01023.x PMID: 18334002

Etxeberria, E., Gonzalez, P., & Pozueta, J. (2009). Evidence for two endocytic transport pathways in plant cells. *Plant Science*, *177*(4), 341–348. DOI: 10.1016/j.plantsci.2009.06.014

Ferrier, D. C., & Honeychurch, K. C. (2021). Carbon Nanotube (CNT)-based biosensors. *Biosensors (Basel)*, *11*(12), 12. Advance online publication. DOI: 10.3390/bios11120486 PMID: 34940243

Golestanipour, A., Nikkhah, M., Aalami, A., & Hosseinkhani, S. (2018). Gene delivery to tobacco root cells with single-walled carbon nanotubes and cell-penetrating fusogenic peptides. *Molecular Biotechnology*, *60*(12), 863–878. DOI: 10.1007/s12033-018-0120-5 PMID: 30203379

González-García, Y., López-Vargas, E. R., Pérez-Álvarez, M., Cadenas-Pliego, G., Benavides-Mendoza, A., Valdés-Reyna, J., Pérez-Labrada, F., & Juárez-Maldonado, A. (2022). Seed priming with carbon nanomaterials improves the bioactive compounds of tomato plants under saline stress. *Plants*, *11*(15), 15. DOI: 10.3390/plants11151984 PMID: 35956461

Haghighi, M., & Teixeira da Silva, J. A. (2014a). The effect of carbon nanotubes on the seed germination and seedling growth of four vegetable species. *Journal of Crop Science and Biotechnology*, *17*(4), 201–208. DOI: 10.1007/s12892-014-0057-6

Haghighi, M., & Teixeira da Silva, J. A. (2014b). The effect of carbon nanotubes on the seed germination and seedling growth of four vegetable species. *Journal of Crop Science and Biotechnology*, *17*(4), 201–208. DOI: 10.1007/s12892-014-0057-6

Hammond, J. L., Formisano, N., Estrela, P., Carrara, S., & Tkac, J. (2016). Electrochemical biosensors and nanobiosensors. *Essays in Biochemistry*, *60*(1), 69–80. DOI: 10.1042/EBC20150008 PMID: 27365037

Hao, Y., Ma, C., Zhang, Z., Song, Y., Cao, W., Guo, J., Zhou, G., Rui, Y., Liu, L., & Xing, B. (2018). Carbon nanomaterials alter plant physiology and soil bacterial community composition in a rice-soil-bacterial ecosystem. *Environmental Pollution*, *232*, 123–136. DOI: 10.1016/j.envpol.2017.09.024 PMID: 28947315

Heydari, S., Masoumi, N., Esmaeeli, E., Ayyoubzadeh, S. M., Ghorbani-Bidkorpeh, F., & Ahmadi, M. (2024). Artificial intelligence in nanotechnology for treatment of diseases. *Journal of Drug Targeting*, *32*(10), 1247–1266. DOI: 10.1080/1061186X.2024.2393417 PMID: 39155708

Hu, Y., Zhang, P., Zhang, X., Liu, Y., Feng, S., Guo, D., Nadezhda, T., Song, Z., & Dang, X. (2021). Multi-wall carbon nanotubes promote the growth of maize (*Zea mays*) by regulating carbon and nitrogen metabolism in leaves. *Journal of Agricultural and Food Chemistry*, *69*(17), 4981–4991. DOI: 10.1021/acs.jafc.1c00733 PMID: 33900073

Hu, Y., Zhang, P., Zhang, X., Liu, Y., Feng, S., Guo, D., Nadezhda, T., Song, Z., & Dang, X. (2021). Multi-wall carbon nanotubes promote the growth of maize (Zea mays) by regulating carbon and nitrogen metabolism in leaves. *Journal of Agricultural and Food Chemistry*, *69*(17), 4981–4991. DOI: 10.1021/acs.jafc.1c00733 PMID: 33900073

James, M., & Owen, L. (2022). Harnessing AI and machine learning for microfluidic drug screening in plant pathogen management. *Revista Española de Documentación Cientifica*, *16*(2), 2.

Joe, B., & Keith, D. (2022). Leveraging artificial intelligence and machine learning in microfluidics for high-throughput drug screening against plant pathogens. *Revista Española de Documentación Cientifica*, *16*(4), 4.

Karthik, P. S., Himaja, A. L., & Singh, S. P. (2014). Carbon-allotropes: Synthesis methods, applications and future perspectives. *Carbon Letters*, *15*(4), 219–237. DOI: 10.5714/CL.2014.15.4.219

Kim, Y. A., Hayashi, T., Endo, M., Kaburagi, Y., Tsukada, T., Shan, J., Osato, K., & Tsuruoka, S. (2005). Synthesis and structural characterization of thin multi-walled carbon nanotubes with a partially facetted cross section by a floating reactant method. *Carbon*, *43*(11), 2243–2250. DOI: 10.1016/j.carbon.2005.03.039

Kumar, V., Sachdev, D., Pasricha, R., Maheshwari, P. H., & Taneja, N. K. (2018). Zinc-supported multiwalled carbon nanotube nanocomposite: A synergism to micronutrient release and a smart distributor to promote the growth of onion seeds in arid conditions. *ACS Applied Materials & Interfaces*, *10*(43), 36733–36745. DOI: 10.1021/acsami.8b13464 PMID: 30286290

Kwak, S.-Y., Lew, T. T. S., Sweeney, C. J., Koman, V. B., Wong, M. H., Bohmert-Tatarev, K., Snell, K. D., Seo, J. S., Chua, N.-H., & Strano, M. S. (2019). Chloroplast-selective gene delivery and expression in planta using chitosan-complexed single-walled carbon nanotube carriers. *Nature Nanotechnology*, *14*(5), 447–455. DOI: 10.1038/s41565-019-0375-4 PMID: 30804482

Li, R., He, Y., Li, Y., Qin, W., Abbas, A., Ji, R., Li, S., Wu, Y., Sun, X., & Yang, J. (2024). Identification of cotton pest and disease based on CFNet- VoV-GCSP -LSKNet-YOLOv8s: A new era of precision agriculture. *Frontiers in Plant Science*, *15*, 1348402. Advance online publication. DOI: 10.3389/fpls.2024.1348402 PMID: 38444536

Li, Y., Kim, W., Zhang, Y., Rolandi, M., Wang, D., & Dai, H. (2001). Growth of single-walled carbon nanotubes from discrete catalytic nanoparticles of various sizes. *The Journal of Physical Chemistry B*, *105*(46), 11424–11431. DOI: 10.1021/jp012085b

Li, Z., Zhou, X., Cheng, Q., Zhai, W., Mao, B., Li, Y., & Chen, Z. (2023). An integrated feature selection approach to high water stress yield prediction. *Frontiers in Plant Science*, *14*, 1289692. Advance online publication. DOI: 10.3389/fpls.2023.1289692 PMID: 38111876

Lin, D., & Xing, B. (2007). Phytotoxicity of nanoparticles: Inhibition of seed germination and root growth. *Environmental Pollution*, *150*(2), 243–250. DOI: 10.1016/j.envpol.2007.01.016 PMID: 17374428

Liu, H., Li, Y., Zhu, M., & Chen, T. (2019). A kind of conical cup-stacked carbon nanotube. *Fullerenes, Nanotubes, and Carbon Nanostructures*, *27*(5), 417–422. DOI: 10.1080/1536383X.2019.1578754

Liu, Q., Chen, B., Wang, Q., Shi, X., Xiao, Z., Lin, J., & Fang, X. (2009). Carbon Nanotubes as Molecular Transporters for Walled Plant Cells. *Nano Letters*, *9*(3), 1007–1010. DOI: 10.1021/nl803083u PMID: 19191500

López-Vargas, E. R., González-García, Y., Pérez-Álvarez, M., Cadenas-Pliego, G., González-Morales, S., Benavides-Mendoza, A., Cabrera, R. I., & Juárez-Maldonado, A. (2020). Seed priming with carbon nanomaterials to modify the germination, growth, and antioxidant status of tomato seedlings. *Agronomy (Basel)*, *10*(5), 5. Advance online publication. DOI: 10.3390/agronomy10050639

Magar, H. S., Ghica, M. E., Abbas, M. N., & Brett, C. M. A. (2017). A novel sensitive amperometric choline biosensor based on multiwalled carbon nanotubes and gold nanoparticles. *Talanta*, *167*, 462–469. DOI: 10.1016/j.talanta.2017.02.048 PMID: 28340746

Majeed, N., Panigrahi, K. C. S., Sukla, L. B., John, R., & Panigrahy, M. (2020a). Application of carbon nanomaterials in plant biotechnology. *Materials Today: Proceedings*, *30*, 340–345. DOI: 10.1016/j.matpr.2020.01.618

Majeed, N., Panigrahi, K. C. S., Sukla, L. B., John, R., & Panigrahy, M. (2020b). Application of carbon nanomaterials in plant biotechnology. *Materials Today: Proceedings*, *30*, 340–345. DOI: 10.1016/j.matpr.2020.01.618

Maksimova, Yu. G. (2019). Microorganisms and carbon nanotubes: Interaction and applications [Review]. *Applied Biochemistry and Microbiology*, *55*(1), 1–12. DOI: 10.1134/S0003683819010101

Martinez-Ballesta, M. C., Chelbi, N., Lopez-Zaplana, A., & Carvajal, M. (2020). Discerning the mechanism of the multiwalled carbon nanotubes effect on root cell water and nutrient transport. *Plant Physiology and Biochemistry*, *146*, 23–30. DOI: 10.1016/j.plaphy.2019.11.008 PMID: 31722266

Mathew, S., Tiwari, D. K., & Tripathi, D. (2021). Interaction of carbon nanotubes with plant system: A review. *Carbon Letters*, *31*(2), 167–176. DOI: 10.1007/s42823-020-00195-1

Mousavi, S. R., Jahandideh Mahjenabadi, V. A., Khoshru, B., & Rezaei, M. (2024). Spatial prediction of winter wheat yield gap: Agro-climatic model and machine learning approaches. *Frontiers in Plant Science*, *14*, 1309171. Advance online publication. DOI: 10.3389/fpls.2023.1309171 PMID: 38264030

Negi, P., & Anand, S. (2024). Plant disease detection, diagnosis, and management: Recent advances and future perspectives. In Pandey, K., Kushwaha, N. L., Pande, C. B., & Singh, K. G. (Eds.), *Artificial Intelligence and Smart Agriculture: Technology and Applications* (pp. 413–436). Springer Nature., DOI: 10.1007/978-981-97-0341-8_20

Niazian, M., & Niedbała, G. (2020). Machine learning for plant breeding and biotechnology. *Agriculture*, *10*(10), 10. Advance online publication. DOI: 10.3390/agriculture10100436

Osama, K., Mishra, B. N., & Somvanshi, P. (2015). Machine learning techniques in plant biology. In Barh, D., Khan, M. S., & Davies, E. (Eds.), *PlantOmics: The Omics of Plant Science* (pp. 731–754). Springer India., DOI: 10.1007/978-81-322-2172-2_26

Oviedo, L. R., Oviedo, V. R., Martins, M. O., Fagan, S. B., & da Silva, W. L. (2022). Nanoarchitectonics: The role of artificial intelligence in the design and application of nanoarchitectures. *Journal of Nanoparticle Research*, *24*(8), 157. DOI: 10.1007/s11051-022-05535-y

Pandey, K., Lahiani, M. H., Hicks, V. K., Hudson, M. K., Green, M. J., & Khodakovskaya, M. (2018). Effects of carbon-based nanomaterials on seed germination, biomass accumulation and salt stress response of bioenergy crops. *PLoS One*, *13*(8), e0202274. DOI: 10.1371/journal.pone.0202274 PMID: 30153261

Patel, A., Tiwari, S., Parihar, P., Singh, R., & Prasad, S. M. (2019). Carbon nanotubes as plant growth regulators: Impacts on growth, reproductive system, and soil microbial community. In Tripathi, D. K., Ahmad, P., Sharma, S., Chauhan, D. K., & Dubey, N. K. (Eds.), *Nanomaterials in Plants, Algae and Microorganisms* (pp. 23–42). Academic Press., DOI: 10.1016/B978-0-12-811488-9.00002-0

Patel, D. K., Kim, H.-B., Dutta, S. D., Ganguly, K., & Lim, K.-T. (2020). Carbon nanotubes-based nanomaterials and their agricultural and biotechnological applications. *Materials (Basel)*, *13*(7), 7. Advance online publication. DOI: 10.3390/ma13071679 PMID: 32260227

Pourhashem, S., Ghasemy, E., Rashidi, A., & Vaezi, M. R. (2020). A review on application of carbon nanostructures as nanofiller in corrosion-resistant organic coatings. *Journal of Coatings Technology and Research*, *17*(1), 19–55. DOI: 10.1007/s11998-019-00275-6

Rahimi, D., Kartoolinejad, D., Nourmohammadi, K., & Naghdi, R. (2016). Increasing drought resistance of *Alnus subcordata* C.A. Mey. Seeds using a nano priming technique with multi-walled carbon nanotubes. *Journal of Forest Science*, *62*(6), 269–278. DOI: 10.17221/15/2016-JFS

Ren, L., Deng, S., Chu, Y., Zhang, Y., Zhao, H., Chen, H., & Zhang, D. (2020). Single-wall carbon nanotubes improve cell survival rate and reduce oxidative injury in cryopreservation of *Agapanthus praecox* embryogenic callus. *Plant Methods*, *16*(1), 130. DOI: 10.1186/s13007-020-00674-6 PMID: 32973916

Rico, C. M., Majumdar, S., Duarte-Gardea, M., Peralta-Videa, J. R., & Gardea-Torresdey, J. L. (2011). Interaction of nanoparticles with edible plants and their possible implications in the food chain. *Journal of Agricultural and Food Chemistry*, *59*(8), 3485–3498. DOI: 10.1021/jf104517j PMID: 21405020

Safdar, M., Kim, W., Park, S., Gwon, Y., Kim, Y.-O., & Kim, J. (2022a). Engineering plants with carbon nanotubes: A sustainable agriculture approach. *Journal of Nanobiotechnology*, *20*(1), 275. DOI: 10.1186/s12951-022-01483-w PMID: 35701848

Safdar, M., Kim, W., Park, S., Gwon, Y., Kim, Y.-O., & Kim, J. (2022b). Engineering plants with carbon nanotubes: A sustainable agriculture approach. *Journal of Nanobiotechnology*, *20*(1), 275. DOI: 10.1186/s12951-022-01483-w PMID: 35701848

Samadi, S., Asgari Lajayer, B., Moghiseh, E., & Rodríguez-Couto, S. (2021). Effect of carbon nanomaterials on cell toxicity, biomass production, nutritional and active compound accumulation in plants. *Environmental Technology & Innovation*, *21*, 101323. DOI: 10.1016/j.eti.2020.101323

Samadi, S., Saharkhiz, M. J., Azizi, M., Samiei, L., Karami, A., & Ghorbanpour, M. (2021). Single-wall carbon nano tubes (SWCNTs) penetrate *Thymus daenensis* Celak. Plant cells and increase secondary metabolite accumulation in vitro. *Industrial Crops and Products*, *165*, 113424. DOI: 10.1016/j.indcrop.2021.113424

Seddighinia, F. S., Iranbakhsh, A., Oraghi Ardebili, Z., Nejad Satari, T., & Soleimanpour, S. (2020). Seed priming with cold plasma and multi-walled carbon nanotubes modified growth, tissue differentiation, anatomy, and yield in bitter melon (*Momordica charantia*). *Journal of Plant Growth Regulation*, *39*(1), 87–98. DOI: 10.1007/s00344-019-09965-2

Serag, M. F., Kaji, N., Habuchi, S., Bianco, A., & Baba, Y. (2013). Nanobiotechnology meets plant cell biology: Carbon nanotubes as organelle targeting nanocarriers. *RSC Advances*, *3*(15), 4856. DOI: 10.1039/c2ra22766e

Sha, R., Zhu, S., Wu, L., Li, X., Zhang, H., Yao, D., Lv, Q., Wang, F., Zhao, F., Li, P., & Yu, K. (2022). Pre-Harvest application of multi-walled carbon nanotubes improves the antioxidant capacity of 'Flame Seedless' grapes during storage. *Sustainability (Basel)*, *14*(15), 15. Advance online publication. DOI: 10.3390/su14159568

Sharma, M. M. M., Kapoor, D., Loyal, A., Kumar, R., & Sharma, P. (2025). Role of carbon-based nanomaterials in biotic stress managements. In *Smart Nanomaterials Technology* (pp. 255–269). Springer Nature Singapore., DOI: 10.1007/978-981-97-5104-4_13

Shojaei, T. R., Salleh, M. A. M., Tabatabaei, M., Mobli, H., Aghbashlo, M., Rashid, S. A., & Tan, T. (2019). Applications of nanotechnology and carbon nanoparticles in agriculture. In Rashid, S. A., Raja Othman, R. N. I., & Hussein, M. Z. (Eds.), *Synthesis, Technology and Applications of Carbon Nanomaterials* (pp. 247–277). Elsevier., DOI: 10.1016/B978-0-12-815757-2.00011-5

Singh, A., Hua Hsu, M., Gupta, N., Khanra, P., Kumar, P., Prakash Verma, V., & Kapoor, M. (2020). Derivatized carbon nanotubes for gene therapy in mammalian and plant cells. *ChemPlusChem*, *85*(3), 466–475. DOI: 10.1002/cplu.201900678 PMID: 32159284

Thakur, A., Bharti, R., & Sharma, R. (2022). Carbon nanotubes: Types, synthesis, cytotoxicity and applications in biomedical. *Materials Today: Proceedings*, *50*, 2256–2268. DOI: 10.1016/j.matpr.2021.10.002

Thavaseelan, D., & Priyadarshana, G. (2021). Nanofertilizer use for sustainable agriculture. *J Res Technol Eng*, *2*(1), 41–59.

Torney, F., Trewyn, B. G., Lin, V. S.-Y., & Wang, K. (2007). Mesoporous silica nanoparticles deliver DNA and chemicals into plants. *Nature Nanotechnology*, *2*(5), 295–300. DOI: 10.1038/nnano.2007.108 PMID: 18654287

Van Dijk, A. D. J., Kootstra, G., Kruijer, W., & De Ridder, D. (2021). Machine learning in plant science and plant breeding. *iScience*, *24*(1), 101890. DOI: 10.1016/j.isci.2020.101890 PMID: 33364579

Vithanage, M., Seneviratne, M., Ahmad, M., Sarkar, B., & Ok, Y. S. (2017). Contrasting effects of engineered carbon nanotubes on plants: A review. *Environmental Geochemistry and Health*, *39*(6), 1421–1439. DOI: 10.1007/s10653-017-9957-y PMID: 28444473

Wang, C., Li, C., Han, Q., Wu, F., & Zou, X. (2023). A performance analysis of a litchi picking robot system for actively removing obstructions, Using an artificial intelligence algorithm. *Agronomy (Basel)*, *13*(11), 2795. DOI: 10.3390/agronomy13112795

Wang, X., Han, H., Liu, X., Gu, X., Chen, K., & Lu, D. (2012). Multi-walled carbon nanotubes can enhance root elongation of wheat (*Triticum aestivum*) plants. *Journal of Nanoparticle Research*, *14*(6), 841. DOI: 10.1007/s11051-012-0841-5

Wu, H., Nißler, R., Morris, V., Herrmann, N., Hu, P., Jeon, S.-J., Kruss, S., & Giraldo, J. P. (2020). Monitoring plant health with near-infrared fluorescent H_2O_2 nanosensors. *Nano Letters*, *20*(4), 2432–2442. DOI: 10.1021/acs.nanolett.9b05159 PMID: 32097014

Xu, F. X., Zhou, X. B., Zhang, L., & Guo, X. Y., M., L., Zhu, Y. C., & Jiang. (2024). Effects of nitrogen application rate on grain yield of hybrid rice and soil nutrient under rice-fish co-culture system in Southeast of Sichuan Province. *Zhiwu Yingyang Yu Feiliao Xuebao*, *30*(5), 909–921.

Yan, H., Zhao, G., Lu, W., Hu, C., Wang, X., Liu, G., Shi, N., Dieu Thuy, N. T., & Zhang, L. (2023). A flexible and wearable paper-based chemiresistive sensor modified with SWCNTs-PdNPs-polystyrene microspheres composite for the sensitive detection of ethylene gas: A new method for the determination of fruit ripeness and corruption. *Analytica Chimica Acta*, *1239*, 340724. DOI: 10.1016/j.aca.2022.340724 PMID: 36628724

Zhang, W., Zeng, Z., Liu, Z., Huang, J., Xiao, R., Shao, B., Liu, Y., Liu, Y., Tang, W., Zeng, G., Gong, J., & He, Q. (2020). Effects of carbon nanotubes on biodegradation of pollutants: Positive or negative? *Ecotoxicology and Environmental Safety*, *189*, 109914. DOI: 10.1016/j.ecoenv.2019.109914 PMID: 31761551

Zhao, G., Zhao, Y., Lou, W., Abdalmegeed, D., Guan, R., & Shen, W. (2020). Multi-walled carbon nanotubes can promote Brassica napus L. and Arabidopsis thaliana L. root hair development through nitric oxide and ethylene pathways. *International Journal of Molecular Sciences*, *21*(23), 9109. DOI: 10.3390/ijms21239109 PMID: 33266061

Zhao, Q., Ma, C., White, J. C., Dhankher, O. P., Zhang, X., Zhang, S., & Xing, B. (2017). Quantitative evaluation of multi-wall carbon nanotube uptake by terrestrial plants. *Carbon*, *114*, 661–670. DOI: 10.1016/j.carbon.2016.12.036

APPENDIX

Key Terms

1. Carbon nanotubes
2. Nanocarriers
3. Functionalization
4. Plant biotechnology
5. Gene delivery
6. Plant uptake mechanisms
7. Biocompatibility
8. Biosensors
9. Agrochemical delivery
10. Abiotic stress
11. Biotic stress
12. Controlled release systems

Abbreviations

1. CNTs - Carbon nanotubes
2. SWCNTs - Single Walled Carbon Nanotubes
3. MWCNTs - Multi Walled Carbon Nanotubes
4. CSCNTs - Conical Stacked cup Carbon Nanotubes
5. o-MWCNTs - oxidized Multi Walled Carbon Nanotubes
6. POD - Peroxidase
7. SOD - Superoxide Dismutase
8. CNM - Carbon Nano Materials
9. siRNA - small interfering RNA
10. Arg- SWCNT - Arginine-Functionalized Single Walled Carbon Nanotubes
11. pDNA-CNT- Plasmid DNA-Conjugated Carbon Nanotube
12. PEI-SWCNT - Polyethylenimine-Functionalized Single-Walled Carbon Nanotube
13. XGBoost - eXtreme Gradient Boosting
14. PCRF-RFE - Pearson Correlation Coefficient and Random Forest
15. ML - Machine Learning
16. ANN - Artificial Neural Networks
17. RF - Random Forests
18. SE - somatic embryogenesis
19. CNN - Convolutional Neural Networks
20. NGS - Next-Generation Sequencing
21. ENMs - Engineered Nanomaterials

Chapter 4
Nanocarriers in Plant Genetic Engineering

Odangowei Inetiminebi Ogidi
https://orcid.org/0000-0002-9961-893X
Bayelsa Medical University, Yenagoa, Nigeria

Ebifanimi Okara
https://orcid.org/0009-0001-6924-3524
Federal Medical Center, Yenagoa, Nigeria

Mukul Machhindra Barwant
https://orcid.org/0000-0002-5639-4251
Sanjivani Arts Commerce and Science College, Kopargaon, India

ABSTRACT

The enhancement of grain yield, the reinforcement of crops against erratic severe weather conditions, and the mitigation of production losses caused by insects and viruses are all instances of contemporary agricultural techniques that are imperative to satisfy the continuously rising global food demand and sustain a burgeoning population. In order to fulfil these requirements, genome editing has been employed on several plant species. An integral part of genome engineering is introducing gene cassettes into plant cells. Traditional methods have a number of drawbacks, including lengthy and painful processes, the possibility of tissue damage, and inefficient conversion. One technological benefit of the nanoparticle-mediated gene delivery system over conventional methods is its high efficacy and flexibility in different plant species. A concise history of plant biomolecular delivery techniques is provided in this chapter, along with an outline of their key features and an analysis of their pros and cons.

INTRODUCTION

Modern agriculture would not exist without plant genetic engineering, which has allowed for the creation of crops with better tolerance to environmental pressures, greater resilience to pests and diseases, and higher yields (Ogoun & Ogidi, 2022). Conventional methods for delivering genetic material into plant cells, such as Agrobacterium-mediated transformation and biolistics, have paved the way for

DOI: 10.4018/979-8-3693-9869-2.ch004

significant breakthroughs. However, these techniques are often associated with low efficiency, high costs, and unintended genetic alterations, limiting their applicability across diverse plant species and genotypes (Altpeter et al., 2016). Nanocarriers, a class of nanotechnology-based delivery systems, have emerged as a promising alternative that addresses these limitations, offering innovative solutions for precise and efficient genetic transformation in plants.

Nanocarriers are nanoscale delivery vehicles engineered to transport nucleic acids, proteins, and small molecules into plant cells with high specificity and minimal off-target effects. These nanomaterials include a diverse range of platforms, such as carbon nanotubes (CNTs), graphene oxide (GO), polymeric nanoparticles, lipid-based systems, and metal-organic frameworks (MOFs) (Zhang et al., 2020a). Due to their small size, high surface area, and customizable functionalization, nanocarriers can effectively penetrate the rigid plant cell wall, which has historically been a formidable barrier in genetic engineering. This capability allows nanocarriers to deliver biomolecules directly into plant cells or organelles, such as chloroplasts or nuclei, with minimal physical damage. A significant advantage of nanocarriers is their versatility in delivering various types of genetic materials. For example, nanocarriers can transport plasmid DNA for transient gene expression, small interfering RNA (siRNA) for gene silencing, or ribonucleoprotein (RNP) complexes for precise genome editing using CRISPR-Cas systems (Demirer et al., 2019a). Unlike traditional methods, nanocarriers enable DNA-free genome editing, where CRISPR components are delivered without integrating foreign DNA into the plant genome. This approach not only enhances the precision of genetic modifications but also addresses public and regulatory concerns related to genetically modified organisms (GMOs) (Wang et al., 2021).

The potential applications of nanocarriers extend beyond genetic transformation. In addition to introducing foreign genes, nanocarriers can be employed to modulate endogenous gene expression, deliver plant hormones for developmental control, or transport agrochemicals such as fertilizers and pesticides in a targeted manner (Ogidi et al., 2023; Lew et al., 2020; Ogidi & Akpan, 2023). These multifunctional capabilities position nanocarriers as transformative tools in plant biotechnology, enabling the development of next-generation crops tailored to address global challenges, including climate change and food security. Moreover, nanocarriers facilitate the genetic engineering of plant species and tissues that have been recalcitrant to conventional transformation methods. Many crop species, particularly monocots such as wheat and maize, are difficult to transform due to their thick cell walls and complex cellular architecture. Nanocarriers offer a solution by bypassing these barriers, enabling efficient gene delivery in a wide range of plant species (Nair et al., 2021). Additionally, nanocarriers can be delivered through non-invasive methods, such as foliar sprays, root immersion, or vascular injection, reducing the need for sophisticated laboratory infrastructure and making genetic engineering more accessible to resource-limited settings (Lew et al., 2020).

Despite their promise, the use of nanocarriers in plant genetic engineering is still at an early stage, and several challenges remain. These include the scalability of nanocarrier production, ensuring biosafety and environmental sustainability, and optimizing delivery efficiency for different plant types and tissues. Additionally, the long-term stability of nanocarrier-mediated modifications and their potential off-target effects require thorough investigation (Nair et al., 2021). Regulatory frameworks must also evolve to address the unique challenges posed by nanotechnology in agriculture, ensuring that its applications are safe, ethical, and widely accepted. Nanocarriers represent a paradigm shift in plant genetic engineering, offering innovative approaches to overcome longstanding challenges in gene delivery and crop improvement. Their potential to deliver diverse biomolecules with high precision and efficiency has positioned them as critical tools for advancing sustainable agriculture. This chapter explores the principles, mechanisms, applications, and challenges associated with nanocarriers in plant genetic engineering, highlighting their transformative role in addressing the demands of a rapidly changing world (Lew et al., 2020).

Objectives of the Chapter

The aims of this chapter are as follows:

(1) To review conventional genetic transformation methods.
(2) To discuss the merits and limitations of traditional delivery methods.
(3) To review the mechanisms and types of nanocarriers in plant genetic engineering.
(4) To evaluate NP-mediated biomolecule delivery to plants.
(5) To discuss nanomaterials for plant genetic engineering.
(6) To X-ray nanocarriers hold promise for nuclease-based plant genome editing.
(7) To discuss the environmental and regulatory considerations for nanoparticle-enabled plant gene editing.
(8) To evaluate various case studies, ethical and public perception of nanocarriers in plant genetic engineering.
(9) To review the future prospect and conclusion.

Organization of the Chapter

The rest of this chapter is organized as follows: Section 2 explores the conventional genetic transformation methods. Section 3 discusses the merits and limitations of traditional delivery methods. Section 4 traverse the mechanisms and types of nanocarriers in plant genetic engineering. Section 5 discusses NP-mediated biomolecule delivery to plants. Section 6 X-ray nanomaterials for plant genetic engineering. Section 7 discusses the nanocarriers hold promise for nuclease-based plant genome editing. Section 8 reviews the environmental and regulatory considerations for nanoparticle-enabled plant gene editing. Section 9 evaluate various case studies, ethical and public perception of nanocarriers in plant genetic engineering. Finally, a review on future prospects and a conclusion summarizes nanocarriers in plant genetic engineering.

CONVENTIONAL GENETIC TRANSFORMATION METHODS

In recent decades, there has been a surge in interest in studying plant genetic engineering. Research in plant biology has entered a new age with the advancements in conventional genetic engineering techniques (Figure 1). Most species have a strong multilayered cell wall, which makes plant regeneration difficult, and these genetic engineering techniques have a number of drawbacks that make them ineffective (Shou et al., 2004).

Figure 1. Conventional genetic transformation methods

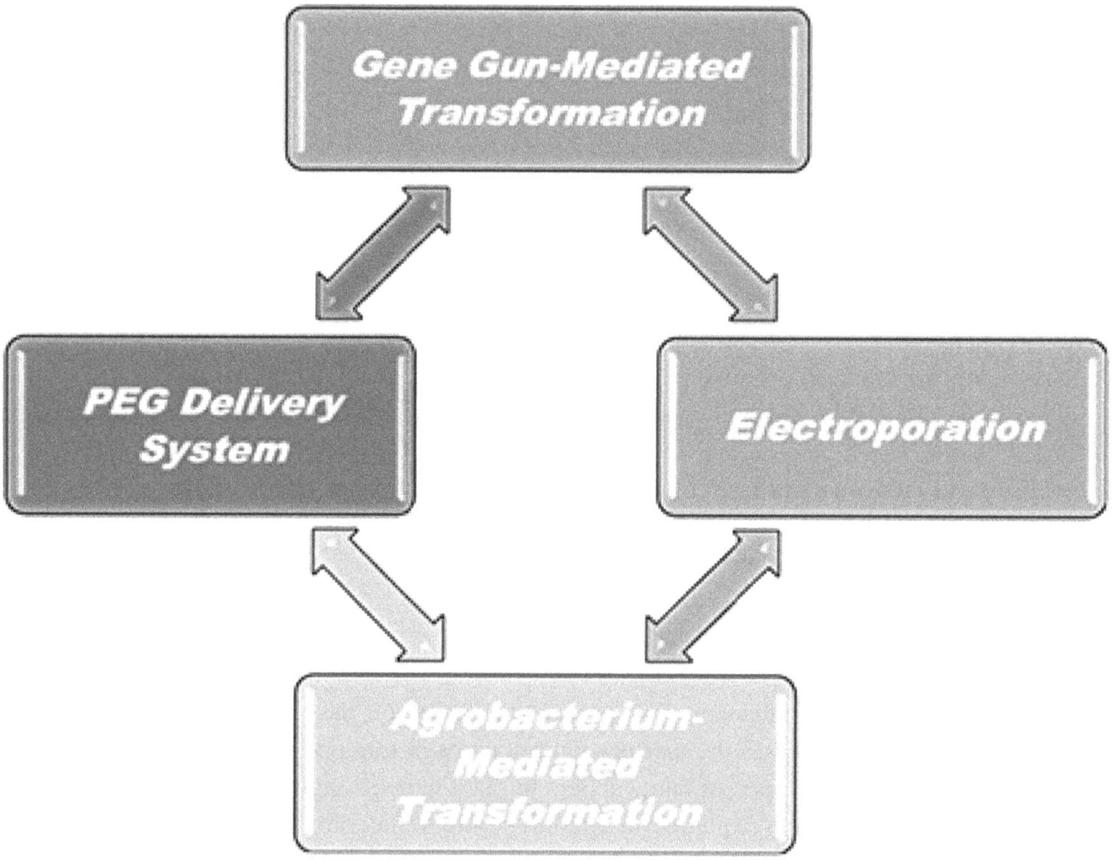

Gene Gun-Mediated Transformation

A biolistic particle delivery technology, typically used in plant genetic engineering, is the basis of gene gun-mediated transformation, also known as particle bombardment. To summarise, a cold gas shock wave is produced by compressing gas that is already inside the bombardment chamber. The tiny gold powder, which contains DNA molecules, is hit by this wave. By passing through the cell wall, membrane, cytoplasm, and other structural layers, DNA-laden gold particles are able to carry out gene transfer under the influence of gas pressure. The restriction that is dependent on the species is successfully handled by the gene-gun transformation (Cunningham et al., 2018). Moreover, in contrast to the Agrobacterium transformation method, the gene cannon is applicable for plasmid biological delivery. Limitations include damage to plant tissues, inconsistent low-level expression, many copies, and ineffective integration during bombardment. The widespread application is additionally constrained by the requirement for specialised facilities and costly materials (Wang et al., 2019). The targeting of the nuclear genome by biolistic delivery could be more prevalent if not for the nonspecific localisation of molecular payloads (Cunningham et al., 2018).

Agrobacterium-Mediated Transformation

The Gram-negative bacterium Agrobacterium is present in soil in every location. Crown galls or hairy roots are the natural consequences of its chemotaxis towards the majority of dicotyledons (Michielse et al., 2005). The Ri plasmid of Agrobacterium rhizogenes and the Ti plasmid of Agrobacterium tumefaciens both include a fragment of T-DNA. Stable transformation can occur when Agrobacterium infects plant cells through incisions and integrates the target gene. Subsequently, the plant genome may incorporate a solitary strand of T-DNA. Agrobacterium-mediated transformation is distinguished from other plant genetic engineering techniques by its exceptional level of efficiency, simplicity of use, and stability. Most of them can be classified as either in vitro systems or in vivo/in planta systems. It is essential to have a diverse selection of sterile explants for in vitro systems. The package includes internodes, segments, petioles, leaves, hypocotyls, cell suspension cultures, calli, and roots. The processes are laborious and meaningless due to the fact that regeneration necessitates only one converted cell. The in planta approach enables the in situ examination of a variety of reproductive components, such as the axillary meristem and pollen tube. This technology has been extensively employed by scientists since the pollen tube's mechanism was elucidated in 1983 (Ahmar et al., 2021). Following the pollination process, the ovary is injected with a solution that contains the DNA of the desired gene. The channels of the pollen tube are employed to facilitate the delivery and integration of exogenous DNA into the genome of the cells that will be receiving it. The fertilised egg endures a transformation into a transgenic organism during its development. Foliar spraying is a successful technique of gene editing that introduces plasmids into plant cells using Cas9/sgRNAs (Ahmar et al., 2021).

The tissue culture-independent pollen tube technique is successful, quick, and simple to use; however, it is limited to flowering plants, particularly Arabidopsis, due to the fact that it generates transformations that are inconsistent with one another (Ali et al., 2014). Meristems can now be regenerated from the ground up. This method has effectively produced gene-edited dicotyledonous plants by eliminating the necessity for in vitro tissue culture (Maher et al., 2020). Nevertheless, the Agrobacterium-based approach is characterised by two significant drawbacks. Nevertheless, it is restricted to DNA transport and possesses host-constrained properties (Zhang et al., 2019a, 2020b; Landry & Mitter, 2019). Nevertheless, the potential for undesirable characteristics to arise through endogenous gene disruption and random DNA integration is present in Agrobacterium-mediated transport (Landry & Mitter, 2019; Yan et al., 2022). The use of Agrobacterium to produce GMOs for the delivery of biomolecules has increased public and regulatory concerns. The USDA regulates plants in the United States that have been transformed by Agrobacterium. Until 2018, only a small number of EU member states cultivated genetically modified (GMO) crops (Yan et al., 2022).

Electroporation

Electroporation can instantly enhance the permeability of the cell membrane by exposing it to a high-intensity electric field (Zhang & Hu, 2014). This facilitates the diffusion of extracellular substances into the cell. This approach may transduce both prokaryotic and eukaryotic cells with nucleotides, RNA, proteins, dyes, and even viral particles. Electroporation is extensively utilised in medicinal and food biotechnology owing to its broad applicability, controlled dosage, and high efficiency (Dong & Chang, 2021). Research utilising electroporation for intracellular delivery has expanded from in vitro investigations to in vivo studies (Ozyigit, 2020; Dong & Chang, 2021). The robust cell walls of plants hindered the widespread application of electroporation for gene transfer, in contrast to its use in mammals. Moreover, electroporation can result in compromised target tissues and impaired cellular function (Yan et al., 2022).

PEG Delivery System

In the presence of divalent cations (e.g., Mg^{2+}, Ca^{2+}), the cell membrane can absorb DNA particles through endophagocytosis, and PEG successfully induces DNA to precipitate granularly. Extensive use of PEG for protoplast transformation has been shown (Wu et al., 2020). Many benefits, including low cost, ease of operation, mild reaction conditions, and high conversion efficiency, make the PEG delivery method a favourite. The PEG delivery approach has genotype limits and must be applied to naked protoplasts, which are typically used for transient gene expression (Zhang et al., 2022a).

THE MERITS AND LIMITATIONS OF TRADITIONAL DELIVERY METHODS

Numerous individuals are of the opinion that genome editing will be necessary for future crop breeding initiatives that are designed to promote sustainable development and enhancement (Steinwand & Ronald, 2020). Transfection by viruses, particle bombardment, Agrobacterium-mediated infection, and PEG-induced transfection are among the most frequently employed engineering techniques in the past century (Behzadi et al., 2017). The majority of methods for introducing genetic modifications to plants involve the use of Agrobacterium infection (Abdallah et al., 2015). After being integrated into the plant genome, the binary vectors release transfer DNA (T-DNA). They are produced from the Agrobacterium tumor-inducing (Ti) plasmid (Gordon & Christie, 2014). Agrobacterium infections in dicots are simpler to regulate than in monocots due to the fact that dicots are the bacterium's natural hosts. One potential explanation is that monocotyledonous plants do not respond to damage in the same manner as dicotyledonous plants; rather, dicotyledonous plants endure transformation through the production of Vir genes. The expression of the Vir gene is induced by the addition of exogenous acetosyringone (AS) when transfecting monocot plants with Agrobacterium. An alternative approach is offered by constitutive strains, which consistently induce the expression of Vir genes (Gelvin, 2021; Wu et al., 2018).

Agrobacterium is currently mediating transformations that are both successful and long-lasting in a number of model monocot species. Genetic transformation by Agrobacterium remains challenging in numerous cultivars, regardless of whether they are monocots or dicots. Only when specific conditions are satisfied are Agrobacterium-mediated transformations effective (Gelvin, 2017). However, there is a significant amount of variation among genotype-dependent plant varieties, even within the same species. Responses to Agrobacterium infection are subject to variation depending on metabolic states, environmental factors, and genetic makeups (Das et al., 2002). The ability to regenerate varies among cultivars of nearly every plant species. During regeneration, plants activate their disease resistance, resulting in responses that are highly variable among different species. Recent research has demonstrated that developmental regulators, such as BBM, WUS, GRF5, and GRF4-GIF1, can be overexpressed in plants, either temporarily or permanently, thereby enabling effective genetic modifications (Aesaert et al., 2022; Debernardi et al., 2020; Kong et al., 2020; Lowe et al., 2016; Qiu et al., 2022).

Nevertheless, Agrobacterium's physical and metabolic characteristics render it an unsuitable candidate for the introduction of foreign DNAs into certain plant species (Gelvin, 2017). Agrobacterium's gene transfer process is impeded by a variety of specialised metabolites, such as ethylene, salicylic acid, and γ-aminobutyric acid (GABA), as indicated by Ciardi et al. (2000), Nonaka & Ezura, (2014) and Nonaka et al. (2017). The plants may exhibit symptoms of illness when the genetic material of certain host plants is incompatible with Agrobacterium (Baker et al., 2005; Ogidi, 2024). Furthermore, when assessing various cultivars, it is necessary to take into account a variety of tissue culture conditions, including the density of Agrobacterium, the concentration of AS, the duration of the preculture, the co-inoculation interval, the types and concentrations of hormones, and so forth. It is impossible for Agrobacterium to transmit RNA, RNPs, or oligonucleotides; this is the sole variety of transformation material that it can transmit. These materials are also essential for the development of cultivars that are precisely modified and transgene-free in crop genome editing (Ogidi & Emaikwu, 2024; Gu et al., 2021; Baker et al., 2005; Park et al., 2019).

Agrobacterium-mediated transformation is not suitable for a wide range of plant taxa or applications, despite its ease of implementation and minimal resource requirements. A gene cannon, which is a biolistic delivery system, was developed in 1987 to execute particle bombardment. When exposed to exogenous biomolecules (e.g., RNA, proteins, and peptides) affixed to gold particles, they can enter the nucleus and other organelles of plants when they are subjected to intense pressure from external sources (Jinturkar et al., 2011; Liang et al., 2017; Zhang et al., 2016). This method is considered highly promising for plant genetic engineering, particularly for transformations that involve chloroplasts and mitochondria, due to the absence of restrictions on the available genetic materials and tissue types (Baltes et al., 2017). The frequent chromosome deletions, translocations, and inversions, as well as the random and multiple-copy insertions of target DNA caused by highly pressurised particles, severely restrict its use in commercial breeding. Protoplast transfection, which was facilitated by PEG, was one of the primary methods for altering cereal crops. Exogenous genes can rapidly penetrate the nucleus when exposed to a PEG solution, which significantly alters the permeability of the cell membrane (Yoo et al., 2007).

This method has been effectively employed to deliver plasmids, RNA, and RNPs to a diverse array of plant species. This has led to its widespread utilisation for the purpose of identifying genomic editing targets, assessing their efficacy, and investigating gene localisation and function in the short term. The process of transforming transfected protoplasts into fully grown plants is both technically challenging and time-consuming. The viral replication cycle offers an optimal environment for the encapsulation and replication of target genes, rendering gene delivery through plant viruses a viable alternative (Ji et al., 2015). The VIGS technique, which employs viral vectors to investigate gene function, has been employed by numerous individuals (Dinesh-Kumar et al., 2003). Last year, it was proposed to enhance the quality of crops by applying Agrobacterium with viral vectors to the foliage (Torti et al., 2021). Using viral vectors to transport single guide RNAs (sgRNAs), transgenic plants such as cotton, tobacco, Arabidopsis thaliana, tomato, wheat, and maize have been recently engineered to overexpress Cas9. The precise modification of the target location of the genome was significantly simplified after the Cas9-sgRNA complex was assembled. Genome editing has the potential to generate organisms that are virus-free in the future (Ma et al., 2020).

However, viruses presented numerous challenges. There are numerous safety concerns associated with plant viruses. There are numerous viruses, some of which are transmitted through seeds and others that cause severe developmental issues. Despite the presence of modified viral vectors that reduce their toxicity, these viruses can still cause growth problems. Tobacco is the primary material employed in research on viruses and their vectors due to its substantially reduced endogenous gene silencing. Endogenous gene silencing is a significant impediment to the efficacy of viral vectors. Another constraint is the extensive array of organisms that can be infected by viruses. Tobacco rattle virus (TRV) vectors are typically employed by dicot and monocot plants. Nevertheless, TRV is incompatible with nonhost species, such as rice. The systematic distribution of virus vectors necessitates a significant amount of knowledge and effort (Ratcliff et al., 2001). Table 1 showing merits and limitations of traditional delivery methods for nanocarriers in plant genetic engineering.

Table 1. Merits and limitations of traditional delivery methods for nanocarriers in plant genetic engineering

Delivery Method	Merits	Limitations	Examples of Use
Agrobacterium-Mediated Transformation	- High efficiency in delivering genetic material to dicotyledonous plants.	- Limited to certain plant species (e.g., dicots); low efficiency in monocots.	Used for delivering CRISPR/Cas9 constructs in Arabidopsis and tobacco.
	- Low cost and well-established protocol.	- Potential biosafety concerns due to bacterial contamination.	
Biolistic (Gene Gun) Method	- Capable of delivering DNA, RNA, and nanoparticles to a wide range of plant species, including monocots.	- Expensive equipment and high operational costs.	Delivery of plasmid DNA into maize, rice, and wheat using gold or tungsten particles.
	- No species-specific limitations.	- Can cause significant physical damage to plant tissues.	
Microinjection	- Allows precise delivery of genetic material into specific cells or organelles.	- Labor-intensive and requires technical expertise.	Used for direct delivery of DNA into tobacco protoplasts.
	- High success rate in controlled environments.	- Limited to large, accessible cells; not suitable for high-throughput applications.	
Electroporation	- Efficient for delivering DNA or RNA into protoplasts and some intact cells.	- Requires the production of protoplasts, which can be labor-intensive and time-consuming.	Used in rice and wheat for delivering CRISPR components.
	- Relatively simple and low cost compared to biolistics.	- Limited to specific cell types and may reduce cell viability due to high voltage.	
PEG-Mediated Transformation	- Simple and cost-effective for delivering plasmids and nanoparticles into protoplasts.	- Limited to protoplasts, making it challenging for whole-plant transformation.	Delivery of plasmid DNA in rice and Arabidopsis protoplasts.
	- No need for specialized equipment.	- Inefficient for stable transformation; primarily used for transient gene expression studies.	
Silicon Carbide Whisker-Mediated Transformation	- Suitable for delivering DNA into both monocots and dicots.	- Risk of damaging plant cells due to the abrasive nature of whiskers.	Used in maize and soybean for genetic transformation.
	- No requirement for complex equipment.	- Low efficiency compared to Agrobacterium and biolistics.	

NP DELIVERY MECHANISMS IN PLANT GENETIC ENGINEERING

Nanoparticle (NP) delivery systems have emerged as transformative tools in plant genetic engineering, offering non-invasive, efficient alternatives to traditional methods like Agrobacterium-mediated transformation and biolistics. NPs can deliver DNA, RNA, proteins, or gene-editing tools such as CRISPR/Cas systems directly into plant cells and tissues with high precision and minimal damage. One of the primary mechanisms is passive diffusion or endocytosis, where NPs enter plant cells through natural openings like stomata or are absorbed across cell membranes depending on size, charge, and surface chemistry. Mesoporous silica nanoparticles (MSNs), carbon nanotubes (CNTs), and chitosan-based NPs have shown effective uptake and transport into cells, bypassing the rigid plant cell wall under optimized conditions (Demirer et al., 2019; Wang et al., 2022). These NPs can be designed to release genetic cargo in response to environmental triggers such as pH or redox changes, ensuring targeted and controlled gene delivery.

Another emerging strategy involves active targeting and endosomal escape, where NPs are functionalized with ligands or peptides that recognize and bind to specific receptors on plant cell surfaces. This facilitates receptor-mediated endocytosis, improving transformation efficiency. Lipid-based nanoparticles and gold nanoparticles have demonstrated success in transporting CRISPR components and siRNAs to induce gene silencing or genome editing in various plant species (Kwak et al., 2019). Additionally, magnetofection and electrostatic interactions are being explored to enhance delivery efficiency without causing tissue damage. These mechanisms not only enhance transformation rates but also reduce the need for tissue culture, paving the way for high-throughput, species-independent gene editing (Zhang et al., 2021). Types of nanoparticle delivery mechanisms in plant genetic engineering is shown in Figure 2.

Figure 2. Types of nanoparticle delivery mechanisms in plant genetic engineering

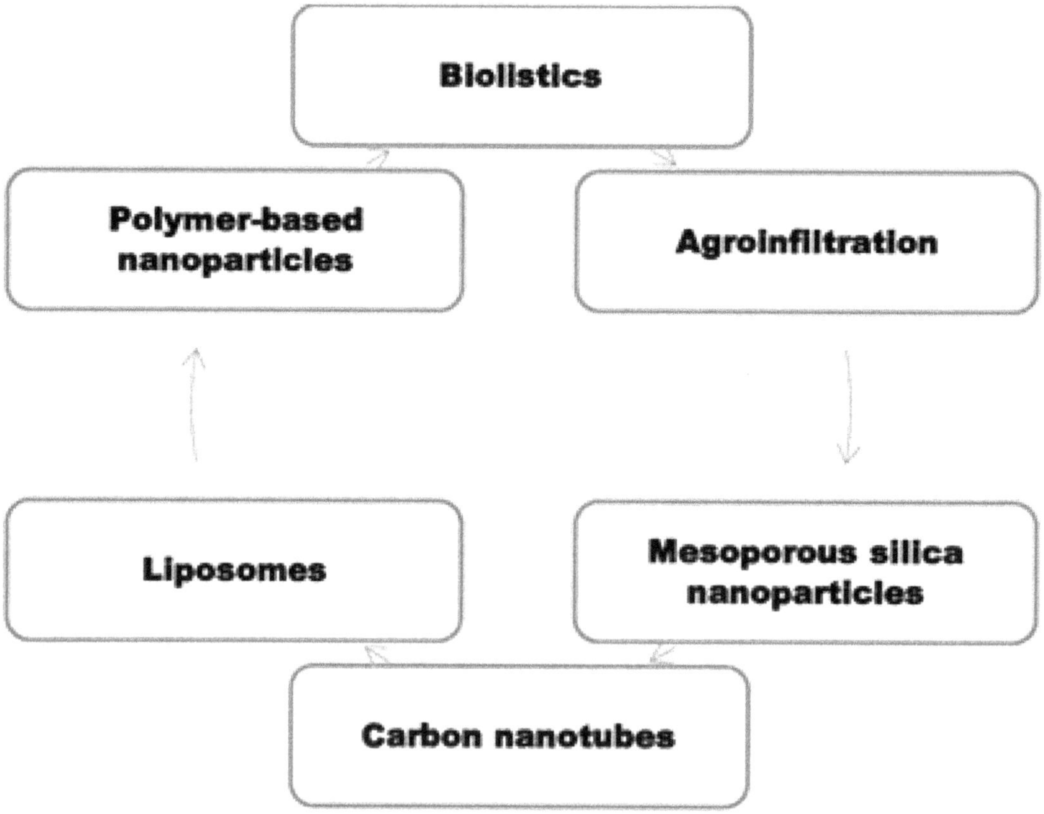

TYPES OF NANOCARRIERS IN PLANT GENETIC ENGINEERING

The ability of nanoparticles (NPs) from various sources to transfer genetic materials to plants has been the subject of much research. Carbon nanotubes (Demirer et al., 2019b), carbon dots (Schwartz et al., 2020), liposomes (Liu et al., 2020), conjugated polymeric nanoparticles (Silva et al., 2010), and metallic nanoparticles (Zhang et al., 2022b; Zhao et al., 2020) are all part of this category. Some nanocarriers used in plant genetic engineering is illustrated in Figure 3 and Table 2.

Table 2. Nanocarriers in plant genetic engineering

Nanocarrier Type	Material Composition	Application in Plant Genetic Engineering	Advantages	Examples/Studies	Reference
Lipid-Based Nanocarriers	Liposomes, solid lipid nanoparticles	Delivery of plasmids, RNA, or CRISPR components to plant cells.	Biocompatibility, high encapsulation efficiency.	Liposome-mediated delivery of CRISPR/Cas9 for genome editing in maize.	(Wang et al., 2020).
Polymeric Nanoparticles	Chitosan, polyethylene glycol (PEG)	Transport of siRNA and DNA for gene silencing and expression.	Controlled release, stability under physiological conditions.	Chitosan nanoparticles used for siRNA delivery in tomato for disease resistance.	(Subiksha et al., 2019).
Carbon-Based Nanocarriers	Carbon nanotubes, graphene oxide	Penetration of plant cell walls to deliver DNA and RNA molecules.	High strength, efficient cellular uptake.	Carbon nanotubes used to deliver plasmid DNA to tobacco cells.	(Demirer et al., 2019a).
Metal and Metal Oxide NPs	Gold nanoparticles, mesoporous silica	Delivery of CRISPR components, DNA, or RNA directly into plant tissues.	High precision, minimal off-target effects.	Gold nanoparticle-mediated delivery of DNA in Arabidopsis.	(Zhang et al., 2021a).
Silicon-Based Nanocarriers	Mesoporous silica nanoparticles	Delivery of dsRNA for pest and pathogen resistance in crops.	High loading capacity, protection from degradation.	Mesoporous silica nanoparticles for dsRNA delivery in rice to suppress fungal infections.	(Mitter et al., 2017).
Magnetic Nanoparticles	Iron oxide, cobalt ferrite	Gene delivery facilitated by magnetic fields (magnetofection).	Targeted delivery, non-invasive.	Iron oxide nanoparticles used for DNA delivery in wheat.	(Alidoust & Kaviani, 2020).
Dendrimers	Polyamidoamine (PAMAM)	Transfer of genetic material for gene expression and regulation.	Precise targeting, high transfection efficiency.	PAMAM dendrimers for transgene delivery in Arabidopsis.	(Pathak et al., 2019).

Figure 3. Types of nanoparticles

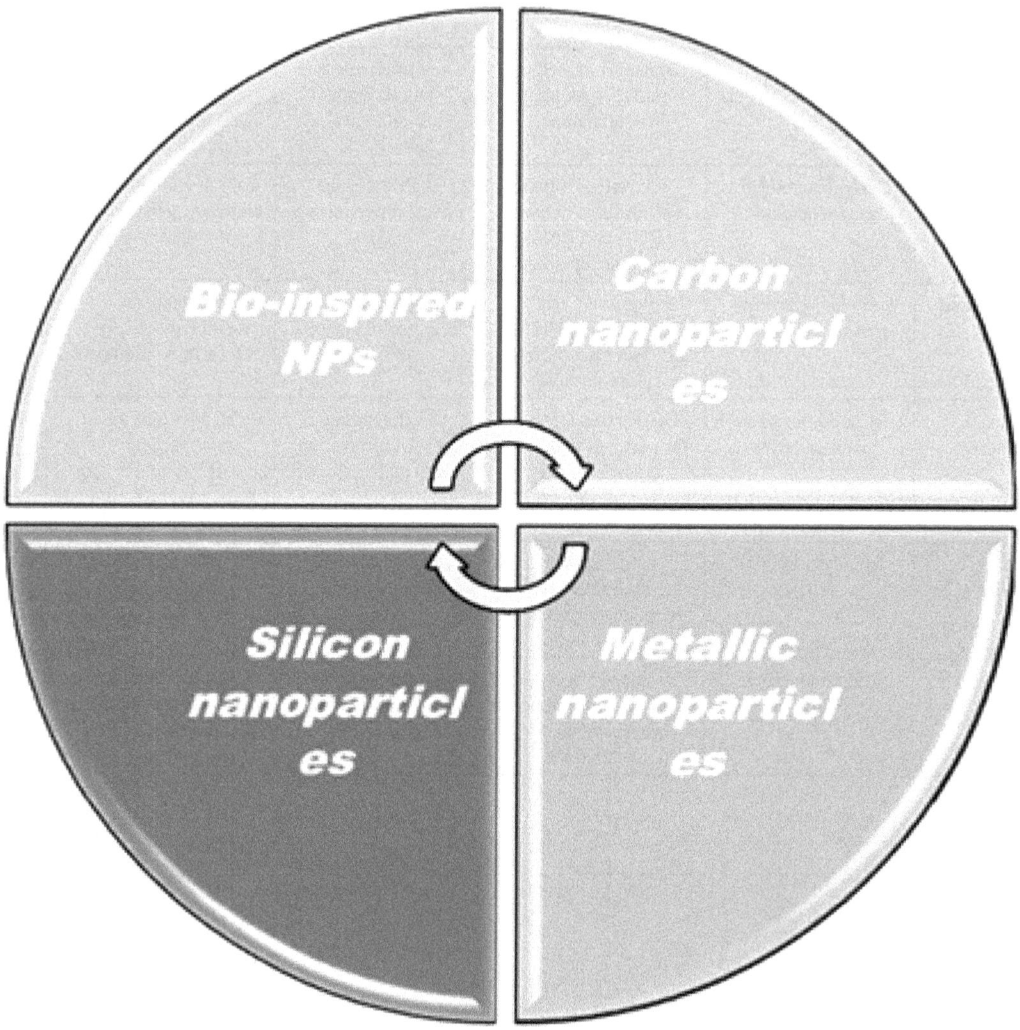

Carbon Nanoparticles

Despite the fact that carbon nanomaterial internalisation was demonstrated in 2009 (Liu et al., 2009), it was not utilised for delivery purposes until 2019. Carbon nanotubes (CNTs) and carbon dots are capable of transmitting genetic elements from plants to other organisms due to their robust tensile strength and diminutive size. A 4.2 kb DNA plasmid can be transported into N. by polyethylenimine (PEI)-linked carbon nanotubes (CNTs) to momentarily express GFP23 in a variety of plant species, including cotton, rocket, wheat, Nicotiana benthamiana, and others (Demirer et al., 2019b). In order to address the challenge of delivering larger plasmids (≥10 kb) using PEI CNTs, it is essential to modify the plasmid-to-carrier ratios and surface chemistry. The significance of NP structure-function design concepts is underscored by the significant impact of the affixed polymer 108 on the biocompatibility and toxicity of CNT carriers. Carbon particles functionalised with PEI are capable of transporting any type of cargo. This encompasses small interfering RNA (siRNA). The distribution of carbon atoms is significantly influenced by the size of the carrier, in contrast to CNTs. Multi-walled CNTs that are at least 12.1 nm in size have the potential to transport plasmid DNA, whereas carbon dots that are larger than 8.7 nm have difficulty disseminating siRNA (Demirer et al., 2019b; Schwartz et al., 2020). This variation is due to the fact that the dimensions and capacities of various carriers for goods differ. Carbon nanoparticles are employed in both nuclear and non-nuclear genome editing techniques. Before employing LEEP model design heuristics to modify CNTs for chloroplast targeting, the carrier surface was functionalised with chitosan and plasmid DNA (Kwak et al., 2019). In accordance with the LEEP model's predictions, this surface functionalisation optimised the zeta potential (the electrical potential at the slipping plane of a particle suspended in a fluid) to 35 mV for chloroplast internalisation and reduced the size of the CNT carrier to a minimal dimension of 10 nm. This strategy was effective in modifying chloroplast genomes across multiple species, as demonstrated by fluorescence microscopy with a GFP vector (Kwak et al., 2019). This emphasises the effectiveness of rational design, which is predicated on fundamental structure-function principles. It is uncertain whether CNTs are effective in expediting plastid transformation in plants, as they were not compared to biolistic delivery techniques in this study. Peptide sequences that bind to CNT carriers and specifically target the mitochondria are another method for controlling mitochondria. The thirty-fold increase in Arabidopsis efficiency when targeting peptide sequences is used in conjunction with CNT conjugation rather than alone is likely due to cargo protection. The species independence of this mitochondrial targeting technique is still a topic of debate (Kwak et al., 2019).

Metallic Nanoparticles

In the past, many metallic delivery techniques have been used to transfer genetic material to animal systems. Of them, gold nanoparticles have been the most extensively studied and used to transport biomolecules. Biolistic chemical transfer using micro-sized gold has been practiced by plants for many years. Because of its ability to prevent cargos from being degraded by nucleases, allow for accurate control of cargo quantity, and allow thiolated nucleic acids to be easily attached to gold carriers, gold is widely used for nucleic acid delivery (Vijayakumar et al., 2010). Vijayakumar et al. (2010) detailed the first studies on using gold nanoparticles to transfer plasmid DNA to plants. When it came to delivering DNA, the biolistics of gold nanoparticles were about 40% more efficient and caused about 40% less harm to plant tissues than micro-gold particles. While some research has shown success in siRNA distribution using gold nanoparticles for DNA transport, the majority of this work has relied on biolistics. One example is the delivery of siRNA to adult plants by means of polydisperse spherical gold nanoparticles, which can specifically target Arabidopsis protoplasts. This results in a staggering 80% reduction in gene expression (Lei et al., 2020). It is also possible to target mature N. with siRNA that uses gold nanoclusters. Assimilation of carrier by benthamiana leaves in plants (Zhang et al., 2021b).

The delivery of siRNA using gold nanoparticles does not require the internalisation of carriers. Gold nanoparticles, according to a detailed analysis of their size and shape, do not penetrate cells but instead become a part of the plant cell wall (Zhang et al., 2022a). But even if they don't absorb it, these carriers nonetheless manage to silence genes by transmitting siRNA. In addition, although 5 nm spherical gold nanoparticles cannot enter plant cells, 13 nm and smaller gold nanorods can, perhaps because of their higher aspect ratio. Research shows that shape is a key factor in understanding NP transport in plant tissues and cells (Lei et al., 2020). In animal systems, cylindrical nanoparticles have an advantage over spherical ones when it comes to cellular internalisation and tissue transit since their smallest dimension is parallel to the surface of the cell membrane. There have also been reports about magnetic nanoparticle (MNP) delivery. Magnetofection, which involves directly modifying pollen using magnetic field-driven MNPs, can be used to alter many dicot and monocot species (Zhang et al., 2021a).

Transforming pollen grains allows them to undergo genetic alterations that are passed down to their offspring, effectively avoiding regeneration. During the initial magnetofection experiment, the staining of treated pollen was used to verify the transfer of plasmid DNA encoding the β-glucuronidase (GUS) protein to pollen from several species. Subsequent studies were unable to reproduce the original finding while magnetofecting pollen, and this was in part due to the reliability of the GUS protein reporter system. With the use of a more dependable GFP reporter system, we were able to confirm that MNP pollen transformed maize. The scientists believe that variations in pollen structure, namely the pollen aperture, the pliable, softer parts of the pollen grain that allowed the nanoparticles to flow through which are to blame for the previously noted repeatability problem (Lei et al., 2020). Pollen grains have an inherent charge that can impede transformation by electrostatically drawing or repelling nanoparticles. Since nanoparticles are often charged, this makes it harder for them to pass through pollen apertures. The method of pollen magnetofection is controversial and only works on a small number of species. Thorough research on NP-pollen interactions across various taxa is necessary to understand the factors impacting NP-mediated pollen transformation (Zhang et al., 2021a,b).

Silicon Nanoparticles

The initial application of silicon-based delivery systems in plants was the transfer of DNA plasmids to intact N using 100-200 nm diameter mesoporous silica nanoparticles (MSNs) (Knewer et al., 2000), following extensive reporting of these systems in animal systems. Applying the principles of biology to the cultivation of sweet potatoes. It is important to remember that gold nanoparticle coatings are necessary for MSNs to distribute plasmids. This method achieved the remarkable feat of genetic manipulation by directly applying the site-specific recombinase Cre (Radu et al., 2004). This is especially noteworthy in light of the scarcity of literature on the delivery of DNA-free genome alteration proteins to plants. Although this method has demonstrated a 20% increase in delivery efficacy on specific tissues (e.g., onion epidermal and immature maize), it has not yet been validated for complete tissues. The delayed recognition of genetic alteration mediated by silicon since 2015 and the effectiveness of delivery systems based on carbon and metals, which do not rely on biolistics, are likely to be the culprits (Bharali et al., 2005).

Bio-Inspired NPs

To date, only protoplasts and other undeveloped plant tissues have successfully received genetic material transferred to plant cells via bioinspired nanoparticles such as liposomes or vesicles. The DNA plasmid encoding the site-specific nuclease Cas9 (Mahmoud et al., 2022) and its facilitation in direct administration can be administered by liposome-based protoplast transfection (Liu et al., 2020). The editing efficiencies were 6% with DNAfree direct delivery of Cas9 ribonucleoprotein (RNP) utilising the commercial liposome transfection agent Lipofectamine 3000, compared to 3% with macro-gold biolistic delivery (Liu et al., 2020). In a similar manner, Cas9 RNP was delivered to the Arabidopsis callus via a cell-penetrating peptide-decorated vesicle approach; nevertheless, the editing efficiency was less than 1%. Liposomes can facilitate the delivery of iron and other nutrients to the cells of entire plants. Liposomes originating from lecithin, including a backbone of 16–18 carbon atoms, were translocated to the roots and distal leaves following topical application (Knewer et al., 2000). While further investigation in mature plants is required, these results indicate that liposomes may serve as a viable medium for genetic transfer to both protoplasts and whole plants. The efficiency of liposomal delivery will likely be determined by its composition, highlighting the importance of study for informed design. A prospective basis for forthcoming research in liposomal-based delivery may use the endogenous vesicle trafficking systems implicated in plant-fungal interactions (Bharali et al., 2005).

NP-MEDIATED BIOMOLECULE DELIVERY TO PLANTS

Nanocarriers, which are also referred to as nanoparticles (NPs), are one approach to injecting biomolecules such as DNA, proteins, or other components into plant cells. This technique has become an invaluable resource for plant genetic engineering by surpassing the limitations of more conventional delivery methods, including biolistics and Agrobacterium-mediated transformation. Nanocarriers are capable of effectively traversing plant cell walls and membranes, thereby circumventing physical obstacles that typically obstruct biomolecule absorption. Key features of NP-mediated biomolecules delivery to plants is shown in Figure 4.

Figure 4. Key features of NP-mediated biomolecules delivery to plants

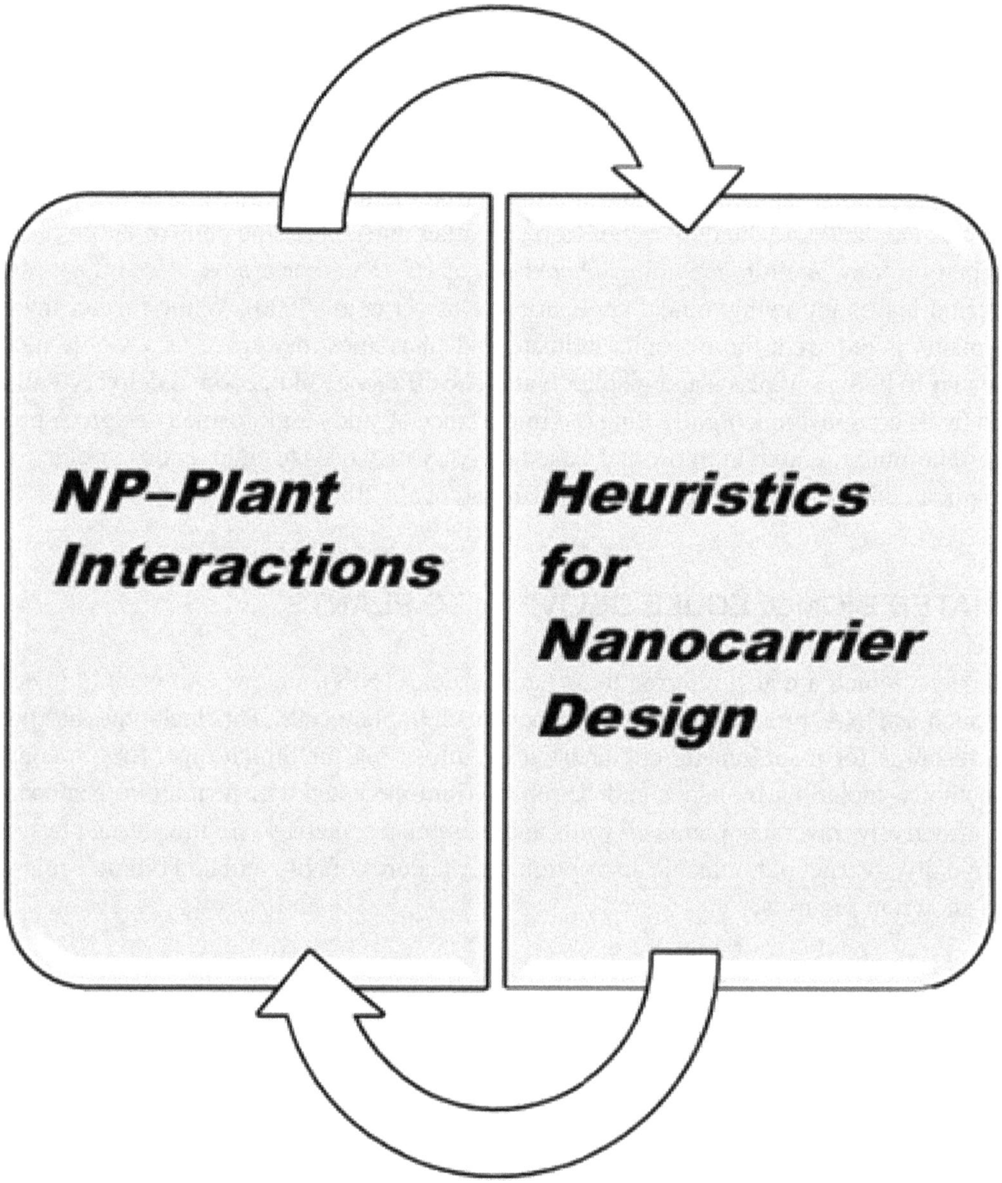

NP–Plant Interactions

The majority of research on NP-plant systems has focused on three primary domains: the synthesis of metallic nanoparticles on plants (Vijayaraghavan & Ashokkumar, 2017), the transport of agrochemicals (Baker et al., 2016), and the uptake of NPs, which has demonstrated both advantageous and detrimental effects on plant growth (Tripathi et al., 2017; Zuverza-Mena et al., 2017). Diverse types of nanoparticles, including metal/metal oxide nanoparticles (Kurepa et al., 2010; Gonzalez-Melendi et al., 2008;Larue et al., 2012), quantum dots (Koo et al., 2015), carbon nanotubes (CNTs) (Liu et al., 2009), and multi-surface nanostructures (MSNs) (Hussain et al., 2013), can be directly assimilated to differing extents by dicotyledonous and monocotyledonous plants. Certain nanoparticles induce phytotoxicity by vascular constriction, oxidative stress, or DNA damage upon absorption (Tripathi et al., 2017). Evidence indicates that NPs exert an antagonistic effect, promoting root and leaf growth along with chloroplast synthesis (Zuverza-Mena et al., 2017). Further investigation focussing on the physical and chemical properties of nanoparticles is essential to elucidate the trade-offs between phytotoxicity and growth enhancement, contingent upon species, growing conditions, dosage, and nanoparticle characteristics. Given the significant potential for analogous breakthroughs in plants, it is imperative to address the information gap concerning the physiological responses of plants to nanoparticle uptake while simultaneously enhancing plant science through manufactured nanomaterials. This is due to a "nanorevolution" in targeted delivery systems for animals (Tripathi et al., 2017).

Heuristics for Nanocarrier Design

The understanding of the structure-function landscape is restricted in terms of the chemical and physical properties of nanoparticles that influence payload loading and cellular internalisation, despite the fact that it is advantageous to design nanocarriers using a heuristic approach. The size exclusion limits (SELs) for various plant tissues and organs are determined by pore diameters, which have been extensively investigated in the literature (Wang et al., 2016; Eichert et al., 2008; Eichert & Goldbach, 2008; Schwab et al., 2015; Ma et al., 2010; Larue et al., 2014). The absorption and distribution of NPs by plant tissues are regulated by these SELs. Contrary to the widely held belief that the cell wall is incapable of allowing particles larger than 5-20 nm to pass through the cell membrane, recent investigations have demonstrated that nanoparticles as large as 50 nm can do so. Nevertheless, the precise mechanisms that underlie this phenomenon remain unknown (Wang et al., 2016; Schwab et al., 2015). The plasma and nuclear membranes in genetic engineering present additional delivery challenges. This is particularly true when gene function is influenced by cytosolic or nuclear localisation. The cell wall, which has a SEL of 500 nm, has practical applications (Wang et al., 2016). These parameters are optimised for use in nanocarriers due to their substantial influence on the translocation of cell membrane charge and nanoparticle form (Albanese et al., 2012). The processes of energy-dependent endocytosis and energy-independent direct penetration, which are the two methods by which plants absorb cells, are still inadequately understood. The prevailing opinion is that cationic nanoparticles are more easily attached and internalised by negatively charged cell membranes than anionic ones (Albanese et al., 2012). This predilection for charges has been demonstrated in protoplasts and wall-bearing cells of plants (Torney et al., 2007; Zhu et al., 2012).

Nanoparticles encased in vesicles must be able to evade the endosome in order to be able to distribute subcellularly. This will prevent them from being further processed if they are designated for destruction or exocytosis. The requirement for endosomal vesicle synthesis is eliminated by direct cellular entrance, regardless of the impact of endocytic proteins and vesicle cargo on the endosome's fate. Nevertheless, the precise subcellular localisation of nanoparticles in plants remains unknown. Serag and colleagues have confirmed that high aspect ratio nanoparticles facilitate vesicle-free internalisation in mammalian cells by demonstrating carbon nanotube internalisation in protoplasts through direct penetration and endocytosis (Serag et al., 2013; Serag et al., 2011). Further research is necessary to determine the mechanism by which high aspect ratio nanoparticles penetrate plant cell walls, as Serag and colleagues exclusively observed direct penetration in protoplasts for cell wall-impermeable multiwalled carbon nanotubes (Serag et al., 2013; Serag et al., 2011). In their investigation, Wong and colleagues discovered that excised chloroplasts passively internalise single-walled carbon nanotubes through a process that is contingent upon the nanoparticles' size and zeta potential (Wong et al., 2016). It is widely recognised that cationic, pH-buffering polymers can function as ligands that facilitate endosomal escape (Selby et al., 2017). Chang et al. (2013), discovered that spherical MSNs that had been organically functionalised could passively and energy-freely penetrate impermeable root cells (Chang et al., 2013).

Plants typically transport endocytosed single-walled carbon nanotubes to vacuoles; however, they may wind up in the cytosol when they bind with DNA (Liu et al., 2009; Serag et al., 2013). Adsorption on surfaces, which is frequently referred to as physisorption, is a straightforward method for connecting biomolecules to the majority of nanoparticles. Electrostatic interactions are the preferred method for noncovalent cargo loading, as physisorption may be unstable with specific nanoparticles and payloads (Zhao et al., 2016). Cationic surfaces are capable of electrostatically loading genetic material due to their ability to attract negatively charged DNA and RNA, which enhances endocytic absorption and release. Some 'click' chemistries are frequently employed to functionalise the surface of covalent nanoparticles (Lallang et al., 2012). The successful covalent bonding of thiolated DNA and proteins to gold nanoparticles is a significant advancement in covalent bioconjugation, particularly for the application in plants (Lee et al., 2017). Nevertheless, there are still opportunities for innovation in this discipline. Macromolecules or small compounds can be mounted onto porous nanoparticles, such as mesoporous silica nanoparticles (MSNs), for controlled intracellular release, rather than relying on surface functionalisation (Slowing et al., 2008). The initial step in selecting the optimal ligand, cargo, and nanoparticle (NP) for a specific application is to identify NPs that possess at least one of the aforementioned qualities. These NPs have been demonstrated to be beneficial in the transport of biomolecules within plants. These heuristics will likely need to be modified for use in other systems in order to achieve success in nanocarrier design, a challenging multivariable optimisation process, until a complete NP structure-function link for plant systems is established (Zhao et al., 2016).

NANOMATERIALS FOR PLANT GENETIC ENGINEERING

Nanoparticles are advantageous for intracellular biomolecule delivery due to their ability to traverse biological membranes, contain and release a variety of payloads, and endure physical and chemical modifications to facilitate multidimensional targeting. The landscape of targeted delivery and controlled release in mammals has been entirely transformed by the distinctive characteristics of nanoparticles. Regrettably, the majority of nanocarrier delivery in plants remains unexplored as a result of the widespread use of chemical and mechanical methods to penetrate cell membranes (Lee et al., 2017). Passive biomolecule delivery to plants is a promising approach for in vivo plant genetic engineering, particularly for transient expression in somatic tissue. This method is less intrusive and species-independent. One of the numerous challenges associated with in vitro plant studies is the significant variation in regeneration capacity that occurs across species, genotypes, and even within a single plant, contingent upon the developmental age of the donor tissue (Birnbaum & Sanchez-Alvarado, 2008). This emphasises the potential of nanoparticle-based plant delivery strategies. Long-term transformation necessitates the regeneration of progeny from embryogenic calli, irrespective of the delivery method. Consequently, it is imperative to optimise both supply and regeneration simultaneously in order to enhance efficiency and expand the potential for consistent transformation in all plant species (Zhao et al., 2016).

Recent research has demonstrated that NPs can facilitate passive transport in plants, both in vitro (Naqvi et al., 2012; Pasupathy et al., 2008; Burlaka et al., 2015) and in vivo (Chang et al., 2013; Jiang et al., 2014) in N. Sativa and other plants. Demirer et al. (2019a) have recently demonstrated the passive delivery of DNA plasmids and protected siRNA using functionalised carbon nanotube (CNT) nanoparticles. This method was employed to intermittently express GFP in Eruca sativa (arugula) leaves and to temporally suppress the constitutive production of GFP in transgenic Nicotiana benthamiana leaves (Demirer et al., 2019b). The study demonstrates that passive NP delivery in model and crop species can be successful and minimally deleterious by utilising CNT-mediated transient GFP expression in Triticum aestivum (wheat) (Demirer et al., 2019a). Despite the fact that further research is required to refine the characteristics and functionalisation of nanoparticles, these initial results indicate that nanoparticles have the potential to serve as a means of transporting biomolecules to plants, potentially resolving certain issues with existing methods. It is imperative to investigate NP-based transport techniques for various biomolecular payloads in order to enhance the delivery of nuclease-based gene editing technologies, which are instruments for revolutionary genome engineering (Demirer et al., 2019a).

NANOCARRIERS HOLD PROMISE FOR NUCLEASE-BASED PLANT GENOME EDITING

Regulatory monitoring of transgenic plants may require modification due to the potential for genome editing techniques to expedite molecular biology and genetic research in plants. Among the numerous variables that influence this likelihood are the species, susceptible tissues, delivery route, expression strategy (DNA, RNA, or protein), and the capacity of a host genome to incorporate a transgene. Regulatory agencies and other entities are demonstrating a significant preference for genome editing techniques that are "DNA-free" in order to eliminate the prospect of transgenic insertion. New evidence indicates that RNPs are administered simultaneously in A. o. thaliana. Consequently, as a consequence of PEG-transfection (Woo et al., 2015). These two investigations are severely restricted in their approach due to issues with progeny regeneration, which in turn limits their output. Reliable and efficient gene editing in plants is impeded by two primary obstacles. Embryogenic calli must be employed to regenerate plants from seed, as there is currently no method to modify plant germline cells, with the exception of the Arabidopsis floral dip. The second issue is that the cell wall serves as a formidable transport barrier, rendering conventional methods of delivering biomolecules to plants either inefficient, detrimental, or even host-specific (Liu et al., 2017).

The primary impediments to the widespread application of plant genome editing are the absence of a bio-cargo delivery mechanism that is effective across species and does not target germline cells. Although enhanced nuclease systems exhibit significant potential for plant genome engineering, it is imperative to develop new carriers in order to circumvent the constraints of existing delivery methods. Plants may be capable of directly modifying their germlines or utilising these carriers to more effectively regenerate their progeny. The use of nanoparticles to transport RNA, ribonucleoproteins, and plasmids has begun to enhance genome editing. In mammalian cells, nanoparticles are frequently employed to facilitate the direct cytosolic/nuclear transport of Cas-RNPs to a variety of cell types. Research has shown that the administration of RNPs substantially reduces off-target effects in comparison to plasmid-based CRISPR systems (Mout et al., 2017). However, the cell wall has thus far impeded the development of a comparable system in plants, which could potentially facilitate the passive transmission of genome editing materials to mature plants and even between species. This implies that there is a significant amount of potential for advancement in the development of NP carriers that possess the appropriate shape and chemistry to autonomously traverse the cell walls and membranes of dense plant tissues. Some nanoparticle formulations have the capacity to passively incorporate DNA, RNA, or proteins in plants, as indicated by prior research (Chang et al., 2013; Jiang et al., 2014). Additional investigation is warranted in the context of the integration of engineered nuclease systems into plants, specifically these nanoparticle scaffolds: polymeric nanoparticles, carbon nanotubes, and mesoporous silica nanoparticles (Demirer et al., 2019b).

ENVIRONMENTAL AND REGULATORY CONSIDERATIONS FOR NANOPARTICLE-ENABLED PLANT GENE EDITING

Conventional genetically modified (GMO) commodities are subject to stringent regulatory systems due to the presence of foreign DNA. A potential alternative to traditional transgenic methodologies is the transient expression of CRISPR reagents in genome-edited crops through nanoparticle-mediated delivery (Gravely et al., 2019). This would enable the cultivation of cereals without transgenes. This has the potential to decrease the costs associated with regulatory procedures for genome-edited crops and prevent the use of GMO labels in numerous countries. Consequently, it has the potential to foster creativity, provide access to affordable resources, and ultimately result in their monetisation. Nevertheless, the regulatory frameworks for organisms that have been modified through gene editing are inconsistent across the globe. The recent decisions of the European Court of Justice to classify genome-edited crops as conventional GMOs may impede the progress of plant genome editing (Zaidi et al., 2019). The relaxation of restrictions by Japan and the United States may lead to a more favourable climate, which could result in less stringent regulatory policies on a global scale (Waltz, 2018). Navigating the safety and environmental evaluations conducted by regulatory authorities may prove challenging due to our inadequate comprehension of nanoparticle ecotoxicity, exposure pathways, and lifecycle impact assessments. The use of nanoparticles is influenced by a variety of factors, including politics, consumer acceptance, economics, ethics, and the law (Yang & Duncan, 2021).

Therefore, prior to employing nanoparticles for plant gene editing outside of regulated laboratory environments, public health organisations, consumers, and manufacturers must evaluate the advantages and disadvantages, including environmental impacts, cost, and effectiveness. Consequently, the regulatory framework that governs commodities that have undergone genome modifications will become considerably more complex. While sentinel field data can be collected at multiple test sites to assist in decision-making, this necessitates long-term financing options that are specifically designed for this purpose. Nanotechnology provides researchers in the field of plant science with a distinctive instrument that enables them to investigate fundamental plant biology by employing efficient and species-independent genetic modifications (Bartczak et al., 2012). Advancements in next-generation omics technologies, including genomics, transcriptomics, proteomics, and metabolomics, will enhance our comprehension of the impact of these nanoparticles on the plant metabolic network and transcriptome. Our findings suggest that the design of nanoparticles may be enhanced to mitigate concerns regarding contamination and toxicity by expediting the disintegration, exocytosis, and excretion of genetic payload from plants (Oh & Park (2014). Peptide and chemical functionalisation techniques have been employed by scientists to regulate the assimilation of nanoparticles by humans. A comparable approach may be implemented to regulate the manner in which nanoparticles are processed by plant cells (Kim et al., 2015).

CASE STUDIES OF NANOCARRIERS IN PLANT GENETIC ENGINEERING

Nanocarriers have emerged as pivotal tools in plant genetic engineering, facilitating the precise delivery of genetic material into plant cells. Table 3 shows notable case studies illustrating their application. These case studies demonstrate the diverse applications and advantages of nanocarriers in advancing plant genetic engineering.

Table 3. Case studies of nanocarriers in plant genetic engineering

Nanocarrier Type	Case Study	Application	Key Outcomes	Reference
Carbon Nanotubes (CNTs)	Delivery of DNA and RNA using single-walled carbon nanotubes (SWNTs)	Transient gene expression and gene silencing in mature plants	Enabled efficient genetic modification without transgene integration, reducing tissue damage.	(Demirer et al., 2019a).
Mesoporous Silica Nanoparticles (MSNs)	Functionalized MSNs for DNA delivery in plant cells	Enhanced transformation efficiency	Successfully delivered DNA to plant tissues, showing high biocompatibility and efficiency.	(Hussain et al., 2013)
Gold Nanoparticles (AuNPs)	Gold nanoparticle-mediated DNA delivery in Arabidopsis	Delivery of plasmid DNA for genome editing	Achieved precise gene expression with minimal toxicity to plant cells.	(Zhang et al., 2021a)
Liposomes	Liposome-based delivery of CRISPR/Cas9 components in maize	Genome editing	Successfully delivered CRISPR/Cas9 complexes, enabling precise genetic modifications.	(Wang et al., 2020)
DNA Nanostructures	DNA nanostructures for siRNA delivery in mature plants	Gene silencing	Achieved targeted gene silencing with high efficiency and minimal off-target effects.	(Zhang et al., 2019b)

ETHICS AND PUBLIC PERCEPTION OF NANOTECHNOLOGY IN PLANT GENETIC ENGINEERING

The integration of nanotechnology into plant genetic engineering presents significant ethical considerations and public perception challenges. While nanomaterials offer unprecedented precision in gene delivery and transformation efficiency, their application in genetically modified (GM) crops raises concerns about biosafety, environmental impact, and long-term sustainability. One ethical dilemma involves the unintended consequences of nanoparticle use in open agricultural environments, where nanoparticles may accumulate in soil and water systems, potentially disrupting microbial communities or entering the food chain (Rajput et al., 2022). Moreover, the use of synthetic materials or gene-editing tools enabled by nanocarriers might invoke bioethical debates similar to those surrounding traditional GMOs, particularly regarding transparency, consent, and potential ecological risks (Shew et al., 2018).

Public perception plays a critical role in the adoption and regulation of nanotechnology-enhanced genetically engineered crops. Studies have shown that public acceptance is influenced by factors such as trust in scientific institutions, perceived naturalness of the technology, and the level of understanding about its benefits and risks (Zhang et al., 2020). In many regions, skepticism towards GMOs persists, and the term "nano-enabled" genetic modification may compound these fears if not adequately addressed through public engagement and science communication. Ethical frameworks that incorporate stakeholder participation and respect for cultural values are essential in promoting informed dialogue and responsible innovation in this field (Fisher et al., 2015).

To ensure socially responsible development, policymakers, scientists, and regulatory bodies must collaboratively establish guidelines that balance innovation with ethical accountability. This includes conducting rigorous risk assessments, promoting transparency in research practices, and developing labeling systems that inform consumers about nano-enabled plant products. Educational initiatives that demystify nanotechnology and highlight its potential benefits—such as reducing pesticide use or enhancing food security—can also help build public trust (Groves et al., 2021). Ultimately, integrating ethical deliberation and public perception into the scientific and regulatory process is vital for the sustainable advancement of nanotechnology in plant genetic engineering.

CONCLUSION AND FUTURE PROSPECT

Nanocarriers have emerged as a transformative innovation in plant genetic engineering, offering precise, efficient, and versatile solutions for delivering genetic material and biomolecules to plants. These nanoscale delivery vehicles have successfully addressed several limitations of traditional methods, such as Agrobacterium-mediated transformation and biolistics, which have long struggled with issues like tissue specificity, low transformation efficiency, and limited applicability across diverse plant species. By leveraging the unique physicochemical properties of nanocarriers, including their high surface area, biocompatibility, and tunable functionality, researchers have opened new avenues for plant biotechnology, with implications for sustainable agriculture and global food security. The fact that nanocarriers can penetrate the extremely strong cell walls of plants is one of their greatest strengths. Cellular injury and decreased viability are common outcomes of the invasive or disruptive procedures used by conventional approaches to access these structures. In contrast, nanocarriers may transport genetic material with little physical harm, leading to improved transformation efficiency while protecting plant tissues. Their versatility as tools for both transient expression and stable genetic transformation is further enhanced by their capacity to encapsulate and protect different payloads, including DNA, RNA, CRISPR-Cas components, and agrochemicals.

Their potential has been further enhanced by combining nanocarriers with state-of-the-art genome editing technologies like CRISPR-Cas. Nanocarriers alleviate regulatory worries and boost public support for GMO crops by allowing DNA-free genome editing, which allows for targeted alterations without the introduction of foreign genetic material. When designing crops to withstand abiotic stresses like drought and salinity, as well as to have higher nutritional profiles and disease resistance, this skill becomes very important. Furthermore, genetic engineering becomes more scalable and accessible due to the non-invasive administration methods of nanocarriers, which include root immersion and foliar sprays. These methods simplify the delivery process and lessen reliance on specialised laboratory infrastructure. A number of obstacles persist, even with these improvements. To encourage their broad use in farming, nanocarrier production needs to be made more scalable and cost-effective.

Another important aspect that needs to be thoroughly investigated is the biosafety and environmental safety of nanocarriers. This is especially important when thinking about how long they can stay in ecosystems and what impact they could have on non-target creatures. Because plant genetic engineering has to meet the varied needs of farmers around the world, future studies should concentrate on finding ways to maximise delivery efficiency across many different types of plants and tissues. Because nanotechnology in agriculture is so novel, regulatory frameworks will need to change to reflect this. Policies should strike a balance between requiring thorough safety evaluations and allowing room for innovation and acceptance. In order to build trust and acceptance, it is crucial to communicate openly about the risks, advantages, and ethical implications of nanocarrier-based technology. Nanotechnology and plant genetic engineering, when used together, have the potential to solve some of the most important problems of our day.

Nanocarriers provide a mechanism to create robust, high-yielding crops that can sustainably meet the demands of a world food system that is under increasing pressure from factors like population increase, climate change, and limited resources. New capabilities and applications will be made possible by the ongoing evolution of nanocarrier technologies, which are driven by advances in molecular biology, bioinformatics, and material science. To sum up, nanocarriers open the door between theoretical advancements in plant genetic engineering and their real-world agricultural applications. Nanocarriers have the potential to revolutionise genetic transformation—improving its accuracy, efficiency, and reach—and hence have far-reaching effects on global food security, environmental preservation, and economic growth. Nanocarriers have the potential to revolutionise agricultural biotechnology by turning obstacles into possibilities for future resilience and sustainability through ongoing research, innovation, and collaboration.

REFERENCES

Abdallah, N. A., Prakash, C. S., & McHughen, A. G. (2015). Genome editing for crop improvement: Challenges and opportunities. *GM Crops and Food: Biotechnology in Agriculture and the Food Chain*, *6*(4), 183–205. DOI: 10.1080/21645698.2015.1129937 PMID: 26930114

Aesaert, S., Impens, L., Coussens, G., Van Lerberge, E., Vanderhaeghen, R., Desmet, L., Vanhevel, Y., Bossuyt, S., Wambua, A. N., Van Lijsebettens, M., Inzé, D., De Keyser, E., Jacobs, T. B., Karimi, M., & Pauwels, L. (2022). Optimized Transformation and Gene Editing of the B104 Public Maize Inbred by Improved Tissue Culture and Use of Morphogenic Regulators. *Frontiers in Plant Science*, *13*, 883847. DOI: 10.3389/fpls.2022.883847 PMID: 35528934

Ahmar, S., Mahmood, T., Fiaz, S., Mora-Poblete, F., Shafique, M. S., Chattha, M. S., & Jung, K. H. (2021). Advantage of Nanotechnology-Based Genome Editing System and Its Application in Crop Improvement. *Frontiers in Plant Science*, *12*, 663849. DOI: 10.3389/fpls.2021.663849 PMID: 34122485

Albanese, A., Tang, P. S., & Chan, W. C. (2012). The effect of nanoparticle size, shape, and surface chemistry on biological systems. *Annual Review of Biomedical Engineering*, *14*(1), 1–16. DOI: 10.1146/annurev-bioeng-071811-150124 PMID: 22524388

Ali, A., Bang, S. W., Chung, S. M., & Staub, J. E. (2014). Plant transformation via pollen tube-mediated gene transfer. Plant. *Plant Molecular Biology Reporter*, *33*(3), 742–747. DOI: 10.1007/s11105-014-0839-5

Altpeter, F., Springer, N. M., Bartley, L. E., Blechl, A., Brutnell, T. P., Citovsky, V., & Voytas, D. F. (2016). Advancing crop transformation in the era of genome editing. *The Plant Cell*, *28*(7), 1510–1520. DOI: 10.1105/tpc.16.00196 PMID: 27335450

Baker, C. J., Mock, N. M., Whitaker, B. D., Roberts, D. P., Rice, C. P., Deahl, K. L., & Aver'yanov, A. A. (2005). Involvement of acetosyringone in plant-pathogen recognition. *Biochemical and Biophysical Research Communications*, *328*(1), 130–136. DOI: 10.1016/j.bbrc.2004.12.153 PMID: 15670760

Baker, S., Volova, T., Prudnikova, S. V., Satish, S., & Prasad, M. N. N. (2017). Nanoagroparticles emerging trends and future prospect in modern agriculture system. *Environmental Toxicology and Pharmacology*, *53*, 10–17. DOI: 10.1016/j.etap.2017.04.012 PMID: 28499265

Baltes, N. J., Gil-Humanes, J., & Voytas, D. F. (2017). Genome Engineering and Agriculture: Opportunities and Challenges. *Progress in Molecular Biology and Translational Science*, *149*, 1–26. DOI: 10.1016/bs.pmbts.2017.03.011 PMID: 28712492

Bartczak, D., Nitti, S., Millar, T. M., & Kanaras, A. G. (2012). Exocytosis of Peptide Functionalized Gold Nanoparticles in Endothelial Cells. *Nanoscale*, *4*(15), 4470–4472. DOI: 10.1039/c2nr31064c PMID: 22743818

Behzadi, S., Serpooshan, V., Tao, W., Hamaly, M. A., Alkawareek, M. Y., Dreaden, E. C., Brown, D., Alkilany, A. M., Farokhzad, O. C., & Mahmoudi, M. (2017). Cellular uptake of nanoparticles: Journey inside the cell. *Chemical Society Reviews*, *46*(14), 4218–4244. DOI: 10.1039/C6CS00636A PMID: 28585944

Bharali, D. J., Klejbor, I., Stachowiak, E. K., Dutta, P., Roy, I., Kaur, N., Bergey, E. J., Prasad, P. N., & Stachowiak, M. K. (2005). Organically modified silica nanoparticles: A nonviral vector for in vivo gene delivery and expression in the brain. *Proceedings of the National Academy of Sciences of the United States of America*, *102*(32), 11539–11544. DOI: 10.1073/pnas.0504926102 PMID: 16051701

Birnbaum, K. D., & Sánchez Alvarado, A. (2008). Slicing across kingdoms: Regeneration in plants and animals. *Cell*, *132*(4), 697–710. DOI: 10.1016/j.cell.2008.01.040 PMID: 18295584

Burlaka, O. M., Pirko, Y. V., Yemets, A. I., & Blume, Y. B. (2015).. . *Tsitologiia i Genetika*, *49*(6), 3–12. PMID: 26841488

Chang, F. P., Lin-Yun, K., Chia-An, H., Wann-Neng, J., Yann, H., Yueie, C. H., & Chung-Yuan, M. (2013). A simple plant gene delivery system using mesoporous silica nanoparticles as carriers. *Journal of Materials Chemistry. B*, *1*(39), 5279. DOI: 10.1039/c3tb20529k PMID: 32263331

Ciardi, J. A., Tieman, D. M., Lund, S. T., Jones, J. B., Stall, R. E., & Klee, H. J. (2000). Response to Xanthomonas campestris pv. vesicatoria in tomato involves regulation of ethylene receptor gene expression. *Plant Physiology*, *123*(1), 81–92. DOI: 10.1104/pp.123.1.81 PMID: 10806227

Cunningham, F. J., Goh, N. S., Demirer, G. S., Matos, J. L., & Landry, M. P. (2018). Nanoparticle-Mediated Delivery towards Advancing Plant Genetic Engineering. *Trends in Biotechnology*, *36*(9), 882–897. DOI: 10.1016/j.tibtech.2018.03.009 PMID: 29703583

Das, D., Reddy, M., Upadhyaya, K., & Sopory, S. (2002). An efficient leaf-disc culture method for the regeneration via somatic embryogenesis and transformation of grape (Vitis vinifera L.). *Plant Cell Reports*, *20*(11), 999–1005. DOI: 10.1007/s00299-002-0441-4

Debernardi, J. M., Tricoli, D. M., Ercoli, M. F., Hayta, S., Ronald, P., Palatnik, J. F., & Dubcovsky, J. (2020). A GRF-GIF chimeric protein improves the regeneration efficiency of transgenic plants. *Nature Biotechnology*, *38*(11), 1274–1279. DOI: 10.1038/s41587-020-0703-0 PMID: 33046875

Demirer, G. S., Zhang, H., Goh, N. S., González-Grandío, E., & Landry, M. P. (2019). Carbon nanotube-mediated DNA delivery without transgene integration in intact plants. *Nature Protocols*, *14*(10), 2954–2971. DOI: 10.1038/s41596-019-0208-9 PMID: 31534231

Demirer, G. S., Zhang, H., Goh, N. S., Pinals, R. L., Chang, R., & Landry, M. P. (2019a). Carbon nanotubes deliver siRNA and plasmid DNA into plant cells without transgene integration. *Nature Nanotechnology*, *14*(5), 456–464. DOI: 10.1038/s41565-019-0382-5 PMID: 30804481

Demirer, G. S., Zhang, H., Matos, J. L., Goh, N. S., Cunningham, F. J., Sung, Y., Chang, R., Aditham, A. J., Chio, L., Cho, M. J., Staskawicz, B., & Landry, M. P. (2019b). High aspect ratio nanomaterials enable delivery of functional genetic material without DNA integration in mature plants. *Nature Nanotechnology*, *14*(5), 456–464. DOI: 10.1038/s41565-019-0382-5 PMID: 30804481

Dinesh-Kumar, S. P., Anandalakshmi, R., Marathe, R., Schiff, M., & Liu, Y. (2003). Virus-induced gene silencing. *Methods in Molecular Biology (Clifton, N.J.)*, *236*, 287–294. DOI: 10.1385/1-59259-413-1:287 PMID: 14501071

Dong, Z. Z., & Chang, L. Q. (2021). Recent electroporation-based systems for intracellular molecule delivery. *Nanotechnology and Precision Engineering*, *4*(4), 045001. DOI: 10.1063/10.0005649

Eichert, T., & Goldbach, H. E. (2008). Equivalent pore radii of hydrophilic foliar uptake routes in stomatous and astomatous leaf surfaces—Further evidence for a stomatal pathway. *Physiologia Plantarum*, *132*(4), 491–502. DOI: 10.1111/j.1399-3054.2007.01023.x PMID: 18334002

Eichert, T., Kurtz, A., Steiner, U., & Goldbach, H. E. (2008). Size exclusion limits and lateral heterogeneity of the stomatal foliar uptake pathway for aqueous solutes and water-suspended nanoparticles. *Physiologia Plantarum*, *134*(1), 151–160. DOI: 10.1111/j.1399-3054.2008.01135.x PMID: 18494856

Fisher, E., Mahajan, R. L., & Mitcham, C. (2015). Midstream modulation of technology: Governance from within. *Bulletin of Science, Technology & Society*, *26*(6), 485–496. DOI: 10.1177/0270467606295402

Gelvin, S. B. (2017). Integration of Agrobacterium T-DNA into the Plant Genome. *Annual Review of Genetics*, *51*(1), 195–217. DOI: 10.1146/annurev-genet-120215-035320 PMID: 28853920

Gelvin, S. B. (2021). Plant DNA Repair and *Agrobacterium* T-DNA Integration. *International Journal of Molecular Sciences*, *22*(16), 8458. DOI: 10.3390/ijms22168458 PMID: 34445162

González-Melendi, P., Fernández-Pacheco, R., Coronado, M. J., Corredor, E., Testillano, P. S., Risueño, M. C., Marquina, C., Ibarra, M. R., Rubiales, D., & Pérez-de-Luque, A. (2008). Nanoparticles as smart treatment-delivery systems in plants: Assessment of different techniques of microscopy for their visualization in plant tissues. *Annals of Botany*, *101*(1), 187–195. DOI: 10.1093/aob/mcm283 PMID: 17998213

Gordon, J. E., & Christie, P. J. (2014). The Agrobacterium Ti Plasmids. *Microbiology spectrum, 2*(6), . https://doi.org/DOI: 10.1128/microbiolspec.PLAS-0010-2013

Gravely, M., Safaee, M. M., & Roxbury, D. (2019). Biomolecular Functionalization of a Nanomaterial to Control Stability and Retention within Live Cells. *Nano Letters*, *19*(9), 6203–6212. DOI: 10.1021/acs.nanolett.9b02267 PMID: 31424226

Groves, C., Frater, L., Lee, R. G., & Stokes, E. (2021). Is there room at the bottom for CSR? Corporate social responsibility and nanotechnology in the UK. *Journal of Business Ethics*, *101*(4), 525–552. DOI: 10.1007/s10551-010-0731-7

Gu, X., Liu, L., & Zhang, H. (2021). Transgene-free Genome Editing in Plants. *Frontiers in Genome Editing*, *3*, 805317. DOI: 10.3389/fgeed.2021.805317 PMID: 34927134

Hussain, H. I., Yi, Z., Rookes, J. E., Kong, L. X., & Cahill, D. M. (2013). Mesoporous silica nanoparticles as a biomolecule delivery vehicle in plants. *Journal of Nanoparticle Research*, *15*(6), 1676. DOI: 10.1007/s11051-013-1676-4

Ji, X., Zhang, H., Zhang, Y., Wang, Y., & Gao, C. (2015). Establishing a CRISPR-Cas-like immune system conferring DNA virus resistance in plants. *Nature Plants*, *1*(10), 15144. DOI: 10.1038/nplants.2015.144 PMID: 27251395

Jiang, L., Ding, L., He, B., Shen, J., Xu, Z., Yin, M., & Zhang, X. (2014). Systemic gene silencing in plants triggered by fluorescent nanoparticle-delivered double-stranded RNA. *Nanoscale*, *6*(17), 9965–9969. DOI: 10.1039/C4NR03481C PMID: 25068243

Jinturkar, K. A., Rathi, M. N., & Misra, A. (2011). 3—Gene Delivery Using Physical Methods. In Misra, A. (Ed.), *Challenges in Delivery of Therapeutic Genomics and Proteomics* (pp. 83–126). Elsevier. DOI: 10.1016/B978-0-12-384964-9.00003-7

Kim, C., Tonga, G. Y., Yan, B., Kim, C. S., Kim, S. T., Park, M. H., Zhu, Z., Duncan, B., Creran, B., & Rotello, V. M. (2015). Regulating Exocytosis of Nanoparticles via Host-Guest Chemistry. *Organic & Biomolecular Chemistry*, *13*(8), 2474–2479. DOI: 10.1039/C4OB02433H PMID: 25569869

Kneuer, C., Sameti, M., Bakowsky, U., Schiestel, T., Schirra, H., Schmidt, H., & Lehr, C. M. (2000). A nonviral DNA delivery system based on surface modified silica-nanoparticles can efficiently transfect cells in vitro. *Bioconjugate Chemistry*, *11*(6), 926–932. DOI: 10.1021/bc0000637 PMID: 11087343

Kong, J., Martin-Ortigosa, S., Finer, J., Orchard, N., Gunadi, A., Batts, L. A., Thakare, D., Rush, B., Schmitz, O., Stuiver, M., Olhoft, P., & Pacheco-Villalobos, D. (2020). Overexpression of the Transcription Factor *GROWTH-REGULATING FACTOR5* Improves Transformation of Dicot and Monocot Species. *Frontiers in Plant Science*, *11*, 572319. DOI: 10.3389/fpls.2020.572319 PMID: 33154762

Koo, Y., Wang, J., Zhang, Q., Zhu, H., Chehab, E. W., Colvin, V. L., Alvarez, P. J., & Braam, J. (2015). Fluorescence reports intact quantum dot uptake into roots and translocation to leaves of Arabidopsis thaliana and subsequent ingestion by insect herbivores. *Environmental Science & Technology*, *49*(1), 626–632. DOI: 10.1021/es5050562 PMID: 25437125

Kurepa, J., Paunesku, T., Vogt, S., Arora, H., Rabatic, B. M., Lu, J., Wanzer, M. B., Woloschak, G. E., & Smalle, J. A. (2010). Uptake and distribution of ultrasmall anatase TiO2 Alizarin red S nanoconjugates in Arabidopsis thaliana. *Nano Letters*, *10*(7), 2296–2302. DOI: 10.1021/nl903518f PMID: 20218662

Kwak, S. Y., Lew, T. T. S., Sweeney, C. J., Koman, V. B., Wong, M. H., Bohmert-Tatarev, K., Snell, K. D., Seo, J. S., Chua, N. H., & Strano, M. S. (2019). Chloroplast-selective gene delivery and expression in planta using chitosan-complexed single-walled carbon nanotube carriers. *Nature Nanotechnology*, *14*(5), 447–455. DOI: 10.1038/s41565-019-0375-4 PMID: 30804482

Kwak, S. Y., Wong, M. H., Lew, T. T. S., Bisker, G., Lee, M. A., Kaplan, A., & Strano, M. S. (2019). Nanosensor technology applied to living plant systems. *Annual Review of Analytical Chemistry (Palo Alto, Calif.)*, *12*, 113–140. DOI: 10.1146/annurev-anchem-061318-114910 PMID: 28605605

Lallana, E., Sousa-Herves, A., Fernandez-Trillo, F., Riguera, R., & Fernandez-Megia, E. (2012). Click chemistry for drug delivery nanosystems. *Pharmaceutical Research*, *29*(1), 1–34. DOI: 10.1007/s11095-011-0568-5 PMID: 21913032

Landry, M. P., & Mitter, N. (2019). How nanocarriers delivering cargos in plants can change the GMO landscape. *Nature Nanotechnology*, *14*(6), 512–514. DOI: 10.1038/s41565-019-0463-5 PMID: 31168070

Larue, C., Castillo-Michel, H., Sobanska, S., Cécillon, L., Bureau, S., Barthès, V., Ouerdane, L., Carrière, M., & Sarret, G. (2014). Foliar exposure of the crop Lactuca sativa to silver nanoparticles: Evidence for internalization and changes in Ag speciation. *Journal of Hazardous Materials*, *264*, 98–106. DOI: 10.1016/j.jhazmat.2013.10.053 PMID: 24275476

Larue, C., Laurette, J., Herlin-Boime, N., Khodja, H., Fayard, B., Flank, A. M., Brisset, F., & Carriere, M. (2012). Accumulation, translocation and impact of TiO2 nanoparticles in wheat (Triticum aestivum spp.): Influence of diameter and crystal phase. *The Science of the Total Environment*, *431*, 197–208. DOI: 10.1016/j.scitotenv.2012.04.073 PMID: 22684121

Lee, K., Conboy, M., Park, H. M., Jiang, F., Kim, H. J., Dewitt, M. A., Mackley, V. A., Chang, K., Rao, A., Skinner, C., Shobha, T., Mehdipour, M., Liu, H., Huang, W. C., Lan, F., Bray, N. L., Li, S., Corn, J. E., Kataoka, K., & Murthy, N. (2017). Nanoparticle delivery of Cas9 ribonucleoprotein and donor DNA *in vivo* induces homology-directed DNA repair. *Nature Biomedical Engineering*, *1*(11), 889–901. DOI: 10.1038/s41551-017-0137-2 PMID: 29805845

Lei, W. X., An, Z., Zhang, B., Wu, Q., Gong, W., Li, J., & Chen, W. (2020). Construction of gold-SiRNA NPR1 nanoparticles for effective and quick silencing of NPR1 in Arabidopsis thaliana. *RSC Advances*, *10*(33), 19300–19308. DOI: 10.1039/D0RA02156C PMID: 35515443

Lew, T. T. S., Koman, V. B., & Strano, M. S. (2020). The emergence of plant nanobionics and living plants as technology. *Nature Plants*, *6*(5), 557–569. DOI: 10.1038/s41477-020-0655-6

Liang, Z., Chen, K., Li, T., Zhang, Y., Wang, Y., Zhao, Q., Liu, J., Zhang, H., Liu, C., Ran, Y., & Gao, C. (2017). Efficient DNA-free genome editing of bread wheat using CRISPR/Cas9 ribonucleoprotein complexes. *Nature Communications*, *8*(1), 14261. DOI: 10.1038/ncomms14261 PMID: 28098143

Liu, C., Zhang, L., Liu, H., & Cheng, K. (2017). Delivery strategies of the CRISPR-Cas9 gene-editing system for therapeutic applications. *Journal of controlled release: official journal of the Controlled Release Society, 266*, 17–26. https://doi.org/DOI: 10.1016/j.jconrel.2017.09.012

Liu, Q., Chen, B., Wang, Q., Shi, X., Xiao, Z., Lin, J., & Fang, X. (2009). Carbon nanotubes as molecular transporters for walled plant cells. *Nano Letters*, *9*(3), 1007–1010. DOI: 10.1021/nl803083u PMID: 19191500

Liu, W., Rudis, M. R., Cheplick, M. H., Millwood, R. J., Yang, J. P., Ondzighi-Assoume, C. A., Montgomery, G. A., Burris, K. P., Mazarei, M., Chesnut, J. D., & Stewart, C. N.Jr. (2020). Lipofection-mediated genome editing using DNA-free delivery of the Cas9/gRNA ribonucleoprotein into plant cells. *Plant Cell Reports*, *39*(2), 245–257. DOI: 10.1007/s00299-019-02488-w PMID: 31728703

Lowe, K., Wu, E., Wang, N., Hoerster, G., Hastings, C., Cho, M. J., Scelonge, C., Lenderts, B., Chamberlin, M., Cushatt, J., Wang, L., Ryan, L., Khan, T., Chow-Yiu, J., Hua, W., Yu, M., Banh, J., Bao, Z., Brink, K., & Gordon-Kamm, W. (2016). Morphogenic Regulators *Baby boom* and *Wuschel* Improve Monocot Transformation. *The Plant Cell*, *28*(9), 1998–2015. DOI: 10.1105/tpc.16.00124 PMID: 27600536

Ma, X., Geisler-Lee, J., Deng, Y., & Kolmakov, A. (2010). Interactions between engineered nanoparticles (ENPs) and plants: Phytotoxicity, uptake and accumulation. *The Science of the Total Environment*, *408*(16), 3053–3061. DOI: 10.1016/j.scitotenv.2010.03.031 PMID: 20435342

Ma, X., Zhang, X., Liu, H., & Li, Z. (2020). Highly efficient DNA-free plant genome editing using virally delivered CRISPR-Cas9. *Nature Plants*, *6*(7), 773–779. DOI: 10.1038/s41477-020-0704-5 PMID: 32601419

Maher, M. F., Nasti, R. A., Vollbrecht, M., Starker, C. G., Clark, M. D., & Voytas, D. F. (2020). Plant gene editing through de novo induction of meristems. *Nature Biotechnology*, *38*(1), 84–89. DOI: 10.1038/s41587-019-0337-2 PMID: 31844292

Mahmoud, L. M., Kaur, P., Stanton, D., Grosser, J. W., & Dutt, M. (2022). A cationic lipid mediated CRISPR/Cas9 technique for the production of stable genome edited citrus plants. *Plant Methods*, *18*(1), 33. DOI: 10.1186/s13007-022-00870-6 PMID: 35303912

Michielse, C. B., Hooykaas, P. J., van den Hondel, C. A., & Ram, A. F. (2005). Agrobacterium-mediated transformation as a tool for functional genomics in fungi. *Current Genetics*, *48*(1), 1–17. DOI: 10.1007/s00294-005-0578-0 PMID: 15889258

Mitter, N., Worrall, E. A., Robinson, K. E., Li, P., Jain, R. G., Taochy, C., Fletcher, S. J., Carroll, B. J., Lu, G. Q., & Xu, Z. P. (2017). Clay nanosheets for topical delivery of RNAi for sustained protection against plant viruses. *Nature Plants*, *3*(3), 16207. DOI: 10.1038/nplants.2016.207 PMID: 28067898

Mout, R., Ray, M., Yesilbag Tonga, G., Lee, Y. W., Tay, T., Sasaki, K., & Rotello, V. M. (2017). Direct Cytosolic Delivery of CRISPR/Cas9-Ribonucleoprotein for Efficient Gene Editing. *ACS Nano*, *11*(3), 2452–2458. DOI: 10.1021/acsnano.6b07600 PMID: 28129503

Nair, R., Tomar, M., & Thevuthasan, S. (2021). Nanotechnology in plant biotechnology and agriculture: Prospects and challenges. *Nanotechnology Reviews*, *10*(1), 259–274. DOI: 10.1515/ntrev-2020-0105

Naqvi, S., Maitra, A. N., Abdin, M. Z., Akmal, M., Arora, I., & Samim, M. (2012). Calcium phosphate nanoparticle mediated genetic transformation in plants. *Journal of Materials Chemistry*, *22*(8), 3500. DOI: 10.1039/c2jm11739h

Nonaka, S., & Ezura, H. (2014). Plant-Agrobacterium interaction mediated by ethylene and super-Agrobacterium conferring efficient gene transfer. *Frontiers in Plant Science*, *5*, 681. DOI: 10.3389/fpls.2014.00681 PMID: 25520733

Nonaka, S., Someya, T., Zhou, S., Takayama, M., Nakamura, K., & Ezura, H. (2017). An Agrobacterium tumefaciens Strain with Gamma Aminobutyric Acid Transaminase Activity Shows an Enhanced Genetic Transformation Ability in Plants. *Scientific Reports*, *7*(1), 42649. DOI: 10.1038/srep42649 PMID: 28220841

Ogidi, O. I. (2024). Recent advances in anticancer activity and Bioinformatics Approach from Potential Plants. In: Computational Approaches in Biomaterials, Bioinformatics and Biomedical Engineering Applications. *CRC Press, Taylor and Francis Group, Florida, USA*. DOI: 10.1201/9781003354437-3

Ogidi, O. I., & Akpan, U. M. (2023). Impacts of Chemical Use in Agricultural Practices: Perspectives of Soil Microorganism and Vegetation. In *One Health implications of Agrochemicals and their sustainable Alternative*. Springer Nature Singapore Pte Ltd. DOI: 10.1007/978-981-99-3439-3_27

Ogidi, O. I., & Emaikwu, N. G. (2024). Plant Phenolic compound isolation and its Bioinformatics Approaches of Molecular Mechanisms in Antimicrobial activities and Resistance. In: Computational Approaches in Biomaterials, Bioinformatics and Biomedical Engineering Applications. *CRC Press, Taylor and Francis Group, Florida, USA*. DOI: 10.1201/9781003354437-8

Ogidi, O. I., Izah, S. C., & Akpan, U. M. (2023). Food Quality and Agrochemical Use: Integrated Monitoring, Assessment, and Management Policies. In: One Health implications of Agrochemicals and their sustainable Alternative. *Springer Nature Singapore Pte Ltd.*

Ogoun, T. R., & Ogidi, O. I. (2022). Evaluation of natural and Artificial ripened Pineapple (*Ananascosmosus* L. Merr.) fruits on Biochemical parameters of wistar rats. *World Journal of Pharmaceutical and Medical Research,*8(8):41-46. www.wjpmr.com

Oh, N., & Park, J. H. (2014). Surface Chemistry of Gold Nanoparticles Mediates Their Exocytosis in Macrophages. *ACS Nano*, *8*(6), 6232–6241. DOI: 10.1021/nn501668a PMID: 24836308

Ozyigit, I. I. (2020). Gene transfer to plants by electroporation: Methods and applications. *Molecular Biology Reports*, *47*(4), 3195–3210. DOI: 10.1007/s11033-020-05343-4 PMID: 32242300

Park, J., Choi, S., Park, S., Yoon, J., Park, A. Y., & Choe, S. (2019). DNA-Free Genome Editing via Ribonucleoprotein (RNP) Delivery of CRISPR/Cas in Lettuce. *Methods in Molecular Biology (Clifton, N.J.)*, *1917*, 337–354. DOI: 10.1007/978-1-4939-8991-1_25 PMID: 30610648

Pasupathy, K., Lin, S., Hu, Q., Luo, H., & Ke, P. C. (2008). Direct plant gene delivery with a poly(amidoamine) dendrimer. *Biotechnology Journal*, *3*(8), 1078–1082. DOI: 10.1002/biot.200800021 PMID: 18543240

Qiu, F., Xing, S., Xue, C., Liu, J., Chen, K., Chai, T., & Gao, C. (2022). Transient expression of a TaGRF4-TaGIF1 complex stimulates wheat regeneration and improves genome editing. *Science China. Life Sciences*, *65*(4), 731–738. DOI: 10.1007/s11427-021-1949-9 PMID: 34406572

Radu, D. R., Lai, C. Y., Jeftinija, K., Rowe, E. W., Jeftinija, S., & Lin, V. S. (2004). A polyamidoamine dendrimer-capped mesoporous silica nanosphere-based gene transfection reagent. *Journal of the American Chemical Society*, *126*(41), 13216–13217. DOI: 10.1021/ja046275m PMID: 15479063

Rajput, V. D., Minkina, T., Sushkova, S., & Mandzhieva, S. (2022). Nanotechnology in agriculture: Opportunities, toxicological risks, and ethical concerns. *Environmental Geochemistry and Health*, *44*, 247–262. DOI: 10.1007/s10653-021-01007-5

Ratcliff, F., Martin-Hernandez, A. M., & Baulcombe, D. C. (2001). Technical Advance. Tobacco rattle virus as a vector for analysis of gene function by silencing. *The Plant journal: for cell and molecular biology, 25*(2), 237–245. https://doi.org/DOI: 10.1046/j.0960-7412.2000.00942.x

Schwab, F., Zhai, G., Kern, M., Turner, A., Schnoor, J. L., & Wiesner, M. R. (2016). Barriers, pathways and processes for uptake, translocation and accumulation of nanomaterials in plants—Critical review. *Nanotoxicology*, *10*(3), 257–278. DOI: 10.3109/17435390.2015.1048326 PMID: 26067571

Schwartz, S. H., Hendrix, B., Hofer, P., Sanders, R. A., & Zheng, W. (2020). Carbon dots for eficient small interfering RNA delivery and gene silencing in plants. *Plant Physiology*, *184*(2), 647–657. DOI: 10.1104/pp.20.00733 PMID: 32764133

Selby, L. I., Cortez-Jugo, C. M., Such, G. K., & Johnston, A. P. R. (2017). Nanoescapology: progress toward understanding the endosomal escape of polymeric nanoparticles. *Wiley interdisciplinary reviews. Nanomedicine and nanobiotechnology, 9*(5), . https://doi.org/DOI: 10.1002/wnan.1452

Serag, M. F., Kaji, N., Gaillard, C., Okamoto, Y., Terasaka, K., Jabasini, M., Tokeshi, M., Mizukami, H., Bianco, A., & Baba, Y. (2011). Trafficking and subcellular localization of multiwalled carbon nanotubes in plant cells. *ACS Nano*, *5*(1), 493–499. DOI: 10.1021/nn102344t PMID: 21141871

Serag, M. F., Noritada, K., Satoshi, H., Alberto, B., & Yoshinobu, B. (2013). Nanobiotechnology meets plant cell biology: Carbon nanotubes as organelle targeting nanocarriers. *RSC Advances*, *3*(15), 4856. DOI: 10.1039/c2ra22766e

Shew, A. M., Nalley, L. L., Snell, H. A., Nayga, R. M.Jr, & Dixon, B. L. (2018). CRISPR versus GMOs: Public acceptance and valuation. *Global Food Security*, *19*, 71–80. DOI: 10.1016/j.gfs.2018.10.005

Shou, H., Frame, B. R., Whitham, S. A., & Wang, K. (2004). Assessment of transgenic maize events produced by particle bombardment or Agrobacterium-mediated transformation. *Molecular Breeding*, *13*(2), 201–208. DOI: 10.1023/B:MOLB.0000018767.64586.53

Silva, A. T., Nguyen, A., Ye, C., Verchot, J., & Moon, J. H. (2010). Conjugated polymer nanoparticles for effective siRNA delivery to tobacco BY-2 protoplasts. *BMC Plant Biology*, *10*(1), 291. DOI: 10.1186/1471-2229-10-291 PMID: 21192827

Slowing, I. I., Vivero-Escoto, J. L., Wu, C. W., & Lin, V. S. (2008). Mesoporous silica nanoparticles as controlled release drug delivery and gene transfection carriers. *Advanced Drug Delivery Reviews*, *60*(11), 1278–1288. DOI: 10.1016/j.addr.2008.03.012 PMID: 18514969

Steinwand, M. A., & Ronald, P. C. (2020). Crop biotechnology and the future of food. *Nature Food*, *1*(5), 273–283. DOI: 10.1038/s43016-020-0072-3

Torney, F., Trewyn, B. G., Lin, V. S., & Wang, K. (2007). Mesoporous silica nanoparticles deliver DNA and chemicals into plants. *Nature Nanotechnology*, *2*(5), 295–300. DOI: 10.1038/nnano.2007.108 PMID: 18654287

Torti, S., Schlesier, R., Thümmler, A., Bartels, D., Römer, P., Koch, B., Werner, S., Panwar, V., Kanyuka, K., Wirén, N. V., Jones, J. D. G., Hause, G., Giritch, A., & Gleba, Y. (2021). Transient reprogramming of crop plants for agronomic performance. *Nature Plants*, *7*(2), 159–171. DOI: 10.1038/s41477-021-00851-y PMID: 33594264

Tripathi, D. K., Shweta, Singh, S., Singh, S., Pandey, R., Singh, V. P., Sharma, N. C., Prasad, S. M., Dubey, N. K., & Chauhan, D. K. (2017). An overview on manufactured nanoparticles in plants: Uptake, translocation, accumulation and phytotoxicity. *Plant physiology and biochemistry: PPB, 110*, 2–12. https://doi.org/DOI: 10.1016/j.plaphy.2016.07.030

Vijayakumar, P. S., Abhilash, O. U., Khan, B. M., & Prasad, B. L. V. (2010). Nanogold-loaded sharp edged carbon bullets as plant-gene carriers. *Advanced Functional Materials*, *20*(15), 2416–2423. DOI: 10.1002/adfm.200901883

Vijayaraghavan, K., & Ashokkumar, T. (2017). Plant-mediated biosynthesis of metallic nanoparticles: A review of literature, factors affecting synthesis, characterization techniques and applications. *Journal of Environmental Chemical Engineering*, *5*(5), 4866–4883. DOI: 10.1016/j.jece.2017.09.026

Waltz, E. (2018). With a free pass, CRISPR-edited plants reach market in record time. *Nature Biotechnology*, *36*(1), 6–7. DOI: 10.1038/nbt0118-6b PMID: 29319694

Wang, J. W., Grandio, E. G., Newkirk, G. M., Demirer, G. S., Butrus, S., Giraldo, J. P., & Landry, M. P. (2019). Nanoparticle-mediated genetic engineering of plants. *Molecular Plant*, *12*(8), 1037–1040. DOI: 10.1016/j.molp.2019.06.010 PMID: 31283997

Wang, P., Lombi, E., Zhao, F. J., & Kopittke, P. M. (2016). Nanotechnology: A New Opportunity in Plant Sciences. *Trends in Plant Science*, *21*(8), 699–712. DOI: 10.1016/j.tplants.2016.04.005 PMID: 27130471

Wang, P., Lombi, E., Zhao, F. J., & Kopittke, P. M. (2021). Nanotechnology for plant nutrition: A review. *Nature Nanotechnology*, *16*(3), 247–260. DOI: 10.1038/s41565-021-00893-3

Wang, P., Lombi, E., Zhao, F. J., & Kopittke, P. M. (2022). Nanotechnology: A new opportunity in plant sciences. *Trends in Plant Science*, *27*(1), 33–43. DOI: 10.1016/j.tplants.2021.08.008 PMID: 27130471

Wang, P., Zhang, L., Zheng, W., & Guo, Z. (2020). Lipid nanoparticles for nucleic acid delivery in plants. *Journal of Controlled Release*, *323*, 628–649. DOI: 10.1016/j.jconrel.2020.05.007

Wong, M. H., Misra, R. P., Giraldo, J. P., Kwak, S. Y., Son, Y., Landry, M. P., Swan, J. W., Blankschtein, D., & Strano, M. S. (2016). Lipid Exchange Envelope Penetration (LEEP) of Nanoparticles for Plant Engineering: A Universal Localization Mechanism. *Nano Letters*, *16*(2), 1161–1172. DOI: 10.1021/acs.nanolett.5b04467 PMID: 26760228

Woo, J. W., Kim, J., Kwon, S. I., Corvalán, C., Cho, S. W., Kim, H., Kim, S. G., Kim, S. T., Choe, S., & Kim, J. S. (2015). DNA-free genome editing in plants with preassembled CRISPR-Cas9 ribonucleoproteins. *Nature Biotechnology*, *33*(11), 1162–1164. DOI: 10.1038/nbt.3389 PMID: 26479191

Wu, C. F., Smith, D. A., Lai, E. M., & Chang, J. H. (2018). The Agrobacterium Type VI Secretion System: A Contractile Nanomachine for Interbacterial Competition. *Current Topics in Microbiology and Immunology*, *418*, 215–231. DOI: 10.1007/82_2018_99 PMID: 29992360

Wu, S., Zhu, H., Liu, J., Yang, Q., Shao, X., Bi, F., Hu, C., Huo, H., Chen, K., & Yi, G. (2020). Establishment of a PEG-mediated protoplast transformation system based on DNA and CRISPR/Cas9 ribonucleoprotein complexes for banana. *BMC Plant Biology*, *20*(1), 425. DOI: 10.1186/s12870-020-02609-8 PMID: 32933485

Yan, Y., Zhu, X., Yu, Y., Li, C., Zhang, Z., & Wang, F. (2022). Nanotechnology Strategies for Plant Genetic Engineering. *Advanced Materials*, *34*(7), e2106945. DOI: 10.1002/adma.202106945 PMID: 34699644

Yang, T., & Duncan, T. V. (2021). Challenges and Potential Solutions for Nanosensors Intended for Use with Foods. *Nature Nanotechnology*, *163*(16), 251–265. DOI: 10.1038/s41565-021-00867-7 PMID: 33712739

Yoo, S. D., Cho, Y. H., & Sheen, J. (2007). Arabidopsis mesophyll protoplasts: A versatile cell system for transient gene expression analysis. *Nature Protocols*, *2*(7), 1565–1572. DOI: 10.1038/nprot.2007.199 PMID: 17585298

Zaidi, S. S., Vanderschuren, H., Qaim, M., Mahfouz, M. M., Kohli, A., Mansoor, S., & Tester, M. (2019). New plant breeding technologies for food security. *Science*, *363*(6434), 1390–1391. DOI: 10.1126/science.aav6316 PMID: 30923209

Zhang, C., & Hu, H. (2014). High-efficiency nuclear transformation of the diatom Phaeodactylum tricornutum by electroporation. *Marine Genomics*, *16*, 63–66. DOI: 10.1016/j.margen.2013.10.003 PMID: 24269346

Zhang, D., Yao, X., Wang, J., & Wu, Y. (2021). Advances in nanotechnology for plant genetic engineering. *Biotechnology Advances*, *48*, 107712. DOI: 10.1016/j.biotechadv.2021.107712

Zhang, H., Cao, Y., Xu, D., Goh, N. S., Demirer, G. S., Cestellos-Blanco, S., Chen, Y., Landry, M. P., & Yang, P. (2021a). Gold-Nanocluster-Mediated Delivery of siRNA to Intact Plant Cells for Efficient Gene Knockdown. *Nano Letters*, *21*(13), 5859–5866. DOI: 10.1021/acs.nanolett.1c01792 PMID: 34152779

Zhang, H., Demirer, G. S., Zhang, D., Ye, T., Goh, N. S., & Landry, M. P. (2019). DNA nanostructures coordinate gene silencing. *Proceedings of the National Academy of Sciences of the United States of America*, *116*(15), 7543–7548. DOI: 10.1073/pnas.1818290116 PMID: 30910954

Zhang, H., Demirer, G. S., Zhang, D., Ye, T., Goh, N. S., & Landry, M. P. (2021c). Engineering DNA nanostructures for delivery of genetic material into plants. *Nature Protocols*, *16*(3), 1704–1722. DOI: 10.1038/s41596-020-00506-1

Zhang, H., Demirer, G. S., Zhang, D., Ye, T., & Landry, M. P. (2021b). Gold nanoparticle applications in plant biology. *Nature Protocols*, *16*(3), 1704–1722. DOI: 10.1038/s41596-020-00506-1

Zhang, H., Demirer, G. S., Zhang, H., Ye, T., Goh, N. S., Aditham, A. J., Cunningham, F. J., Fan, C., & Landry, M. P. (2020a). DNA nanostructures coordinate gene silencing in mature plants. *Proceedings of the National Academy of Sciences of the United States of America*, *116*(15), 7543–7548. DOI: 10.1073/pnas.1818290116 PMID: 30910954

Zhang, H., Goh, N. S., Wang, J. W., Pinals, R. L., González-Grandío, E., Demirer, G. S., Butrus, S., Fakra, S. C., Del Rio Flores, A., Zhai, R., Zhao, B., Park, S. J., & Landry, M. P. (2022a). Nanoparticle cellular internalization is not required for RNA delivery to mature plant leaves. *Nature Nanotechnology*, *17*(2), 197–205. DOI: 10.1038/s41565-021-01018-8 PMID: 34811553

Zhang, H., Zhang, H., Demirer, G. S., González-Grandío, E., Fan, C., & Landry, M. P. (2020b). Engineering DNA nanostructures for siRNA delivery in plants. *Nature Protocols*, *15*(9), 3064–3087. DOI: 10.1038/s41596-020-0370-0 PMID: 32807907

Zhang, Q., Zhao, L., Shen, M., Liu, J., Li, Y., Xu, S., Chen, L., Shi, G., & Ding, Z. (2022b). Establishment of an Efficient Polyethylene Glycol (PEG)-Mediated Transformation System in *Pleurotus eryngii* var. *ferulae* Using Comprehensive Optimization and Multiple Endogenous Promoters. *Journal of Fungi (Basel, Switzerland)*, *8*(2), 186. DOI: 10.3390/jof8020186 PMID: 35205941

Zhang, Y., Liang, Z., Zong, Y., Wang, Y., Liu, J., Chen, K., Qiu, J. L., & Gao, C. (2016). Efficient and transgene-free genome editing in wheat through transient expression of CRISPR/Cas9 DNA or RNA. *Nature Communications*, *7*(1), 12617. DOI: 10.1038/ncomms12617 PMID: 27558837

Zhang, Y., Zhang, Y., & Jin, S. (2020). Public acceptance of genetically modified food: Socio-demographic variables, general attitudes, and information framing. *Science Communication*, *42*(2), 248–273. DOI: 10.1177/1075547019898143

Zhao, H., Zhi, Y. L., Lara, Y., Arvind, D., Xin, Z., & Jun, W. (2016). Polymer-based nanoparticles for protein delivery: Design, strategies and applications. *Journal of Materials Chemistry. B*, *4060*(23), 4060–4071. DOI: 10.1039/C6TB00308G PMID: 32264607

Zhao, X., Meng, Z., Wang, Y., Chen, W., Sun, C., Cui, B., Cui, J., Yu, M., Zeng, Z., Guo, S., Luo, D., Cheng, J. Q., Zhang, R., & Cui, H. (2017). Pollen magnetofection for genetic modification with magnetic nanoparticles as gene carriers. *Nature Plants*, *3*(12), 956–964. DOI: 10.1038/s41477-017-0063-z PMID: 29180813

Zhu, Z. J., Wang, H., Yan, B., Zheng, H., Jiang, Y., Miranda, O. R., Rotello, V. M., Xing, B., & Vachet, R. W. (2012). Effect of surface charge on the uptake and distribution of gold nanoparticles in four plant species. *Environmental Science & Technology*, *46*(22), 12391–12398. DOI: 10.1021/es301977w PMID: 23102049

Zuverza-Mena, N., Martínez-Fernández, D., Du, W., Hernandez-Viezcas, J. A., Bonilla-Bird, N., López-Moreno, M. L., Komárek, M., Peralta-Videa, J. R., & Gardea-Torresdey, J. L. (2017). Exposure of engineered nanomaterials to plants: Insights into the physiological and biochemical responses-A review. *Plant physiology and biochemistry. PPB*, *110*, 236–264. PMID: 27289187

APPENDIX

Abbreviations

CNTs carbon nanotubes
GO graphene oxide
MOFs metal-organic frameworks
RNP ribonucleoprotein
siRNA small interfering ribonucleic acid
DNA Dioxoribonucleic acid
CRISPR Clustered regularly interspaced short palindromic repearts
GMOs genetically modified organisms
T-DNA transfer deoxoribonucleic acid
USDA United State Department of Agriculture
EU European Union
Mg^{2+} Magnesium ion
Ca2+ Calcium ion
PEG polyethylene glycol
AS acetosyringone
BBM Baby Boom
GABA gamma-aminobutyric acid
sgRNAs single guide ribonucleic acids
TRV Tobacco rattle virus
NPs nanoparticles
dsRNA double stranded ribonucleic acid
PAMAM Polyamidoamine
CNTs Carbon nanotubes
PEI polyethylenimine
nm nanometer
MNP magnetic nanoparticle
GUS β-glucuronidase
GFP green fluorescent protein
MSNs mesoporous silica nanoparticles
SELs size exclusion limits

Chapter 5
Nanocarriers for Targeted Delivering and Translocation of Chemical Cargoes in Plant Cells

Preethi Rajesh
https://orcid.org/0009-0001-8077-2822
Garden City University, India

ABSTRACT

Nanocarriers offer a transformative approach to agrochemical delivery, revolutionizing modern agriculture. These nanoscale vehicles enhance bioavailability, control release, and improve target specificity of diverse cargoes, including pesticides, herbicides, nutrients, PGRs, genetic material, and biostimulants. Through passive (root absorption, stomatal penetration, cuticular diffusion) and active (endocytosis, plasmodesmata transport) uptake mechanisms, nanocarriers deliver cargo directly to plant tissues. Surface functionalization optimizes interaction with plant cells, enhancing adhesion, permeability, and controlled release. This technology improves nutrient uptake, pest control, and stress tolerance, while minimizing environmental impact. However, safety and environmental concerns necessitate rigorous research and regulatory frameworks. Future directions include bioengineered nanocarriers, AI integration, and scalable production for sustainable agriculture.

1.0 INTRODUCTION

1.1 Overview of Nanotechnology in Agriculture

Nanotechnology, the manipulation of materials at the nanoscale (1-100 nm), has gained increasing attention in agriculture due to its ability to enhance the efficiency of agrochemical delivery systems. By using nanocarriers, researchers and agricultural scientists aim to improve the delivery, uptake, and translocation of chemical cargoes such as pesticides, fertilizers, growth regulators, and genetic material in plant cells (Kah et al., 2019).

Traditional methods of agrochemical application, such as foliar spraying and soil incorporation, suffer from inefficiencies, including volatilization, photodegradation, leaching, and poor bioavailability (Jampilek et al., 2018). Nanocarriers offer a promising solution to these issues by enabling the controlled

DOI: 10.4018/979-8-3693-9869-2.ch005

and targeted release of active compounds, thereby reducing environmental impact and improving plant health. These nanoformulations help minimize chemical wastage, enhance plant absorption, and reduce toxicity risks to non-target organisms (Torres et al., 2020).

Nanocarriers, which include polymeric nanoparticles, liposomes, silica nanoparticles, metal and metal oxide nanoparticles, and carbon-based materials, provide a novel mechanism for agrochemical delivery. Their small size, large surface area, and surface functionalization capabilities allow them to penetrate plant barriers efficiently and translocate within tissues, reaching target sites with minimal losses (Wang et al., 2021). As the field of nanotechnology in agriculture advances, nanocarrier-based formulations are expected to play a critical role in the development of smart and sustainable agricultural practices.

1.2 Importance of Targeted Delivery in Plant Cells

Targeted delivery is essential for optimizing the efficacy of agrochemicals, reducing environmental contamination, and increasing agricultural productivity. Unlike conventional application methods that result in non-specific distribution, nanocarriers facilitate precise transport and controlled release of chemical cargoes to specific plant tissues, organelles, or cellular compartments (Ghormade et al., 2011).

Advantages of Nanocarrier-Based Targeted Delivery:

Improved Bioavailability: Nanocarriers enhance the solubility and stability of poorly water-soluble agrochemicals, ensuring better plant uptake and utilization (Liu et al., 2020). Many agrochemicals, like pesticides and fertilizers, aren't easily absorbed by plants. They might be poorly soluble in water, making it difficult for plants to take them up. Nanocarriers solve this problem by:

- **Encapsulation:** They surround the agrochemical, protecting it from degradation and improving its solubility. Think of it like wrapping the active ingredient in a protective, water-friendly package.
- **Enhanced Delivery:** This improved solubility and protection allows the nanocarrier to effectively deliver the agrochemical to the plant's surface or even inside the plant tissue, where it can be utilized. This leads to better absorption and, therefore, greater effectiveness of the agrochemical.

Reduced Environmental Contamination: Targeted delivery prevents excessive leaching and runoff, minimizing the adverse effects of agrochemicals on soil and water ecosystems (De Oliveira et al., 2019). Traditional spraying methods often lead to significant waste. Agrochemicals can drift away, run off into waterways, or leach into the soil, harming non-target organisms and polluting the environment. Targeted delivery with nanocarriers minimizes these risks by:

- **Precise Placement:** Nanocarriers can be designed to deliver the agrochemical directly to the target area (e.g., the leaf surface where a pest is present, or the root zone). This reduces the amount of chemical that comes into contact with the surrounding environment.
- **Reduced Runoff:** By improving the adherence of the agrochemical to the plant and preventing leaching, nanocarriers decrease the amount of chemical that can be washed away by rain or irrigation.

Controlled and Sustained Release: Nanocarriers enable time-dependent and stimuli-responsive release of active compounds, reducing the need for frequent applications (Franci et al., 2015). Imagine a fertilizer that releases all its nutrients at once. The plant might not be able to use them all immediately, and some could be lost. Nanocarriers offer a better approach:

- **Timed Release:** They can be designed to release the agrochemical slowly and steadily over time, matching the plant's needs and reducing the risk of nutrient overload or toxicity.
- **Stimuli-Responsive Release:** Some nanocarriers can be triggered to release their cargo by specific environmental cues, such as changes in pH, temperature, or the presence of certain enzymes. This allows for even more precise control over the delivery of the agrochemical. For example, a nanocarrier could be designed to release a pesticide only when a specific pest is detected.

Enhanced Absorption and Uptake: Nanoparticles can penetrate plant cell walls, bypass physiological barriers, and facilitate intracellular transport, improving the efficiency of agrochemical delivery (Ederli et al., 2021). Plant cell walls are a barrier to the uptake of many substances. Nanocarriers can help overcome this by:

- **Penetration:** Some nanoparticles are small enough to penetrate plant cell walls, allowing for direct delivery of the agrochemical into the plant cells.
- **Improved Transport:** Nanocarriers can facilitate the transport of the agrochemical within the plant, ensuring that it reaches the target site of action.

Precision Agriculture Applications: Smart nanocarrier systems can respond to specific environmental stimuli such as pH, temperature, and enzymatic activity to release cargo at optimal conditions (Sujatha et al., 2022). Precision agriculture aims to optimize resource use and minimize environmental impact. Smart nanocarrier systems are a perfect fit for this approach:

- **Site-Specific Delivery:** Nanocarriers can be tailored to deliver different agrochemicals to different parts of the field, based on the specific needs of the plants in each area.
- **Real-Time Response:** Sensors can be integrated with nanocarrier systems to monitor environmental conditions and trigger the release of agrochemicals only when needed. This reduces the overall amount of chemicals used and minimizes environmental impact. For example, a sensor could detect a nutrient deficiency and trigger the release of fertilizer from the nanocarrier.

Plant cells are protected by a rigid cell wall composed of cellulose, hemicellulose, and pectin, which makes the direct uptake of agrochemicals challenging (Nair et al., 2010). Nanocarriers address this challenge by utilizing various transport mechanisms, including passive diffusion, endocytosis, and plasmodesmata-mediated movement (Ke et al., 2018). Functionalization with targeting ligands, such as peptides, proteins, or aptamers, can further improve specificity and efficiency in plant systems (Shang et al., 2022).

1.3 Challenges in Conventional Chemical Delivery Methods

Traditional agrochemical delivery methods have several limitations that hinder their effectiveness. These include:

1.3.1 Low Efficiency and Chemical Losses

Environmental Degradation

Agrochemicals, whether sprayed on leaves or applied to the soil, are vulnerable to breakdown by environmental factors. Sunlight (UV radiation), naturally occurring microbes in the soil, leaching and even rainfall can degrade these chemicals, reducing their effectiveness before they even reach the target. (Kookana et al., 2014).

- **Photodegradation (UV Radiation):** Sunlight, particularly the ultraviolet (UV) component, can break down chemical bonds in agrochemicals. This process, called photodegradation, can alter the chemical structure of the active ingredient, rendering it ineffective or even creating harmful byproducts. The rate of photodegradation depends on factors like the chemical's sensitivity to UV light, the intensity of sunlight, and the duration of exposure.
- **Microbial Degradation:** Soil is teeming with microorganisms (bacteria, fungi, etc.) that can break down organic compounds, including agrochemicals. These microbes use the chemicals as a food source, effectively decomposing them. The rate of microbial degradation depends on the type of chemical, the types of microbes present in the soil, soil temperature, moisture, and other environmental conditions.
- **Chemical Degradation (Hydrolysis, Oxidation):** Agrochemicals can also undergo chemical reactions in the environment, such as hydrolysis (reaction with water) or oxidation (reaction with oxygen). These reactions can alter the chemical structure and reduce effectiveness.
- **Leaching:** Rainfall or irrigation can cause agrochemicals to leach (move) through the soil profile. This can carry them away from the target area (e.g., the root zone) and potentially into groundwater or surface water. Leaching is influenced by soil type, rainfall intensity, and the chemical's solubility.
- **Volatilization:** Some agrochemicals can evaporate (volatilize) from the soil or plant surface into the atmosphere. This not only reduces their effectiveness but can also contribute to air pollution.

Off-Target Application

A significant portion of applied chemicals often misses the intended target. Spray drift, runoff, and uneven distribution contribute to this loss. This not only wastes valuable resources but also increases the risk of environmental contamination.

The problem of Off-Target Application is as follows:

- **Wasted Resources:** When agrochemicals miss their intended target (the pests, weeds, or specific plant parts), it's a direct waste of resources. Farmers have to use more chemicals than necessary, increasing costs and potentially reducing profitability.
- **Environmental Contamination:** Off-target chemicals can end up in the soil, water, and air, leading to a range of environmental problems:
- **Water Pollution:** Runoff can carry chemicals into rivers, lakes, and groundwater, harming aquatic life and potentially contaminating drinking water sources.
- **Soil Contamination:** Chemicals can accumulate in the soil, affecting soil health and potentially harming beneficial organisms.
- **Air Pollution:** Spray drift can carry chemicals long distances, affecting air quality and potentially exposing people and wildlife to harmful substances.

Factors Contributing to Off-Target Application

- **Spray Drift:** Wind can carry tiny droplets of pesticides or other agrochemicals far from the intended target. This is especially problematic with aerial spraying or when using fine sprays.
- **Runoff:** Rain or irrigation can wash away chemicals from the soil surface, carrying them into waterways. This is more likely to occur on sloped land or in areas with heavy rainfall.
- **Uneven Distribution:** Traditional application methods, like spraying or broadcasting granules, can result in uneven distribution of chemicals. Some areas may receive too much, while others receive too little.

Poor Uptake

Even when the agrochemical reaches the plant, it may not be readily absorbed. Plants have natural barriers, like the waxy cuticle on leaves, that can prevent or limit the uptake of these chemicals. The majority of applied chemicals do not reach their intended target due to poor uptake efficiency, necessitating repeated applications (Chen & Yada, 2011).

Factors contributing to poor uptake

- **The Cuticle:** Plant leaves are covered with a waxy layer called the cuticle. This layer is designed to protect the plant from water loss and other environmental stressors. However, it also acts as a barrier to many agrochemicals, especially those that are not lipophilic (fat-soluble). The cuticle is particularly effective at preventing the uptake of water-soluble or large molecules.
- **Cell Walls:** Plant cells are surrounded by rigid cell walls made primarily of cellulose. While porous, cell walls can still limit the passage of some agrochemicals, especially larger molecules or those with specific chemical properties that don't interact favorably with the cell wall components.
- **Cell Membranes:** The plasma membrane surrounding the cell's interior is another selective barrier. It controls what enters and exits the cell. Getting an agrochemical across the cell membrane often requires specific transport mechanisms, which may not exist for all chemicals.

Consequences of Poor Uptake:

- **Reduced Effectiveness:** If the agrochemical can't get inside the plant cells or tissues where it needs to act (e.g., inside a pest's body or within the plant's vascular system), it won't be effective.
- **Wasted Chemicals:** Poor uptake means that a large proportion of the applied agrochemical is wasted. This is not only costly for the farmer but also increases the risk of environmental contamination.
- **Repeated Applications:** To compensate for poor uptake, farmers often have to apply agrochemicals more frequently. This increases costs, labor, and the potential for environmental and health problems.
- **Development of Resistance:** Repeated applications can also contribute to the development of resistance in pests and pathogens. The more a pest is exposed to a chemical, the greater the chance that it will evolve resistance mechanisms.

1.3.2 Environmental and Health Concerns

Soil and Water Contamination Leading to Ecological Imbalances

Overuse of chemical fertilizers and pesticides contributes to soil and water contamination, leading to ecological imbalances (Khan et al., 2019).

Soil Contamination:

- **Chemical Residues:** Pesticides and fertilizers can leave behind residues in the soil. These residues can persist for long periods and may be toxic to beneficial soil organisms (like earthworms and microbes) that are essential for soil health.
- **Nutrient Imbalance:** Overuse of fertilizers can disrupt the natural nutrient balance in the soil. Excess nitrogen, for example, can lead to soil acidification and can affect the availability of other essential nutrients for plants.
- **Heavy Metal Accumulation:** Some fertilizers may contain heavy metals (like cadmium or arsenic) as impurities. These metals can accumulate in the soil over time, posing risks to plant health and potentially entering the food chain.

Water Contamination:

- **Runoff:** Excess fertilizers and pesticides can be carried away by runoff into rivers, lakes, and streams. This can contaminate water sources and harm aquatic life.
- **Leaching:** Chemicals can also leach through the soil and contaminate groundwater, which is a major source of drinking water for many communities.

Ecological Imbalances:

- **Loss of Biodiversity:** Contamination from fertilizers and pesticides can harm or kill non-target organisms, reducing biodiversity in both terrestrial and aquatic ecosystems.
- **Disruption of Food Webs:** The loss of key species can disrupt food webs and have cascading effects throughout the ecosystem.
- **Pest Resurgence:** The overuse of pesticides can lead to the development of resistance in pests, making them harder to control and potentially leading to pest resurgences.

Runoff Leading to Eutrophication and Groundwater Pollution

Runoff from agricultural fields can result in eutrophication and groundwater pollution, posing serious risks to aquatic ecosystems and human health (Fernandez-Cornejo et al., 2014).

Eutrophication:

- **Nutrient Enrichment:** Runoff from agricultural fields often contains high levels of nitrogen and phosphorus from fertilizers. When this runoff enters water bodies, it can cause an overgrowth of algae, known as an algal bloom.
- **Oxygen Depletion:** As the algae die and decompose, they consume large amounts of oxygen in the water. This oxygen depletion (hypoxia) can kill fish and other aquatic animals.
- **Dead Zones:** In severe cases, eutrophication can create "dead zones" in water bodies where oxygen levels are so low that most aquatic life cannot survive.

Groundwater Pollution:

- **Nitrate Contamination:** Nitrate from fertilizers is a common groundwater contaminant. High levels of nitrate in drinking water can pose health risks, particularly to infants.
- **Pesticide Contamination:** Pesticides can also leach into groundwater, contaminating drinking water sources and posing health risks.

Risks to Aquatic Ecosystems:

- **Loss of Biodiversity:** Eutrophication and pollution can lead to the loss of fish, invertebrates, and other aquatic species.

- **Habitat Degradation:** Algal blooms can reduce light penetration in water bodies, harming submerged plants and affecting aquatic habitats.

Risks to Human Health:

- **Contaminated Drinking Water:** Nitrate and pesticide contamination of groundwater can pose health risks to people who rely on groundwater for drinking.
- **Harmful Algal Blooms:** Some algal blooms can produce toxins that are harmful to humans and animals. Contact with these toxins can cause skin irritation, respiratory problems, and even more serious health issues.

1.3.3 Limited Uptake and Translocation

Limited Absorption and Inefficient Translocation (Nair et al., 2010):

Many agrochemicals are unable to cross the cuticle and cell walls effectively, leading to limited absorption and inefficient translocation within plant tissues (Nair et al., 2010).

- **The Cuticle Barrier:** Plants are covered with a waxy, protective layer called the cuticle, primarily composed of cutin. This layer's main function is to prevent water loss, but it also acts as a significant barrier to the penetration of many agrochemicals. The cuticle is particularly effective at blocking water-soluble (hydrophilic) substances. Agrochemicals must be able to penetrate this waxy layer to reach the plant's interior.
- **Cell Wall Challenges:** Once past the cuticle, agrochemicals encounter the plant cell wall, a rigid structure composed mainly of cellulose, hemicellulose, and pectin. While porous, the cell wall can still hinder the passage of larger molecules or those with specific chemical properties. It acts as a filter, limiting what can reach the cell membrane.
- **Translocation Issues:** Even if an agrochemical is absorbed into the plant, it needs to be transported (translocated) to its target site of action. This often involves movement through the plant's vascular system (xylem and phloem). Many agrochemicals struggle to move effectively through these tissues, especially if they are not readily water-soluble or if they are rapidly metabolized by the plant. Systemic movement is essential for many pesticides and fungicides that need to protect the entire plant.

Low Bioavailability of Large Molecular Weight Compounds and Hydrophobic Molecules

Large molecular weight compounds and hydrophobic molecules exhibit low bioavailability, further reducing their effectiveness (Schwab et al., 2015).

- **Large Molecular Weight:** Large molecules (like some proteins or polymers) often have difficulty crossing the cuticle and cell walls. The plant's transport systems are often designed to handle smaller molecules. Large molecules may simply be too big to pass through the pores in the cell wall or to be transported across cell membranes.
- **Hydrophobic Molecules:** Hydrophobic (water-repelling) molecules tend to have low bioavailability because they don't interact well with the aqueous environment within the plant. The cuticle, being waxy, is more permeable to hydrophobic substances, but these molecules may then struggle to move through the more water-based environment of the cell wall and within the plant's interior.

This makes it hard for them to be transported effectively. They may also have difficulty interacting with the enzymes and target sites within the plant, which are often in aqueous environments.

1.3.4 Lack of Controlled and Targeted Release

Non-Specific Distribution and Increased Wastage

Conventional methods rely on bulk application, resulting in non-specific distribution and increased wastage (Perez-de-Luque & Rubiales, 2009).

- **Bulk Application:** Traditional methods like spraying, broadcasting granules, or seed treatments often involve applying the agrochemical over a large area, regardless of the precise location of the pest, weed, or nutrient deficiency. This "blanket approach" results in non-specific distribution, meaning the chemical is applied where it's needed, *and* where it's not.
- **Wastage:** A significant portion of the applied chemical may miss the target altogether. It can drift away in the wind (spray drift), run off into waterways, leach into the soil, or simply degrade before it can be used by the plant. This off-target application represents a significant waste of resources and increases the risk of environmental contamination.
- **Uneven Coverage:** Bulk application can also lead to uneven coverage. Some areas may receive too much of the chemical, potentially harming plants or beneficial organisms, while other areas may receive too little, leaving them vulnerable to pests or nutrient deficiencies.

Frequent Reapplication and Increased Costs

The absence of controlled release mechanisms necessitates frequent reapplication, increasing labor costs and input expenses (Santos et al., 2020).

- **Lack of Controlled Release:** Conventional agrochemicals are typically released immediately upon application. This can lead to rapid degradation or loss of the chemical before the plant can fully utilize it. The lack of controlled release mechanisms means that the chemical is not available to the plant over an extended period.
- **Frequent Reapplication:** Because of rapid degradation or loss, and the lack of sustained availability, farmers often need to reapply agrochemicals frequently to maintain adequate levels of protection or nutrition. This increases labor costs, the amount of chemical used, and the overall input expenses for the farmer.
- **Environmental Impact:** Frequent applications also increase the environmental burden. More chemicals are released into the environment, increasing the risk of pollution and harming non-target organisms.
- **Resistance Development:** Repeated applications of pesticides can contribute to the development of resistance in pests and pathogens. The more frequently a pest is exposed to a pesticide, the greater the chance that it will evolve resistance mechanisms, making the pesticide less effective over time.

1.3.5 Development of Resistance in Pests and Pathogens

Development of Resistance in Pests and Pathogens

Repeated use of chemical pesticides contributes to the development of resistance in pests and pathogens, reducing the long-term efficacy of these compounds (Jayaraj et al., 2019).

- **Evolutionary Pressure:** When a pesticide is applied, it kills most of the susceptible pests or pathogens. However, some individuals may have slight genetic variations that make them less susceptible (more resistant). These resistant individuals survive and reproduce, passing on their resistance genes to their offspring.
- **Selection and Proliferation:** With repeated use of the same pesticide, the selection pressure favors the resistant individuals. Over time, the proportion of resistant individuals in the population increases, while the susceptible individuals become less common. This is how pesticide resistance develops.
- **Decreased Efficacy:** As resistance spreads, the pesticide becomes less and less effective. Higher doses or more frequent applications are needed to achieve the same level of pest control. Eventually, the pesticide may become virtually useless against the resistant population.
- **Cross-Resistance:** Pests can also develop cross-resistance, meaning resistance to multiple pesticides, even those with different modes of action. This makes it even more challenging to manage pests effectively.

Mitigating Resistance with Controlled-Release Nanoformulations

Controlled-release nanoformulations help mitigate resistance issues by ensuring prolonged activity at lower dosages (Oberdörster et al., 2005).

- **Sustained Exposure:** Controlled-release nanoformulations release the pesticide slowly and steadily over time. This provides sustained exposure to the pest or pathogen, rather than a large initial dose followed by a rapid decline in concentration. This sustained exposure can be more effective at killing the pest, even those that might have some initial level of tolerance.
- **Lower Dosages:** Because the pesticide is released in a controlled manner, lower overall dosages may be required to achieve effective pest control. This reduces the selection pressure for resistance development, as pests are exposed to lower concentrations of the pesticide for longer periods.
- **Reduced Frequency of Application:** With controlled release, fewer applications may be needed, further reducing the selection pressure and the risk of resistance development.
- **Targeted Delivery:** Nanoformulations can also improve the targeted delivery of pesticides, reducing the exposure of non-target organisms and minimizing environmental contamination. This can also help to preserve beneficial insects that may help to control pests naturally.

Some nanoformulations can also enhance the mechanism of action of the pesticide, making it more effective at killing the pest. For example, they may improve the penetration of the pesticide into the pest's body or increase its binding to its target site.

1.4 Nanocarriers as an Advanced Agrochemical Delivery System

Nanocarriers offer a transformative solution for overcoming the limitations of conventional agrochemical delivery systems. Their advanced physicochemical properties allow for enhanced transport, controlled release, and site-specific action within plant tissues (Raliya et al., 2016).

Key Features of Nano carriers:

1. Nano-sized Particles (Torney et al., 2007):
    - **Enhanced Interaction:** The incredibly small size of nanoparticles (typically measured in nanometers, billionths of a meter) allows for a much greater surface area to volume ratio compared to larger particles. This increased surface area means more contact points with plant surfaces (leaves, roots, etc.), leading to better adhesion and interaction.
    - **Intracellular Transport:** The tiny dimensions of nanoparticles enable them to penetrate plant cell walls and even cell membranes, facilitating the delivery of agrochemicals directly into plant cells. This is particularly important for substances that need to reach specific cellular targets to be effective. This intracellular delivery can bypass some of the plant's natural defense mechanisms and improve the efficacy of the delivered substance.
    - **Improved Distribution:** Nanoparticles can be more evenly distributed within the plant tissues compared to larger particles, ensuring that the agrochemical reaches all the necessary sites of action.
2. Surface Functionalization (Xing et al., 2018):
    - **Targeting:** The surface of nanocarriers can be chemically modified (functionalized) with specific molecules that can target particular plant tissues, cells, or even specific organelles within cells. This targeted delivery reduces off-target effects and minimizes the amount of chemical needed. Think of it like adding a "homing device" to the nanocarrier.
    - **Enhanced Bioavailability:** Surface modifications can also improve the bioavailability of agrochemicals. For example, adding hydrophilic (water-loving) groups to the surface of a nanocarrier can make it easier for the nanocarrier to interact with the aqueous environment within the plant, improving its uptake and transport.
    - **Improved Adhesion:** Functionalization can also enhance the adhesion of the nanocarrier to plant surfaces, preventing it from being washed away by rain or irrigation.
3. Controlled and Stimuli-Responsive Release (Ghormade et al., 2011):
    - **Controlled Release:** Nanocarriers can be designed to release their cargo (the agrochemical) in a controlled manner, either slowly over time or in response to specific triggers. This controlled release minimizes waste, reduces the risk of toxicity, and ensures that the agrochemical is available to the plant when and where it's needed.
    - **Stimuli-Responsive Release:** Some "smart" nanocarriers can be designed to release their cargo in response to specific environmental stimuli, such as changes in pH, temperature, light, or the presence of certain enzymes. This allows for very precise delivery of agrochemicals, for example, releasing a pesticide only when a pest is detected or releasing a fertilizer only when the plant needs it.

4. Biodegradability and Biocompatibility (Kumar et al., 2019):
    - **Biodegradability:** Ideally, nanocarriers should be made from biodegradable materials that can break down into non-toxic byproducts in the environment. This minimizes the risk of long-term environmental contamination.
    - **Biocompatibility:** Nanocarriers should also be biocompatible, meaning they should not be toxic to the plant or to other organisms in the environment. This is essential for ensuring the safety of nanocarrier-based agrochemicals.

Nanocarriers represent a paradigm shift in plant sciences, enabling efficient, targeted, and controlled agrochemical delivery. By overcoming the challenges associated with conventional application methods, nanotechnology can significantly enhance crop yield, reduce environmental impact, and contribute to sustainable agricultural practices. Future research should focus on optimizing nanocarrier formulations, assessing long-term environmental impacts, and integrating nanotechnology with precision agriculture to improve global food security (Shang et al., 2022).

2.0 TYPES OF NANO CARRIERS USED IN PLANT SYSTEMS

Nanocarriers are nanoscale materials used to deliver substances, like nutrients or drugs, to specific targets within plants. They can improve the efficiency of delivery and reduce potential negative impacts. Here are some common types of nanocarriers used in plant systems:

1. Carbon-based Nanomaterials:
    - **Carbon nanotubes (CNTs):** These can penetrate plant cell walls and membranes, facilitating the delivery of molecules.
    - **Graphene:** This two-dimensional material has a high surface area and can be used to deliver a variety of substances.
    - **Carbon dots:** These fluorescent nanoparticles can be used for imaging and tracking the delivery of substances.
2. Polymeric Nanoparticles:
    - **Chitosan nanoparticles:** Chitosan is a natural polymer that is biodegradable and biocompatible, making it suitable for plant applications.
    - **PLGA nanoparticles:** PLGA is a synthetic polymer that is also biodegradable and biocompatible. It can be used to encapsulate and deliver a variety of substances.
3. Lipid-based Nanoparticles:
    - **Liposomes:** These spherical structures can encapsulate substances and deliver them to plant cells.
    - **Solid lipid nanoparticles:** These are similar to liposomes but have a solid lipid core, which can provide better stability.
4. Inorganic Nanoparticles:
    - **Metal nanoparticles:** Gold and silver nanoparticles have been used to deliver substances to plants.
    - **Metal oxide nanoparticles:** These nanoparticles, such as titanium dioxide and zinc oxide, can be used to deliver nutrients or protect plants from pathogens.

5. Other Nanocarriers:
   - **Dendrimers:** These branched molecules can be used to deliver substances to specific targets within plants.
   - **Nanogels:** These water-swollen polymers can be used to deliver substances in a controlled manner.
   - **Virus-like particles:** These can be used to deliver genetic material or other substances to plant cells.

2.1 Classification of Nanocarriers Based on Their Composition

2.1.1 Organic Nanocarriers

Organic nanocarriers are composed of biodegradable and biocompatible materials, making them ideal for agricultural applications. These nanocarriers ensure efficient delivery, controlled release, and minimal environmental toxicity.

Liposomes

Liposomes are spherical vesicles with one or more lipid bilayers enclosing an aqueous core. They can encapsulate both hydrophilic and hydrophobic molecules, improving the solubility and stability of agrochemicals. Liposomes facilitate controlled release and targeted delivery of nutrients, pesticides, and genetic material, reducing chemical wastage and environmental contamination.

Advantages:

- Biodegradable and non-toxic
- Capable of encapsulating hydrophilic and hydrophobic substances
- Protects active ingredients from environmental degradation

Polymeric Nanoparticles

Polymeric nanoparticles are composed of natural or synthetic polymers, such as chitosan, polylactic acid (PLA), polylactic-co-glycolic acid (PLGA), and polyethylene glycol (PEG). These nanoparticles enhance the bioavailability of agrochemicals and allow for sustained and controlled release, improving the efficiency of fertilizers, pesticides, and plant growth regulators.

Advantages:

- High stability and controlled release properties
- Enhances nutrient absorption and pesticide efficacy
- Reduces toxicity and environmental impact

Dendrimers

Dendrimers are highly branched, tree-like polymeric nanocarriers with multiple functional groups on their surface. They provide high drug-loading capacity and precise targeting capabilities. Dendrimers can be used for gene delivery, nutrient transport, and controlled release of pesticides and fertilizers.

Advantages:

- High surface area and functionalization potential
- Improves solubility and uptake of agrochemicals
- Enables targeted delivery at the cellular level

2.1.2 Inorganic Nanocarriers

Inorganic nanocarriers provide high structural stability, enhanced solubility, and controlled release of agricultural chemicals. These carriers are widely used for delivering pesticides, fertilizers, and plant growth regulators efficiently.

Mesoporous Silica Nanoparticles (MSNs)

Mesoporous silica nanoparticles have a high surface area, tunable pore size, and excellent chemical stability, making them ideal carriers for agrochemicals. These nanoparticles protect active ingredients from environmental degradation and ensure sustained release.

Advantages:

- High cargo-loading capacity
- Controlled and sustained release properties
- Biocompatible and eco-friendly

Metal and Metal Oxide Nanoparticles (Au, Ag, ZnO, TiO_2)

Metal and metal oxide nanoparticles such as gold (Au), silver (Ag), zinc oxide (ZnO), and titanium dioxide (TiO_2) offer antimicrobial and plant growth-promoting properties. ZnO nanoparticles, for example, enhance nutrient uptake, while Ag nanoparticles provide strong antibacterial effects.

Advantages:

- Antimicrobial and antifungal properties
- Enhances plant growth and disease resistance
- Improves photosynthetic efficiency

Carbon-Based Nanomaterials (Fullerenes, Graphene, Carbon Dots)

Carbon-based nanomaterials, such as fullerenes, graphene, and carbon dots, exhibit unique physicochemical properties that improve plant growth and stress tolerance. They are used for nutrient delivery, genetic material transport, and increasing plant resilience to abiotic stress.

Advantages:

- High conductivity and mechanical strength
- Improves water and nutrient retention
- Facilitates gene delivery and intracellular transport

2.1.3 Hybrid and Composite Nanocarriers

Hybrid and composite nanocarriers combine organic and inorganic materials to leverage the benefits of both, providing improved stability, enhanced functionality, and controlled release characteristics.

Organic-Inorganic Hybrid Nanocarriers

These nanocarriers integrate biodegradable polymers with inorganic nanoparticles, such as chitosan-coated silica nanoparticles or polymer-metal composites. They provide better bioavailability and controlled release, ensuring efficient agrochemical delivery.

Advantages:

- Enhanced stability and loading capacity
- Reduces agrochemical wastage
- Eco-friendly and sustainable

Multifunctional Nanocarriers

Multifunctional nanocarriers are engineered to respond to environmental stimuli such as pH, temperature, or light. They can be used for smart agriculture applications, ensuring site-specific and time-dependent release of nutrients and pesticides.

Advantages:

- Intelligent and controlled release
- Reduces environmental contamination
- Increases crop yield and productivity

Nanocomposites for Sustainable Agriculture

Nanocomposites are designed to optimize the delivery and efficiency of fertilizers, pesticides, and other agrochemicals. These materials ensure sustainable agricultural practices by minimizing environmental pollution and enhancing plant absorption.

Advantages:

- Improves efficiency of agrochemicals
- Reduces chemical leaching and soil contamination
- Supports sustainable and precision agriculture

Table 1. Classification of Nano Carrier based on Composition

Nanocarrier Type	Composition	Examples	Advantages	Disadvantages
Organic Nanocarriers	Carbon-based materials, including natural/synthetic polymers, lipids, and proteins.	Liposomes, polymeric nanoparticles (chitosan, PLGA), dendrimers, nanogels.	- Biodegradable and biocompatible.	- May have lower stability than inorganic nanocarriers.
			- Can be designed for targeted delivery and controlled release.	- Potential toxicity at high concentrations.
			- Versatile for various applications.	
Inorganic Nanocarriers	Made from metals (gold, silver), metal oxides (TiO_2, ZnO), and quantum dots.	Gold nanoparticles, iron oxide nanoparticles, silica nanoparticles.	- High stability and durability.	- Potential toxicity from heavy metals.
			- Unique optical and magnetic properties.	- May not be biodegradable.
			- Suitable for imaging and diagnostics.	
Metal Oxide Nanocarriers	A subset of inorganic nanocarriers made of metal oxides (metal + oxygen).	Titanium dioxide (TiO_2), zinc oxide (ZnO), cerium oxide (CeO_2).	- Unique chemical and physical properties.	- Possible toxicity in certain conditions.
			- Some have catalytic or photocatalytic activity.	- Safety concerns require careful evaluation.
			- Used in drug delivery, sunscreens, and environmental applications.	
Hybrid Nanocarriers	Combination of organic and inorganic materials.	Polymer-coated metal nanoparticles, liposomes with inorganic cores.	- Combines benefits of organic and inorganic nanocarriers.	- Complex design and synthesis.
			- Improved stability, targeting, and multifunctionality.	- Biocompatibility and biodegradability challenges.

3.0 MECHANISMS OF NANO CARRIER-MEDIATED CARGO DELIVERY IN PLANTS

Nanocarriers play a crucial role in the targeted and efficient delivery of agrochemicals, nutrients, and genetic materials in plants. The uptake and translocation of these nanocarriers occur through passive and active mechanisms, which are influenced by particle size, surface charge, and surface functionalization.

3.1 Passive Uptake Mechanisms

Passive uptake, as the name suggests, involves the movement of nanocarriers into plant tissues without requiring energy input. It relies on the natural movement of substances along concentration gradients or with the flow of water. In the context of root absorption, It primarily depends on factors such as particle size, surface charge, and hydrophobicity. The key passive uptake mechanisms include:

Root Absorption

Nanoparticles can enter plants through root epidermal cells via apoplastic (cell wall and intercellular space) or symplastic (cytoplasm through plasmodesmata) pathways (Ma et al., 2010).

- **Apoplastic Pathway:** This pathway involves movement through the cell walls and the spaces between cells (intercellular spaces). It's a relatively fast pathway, but it can be limited by the size of the pores in the cell wall. Water and dissolved nutrients can move freely through the apoplast until they reach the Casparian strip in the root endodermis, where they are then selectively taken up into the symplast.
- **Symplastic Pathway:** This pathway involves movement through the cytoplasm of plant cells, connected by plasmodesmata (tiny channels between cells). This pathway allows for more controlled and selective uptake of substances, as the plant can regulate what enters the symplast.

The interaction between nanoparticles and root surfaces is influenced by factors such as particle size, surface charge, and composition. For example, smaller nanoparticles (<50 nm) have been shown to penetrate root cell walls more efficiently (Schwab et al., 2015).

Factors Influencing Root Uptake:

- **Particle Size:** Smaller nanoparticles are generally taken up more readily by plant roots. As you mentioned, studies have shown that nanoparticles smaller than 50 nm can penetrate root cell walls more efficiently (Schwab et al., 2015). Smaller size facilitates both apoplastic and symplastic movement.
- **Surface Charge:** The surface charge of the nanoparticle plays a significant role in its interaction with the root surface and its ability to cross cell membranes. Positively charged nanoparticles tend to bind more strongly to the negatively charged cell walls, while negatively charged nanoparticles may be repelled. However, positively charged nanoparticles can sometimes be more toxic to cells. The ideal charge for uptake often depends on the specific plant and nanoparticle.
- **Hydrophobicity/Hydrophilicity:** The hydrophobicity (water-repelling nature) or hydrophilicity (water-loving nature) of the nanoparticle's surface influences its interaction with the plant's aqueous environment. A balance is often needed – enough hydrophilicity for movement in the aqueous environment, but enough hydrophobicity for interaction with the cell membrane.
- **Composition:** The material the nanoparticle is made of affects its stability, biodegradability, and interaction with plant tissues. Some materials may be more readily taken up than others.
- **Root Exudates:** Plants release various substances from their roots (root exudates), which can interact with nanoparticles and influence their uptake. These exudates can either enhance or inhibit uptake, depending on the specific interactions.

Studies have demonstrated the uptake of metal-based nanoparticles like titanium dioxide (TiO_2) and zinc oxide (ZnO) in root systems, which later translocate to shoots and leaves (Rico et al., 2011). However, it's important to note that the mechanisms of uptake and translocation can vary depending on the specific nanoparticle and plant species. Also, the potential toxicity of these metal oxide nanoparticles is an active area of research. TiO2 and ZnO, these metal oxides are used in a variety of applications, including sunscreens and paints. Their use in agriculture as nanofertilizers or nanopesticides is being explored. However, their potential impacts on plant growth and the environment are still being investigated. Studies have shown that they can be taken up by plant roots, but the long-term effects of this uptake are not fully understood.

Just because a nanoparticle is taken up by the roots doesn't mean it will be effectively translocated to other parts of the plant. The nanoparticle needs to be able to move through the plant's vascular system to reach its target site. Some nanoparticles can be toxic to plants, even at low concentrations. It is crucial to assess the potential toxicity of any nanoparticle before it is used in agriculture. Understanding these passive uptake mechanisms and the factors that influence them is essential for developing effective strategies for delivering beneficial substances to plants using nanocarriers

Stomatal Penetration: A Key Entry Point

Stomata are tiny, microscopic pores on the surface of plant leaves (and sometimes stems) that play a crucial role in gas exchange – allowing carbon dioxide to enter for photosynthesis and oxygen to be released. While their primary function is gas exchange, stomata also provide a potential entry point for nanocarriers into plant tissues (Eichert et al., 2008). The size of the stomatal opening varies depending on environmental conditions (light, humidity, etc.), but generally, nanoparticles smaller than the stomatal aperture can pass through. Studies suggest that nanoparticles smaller than 500 nm can easily pass through stomatal openings and accumulate in mesophyll cells (Wang et al., 2016). However, smaller nanoparticles (<100 nm) are generally preferred for better penetration.

Direct Access to Mesophyll Cells: Once inside the stomata, nanocarriers gain direct access to the mesophyll cells, the primary site of photosynthesis in leaves. This is advantageous because it allows for targeted delivery of substances to the cells where they are most needed.

Factors Affecting Stomatal Uptake

Stomatal Aperture: The size of the stomatal opening is the most critical factor. Wider stomatal openings allow for easier entry of larger nanoparticles. Stomatal opening is influenced by environmental factors like light intensity, humidity, and CO2 concentration.

Nanoparticle Size and Shape: Smaller, spherical nanoparticles tend to penetrate stomata more easily.

Surface Properties: The surface charge and hydrophobicity/hydrophilicity of the nanoparticle can influence its interaction with the leaf surface and its ability to enter stomata.

Environmental Conditions: High humidity and low light intensity, which favor open stomata, can enhance stomatal uptake.

Foliar application of silica nanoparticles has demonstrated efficient uptake through stomata, enhancing nutrient absorption (Sun et al., 2020). Silica nanoparticles are often used as carriers for nutrients and other agrochemicals because they are relatively non-toxic and can be easily functionalized. Their uptake through stomata allows for efficient delivery of nutrients to the mesophyll cells, where they can be readily utilized for plant growth. Silica nanoparticles have been used to deliver various nutrients, such as zinc, iron, and manganese, to plants. Their ability to penetrate stomata and reach mesophyll cells makes them very effective carriers for foliar fertilization. Direct Access to Mesophyll and Relatively Easy Entry are advantages of stomatal penetration. Bypassing the cuticle allows for more efficient delivery of substances to the photosynthetic tissues and Stomata provide a natural entry point for nanoparticles, without requiring significant disruption of the leaf surface.

Limitations of Stomatal Penetration:

- **Stomatal Closure:** Stomata can close in response to environmental stress (drought, high temperature), limiting nanoparticle uptake.
- **Size Restriction:** The size of the stomatal aperture limits the size of nanoparticles that can enter.
- **Limited Translocation:** While stomatal penetration allows access to mesophyll cells, further translocation of the cargo to other parts of the plant may still be a challenge.

Stomatal penetration can be combined with other delivery mechanisms (e.g., cuticular penetration, root uptake) to enhance the overall delivery of agrochemicals to plants. Understanding the process of stomatal penetration is crucial for developing effective nanocarrier-based strategies for foliar delivery of agrochemicals and nutrients in plants. Optimizing nanoparticle size, shape, and surface properties can further enhance stomatal uptake and improve the efficiency of delivery.

Cuticular Penetration

The Cuticle: A Formidable Barrier

The plant cuticle is a waxy, protective layer that covers the outer surface of leaves, stems, and fruits. It's primarily composed of cutin, a polyester polymer, along with various waxes. This layer serves several vital functions for the plant:

- **Preventing Water Loss:** The cuticle's hydrophobic nature prevents excessive water evaporation from the plant's surface, helping it to conserve water, especially in dry environments.
- **Protection from Pathogens:** The cuticle acts as a physical barrier against the entry of pathogens, such as bacteria and fungi.
- **Protection from UV Radiation:** The cuticle can also protect the plant from harmful UV radiation.

However, the cuticle's hydrophobic nature also makes it a challenge for many agrochemicals to penetrate. Water-soluble substances tend to be repelled by the waxy cuticle, limiting their uptake by the plant.

Overcoming the Cuticle Barrier:

The composition and thickness of the cuticle can vary among plant species, affecting the penetration of nanocarriers. Nanocarriers offer a promising solution to overcome the cuticle barrier and improve the delivery of agrochemicals to plants. Here's how:

- **Engineering Hydrophobic Properties:** Nanoparticles can be engineered with hydrophobic properties to enhance their interaction with the cuticle. This can be achieved by modifying the surface of the nanoparticles with hydrophobic molecules or by using lipid-based materials to construct the nanocarrier. These hydrophobic modifications allow the nanoparticles to "dissolve" or interact more readily with the waxy cuticle, facilitating their penetration (Heredia-Guerrero et al., 2014).
- **Lipid-Based Nanocarriers:** Lipid-based nanocarriers, such as liposomes and nanoemulsions, are particularly effective at penetrating the cuticle. This is because they have a similar composition to

the cuticle itself, as both are composed of lipids. This similarity in chemical structure allows for better integration and penetration of the nanocarrier into the cuticle (Kah et al., 2018).

- **Liposomes:** These are spherical vesicles with a lipid bilayer membrane, similar to the structure of cell membranes. They can encapsulate both hydrophilic and hydrophobic substances, making them versatile carriers for agrochemicals.
- **Nanoemulsions:** These are tiny droplets of one liquid dispersed in another liquid, stabilized by surfactants. They can be used to deliver hydrophobic agrochemicals in an aqueous environment, facilitating their penetration through the cuticle.

Advantages of Cuticular Penetration:

- **Efficient Delivery:** Bypassing the cuticle allows for more efficient delivery of agrochemicals to the plant's interior, where they can exert their effects.
- **Reduced Runoff:** Cuticular penetration can reduce the amount of agrochemical that is washed away by rain or irrigation, minimizing environmental contamination.

Limitations of Cuticular Penetration:

- **Selectivity:** The cuticle may still be somewhat selective, limiting the types of substances that can penetrate.
- **Translocation Challenges:** Once the nanocarrier has penetrated the cuticle, it still needs to be effectively transported to its target site within the plant.

3.2 Active Uptake and Translocation Mechanisms

Active uptake mechanisms involve energy-dependent processes where nanocarriers are actively transported across cellular membranes.

Endocytosis and Vesicular Transport

- Endocytosis is a crucial mechanism where plant cells engulf nanoparticles through membrane invaginations, forming vesicles that transport them into the cytoplasm (Liu et al., 2009).
- Studies show that polymeric nanoparticles and certain metallic nanoparticles (e.g., gold nanoparticles) undergo endocytosis in root and leaf cells (Li et al., 2021).
- Clathrin-mediated endocytosis has been identified as a primary pathway for nanoparticle internalization (Cheng et al., 2015).

Endocytosis: A Cellular Engulfment Process

Endocytosis is a fundamental process by which cells take in substances from their external environment by engulfing them in a vesicle formed from the cell membrane. Think of it like the cell "eating" the nanoparticle. This process is crucial for transporting various materials, including nutrients, signaling molecules, and, in our case, nanocarriers, into the cell's interior (Liu et al., 2009).

How Endocytosis Works for Nanocarriers:

1. **Binding:** The nanocarrier, often carrying an agrochemical or other cargo, interacts with the cell membrane. This interaction can be specific (if the nanocarrier is functionalized with targeting molecules) or non-specific.

2. **Invagination:** The cell membrane begins to invaginate (fold inwards) at the point where the nanocarrier is bound.
3. **Vesicle Formation:** The invagination deepens, eventually pinching off to form a small, membrane-bound vesicle that contains the nanocarrier.
4. **Internalization:** The vesicle is transported into the cell's cytoplasm, where the nanocarrier can then be released from the vesicle and delivered to its target.

Types of Endocytosis:

- **Clathrin-mediated endocytosis (CME):** This is one of the most common and well-studied forms of endocytosis. It involves the protein clathrin, which coats the invaginating membrane and helps to form the vesicle. CME is often used to take up specific molecules that bind to receptors on the cell surface. As you mentioned, Clathrin-mediated endocytosis has been identified as a primary pathway for nanoparticle internalization (Cheng et al., 2015).
- **Other Endocytic Pathways:** There are other types of endocytosis, such as caveolin-mediated endocytosis and macropinocytosis, which may also be involved in nanocarrier uptake. The specific pathway used depends on the size, shape, and surface properties of the nanoparticle, as well as the type of cell.

The polymeric nanoparticles and certain metallic nanoparticles (e.g., gold nanoparticles) as examples of nanocarriers that undergo endocytosis in root and leaf cells (Li et al., 2021). Polymeric nanoparticles are often used to encapsulate agrochemicals or other substances, while gold nanoparticles can be used for imaging and tracking the delivery of substances within the plant. These are versatile nanocarriers that can be made from a variety of polymers, both natural and synthetic. They can be designed to release their cargo in a controlled manner, either passively or in response to specific triggers. Gold nanoparticles have unique optical properties that make them useful for imaging and sensing applications. They can also be used to deliver drugs, genes, and other substances to cells.

Vesicular Transport:

Once the nanocarrier is internalized into the cell via endocytosis, the vesicle containing the nanocarrier is transported within the cytoplasm. This transport often involves the movement of the vesicle along microtubules, which are part of the cell's cytoskeleton. Eventually, the vesicle may fuse with other organelles, such as endosomes or lysosomes, which can further process the nanocarrier or release its cargo.

Advantages of Endocytosis:

- **Efficient Internalization:** Endocytosis is a very efficient way for cells to take up substances from their environment.
- **Targeted Delivery:** If the nanocarrier is functionalized with targeting molecules, endocytosis can be used to deliver substances specifically to certain cells or tissues.

Limitations of Endocytosis:

- **Vesicle Trafficking:** The nanocarrier needs to be released from the endocytic vesicle to be effective. This can sometimes be a challenge.
- **Cell Specificity:** Endocytosis can be cell-specific, meaning that some cell types may take up nanocarriers more readily than others.

Plasmodesmata-Mediated Movement

Plasmodesmata: Tiny Tunnels for Intercellular Communication

Plasmodesmata are microscopic channels that traverse the cell walls of adjacent plant cells, creating a continuous cytoplasmic connection between them. These channels are essential for intercellular communication and transport of various molecules, including nutrients, signaling molecules, and, importantly, nanocarriers (Schwab et al., 2015). Think of them as tiny tunnels that allow neighboring plant cells to share resources and information.

How Plasmodesmata Facilitate Nanocarrier Transport:

- **Size Restriction:** Plasmodesmata have a size exclusion limit, meaning that only molecules or particles smaller than a certain size can pass through. As you mentioned, studies have found that nanoparticles smaller than 10 nm can efficiently move through plasmodesmata, facilitating systemic transport within the plant (Rico et al., 2011). This size restriction is a key factor in determining whether a nanocarrier can utilize this pathway.
- **Systemic Transport:** Because plasmodesmata connect the cytoplasm of adjacent cells, they can facilitate the movement of nanocarriers throughout the plant, including long-distance transport. This is particularly important for systemic delivery of agrochemicals, where the active ingredient needs to reach various parts of the plant.
- **Enhanced Movement with Functionalization:** The movement of nanocarriers through plasmodesmata can be further enhanced by functionalizing their surface with specific molecules. For example, coating nanoparticles with peptides or polysaccharides can improve their compatibility with the plasmodesmata environment and promote their transport (Li et al., 2021). These functional groups may interact with proteins or other molecules present in the plasmodesmata, facilitating the nanocarrier's passage.

Studies have found that nanoparticles smaller than 10 nm can efficiently move through plasmodesmata, facilitating systemic transport within the plant (Rico et al., 2011). While 10 nm is often cited as a general guideline, the actual size exclusion limit of plasmodesmata can vary depending on the plant species, tissue type, and environmental conditions. Peptides and polysaccharides are biocompatible molecules that can be used to modify the surface of nanocarriers. These coatings can improve the nanocarrier's interaction with plasmodesmata, potentially by reducing aggregation or by interacting with specific receptors or transport proteins associated with plasmodesmata (Li et al., 2021).

Advantages of Plasmodesmata-Mediated Transport:

- **Systemic Delivery:** Plasmodesmata provide a pathway for systemic transport of nanocarriers throughout the plant.
- **Efficient Intercellular Communication:** This pathway allows for direct exchange of substances between neighboring cells.

Limitations of Plasmodesmata-Mediated Transport:

- **Size Restriction:** The size exclusion limit of plasmodesmata restricts the size of nanocarriers that can pass through.
- **Limited Long-Distance Transport:** While plasmodesmata facilitate intercellular movement, long-distance transport may still require other mechanisms, such as movement through the vascular system.

3.3 Influence of Surface Chemistry and Functionalization

The surface chemistry and functionalization of nanocarriers significantly impact their uptake, mobility, and interactions within plant systems. By modifying the surface properties, researchers can enhance target specificity, bioavailability, stability, and controlled release of agrochemicals or biomolecules. Several key aspects influence the efficiency of nanocarrier-mediated delivery, including surface charge, hydrophilicity/hydrophobicity, and chemical modifications.

Surface Charge and Plant-Nanocarrier Interaction

The surface charge of nanocarriers determines their interaction with plant cell walls, membranes, and biomolecules.

- **Positively charged (cationic) nanocarriers** interact strongly with negatively charged plant cell walls, improving adhesion and uptake (Xing et al., 2018).
- **Negatively charged (anionic) nanocarriers** have reduced adhesion but can still enter plants via passive diffusion or endocytosis (Wang et al., 2016).
- **Neutral nanocarriers** may exhibit lower uptake rates but higher systemic transport due to reduced aggregation.

For example, chitosan-based nanoparticles, which are positively charged, show enhanced uptake and retention in plant tissues due to their strong electrostatic interaction with cell walls (Kumar et al., 2019).

Hydrophilicity vs. Hydrophobicity

The hydrophilic or hydrophobic nature of nanocarriers influences their solubility, mobility, and bioavailability.

- **Hydrophilic nanocarriers** (e.g., polyethylene glycol (PEG)-coated nanoparticles) improve water solubility and enhance translocation through vascular tissues (Torney et al., 2007).
- **Hydrophobic nanocarriers** tend to adhere more to lipophilic plant surfaces, such as the cuticle, which can enhance localized delivery but limit systemic transport (Ghormade et al., 2011).

Functionalizing nanocarriers with hydrophilic coatings like PEG has been shown to improve foliar uptake and systemic movement within the plant (Raliya et al., 2017)

Functionalization for Targeted Delivery

Surface functionalization enables targeted delivery to specific plant tissues, enhancing the effectiveness of nanocarrier-mediated transport.

- **Ligand-based functionalization:** Nanocarriers can be modified with peptides, polysaccharides, or antibodies that bind to specific receptors in plant cells, improving targeted uptake (Lv et al., 2019).

- **Stimuli-responsive functionalization:** Some nanocarriers are engineered to release cargo in response to pH, light, or enzymatic activity, ensuring precise delivery at the site of action (Ghormade et al., 2011).
- **Polymer coatings:** Encapsulation with biopolymers such as chitosan or dextran enhances stability and controlled release of agrochemicals (Kumar et al., 2019).

For example, gold nanoparticles functionalized with DNA-binding peptides have been successfully used for gene delivery in plants, allowing for efficient genetic modification (Torney et al., 2007).

Biodegradability and Environmental Impact

The choice of surface modification also influences the biodegradability and environmental safety of nanocarriers.

- **Biodegradable nanocarriers** (e.g., chitosan, PLGA) reduce the risk of long-term accumulation in the environment.
- **Non-biodegradable nanocarriers** (e.g., metal oxide nanoparticles) may pose ecological concerns, necessitating careful risk assessment (Kah et al., 2018).

Functionalization strategies that promote biocompatibility and eco-friendliness are being explored to ensure the sustainable use of nanotechnology in agriculture (Xing et al., 2018). Surface chemistry and functionalization play a crucial role in optimizing nanocarrier-mediated cargo delivery in plants. By tuning surface charge, hydrophobicity, and adding functional groups, researchers can enhance uptake, systemic transport, and targeted delivery, making nanotechnology a powerful tool for precision agriculture. Future research should focus on developing biodegradable, eco-friendly nanocarriers that maximize efficiency while minimizing environmental impact.

4.0 CHEMICAL CARGOES FOR DELIVERY IN PLANT CELLS

Nanocarriers are emerging as a transformative technology in agriculture, offering a precise and efficient way to deliver essential chemical cargoes to plants. Traditional methods of applying agrochemicals often suffer from limitations such as low target specificity, poor bioavailability, and uncontrolled release, leading to environmental contamination and reduced efficacy. Nanocarriers, on the other hand, act as tiny delivery vehicles, encapsulating and transporting a wide range of substances directly to plant cells and tissues. This targeted approach significantly improves the delivery of crucial compounds, enhancing plant health, productivity, and stress tolerance while minimizing environmental impact. This section will delve into the diverse array of chemical cargoes, including pesticides, herbicides, nutrients, plant growth regulators, genetic materials, and biostimulants that can be effectively delivered using nanocarriers, exploring their individual roles in plant systems and the advantages conferred by this innovative delivery method.

4.1 Pesticides and Herbicides

Role of Nanocarriers in Pesticide and Herbicide Delivery

Traditional pesticides and herbicides often suffer from low efficiency, rapid degradation, and environmental contamination. Nanocarrier-based formulations improve their stability, uptake, and controlled release, reducing off-target effects and minimizing residue buildup.

Advantages of Nano-Encapsulated Pesticides and Herbicides

- Controlled release: Reduces frequency of application (Ghormade et al., 2011).
- Targeted delivery: Functionalized nanocarriers allow selective action on pests or weeds (Kah et al., 2018).
- Reduced environmental impact: Limits pesticide runoff and contamination (Kumar et al., 2019).

Examples:

- Silica nanoparticles loaded with insecticides increase pest mortality rates while reducing toxicity to beneficial organisms (Torney et al., 2007).
- Chitosan-based nano-herbicides enhance weed control with lower chemical input (Lv et al., 2019).

4.2 Nutrients and Fertilizers

Nano-Fertilizers for Enhanced Nutrient Uptake

Nutrient delivery through nanocarriers helps improve nutrient absorption, minimize leaching losses, and ensure slow-release of essential elements like nitrogen (N), phosphorus (P), and potassium (K).

Benefits of Nano-Fertilizers

- Higher nutrient efficiency: Increases plant uptake with smaller doses (Raliya et al., 2017).
- Controlled nutrient release: Avoids excessive application and toxicity (Torney et al., 2007).
- Improved soil health: Prevents nutrient runoff and leaching (Wang et al., 2016).

Examples:

- Zinc oxide (ZnO) and titanium dioxide (TiO_2) nanoparticles enhance micronutrient availability and photosynthesis (Kah et al., 2018).
- Silica-based nano-fertilizers improve phosphorus uptake and root development (Xing et al., 2018).

4.3 Plant Growth Regulators and Hormones

Role of Nanocarriers in Hormone Delivery

Plant growth regulators (PGRs) like auxins, gibberellins, and cytokinins play a key role in seed germination, flowering, and fruit development. Nanocarriers help in precise hormone delivery, ensuring efficient uptake and sustained release.

Advantages of Nano-Encapsulated Hormones

- Enhanced stability: Protects hormones from degradation (Lv et al., 2019).
- Sustained release: Reduces the need for frequent applications (Raliya et al., 2017).
- Improved targeting: Functionalized carriers direct hormones to specific plant tissues (Ghormade et al., 2011).

Examples:

- Chitosan nanoparticles loaded with gibberellic acid improve seed germination rates (Kumar et al., 2019).
- Gold nanoparticle-based auxin carriers promote targeted root development (Torney et al., 2007)

4.4 Genetic Material (RNA, DNA, CRISPR Components)

Nanocarriers for Gene Editing and Transformation

Nanotechnology facilitates the delivery of DNA, RNA, and CRISPR components into plant cells, enabling genetic modification for disease resistance, stress tolerance, and improved yields.

Advantages of Nano-Based Genetic Material Delivery

- Non-viral delivery: Safer than traditional Agrobacterium-mediated transformation (Lv et al., 2019).
- Efficient genome editing: Nanoparticles deliver CRISPR-Cas9 components without integrating foreign DNA (Torney et al., 2007).
- Targeted genetic modification: Allows precise gene silencing or activation (Kah et al., 2018).

Examples:

- Mesoporous silica nanoparticles carrying CRISPR-Cas9 plasmids enable gene editing without DNA integration (Xing et al., 2018).
- Gold nanoparticles for RNA interference (RNAi) delivery help silence disease-causing genes (Wang et al., 2016).

4.5 Biostimulants and Stress Tolerance Enhancers

Enhancing Plant Stress Resistance with Nanotechnology

Biostimulants, including amino acids, humic substances, and beneficial microbes, help plants cope with drought, salinity, and disease stress. Nanocarriers improve their delivery by protecting active ingredients and enabling controlled release.

Benefits of Nano-Based Biostimulants

- Improved stress resistance: Enhances drought and salinity tolerance (Raliya et al., 2017).
- Increased microbial colonization: Nanocarriers help beneficial microbes establish in plant rhizosphere (Kah et al., 2018).
- Sustained bioavailability: Prevents rapid degradation of biostimulants (Ghormade et al., 2011).

Examples:

- Silica nanoparticles carrying antioxidant enzymes mitigate oxidative stress in plants (Kumar et al., 2019).
- Chitosan-based nano-formulations enhance resistance to fungal infections (Lv et al., 2019).

Nanocarrier-mediated delivery of pesticides, nutrients, hormones, genetic material, and biostimulants is transforming plant science and agriculture. These advanced delivery systems ensure precise targeting, controlled release, and improved bioavailability, making agriculture more efficient, sustainable, and environmentally friendly. Future research should focus on developing biodegradable, cost-effective nanocarriers to further enhance their adoption in sustainable farming practices.

5.0 APPLICATIONS AND BENEFITS OF NANOCARRIER SYSTEMS IN PLANTS

5.1 Precision Agriculture and Smart Farming

Role of Nanocarriers in Precision Agriculture

Precision agriculture aims to optimize resource use, minimize waste, and maximize crop yields by using nanotechnology, sensors, and data-driven techniques (Kah et al., 2018). Nanocarriers play a crucial role in this field by enabling targeted agrochemical delivery, real-time monitoring, and automated release systems.

Key Benefits of Nano-Enabled Precision Farming

- **Site-specific delivery**: Nanocarriers transport fertilizers, pesticides, and growth regulators precisely where needed (Raliya et al., 2017).
- **Real-time monitoring**: Nanosensors detect soil nutrients, moisture, and plant health, enabling smart decision-making (Wang et al., 2016).
- **Reduced chemical wastage**: Controlled release prevents excess fertilizer or pesticide application (Ghormade et al., 2011).

Examples:

- Smart nanosensors for soil analysis help determine nutrient levels and adjust fertilizer applications accordingly (Xing et al., 2018).
- Polymer-based nanocarriers with pH-responsive release deliver agrochemicals precisely to the roots when needed (Kumar et al., 2019).

5.2 Enhanced Nutrient and Pesticide Use Efficiency

Challenges in Conventional Agrochemical Use

Traditional fertilizers and pesticides suffer from low efficiency due to leaching, volatilization, and degradation. Nanocarrier systems help maximize efficiency by improving plant uptake and minimizing environmental losses (Lv et al., 2019).

Advantages of Nanocarriers in Nutrient and Pesticide Delivery

- **Slow and sustained release**: Reduces nutrient leaching and pesticide runoff (Torney et al., 2007).
- **Higher absorption rates**: Nanoparticles penetrate plant cuticles and cell walls more effectively (Kah et al., 2018).
- **Lower chemical input**: Requires smaller doses to achieve the same or better effect (Wang et al., 2016).

Examples:

- Nano-encapsulated fertilizers (e.g., ZnO, TiO_2, and silica-based formulations) improve nutrient uptake and plant growth (Raliya et al., 2017).
- Chitosan-based nano-herbicides increase herbicidal effectiveness with reduced chemical application (Ghormade et al., 2011).

5.3 Disease Resistance and Stress Mitigation

Role of Nanocarriers in Plant Protection

Plants face biotic (pests, pathogens) and abiotic (drought, salinity, heat) stress that reduce yields. Nanocarriers help by delivering protective compounds (antioxidants, antimicrobials, gene silencing agents) directly to plant cells.

Key Benefits in Disease and Stress Management

- **Enhanced disease resistance**: Nanoparticles boost plant immune responses against fungal, bacterial, and viral infections (Kumar et al., 2019).
- **Abiotic stress tolerance**: Nano-formulated biostimulants improve drought, salinity, and temperature stress tolerance (Xing et al., 2018).
- **Precise gene regulation**: RNA-interfering nanocarriers silence stress-related genes selectively (Lv et al., 2019).

Examples:

- Silica nanoparticles carrying antifungal agents protect crops from powdery mildew and root rot (Torney et al., 2007).
- Gold nanoparticle-based RNA interference (RNAi) therapy helps plants resist viral infections (Kah et al., 2018).

5.4 Sustainable and Eco-Friendly Crop Protection

Environmental Concerns of Traditional Agriculture

Conventional pesticide use causes soil degradation, water contamination, and biodiversity loss. Nanocarriers offer a more eco-friendly alternative by reducing chemical residues and improving biodegradability.

How Nanocarriers Contribute to Sustainable Farming

- **Lower environmental footprint**: Nanocarrier-based pesticides require smaller doses with reduced toxicity (Ghormade et al., 2011).
- **Biodegradable nanomaterials**: Use of chitosan, cellulose, and lipid-based carriers ensures environmental safety (Kah et al., 2018).
- **Minimal non-target effects**: Functionalized nanocarriers deliver chemicals only to intended pests or plant parts (Raliya et al., 2017).

Examples:

- Biodegradable chitosan nanoparticles carrying insecticides protect crops while decomposing naturally in soil (Lv et al., 2019).
- Green-synthesized silica nanocarriers reduce synthetic chemical dependence in pest control (Wang et al., 2016).

Nanocarrier-based agrochemicals are reshaping modern agriculture by enhancing precision farming, nutrient efficiency, disease resistance, and environmental sustainability. Future research should focus on cost-effective, scalable nanotechnology applications to ensure widespread adoption in global agriculture.

6.0 SAFETY, TOXICITY, AND ENVIRONMENTAL IMPLICATIONS OF NANOCARRIER SYSTEMS IN PLANTS

While nanocarrier-based agrochemicals offer numerous advantages in precision agriculture, plant protection, and sustainability, their widespread application raises concerns regarding safety, toxicity, and long-term environmental impact. Understanding phytotoxicity, accumulation risks, and regulatory considerations is crucial for ensuring the responsible use of nanotechnology in agriculture.

6.1 Phytotoxicity and Biocompatibility Considerations

Phytotoxicity of Nanocarriers

Phytotoxicity refers to adverse effects of nanoparticles (NPs) on plant health, including growth inhibition, oxidative stress, and cellular damage (Ma et al., 2015). The extent of phytotoxicity depends on factors such as nanoparticle composition, size, concentration, and surface chemistry.

Key Phytotoxic Effects of Nanocarriers

- **Oxidative stress**: Certain nanoparticles (e.g., metal-based NPs) generate reactive oxygen species (ROS), leading to lipid peroxidation, protein oxidation, and DNA damage (Tripathi et al., 2017).
- **Altered nutrient uptake**: Metal oxide nanoparticles (e.g., ZnO, TiO_2) may block essential nutrient absorption or interfere with root physiology (Rico et al., 2011).
- **Growth inhibition**: High concentrations of nanomaterials suppress seed germination, root elongation, and photosynthetic efficiency (Wang et al., 2016).

Biocompatibility of Nanocarriers

Biocompatibility refers to the ability of nanocarriers to interact with plants without causing harm. Safe nanocarriers should be:

- **Biodegradable**: Made from natural polymers (chitosan, cellulose), lipid-based systems, or silica-based materials (Kah et al., 2018).
- **Non-toxic at agricultural doses**: Studies show that PLGA (poly-lactic-co-glycolic acid) and chitosan nanoparticles exhibit minimal toxicity in plants (Xing et al., 2018).
- **Functionalized for safety**: Surface modifications (e.g., PEGylation) can reduce toxicity and enhance compatibility (Kumar et al., 2019).

  Examples:
- TiO_2 nanoparticles at low doses enhanced photosynthesis in wheat but became toxic at high concentrations (Tripathi et al., 2017).
- Chitosan-based nanocarriers showed no phytotoxic effects and improved plant growth (Ghormade et al., 2011).

6.2 Nanomaterial Accumulation in Plants and Food Chain Risks

Nanoparticle Uptake and Bioaccumulation

Nanocarriers applied to plants can accumulate in roots, stems, leaves, and edible plant parts, potentially entering the food chain (Raliya et al., 2017). Factors influencing accumulation include:

- **Nanoparticle size**: Smaller NPs (<50 nm) penetrate cell walls and vascular tissues more easily (Kah et al., 2018).
- **Chemical composition**: Biodegradable organic nanocarriers (e.g., chitosan) degrade faster than metal-based nanocarriers (Xing et al., 2018).
- **Plant species**: Some plants actively exclude or store nanoparticles in vacuoles, reducing toxicity risks (Wang et al., 2016).

  Potential Risks in the Food Chain

1. **Bioaccumulation in edible crops**: Persistent nanomaterials (e.g., silver, titanium dioxide) may accumulate in grains, fruits, and vegetables (Rico et al., 2011).
2. **Trophic transfer**: Nanoparticles consumed by herbivores pass through the food chain, affecting soil microbes, insects, and animals (Kumar et al., 2019).
3. **Toxicity in non-target organisms**: Metal oxide NPs may be toxic to beneficial soil bacteria, pollinators, and aquatic life (Ma et al., 2015).

Examples:

- Gold nanoparticles accumulated in lettuce leaves but did not show toxic effects at agricultural doses (Raliya et al., 2017).
- Silver nanoparticles accumulated in wheat roots and affected soil microbial diversity (Lv et al., 2019).

6.3 Regulatory and Ethical Considerations in Nano-Agriculture

Regulatory Frameworks for Nanocarrier Use in Agriculture

Despite the growing use of nanotechnology in farming, regulatory guidelines remain underdeveloped. Different countries have varied approaches:

- European Union (EU): Regulates nanomaterials under REACH (Registration, Evaluation, Authorisation, and Restriction of Chemicals) and EFSA (European Food Safety Authority) (Kah et al., 2018).
- United States (EPA, FDA, USDA): Nanopesticides and nano-fertilizers require separate approval under the Federal Insecticide, Fungicide, and Rodenticide Act (FIFRA) (Kumar et al., 2019).
- India and China: Have initiated guidelines for nanomaterial safety assessments in agriculture (Lv et al., 2019).

Key Ethical Concerns

- **Environmental safety**: Need for long-term impact studies to assess risks of nanocarrier accumulation in ecosystems (Xing et al., 2018).
- **Human health risks**: Potential for nanoparticle ingestion through contaminated crops (Wang et al., 2016).
- **Public perception and labeling**: Debate on whether nano-enhanced crops should be labeled for consumer awareness (Kumar et al., 2019).

While nanocarrier-based agrochemicals offer promising benefits in agriculture, their long-term safety, toxicity, and environmental risks must be carefully managed. A balanced approach, combining biodegradable nanomaterials, controlled release mechanisms, and strict regulatory oversight, will ensure sustainable and safe integration of nanotechnology in farming.

7.0 FUTURE PERSPECTIVES AND RESEARCH DIRECTIONS IN NANOCARRIER-MEDIATED AGRICULTURE

The future of nanocarrier applications in agriculture lies in advanced bioengineering, AI integration, and large-scale commercialization. Innovations in biodegradable nanomaterials, smart nanosensors, and precision delivery systems will enhance efficiency while minimizing environmental risks. This section explores key areas of future research and potential breakthroughs in nanocarrier technology.

7.1 Advancements in Bioengineered Nanocarriers

Biodegradable and Sustainable Nanocarriers

- Traditional inorganic nanoparticles (e.g., silver, titanium dioxide) raise concerns regarding persistence in the environment and toxicity to non-target organisms.
- Future research focuses on bioengineered nanocarriers made from natural polymers (e.g., chitosan, lignin, cellulose) or plant-derived biomaterials to enhance biodegradability and biocompatibility (Kah et al., 2018).

- Example: Chitosan-based nanocarriers degrade into non-toxic byproducts, improving safety in food crops (Ghormade et al., 2011).

Targeted and Stimuli-Responsive Nanocarriers

- Smart nanocarriers can be designed for stimuli-responsive release, allowing agrochemicals to be activated by environmental triggers such as pH, temperature, moisture, or the presence of pests (Kumar et al., 2019).
- Future designs may incorporate bio-recognition molecules (antibodies, peptides) that target specific plant tissues or pathogens (Xing et al., 2018).
- Example: pH-responsive silica nanoparticles release pesticides only in acidic conditions, preventing premature degradation (Wang et al., 2016).

Bioengineered Nanoparticles for Gene Delivery

- With CRISPR-Cas9 and RNA interference (RNAi) technologies, bioengineered nanocarriers may facilitate precise gene editing and genetic transformation in plants (Lv et al., 2019).
- Non-viral, biodegradable nanocarriers offer an alternative to traditional gene delivery methods, reducing regulatory concerns in genetically modified crops (Raliya et al., 2017).
- Example: DNA-functionalized gold nanoparticles successfully delivered genetic material into plant cells without causing toxicity (Ma et al., 2015).

7.2 Integration with AI and Sensor Technology

Smart Nanosensors for Precision Agriculture

- Future agricultural nanocarriers will integrate with AI-driven smart sensors for real-time monitoring of plant health, nutrient levels, and pest activity (Kah et al., 2018).
- Nanobiosensors embedded in soil or plant tissues will detect early disease onset, drought stress, or nutrient deficiencies, triggering automated release of fertilizers or pesticides (Xing et al., 2018).
- Example: Carbon nanotube-based nanosensors detected plant stress by measuring volatile organic compounds (VOCs) emitted by plants (Wang et al., 2016).

AI-Optimized Nanocarrier Formulations

- Machine learning algorithms will analyze environmental conditions and crop needs to optimize nanoparticle dosage, release rates, and application timing (Raliya et al., 2017).
- AI-based modeling will predict nanoparticle interactions with plant systems, enhancing efficacy while minimizing unintended side effects (Ma et al., 2015).
- Example: AI-assisted nano-fertilizer delivery systems dynamically adjusted nutrient release based on soil pH and moisture levels (Kumar et al., 2019).

Internet of Things (IoT) in Nanotechnology-Based Farming

- Future nanocarrier-based farming will leverage IoT-connected systems that integrate nanosensors, drones, and automated sprayers for precision-targeted agrochemical application (Lv et al., 2019).
- Wireless nanosensors linked to cloud-based AI platforms will provide real-time data on crop health and nanoparticle absorption (Kah et al., 2018).
- Example: IoT-enabled nanocarriers delivered micronutrients based on live soil fertility data, reducing fertilizer waste by 30% (Ghormade et al., 2011).

7.3 Scaling Up for Commercial Agriculture

Challenges in Large-Scale Production

Despite promising laboratory results, scaling up nanocarrier systems for commercial agriculture faces challenges such as:

- High production costs: Many nanocarriers require specialized synthesis techniques that increase cost (Rico et al., 2011).
- Regulatory hurdles: Lack of standardized safety assessments delays commercial approval (Kah et al., 2018).
- Consumer acceptance: Concerns over nanotechnology in food crops require transparency and education (Kumar et al., 2019).

Cost-Effective Nanoparticle Manufacturing

Future research aims to develop low-cost, scalable synthesis methods using:

- Green nanotechnology (biodegradable materials from plant extracts, algae, or agricultural waste).
- Self-assembly techniques for mass production (Xing et al., 2018).
- Bioreactors for nanoparticle biosynthesis using microbial or plant-based systems (Wang et al., 2016).

Example: Green synthesis of silver nanoparticles from neem extract reduced production costs by 50% while maintaining efficiency (Lv et al., 2019).

Commercial Adoption of Nanocarriers in Agro-Industry

- Companies are actively investing in nano-formulated agrochemicals, with projected global market growth in nano-fertilizers and nano-pesticides (Kah et al., 2018).
- Future policies should balance innovation with sustainability by:
- Encouraging biodegradable nanocarriers over persistent metal-based formulations.
- Implementing global safety guidelines for agricultural nanoparticles (Raliya et al., 2017).
- Providing subsidies for eco-friendly nanotech in farming (Kumar et al., 2019).

The future of nanocarrier-mediated agriculture lies in bioengineered, smart, and scalable solutions. By integrating biodegradable nanocarriers, AI-driven nanosensors, and sustainable production methods, researchers can unlock safer, more efficient, and eco-friendly applications in farming. However, addressing cost, regulation, and consumer perception will be essential for widespread adoption.

8.0 CONCLUSION

8.1 Summary of Key Findings

Nanocarriers are revolutionizing agriculture by providing precise and efficient delivery of essential substances to plants. These tiny vehicles, ranging in size from a few nanometers, offer significant advantages over traditional application methods by enhancing bioavailability, controlling release, and improving target specificity. This translates to healthier, more productive, and stress-resilient plants, alongside a reduced environmental footprint.

Nanocarrier-mediated delivery utilizes both passive and active uptake mechanisms. Passive uptake involves processes like root absorption, stomatal penetration, and cuticular diffusion. Through root absorption, nanocarriers enter the plant via root hairs and epidermal cells, eventually reaching the vascular system for distribution throughout the plant. Stomatal penetration allows entry through leaf pores, leveraging transpiration to reach mesophyll cells, the primary sites of photosynthesis. Cuticular diffusion, though more challenging due to the waxy nature of the cuticle, can be achieved with surface-modified nanocarriers.

Active uptake mechanisms, on the other hand, rely on the plant's biological processes. Endocytosis involves the plant cell engulfing the nanocarrier, forming a vesicle that transports it into the cytoplasm. Clathrin-mediated endocytosis is a common pathway for this process. Plasmodesmata, the microscopic channels connecting plant cells, offer another avenue for active transport, allowing nanocarriers to move symplastically, bypassing cell walls and reaching distant tissues.

The efficacy of nanocarrier delivery is heavily influenced by their surface chemistry and functionalization. Modifying the nanoparticle surface with molecules like polyethylene glycol (PEG), chitosan, or ligands optimizes interaction with plant cell membranes. These modifications enhance adhesion, permeability, and cellular recognition, improving uptake and translocation. Furthermore, surface functionalization can protect the cargo from degradation and enable controlled or stimuli-responsive release, triggered by environmental cues like pH, light, or enzymes.

The range of chemical cargoes delivered by nanocarriers is vast, encompassing pesticides, herbicides, nutrients, plant growth regulators, genetic materials, and biostimulants. Nano-encapsulation of pesticides and herbicides allows for targeted pest control, minimizing environmental toxicity by preventing leaching and volatilization. Nutrient-loaded nanocarriers enhance plant nutrition by improving the solubility, stability, and uptake of essential macro and micronutrients. Delivering plant growth regulators and hormones via nanocarriers allows for precise modulation of physiological processes, improving crop yield and quality. Genetic materials, including DNA, RNA, and CRISPR-Cas components, can be efficiently delivered using nanoscale vectors, enabling advanced genetic engineering. Finally, biostimulants and stress tolerance enhancers encapsulated in nanocarriers promote plant resilience to various stressors.

The benefits of nanocarrier systems extend to precision agriculture and smart farming. Nanosensors can monitor plant health and nutrient status in real-time, facilitating data-driven decisions for optimized input application. Enhanced nutrient and pesticide use efficiency reduces waste and environmental contamination. Nanocarrier-mediated delivery of antimicrobials strengthens plant defense mechanisms, reducing reliance on chemical pesticides. Eco-friendly nanocarriers, like biodegradable polymers, minimize the ecological footprint of agricultural practices.

Despite the numerous advantages, concerns regarding the safety and environmental implications of nanocarrier applications remain. Phytotoxicity and biocompatibility assessments are crucial to understand nanoparticle impacts on plant physiology. The potential accumulation of nanomaterials in edible tissues raises concerns about human health risks. Robust regulatory frameworks and ethical guidelines are essential to ensure the safe and responsible deployment of nanotechnology in agriculture.

Future research focuses on developing bioengineered nanocarriers with enhanced targeting, biocompatibility, and controlled release. Integrating nanotechnology with artificial intelligence and sensor technology can revolutionize agricultural practices through automated and precision-controlled input delivery. Scaling up nanocarrier production for commercial agriculture requires advancements in cost-effective synthesis, large-scale formulation, and regulatory approvals.

In conclusion, nanocarriers represent a significant advancement in plant science, offering innovative solutions for targeted delivery and translocation of chemical cargoes. By improving efficiency, reducing environmental impact, and enhancing plant resilience, nanocarrier technology has the potential to transform modern agriculture towards a more sustainable and productive future.

8.2 Future Outlook for Nanotechnology in Plant Sciences

Nanotechnology is poised to revolutionize plant sciences, offering groundbreaking solutions for sustainable agriculture, improved crop productivity, and enhanced stress resilience. As research advances, the integration of nanotechnology into plant systems will lead to more precise, efficient, and environmentally friendly agricultural practices. The future outlook for nanotechnology in plant sciences encompasses key developments in bioengineered nanocarriers, integration with artificial intelligence (AI) and sensor technologies, and large-scale commercial applications, all of which hold the potential to transform global food security and sustainability.

One of the most promising directions in nanotechnology for plant sciences is the development of bioengineered nanocarriers with enhanced targeting and delivery capabilities. Future nanocarriers will be designed to exhibit high biocompatibility, ensuring minimal toxicity to plants, soil microorganisms, and humans. These bioengineered carriers will incorporate stimuli-responsive materials that release agrochemicals based on environmental conditions such as temperature, humidity, or soil nutrient levels. Smart nanocarriers that respond to plant biochemical signals, such as specific enzymes or pH changes, will enable ultra-precise delivery of fertilizers, pesticides, and growth regulators, reducing waste and maximizing effectiveness. Furthermore, advancements in bio-nanotechnology will allow for the use of plant-derived or biodegradable nanomaterials, ensuring sustainability and minimizing ecological impact.

The integration of nanotechnology with AI and sensor technologies will usher in a new era of precision agriculture. Nano-sensors embedded in plants or soil will provide real-time data on nutrient status, disease outbreaks, and environmental stressors. AI-powered analytics will process this data to optimize input application, reducing overuse of fertilizers and pesticides while enhancing crop yields. The development of nanoscale biosensors capable of detecting pathogens at early infection stages will improve plant disease management, allowing farmers to take preventive measures before significant damage occurs. Additionally, nanotechnology-driven smart farming systems will enable automated and remote-controlled agricultural practices, improving efficiency and reducing labor costs.

Scaling up nanotechnology for commercial agriculture remains a crucial challenge but also presents vast opportunities. Cost-effective production methods for nanocarriers must be developed to make the technology accessible to farmers worldwide. Governments and regulatory bodies will need to establish clear guidelines for the safe use of nanomaterials in agriculture, addressing concerns about toxicity, environmental persistence, and food safety. Collaboration between researchers, policymakers, and industry stakeholders will be essential in ensuring that nanotechnology-based agricultural solutions are both effective and widely adopted.

Looking ahead, nanotechnology has the potential to play a vital role in addressing global agricultural challenges such as climate change, soil degradation, and food security. By improving nutrient use efficiency, reducing dependency on chemical pesticides, and enhancing plant resilience to environmental stressors, nanotechnology can contribute to more sustainable and resilient farming systems. With continued research and innovation, nanotechnology will not only enhance the productivity of current agricultural practices but also pave the way for future advancements in plant sciences, ultimately shaping a more efficient, sustainable, and environmentally responsible agricultural landscape.

REFERENCES

Chen, H., & Yada, R. (2011). Nanotechnologies in agriculture: New tools for sustainable development. *Trends in Food Science & Technology*, *22*(11), 585–594. DOI: 10.1016/j.tifs.2011.09.004

Cheng, B.. (2015). Clathrin-mediated endocytosis is involved in nanoparticle uptake in plants. *Nature Nanotechnology*, *10*(9), 791–798. PMID: 26237346

De Oliveira, J. L., Campos, E. V. R., & Fraceto, L. F. (2019). Recent developments on nanopesticides and nanofertilizers: A path to the future of sustainable agriculture. *Nano Today*, *24*, 100190.

Ederli, L., Varone, L., & De Gara, L. (2021). Nanotechnology applications in plant nutrition and crop protection: Prospects and challenges. *Plant Physiology and Biochemistry*, *167*, 245–256.

Eichert, T.. (2008). Stomatal uptake of nanoparticle-based agrochemicals. *Plant Physiology*, *148*(1), 31–38.

Fernandez-Cornejo, J., Jans, S., & Smith, M. (2014). Issues in the economics of pesticide use in agriculture: A review of the empirical evidence. *Agricultural Economics*, *30*(2), 123–132.

Franci, G., Falanga, A., Galdiero, S., Palomba, L., Rai, M., Morelli, G., & Galdiero, M. (2015). Silver nanoparticles as potential antibacterial agents. *Molecules (Basel, Switzerland)*, *20*(5), 8856–8874. DOI: 10.3390/molecules20058856 PMID: 25993417

Ghormade, V., Deshpande, M. V., & Paknikar, K. M. (2011). Perspectives for nano-biotechnology enabled protection and nutrition of plants. *Biotechnology Advances*, *29*(6), 792–803. DOI: 10.1016/j.biotechadv.2011.06.007 PMID: 21729746

Heredia-Guerrero, J. A.. (2014). Plant cuticle penetration by nanomaterials. *Advanced Functional Materials*, *24*(18), 2805–2814.

Jayaraj, J., Megha, P., & Sreedev, P. (2019). Organochlorine pesticides, their toxic effects on living organisms and their fate in the environment. *Interdisciplinary Toxicology*, *9*(3-4), 90–100. DOI: 10.1515/intox-2016-0012 PMID: 28652852

Kah, M.. (2018). Foliar uptake of nanopesticides: Current state and future challenges. *Environmental Science. Nano*, *5*(1), 6–20.

Kah, M., Hofmann, T., & Nanosciences, A. (2018). Nanopesticides and nanofertilizers: Emerging contaminants or opportunities for risk mitigation? *Frontiers in Environmental Science*, *6*, 6.

Kah, M., Hofmann, T., & Nanosciences, A. (2018). Nanopesticides and nanofertilizers: Emerging contaminants or opportunities for risk mitigation? *Frontiers in Environmental Science*, *6*, 6.

Kah, M., Kookana, R. S., Gogos, A., & Bucheli, T. D. (2019). A critical evaluation of nanopesticides and nanofertilizers against their conventional analogues. *Nature Nanotechnology*, *14*(6), 532–540. DOI: 10.1038/s41565-019-0439-5 PMID: 31168071

Ke, P. C., Lin, S., Parakhonskiy, B., & Chirth, C. (2018). Nanoparticles for plant imaging and detection. *Current Opinion in Biotechnology*, *49*, 166–172.

Khan, M. A., Naeem, M., Akram, W., & Hussain, I. (2019). Impacts of pesticide exposure on human health and ecosystem. *Pesticide Science and Technology*, *17*(3), 312–328.

Kookana, R. S., Drechsel, P., Jamwal, P., & Vanderzalm, J. (2014). Urban wastewater: A valuable resource turned a challenge. *Environmental Science & Technology*, *48*(14), 7467–7475.

Kumar, D., Pathak, M., & Chaturvedi, A. (2019). Recent advances in nanoformulations for agrochemical delivery: Current status and future perspectives. *Journal of Nanotechnology Research*, *5*(3), 45–58.

Kumar, S., Nehra, M., Dilbaghi, N., Marrazza, G., Hassan, A. A., & Kim, K. H. (2019). Nano-based smart pesticide formulations: Emerging opportunities for agriculture. *Journal of Controlled Release*, *294*, 131–153. DOI: 10.1016/j.jconrel.2018.12.012 PMID: 30552953

Li, J.. (2021). Role of endocytosis in nanoparticle uptake in plants. *Journal of Nanobiotechnology*, *19*(1), 32–45. PMID: 33499885

Liu, Q.. (2009). Mechanisms of nanoparticle uptake and translocation in plants. *Environmental Toxicology and Chemistry*, *28*(5), 994–1002.

Liu, R., Lal, R., & Potapov, A. (2020). Nanomaterials for sustainable agricultural practices: An overview. *Environmental Chemistry Letters*, *18*, 523–530.

Lv, J., Christie, P., Zhang, S., & Lu, Y. (2019). Functionalized nanoparticles as advanced agrochemicals for sustainable agriculture: Current status and future trends. *Journal of Agricultural and Food Chemistry*, *67*(13), 3281–3291.

Ma, X.. (2010). Uptake and transport of engineered nanoparticles in plants. *Journal of Environmental Quality*, *39*(6), 1893–1903. PMID: 21284305

Ma, X., Geisler-Lee, J., Deng, Y., & Kolmakov, A. (2015). Interactions between engineered nanoparticles and plants: Phytotoxicity and plant uptake. *Environmental Science & Technology*, *49*(5), 2757–2771. PMID: 25664567

Nair, R., Varghese, S. H., Nair, B. G., Maekawa, T., Yoshida, Y., & Kumar, D. S. (2010). Nanoparticulate material delivery to plants. *Plant Science*, *179*(3), 154–163. DOI: 10.1016/j.plantsci.2010.04.012

Oberdörster, G., Oberdörster, E., & Oberdörster, J. (2005). Nanotoxicology: An emerging discipline evolving from studies of ultrafine particles. *Environmental Health Perspectives*, *113*(7), 823–839. DOI: 10.1289/ehp.7339 PMID: 16002369

Perez-de-Luque, A., & Rubiales, D. (2009). Nanotechnology for parasitic plant control. *Pest Management Science*, *65*(5), 540–545. DOI: 10.1002/ps.1732 PMID: 19255973

Raliya, R., Nair, R., Chavalmane, S., Wang, W. N., & Biswas, P. (2016). Mechanistic evaluation of translocation and physiological impact of titanium dioxide and zinc oxide nanoparticles on wheat. *Scientific Reports*, *6*, 22936.

Raliya, R., Nair, R., Chavalmane, S., Wang, W. N., & Biswas, P. (2017). Mechanistic evaluation of translocation and physiological impact of titanium dioxide and zinc oxide nanoparticles on the tomato (*Solanum lycopersicum L.*) plant. *Metallomics*, *9*(7), 964–978. PMID: 26463441

Rico, C. M.. (2011). Effect of nanoparticles on plant growth and development. *Journal of Nanoparticle Research*, *13*(11), 4001–4014.

Rico, C. M., Majumdar, S., Duarte-Gardea, M., Peralta-Videa, J. R., & Gardea-Torresdey, J. L. (2011). Interaction of nanoparticles with edible plants and their possible implications in the food chain. *Journal of Agricultural and Food Chemistry*, *59*(8), 3485–3498. DOI: 10.1021/jf104517j PMID: 21405020

Santos, A., Aguilar, M., Torres, F., & Ruiz, L. (2020). Advances in nanotechnology for agricultural sustainability: Challenges and future prospects. *Journal of Agricultural Science and Technology*, *22*(2), 89–103.

Schwab, F.. (2015). Nanoparticle interactions with plants: Uptake and mechanisms. *Journal of Agricultural and Food Chemistry*, *63*(14), 3839–3850.

Schwab, F., Zhai, G., Kern, M., Turner, A., Schnoor, J. L., & Wiesner, M. R. (2015). Barriers, pathways and processes for uptake, translocation and accumulation of nanomaterials in plants—Critical review. *Nanotoxicology*, *10*(3), 257–278. DOI: 10.3109/17435390.2015.1048326 PMID: 26067571

Shang, Y., Hasan, A., & Ahammed, G. (2022). Nanocarrier-mediated plant biostimulation and defense enhancement: New frontiers for agricultural sustainability. *Advances in Plant Biology*, *16*, 75–92.

Sujatha, P., Karunakaran, G., & Rajendran, S. (2022). Smart nanocarrier-based targeted delivery of pesticides: A review. *Environmental Nanotechnology, Monitoring & Management*, *17*, 100628.

Sun, D.. (2020). Foliar uptake and distribution of silica nanoparticles in plants. *Scientific Reports*, *10*(1), 12345. PMID: 32704070

Torney, F., Trewyn, B. G., Lin, V. S. Y., & Wang, K. (2007). Mesoporous silica nanoparticles deliver DNA and chemicals into plants. *Nature Nanotechnology*, *2*(5), 295–300. DOI: 10.1038/nnano.2007.108 PMID: 18654287

Torres, J. F., Rubilar, O., Tortella, G., Diez, M. C., & Carrión, O. (2020). Nanotechnology and plant biotechnology: Advances and challenges in agroindustry. *Plant Science*, *297*, 110503.

Tripathi, D. K., Shweta, , Singh, S., Singh, S., Pandey, R., Singh, V. P., Sharma, N. C., Prasad, S. M., Dubey, N. K., & Chauhan, D. K. (2017). An overview on manufactured nanoparticles in plants: Uptake, translocation, accumulation, phytotoxicity and biotransformation. *Plant Physiology and Biochemistry*, *110*, 2–12. DOI: 10.1016/j.plaphy.2016.07.030 PMID: 27601425

Wang, P.. (2016). Transport and fate of nanoparticles in plants: Current knowledge and future perspectives. *Environmental Science & Technology*, *50*(10), 6001–6014.

Wang, P., Lombi, E., Zhao, F. J., & Kopittke, P. M. (2021). Nanotechnology for plant nutrition: Opportunities and challenges. *Trends in Plant Science*, *26*(6), 590–600.

Wang, W., Raliya, R., Sahle-Demessie, E., & Biswas, P. (2016). Nanoparticle transport in plants: Current knowledge and future perspectives. *Environmental Science. Nano*, *3*(6), 1044–1060.

Wang, W., Raliya, R., Sahle-Demessie, E., & Biswas, P. (2016). Nanoparticle transport in plants: Current knowledge and future perspectives. *Environmental Science. Nano*, *3*(6), 1044–1060.

Wang, W. N., Tarafdar, J. C., & Biswas, P. (2016). Nanoparticle synthesis and their applications in agriculture. *Journal of Nanobiotechnology*, *14*(1), 40. PMID: 27278090

Xing, M., Liu, Y., Guo, Z., & Tang, Y. (2018). Engineered nanomaterials in sustainable agriculture: From design to applications. *Environmental Science & Technology*, *52*(12), 7361–7375.

Xing, R., Liu, S., Guo, Z., Yu, H., Wang, P., Li, C., & Li, P. (2018). Biofunctionalization of nanoparticles for plant imaging and targeted pesticide delivery: A review. *Journal of Agricultural and Food Chemistry*, *66*(28), 7066–7073.

Chapter 6
Nanocarriers for Delivering Biomolecules in Plants: Mechanisms and Applications

Junaid Shehzad
Quaid-i-Azam University, Islamabad, Pakistan

Sobia Bibi
Quaid-i-Azam University, Islamabad, Pakistan

Ilham Khan
Quaid-i-Azam University, Islamabad, Pakistan

Murtaza Hasan
The Islamia University of Bahawalpur, Pakistan

Ghazala Mustafa
https://orcid.org/0000-0003-1086-4462
Department of Plant Sciences, Quaid-i-Azam University, Islamabad, Pakistan

ABSTRACT

Delivering functional biomolecules, such as RNA, DNA, and proteins, into plants offers a valuable opportunity to manipulate plant traits and address environmental sustainability challenges. Existing methods for biomolecule delivery face significant hurdles. Techniques like gene guns and Agrobacterium-mediated transformation are often inefficient. Nanocarriers have emerged as a promising alternative, offering precise, efficient, and versatile delivery systems. They can bypass the cell wall of plants without causing damage and enable broader applications in plant genetic engineering and biotechnology. This chapter examines the types of nanocarriers used for DNA, RNA, and protein delivery, exploring their uptake and transport mechanisms in plants. Nanocarrier-based delivery, system has improved efficiency, targeted delivery, and controlled release, alongside potential limitations like toxicity. The chapter summarizes future directions for optimizing nanocarriers, focusing on advancing plant genetic engineering and promoting sustainable agricultural practices.

DOI: 10.4018/979-8-3693-9869-2.ch006

1. INTRODUCTION

Nanocarriers (NCs) have emerged as a new and promising technology for delivering biomolecules in plants to provide a new approach to enhancing plant biotechnological processes and plant genetic improvement. The application of biomolecules such as DNA, RNA, and protein in plant cells plays a crucial part in gene modification by allowing scientists to modify cellular processes, introduce advantageous traits, and improve plant adaptation to environmental stresses (Molla et al., 2021; Chung et al., 2021). Agrobacterium-mediated transformation (AMT), biolistic gene transfer, and electroporation have found broad applications in plant biotechnology for conventional gene delivery processes (Kant and Dasgupta, 2017; Men et al., 2020). Agroinfiltration has numerous limitations, such as low transformation in specific plants, expensive implementation costs, the potential for causing damage to tissues, and random integration of transgenes that induce genomic changes in an undesirable manner (Altpeter et al., 2016). To counter these limitations, technology-based delivery mechanisms have been found to be significant in research work. Various advantages of NCs include greater bioavailability, higher cellular permeability, target-specific localization, and low toxicity (Venugopal and Dively, 2017; Mora et al., 2022). Carriers in these situations transport biomolecules through biological interfaces to increase transformation efficiency and gene expression in plants. Inspired by their immense potential in mammalian drug delivery technology, NCs have found their way into plant biotechnology to provide a potential alternative to conventional methods of gene delivery (Su et al., 2023).

NCs can be categorized as inorganic or organic in origin. Inorganic NCs such as superparamagnetic iron oxide nanoparticles, mesoporous silica nanoparticles, and silver or gold nanoparticles have been explored for delivering biomolecules (Molla et al., 2021). They have not found much utility in plants due to their biocompatibility concerns as well as their potential for toxicity (Davies et al., 2013). Alternatively, organic carriers in the form of lipid-based nanocarriers such as liposomes, solid lipid nanoparticles, and niosomes have demonstrated higher biocompatibility, lower toxicity, and good loading of biomolecules (Kumar et al., 2013; Ullah et al., 2024). They increase gene stability, protect biomolecules from degradation, and facilitate sustained release in plants' cells. The uses of NC-delivery of biomolecules extend far beyond their potential in gene transformation. For instance, NCs have found utility in RNA interference (RNAi)-induced gene silencing by allowing specific regulation of gene expression in the absence of integration of stable DNA (Doudna and Charpentier, 2014). Additionally, the use of ribonucleoprotein complexes such as CRISPR-Cas9 in combination with NCs has presented a method for precise genome editing in plants in the absence of transgenes (Miller et al., 2021; Hamada et al., 2018). All these potential developments notwithstanding, much work has to be achieved in designing NCs to further improve their delivery efficiency, curtail off-target effects, and facilitate their delivered biomolecules' long-term stability (Zhang et al., 2015).

The chapter aims to explore mechanisms of and potential applications for NCs in plant biotechnology in their ability to transport DNA, RNA, and protein for genome editing and genetic transformation. Through the presentation of challenges of significance and suggestion of potential approaches to simplifying NC-based delivery systems, this chapter provides insight into future plant genetic modification and green agricultural practice.

2. BIOMOLECULE DELIVERY METHODS FOR GENETIC ENGINEERING

Delivery of different exogenous macromolecular materials, such as drugs, genetic materials, or peptides, into plant cells is crucial and also significant for many plant science objectives. Intracellular delivery ranges from basic research in biology to functional research for crop yield increase, crop improvement in quality, or disease or stress resistance. Intracellular delivery is still challenging to a greater extent in that the plant cell wall is thought to play the role of a hindrance to delivering biomolecules (Zhang et al., 2024). The DNA delivery to plant cells is of crucial significance in plant genetic engineering with far-reaching implications for basic plant biology as well as for plant biotechnology in agricultural objectives. The target of choice for plant transformation is the nuclear genome, while the organellar genomes in mitochondria and chloroplasts serve as targets for delivering beneficial agricultural characteristics as well as for research on inter-organelle communications. Polyethylene glycol mediated protoplast transfection (Wang et al., 2022), particle bombardment (Ozyigit and Yucebilgili, 2020), and Agrobacterium-mediated transformation (Florencio et al., 2018) are used widely for delivering DNA. Agrobacterium-mediated transformation involves Agrobacterium tumefaciens' infection, by depends on Agrobacterium tumefaciens' ability to insert its DNA (T-DNA) into plants' nuclear genome to play the leading method for transforming the nucleus (Ziemienowicz, 2014). Agrobacterium-mediated transformation cannot, however, apply to many plant crops, in particular monocotyledonous plants, nor to delivering DNA to chloroplasts or to delivering to mitochondria in their initial step of delivering DNA in their target cells or tissues. Recently, Agrobacterium-mediated transformation has found utility in an effective mutation of plastid genomes and of mitochondrial genomes through mediation by a TALEN DNA-binding domain linked to cytidine deaminase and to a plastid-transit signal, and nuclear encoded transcription activator like effector nuclease linked to a mitochondria-targeting signal and by respectively (Kazama et al., 2019; Nakazato et al., 2021). Gene insertion in organelle genomes is still extremely challenging. Gene bombardment by particle bombardment is yet another method that is compatible and effective for many plant crops. It facilitates the delivery of DNA to intact cells of plants by allowing for concurrent multi-gene transfers with no biological restrictions or host limitations (Mookkan, 2018). It involves physically accelerating micro-particles coated with DNA onto target tissues by high-pressure gas (Liu et al., 2019). The physical introduction of DNA by this method facilitates the species-independent introduction of DNA and transformation of the nucleus as well as of plastids but has low transformation proficiency (commonly less than 1% for plastids and around 1% for the nucleus), tissue damage, and transgene insertion at arbitrary positions in the plant genome that results in gene silencing (Ahmad et al., 2021). PEG or PEG derivatives have found immense utility as carriers of drugs or conjugated to protein, peptides, enzymes, or small molecule agents for the improvement of the proficiency of drug delivery and curative effects (Zou et al., 2024). PEG-mediated delivery involves protoplasts that need to be produced by the removal of the plant cell wall by enzymes. Only in a few plants has the regeneration of complete plants from the protoplasts been achieved.

While these approaches have found their way to transforming plants with DNA, several species have remained recalcitrant to stable transformation of their genomes. The transformation in the chloroplast genome has not yet been successfully attained in monocotyledons such as crop plants of economic significance, nor have stable mitochondria transformants ever been isolated in any species of plants (Maliga, 2022). Aside from these limitations of the approaches described above, tissue culture and regeneration steps offer severe difficulties in obtaining stable lines of transgenic plants (Squire et al., 2023). Induction of callus (a mass of unorganized cells), transformed tissue selection, followed by differentiation to yield mature plants, has not yet been developed in several crops. Direct manipulation of the germline genome by plant transformation is one of the possibilities to counter these limitations. Aside from the approach of floral dip with AMT in *Arabidopsis* buds of flowers, (Rao, 2019), in planta transformation has remained a low-throughput approach that has not been easy to replicate due to poor delivery of foreign genes to germlines by methods in practice so far.

3. NANOCARRIERS FOR DELIVERING BIOMOLECULES TO PLANTS

Nanoparticles are described as small-size particles with less than of 100 nm in diameters. Several amendment possibilities make them capable of delivering cargoes of varying materials like proteins, drugs, and nucleic acids within cells (Wang et al., 2017; Deng et al., 2014; Yin and Zhong, 2020). Transforms mediated by NP make it easy to modify plant genomes to include beneficial characteristics to improve crop yield and offer tolerance to varying stress levels (Yong et al., 2024). Over the past ten years, NC mediated delivery of DNA in plants has undergone swift developments and shown a potential to surpass the limitations of existing approaches. Particularly, recent research work has shown success in demonstrating NC-mediated DNA delivery to specific plant cell organelles like mitochondria and chloroplasts (Santana et al., 2020; Miyamoto et al., 2024), opening doors to plant organelle transformation. Additionally, modification of the germline genome has been achieved through the employment of NCs for modification of walled cells in plants in planta (Zhao et al., 2017; Wang et al., 2022). Although several NCs have succeeded in their development for the mammalian systems, their employment in plants has remained low, in all likelihood due to dissimilarity in both systems, characterized by the absence or presence of cellular walls. The cationic liposomes that have served as the gold standard for mammalian cells' transfection have found success in their employment in delivering DNA in protoplasts but have found poor performance in walled cells of plants (Miyamoto et al., 2024), proposing defensive capability of cellular walls to cells' invasion by foreign materials. NCs suitable for delivering DNA in plants are presented in the succeeding sections.

4. MECHANISMS OF NANOCARRIER UPTAKE AND TRANSPORT IN PLANTS

Plant absorption, or movement of materials to plants, is vital for plant growth and development. Plants need a variety of necessary nutrients, gases, and water for their survival and growth. Two of the primary processes through which materials get to plants include the assimilation of nutrients and water uptake. Plants need water as it makes several physiological processes in plants possible while serving as a solvent for nutrients (Ahanger et al., 2017). Aside from conventional nutrient and water uptake, plants have the potential to absorb other nanoparticles, in addition to other foreign materials, other than water, nutrients, and gas. Double-stranded RNA and other genetic materials fit in this category. Depending on their type and their use, their transport to plants has numerous effects on plants, ranging from beneficial to perhaps alarming. According to Khan et al. (2019), nanoparticles have their classification based on their size and structure. Chemical structure dictates the inherent optical, electrical, magnetic, and catalytic characteristics of nanoparticles (Dontsova et al., 2019). Owing to their unique electrical, optical, and catalytic properties, their effects on plant physiologies vary in kind (Dolai et al., 2020).

4.1 Pathways of Nanocarriers Entry

NCs enter plants through structures that are subterranean as well as aerial in nature. Most of these subterranean or aerial structures of cells that these NCs enter have varying physiological and morphological effects on plants. Plant variety, the growth cycle of plants, conditions of plant growth, method of application of these chemicals, dose of these chemicals, and their exposure time all affect these chemicals' effects on plants (Rizwan et al., 2017). NCs that enter through stomata get delivered by phloem in plants **(Figure 1)**. As per mass flow or pressure flow theory, NCs move by pressure gradient from roots to leaves (Yıldız et al., 2018). Once these nanoparticles are introduced to the plant cell, they move from cell to cell by plasmodesmata through the apoplast or the symplast (Usman et al., 2020). Pores in cell wall make it easy or difficult for these nanoparticles to move through it, or not at all. Although larger nanoparticles move through hydathodes and stomata (Hossain et al., 2016), these small nanoparticles move through the cell wall so easily (Zaheer et al., 2022). The ability of these nanoparticles to move through natural obstacles like the endodermis further affects their movement in plant tissues.

Figure 1. Nanocarriers' uptake, transport, and cell internalization mechanisms in plants

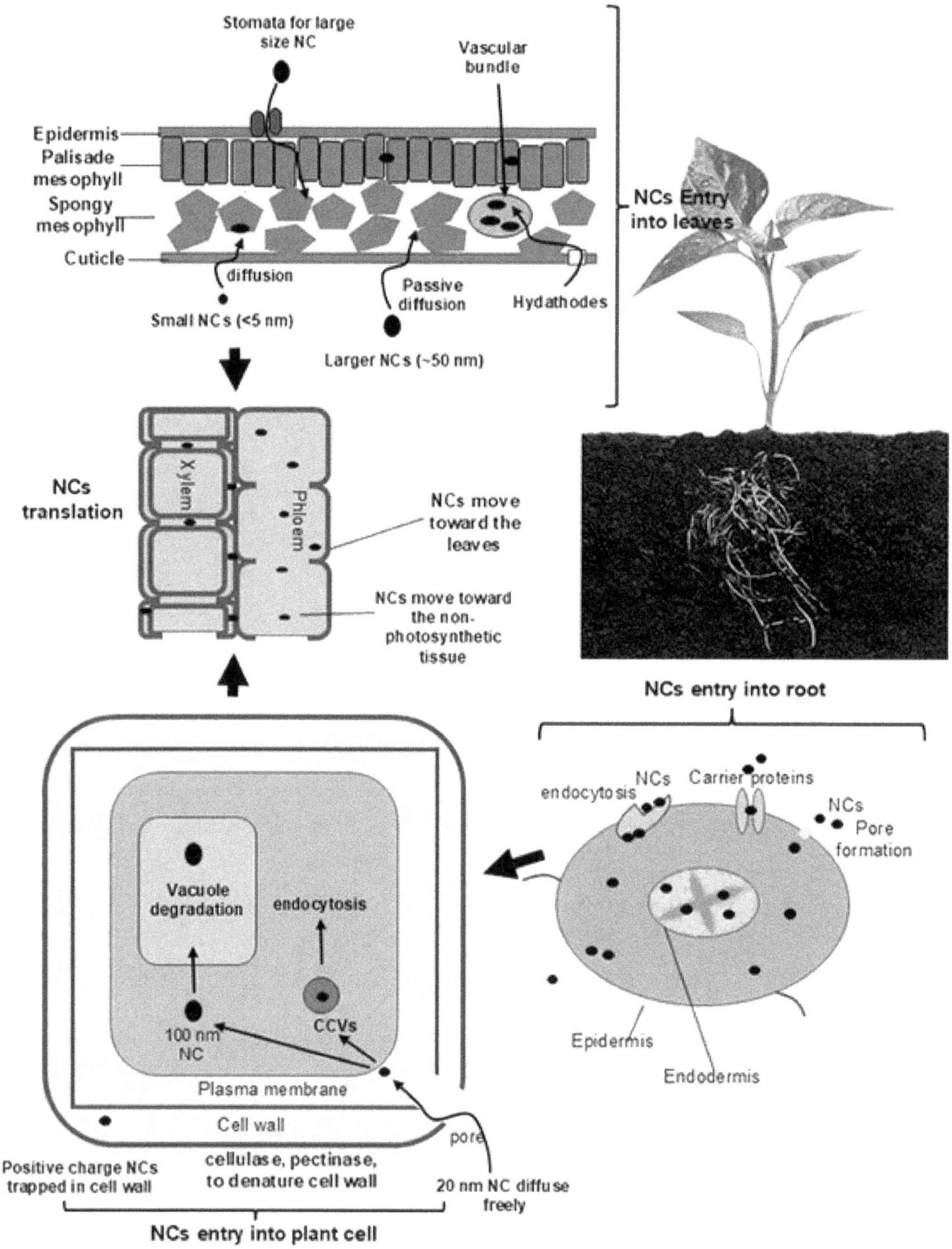

4.2 Intracellular Transport and Fate

After nanoparticles have penetrated plant cells, they take specific tracks within cells that determine their fates. Most of the nanoparticles start to concentrate in other cellular compartments, such as vacuoles and cytoplasm, while others get sequestered in certain tissues, such as roots. Studies have established that small nanoparticles have greater internalization efficiency than larger ones and possess greater penetrability in plants (Hossain et al., 2016; Ullah et al., 2023). After entering cells, nanoparticles may get incorporated into cellular metabolism or be channeled to degradation processes. Movement within cells in plants is regulated by electrostatic force attraction, lipid membrane binding, and intracellular transport networks. Endocytosis aids transport of nanoparticles to ascertain if they receive lysosome fusion for degradation or escape to the cytoplasm for further transport to other cells (Tripathi et al., 2017). Nanoparticles that escape degradation may reach cellular components such as nucleus, chloroplasts, and mitochondria, where they exert effects on cellular processes (Yan et al., 2022).

5. NANOCARRIERS FOR DNA DELIVERY

The NC mediated DNA delivery in plant genome has evolved at a quick rate in the last ten years and has demonstrated its capability to surpass the limitations of traditional approaches (Law et al., 2022). Notably, in novel research work, NC mediated DNA delivery to specific target organelles like chloroplasts and mitochondria has been shown to succeed (Santana et al., 2022; Miyamoto et al., 2022), opening doors to plant organelle transformation. Modification of the germline genome has been attained through modification by NCs in a direct manner (Wang et al., 2022; Zhao et al., 2017) which has the potential to make planta transformation of recalcitrant plant crops possible. Although numerous NCs have been reported to succeed in working in mammalian systems, their utility in plants has remained low, in all likelihood due to dissimilarity in both systems represented by the absence or presence of cell walls (Khan et al., 2022). For instance, cationic liposomes that have remained the gold standard for transfecting mammalian cells have found success in delivering DNA to protoplasts but have demonstrated low transfection in walled cells of plants (Miyamoto et al., 2020; Khan et al., 2022) implying defensive property of cell walls to repel foreign molecule entry in cells. NCs employed for delivering DNA in plants have been described in (**Figure 2**).

Figure 2. Nanocarriers mediated DNA delivery in plants. Delivery of cargo in plants consists of three major steps, cargo loading on nanocarriers, nanocarriers application on plants, and nanocarriers entry into plant cells

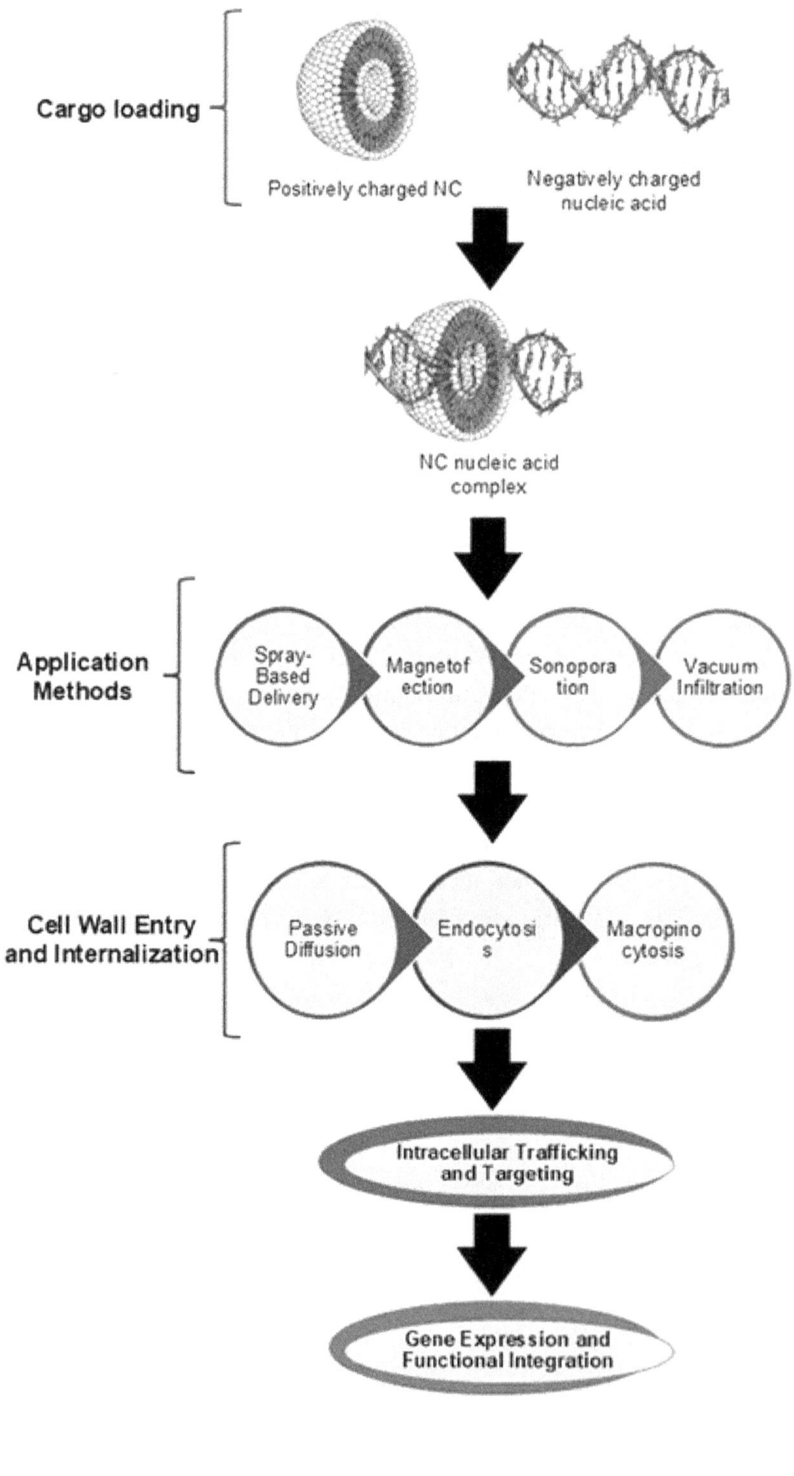

For the last few decades, several categories of nanoparticles have found their application in sustained release and controlled delivery of cargo DNA in carbon carriers (carbon dots, multi and single-walled carbon nanotubes), metal carriers (gold rods, gold clusters, gold nanosphere, iron oxide clusters), bio-inspired and silicon carriers (Cunningham et al., 2018). The carbon dots are microscopic carbon-based nanoparticles of size less than 10 nm. Carbon dots have three categories on a chemical structure basis (1) carbon nanodots or CNDs, (2) graphene quantum dots or GQDs, and (3) polymer dots or PDs. Carbon nanodots have two further subgroups: carbon quantum dots and carbon nanoparticles (Lopes et al., 2022). Owing to their excellent tensile strength and ultra-small size, carbon dots have found their utility as carriers for the transfer of genetic cargo, reported in plants capable of evading the plant cell wall. Carbon dots have found their attachment to cationic charge polymers by covalent bonding that aids in anchoring to exogenous DNA of negative charge for delivery to integral plants (Nayak et al., 2021). Carbon dots have found their utility as effective carriers for several exogenous nucleic acids like siRNA, plasmid DNA, and dsRNA (Schwartz et al., 2020; Martin-Ortigosa et al., 2014; Wang et al., 2020) to engineer plant organellar and nuclear genomes. The theoretical model of lipid exchange envelope penetration has been established recently, which predicts the target of NPs within a cell to reach the mitochondria, chloroplasts, or remain in the cytosol. The LEEP relies on sufficient surface charge and size of nanoparticles similar to that of the target plasma membrane to create a potential across the membrane, followed by diffusional weakening of the lipid bilayer, followed by pore formation to take in cargoes (Wong and Wright, 2016; Khan et al., 2022). Owing to their small size, Carbon dots were found to travel through the plant through vascular tissues. The carbon dots plasmid-coated method was employed to modify *Triticum aestivum* chloroplast and nuclear with success, confirmed by GFP-SPO11 vector fluorescence microscopy (Doyle et al., 2019). Carbon dots function through a combination of electrostatic interaction with plasmid DNA and efficient cellular uptake, likely via passive diffusion or stomatal penetration, aided by their small size (<10 nm). Once inside, plasmids may be released into the cytoplasm, where transcription and expression begin. This mechanism contrasts with other nanocarriers such as mesoporous silica nanoparticles, which require endosomal uptake and often need surface modification for targeting. In addition to gene manipulation research work, Carbon dots have found extensive utility in biomolecule sensing and the improvement of the nutritional status of plants through biofortification (Parra-Torrejón et al., 2023).

Carbon nano tubes are a tiny, high aspect ratio carbon nanomaterial and offer a new platform for the delivery of nucleic acid delivery plants (Squire et al., 2023). In 2009, it was confirmed that single-walled carbon nano tubes containing short single-stranded DNA can easily cross the plant cell membranes and cell walls to enter into cells (Liu et al., 2009). Similarly, Khodakovskaya et al. (2009) reported that multi-walled carbon nano tubes can easily enter the plant root and seed systems, facilitating the delivery of genetic material at early developmental stages. More recently, the PEI-conjugated carbon nano tubes have been used for transient transgene expression and delivery of DNA in numerous plant species, including *Eruca sativa*, *N. benthamiana*, *Gossypium hirsutum,* and *Triticum aestivum* (Demirer et al., 2019). Unlike MNPs and MSNs, carbon nano tubes can easily cross the plasma membranes and cell walls without the use of any external force in various plant species, because of their high aspect ratio. Additionally, by using the lipid exchange envelope penetration model (Lew et al., 2018), scientists have predicted the localization of nanocarrier based on their sizes and zeta potentials. Apart from this chitosan-conjugated carbon nano tubes have been designed to deliver DNA into the chloroplast. Despite the tremendous potential of carbon nano tubes as a carrier for plant genetic material, their use in plant species poses several challenges. For example, carbon nano tubes are highly cytotoxic within plant cells, due to their ability to generate reactive oxygen species and disrupt the biological membranes (Santana et al., 2022), which limits the applicable dose of carbon nano tubes and their cargo DNA. The selection of the polymer conjugated to DNA for binding is critical as it affects the biocompatibility and toxicity of carbon nano tubes (González-Grandío et al., 2021). Mechanisms of carbon nano tubes entry within plant cells and their organelles (e.g., in the chloroplast, and within the nucleus) are not yet clear (Serag et al., 2011). Although carbon nano tubes alone have been predicted to permeate through chloroplast envelope as well as through the plasma membranes by energy-independent direct permeation (Lew et al., 2018), their entry within carbon nano tubes/DNA complexes within cells and their organelles has to be further explored. Additionally, even if carbon nano tubes' complexation with DNA protects them from degradation by enzymes within cells, the release of DNA from carbon nano tubes at their target location is still challenging in most carbon nano tube designs. Overall, to realize the full potential of carbon nano tubes in permeating through cell walls as well as through their membranes, their concerns, like their biodegradability, cytotoxicity, release mechanisms of their cargo, and interactions with plant cells, need to be tackled by scientists to realize their full potential in plant genetic modification.

Biomimetic nanoparticles or bio-inspired nanoparticles are synthesized through biological materials such as viruses, bacteria, biomolecules, and plant extracts (Keum et al., 2022). Metal or organic structures of biNPs have biological materials coating compatible with an adequate magnitude of charge in parallel to the cellular membrane to offer adequate potential difference to transport cargoes to plant cells. Due to their adjustable characteristics, size, structure, shape, surface-marge moiety, and surface-marge biNPs are used as nucleic acid and drug carriers in delivering cargoes efficiently for research or therapeutic purposes. Closely vesicle-shaped structures of liposomes or vesicles have at least one lipid bilayer with an internal space or cavity in their structure. Cationic liposomes and polymeric-based nanoparticles, such as liposomes, transport molecular cargo such as siRNA and plasmid DNA. These liposomal nanocarriers can fuse with plant cell membranes or be internalized via endocytosis, facilitating the intracellular release of their molecular cargo. Similarly, polymeric-based nanoparticles engineered from biodegradable polymers can be tailored for controlled release and targeted delivery of genetic material. The biocompatibility, customizable surface chemistry, and structural versatility of liposomes and polymeric nanoparticles make them promising vehicles for non-invasive, efficient gene delivery in plants. (Cunningham et al., 2020). Varying structures of cargoes, such as plasmid DNA, and ribonucleoprotein complexes of RNP complexes are delivered by these lipoplexes. Efficient gene editing in tobacco plants and citrus plants was demonstrated by free lipofectamine-grafted Cas9 ribonucleoprotein delivery. In this approach, lipofectamine a cationic lipid-based transfection reagent, facilitates the direct transport of preassembled Cas9 protein and guide RNA into plant cells without the need for DNA vectors. This DNA-free editing strategy minimizes the risk of off-target integration and has shown promising results in achieving targeted genome modifications in dicot plant species (Mahmoud et al., 2022; Liu et al., 2020). Polymeric nanoparticles are solid-shaped nanoparticles made up of macromolecular polymers ranging from diameters of 10 to 1,000 in their size. Depending on their structure, these polymer nanoparticles have been categorized into two classes, namely naturally derived biopolymer materials or biopolymer materials of natural origin, or branched dendrimers, or dendrimers having a branched structure synthesized from chemical materials.

As a novel example of meal-based nanoparticle-facilitated DNA delivery in plants, positively-charged polyethyleneimine coated iron oxide nanoparticles have found application in direct pollen transformation (Zhao et al., 2017). Positively-charged PEI was used to bind to negatively-charged DNA to create an ionic MNP/DNA complex that was guided by an external magnetic field to introduce through pollen through surface openings of diameters ranging from 5 to 10 μm to induce transformation. The method was referred to as magnetoreception and was demonstrated to not only induce transient expression of the transgene but also to integrate the transgene into the genome in the pollen of several dicot and monocot plants. Importantly, magnetometer pollens of cotton gave rise to transformed seeds by way of pollination, to give way to the successful generation of transformed plants from such seeds. The pollen magnetoreception is a potential method of choice for genetic transformation in that it has the potential to sidestep intricate plant generation from tissue culture. Reproducibility of the method is controversial in monocot plants in particular (Vejlupkova et al., 2020), in that MNP-pollen interactions need to be further validated, as well as the release of DNA from NMP to take place before integration to the genome occurs.

6. NANOCARRIERS FOR RNA DELIVERY

Apart from DNA delivery, scientists are considering the RNA delivery system a unique and alternative opportunity for improvements of plant traits (**Table 1**). Specifically, the RNA interference system enables the precise silencing of a gene of interest by using small interfering RNA in plant cells (Das and Sherif, 2020). This process can be started by targeted cleavage of RNA molecules by using an RNA-induced silencing complex. In the last few decades, NCs-mediated RNA delivery has gained attention. This method comprises the direct delivery of two different exogenous RNAi effectors, including dsRNA and siRNA, by NCs. This procedure is potentially applicable to an extensive plant species without any risk of random DNA insertion into the genome. While the NCs used for DNA delivery, such as CPPs and CNTs, are also applicable for RNA delivery. Additionally, numerous other types of NCs have been reported for the delivery of RNAi effectors, facilitating gene silencing in plant cells. While the NCs-mediated RNA delivery has been extensively reported in animals, its potential for plant cells still remains understudied (Barampuram and Zhang, 2011). Numerous previous experiments have taken advantage of NCs to deliver RNA, proteins (Chen et al., 2021), or plasmid DNA (Lang et al., 2021) to intact plant cells. Polymeric nanoparticles have shown potential for siRNA delivery to plant protoplasts without cell walls, but polymeric nanoparticles have not shown any evidence to traverse the cell wall for gene silencing in integral plant cells containing cell walls (Chen et al., 2021). However, a recent study has reported that clay nanosheets can enable the pathogen-specific double-stranded RNA delivery into intact plant cells for virus resistance (Bigeard et al., 2015). These positively charged clay nanosheets electrostatically bind with negatively charged dsRNA, forming stable nanocomplexes that protect the RNA from environmental degradation and washing. Topical application of clay nanosheets allowed the silencing of homologous RNA to increase the viral protection on the leaf surface. Therefore, the use of clay nanosheets is a promising use of nanoparticles for RNAi delivery into plant cells, providing a way toward future developments in the plant bionanotechnology field. Mitter et al. (2017) reported that double-strand RNA can be loaded in LDH clays, which are protected from the activity of nucleases and available in a sustained manner for nearly 30 days after spraying. The dsRNA uptake in plant cells was monitored by using fluorophore-labeled dsRNA after application onto *Arabidopsis* leaves for 48 hours. The absorption in xylem vascular tissue was observed for all treated leaves and even in untreated apical leaves. The application of a β-glucuronidase reporter system in a transgenic *Arabidopsis* strain enabled the detection of decreased GUS activity in seedlings treated with GUS-dsRNA-LDH or GUS-dsRNA. This indicated that dsRNA was released from the clay nanosheets within the plant, leading to post-transcriptional negative regulation of GUS expression. The incorporation of dsRNA into the layered double hydroxide nanocomposite effectively boosted the resistance against viruses in the assays conducted, resulting in a significant reduction in the overall number of necrotic lesions on the leaf surfaces of *N. tabacum* and *V. unguiculata.* Apart from this, its non-toxic nature, ease of use, and compatibility with conventional spraying methods make it a compelling platform for field-level deployment of RNAi-based solutions. Similarly, Worrall et al. (2019) used layered double hydroxide nanoparticles loaded with double-strand RNA directed to the coating protein-coding region that provides effective infection protection in plants against the bean mosaic virus. According to their research, *N. benthamiana* treated with layered double hydroxide containing dsRNA was not affected by BCMV, while 13 out of 24 plants that did not receive the dsRNA treatment tested positive for BCMV. Additionally, LDH-dsRNA formulations provided protective effects in cowpea (*Vigna unguiculata*) plants against aphid-transmitted viruses, underscoring the potential of this delivery system for combating both direct and vector-mediated infections. Demirer et al. (2020) showed the application of SWNTs for delivering RNAi into the cells of *N. benthamiana* leaves, intending to silence green fluorescent protein. The goal was to silence the expression of green fluorescent protein, which was successfully achieved following a 6-hour post-infiltration period. The SWNT-siRNA complexes were passively absorbed into the leaf cells, resulting in a ~95% reduction in GFP transcript levels, effectively achieving gene knockdown. This study highlights the capacity of SWNTs to facilitate efficient RNA delivery without genetic modification or cellular damage, offering a powerful tool for transient gene silencing in plant biotechnology.

7. NANOCARRIERS FOR PROTEIN DELIVERY

The NCs mediated proteins delivery into plant cells still remains a challenge with few manageable solutions. The recent advances in nanomaterial biosynthesis have improved the biomacromolecule delivery method into the cell. For effective delivery of proteins into a plant cell, the nanocarrier must bypass the two main key barriers: the plasma membrane and plant cell wall. Plant cells are protected by a complex network of biopolymers that are connected to form a multilayer cell wall that restricts the biomolecules access to the cell membrane (Sun et al., 2022). The researchers have studied the cell wall size exclusion limit by using a number of methods including electron microscopy (Jiang et al., 2018), uptake of dye-labeled nanoscale materials (Babi et al., 2022), and gas adsorption (Shi and Avramidis, 2019). Some studies have reported the uptake of some large nanomaterials up to 100 nm by plant cell walls (Karny et al., 2018), most NCs pass the cell wall by diffusion method but many researchers also proposed other possibilities, including biotransformation and tissue infiltration (Mad'Atari and Folta, 2019). After passing the cell wall, NCs containing protein cargo enter the symplast by using a variety of mechanisms, including plasmodesmata, or endocytosis. However, the most reported mechanism for NCs cellular uptake is endocytosis, especially the clathrin-mediated endocytosis in plant cells. After entering the cell by endocytosis the delivered protein materials must escape from endosomes. Without the endosome, the endocytosed materials are transferred into different cellular organelles, such as the lysosomes and central vacuole, where they are consequently destroyed (Contento and Bassham, 2012).

After seeing the barriers to protein within the plant cells, we can suggest the necessities for designing an efficient method for protein delivery. Firstly, the size of the cargo protein and carriers should be small enough to cross the cell wall barrier, which limits the choice of NC. Secondly, the protein carrier must have such a structure that enables it to easily cross the cell wall. Many NCs, such as quantum dots and single-walled carbon nanotubes have structures that elicit endocytosis (Yaron et al., 2011). Apart from this, many NCs have pro-endosomolytic moieties that protect them from degradation after entering into the cytoplasm. Plants also contain a third barrier that was fully elaborated: the cuticle can be bypassed in the laboratory setting by applying a carrier or biomolecule solution. Therefore, many variables determine the efficient delivery of protein cargo by NCs.

Different NCs, including cell-penetrating peptides, macropinocytosis-Inducing peptides, and polyion complex vesicles, have been used for delivery of protein in plants. Among these cell-penetrating peptides are promising tools for protein delivery in plants. Earlier, Chang et al. (2005) reported the successful delivery of cargo fluorescent proteins into tomato roots and onion epidermis by conjugating arginine-rich cell-penetrating peptides to the proteins. These cell-penetrating peptides were covalently conjugated to the protein cargo, allowing passive translocation across the plant cell membrane without requiring physical disruption or external energy input. After that, Numata (2015) developed synthetic peptides consisting of a polycationic peptide and a cell-penetrating peptide which enabled the successful delivery of cargo proteins without using protoplasts. However, these NCs are only suitable for proteins with a negative charge and if the protein size is too large, they are trapped within the cell walls. The recent advancement in the field of NCs involves the use of a synthetic peptide called dTat-Sar-EED4, which contains three components: a hydrophobic endosomolytic domain, hexameric sarcosine linker, and a d-arginine-rich domain. The dTat-Sar-EED4 peptide exhibits exceptional efficiency for the delivery of different enzymes, such as tetrameric enzyme and β-glucuronidase in various plant systems. The peptide-based NC system has revolutionized protein delivery in plants but it also has some limitations.

Table 1. Different types of nanocarriers used for transport of biomolecules in plant cell

Cargo	Composition	Type of nanocarrier	Mechanism of delivery	Target Plant	Advantages	References
DNA	2,3-epoxypropyltrimethylammonium chloride	Plant derived biopolymers cellulose nanocrystals	Leaf injection and vacuum infiltration	Nicotiana tabacum	Increases the speed of the transformation	Vinzant et al., 2025
	Polycationic biopolymer, poly(allylamine) hydrochloride coated	(TMGMV)-like nanocarriers	Energy independent mechanisms	Arabidopsis protoplasts	Facile and biocompatible tool for advancing genetic engineering	Islam et al., 2024
	Contain various biocompatible and biodegradable polymers	Polymeric high aspect ratio nanocarriers	Endocytosis	Nicotiana benthamiana	Facilitate soft-material delivery	Zhang et al., 2024
	Silica	Mesoporous silica nanoparticles	Spray and injection	Solanum lycopersicum	Biocompatible	Hajiahmadi et al., 2019
	LDH-lactate nanosheets	Layered double hydroxide lactate nanosheets	Co-incubation	Arabidopsis thaliana	Novel gene carrier without any cytological effects	Bao et al., 2016
	Silica	Mesoporous silica nanoparticles	Endocytosis	Arabidopsis thaliana roots	Stable, High transfection efficiency	Chang et al., 2013

continued on following page

Table 1. Continued

Cargo	Composition	Type of nanocarrier	Mechanism of delivery	Target Plant	Advantages	References
RNA	Layered double hydroxide	Clay nanoparticles	Infiltration	Arabidopsis thaliana	Enhance internalization of nucleic acids	Yong et al., 2022
	Layered double hydroxide	Clay nanoparticles	Energy dependent and energy independent manners	Solanum lycopersicum pollen	Effective nonviral vector for effective delivery	Yong et al., 2021
	Layered double hydroxide	Colloidal LDH nanosheets	Leaves spray, Petiole adsorption; and root dipping	Solanum lycopersicum	Biodegradable nano-delivery systems with high specificity and sustainability	Mosa and Youssef, 2021
	LDH nanosheets	Clay nanosheet	Spraying plant leaves with an atomizer	Solanum lycopersicum	Sustainable and promising tool for begomovirus control	Liu et al., 2020
	Layered double hydroxide	BioClay	Topical application, spraying	Nicotiana tabacum	Biodegradable & Eco-Friendly	Mitter et al., 2017
Protein	Biodegradable Polymers	Polymeric nanocarriers	Spray	Nicotiana benthamiana	Autonomous Cell Wall Penetration	Zhang et al., 2024
	Cationic lipid-based transfection reagent	Lipofectamine 3000	Liposome transfection	Nicotiana tabacum	High Transfection Efficiency	Liu et al., 2020

8. ADVANTAGES OF NANOCARRIERS

A novel approach utilizing nanoparticle-mediated gene transformation can effectively bypass the plant cell wall barrier, enabling precise DNA or RNA delivery into plant cells for either transient or stable transformation. Nanotechnology presents an innovative method that has the potential to revolutionize traditional crop improvement and significantly contribute to the advancement of modern agriculture. As a result, nanoparticles could offer entirely new strategies for plant biotechnology and genetic engineering. A recent technological breakthrough in plant science is genetic engineering, which serves as an important tool for enhancing crop quality and yield, increasing secondary metabolite production in medicinal plants, and for sustainable agriculture growth (Lv et al., 2020). Recently, various nanoparticles have been explored as carriers for delivering biomolecules into intact plant cells. For example, carbon-based nanomaterials, layered double hydroxides, DNA nanostructures, gold clusters, and mesoporous silica nanoparticles have been identified for the delivery of RNA and plasmid DNA into plant tissues to modify the plant genome and gene expression (Zhan et al., 2020). Among nanoparticle carriers, DNA-based nanostructures stand out due to their non-toxic nature, high efficiency, and easy metabolism in plants, microorganisms, and animals (Li et al., 2019). They also exhibit remarkable biological stability, resisting degradation by nucleases (Kim et al., 2020). Furthermore, these nanostructures are highly compatible with target tissues and do not trigger an immune response (Zhan et al., 2020). Their flexible, modular, and straightforward application in plant cells makes them a valuable tool for genetic engineering. Furthermore, peptide-based NCs provide additional advantages for plant applications. Peptides function as highly effective cell-penetrating molecules capable of transporting biomolecular cargoes into the cytoplasm of walled plant cells. These peptides facilitate rapid and efficient intracellular delivery, even in the presence of the rigid plant cell wall (Lv et al., 2020). Similarly, single-walled carbon nanotubes serve as powerful gene-delivery vehicles, capable of penetrating organelle membranes. Their high aspect ratio, biocompatibility, exceptional tensile strength, and large surface area-to-volume ratio make them particularly beneficial for plant genetic engineering (Kwak et al., 2019; Mohanta et al., 2019; Hendler-Neumark and Bisker, 2019). Moreover, recent advancements in molecular payload (MP) technology have streamlined genetic engineering by enabling precise and efficient delivery of plasmids, RNA, and ribonucleoproteins. In mammalian cells, nanoparticles are widely used for direct nuclear and cytosolic delivery of Cas–RNP complexes (Mout et al., 2017). In plants, nanoparticle uptake depends largely on cell wall pore diameter, which varies across tissues and organs. Since the plant cell wall's exclusion size limit is generally up to 50 nm, nanoparticles offer a promising means for gene delivery due to their small size (Cunningham et al., 2018). Overall, this advanced nanotechnology presents an environmentally sustainable approach to enhancing plant traits and improving crop quality without relying on conventional genetic modification techniques.

9. DISADVANTAGES OF NANOCARRIERS

Along with their benefits, NCs have drawbacks. Low encapsulation efficiency, poor water solubility, and poor stability are some of the issues that currently plague zein-based delivery NCs (Elzoghby et al., 2017). When they enter plant cells, some nanoparticles, like specific metal and metal oxide nanoparticles, can produce ROS. Increased ROS levels can harm plant cells or even kill them by causing oxidative stress, damaging cell structures, and interfering with cellular processes (Rastogi et al., 2017; Shehzad and Mustafa, 2023). Plant cell membrane integrity may be compromised by high quantities of nanoparticles (Wang et al., 2013). Cell viability may suffer as a result of the leakage of cellular contents. According to Rossi et al. (2028), NPs may potentially affect plant symplastic processes by changing the integrity of the root cell membrane or ion transport activities.

The uptake and distribution of vital nutrients can be hampered by excessive nanoparticle buildup, which can upset the nutritional balance in plants and result in nutrient deficits or toxicities. The plant's water and nutrient transport systems may also be physically obstructed by high nanoparticle concentrations, which would hinder the flow of materials throughout the plant (Hong et al., 2021). High concentrations of nanoparticles may cause stress reactions in plants, taking resources away from growth and development and lowering the health of the plant as a whole (Rani et al., 2022). The type of nanoparticle, its size, shape, surface characteristics, and the plant species involved are some of the variables that can affect its phytotoxicity. Because it can greatly affect the efficiency and efficacy of nanoparticle transport, choosing the right concentration of nanoparticles is therefore crucial.

Nanoparticles, whether engineered or naturally occurring, can have varying negative effects on plants. The nanoparticles impact on plants depends on various factors such as their concentration, type of nanoparticles, specific plant species, and duration of exposure (Rani et al., 2022). Nanoparticles can show phytotoxic effects, leading to damage being caused to the plant organ tissues, and cells which can result in reduced biomass, stunted growth, and compromised plant health (Yan et al., 2019). Similarly, exposure to certain types of nanoparticles can alter the root morphology and ultra structure. This includes changes in the root hair structure, root length, and surface area, which can impact water and nutrient uptake (Rani et al., 2022). By producing ROS, nanoparticles can cause oxidative pressure in plants. Plant health may be affected by elevated reactive oxygen species levels that cause damage to cellular components like membranes, DNA, and proteins (Kralova et al., 2021; Shehzad and Mustafa, 2023). Plants may be exposed to certain nanoparticles for an extended period if they remain in the environment. Furthermore, there is worry that nanoparticles may bioaccumulate in plant tissues, which could have an impact on creatures further up the food chain (Abbas et al., 2020). NCs in plants face challenges like low encapsulation efficiency, poor stability, and potential toxicity, leading to oxidative stress, nutrient imbalance, and impaired growth. Their phytotoxic effects depend on nanoparticle type, concentration, and exposure, potentially disrupting cellular functions and bioaccumulating in plant tissues.

10. CONCLUSION

Recent developments in NCs mediated biomolecule delivery techniques have significantly improved plant genetic engineering. Traditional methods for genetic engineering like biolistic particle bombardment, Agrobacterium-mediated transformation, and PEG-mediated delivery have some limitations. These techniques often have varying levels of efficiency due to potential tissue damage and dependency on species. These limitations highlight the need for more efficient and versatile delivery systems in plants, especially for genetic modifications. NC based delivery is an alternative and promising tool to address the limitations of conventional methods. These NCs can easily overcome different barriers in plant cells such as the plasma membrane and cell wall to enhance the uptake of biomolecules. The versatility of NCs enhanced the compatibility with a wide range of plant species, facilitating both stable and transient genetic transformations. Mostly, a stable complex is formed between NCs and cargo, but it is critical for NCs to release the cargo at the premeditated destination to ensure its proper function. Extensive studies have been performed to control the release of biomolecules from NCs in plant cells by exploring temperature, pH, and light. However, attaining controlled release of biomolecules such as proteins, RNA and DNA still remains a challenge. As the RNP, proteins, and DNA delivery to plant cells hold substantial potential for numerous future applications, therefore the development of efficient release systems for these biomolecular cargoes is crucial.

REFERENCES

Abbas, Q., Yousaf, B., Ullah, H., Ali, M. U., Ok, Y. S., & Rinklebe, J. (2020). Environmental transformation and nano-toxicity of engineered nano-particles (ENPs) in aquatic and terrestrial organisms. *Critical Reviews in Environmental Science and Technology*, *50*(23), 2523–2581. DOI: 10.1080/10643389.2019.1705721

Ahanger, M. A., & Agarwal, R. M. (2017). Salinity stress induced alterations in antioxidant metabolism and nitrogen assimilation in wheat (Triticum aestivum L) as influenced by potassium supplementation. *Plant Physiology and Biochemistry*, *115*, 449–460. DOI: 10.1016/j.plaphy.2017.04.017 PMID: 28478373

Ahmad, R., Srivastava, S., Ghosh, S., & Khare, S. K. (2021). Phytochemical delivery through nanocarriers: A review. *Colloids and Surfaces. B, Biointerfaces*, *197*, 111389. DOI: 10.1016/j.colsurfb.2020.111389 PMID: 33075659

Altpeter, F., & James, V. A. (2005). Genetic transformation of turf-type bahiagrass (Paspalum notatum Flugge) by biolistic gene transfer. *International Turfgrass Society Research Journal*, *10*, 485–489.

Altpeter, F., Springer, N. M., Bartley, L. E., Blechl, A. E., Brutnell, T. P., Citovsky, V., & Stewart, C. N.Jr. (2016). Advancing crop transformation in the era of genome editing. *The Plant Cell*, *28*(7), 1510–1520. DOI: 10.1105/tpc.16.00196 PMID: 27335450

Babi, M., Fatona, A., Li, X., Cerson, C., Jarvis, V. M., Abitbol, T., & Moran-Mirabal, J. M. (2022). Efficient labeling of nanocellulose for high-resolution fluorescence microscopy applications. *Biomacromolecules*, *23*(5), 1981–1994. DOI: 10.1021/acs.biomac.1c01698 PMID: 35442640

Bao, W., Wang, J., Wang, Q., O'Hare, D., & Wan, Y. (2016). Layered double hydroxide nanotransporter for molecule delivery to intact plant cells. *Scientific Reports*, *6*(1), 26738. DOI: 10.1038/srep26738 PMID: 27221055

Barampuram, S., & Zhang, Z. J. (2011). Recent advances in plant transformation. *Plant chromosome engineering: methods and protocols*, 1-35.

Bigeard, J., Colcombet, J., & Hirt, H. (2015). Signaling mechanisms in pattern-triggered immunity (PTI). *Molecular Plant*, *8*(4), 521–539. DOI: 10.1016/j.molp.2014.12.022 PMID: 25744358

Chang, F. P., Kuang, L. Y., Huang, C. A., Jane, W. N., Hung, Y., Hsing, Y. I. C., & Mou, C. Y. (2013). A simple plant gene delivery system using mesoporous silica nanoparticles as carriers. *Journal of Materials Chemistry. B*, *1*(39), 5279–5287. DOI: 10.1039/c3tb20529k PMID: 32263331

Chang, M., Chou, J. C., & Lee, H. J. (2005). Cellular internalization of fluorescent proteins via arginine-rich intracellular delivery peptide in plant cells. *Plant & Cell Physiology*, *46*(3), 482–488. DOI: 10.1093/pcp/pci046 PMID: 15695452

Chen, J., Zhang, X., Millican, R., Sherwood, J., Martin, S., Jo, H., & Jun, H. W. (2021). Recent advances in nanomaterials for therapy and diagnosis for atherosclerosis. *Advanced Drug Delivery Reviews*, *170*, 142–199. DOI: 10.1016/j.addr.2021.01.005 PMID: 33428994

Chung, S. A., Langford, C. A., Maz, M., Abril, A., Gorelik, M., Guyatt, G., & Mustafa, R. A. (2021). 2021 American College of Rheumatology/Vasculitis Foundation guideline for the management of antineutrophil cytoplasmic antibody–associated vasculitis. *Arthritis & Rheumatology (Hoboken, N.J.)*, *73*(8), 1366–1383. DOI: 10.1002/art.41773 PMID: 34235894

Contento, A. L., & Bassham, D. C. (2012). Structure and function of endosomes in plant cells. *Journal of Cell Science*, *125*(15), 3511–3518. DOI: 10.1242/jcs.093559 PMID: 22935651

Cunningham, A. J., Gibson, V. P., Banquy, X., Zhu, X. X., & Jeanne, L. C. (2020). Cholic acid-based mixed micelles as siRNA delivery agents for gene therapy. *International Journal of Pharmaceutics*, *578*, 119078. DOI: 10.1016/j.ijpharm.2020.119078 PMID: 31988037

Cunningham, F. J., Goh, N. S., Demirer, G. S., Matos, J. L., & Landry, M. P. (2018). Nanoparticle-mediated delivery towards advancing plant genetic engineering. *Trends in Biotechnology*, *36*(9), 882–897. DOI: 10.1016/j.tibtech.2018.03.009 PMID: 29703583

Das, P. R., & Sherif, S. M. (2020). Application of exogenous dsRNAs-induced RNAi in agriculture: Challenges and triumphs. *Frontiers in Plant Science*, *11*, 946. DOI: 10.3389/fpls.2020.00946 PMID: 32670336

Davies, M. J., Badcock, T. J., Dawson, P., Kappers, M. J., Oliver, R. A., & Humphreys, C. J. (2013). High excitation carrier density recombination dynamics of InGaN/GaN quantum well structures: Possible relevance to efficiency droop. *Applied Physics Letters*, *102*(2), 022106. DOI: 10.1063/1.4781398

Demirer, G. S., Zhang, H., Goh, N. S., Pinals, R. L., Chang, R., & Landry, M. P. (2020). Carbon nanocarriers deliver siRNA to intact plant cells for efficient gene knockdown. *Science Advances*, *6*(26), eaaz0495. DOI: 10.1126/sciadv.aaz0495 PMID: 32637592

Demirer, G. S., Zhang, H., Matos, J. L., Goh, N. S., Cunningham, F. J., Sung, Y., & Landry, M. P. (2019). High aspect ratio nanomaterials enable delivery of functional genetic material without DNA integration in mature plants. *Nature Nanotechnology*, *14*(5), 456–464. DOI: 10.1038/s41565-019-0382-5 PMID: 30804481

Deng, X., Cao, M., Zhang, J., Hu, K., Yin, Z., Zhou, Z., & Zeng, Y. (2014). Hyaluronic acid-chitosan nanoparticles for co-delivery of MiR-34a and doxorubicin in therapy against triple negative breast cancer. *Biomaterials*, *35*(14), 4333–4344. DOI: 10.1016/j.biomaterials.2014.02.006 PMID: 24565525

Dolai, S., Maiti, P., Ghorai, A., Bhunia, R., Paul, P. K., & Ghosh, D. (2020). Exfoliated molybdenum disulfide-wrapped CdS nanoparticles as a nano-heterojunction for photo-electrochemical water splitting. *ACS Applied Materials & Interfaces*, *13*(1), 438–448. DOI: 10.1021/acsami.0c16972 PMID: 33356109

Doudna, J. A., & Charpentier, E. (2014). The new frontier of genome engineering with CRISPR-Cas9. *Science*, *346*(6213), 1258096. DOI: 10.1126/science.1258096 PMID: 25430774

Doyle, C., Higginbottom, K., Swift, T. A., Winfield, M., Bellas, C., Benito-Alifonso, D., & Whitney, H. M. (2019). A simple method for spray-on gene editing in planta. *BioRxiv*, 805036.

Elzoghby, A., Freag, M., Mamdouh, H., & Elkhodairy, K. (2017). Zein-based nanocarriers as potential natural alternatives for drug and gene delivery: Focus on cancer therapy. *Current Pharmaceutical Design*, *23*(35), 5261–5271. PMID: 28641543

Florencio, C. S., Brandão, F. A. S., de Mello Teixeira, M., Bocca, A. L., Felipe, M. S. S., Vicente, V. A., & Fernandes, L. (2018). Genetic manipulation of Fonsecaea pedrosoi using particles bombardment and Agrobacterium mediated transformation. *Microbiological Research*, *207*, 269–279. DOI: 10.1016/j.micres.2018.01.001 PMID: 29458863

González-Grandío, E., Demirer, G. S., Jackson, C. T., Yang, D., Ebert, S., Molawi, K., & Landry, M. P. (2021). Carbon nanotube biocompatibility in plants is determined by their surface chemistry. *Journal of Nanobiotechnology*, *19*(1), 1–15. DOI: 10.1186/s12951-021-01178-8 PMID: 34930290

Hajiahmadi, Z., Shirzadian-Khorramabad, R., Kazemzad, M., & Sohani, M. M. (2019). Enhancement of tomato resistance to Tuta absoluta using a new efficient mesoporous silica nanoparticle-mediated plant transient gene expression approach. *Scientia Horticulturae*, *243*, 367–375. DOI: 10.1016/j.scienta.2018.08.040

Hamada, H., Liu, Y., Nagira, Y., Miki, R., Taoka, N., & Imai, R. (2018). Biolistic-delivery-based transient CRISPR/Cas9 expression enables in planta genome editing in wheat. *Scientific Reports*, *8*(1), 14422. DOI: 10.1038/s41598-018-32714-6 PMID: 30258105

Hendler-Neumark, A., & Bisker, G. (2019). Fluorescent single-walled carbon nanotubes for protein detection. *Sensors (Basel)*, *19*(24), 5403. DOI: 10.3390/s19245403 PMID: 31817932

Hong, J., Wang, C., Wagner, D. C., Gardea-Torresdey, J. L., He, F., & Rico, C. M. (2021). Foliar application of nanoparticles: Mechanisms of absorption, transfer, and multiple impacts. *Environmental Science. Nano*, *8*(5), 1196–1210. DOI: 10.1039/D0EN01129K

Hossain, Z., Mustafa, G., Sakata, K., & Komatsu, S. (2016). Insights into the proteomic response of soybean towards Al2O3, ZnO, and Ag nanoparticles stress. *Journal of Hazardous Materials*, *304*, 291–305. DOI: 10.1016/j.jhazmat.2015.10.071 PMID: 26561753

Islam, M. R., Youngblood, M., Kim, H. I., González-Gamboa, I., Monroy-Borrego, A. G., Caparco, A. A., & Giraldo, J. P. (2024). DNA delivery by virus-like nanocarriers in plant cells. *Nano Letters*, *24*(26), 7833–7842. DOI: 10.1021/acs.nanolett.3c04735 PMID: 38887996

Jiang, Y., Lawrence, M., Ansell, M. P., & Hussain, A. (2018). Cell wall microstructure, pore size distribution and absolute density of hemp shiv. *Royal Society Open Science*, *5*(4), 171945. DOI: 10.1098/rsos.171945 PMID: 29765652

Kant, R., & Dasgupta, I. (2017). Phenotyping of VIGS-mediated gene silencing in rice using a vector derived from a DNA virus. *Plant Cell Reports*, *36*(7), 1159–1170. DOI: 10.1007/s00299-017-2156-6 PMID: 28540496

Karny, A., Zinger, A., Kajal, A., Shainsky-Roitman, J., & Schroeder, A. (2018). Therapeutic nanoparticles penetrate leaves and deliver nutrients to agricultural crops. *Scientific Reports*, *8*(1), 1–10. DOI: 10.1038/s41598-018-25197-y PMID: 29773873

Kazama, T., Okuno, M., Watari, Y., Yanase, S., Koizuka, C., Tsuruta, Y., & Arimura, S. I. (2019). Curing cytoplasmic male sterility via TALEN-mediated mitochondrial genome editing. *Nature Plants*, *5*(7), 722–730. DOI: 10.1038/s41477-019-0459-z PMID: 31285556

Keum, T., Noh, G., Seo, J. E., Bashyal, S., Sohn, D. H., & Lee, S. (2022). Examination of effective buccal absorption of salmon calcitonin using cell-penetrating peptide-conjugated liposomal drug delivery system. *International Journal of Nanomedicine*, *17*, 697–710. DOI: 10.2147/IJN.S335774 PMID: 35210769

Khan, I., Hasan, M., Kausar, R., Shehzad, J., & Mustafa, G. (2022). Plant molecular responses to nanoparticle stress. In *Plant and nanoparticles* (pp. 239–264). Springer Nature Singapore. DOI: 10.1007/978-981-19-2503-0_9

Khan, N. A., Khan, S. U., Ahmed, S., Farooqi, I. H., Dhingra, A., Hussain, A., & Changani, F. (2019). Applications of nanotechnology in water and wastewater treatment: A review. *Asian Journal of Water, Environment and Pollution*, *16*(4), 81–86. DOI: 10.3233/AJW190051

Khodakovskaya, M., Dervishi, E., Mahmood, M., Xu, Y., Li, Z., Watanabe, F., & Biris, A. S. (2009). Carbon nanotubes are able to penetrate plant seed coat and dramatically affect seed germination and plant growth. *ACS Nano*, *3*(10), 3221–3227. DOI: 10.1021/nn900887m PMID: 19772305

Kim, C. J., Park, J. E., Hu, X., Albert, S. K., & Park, S. J. (2020). Peptide-driven shape control of low-dimensional DNA nanostructures. *ACS Nano*, *14*(2), 2276–2284. DOI: 10.1021/acsnano.9b09312 PMID: 31962047

Kralova, K., & Jampilek, J. (2021). Responses of medicinal and aromatic plants to engineered nanoparticles. *Applied Sciences (Basel, Switzerland)*, *11*(4), 1813. DOI: 10.3390/app11041813

Kumar, S., & Randhawa, J. K. (2013). High melting lipid based approach for drug delivery: Solid lipid nanoparticles. *Materials Science and Engineering C*, *33*(4), 1842–1852. DOI: 10.1016/j.msec.2013.01.037 PMID: 23498204

Kwak, S. Y., Lew, T. T. S., Sweeney, C. J., Koman, V. B., Wong, M. H., Bohmert-Tatarev, K., & Strano, M. S. (2019). Chloroplast-selective gene delivery and expression in planta using chitosan-complexed single-walled carbon nanotube carriers. *Nature Nanotechnology*, *14*(5), 447–455. DOI: 10.1038/s41565-019-0375-4 PMID: 30804482

Lang, C., Mission, E. G., Fuaad, A. A. H. A., & Shaalan, M. (2021). Nanoparticle tools to improve and advance precision practices in the Agrifoods Sector towards sustainability-A review. *Journal of Cleaner Production*, *293*, 126063. DOI: 10.1016/j.jclepro.2021.126063

Law, S. S. Y., Liou, G., Nagai, Y., Giménez-Dejoz, J., Tateishi, A., Tsuchiya, K., & Numata, K. (2022). Polymer-coated carbon nanotube hybrids with functional peptides for gene delivery into plant mitochondria. *Nature Communications*, *13*(1), 2417. DOI: 10.1038/s41467-022-30185-y PMID: 35577779

Lew, T. T. S., Wong, M. H., Kwak, S. Y., Sinclair, R., Koman, V. B., & Strano, M. S. (2018). Rational design principles for the transport and subcellular distribution of nanomaterials into plant protoplasts. *Small*, *14*(44), 1802086. DOI: 10.1002/smll.201802086 PMID: 30191658

Li, M., Liu, J., Deng, M., Ge, Z., Afshan, N., Zuo, X., & Li, Q. (2019). Rapid transmembrane transport of DNA nanostructures by chemically anchoring artificial receptors on cell membranes. *ChemPlusChem*, *84*(4), 323–327. DOI: 10.1002/cplu.201900025 PMID: 31939210

Liu, Q., Chen, B., Wang, Q., Shi, X., Xiao, Z., Lin, J., & Fang, X. (2009). Carbon nanotubes as molecular transporters for walled plant cells. *Nano Letters*, *9*(3), 1007–1010. DOI: 10.1021/nl803083u PMID: 19191500

Liu, Q., Fang, L., Yu, G., Wang, D., Xiao, C. L., & Wang, K. (2019). Detection of DNA base modifications by deep recurrent neural network on Oxford Nanopore sequencing data. *Nature Communications*, *10*(1), 2449. DOI: 10.1038/s41467-019-10168-2 PMID: 31164644

Liu, Q., Li, Y., Xu, K., Li, D., Hu, H., Zhou, F., & Li, C. (2020). Clay nanosheet-mediated delivery of recombinant plasmids expressing artificial miRNAs via leaf spray to prevent infection by plant DNA viruses. *Horticulture Research*, *7*(1), 7. DOI: 10.1038/s41438-020-00400-2 PMID: 33328436

Liu, T., Xiao, B., Xiang, F., Tan, J., Chen, Z., Zhang, X., & Deng, J. (2020). Ultrasmall copper-based nanoparticles for reactive oxygen species scavenging and alleviation of inflammation related diseases. *Nature Communications*, *11*(1), 2788. DOI: 10.1038/s41467-020-16544-7 PMID: 32493916

Liu, W., Rudis, M. R., Cheplick, M. H., Millwood, R. J., Yang, J. P., Ondzighi-Assoume, C. A., & Stewart, C. N. (2020). Lipofection-mediated genome editing using DNA-free delivery of the Cas9/gRNA ribonucleoprotein into plant cells. *Plant Cell Reports*, *39*(2), 245–257. DOI: 10.1007/s00299-019-02488-w PMID: 31728703

Lopes, R. C., Rocha, B. G., Maçôas, E. M., Marques, E. F., & Martinho, J. M. (2022). Combining metal nanoclusters and carbon nanomaterials: Opportunities and challenges in advanced nanohybrids. *Advances in Colloid and Interface Science*, *304*, 102667. DOI: 10.1016/j.cis.2022.102667 PMID: 35462268

Lv, Z., Jiang, R., Chen, J., & Chen, W. (2020). Nanoparticle-mediated gene transformation strategies for plant genetic engineering. *The Plant Journal*, *104*(4), 880–891. DOI: 10.1111/tpj.14973 PMID: 32860436

Mad'Atari, M. F. B., & Folta, K. M. (2019). Transformation improvement with the Standardized Pressure A grobacterium Infiltration Device (SPAID). *BMC Research Notes*, *12*, 1–6. PMID: 30876440

Mahmoud, K., Swidan, S., El-Nabarawi, M., & Teaima, M. (2022). Lipid based nanoparticles as a novel treatment modality for hepatocellular carcinoma: A comprehensive review on targeting and recent advances. *Journal of Nanobiotechnology*, *20*(1), 109. DOI: 10.1186/s12951-022-01309-9 PMID: 35248080

Maliga, P. (2022). Engineering the plastid and mitochondrial genomes of flowering plants. *Nature Plants*, *8*(9), 996–1006. DOI: 10.1038/s41477-022-01227-6 PMID: 36038655

Martin-Ortigosa, S., Peterson, D. J., Valenstein, J. S., Lin, V. S. Y., Trewyn, B. G., Lyznik, L. A., & Wang, K. (2014). Mesoporous silica nanoparticle-mediated intracellular Cre protein delivery for maize genome editing via loxP site excision. *Plant Physiology*, *164*(2), 537–547. DOI: 10.1104/pp.113.233650 PMID: 24376280

Men, W., Zhu, P., Dong, S., Liu, W., Zhou, K., Bai, Y., & Zhang, S. (2020). Layer-by-layer pH-sensitive nanoparticles for drug delivery and controlled release with improved therapeutic efficacy in vivo. *Drug Delivery*, *27*(1), 180–190. DOI: 10.1080/10717544.2019.1709922 PMID: 31924103

Miller, E. M., Samec, T. M., & Alexander-Bryant, A. A. (2021). Nanoparticle delivery systems to combat drug resistance in ovarian cancer. *Nanomedicine; Nanotechnology, Biology, and Medicine*, *31*, 102309. DOI: 10.1016/j.nano.2020.102309 PMID: 32992019

Mitter, N., Worrall, E. A., Robinson, K. E., Li, P., Jain, R. G., Taochy, C., & Xu, Z. P. (2017). Clay nanosheets for topical delivery of RNAi for sustained protection against plant viruses. *Nature Plants*, *3*(2), 1–10. DOI: 10.1038/nplants.2016.207 PMID: 28067898

Mitter, N., Worrall, E. A., Robinson, K. E., Xu, Z. P., & Carroll, B. J. (2017). Induction of virus resistance by exogenous application of double-stranded RNA. *Current Opinion in Virology*, *26*, 49–55. DOI: 10.1016/j.coviro.2017.07.009 PMID: 28778033

Miyamoto, N., Sakuragi, M., & Kitade, Y. (2024). Advanced Nanotechnology-Based Nucleic Acid Medicines. *Pharmaceutics*, *16*(11), 1367. DOI: 10.3390/pharmaceutics16111367 PMID: 39598491

Miyamoto, T., Tsuchiya, K., & Numata, K. (2020). Dual peptide-based gene delivery system for the efficient transfection of plant callus cells. *Biomacromolecules*, *21*(7), 2735–2744. DOI: 10.1021/acs.biomac.0c00481 PMID: 32432860

Miyamoto, T., Tsuchiya, K., Toyooka, K., Goto, Y., Tateishi, A., & Numata, K. (2022). Relaxation of the Plant Cell Wall Barrier via Zwitterionic Liquid Pretreatment for Micelle-Complex-Mediated DNA Delivery to Specific Plant Organelles. *Angewandte Chemie*, *134*(32), e202204234. DOI: 10.1002/ange.202204234 PMID: 35670289

Mohanta, D., Patnaik, S., Sood, S., & Das, N. (2019). Carbon nanotubes: Evaluation of toxicity at bio interfaces. *Journal of Pharmaceutical Analysis*, *9*(5), 293–300. DOI: 10.1016/j.jpha.2019.04.003 PMID: 31929938

Molla, K. A., Sretenovic, S., Bansal, K. C., & Qi, Y. (2021). Precise plant genome editing using base editors and prime editors. *Nature Plants*, *7*(9), 1166–1187. DOI: 10.1038/s41477-021-00991-1 PMID: 34518669

Mookkan, M. (2018). Particle bombardment–mediated gene transfer and GFP transient expression in Setaria viridis. *Plant Signaling & Behavior*, *13*(4), e1441657. DOI: 10.1080/15592324.2018.1441657 PMID: 29621423

Mora-Godínez, S., Abril-Martínez, F., & Pacheco, A. (2022). Green synthesis of silver nanoparticles using microalgae acclimated to high CO2. *Materials Today: Proceedings*, *48*, 5–9. DOI: 10.1016/j.matpr.2020.04.761

Mosa, M. A., & Youssef, K. (2021). Topical delivery of host induced RNAi silencing by layered double hydroxide nanosheets: An efficient tool to decipher pathogenicity gene function of Fusarium crown and root rot in tomato. *Physiological and Molecular Plant Pathology*, *115*, 101684. DOI: 10.1016/j.pmpp.2021.101684

Mout, R., Ray, M., Yesilbag Tonga, G., Lee, Y. W., Tay, T., Sasaki, K., & Rotello, V. M. (2017). Direct cytosolic delivery of CRISPR/Cas9-ribonucleoprotein for efficient gene editing. *ACS Nano*, *11*(3), 2452–2458. DOI: 10.1021/acsnano.6b07600 PMID: 28129503

Nakazato, I., Okuno, M., Yamamoto, H., Tamura, Y., Itoh, T., Shikanai, T., & Arimura, S. I. (2021). Targeted base editing in the plastid genome of Arabidopsis thaliana. *Nature Plants*, *7*(7), 906–913. DOI: 10.1038/s41477-021-00954-6 PMID: 34211131

Nayak, S., Das, P., & Singh, M. K. (2021). Carbon dot with aggregation induced emission and pH triggered disintegration. *Colloid and Interface Science Communications*, *45*, 100537. DOI: 10.1016/j.colcom.2021.100537

Numata, K. (2015). Poly (amino acid) s/polypeptides as potential functional and structural materials. *Polymer Journal*, *47*(8), 537–545. DOI: 10.1038/pj.2015.35

Ozyigit, I. I., & Yucebilgili Kurtoglu, K. (2020). Particle bombardment technology and its applications in plants. *Molecular Biology Reports*, *47*(12), 9831–9847. DOI: 10.1007/s11033-020-06001-5 PMID: 33222118

Parra-Torrejón, B., Cáceres, A., Sánchez, M., Sainz, L., Guzmán, M., Bermúdez-Perez, F. J., & Delgado-López, J. M. (2023). Multifunctional nanomaterials for biofortification and protection of tomato plants. *Environmental Science & Technology*, *57*(40), 14950–14960. DOI: 10.1021/acs.est.3c02559 PMID: 37753594

Rani, N., Kumari, K., Sangwan, P., Barala, P., Yadav, J., Vijeta, , Rahul, , & Hooda, V. (2022). Nano-iron and nano-zinc induced growth and metabolic changes in Vigna radiata. *Sustainability (Basel)*, *14*(14), 8251. DOI: 10.3390/su14148251

Rao, N. N. (2019). *Floral Dip Transformation of Camelina Sativa L. With a Myo-Inositol Oxygenase Gene to Potentially Improve Oil Yield.* Arkansas State University.

Rastogi, A., Zivcak, M., Sytar, O., Kalaji, H. M., He, X., Mbarki, S., & Brestic, M. (2017). Impact of metal and metal oxide nanoparticles on plant: A critical review. *Frontiers in Chemistry*, *5*, 78. DOI: 10.3389/fchem.2017.00078 PMID: 29075626

Rizwan, M., Ali, S., Qayyum, M. F., Ok, Y. S., Adrees, M., Ibrahim, M., & Abbas, F. (2017). Effect of metal and metal oxide nanoparticles on growth and physiology of globally important food crops: A critical review. *Journal of Hazardous Materials*, *322*, 2–16. DOI: 10.1016/j.jhazmat.2016.05.061 PMID: 27267650

Rossi, L., Sharifan, H., Zhang, W., Schwab, A. P., & Ma, X. (2018). Mutual effects and in planta accumulation of co-existing cerium oxide nanoparticles and cadmium in hydroponically grown soybean (Glycine max (L.) Merr.). *Environmental Science. Nano*, *5*(1), 150–157. DOI: 10.1039/C7EN00931C

Santana, I., Jeon, S. J., Kim, H. I., Islam, M. R., Castillo, C., Garcia, G. F., & Giraldo, J. P. (2022). Targeted carbon nanostructures for chemical and gene delivery to plant chloroplasts. *ACS Nano*, *16*(8), 12156–12173. DOI: 10.1021/acsnano.2c02714 PMID: 35943045

Santana, I., Wu, H., Hu, P., & Giraldo, J. P. (2020). Targeted delivery of nanomaterials with chemical cargoes in plants enabled by a biorecognition motif. *Nature Communications*, *11*(1), 2045. DOI: 10.1038/s41467-020-15731-w PMID: 32341352

Schwartz-Duval, A. S., Konopka, C. J., Moitra, P., Daza, E. A., Srivastava, I., Johnson, E. V., & Pan, D. (2020). Intratumoral generation of photothermal gold nanoparticles through a vectorized biomineralization of ionic gold. *Nature Communications*, *11*(1), 4530. DOI: 10.1038/s41467-020-17595-6 PMID: 32913195

Serag, M. F., Kaji, N., Gaillard, C., Okamoto, Y., Terasaka, K., Jabasini, M., & Baba, Y. (2011). Trafficking and subcellular localization of multiwalled carbon nanotubes in plant cells. *ACS Nano*, *5*(1), 493–499. DOI: 10.1021/nn102344t PMID: 21141871

Shehzad, J., & Mustafa, G. (2023). Mechanism of reactive oxygen species regulation in plants. In *Reactive Oxygen Species: Prospects in Plant Metabolism* (pp. 17–41). Springer Nature Singapore. DOI: 10.1007/978-981-19-9794-5_2

Shi, J., & Avramidis, S. (2019). Evolution of wood cell wall nanopore size distribution in the hygroscopic range. *Holzforschung*, *73*(10), 899–910. DOI: 10.1515/hf-2018-0198

Squire, H. J., Tomatz, S., Voke, E., González-Grandío, E., & Landry, M. (2023). The emerging role of nanotechnology in plant genetic engineering. *Nature Reviews Bioengineering*, *1*(5), 314–328. DOI: 10.1038/s44222-023-00037-5

Su, Z., Yao, C., Tipper, J., Yang, L., Xu, X., Chen, X., & Zheng, Y. (2023). Nanostrategy of targeting at embryonic trophoblast cells using CuO nanoparticles for female contraception. *ACS Nano*, *17*(24), 25185–25204. DOI: 10.1021/acsnano.3c08267 PMID: 38088330

Sun, H., Cao, Y., Kim, D., & Marelli, B. (2022). Biomaterials technology for agrofood resilience. *Advanced Functional Materials*, *32*(30), 2201930. DOI: 10.1002/adfm.202201930

Tripathi, D. K., Singh, S., Singh, S., Pandey, R., Singh, V. P., Sharma, N. C., & Chauhan, D. K. (2017). An overview on manufactured nanoparticles in plants: Uptake, translocation, accumulation and phytotoxicity. *Plant Physiology and Biochemistry*, *110*, 2–12. DOI: 10.1016/j.plaphy.2016.07.030 PMID: 27601425

Ullah, H., Khan, I., Mustafa, G., Shehzad, J., Hasan, M., Saeed, U., & Chaudhari, S. K. (2023). Molecular characterization and mitigative role of silicon dioxide nanoparticles in Ocimum basilicum under lead (Pb) stress. *Silicon*, *15*(6), 2551–2565. DOI: 10.1007/s12633-022-02178-5

Ullah, J., Gul, A., Khan, I., Shehzad, J., Kausar, R., Ahmed, M. S., & Mustafa, G. (2024). Green synthesized iron oxide nanoparticles as a potential regulator of callus growth, plant physiology, antioxidative and microbial contamination in Oryza sativa L. *BMC Plant Biology*, *24*(1), 939. DOI: 10.1186/s12870-024-05627-y PMID: 39385076

Usman, M., Farooq, M., Wakeel, A., Nawaz, A., Cheema, S. A., Rehman, H., Ashraf, I., & Sanaullah, M. (2020). Nanotechnology in agriculture: Current status, challenges and future opportunities. *The Science of the Total Environment*, *721*, 137778. DOI: 10.1016/j.scitotenv.2020.137778 PMID: 32179352

Vejlupkova, Z., Warman, C., Sharma, R., Scheller, H. V., Mortimer, J. C., & Fowler, J. E. (2020). No evidence for transient transformation via pollen magnetofection in several monocot species. *Nature Plants*, *6*(11), 1323–1324. DOI: 10.1038/s41477-020-00798-6 PMID: 33139861

Venugopal, P. D., & Dively, G. P. (2017). Climate change, transgenic corn adoption and field-evolved resistance in corn earworm. *Royal Society Open Science*, *4*(6), 170210. DOI: 10.1098/rsos.170210 PMID: 28680673

Vinzant, K., Rashid, M., Clouse, D. E., Ghosh, P., Quadir, M., Davis, V. A., & Khodakovskaya, M. V. (2025). From Plants to Plants: Plant-Derived Biological Polymers as Sustainable and Safe Nanocarriers for Direct Delivery of DNA to Plant Cells. *Nano Letters*, *25*(14), 5572–5581. DOI: 10.1021/acs.nanolett.4c05489 PMID: 40139733

Wang, Q., Ebbs, S. D., Chen, Y., & Ma, X. (2013). Trans-generational impact of cerium oxide nanoparticles on tomato plants. *Metallomics*, *5*(6), 753–759. DOI: 10.1039/c3mt00033h PMID: 23689668

Wang, S., Yu, M., Jiang, J., Zhang, W., Guo, X., Chang, S., & Campbell, M. (2017). Evidence aggregation for answer re-ranking in open-domain question answering. *arXiv preprint arXiv:1711.05116.*

Wang, X., & Zhang, W. (2022). The Janus of Protein Corona on nanoparticles for tumor targeting, immunotherapy and diagnosis. *Journal of Controlled Release*, *345*, 832–850. DOI: 10.1016/j.jconrel.2022.03.056 PMID: 35367478

Wang, Y., Li, X., Liu, L., Liu, B., Wang, F., & Chen, C. (2020). Tissue targeting and ultrasound-targeted microbubble destruction delivery of plasmid DNA and transfection in vitro. *Cellular and Molecular Bioengineering*, *13*(1), 99–112. DOI: 10.1007/s12195-019-00597-w PMID: 32030111

Wong, A. C., & Wright, D. W. (2016). Size-dependent cellular uptake of DNA functionalized gold nanoparticles. *Small*, *12*(40), 5592–5600. DOI: 10.1002/smll.201601697 PMID: 27562251

Worrall, E. A., Bravo-Cazar, A., Nilon, A. T., Fletcher, S. J., Robinson, K. E., Carr, J. P., & Mitter, N. (2019). Exogenous application of RNAi-inducing double-stranded RNA inhibits aphid-mediated transmission of a plant virus. *Frontiers in Plant Science*, *10*, 265. DOI: 10.3389/fpls.2019.00265 PMID: 30930914

Yan, A., & Chen, Z. (2019). Impacts of silver nanoparticles on plants: A focus on the phytotoxicity and underlying mechanism. *International Journal of Molecular Sciences*, *20*(5), 1003. DOI: 10.3390/ijms20051003 PMID: 30813508

Yan, Y., Zhu, X., Yu, Y., Li, C., Zhang, Z., & Wang, F. (2022). Nanotechnology strategies for plant genetic engineering. *Advanced Materials*, *34*(7), 2106945. DOI: 10.1002/adma.202106945 PMID: 34699644

Yaron, P. N., Holt, B. D., Short, P. A., Lösche, M., Islam, M. F., & Dahl, K. N. (2011). Single wall carbon nanotubes enter cells by endocytosis and not membrane penetration. *Journal of Nanobiotechnology*, *9*(1), 1–15. DOI: 10.1186/1477-3155-9-45 PMID: 21961562

Yildiz, T., Gu, R., Zauscher, S., & Betancourt, T. (2018). Doxorubicin-loaded protease-activated near-infrared fluorescent polymeric nanoparticles for imaging and therapy of cancer. *International Journal of Nanomedicine*, *13*, 6961–6986. DOI: 10.2147/IJN.S174068 PMID: 30464453

Yin, L., & Zhong, Z. (2020). Nanoparticles. In *Biomaterials science* (pp. 453–483). Academic Press. DOI: 10.1016/B978-0-12-816137-1.00031-3

Yong, J., Wu, M., Zhang, R., Bi, S., Mann, C. W., Mitter, N., & Xu, Z. P. (2022). Clay nanoparticles efficiently deliver small interfering RNA to intact plant leaf cells. *Plant Physiology*, *190*(4), 2187–2202. DOI: 10.1093/plphys/kiac430 PMID: 36135825

Yong, J., Zhang, R., Bi, S., Li, P., Sun, L., Mitter, N., & Xu, Z. P. (2021). Sheet-like clay nanoparticles deliver RNA into developing pollen to efficiently silence a target gene. *Plant Physiology*, *187*(2), 886–899. DOI: 10.1093/plphys/kiab303 PMID: 34608968

Zaheer, S., Mustafa, G., Shehzad, J., Tariq, T., Chaudhari, S. K., & Hasan, M. (2022). Green Synthesis of SiO 2 NPs Promote Vigna radiata L Germination, Membrane Damage, Antioxidant enzymes and Physiological activities.

Zhan, Y., Ma, W., Zhang, Y., Mao, C., Shao, X., Xie, X., & Lin, Y. (2020). Diversity of DNA nanostructures and applications in oncotherapy. *Biotechnology Journal*, *15*(1), 1900094. DOI: 10.1002/biot.201900094 PMID: 31464361

Zhang, X. H., Tee, L. Y., Wang, X. G., Huang, Q. S., & Yang, S. H. (2015). Off-target effects in CRISPR/Cas9-mediated genome engineering. *Molecular Therapy. Nucleic Acids*, *4*, 4. DOI: 10.1038/mtna.2015.37 PMID: 26575098

Zhang, Y., Cao, Y., Jiang, W., Ma, Q., Shin, J., Sun, H., & Marelli, B. (2024). Polymeric Nanocarriers Autonomously Cross the Plant Cell Wall and Enable Protein Delivery for Stress Sensing. *Advanced Materials*, *36*(41), 2409356. DOI: 10.1002/adma.202409356 PMID: 39149770

Zhang, Y., Huang, C., & Xiong, R. (2024). Advanced materials for intracellular delivery of plant cells: Strategies, mechanisms and applications. *Materials Science and Engineering R Reports*, *160*, 100821. DOI: 10.1016/j.mser.2024.100821

Zhang, Y., Shin, J., Sun, H., Chang, H. F., Martinez, M. R., Perkins, L. A., & Lowry, G. V. (2024). High Aspect Ratio Polymer Nanocarriers for Gene Delivery and Expression in Plants. *Nano Letters*, *25*(2), 681–690. DOI: 10.1021/acs.nanolett.4c04704 PMID: 39810730

Zhao, Q., Lin, Y., Han, N., Li, X., Geng, H., Wang, X., & Wang, S. (2017). Mesoporous carbon nanomaterials in drug delivery and biomedical application. *Drug Delivery*, *24*(2), 94–107. DOI: 10.1080/10717544.2017.1399300 PMID: 29124979

Zhao, X., Meng, Z., Wang, Y., Chen, W., Sun, C., Cui, B., Cui, H., Yu, M., Zeng, Z., Guo, S., Luo, D., Cheng, J. Q., Zhang, R., & Cui, H. (2017). Pollen magnetofection for genetic modification with magnetic nanoparticles as gene carriers. *Nature Plants*, *3*(12), 956–964. DOI: 10.1038/s41477-017-0063-z PMID: 29180813

Ziemienowicz, A. (2014). Agrobacterium-mediated plant transformation: Factors, applications and recent advances. *Biocatalysis and Agricultural Biotechnology*, *3*(4), 95–102. DOI: 10.1016/j.bcab.2013.10.004

Zou, L., Cheng, M., Hu, K., Feng, J., & Tu, L. (2024). Vesicular drug delivery systems for oral absorption enhancement. *Chinese Chemical Letters*, *35*(7), 109129. DOI: 10.1016/j.cclet.2023.109129

Chapter 7 Nanocarriers for Delivering Biostimulants in Plants: Mechanisms and Applications

Ansar Mehmood
University of Poonch, Rawalakot, Pakistan

Amna Mustafa
University of Poonch, Rawalakot, Pakistan

Khawaja Shafique Ahmad
https://orcid.org/0000-0001-8814-9878
University of Poonch, Rawalakot, Pakistan

ABSTRACT

The integration of plant biostimulants and nanocarriers in the agri-food industry is gaining attention, yet most research has focused on their independent applications. This chapter explores how combining biostimulant technology with nanotechnology can enhance agricultural sustainability and food quality. By leveraging nanotechnology's capabilities, plant biostimulants—compounds, microbes, or complex mixtures that improve plant growth and nutrition—can be more effectively delivered and optimized for food security, safety, and quality. This chapter provides a tutorial-style overview of nanotechnology's fundamental principles, highlighting nanoencapsulation and other advanced techniques to create nano-biostimulants. It discusses sustainability, including biostimulant production from waste materials, supporting a circular economy. Ultimately, this integration offers a long-term, eco-friendly solution to food insecurity and enhances the value of biostimulants for sustainable agriculture.Keywords: Biostimulants; Plant growth; Nanotechnology; Nanocarriers

1. INTRODUCTION

Biostimulants are substances that, in trace amounts, stimulate plant growth (Zhang et al., 2003). This description clearly distinguishes these chemicals from traditional soil amendments and fertilizers, which solely encourage plant growth in greater amounts. Biofertilizers can also be differentiated from

DOI: 10.4018/979-8-3693-9869-2.ch007

biostimulants since they often contain a variety of living bacteria that may enhance plant nutrient uptake by populating the rhizosphere and facilitating the passage of nutrients to the plant root hairs. Biofertilizers are regarded as a good substitute for the dangerous synthetic fertilizers because of their composition, shown profitability, and environmental friendliness (Aloo et al., 2022). However, as evidenced by the co-inoculation of rhizobia bacteria and biostimulant treatment in beans, it is hard to completely rule out the possibility of a synergistic use of biostimulants and biofertilizer (Karavidas et al., 2022).

A phenomenon known as biostimulation occurs when metabolic processes are altered to enable an organism to use environmental resources more effectively, develop or yield more significantly, and become more resilient to unfavorable environmental conditions (Vygovsky et al., 2003). Biostimulation is a general biological phenomenon that depends on the interactions among cell molecular structures and outside stimuli or impulses. Plant development and stress tolerance are aided by several biostimulants, such as humic compounds, amino acids, microbial inoculants, and protein hydrolysates (du Jardin, 2015). Due to their rapid breakdown, limited water absorption, and weak plant intake, biostimulant administration has several disadvantages. By guaranteeing accurate and regulated, effective biostimulant release, nanocarrier-based delivery methods offer a way around restrictions (Fraceto et al., 2016). The entire method of delivering biostimulants via nanocarriers is examined in this chapter, along with the potential applications for modern agricultural systems.

A series of processes, including transduction, perception, signaling, effector action, and changes in metabolism, gene expression or cellular (and organismal) characteristics, take place following the application of a biostimulating substance. Living organisms can be biostimulated by chemical, physical or biological environmental factors (Cabrera-De et al., 2018; Bell et al., 2014). Biostimulation in cultivated plants is the outcome of applying compounds or microbes that do not directly provide dynamic components, known plant hormones, or molecules having a disease-suppressive impact (Du Jardin, 2015; Brown, 2015). Materials with at least one dimension smaller than 100 nm are known as nanocarriers of semimetals, metals and nonmetals (NPs), as well as nanomaterials (NMs) of C (like carbon nanotubes), organic (like chitosan), and other materials like nanoclays. Because of its small size, this material exhibits different properties than the bulk material with the same composition. These additional characteristics give the material a value addition that can be used in a variety of industries, including the automotive, energy, pharmaceutical, agriculture and medical sectors (Judy et al., 2014).

Nanoparticle and nanocarrier-induced biostimulation appears to take place in two stages. The first is physicochemical in nature, affecting the passage of metabolites and ions to cells in addition to the activity of receptors, transporters, and other proteins through the energy and surface charges of NPs and NCs, which are linked to the surfaces of walls and cell membranes (Zuverza-Mena et al., 2017). Like the reactions to metal ions or organic molecules in their bulk form, the second phase is brought about by biochemical processes that react to the release of the chemical elements from the NP or NC (for instance, Fe^{3+} from Fe NPs or carbon compounds from NCs of chitosan) (Poschenriede et al., 2013; Hillegass et al., 2010). It is possible that almost any type of inorganic (Cu, CuO, silica) or organic (polystyrene, chitosan) NC or NP in particular concentration ranges has this type of biostimulating capacity because the first phase, the interaction of surfaces, in a physicochemical process can be independent from the composition. In fact, it has been suggested that the surface area of the NPs and NCs be utilized as a predictor of the possible biological impact rather than the concentration (Knox, 2014). This will concentrate on the three most researched and lucrative products: protein hydrolysates, seaweed extracts, and humic compounds (Bastos et al., 2021).

1.1 Humic Substances

Degraded organic compounds make up humic substances (HS), which are prevalent in soil and sediments and are heterogeneous in their molecular combinations. They are also leftovers of microbial activity. Mentoring through applied science, the International Humic Substances Society delineates the range of elements composing conventional fulvic humic acids, as follows: C, 50-60%; N, 0.7-5.1%; H, 3.5-4.8%; and O, 31.6-45.5%. The elements that comprise HS have been researched extensively (Ritchie et al., 2008). The average elemental composition, however, varies widely, and the molecular compositions of HS also vary considerably, depending on the nature of microbial degradation and the make-up of the soil (Pizzeghello et al., 2020). Carboxylic and aromatic groups are generally considered the most important functional groups in HS for stability against chemical and biological degradation (Pizzeghello et al., 2020).

Organic acid that has low-molecular-weight present in root exudates facilitate the dispersion of HS for the absorption of these by the root (Nardi et al., 2000). The carboxylic groups render HS more soluble and biologically active (Muscolo et al., 2007). According to their solubilization, HS produce two types of molecules: the low molecular size (LMS, < 3500 Da) and the high molecular size (HMS, > 3500 Da), working together with the root. In laboratory settings, it has been reported that the HMS fraction interacts specifically with root membrane receptors and cell wall components to internally signal the onset of a signal transduction cascade, while LMS, on account of its small size, is more easily absorbed (Nardi et al., 2002). The root is the primary target for HS since it is the organ where the plant responds the most and where elongation and an increase in root biomass are commonly seen.

Since the elongation and differentiation zone of the root contains small, densely meristematic cells that undergo continuous metabolic activity, making it more susceptible to lateral root initiation, it would be considered a zone of HS action for root formation (Pizzeghello et al., 2020). The stimulatory effect is most likely due to a structural modification in the central cylinder that positively mediates water conductance and nutrient flux rate; increased resource storage in the cortex has contributed to the increased diameter (Canellas et al., 2002). HS affects the structural changes in addition to the allocation of hormones within the plant affecting nitric oxide metabolism and auxin polar transport (Jindo et al., 2020). Advanced H&S studies revealed indole-3-acetic acid (IAA) together with other substances having IAA-like action in the HS matrix (Scaglia et al., 2016). The IAA-like-molecule-root cell interaction is characterized by increased activity of root plasma membrane (PM) H^+-ATPase, vacuolar H^+-ATPase, H^+-pyrophosphatase (H^+-PPase), and cross-induction of NO production that accumulates in pericycle cells (Nardi et al., 2021). Activation of this pump then maintains the H+-electrochemical gradient with osmotic pressure directed into the vacuole to enable proper water uptake and turgor maintenance (Nardi et al., 2021). In addition, PM H^+-ATPase activation appears to be critical for LRA induction (Canellas et al., 2002).

1.2 Seaweed

Many seaweeds are abundant in the ocean, which is marine photosynthetic algae. Some hundreds of species can be found in each one of the three large phyla: Chlorophyta (green algae), Rhodophyta (red algae), and Phaeophyceae (brown algae) (Ali et al., 2021). The polysaccharides in the walls of seaweed are the primary agents known to elicit biological activity in plants. In green seaweeds, these would include carrageenans, ulvans, and agarans while alginates, fucans, and the storage polysaccharide laminarin are found in brown seaweeds (Rioux et al., 2007; Martone et al., 2009; Perez et al., 2016). Various sulfate groups modify approximately of these polysaccharides (such as carrageenan, fucoidan, and ulvan), resulting in a variety of distinct structures and characteristics (Gotteland et al., 2020). Seaweeds' cell wall polysaccharides also include a variety of chemical and inorganic compounds that are recognized for supporting their biostimulant characteristics.

1.3 Protein Hydrolysates

Numerous research studies on a variety of horticultural and agronomic crops have demonstrated that biostimulant as protein hydrolysates (PH) are nn innovative technology with potential uses. Mixtures of polypeptides, oligopeptides, and amino acids prepared from protein sources by partial hydrolysis" are known as PHs (Schaafsma et al., 2009). Many amino acids, including Lys, Trp, Ser, Thr, and Cys, are hydrolyzed when proteins undergo chemical hydrolysis by acid or alkaline treatments at high temperatures (121–137°C). This results in the conversion of their L-forms to D-forms, which restricts their metabolic effects and sometimes has additional detrimental effects on plant metabolism (Cerdan et al., 2009). Enzymatic proteolysis, a less energy-intensive method that ensures greater control over the final product, is used to create PHs. Certain proteases optimized for peptide/oligopeptide proportion, yield, and other physicochemical features are chosen since the kind of protease hydrolyzing the starting material has a considerable impact on almost all PH aspects, including biostimulant qualities (Colla et al., 2017).

2. TYPES OF NANOCARRIERS FOR BIOSTIMULANT DELIVERY

2.1 Lipid-Based Nanocarriers

High biocompatibility and efficient biostimulant encapsulation are two benefits of lipid-based nanocarriers, such as liposomes and nanoemulsions. Biostimulants penetrate into the system via the fusion between plant cell membranes and nanocarriers. Efficient encapsulation of lipid-based nanocarriers allows hydrophobic bioactive compounds and microbial biostimulants be released from them in a safe, controlled manner (Sajid et al., 2020).

2.2 Polymeric Nanocarriers

Regulated biostimulants release can, thus, be accomplished by polymeric nanocarriers made from biodegradable agents such as polylactic acid and chitosan. These carriers cause release by activation of a material through enzyme-triggering or pH variation systems these biomaterials developed to be affected in plant tissues. The unmatched effectiveness observed when applying organic acids and amino acids targeted at plants is seen on the carriers (Rai et al, 2018).

2.3 Inorganic Nanocarriers

Three majors inorganic nanocarrier systems have been detailed to be supported to modify the controlled sustained availability of biostimulants. These include carbon-based nanomaterials with metal-organic frameworks and mesoporous silica nanoparticles. Movement inside these carriers occurs because of diffusion-controlled release processes that allow the complete payload to be gradually lost. In this way, the technique will facilitate plant development and metabolism with very high efficiencies in micronutrient and plant growth regulator delivery (Torney et al., 2007).

2.4 Hybrid Nanocarriers

Nanocarriers containing both polymeric and inorganic aspects, give superior possibilities for an efficient biostimulant to be delivered. These delivery systems respond to the stimuli in the environment and control various routes to ensure timely and proper product discharge. The process becomes simpler as compatible biostimulants are mixed with agrochemicals, resulting in enhanced plant growth and health benefits (Ghormade et al., 2011).

3. MECHANISMS OF NANOCARRIER-MEDIATED BIOSTIMULANT UPTAKE

3.1 Foliar Uptake Mechanisms

After application onto the leaves of a plant, nanocarriers exhibit various penetration pathways. Hydrophobic coatings on nanocarriers enable the delivery of biostimulant chemicals into plant tissues by penetrating plant cuticles. Direct entry into plant cells by nanomaterials through stomata is another pathway. Some nanocarriers, Eichert et al. (2008) elaborated, indeed enhance their effectiveness in foliar applications through possible vesicle-mediated endocytosis.

3.2 Root Uptake Mechanism

An assortment of routes for root application are advertised to obtain effective biostimulant distribution from nanocarriers. These nanocarriers find their way into plant tissues using apoplastic and symplastic routes, entering through plasmodesmata and cell walls. Certain nanocarriers cling to root structures within the rhizosphere to enhance biostimulant availability. Assisted by mycorrhizal fungi, they serve vital means for the distribution of biostimulants through the nanocarrier pathways for transport to plants (Liu et al., 2019).

3.3 Controlled Release Mechanisms

Nanocarriers use mechanisms that are especially designed for the regulation of release from such carriers, such that the biostimulants find optimum utilization by plants. Thus, specific specific soil habitats or cellular pH determine the operation of some carriers to a particular delivery mechanism for content release. Certain mechanisms of delivery result in the activation of enzymes among which constituents will be released in specific time periods. Another system for delivery is redox-sensitive, which uses oxidation-reduction reactions for the timing of release of bioactive components within the systems (Shang et al., 2020).

4. PROS OF NANOCARRIERS IN THE DELIVERY OF BIOSTIMULANTS

These are nanoparticulate carriers: they incorporate most of the delivery benefits as mentioned for biostimulants. These systems also improve stability and bioavailability before that of plants takes place. Such methodologies will reduce losses caused by use, danger to the environment, and provide precision of features about timing used in applications. Nanocarriers were successfully used in improving the stress tolerance of plants by strengthening their defense against pathogens, salt, and drought. They therefore provide environmentally friendly agriculture for farmers through reduced chemical inputs (Wang et al., 2022).

4.1 Application of Nanocarrier-Mediated Biostimulant

4.1.1 Nanoencapsulation

Encapsulation is an additional improvement of nanocarriers concerning the optimization of substance deliveries. This has drawn much interest from scientists attempting to enhance the use of encapsulated molecules for overactive control: in the 1980s, starch microparticles were suggested for use as carriers in medicine for the more efficient transport low molecular weighers or proteins to target regions, as well as during injection to prevent their degradation (Artursson et al., 1984). To tested two different types of albumins loaded microcapsule, which proved that the starch degradation of microcapsule lost loaded molecule highlighting the importance of regulating the actual release of molecules on target (Artursson et al., 1984). Over time, with effective application of various nanoplatforms in healthcare, much wider ranges of nano-encapsulation have been developed and used such as emulsification, nanoprecipitation, layer-by-layer nanoparticles and polymerization (Mora-Huertas et al., 2010).

Wang classified the nanocarriers into four categories according to their loading principle: cavity loading systems, matrix loading, surface loading, and molecular level loading (Fig. 1) (Wang et al., 2019). A common technique is molecular-level loading systems, which combine host and retention molecules. Whereas lower molecular weight of guest molecules by an active element like medicines or nucleic acids, retention molecules are made up of peptide, polymer or tiny molecules. These two parties may interact chemically or physically, which could result in complexes. Because of the hydrophobic contact forces that propel the process, by physical attraction complex formed is called self-assembling. The resulting structures are frequently cyclic molecules with a hydrophilic edge and a hydrophobic core cavity; however, dendrimeric complexes are also possible. To determine the polar property the functional group and pore property are found in polymer structure determine the loading capacity. The weak interaction between hydrophilic contact with hydrophobic pores with the functional group of polymers can charge target molecules. Chemically loaded host molecules are frequently employed to create intelligent release systems, but physically loaded nanocomponents are rarely able to achieve controlled release because of the non-specific nature of the contact. When loading by physical interactions, these restrictions are much more noticeable since content loading is more constrained because only particular guest molecules and sizes may be loaded (Wang et al., 2019).

For instance, *Toddalia asiatica* essential oil, which possesses antifungal properties and suppresses aflatoxin B1, is based on nanoencapsulated chitosan (Roshan et al., 2022). To make the NCs of chitosan-based nanoencapsulated oil using the ionic gelation process, Toddalia essential oil was progressively added dropwise to the low molecular weight chitosan solution (Roshan et al., 2022). Highly reactive chitosan-TPP ions were created by electrostatically interacting with positively charged chitosan (NH_3^+ groups) using negatively charged sodium polyphosphate (TPP) ions. These ions then mix with Toddalia essential oil to create oil NCs that are nanoencapsulated and based on chitosan. According to scanning electron microscope (SEM) pictures, the chitosan NCs were spherical, soft, and compact. They also demonstrated a high load capacity and encapsulation effectiveness when combined with essential oils (Roshan et al., 2022).

Figure 1. Nanoencapsulation strategies

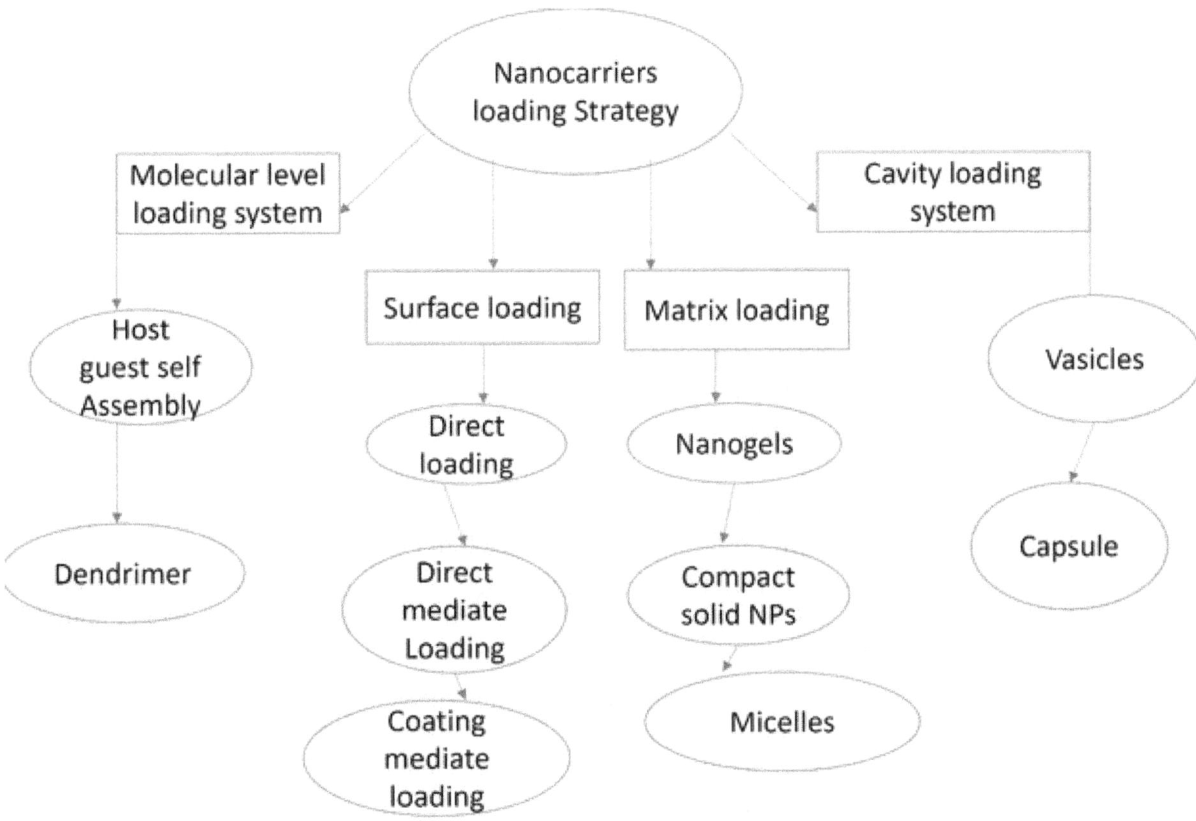

4.1.2 Nano-Biostimulants

Without affecting ecosystems, biostimulants can increase crop production and productivity whereas shielding plants from environmental stresses. Nevertheless, these products' simple degradability is one of their main drawbacks in large-scale applications. Biostimulating agents are often diluted in water and applied topically or to the roots. This stage may expose molecules to biotic and abiotic degradation processes, which may result in a reduction in plant efficacy and period of action. Only a few applications exist in this cutting-edge field, and the possibility of the biostimulant carried in specially designed nanocarriers is still untapped (Jimenez-Arias et al., 2020). Twenty years ago, the first study was carried out in which the self-assembly approach was used to load a-naphthaleneacetate (NAA), a plant growth regulator, into an inorganic matrix of double hydroxide layers of Zn–Al. The pH changes in the solution caused NAA to be released from the matrix (Bin Hussein et al., 2002). Although no studies were carried out on plants using this nanoplatform, the goal of that work was to investigate the ability of the release mechanisms.

The use of nanoencapsulated biostimulant chemicals in plants was a gradual process. It has been shown that the effects of salt stress in maize can be mitigated by using chitosan nanoparticles loaded with the NO donor s-nitrous mercaptosuccinic acid (s-nitrous-MSA) (Oliveira et al., 2016). Compared to plants treated with the free NO precursor mercaptosuccinic acid, salt-stressed maize plants treated with 50 μM and 100 μM S-nitroso-MSA NP chitosan on soil had higher foliar S-nitrosothiol concentrations. The low concentration treatment of S-nitroso-MSA chitosan NPs completely halted the change of photosystem II activity, chlorophyll content, and maize plant development caused by salt stress. These nanoplatforms' 38 nm size and positive zeta potential allowed them to be effectively absorbed and transported into the plants from the roots. Its 38 nm size allowed it to spread to the leaves and penetrate root tissues with ease (Oliveira et al., 2016).

4.1.3 Soil Remediation

A unique remediation platform using specially designed nanocarriers has been developed. The main challenges associated with contaminated sites are complexity and heterogeneity of the soil–water interface, along with the efficient communication among the target pollutant and the remedial agent. High reactivity of NPs and small size may allow them to be good candidates for overcoming these challenges. This adsorptive and oxidative degradation via nanocarrier-assisted processes or alternative combinations with remediation of contaminants may remove a range of different pollutants (Qian et al., 2020).

Adsorption and immobilization are unique methods of remediation to clean up soils contaminated with metals (Zhu et al., 2019). While deploying this technology, given the goal and soil conditions, choosing unique materials or coating agents for the contaminating agents becomes difficult.Various kinds of Nanoparticles, such as metal oxides, carbon NMs, nanocomposite NPs, etc., have been shown to work in immobilizing both organic and inorganic contaminants, with each NP working against a distinct target pollutant. Carbon NMs absorb organic pollutants by π–π interactions and van der Waals forces, while metal oxide NMs, such as FeO_4 and TiO_2 (the most researched), and nanocomposite NPs, such as polysaccharide stabilized Fe–Mn binary oxide, may immobilize heavy metals and organic molecules via surface complexation (Qian et al., 2020).

4.1.4 Nanofertilizers

Several inorganic and organic NCs have been assessed in recent years. Other than their intended use as fertilizer, some of them displayed adverse consequences. For instance, Ti NF has been shown to improve soil characteristics upon application, which has subsequently improved plant growth and biomass production (Pérez-de-Luque et al., 2017). Cu and Ag NPs might be used as an example once more. In soil the plants treated with Cu NPs, *Medicago sativa* displayed superior agronomical performances, such as increased length of roots and shoot and fresh weight (Cota-Ruiz et al., 2020). In contrast, Ag NPs enhanced plant tolerance and growth under heat stress in Triticum aestivum (Ali et al., 2020). These two metal species also exhibit antimicrobial and pesticidal qualities: Phytophthora infestans were effectively controlled in a study involving Cu NPs are used to treat *Solanum lycopersicum* and applied to the leaves (Giannousi et al., 2013), and Ag NPs demonstrated antibacterial activity compared to plant pathogens like *Xanthomonas campestris* and *Xanthomonas axonopodis* (Vanti et al., 2019).

4.1.5 Nanopesticides

Another category of nanopesticides is pesticides which are linked with nanocarriers. When an active molecule physically adheres to a substrate, activity can be controlled. Soft NPs (like lipids, polymers, and nanomatrices) or hard NPs (like silica, nanotubes, or graphene), have the ability to dissolve, adsorb, trap, or enclose pesticides depending on the design of the nanocarriers. By protecting pesticides from photolysis and environmental degradation, the nanocarrier provides excellent stability. Novel nanopesticides using avermectins (AVMs) as the active component supply an interesting example. They include a category of naturally produced medications termed AVMs that don't harm mammals, generated from Streptomyces avermitilis. An increase in chloride ion influx to the cells leads to hyperpolarization and paralysis of the invertebrate neuromuscular system (Pal et al., 2014; Xu et al., 2022). High-energy ultraviolet radiation damages the AVMs, which are extensively used as insecticides, nematicides, anthelminthics, and arachnicides (Batiha et al., 2020; de Souza et al., 2022). AVMs trapped in functional boron nitrile NPs have many good properties, such as good water dispersibility, very high loading capacity, pH-responsive release, gradual/sustained release, good resistance to UV degradation, and immense adhesion to surfaces of plant (Hao et al., 2020).

5. CONCLUSION AND FUTURE PERSPECTIVES

Nanocarriers have transformed agricultural biotechnology with their sophisticated release of biostimulants. Delivery systems for plant growth stimulate progress in specificity and efficiency, enhancing their capacity towards the sustainability of modern farming practices. More studies are needed to reduce safety threats and optimize compositions of nanocarriers fit for mass agricultural deployment. Nanoparticle-based systems for the biotic release of stimulants have a lot of promises for enhancing plant productivity and consequently food security in the world. It is also in this perspective of a green economy that the farming and agriculture sector very well stands to change its mode of farming and disposal of waste from agriculture and other sectors because of the development of nano-biostimulants. By enabling the field deployment of volatile or readily degradable compounds that may otherwise only be used in laboratory experiments, the possibility of finding novel biostimulant sources from any raw material is increased by nanomaterials with integrated biostimulant molecules. Furthermore, because a single substance may have two effects on plants, nanobiostimulants have the potential to completely transform agronomic crop treatments; the encapsulated biostimulant may cause one reaction, and the capsule itself may cause a second. An interdisciplinary team of agronomists, nanomaterial development specialists, and plant physiologists should serve as the foundation for future study in this field.

Using this multidisciplinary approach, the design of nano-biostimulants should concentrate on a better understanding of how NPs and biostimulants behave in plants while promoting the use of inexpensive, environmentally friendly biostimulants and nano-capping materials, perhaps made from waste materials. For instance, lignin and silica that are taken from agricultural waste are promising substances in this respect. When paired with fundamental research, applied studies for field responses are crucial to allowing farmers to use this new, promising technology. A new perspective on agriculture that is centered on the circular economy, lessens the influence of humans on the environment, and is better integrated with the producing region may be facilitated by nano-biostimulants.

REFERENCES

Ali, N., Bilal, M., Khan, A., Ali, F., & Iqbal, H. M. N. (2020). Effective exploitation of anionic, nonionic, and nanoparticle-stabilized surfactant foams for petroleum hydrocarbon contaminated soil remediation. *The Science of the Total Environment, 704*, 135371. DOI: 10.1016/j.scitotenv.2019.135391 PMID: 31806317

Ali, O., Ramsubhag, A., & Jayaraman, J. (2021). Biostimulant properties of seaweed extracts in plants: Implications towards sustainable crop production. *Plants, 10*(3), 531. DOI: 10.3390/plants10030531 PMID: 33808954

Aloo, B. N., Tripathi, V., Makumba, B. A., & Mbega, E. R. (2022). Plant growth-promoting rhizobacterial biofertilizers for crop production: The past, present, and future. *Frontiers in Plant Science, 13*, 1002448. DOI: 10.3389/fpls.2022.1002448 PMID: 36186083

Artursson, P., Edman, P., Laakso, T., & Sjoholm, I. (1984). Characterization of polyacryl starch microparticles as carriers for proteins and drugs. *Journal of Pharmaceutical Sciences, 73*(11), 1507–1513. DOI: 10.1002/jps.2600731103 PMID: 6440978

Bastos, A., Orth, R., Reichstein, M., Ciais, P., Viovy, N., Zaehle, S., Anthoni, P., Arneth, A., Gentine, P., Joetzjer, E., Lienert, S., Loughran, T., McGuire, P. C., O, S., Pongratz, J., & Sitch, S. (2021). Vulnerability of European ecosystems to two compound dry and hot summers in 2018 and 2019. *Earth System Dynamics: ESD, 12*(4), 1015–1035. DOI: 10.5194/esd-12-1015-2021

Batiha, G. E. S., Alqahtani, A., Ilesanmi, O. B., Saati, A. A., El-Mleeh, A., & Hetta, H. F.. (2020). Avermectin derivatives, pharmacokinetics, therapeutic and toxic dosages, mechanism of action, and their biological effects. *Pharmaceuticals (Basel, Switzerland), 13*(8), 196. DOI: 10.3390/ph13080196 PMID: 32824399

Bell, I. R., Ives, J. A., & Jonas, W. B. (2014). Nonlinear effects of nanoparticles: Biological variability from hormetic doses, small particle sizes, and dynamic adaptive interactions. *Dose-Response, 12*(2), 202–232. DOI: 10.2203/dose-response.13-025.Bell PMID: 24910581

Bin Hussein, M. Z., Zainal, Z., Yahaya, A. H., & Foo, D. W. V. (2002). Controlled release of a plant growth regulator, alpha-naphthaleneacetate from the lamella of Zn–Al-layered double hydroxide nanocomposite. *Journal of Controlled Release, 82*(2–3), 417–427. DOI: 10.1016/S0168-3659(02)00172-4 PMID: 12175754

Brown, P., & Saa, S. (2015). Biostimulants in agriculture. *Frontiers in Plant Science, 6*, 671. DOI: 10.3389/fpls.2015.00671 PMID: 26379695

Cabrera-De la Fuente, M., González-Morales, S., Juárez-Maldonado, A., Leija-Martínez, P., & Benavides-Mendoza, A. (2018). Plant nutrition and agronomic management to obtain crops with better nutritional and nutraceutical quality. In Holban, A. M., & Grumezescu, A. M. (Eds.), *Therapeutic Foods* (pp. 99–140). Academic Press. DOI: 10.1016/B978-0-12-811517-6.00004-0

Canellas, L. P., Olivares, F. L., Okorokova-Facanha, A. L., & Facanha, A. R. (2002). Humic acids isolated from earthworm compost enhance root elongation, lateral root emergence, and plasma membrane H+-ATPase activity in maize roots. *Plant Physiology*, *130*(4), 1951–1957. DOI: 10.1104/pp.007088 PMID: 12481077

Cerdan, M., Sanchez-Sanchez, A., Oliver, M., Juarez, M., & Sanchez-Andreu, J. J. (Eds.). Effect of foliar and root applications of amino acids on iron uptake by tomato plants. In: 4th Balkan symposium on vegetables and potatoes; 2009 June 17; Plovdiv, BULGARIA2009

Colla, G., Hoagland, L., Ruzzi, M., Cardarelli, M., Bonini, P., Canaguier, R., & Rouphael, Y. (2017). Biostimulant action of protein hydrolysates: Unraveling their effects on plant physiology and microbiome. *Frontiers in Plant Science*, *8*, 2202. DOI: 10.3389/fpls.2017.02202 PMID: 29312427

Cota-Ruiz, K., Ye, Y. Q., Valdes, C., Deng, C. Y., Wang, Y., Hernandez-Viezcas, J. A., Duarte-Gardea, M., & Gardea-Torresdey, J. L. (2020). Copper nanowires as nanofertilizers for alfalfa plants: Understanding nano-bio systems interactions from microbial genomics, plant molecular responses and spectroscopic studies. *The Science of the Total Environment*, *742*, 140572. DOI: 10.1016/j.scitotenv.2020.140572 PMID: 32623177

de Souza, R. B., & Guimaraes, J. R. (2022). Effects of avermectins on the environment based on their toxicity to plants and soil invertebrates—A review. *Water, Air, and Soil Pollution*, *233*(7), 259. DOI: 10.1007/s11270-022-05744-0 PMID: 35789787

Du Jardin, P. (2015). Plant biostimulants: Definition, concept, main categories and regulation. *Scientia Horticulturae*, *196*, 3–14. DOI: 10.1016/j.scienta.2015.09.021

Eichert, T., Kurtz, A., Steiner, U., & Goldbach, H. E. (2008). Size exclusion limits and lateral heterogeneity of the stomatal foliar uptake pathway for aqueous solutes and watersuspended nanoparticles. *Physiologia Plantarum*, *134*(1), 151–160. DOI: 10.1111/j.1399-3054.2008.01135.x PMID: 18494856

Fraceto, L. F., Grillo, R., Medeiros, G. A., Scognamiglio, V., de Melo, N. F. S., & Rosa, A. H. (2016). Nanotechnology in agriculture: Which innovation potential does it have? *Frontiers in Environmental Science*, *4*, 20. DOI: 10.3389/fenvs.2016.00020

Ghormade, V., Deshpande, M. V., & Paknikar, K. M. (2011). Perspectives for nanobiotechnology-enabled protection and nutrition of plants. *Biotechnology Advances*, *29*(6), 792–803. DOI: 10.1016/j.biotechadv.2011.06.007 PMID: 21729746

Giannousi, K., Avramidis, I., & Dendrinou-Samara, C. (2013). Synthesis, characterization and evaluation of copper based nanoparticles as agrochemicals against *Phytophthora infestans.RSC Advances*, *3*(44), 21743–21752. DOI: 10.1039/c3ra42118j

Gotteland, M., Riveros, K., Gasaly, N., Carcamo, C., Magne, F., Liabeuf, G., Beattie, A., & Rosenfeld, S. (2020). The pros and cons of using algal polysaccharides as prebiotics. *Frontiers in Nutrition*, *7*, 163. DOI: 10.3389/fnut.2020.00163 PMID: 33072794

Hao, L., Gong, L., Chen, L., Guan, M., Zhou, H., Qiu, S., Wen, H., Chen, H., Zhou, X., & Akbulut, M. (2020). Composite pesticide nanocarriers involving functionalized boron nitride nanoplatelets for pH-responsive release and enhanced UV stability. *Chemical Engineering Journal*, *396*, 125233. DOI: 10.1016/j.cej.2020.125233

Hillegass, J. M., Shukla, A., Lathrop, S. A., MacPherson, M. B., Fukagawa, N. K., & Mossman, B. T. (2010). Assessing nanotoxicity in cells in vitro. *Wiley Interdisciplinary Reviews. Nanomedicine and Nanobiotechnology*, *2*(3), 219–231. DOI: 10.1002/wnan.54 PMID: 20063369

Jimenez-Arias, D., Morales-Sierra, S., Borges, A. A., & Diaz, D. D. (2020). Biostimulant nanoencapsulation: The new keystone to fight hunger. *Journal of Agricultural and Food Chemistry*, *68*(27), 7083–7085. DOI: 10.1021/acs.jafc.0c02872 PMID: 32588622

Jindo, K., Canellas, L. P., Albacete, A., dos Santos, L. F., Rocha, R. L. F., & Baia, D. C.. (2020). Interaction between humic substances and plant hormones for phosphorous acquisition. *Agronomy (Basel)*, *10*(5), 640. DOI: 10.3390/agronomy10050640

Judy, J. D., & Bertsch, P. M. (2014). Bioavailability, toxicity, and fate of manufactured nanomaterials in terrestrial ecosystems. *Advances in Agronomy*, *123*, 1–64. DOI: 10.1016/B978-0-12-420225-2.00001-7

Karavidas, I., Ntatsi, G., Vougeleka, V., Karkanis, A., Ntanasi, T., Saitanis, C., Agathokleous, E., Ropokis, A., Sabatino, L., Tran, F., Iannetta, P. P. M., & Savvas, D. (2022). Agronomic practices to increase the yield and quality of common bean (*Phaseolus vulgaris* L.): A systematic review. *Agronomy (Basel)*, *12*(2), 271. DOI: 10.3390/agronomy12020271

Khot, L. R., Sankaran, S., Maja, J. M., Ehsani, R., & Schuster, E. W. (2012). Applications of nanomaterials in agricultural production and crop protection: A review. *Crop Protection (Guildford, Surrey)*, *35*, 64–70. DOI: 10.1016/j.cropro.2012.01.007

Knox, J. P., & Benitez-Alfonso, Y. (2014). Roles and regulation of plant cell walls surrounding plasmodesmata. *Current Opinion in Plant Biology*, *22*, 93–100. DOI: 10.1016/j.pbi.2014.09.009 PMID: 25286000

Liu, R., Zhang, H., & Lal, R. (2019). Effects of stabilized nanoparticles on soil microbial communities and plant growth. *The Science of the Total Environment*, *662*, 296–303.

Martone, P. T., Estevez, J. M., Lu, F. C., Ruel, K., Denny, M. W., Somerville, C., & Ralph, J. (2009). Discovery of lignin in seaweed reveals convergent evolution of cell wall architecture. *Current Biology*, *19*(2), 169–175. DOI: 10.1016/j.cub.2008.12.031 PMID: 19167225

Mora-Huertas, C. E., Fessi, H., & Elaissari, A. (2010). Polymer-based nanocapsules for drug delivery. *International Journal of Pharmaceutics*, *385*(1–2), 113–142. DOI: 10.1016/j.ijpharm.2009.10.018 PMID: 19825408

Muscolo, A., Sidari, M., Francioso, O., Tugnoli, V., & Nardi, S. (2007). The auxin-like activity of humic substances is related to membrane interactions in carrot cell cultures. *Journal of Chemical Ecology*, *33*(1), 115–129. DOI: 10.1007/s10886-006-9206-9 PMID: 17111220

Nardi, S., Concheri, G., Pizzeghello, D., Sturaro, A., Rella, R., & Parvoli, G. (2000). Soil organic matter mobilization by root exudates. *Chemosphere*, *41*(5), 653–658. DOI: 10.1016/S0045-6535(99)00488-9 PMID: 10834364

Nardi, S., Pizzeghello, D., Muscolo, A., & Vianello, A. (2002). Physiological effects of humic substances on higher plants. *Soil Biology & Biochemistry*, *34*(11), 1527–1536. DOI: 10.1016/S0038-0717(02)00174-8

Nardi, S., Schiavon, M., & Francioso, O. (2021). Chemical structure and biological activity of humic substances define their role as plant growth promoters. *Molecules (Basel, Switzerland)*, *26*(8), 2256. DOI: 10.3390/molecules26082256 PMID: 33924700

Oliveira, H. C., Gomes, B. C. R., Pelegrino, M. T., & Seabra, A. B. (2016). Nitric oxide-releasing chitosan nanoparticles alleviate the effects of salt stress in maize plants. *Nitric Oxide*, *61*, 10–19. DOI: 10.1016/j.niox.2016.09.010 PMID: 27693703

Pal, S., Sharma, M. K., Danielsson, B., Willander, M., Chatterjee, R., & Bhand, S. (2014). A miniaturized nanobiosensor for choline analysis. *Biosensors & Bioelectronics*, *54*, 558–564. DOI: 10.1016/j.bios.2013.11.057 PMID: 24325984

Perez, M. J., Falque, E., & Dominguez, H. (2016). Antimicrobial action of compounds from marine seaweed. *Marine Drugs*, *14*(3), 52. DOI: 10.3390/md14030052 PMID: 27005637

Pérez-de-Luque, A. (2017). Interaction of nanomaterials with plants: What do we need for real applications in agriculture? *Frontiers in Environmental Science*, *5*(Apr), 12. DOI: 10.3389/fenvs.2017.00012

Pizzeghello, D., Schiavon, M., Francioso, O., Dalla Vecchia, F., Ertani, A., & Nardi, S. (2020). Bioactivity of size-fractionated and unfractionated humic substances from two forest soils and comparative effects on N and S metabolism, nutrition, and root anatomy of *Allium sativum* L. *Frontiers in Plant Science*, *11*, 1203. DOI: 10.3389/fpls.2020.01203 PMID: 32922415

Poschenrieder, C., Cabot, C., Martos, S., Gallego, B., & Barceló, J. (2013). Do toxic ions induce hormesis in plants? *Plant Science*, *212*, 15–25. DOI: 10.1016/j.plantsci.2013.07.012 PMID: 24094050

Qian, Y. T., Qin, C. D., Chen, M. M., & Lin, S. J. (2020). Nanotechnology in soil remediation— Applications vs. implications. *Ecotoxicology and Environmental Safety*, *201*, 110815. DOI: 10.1016/j.ecoenv.2020.110815 PMID: 32559688

Rai, M., Ingle, A. P., Pandit, R., Paralikar, P., Anasane, N., Santos, C. A., & Duran, N. (2018). Nanotechnology for agriculture: Impact and applications. *The Science of the Total Environment*, *636*, 741–755.

Rioux, L. E., Turgeon, S. L., & Beaulieu, M. (2007). Characterization of polysaccharides extracted from brown seaweeds. *Carbohydrate Polymers*, *69*(3), 530–537. DOI: 10.1016/j.carbpol.2007.01.009

Ritchie, J. D., & Perdue, E. M. (2008). Analytical constraints on acidic functional groups in humic substances. *Organic Geochemistry*, *39*(6), 783–799. DOI: 10.1016/j.orggeochem.2008.03.003

Roshan, A. B., Venkatesh, H. N., Dubery, N. K., & Mohana, D. C. (2022). Chitosan-based nanoencapsulation of *Toddalia asiatica* (L.) Lam essential oil to enhance antifungal and aflatoxin B-1 inhibitory activities for safe storage of maize. *International Journal of Biological Macromolecules*, *204*, 476–484. DOI: 10.1016/j.ijbiomac.2022.02.026 PMID: 35151710

Sajid, M., Płotka-Wasylka, J., & Andruch, V. (2020). Nanomaterials as a sustainable choice for enhancing agricultural productivity. *Environmental Science. Nano*, *7*(12), 3545–3568.

Scaglia, B., Nunes, R. R., Rezende, M. O. O., Tambone, F., & Adani, F. (2016). Investigating organic molecules responsible of auxin-like activity of humic acid fraction extracted from vermicompost. *The Science of the Total Environment*, *562*, 289–295. DOI: 10.1016/j.scitotenv.2016.03.212 PMID: 27100009

Schaafsma, G. (2009). Safety of protein hydrolysates, fractions thereof and bioactive peptides in human nutrition. *European Journal of Clinical Nutrition*, *63*(10), 1161–1168. DOI: 10.1038/ejcn.2009.56 PMID: 19623200

Shang, Y., Hasan, M. K., & Ahammed, G. J. (2020). Applications of nanotechnology in plant growth and crop protection: A mini-review. *Journal of Plant Growth Regulation*, *39*(4), 1381–1393.

Torney, F., Trewyn, B. G., Lin, V. S. Y., & Wang, K. (2007). Mesoporous silica nanoparticles deliver DNA and chemicals into plants. *Nature Nanotechnology*, *2*(5), 295–300. DOI: 10.1038/nnano.2007.108 PMID: 18654287

Vanti, G. L., Nargund, V. B., Basavesha, K. N., Vanarchi, R., Kurjogi, M., & Mulla, S. I.. (2019). Synthesis of *Gossypium hirsutum*-derived silver nanoparticles and their antibacterial efficacy against plant pathogens. *Applied Organometallic Chemistry*, *33*(1), e4630. DOI: 10.1002/aoc.4630

Vygovsky, Y. N., & Malov, A. N. (2003). *The Physics of Laser Biostimulation*. JSC MILTA-PKP GIT.

Wang, N., Cheng, X. J., Li, N., Wang, H., & Chen, H. Y. (2019). Nanocarriers and their loading strategies. *Advanced Healthcare Materials*, *8*(6), 1801002. DOI: 10.1002/adhm.201801002 PMID: 30450761

Xu, Z. L., Tang, T., Lin, Q., Yu, J. Z., Zhang, C. P., Zhao, X. P., Kah, M., & Li, L. (2022). Environmental risks and the potential benefits of nanopesticides: A review. *Environmental Chemistry Letters*, *20*(3), 2097–2108. DOI: 10.1007/s10311-021-01338-0

Zhang, X. Z., Ervin, E. H., & Schmidt, R. E. (2003). Physiological effects of liquid applications of a seaweed extract and a humic acid on creeping bentgrass. *Journal of the American Society for Horticultural Science*, *128*(4), 492–496. DOI: 10.21273/JASHS.128.4.0492

Zhu, Y., Liu, X., Hu, Y., Wang, R., Chen, M., Wu, J., Wang, Y., Kang, S., Sun, Y., & Zhu, M. (2019). Behavior, remediation effect and toxicity of nanomaterials in water environments. *Environmental Research*, *174*, 54–60. DOI: 10.1016/j.envres.2019.04.014 PMID: 31029942

Zuverza-Mena, N., Martínez-Fernández, D., Du, W., Hernandez-Viezcas, J. A., Bonilla-Bird, N., López-Moreno, M. L., Komárek, M., Peralta-Videa, J. R., & Gardea-Torresdey, J. L. (2017). Exposure of engineered nanomaterials to plants: Insights into the physiological and biochemical responses-A review. *Plant Physiology and Biochemistry*, *110*, 236–264. DOI: 10.1016/j.plaphy.2016.05.037 PMID: 27289187

Chapter 8
Nanocarriers for Delivering Nucleic Acids in Plants:
Mechanisms and Applications

Bhanuprakash Ananthakumar
https://orcid.org/0009-0001-5517-7933
Sri Sathya Sai University for Human Excellence, India

Nagashri Nanjundeshwara
Sri Sathya Sai University for Human Excellence, India

Archana Lakshmaiah
Sri Sathya Sai University for Human Excellence, India

Ramya Raghavan
https://orcid.org/0000-0002-9953-543X
Sri Sathya Sai University for Human Excellence, India

ABSTRACT

Nucleic acid nanocarriers represent a transformative approach in plant biotechnology, facilitating the efficient delivery of genetic material into plant cells. This review explores the various types of nanocarriers, including gold nanoparticles, silica nanoparticles, and carbon nanotubes, highlighting their unique advantages, such as enhanced cellular uptake, targeted delivery, and controlled release. By overcoming the physical barriers inherent in plant cell walls, these nanocarriers enable precise genetic modifications and the introduction of traits that enhance growth, stress tolerance, and metabolic profiles. Furthermore, the application of nucleic acid therapeutics, including RNA interference and genome editing technologies, is significantly advanced through the use of nanocarriers. The future of plant biotechnology is poised for innovative breakthroughs as researchers continue to refine these delivery systems, unlocking new possibilities for crop improvement and sustainable agricultural practices.

DOI: 10.4018/979-8-3693-9869-2.ch008

INTRODUCTION

Genetic engineering can enhance crop yields by developing climate-resilient crop varieties. Nanobiotechnology is crucial in delivering genetic materials, nutrients, and stress-responsive agents into plant cells. Nucleic acid delivery introduces desirable crop traits, such as increased yield, pest resistance, and improved nutritional content. Gene silencing is a natural mechanism found in eukaryotic cells that regulates gene expression at the posttranscriptional level, effectively managing the levels of specific cell transcripts. Exogenous RNA interference (RNAi) technology is a nontransgenic method for improving crop resilience against pests and pathogens, addressing the need for sustainable agricultural practices (Septiani et al., 2025). This process, known as RNA interference (RNAi), was identified in the early 2000s. Identifying RNA interference (RNAi) as a gene-silencing mechanism in eukaryotic organisms, conducted by Prof. Andrew Fire and Prof. Craig Mello in 1998, is regarded as one of the most significant breakthroughs in modern biology. Their pioneering work, which demonstrated how RNA could regulate gene expression in the nematode worm Caenorhabditis elegans, earned them the Nobel Prize in Physiology or Medicine in 2006. Understanding gene function and regulation in plants often requires precise delivery of DNA or RNA. It enables researchers to study gene functions, pathways, and interactions, contributing to advancements in plant biotechnology. RNA interference (RNAi) has emerged as a promising biotechnological tool to mitigate risks associated with conventional transgenic methods, enabling the silencing of specific genes to enhance disease resistance in plants.

Figure 1. Research in nanotechnology begins, laying the foundation for future applications in agriculture. History and Development of Technologies to Deliver Nucleic Acids in Plants

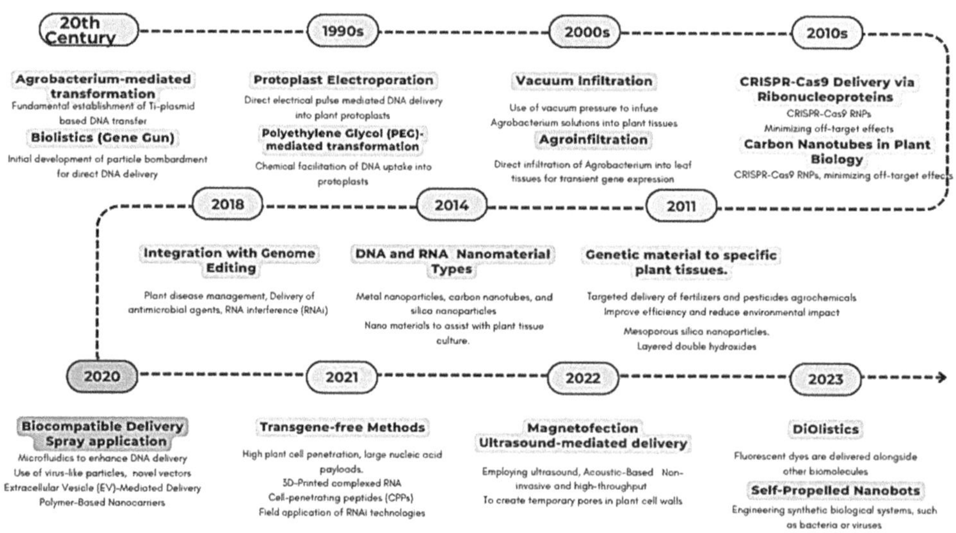

The Giraldo et al. (2014) study demonstrated that single-walled carbon nanotubes (SWCNTs) can Passively transport and localize within chloroplast lipid envelopes to augment photosynthetic activity. It Enhanced light capture and energy conversion. RNAi is triggered by long double-stranded RNA (dsRNA) molecules, which can originate from external or internal sources. Specific proteins process these dsRNAs into smaller molecules called small interfering RNAs (siRNAs). The siRNAs then target the cell's native transcripts through complementary base pairing, resulting in the degradation of the target transcript by a protein complex containing endonucleases. Since the late 1980s, RNAi has been combined with genetic engineering techniques. Efficient nucleic acid delivery in plants supports genetic enhancement, functional research, and the development of resilient crops, ultimately contributing to food security and sustainability in agriculture. Foundational limitations of traditional methods like Agrobacterium-mediated transformation and biolistics, are off-target effects, species and payload limitations, lack of precise delivery and multiple cargos.

Biolistics can cause significant physical damage to plant cells, and genetic modifications.

Agrobacterium has host-range limitations, often resulting in random integration of genetic material, hindering targeted gene editing. Recent studies demonstrate that double-stranded RNA (dsRNA) sprays and transgenic plants with RNAi vectors effectively control plant diseases and contribute to sustainable agricultural practices (Ali et al., 2024).

Nanocarriers, defined by their size ranging from 1 to 1000 nanometers and their specific function, serve as transport and encapsulation systems for active ingredients. These systems are designed to protect their payload from degradation, improve its dispersibility, enable targeted delivery to specific sites of action within the plant, and facilitate controlled release of the active compound. These attributes are particularly valuable in plant biotechnology, where challenges such as the plant cell wall barrier and the need for precise delivery of agrochemicals and genetic material often limit the effectiveness of conventional methods. The effectiveness of nanocarriers depends on their interaction with plant cells, involving:

- Complex uptake mechanisms through passive diffusion, endocytosis, or transport proteins
- Translocation through apoplastic and symplastic pathways
- Controlled release of cargo based on environmental triggers

Cuticular Pathway

The cuticle is the outermost protective layer of plant leaves and stems. It is composed of a hydrophobic waxy surface that prevents water loss and provides a barrier against microbes and environmental stressors. Nanocarriers must breach this barrier to access the underlying tissues (Gao et al., 2025).

Small nanoparticles (typically <50 nm) can diffuse through the cuticle's nano-scale pores. At the same time, larger particles may require mechanical disruption or chemical pre-treatment to penetrate. The hydrophobic nature of the cuticle can repel hydrophilic nanoparticles. The thickness and chemical composition of the cuticle vary across plant species, tissues, and even developmental stages, affecting penetration efficiency. Hydrophobic nanoparticles or those coated with surfactants may traverse the waxy cuticle more quickly due to increased compatibility with the hydrophobic surface (Gandhi et al., 2024). Delivery to the cuticle is often used for foliar applications, such as spraying nanoparticles for genetic modification or disease resistance.

Stomatal Pathway

Stomata are microscopic pores on the surface of leaves that regulate gas exchange and water vapor loss. These natural openings can serve as direct entry points for nanocarriers. Stomatal pathways are often utilized for the foliar delivery of nanocarriers in controlled environments or when stomata are naturally open (Miyamoto & Numata, 2023).

The stomatal pathway bypasses the hydrophobic cuticle, making it an efficient route for nanocarrier uptake. It allows for easier entry of larger nanoparticles compared to the cuticular pathway. Nanocarriers can penetrate through the stomatal pores (typically 3–12 μm in size) and enter the sub-stomatal cavity. From there, nanocarriers can move into the mesophyll tissue and vascular system for systemic distribution. Stomatal opening is regulated by environmental factors such as light, humidity, and CO_2 concentration, limiting penetration efficiency under certain conditions. Excessive nanoparticle loading may clog stomatal pores, affecting plant physiology (Rani et al., 2025).

Root Uptake

Roots are in constant contact with the soil and play a critical role in water uptake, nutrients, and other compounds. Nanocarriers can exploit this natural absorption function to deliver compounds to plants. Root-based delivery is often explored for soil-applied nanocarriers aimed at improving nutrient use efficiency, stress tolerance, or delivery of genetic materials.

Nanocarriers in the rhizosphere (soil surrounding the roots) can be absorbed through the root epidermis via apoplastic pathways (movement between cell walls) or symplastic pathways (movement through cytoplasm via plasmodesmata). Root hair cells and lateral root tips, which lack thick lignified cell walls, serve as hotspots for nanoparticle penetration. Root uptake provides systemic delivery, allowing nanocarriers to travel through the plant's vascular system (xylem and phloem). Root pathway enables sustained release and long-term effects.

Nanoparticle size plays a crucial role; smaller nanoparticles (<20 nm) are more likely to be taken up. Surface charge is also key; cationic nanoparticles have enhanced interactions with the negatively charged root cell walls, promoting adhesion and uptake (Bahl et al., 2024). Soil composition, pH, and organic matter can influence nanoparticle stability and availability for uptake. Potential environmental risks, such as nanoparticle leaching into groundwater, must be carefully managed (Iqbal et al., 2024).

Initial experiments with transgenic plants incorporating RNAi successfully demonstrated gene suppression and silencing, showcasing its potential in plant biotechnology. Better cellular uptake and release kinetics often induce the RNAi mechanism by introducing dsRNA using virus-induced gene silencing (VIGS), a viral method, or through a plasmid bacterial system utilizing *Agrobacterium tumefaciens*. Researchers have leveraged RNAi to investigate gene functions, control insects, invasive plants, and fungal inhibition and resistance against viruses, insects, and nematodes (Peyret et al., 2025). These carriers can be introduced through multiple delivery methods: Foliar application (leaf-based), Soil application, Seed treatment, Biolistic delivery.

Table 1. Comparison of different modes of nanocarrier treatments

	Stomatal Delivery (Foliar Sprays)	Cuticular Penetration (Foliar Sprays)	Root Uptake (Soil Application)	Vascular Transport (Systemic Delivery)
Mechanism	Diffusion through stomatal pores	Diffusion across the cuticle, potentially aided by cracks or defects	Passive diffusion or active transport across root epidermal cells	Movement through xylem and phloem tissues
Application	Agrochemicals, nutrients, biostimulants directly to leaves	Agrochemicals, nutrients, biostimulants to leaf tissues	Nutrients, water, and potentially systemic pesticides	Systemic delivery of therapeutics, signaling molecules, or nutrients
Nanoparticle Size	< 200 nm (ideally < 100 nm for efficient entry)	< 10 nm (for penetration through waxy layers) to < 100 nm (through defects)	< 100 nm for efficient uptake, but larger sizes can also be taken up	Variable, depending on the vascular system and application method
Mode of Action	Surface modifications for enhanced stomatal adhesion and entry	Strategies to disrupt the cuticle barrier (e.g., surfactants, co-solvents)	Understanding root exudate interactions with nanoparticles	Developing nanoparticle tracking methods within vascular tissues

Adapted from (Gao et al., 2025), (Miyamoto & Numata, 2023), (Peyret et al., 2025).

NEED FOR NANOPARTICLES AND NANOCARRIERS

The first documented instance of applying exogenous RNA to trigger the RNA interference (RNAi) mechanism in plants occurred in 2011, as described in a Monsanto patent. This involved spraying double-stranded RNA (dsRNA) on *Nicotiana benthamiana* (Patent No. US20110296556A1). which produced small interfering RNA (siRNA) targeting the phytoene desaturase (PDS) gene. RNAi has since been utilized in nontransformation systems to protect various crop species from insects, nematodes, and pathogens. Delivery systems can also integrate genes that promote beneficial interactions, enhancing plant growth and resilience against pests and pathogens. Altering the genetic makeup of many plant species, particularly their plastids, is difficult due to the presence of a cell wall, which hinders the adequate transportation of biomolecules.

Nanocarriers are systems designed to transport cargo, such as nucleic acids, and typically range in size from 10 to 1000 nm. They have been extensively studied for use in animal cells but face significant challenges in plant cells due to additional barriers like the cuticle and cell wall. Techniques like Agrobacterium-mediated transformation have been less preferred in recent times. The shift away from virus-based vectors for nucleic acid delivery is driven by factors related to safety, efficacy, and practicality (Rani et al., 2025). The effectiveness of nanocarriers depends on their interaction with plant cells, involving:

- Complex uptake through passive diffusion, endocytosis, or transport proteins
- Translocation through apoplastic and symplastic pathways
- Controlled release of cargo based on environmental triggers

Figure 2. Agriculture applications of nanoparticles and nanocarriers

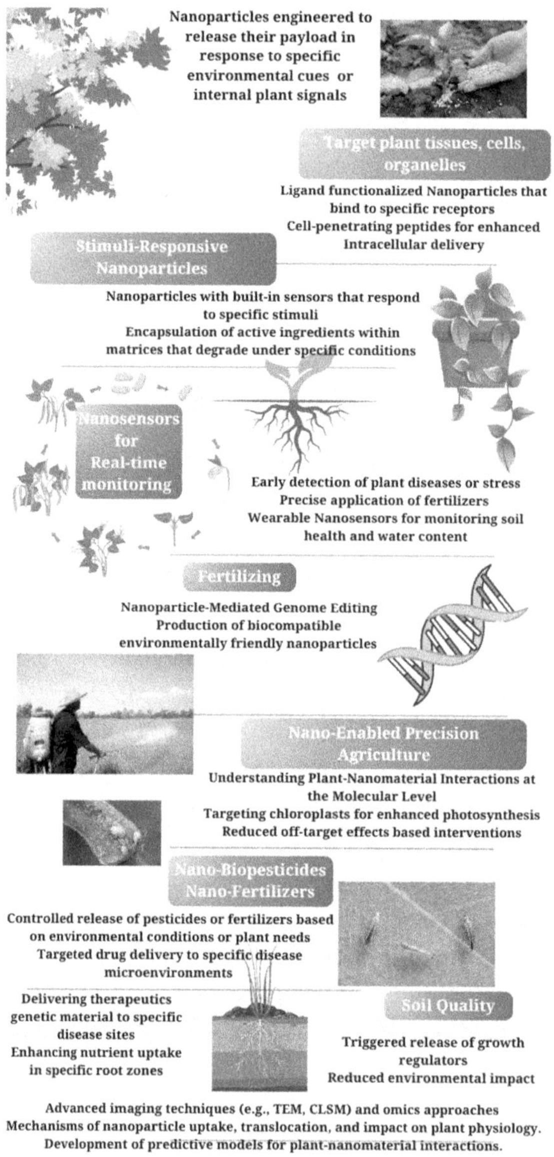

Efficient delivery of nanocarriers carrying nucleic acids into plants requires overcoming the major physical barriers posed by the plant structure. The penetration of nanocarriers into plant tissues primarily occurs through three main pathways: the cuticular pathway, the stomatal pathway, and root uptake. Each pathway offers unique mechanisms and challenges for the internalization of nanoparticles (Table 1).

Based on their surface charge, nanocarriers can be broadly classified into cationic and anionic types. This fundamental property dictates their interactions with plant tissues and cellular components, which are often negatively charged. The surface charge plays a pivotal role in the initial adhesion of nanoparticles to plant surfaces, subsequent uptake into plant cells, transport within the plant vascular system, and interaction with intracellular biological molecules.

Monsanto's RNAi patents for crop protection (US20110296556A1) foundational research provides crucial insights into their potential. Demirer et al. (2019) showcased carbon nanotube-mediated DNA delivery without transgene integration (Demirer et al., 2019). Furthermore, field trials of nano-pesticides in India, potentially through institutions like ICAR, are exploring real-world efficacy. Ethical and regulatory frameworks, such as the OECD Guidelines for Nanoparticle Testing (2020) and EPA Reports on Nano-Agrochemical Safety (2022), are crucial for responsible development. Case studies, such as applying layered double hydroxides (LDHs) in African maize farms, could illustrate scalability, and cost-benefit analyses (Khatoon & Velidandi, 2025).

Immunogenicity

Existing methods for gene introduction, such as Agrobacterium-mediated transformation, biolistic particle delivery, electroporation, and PEG transformation, have limitations, including being labor-intensive, time-consuming, and potentially damaging to plant tissues. Virus-based vectors can elicit strong immune responses in patients, leading to inflammation and other immune-related complications. This immunogenicity may pose significant risks, particularly in therapeutic applications. The plant's immunity to specific viral vectors (e.g., adenoviruses) can reduce the effectiveness of the delivery system upon repeated administration. Nanotechnology is beginning to play a role in genetic engineering, with the cuticle's permeability being influenced by the size and charge of molecules. Nanoparticles serve as potential delivery vehicles for biomolecules. Still, the cell wall's structure varies, affecting intracellular cargo delivery for agriculture-based applications. Emerging virology findings indicate that RNA replication is likely the primary factor in selectivity. However, packaging signals may enhance the efficiency of RNA incorporation into viral particles (Peyret et al., 2025).

Cytotoxicity

Nanotechnology, particularly functionalized carbon-based or green nanomaterials, is synthesized through environmentally friendly methods. Functionalization of carbon nanocarriers can reduce cytotoxicity and improve delivery outcomes. Single-walled carbon nanotubes (SWNTs) can target specific organelles, including chloroplasts and mitochondria. They have potential for RNA interference applications in crop protection (Ashfaq et al., 2025)., and offer unique properties that make them suitable for targeted plant disease management. The green nanoparticles treat plant diseases caused by bacteria, viruses, and fungi, promoting sustainable and less hazardous disease management strategies (Wasule et al., 2024). The cuticle, a lipophilic layer covering plant leaves, and the rigid cell wall present significant mechanical barriers to nucleic acid delivery. The size and charge of molecules influence the cuticle's permeability, while the cell wall's thickness and pore size also affect delivery. The viral vectors can be cytotoxic, causing damage to host cells or tissues. This cytotoxicity can complicate their use in gene therapy and may limit the therapeutic window. Further, incorporating star polycation enhances the efficacy of seboctylamine in inhibiting viral accumulation in soybean plants, resulting in improved agronomic traits and effective control of the Soybean mosaic virus (Gao et al., 2025).

Limited Cargo Capacity

Many viral vectors have restricted packaging capacities, which limits the size of the nucleic acids that can be delivered. Environmental factors like UV radiation and rain degrade naked nucleic acids. This constraint makes it challenging to produce more prominent or multiple genes simultaneously. Nanotechnology offers new approaches for payload delivery in plants to enable their engineering. Transgenic tomato plants containing the RNAi construct exhibited reduced virus titer and symptom severity upon artificial inoculation with groundnut bud necrosis virus (GBNV), suggesting that this gene-silencing mechanism could be a viable strategy for managing GBNV in tomato crops (Gandhi et al., 2024). Mesoporous silica nanoparticles (MSNs) can package siRNA within mesopores, protecting it and allowing gradual release. MSNs can penetrate protoplasts and have shown root uptake in plants. They can deliver cargo throughout the plant, including the vascular system.

Production Challenges

The production of viral vectors can be complex and costly, requiring specialized facilities and techniques. Moreover, achieving higher preparations can be labor-intensive and time-consuming. Viral vectors may show inconsistent biodistribution, leading to uncontrolled targeting of nondesired tissues, which can cause off-target effects and adverse reactions. Foliar treatment with dsRNA/siRNA can induce transient phenotypic changes in plants. Functionalized green nanomaterials, synthesized through eco-friendly methods, are better at managing plant diseases caused by various pathogens. It highlights the potential of nanotechnology to enhance agricultural productivity while reducing reliance on harmful synthetic agrochemicals and promoting sustainable and environmentally friendly disease management strategies (Wasule et al., 2024).

Regulatory Issues

The use of viral vectors in clinical settings often faces stringent regulatory hurdles due to safety concerns. The potential for residual pathogenicity in attenuated viral vectors is a concern. While efforts are made to reduce this risk, residual virulence can threaten patient safety. Mesoporous silica nanoparticles (MSNs) are biodegradable and non-toxic, making them suitable for agricultural applications. Nanomaterials enhance plant immunomodulation and defense mechanisms against biotic and abiotic stresses, which are increasingly challenging due to climate change. It discusses how nanomaterials can regulate physiological and biochemical processes, improve antioxidative systems, and upregulate defense-related gene expression while addressing their toxicity and future developmental prospects (Singh et al., 2024). Given these challenges, researchers are exploring alternative delivery mechanisms, such as bacteria-based vectors, nanoparticles, and other nonviral methods. Field trials have shown promising results in Indian field trials demonstrated reduced active ingredient requirements in nano-pesticides. African agriculture has benefited from Layered Double Hydroxides (LDHs) in maize farming, showing improved nutrient efficiency and crop yields.

Table 2. Examples of nanocarriers: Mechanism for delivering nucleic acids in plants

Nanoparticle Type	Overcoming Cuticle	Targeting Specific Compartments
Lipid-Based Nanoparticles (Liposomes, Lipid Nanoparticles)	Surfactant properties aid penetration	Lipid modification with targeting ligands
Polymer-Based Nanoparticles	Cationic or amphiphilic polymers enhance adhesion	Polymer modification with targeting ligands
Carbon-Based Nanomaterials (Carbon Nanotubes, Graphene Oxide)	High surface area and functional groups promote adhesion	Functionalization with targeting ligands
Metal and Metal Oxide Nanoparticles	Cationic or amphiphilic surface properties enhance adhesion	Surface modification with targeting ligands
Silicate Nanoparticles	High surface area and porous structure for cargo loading	Surface modification with targeting ligands

PROPERTIES OF NANOCARRIERS FOR DELIVERING NUCLEIC ACIDS

In plant biology, nanocarriers encompass various materials, including metal, polymeric, lipid-based, and silica nanoparticles. Each offers unique properties for specific applications.

Advancements in polymeric nanocarriers for enhancing the delivery of dsRNA and siRNA in agricultural applications present a sustainable alternative to traditional pest control methods. This paper highlights the progress in designing and applying these nanocarriers while addressing environmental and health challenges and emphasizing the importance of green chemistry in their development (Silva et al., 2024). Nanocarriers were developed for the precise delivery of nutrients, pesticides, or genes to plants.

- Polymeric Nanocarriers for Nutrient Delivery: Nanocarriers made from polymers that improve nutrient uptake by plants.
- Gold Nanoparticles for Gene Delivery: Gold nanoparticles used for the targeted delivery of genetic material to plants.
- Polymeric Micelles for Drug Delivery: Polymeric micelles as carriers for plant therapeutic agents or growth hormones.
- Carbon-Based Nanocarriers for Pesticide Delivery: Carbonbased nanocarriers for the effective and controlled release of pesticides.
- Peptide-Based Nanocarriers for Disease Resistance: Peptidemodified nanocarriers to target and enhance plant resistance to diseases.
- Stimuli-Responsive Nanocarriers for Controlled Release: Nanocarriers that release their payload in response to specific stimuli like pH or temperature.

Synthetic biology allows overcoming biological barriers to ensure efficient uptake and expression of genetic material. CRISPR/Cas tools enhance environmental sustainability and address nonspecific delivery and insertional mutagenesis challenges. It highlights advancements in nanotechnology that improve precision targeting and cost-effectiveness, facilitating applications in ecological remediation, healthcare, and agriculture (Bahl et al., 2024).

Size and Surface Charge

Smaller nanoparticles with positive surface charges can penetrate plant cell walls and membranes more easily. RNA viruses, specifically positive-strand RNA plant viruses, select RNA for packaging within viral particles (Peyret et al., 2025). Dendrimers are highly branched, star-shaped macromolecules with numerous surface functional groups. Dendrimers are highly tunable in size and surface properties, allowing for specific interaction with plant cells. The zeta potential of the casein nanoparticles is optimized to enhance the success of DNA delivery into plant cells. Guy Mechrez and co-workers present gene expression by protein nanoparticles after DNA delivery into intact plant cells. DNA delivery and expression effectiveness are validated through confocal fluorescence microscopy, RTPCR, and qRTPCR techniques. However, challenges remain in using RNAi for weed control, particularly concerning the carriers' compositions, morphologies, sizes, and chemical properties (Ben-Haim et al., 2024).

Protective Coatings

Encapsulating the cargo in protective coatings can shield it from environmental factors and degradation. Liposomes are spherical vesicles made of phospholipid bilayers that can encapsulate nucleic acids. Liposomes are biocompatible, can protect nucleic acids from degradation, and can be engineered for targeted delivery. Autonomous penetration by casein and other protein nanocarriers can autonomously cross plant cell walls when their size is tailored to be below approximately 14 nm, facilitating the effective delivery of proteins into plant cells without significant barriers (Ben-Haim et al., 2024).

Cell Penetrating Peptides (CPPs)

CPPs can enhance cellular uptake by disrupting cell membranes or interacting with cell surface receptors. Carbon-based Nanomaterials include graphene oxide and carbon nanotubes, which can adsorb nucleic acids on their surface. Carbon nanoparticles have a high surface area for loading and the potential for electrical conductivity in gene delivery applications. The selfassembly process between star polycation and seboctylamine is facilitated by hydrogen bonds and van der Waals forces, leading to a reduction in particle size and the formation of a spherical complex.

The star polycation improves the wetting properties and adhesion performance of seboctylamine on leaf surfaces, as indicated by a decrease in contact angle and increased retention (Gao et al., 2025).

Endosomal Escape

Endocytosis enhancement is engineering nanocarriers to mimic natural plant viruses that can promote their internalization through endocytosis, facilitating nucleic acid delivery (Islam et al., 2024). pH-sensitive or enzyme-degradable materials can help release the cargo from endosomes. Silica Nanoparticles comprise porous silica structures that can encapsulate and protect nucleic acids. Silica has excellent biocompatibility, significant loading capacity, and can be designed for controlled release. Lysozyme-coated layered double hydroxide (LDH) nanosheets facilitate the active uptake of nucleic acids and proteins in plants through endocytosis, targeting root tips, root hairs, and lateral root junctions (Yong et al., 2025).

Controlled Release

Designing nanocarriers that respond to specific environmental stimuli (like pH or temperature changes) allows for the controlled release of nucleic acids, improving the effectiveness of the delivery. Stimuli-responsive materials can release the cargo at specific times or locations. Metal-based carriers such as Gold Nanoparticles (AuNPs), nanosized particles of gold, can be modified to attach nucleic acids for delivery. Copper, silver, or gold-based nanoparticles are easy to synthesize and functionalize, with high stability and potential for photothermal therapy applications.

Coating and Surface Modification for Targeted Delivery

Ligands or aptamers can target specific cell types or organelles. Modifying the surface of nanocarriers by targeting ligands or polymers enhances cell membrane permeability and improves uptake by plant cells. A self-assembled nanoantiviral pesticide combining star polycation (SPc) and seboctylamine is efficient in controlling the Soybean mosaic virus (SMV), a significant threat to soybean crops (Gao et al., 2025). Electroporation or microinjection can assist nanocarrier entry by temporarily permeabilizing the cell membrane. Modifying nanocarriers to target specific plant tissues, such as roots or leaves, enhances the likelihood of successful nucleic acid uptake.

Stability and Activity of dsRNA/siRNA Molecules

The stability and activity of dsRNA and siRNA molecules are crucial for their effectiveness in gene silencing applications. Polymeric Nanoparticles are nanosized particles made from synthetic or natural polymers that can carry and protect nucleic acids. Polymers offer versatile designs, controllable release profiles, and stability in different environments. Comparing nanocarriers to traditional chemical pesticides are essential to assess economic viability. Addressing ethical and socio-economic gaps, particularly through a focus on global equity, is required to ensure accessibility for smallholder farmers. Evaluating polymeric nanoparticle bioaccumulation risks and implementing long-term monitoring strategies will be critical for sustainable and equitable implementation (Khatoon & Velidandi, 2025).

Use of Adequate Solvents

The choice of solvent can significantly impact the stability of RNA molecules. Iron oxide nanoparticles (Fe2O3 NPs) were synthesized using the seed extract of *Trachyspermum ammi*. They were characterized using FTIR, XRD, SEM, and EDX techniques. Using appropriate solvents helps maintain the integrity of dsRNA/siRNA, preventing degradation and ensuring that the molecules remain functional for their intended applications. The application of Fe2O3 NPs improved physiological and biochemical attributes in rose leaves, including increased activities of antioxidant enzymes (SOD, POD, and CAT) and higher expression of the defense-related gene RcWRKY29, suggesting their role in enhancing plant defense mechanisms against blight disease.

Elimination of RNase Enzymes

Ribonucleases (RNases) are enzymes that degrade RNA, posing a significant threat to the stability of dsRNA and siRNA. To protect RNA molecules, it is essential to eliminate RNase activity during preparation and storage. Creating protoplasts (cells without cell walls) can improve nucleic acid delivery, as they are more permeable to nanocarriers. This can be achieved using RNase-free reagents and maintaining a controlled environment to minimize contamination.

Acid-Base Balance

The pH of the environment in which dsRNA/siRNA is handled can also affect their stability. RNA molecules are sensitive to extreme pH levels, which can lead to hydrolysis and degradation. Maintaining an optimal pH range is vital for preserving the integrity and activity of RNA molecules. Delivering nucleic acids into plant cells is difficult due to physical and mechanical barriers such as the cuticle and cell wall (Komarova et al., 2023). Advances in biotechnology have led to the development of polymer-based nanocarriers that enhance the stability and delivery efficiency of RNA molecules into plant cells. However, implementation faces cost barriers for small-scale farmers—the need for technical knowledge and training due to limited access to advanced technologies in developing regions.

Table 3. Cationic vs. anionic nanocarriers

	Cationic Nanocarriers	Anionic Nanocarriers
Surface Charge	Positive	Negative
Primary Interaction	Negatively charged cell walls/membranes (electrostatic)	Potentially via specific interactions, size-dependent entry
Uptake Mechanisms	Electrostatic attraction, endocytosis, direct translocation, stomatal/cuticular entry, transporter interaction	Macropinocytosis (in some cell types), hydrophilic pathway, stomatal/cuticular entry, endocytosis, root uptake at specific regions
Translocation	It may be hindered by strong cell wall binding	Potentially higher root-to-shoot translocation
Examples	Chitosan, PEI, PLL, CNTs, protein NPs	LDH, carbon dots, PGA, clay nanosheets

CATIONIC AND ANIONIC NANOCARRIERS IN PLANT SYSTEMS

Cationic and anionic nanocarriers, distinguished by the nature of their surface charge, exhibit distinct properties and mechanisms of action when interacting with plant systems. Understanding these fundamental differences is crucial for selecting the appropriate nanocarrier for a specific application in plant biotechnology (Table 3).

Cationic polymer-based nanocarriers can form biodegradable complexes known as polyplexes, capable of sustaining the release of dsRNA over time. This approach addresses the challenges associated with the stability and delivery of RNA in plant systems, ultimately enhancing the effectiveness of RNAi applications in agriculture. Cost-benefit analyses reveal several advantages due to enhanced crop yield and reduced chemical input requirements. Improved efficiency in targeted delivery systems combats initial higher costs offset by long-term benefits. There is a potential for significant economic impact across various agricultural sectors.

Cationic nanocarriers are characterized by a positive surface charge by the protonated amine groups or other cationic functional groups on their surface. Gold nanoparticles can be modified with 2-aminoethanethiol, an amine-containing compound, which alters their surface charge from negative to positive. Examples of cationic nanocarriers commonly explored in plant biotechnology include chitosan nanoparticles, polyethylenimine (PEI), poly-L-lysine (PLL), guanidinium-containing nanoparticles, protein-based nanoparticles, and carbon nanotubes (CNTs). Anionic nanocarriers, with a negative surface charge, with carboxylate, phosphate, or sulfonate groups on their surface. Polyglutamic acid (PGA), an anionic polymer frequently used in biological applications. Examples of anionic nanocarriers commonly used in plant biotechnology include layered double hydroxide (LDH) nanoparticles, carbon dots, anionic polymers like polyglutamic acid (PGA), and clay nanosheets.

NANOCARRIERS TESTED FOR DELIVERING NUCLEIC ACIDS IN PLANTS AND THEIR PROPERTIES

The development of nanotechnology enabled the synthesis of nanomaterials with high surface area, tunable size and shape that has the ability to targeted and responsive delivery. Nanocarriers have emerged as a transformative tool for effectively delivering nucleic acids in plants, enabling advancements in genetic engineering, crop improvement, and plant disease management (Ahmed et al., 2024). Several factors determine the efficiency of nanocarrier penetration through the above pathways. These include the physicochemical properties of the nanoparticles, plant anatomy, and environmental conditions. Nanocarriers can penetrate physical barriers like the plant cell wall and cuticle, enhancing the internalization of DNA and RNA (Komarova et al., 2023). Nanomaterials like CNTs offer a large surface area for carrying nucleic acids. CNTs, in particular, have demonstrated the ability to penetrate plant cell walls, facilitating intracellular delivery. Nanoparticles can be engineered to specific sizes and shapes, allowing for controlled cellular uptake. (Miyamoto & Numata, 2023). Nanocarrier systems can be engineered for controlled release, ensuring that nucleic acids are delivered at the right time and location within the plant(Yong et al., 2024)(Miyamoto & Numata, 2023). The unique properties of nanocarriers, particularly CNTs, led to a paradigm shift in plant nucleic acid delivery due to their unique properties (Demirer et al., 2019).

- Reduced off-target effects: Nanocarriers can minimize off-target effects by delivering nucleic acids directly to target cells, reducing unintended genetic modifications.
- Ability to deliver a wider range of cargo: Nanocarriers allow for the delivery of RNA, and other delicate cargo, that is harder to deliver with older methods.
- Increased sustainability: Nanocarriers are opening the door to the ability to deliver nucleic acids via foliar spray, reducing the need for tissue culture.

Various nanocarriers, including metal, carbon-based, and polymeric nanoparticles, metal-organic frameworks (MOFs), Tobacco mild green mosaic virus-like nanocarriers, and Graphene oxide (GO), are utilized to transport nucleic acids into plant cells(Iqbal et al., 2024)(Yong et al., 2024).

- Peptide-Based Nanocarriers: Targeted disease resistance but have scalability concerns.

- Stimuli-Responsive Nanocarriers: Offer precision but require complex synthesis.

Carbon-Based Nanoplatforms

Carbon nanomaterials include various forms like nanotubes, fullerenes, and nanoparticles. They are versatile and can deliver nucleic acids into plant cells. Carbon dots (CDs) are small, biocompatible, and stable cargo-delivery nanomaterials. CDs can enhance plant growth and productivity without toxicity. They can be functionalized to bind nucleic acids and improve cellular uptake. They have shown positive effects on photosynthesis and root growth in plants. CDs are suitable for various plant species and can deliver biologically active cargo.

Carbon Nanotubes are Single-walled carbon nanotubes (SWNTs) that can shuttle various molecular cargo into cells. SWNTs possess unique properties, including fluorescence for imaging in biological systems. SWNTs can protect nucleic acids from degradation and enhance delivery efficiency. They have been shown to deliver pDNA and siRNA into various plant species. Carbon nanofibers and other CB-NMs improve crop yield and nutritional quality while addressing the environmental concerns associated with the excessive use of chemical fertilizers (Ashfaq et al., 2025).

Chitosan-Based Nanocarriers

Chitosan is a biocompatible polymer used for nucleic acid delivery in plants. Its properties depend on molecular weight and degree of acetylation. Chitosan nanoparticles can provide a controlled release of cargo and enhance plant resistance. Various studies have used them to deliver dsRNA and pDNA effectively. Chitosan formulations have shown antiviral activity and increased plant productivity. Limitations include variability in nanoparticle preparation and potential adverse effects (Komarova et al., 2023).

Cell Penetrating Peptides

Cell-penetrating peptides (CPPs) facilitate cargo delivery into plant cells. They can overcome mechanical barriers and enhance endosomal escape. The lysozyme coating enhances the uptake of nanosheets by loosening the plant cell wall and stimulating the expression of genes related to endocytosis and membrane trafficking (Yong et al., 2025). Natural and synthetic CPPs have been identified for effective nucleic acid delivery. CPPs can be applied via spraying, infiltration, or coincubation with cargo. They have shown potential for organelle-specific delivery, including chloroplasts and mitochondria. Future studies should focus on improving CPP delivery efficiency and understanding mechanisms (Komarova et al., 2023).

Table 4. Nanocarriers for delivering nucleic acids in plants. Each has its unique advantages in plant biology applications

	Lipid Nanoparticles (LNPs)	Polymeric Nanoparticles	Carbon Nanomaterials (CNTs, GO)	Inorganic Nanoparticles (Gold, Silica)
Cyto toxicity	Generally low if well-designed	Variable, depends on polymer	Potential for higher cytotoxicity	Generally low if biocompatible coatings are used
Cyto toxicity Mitigation	Surface modifications, PEGylation	Use of biocompatible polymers, surface modifications	Functionalization with biocompatible groups, dosage control	Biocompatible coatings (e.g., silica, PEG), size control
Regulatory Hurdles	Moderate, depending on lipid composition	Moderate, polymer safety assessment	High, concerns about environmental impact	Moderate, potential for metal accumulation
Green Nano materials	Potential with bio-derived lipids	Yes, many options (e.g., chitosan, starch)	Emerging research using bio-derived carbon sources	Limited direct use, but coatings can be green
Stimuli-Responsive Carriers	Yes, pH, temperature, light	Yes, pH, temperature, light	Yes, pH, light, redox	Yes, pH, light

INNOVATIONS IN NANOCARRIER DELIVERY

Nanocarriers have emerged as a transformative technology for delivering nucleic acids in plants, addressing longstanding challenges in plant biotechnology. These systems enhance gene transfer efficiency, facilitate RNA interference, and enable genome editing, thereby improving crop resilience and yield.

Engineering Attenuated Bacterial Strains

Modifying bacterial strains to reduce their pathogenicity while retaining their ability to invade host cells. This ensures safety while maintaining delivery efficacy. Nanocarriers transport nucleic acids into plant cells, including metal, carbon-based, and polymeric nanoparticles (Iqbal et al., 2024). The innovative methods for delivering nucleic acids using bacteria, focusing on their potential in the development of Genetic Circuits and gene therapy

Designing genetic circuits that can regulate the expression of genes crucial for nucleic acid delivery processes (Biradar et al., 2025). This can include circuits that facilitate the escape from vacuoles or programmed lysis to release the nucleic acid cargo at the desired time. Bacterial vectors show significant promise because of their safety and ability to carry larger cargo.

Nanoparticle (NP) Coating

Coating bacteria with nanoparticles that carry nucleic acids. This method leverages the invasive properties of bacteria to promote efficient intracellular delivery and protect nucleic acids from degradation. The lysozyme-coated nanosheets can deliver various types of nucleic acids, including synthetic mRNA, double-stranded RNA, small interfering RNA, and plasmid DNA, into tobacco roots and other plant tissues. The lysozyme-coated LDH nanoparticles are a versatile and efficient tool for delivering functional nucleic acids in dicot and monocot plant species (Yong et al., 2025). *Bacteria produce Outer Membrane Vesicles (OMVs)* as delivery vehicles. OMVs can encapsulate nucleic acids and facilitate their transfer to host cells. The selective packaging of DNA sequences into OMVs can also be engineered for specificity.

Controlled Release Mechanisms

Developing systems that can control the release of nucleic acids from bacterial carriers based on environmental triggers, ensuring that the genetic material is released at the optimal time and location for maximum effectiveness. These systems can be engineered for controlled release, ensuring that nucleic acids are delivered within the plant at the right time and location (Yong et al., 2024). It also introduces plant-derived exosome-like nanoparticles as a promising platform for CRISPR/Cas9 genome editing in plants, aiming to enhance crop quality and resilience (Miyamoto & Numata, 2023).

Targeting Specific Cell Types

Engineering bacterial vectors to express ligands or antibodies on their surface that target specific cell types within the host enhances the specificity and efficiency of nucleic acid delivery. Plant roots can actively take up lysozyme-coated double hydroxide nanoparticles and effectively deliver various types of nucleic acids, including synthetic mRNA and plasmid DNA, into different plant tissues. This method enhances uptake by loosening cell walls and stimulating relevant gene expression, making these nanoparticles a versatile tool for genetic modification in crops. Several methods of application, experimental systems, and approaches are used to assess the nanocarrier formulations' efficiency in laboratory and field settings (Miyamoto & Numata, 2023).

Combining Bacterial and Viral Delivery Mechanisms

Exploring hybrid systems that combine the advantages of bacterial and viral vectors to improve nucleic acid transfer efficiency and reduce potential immunogenic responses. Evidence from both in vitro and in vivo studies indicates that the selectivity of RNA packaging is likely more influenced by the replication of viral RNAs than by specific RNA sequences. High-affinity packaging signals for the viral coat protein may enhance the efficient incorporation of RNA into viral particles, contributing to the assembly of infectious particles (Peyret et al., 2025). By combining these strategies, nanoparticles can effectively deliver cargo to plant cells, overcoming various barriers and improving the efficiency of plant transformation and genetic engineering. Environmental impact studies have identified several concerns due to potential accumulation of nanoparticles in aquatic systems. The risk of biomagnification through food chains can lead to long-term health implications for ecosystems and humans. Regulatory frameworks like OECD guidelines for nanomaterial safety assessment have been established to ensure safety. In USA, environment protection agency (EPA) regulations are under the Toxic Substances Control Act that seeks international cooperation for standardized testing and assessment.

Table 5. Comparison of nanocarrier types tested for plant biotechnology

	Polymeric Nanoparticles	Lipid-Based Nanoparticles	Carbon Nanomaterials	Metal/Metal Oxide Nanoparticles	Dendrimers
Composition	Natural or synthetic polymers (e.g., chitosan, PLGA)	Lipids (e.g., liposomes, solid lipid nanoparticles)	Carbon nanotubes (CNTs), graphene, graphene oxide (GO)	Metals (e.g., silver, gold) or metal oxides (e.g., TiO2, ZnO)	Highly branched, synthetic macromolecules with well-defined structures
Targeted Delivery	Agrochemical, DNA/ RNA for gene editing, controlled release of nutrients	Hydrophobic drugs, nucleic acids, Penetration through cell membranes	DNA/RNA, enhancing photosynthesis (CNTs), Antimicrobial applications (GO)	Antimicrobial applications, Micronutrients, Stress tolerance	Gene editing, Drugs, small molecules
Applications in Agriculture	Stimuli-responsive polymers for controlled release based on environmental cues (pH, temperature, light).	Fusion with plant cell membranes for efficient delivery, co-delivery of multiple compounds.	Targeted delivery to specific organelles, Biosensors and plant health monitoring.	Development of nanofertilizers with controlled release of essential micronutrients, Plant disease diagnostics.	Targeted delivery to specific plant cells or organelles.
Advantages	Biodegradable, Tunable release properties, Good encapsulation efficiency	Biocompatibility cellular uptake, Encapsulates diverse payloads	High surface area, plant stress response, physiological processes	Antimicrobial properties, plant growth and stress tolerance	Highly customizable structure, high drug loading
Limitations	Potential toxicity depending on the polymer, stability issues	Can be less stable than polymeric nanoparticles, potential for lipid oxidation	Potential toxicity concerns, aggregation in the environment,	Potential for phytotoxicity and accumulation in the food chain, environmental concerns	Expensive to synthesize, toxicity depending on surface chemistry

APPLICATIONS IN PLANT BIOTECHNOLOGY

Biotic stress significantly affects agricultural productivity, and the increasing global food demand necessitates enhanced crop productivity to address these challenges (Nagashri et al., 2024). The indiscriminate use of synthetic agrochemicals for managing biotic stresses has detrimental effects on the food chain and biosphere, highlighting the need for eco-friendly alternatives. Nanocarriermediated delivery offers a more universal and precise targeting approach (Wasule et al., 2024).

Nanomaterials can protect nucleic acids from environmental factors and enhance cell penetration. Various experimental systems, including protoplasts and suspension cultures, assess nanocarrier suitability. Functional tests confirm successful delivery through reporter gene expression and phenotypic changes.

Gene Editing

Nanocarriers facilitate advanced gene editing techniques, such as CRISPR/Cas, improving the precision of genetic modifications(Iqbal et al., 2024). Efficient delivery systems allow researchers to develop and test new functionalities, such as biosensors, and explore novel pathways involved in plant development and responses to environmental stressors. Nanocarriers are nanoscale platforms transporting various substances, including drugs, dsRNA, and genes, to targeted biological sites. Advances in nanocarrier technology can improve crop protection and biotechnology applications.

Stress Tolerance

With climate change and abiotic stressors becoming more prominent, delivering nucleic acids can enhance a plant's ability to cope by expressing specific stress-related genes. Nanocarriers can enhance plant tolerance to environmental stresses by providing particular genes, contributing to food security(Miyamoto & Numata, 2023). Polyethylene, polyester, starch, alginate, hyaluronic acid, polyamide, cellulose, polyurethane, and polylactic acid-based carriers are engineered to deliver dsRNA via the plant leaf effectively. For example, researchers have explored alternative RNA delivery methods, including breaking the cuticle structure through abrasion, nontransgenic SIGS (SprayInduced Gene Silencing) approach using surfactants or chemical changes, foliar infiltration, and irrigation methods for roots to absorb RNA. Recent studies have demonstrated the practical application of nanocarriers in agricultural settings, showcasing their potential for large-scale use(Komarova et al., 2023).

Genetic Transformation

Bacterial vectors, especially Agrobacterium tumefaciens, are widely used for the genetic transformation of plants. This method allows for the stable insertion of foreign DNA into the plant genome, facilitating the development of genetically modified plants with desirable traits. Bacterial delivery systems can introduce gene editing tools like CRISPR/Cas9 into plant cells. This application allows for precise modifications of plant genomes to enhance traits like drought tolerance, disease resistance, or nutritional content. While the advancements in nanocarrier technology present significant opportunities for improving plant genetic engineering, challenges remain, such as ensuring biosafety and optimizing delivery efficiency across diverse plant species(Miyamoto & Numata, 2023)(Qin et al., 2023). There is a need for further exploration of the interactions between nanomaterials and plants, particularly regarding their antioxidative and immunomodulatory effects, as well as discussing the toxicity of nanomaterials and future developmental prospects in this field (Singh et al., 2024).

Disease Resistance

Delivery of genes associated with disease resistance through bacterial vectors can enhance the plant's immune response to pathogens. The use of an RNA silencing mechanism targeting the replicase gene to confer resistance against the groundnut bud necrosis virus (GBNV) in tomato plants (Gandhi et al., 2024). Layered double hydroxides (LDHs) protect nucleic acids from environmental challenges. LDH sheets release cargo at an acidic pH, typical of plant cells. LDH/dsRNA complexes can remain on leaf surfaces for extended periods. Foliar spraying of LDH/dsRNA induces silencing of essential genes in plants. LDHs have been used to combat viral infections by delivering pDNA-encoding microRNAs. This approach is necessary for developing crops that can withstand various biotic stresses.

Trait Enhancement

Bacterial vectors can introduce genes that enhance specific traits, such as increased yield, improved quality, or resistance to environmental stresses (e.g., heat, salinity). Layered double hydroxides and lysosome-coated complexes can enter chloroplasts, enhancing cargo delivery efficiency. LDHs are characterized by biosafety and environmental friendliness. They provide a gradual release of cargo, ensuring prolonged effects in plants (Yong et al., 2025). Thus, they contribute to food security and agricultural sustainability.

Metabolic Engineering

Scientists can manipulate plant metabolic pathways by introducing or knocking out specific genes using bacterial vectors. This can directly produce valuable secondary metabolites, such as pharmaceuticals, flavors, or fragrances, in plants. Selected nanocarriers are analyzed based on biosafety of carbon quantum dots (FMCQDs) synthesized from the marine fungus Aspergillus flavus on the growth and curcumin production in Curcuma longa L.FMCQDs were characterized using UVVis spectroscopy and HRTEM, revealing a size of approximately 7.4 nm and a spherical shape, and were found to enhance plant height significantly compared to control groups. Applying FMCQDs and fungal crude extract (FCE) increased photosynthetic pigments in C. longa, with total chlorophyll and carotenoids showing notable improvements while confirming no toxic effects on the plants. FMCQD treatment increased curcumin content (up to 0.060 mg /g). It significantly upregulated the expression of the ClCURS1 gene associated with curcumin production (Vasanthkumar et al., 2024).

Synthetic Biology Applications

Bacterial vectors can be engineered to deliver genetic constructs for creating synthetic biological pathways in plants, allowing for the production of novel compounds or materials not typically found in nature. Engineering attenuated strains to reduce pathogenicity while preserving immunogenic properties (Ananthakumar et al., 2024). Development of genetic circuits for improved delivery mechanisms. Use bacterial conjugation systems and the Type IV Secretion System for nucleic acid delivery. Implementation of nanoparticle coatings and outer membrane vesicles (OMVs) for enhanced targeting and delivery. enetic bioaugmentation as a method to improve bioremediation by delivering necessary genes to resident bacteria through horizontal gene transfer, addressing the issue of missing genes in microbial communities for pollutant breakdown (Marquiegui – Alvaro et al., 2025).

Microbiome Engineering

Highlighting the potential of bacteria as vectors for nucleic acid delivery due to their safety and cargo capacity. Bacterial vectors can modify the plant-associated microbiome, enhancing plant health and growth by introducing beneficial microbial genes that promote nutrient uptake or disease resistance. In plant tissue culture, bacterial vectors can introduce new genes into cultured cells, enabling the regeneration of whole plants with desired genetic modifications. A new method for delivering proteins into plant cells using specially designed polymeric nanocarriers (PNCs) (da Silva et al., 2024). Cationic PNCs can autonomously cross plant cell walls when smaller than 14 nm.

It uses a reductionoxidation sensitive green fluorescent protein (roGFP) to monitor plant responses to stress. This technique offers a way to overcome existing limitations in protein delivery, allowing for more efficient plant engineering across various species.

Bioremediation

Plants can be genetically engineered using bacterial delivery methods to enhance their ability to uptake and detoxify pollutants from the soil, contributing to bioremediation efforts in contaminated environments. The shift from viral to bacterial vectors aims to address issues like immunogenicity and increase cost-effectiveness. (Yong et al., 2024).

Techniques like engineering attenuated strains and nanoparticle coatings are explored to enhance delivery. *Utilizing the Type IV Secretion System (T4SS) allows e*xploiting T4SS to mediate the transfer of nucleic acids into host cells. Understanding and optimizing the mechanisms of T4SS can lead to enhanced efficiency in delivery. The impact of dinotefuran (DIN) on the abundance and transfer mechanisms of antibiotic resistance genes (ARGs) and virulence factor genes (VFGs). The plasmid, which carries synthetic genes for terephthalate degradation, was successfully maintained by P. putida and enabled complete breakdown of the pollutant within 8 days, showcasing its potential for enhancing microbial communities in environmental clean-up efforts (Marquiegui – Alvaro et al., 2025).

Table 6. Upcoming nanocarriers that are showing promising results for plant biotechnology and agriculture applications

	Green Nanomaterials (Cellulose, Chitosan-based)	Stimuli-Responsive Carriers	Virus-Like Particles (VLPs)	Peptide Carriers
Cytotoxicity	Generally low	Variable, depends on material	Low	Variable
Cytotoxicity Mitigation	Use of inherently biocompatible materials, surface modifications	Encapsulation, careful selection of stimuli and materials	Surface modifications, choice of plant-derived viral components	Conjugation with biocompatible linkers
Regulatory Hurdles	Lower, perceived as more environmentally friendly	Moderate assessment of responsiveness and degradation of products	Moderate, biosafety considerations	Moderate, peptide safety and stability
PEGylation	Can be incorporated	Often used to enhance stability and reduce non-specific interactions	Can improve stability and reduce immune response in plants	Can improve stability and reduce degradation

UPCOMING NANOTECHNOLOGY INNOVATIONS FOR AGRICULTURE AND PLANT BIOTECHNOLOGY

Nanocarriers can penetrate physical barriers like the plant cell wall and cuticle, enhancing the internalization of DNA and RNA(Komarova et al., 2023). Spray-induced gene silencing offers a scalable method for crop treatment. Understanding the fate of nanocarriers in plants and the environment is crucial. Understanding and optimizing these pathways and influencing factors is crucial for designing effective nanocarrier-based delivery systems in plant biotechnology. Future research should focus on tailoring nanocarrier properties to specific plant species, tissues, and environmental conditions, ensuring efficient and precise delivery for applications such as genetic modification, stress tolerance, and crop improvement (Ashfaq et al., 2025).

More examples of the nanocarriers that are showing promising results:

Liposomes: Excellent for encapsulating hydrophobic drugs, biocompatible, biodegradable, and provides good control over release. Size-dependent toxicity and short circulation time in vivo.

Mesoporous Silica Nanoparticles (MSNs): High surface area for loading molecules, good stability in biological environments, enhances controlled drug delivery. Limited biodegradability and a complex synthesis process.
Chitosan Nanoparticles: Biodegradable, nontoxic, good for encapsulating a variety of molecules, and improves the stability and solubility of loaded drugs. Limited release control and requires modification for targeted delivery.
Quantum Dots: Unique optical properties, effective for gene and drug delivery, and highly stable. Potential toxicity concerns and expensive synthesis process.
Polymeric Nanocarriers: Good for controlled release and surface modifications.
Gold Nanoparticles: Known for gene delivery but expensive.
Carbon-Based Nanocarriers: Excellent for penetration and drug loading but require surface modifications.

Nanosensors, Nanogenerators, and Nano Enabled Seed Treatments

Nanosensors, Nanogenerators, and Nano enabled Seed Treatment highlight the versatility and potential of bacterial vectors in advancing plant biotechnology, ultimately contributing to improved agricultural practices and sustainable food production. Progress in nanomaterials enhances nucleic acid delivery efficiency in plants. Nanocarriers protect nucleic acids from environmental factors and improve stability. Addressing biosafety issues is essential for transferring technology to field applications (Miyamoto & Numata, 2023).

- Smart Nanoparticles: Stimuli-responsive nanoparticles that open at stomatal pores.
- Bio-inspired Nanocarriers: Mimicking pollen grain adhesion for targeted delivery.
- Phloem-Mobile Nanoparticles: Engineering nanoparticles for efficient transport in the phloem for delivery to sink tissues.
- Nanobiosensors for Plant Health: Integrating nanosensors within the vascular system for real-time monitoring.

Nanofertilizers

Polymeric nanocarriers enhance the delivery of dsRNA and siRNA in agricultural applications, presenting a sustainable alternative to traditional pest control methods (da Silva et al., 2024). Biodegradable and safe nanomaterials suitable for nucleic acid delivery include mesoporous silica nanoparticles, layered double hydroxides, carbon-based materials, chitosan, and cell-penetrating peptides. Progress in designing and applying these nanocarriers addresses environmental and health challenges and emphasizes the importance of green chemistry in their development.

- Penetration Enhancers: Developing biodegradable nano-adjuvants for cuticle permeation.
- Layer-by-Layer Nanocoatings: For controlled release and enhanced penetration.
- Rhizosphere Engineering: Designing nanoparticles to interact with beneficial soil microbes for enhanced nutrient uptake.
- Nanoparticle-Mediated Gene Delivery: Targeting root cells for genetic modification.

Nanopesticides

A self-assembled nano antiviral pesticide combining star polycation (SPc) and seboctylamine for effective control of the destructive Soybean mosaic virus (SMV). The study demonstrates that SPc enhances the adhesion and wetting properties of seboctylamine, leading to significant inhibition of viral accumulation and improved agronomic traits in soybean plants (Gao et al., 2025).

Mesoporous Silica Nanoparticles

Mesoporous Silica Nanoparticles are nontoxic in protoplasts at specific concentrations but can affect seed germination. They have been tested on various plant species and parts, showing versatility. MSNs are primarily used for biolistic transformation and direct injection methods. They have the potential to deliver active enzymes and proteins into plant cells. In addition, nanoencapsulation using inorganic nanocarriers, such as clay silica, minerals, carbon derivatives, and metallic nanoparticles, improves the stability of dsRNA/siRNA and the effective delivery into plant cells to reduce the potential for off target effects. Using exogenous RNA, specifically dsRNA and siRNA, for gene silencing strategies in agriculture presents an efficient and eco-friendly alternative to traditional plant protection methods.

Evaluation Criteria include adjustability for specific cargo and tasks, penetration efficiency, protection of nucleic acids from environmental factors, and the ability to mediate gradual release.

Gelatin Nanoparticles (GNPs)

Gelatin nanoparticles (GNPs) are biocompatible nano fertilizers designed to improve pesticide delivery and performance in agricultural practices. GNPs have a small size (approximately 150 nm) and a positively charged surface (approximately 30 mV), facilitating unrestricted nutrient absorption on plant root surfaces. Gelatin nanoparticles (GNPs) are developed as biocompatible nano fertilizers that enhance pesticide delivery and improve plant growth, particularly root development. They demonstrate significant improvements in insecticidal and weedkilling effects while maintaining pesticide efficacy without the need for additional fertilization and enhancing the effectiveness of agrochemical agents (Park et al., 2024).

Nanoparticles for Human Disease Control

Exosomes, a category of extracellular vesicle (EV), are phospholipid bilayer structures ranging from 30 to 150 nm, produced by various organisms through the endosomal pathway. Recent studies have established the utilization of exosomes as nanocarriers for drug distribution across different therapeutic areas, including cancer, acute liver injury, neuroprotection, oxidative stress, inflammation, etc (Korrapati et al., 2024). The importance of plant-derived exosomes and exosome vesicles derived from mammalian cells or milk loaded with potent plant bioactives (Kathait et al., 2024).

Nanotechnology for Abiotic Stress Tolerance

Developing nanomaterials that can have immunomodulatory effects in plants enhances their resistance to biotic and abiotic stresses caused by environmental challenges, particularly climate change. Plants face significant reductions in crop yield (50–88% worldwide) due to various environmental stresses. These stresses hinder their development and metabolism, prompting the need for effective defense mechanisms. Nanomaterials are identified as regulatory molecules that can modulate a wide range of physiological and biochemical processes in plants, including the immune system, stress-related gene expression, hormonal regulation, and activation of antioxidative defense systems (Singh et al., 2024).

Figure 3. Integration of computational biology and nanotechnology for delivering nucleic acids in plants

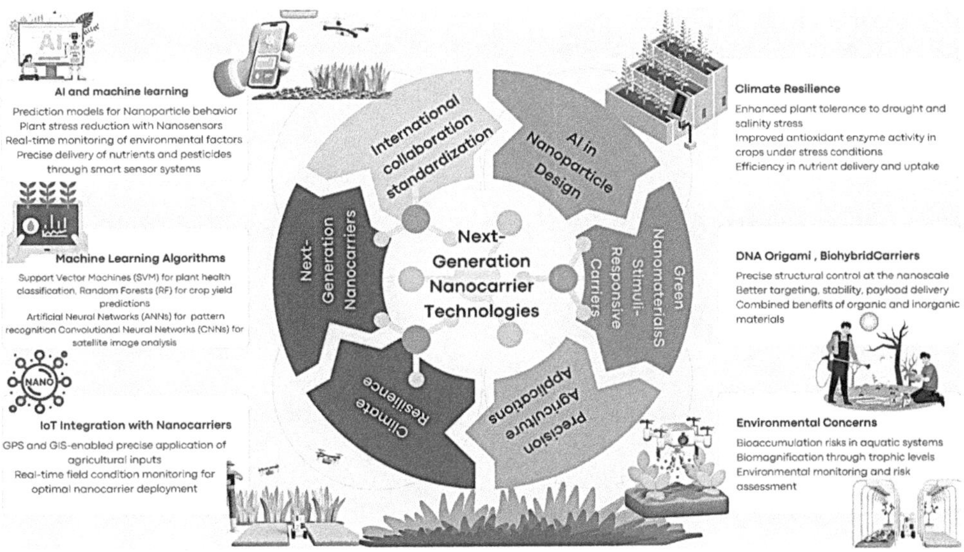

CONCLUSION

Nanocarrier-based delivery systems hold immense potential to revolutionize plant biotechnology by addressing the challenges associated with delivering nucleic acids into plant cells. Recent advancements in nanocarrier design, particularly surface modifications and functionalization, have significantly enhanced their efficacy. Modifications with surfactants, polymers, peptides, or other molecules improve adhesion, penetration, and targeting, allowing for more precise delivery of genetic material. Additionally, the size and shape of nanoparticles play a critical role, as smaller nanoparticles and optimized geometries enhance penetration and cellular uptake.

Nanocarriers have emerged as a transformative technology for delivering nucleic acids in plants, addressing longstanding challenges in plant biotechnology. The charge of nanocarriers is another crucial factor; cationic nanoparticles, for example, interact effectively with the negatively charged surfaces of plant cells, promoting adhesion and internalization. Furthermore, stimuli-responsive materials provide a cutting-edge approach by enabling the controlled release of cargo in response to environmental triggers such as pH, temperature, or redox conditions. Incorporating targeting ligands, such as peptides or aptamers, further enhances the specificity of nanocarriers, enabling delivery to particular cell types or organelles within the plant. These systems enhance the efficiency of gene transfer, facilitate RNA interference, and enable genome editing, thereby improving crop resilience and yield.

As research progresses, the integration of these advanced features will pave the way for more efficient and precise delivery systems. This will unlock new possibilities in plant genetic engineering, trait enhancement, and sustainable agriculture, ultimately contributing to global food security and environmental resilience. Targeting ligands Specific molecules, such as peptides or aptamers, can be attached to nanoparticles to target specific cell types or organelles. While the advancements in nanocarrier technology present significant opportunities for enhancing plant genetic engineering, challenges remain, such as ensuring biosafety and optimizing delivery efficiency across diverse plant species(Miyamoto & Numata, 2023)(Qin et al., 2023).

REFERENCES

Ahmed, T., Noman, M., Qi, Y., White, J. C., & Qi, X. (2024). Protein nanocarrier-enabled plant genetic engineering systems. *Trends in Biotechnology*, *0*(0). Advance online publication. DOI: 10.1016/j.tibtech.2024.08.009 PMID: 39277495

Ali, A., Shahbaz, M., Ölmez, F., Fatima, N., Umar, U. U. D., Ali, M., Akram, M., Seelan, J. S. S., & Baloch, F. S. (2024). RNA interference: A promising biotechnological approach to combat plant pathogens, mechanism and future prospects. *World Journal of Microbiology & Biotechnology*, *40*(11), 339. DOI: 10.1007/s11274-024-04143-3 PMID: 39358476

Ananthakumar, B., Nanjundeshwara, N., Lakshmaiah, A., & Raghavan, R. (2024). Synthetic Gene Circuits as a Promising Approach in Cancer Immunotherapy. *Critical Developments in Cancer Immunotherapy*, 391–418. DOI: 10.4018/979-8-3693-3976-3.ch011

Ashfaq, M., Gupta, G., & Verma, N. (2025). Carbon-based nanocarriers for plant growth promotion: Fuelling when needed. *Nanoscale*, *17*(2), 616–634. DOI: 10.1039/D4NR03268C PMID: 39575969

Bahl, E., Jyoti, A., Singh, A., Siddqui, A., Upadhyay, S. K., Jain, D., Shah, M. P., & Saxena, J. (2024). Nanomaterials for intelligent CRISPR-Cas tools: Improving environment sustainability. *Environmental Science and Pollution Research International*, *31*(60), 67479–67495. Advance online publication. DOI: 10.1007/s11356-024-32101-x PMID: 38291210

Ben-Haim, A. E., Feldbaum, R. A., Belausov, E., Zelinger, E., Maria, R., Nativ-Roth, E., Mani, K. A., Barda, O., Sionov, E., & Mechrez, G. (2024). DNA Delivery to Intact Plant Cells by Casein Nanoparticles with Confirmed Gene Expression (Adv. Funct. Mater. 16/2024). *Advanced Functional Materials*, *34*(16), 2470089. DOI: 10.1002/adfm.202470089

Biradar, S., Korrapati, C., Krishna, R. J., & Nanjundeshwara, N. (2025). Spatial Omics and Gene Circuits: The Future in Cancer Biology. *Spatially Variable Genes in Cancer: Development, Progression, and Treatment Response*, 383–408. DOI: 10.4018/979-8-3693-7728-4.ch014

da Silva, R., Viana, V. E., Avila, L. A., Zotti, M. J., Smagghe, G., Junior, A. M., Camargo, E. R., & Fajardo, A. R. (2024). Advances on polymeric nanocarriers for sustainable agriculture: Enhancing dsRNA/siRNA delivery to combat agricultural pests. *International Journal of Biological Macromolecules*, *282*, 137000. DOI: 10.1016/j.ijbiomac.2024.137000 PMID: 39476891

da Silva, R., Viana, V. E., Avila, L. A., Zotti, M. J., Smagghe, G., Junior, A. M., Camargo, E. R., & Fajardo, A. R. (2024). Advances on polymeric nanocarriers for sustainable agriculture: Enhancing dsRNA/siRNA delivery to combat agricultural pests. *International Journal of Biological Macromolecules*, *282*(Pt 3), 137000. DOI: 10.1016/j.ijbiomac.2024.137000 PMID: 39476891

Demirer, G. S., Zhang, H., Goh, N. S., González-Grandío, E., & Landry, M. P. (2019). Carbon nanotube-mediated DNA delivery without transgene integration in intact plants. *Nature Protocols*, *14*(10), 2954–2971. DOI: 10.1038/s41596-019-0208-9 PMID: 31534231

Gandhi, K., Murugesan, S., & Suppaiah, R. (2024). Replicase gene mediated RNA silencing mechanism confers resistance against groundnut bud necrosis virus in tomato (Lycopersicon esculentum Mill.). *European Journal of Plant Pathology*, *170*(2), 405–416. DOI: 10.1007/s10658-024-02908-y

Gao, L., Wei, Y., Wang, T., Zhu, Q., Liu, Y., Zhang, Z., Hu, Z., Zhou, B., Zhang, H., Jin, N., Jiao, D., Yin, M., Shen, J., & Yan, S. (2025). Preparation of a Self-Assembled Nanoantiviral Pesticide with Strong Adhesion Performance for Efficiency Control of Destructive Soybean Mosaic Virus. *Journal of Agricultural and Food Chemistry*, *73*(4), 2342–2351. DOI: 10.1021/acs.jafc.4c11405 PMID: 39817843

Islam, M. R., Youngblood, M., Kim, H.-I., González-Gamboa, I., Monroy-Borrego, A. G., Caparco, A. A., Lowry, G. V., Steinmetz, N. F., & Giraldo, J. P. (2024). DNA Delivery by Virus-Like Nanocarriers in Plant Cells. *Nano Letters*, *24*(26), 7833–7842. DOI: 10.1021/acs.nanolett.3c04735 PMID: 38887996

Kathait, P., Patel, P. K., & Sahu, A. N. (2024). Harnessing exosomes and plant-derived exosomes as nanocarriers for the efficient delivery of plant bioactives. *Nanomedicine*. https://www.tandfonline.com/doi/abs/10.1080/17435889.2024.2354159

Khatoon, U. T., & Velidandi, A. (2025). An Overview on the Role of Government Initiatives in Nanotechnology Innovation for Sustainable Economic Development and Research Progress. *Sustainability (Basel)*, *17*(3), 3. Advance online publication. DOI: 10.3390/su17031250

Komarova, T., Ilina, I., Taliansky, M., & Ershova, N. (2023). Nanoplatforms for the Delivery of Nucleic Acids into Plant Cells. *International Journal of Molecular Sciences*, *24*(23), 23. Advance online publication. DOI: 10.3390/ijms242316665 PMID: 38068987

Korrapati, C., Yaraladdi, A. B., Vaddi, H., & Raghavan, R. (2024). Applications of the Saffron European Kitchen Spice Crocus sativus. In *Therapeutic and Pharmacological Applications of Ethnobotany* (pp. 98–127). IGI Global., DOI: 10.4018/979-8-3693-1986-4.ch005

Marquiegui - Alvaro, A., Kottara, A., Chacón, M., Cliffe, L., Brockhurst, M., & Dixon, N.Marquiegui – Alvaro. (2025). Genetic Bioaugmentation-Mediated Bioremediation of Terephthalate in Soil Microcosms Using an Engineered Environmental Plasmid. *Microbial Biotechnology*, *18*(1), e70071. DOI: 10.1111/1751-7915.70071 PMID: 39801293

Nagashri, N., Archana, L., & Raghavan, R. (2024). Microbial Volatile Compounds (MVOCs) in Food Industries and Food Safety Applications. In *Innovations, Securities, and Case Studies Across Healthcare, Business, and Technology* (pp. 75–96). DOI: 10.4018/979-8-3693-1906-2.ch005

Park, S., Safdar, M., Kim, W., Seol, J., Kim, D., Lee, K.-H., Son, H. I., & Kim, J. (2024). Gelatin Nanoparticles can Improve Pesticide Delivery Performance to Plants. *Small*, *20*(42), 2402899. DOI: 10.1002/smll.202402899 PMID: 38949406

Peyret, H., Shah, S. N., Meshcheriakova, Y., Saunders, K., & Lomonossoff, G. P. (2025). How do RNA viruses select which RNA to package? The plant virus experience. *Virology*, *604*, 110435. DOI: 10.1016/j.virol.2025.110435 PMID: 39893746

Rani, N., Kumari, K., & Hooda, V. (2025). The role of nanoparticles in transforming plant genetic engineering: Advancements, challenges and future prospects. *Functional & Integrative Genomics*, *25*(1), 23. DOI: 10.1007/s10142-025-01528-x PMID: 39841261

Septiani, P., Pramesti, Y., Ghildan, M., Aprilia, K. Z., Awaludin, R., Medina, S., Subandiyah, S., & Meitha, K. (2025). RNAi-based biocontrol for crops: A revised expectation for a non-recent technology. *Planta*, *261*(2), 44. DOI: 10.1007/s00425-025-04625-0 PMID: 39862243

Singh, R., Choudhary, P., Kumar, S., & Daima, H. K. (2024). Mechanistic approaches for crosstalk between nanomaterials and plants: Plant immunomodulation, defense mechanisms, stress resilience, toxicity, and perspectives. *Environmental Science. Nano*, *11*(6), 2324–2351. DOI: 10.1039/D4EN00053F

Vasanthkumar, R., Baskar, V., Vinoth, S., Roshna, K., Mary, T. N., Alagupandi, R., Saravanan, K., Radhakrishnan, R., Arun, M., & Gurusaravanan, P. (2024). Biogenic carbon quantum dots from marine endophytic fungi (*Aspergillus flavus*) to enhance the curcumin production and growth in *Curcuma longa* L. *Plant Physiology and Biochemistry*, *211*, 108644. DOI: 10.1016/j.plaphy.2024.108644 PMID: 38710114

Wasule, D. L., Shingote, P. R., & Saxena, S. (2024). Exploitation of functionalized green nanomaterials for plant disease management. *Discover Nano*, *19*(1), 118. DOI: 10.1186/s11671-024-04063-z PMID: 39023655

Yong, J., Xu, W., Wu, M., Zhang, R., Mann, C. W. G., Liu, G., Brosnan, C. A., Mitter, N., Carroll, B. J., & Xu, Z. P. (2025). Lysozyme-coated nanoparticles for active uptake and delivery of synthetic RNA and plasmid-encoded genes in plants. *Nature Plants*, *11*(1), 131–144. DOI: 10.1038/s41477-024-01882-x PMID: 39747606

Chapter 9
Nanocarriers for Delivering Plant Growth Promoting Microorganisms

Ravi Kant
Guru Gobind Singh Indraprastha University, Dwarka, India

Komal
https://orcid.org/0009-0006-6419-1186
Guru Gobind Singh Indraprastha University, Dwarka, India

Deepak
https://orcid.org/0009-0006-8008-8760
Guru Gobind Singh Indraprastha University, Dwarka, India

Divya Rajesh Shukla
https://orcid.org/0009-0006-4996-0354
Guru Gobind Singh Indraprastha University, Dwarka, India

Sayanti Mandal
https://orcid.org/0000-0003-4054-833X
Sharda School of Engineering and Sciences, Sharda University, India

ABSTRACT

The increasing world population and development have led to a heightened demand for food production, often met through the extensive application of chemical fertilizers and pesticides. To mitigate these adverse effects, sustainable agricultural practices are essential, emphasizing the need for alternatives to chemical inputs. Plant growth-promoting microorganisms (PGPM), particularly plant growth-promoting bacteria (PGPBs), have emerged as sustainable solutions to enhance plant growth. Despite their potential, the practical application of PGPMs is often compromised by issues such as instability, reduced efficacy, and environmental sensitivity. Nanocarriers such as biochar, clay, alginate, starch, chitosan, and gelatin have shown a potential to increase the viability and beneficial properties of PGPMs. This chapter reviews the potential of nanocarriers in delivering PGPMs, highlighting the advancements and benefits of integrating nanotechnology in sustainable agriculture.

DOI: 10.4018/979-8-3693-9869-2.ch009

1. INTRODUCTION

Population growth has led to a heightened demand for food (Ma et al., 2018) As a result, the extensive application of pesticides and fertilizers to increase agricultural productivity has led to biodiversity loss, soil and water contamination, disturbance of ecosystem functioning, and the depletion of soil microbial populations (Vance, 2001; Vessey, 2003). There has never been a greater pressing demand for alternatives that can partly or even completely replace chemical application in agriculture due to environmental deterioration and finite land resources. Sustainable agriculture requires tackling the challenges related to both agricultural practices and environmental health (Tilman et al., 2002).

Crop productivity with less damage to environment can be achieved by using biofertilizers, biopesticides, and eco-friendly water and soil management practices. One sustainable method is the use of PGPB. These bacteria are found in the rhizosphere, which is the soil, and they create symbiotic connections with the roots of plants. Rhizospheric bacteria live freely in the soil and form symbiotic relationships with plant roots, while endophytic PGPB (PGPB) colonize leaves, flowers, or inner tissues of plants (Malgioglio et al., 2022). Both types of bacteria help plants grow by improving nutrient uptake, increasing stress resistance, and protecting against diseases (Castro et al., 2018; Malgioglio et al., 2022; Ryan et al., 2008; Waller et al., 2005). As a result, biofertilizers that use these microorganisms are becoming more popular in organic farming. Conventional agrochemicals are not very efficient and have high environmental impacts, leading to the need for new strategies that can provide plants with nutrients and protection while reducing negative effects on the environment. Common fertilizers are inefficient because of nutrient loss through leaching, volatilization, or precipitation. Additionally, they lead to environmental problems that might impact crop quality, such as eutrophication, elevated soil salinity, and air pollution from nitrate accumulation.

Nanomaterial, sized between 1-100 nm, have various beneficial applications in agriculture (Narang et al., 2024); Abdelaziz et al., 2021; Guo et al., 2018; Paramo et al., 2020; Younas et al., 2024). Unlike conventional agrochemicals, nanoparticles can be applied in smaller quantities and provide a slow, controlled release of nutrients, thereby reducing environmental impact. Despite these advantages, questions about their safety and sustainability remain. While nanoparticles have been used in medicine for drug delivery, their agricultural applications are still emerging (González-Melendi et al., 2008; Sampathkumar et al., 2020; Mahanta et al., 2023; Pattnaik et al., 2024). To create a sustainable agricultural system, it's essential to enhance crop yields while maintaining soil health and ecosystem services. Plant production and health are increased when PGPB is used as an agricultural input, providing a sustainable and environmentally beneficial substitute for toxic chemical fertilizers, herbicides, fungicides, and insects. In the past few years, research in agricultural biotechnology has focused on developing microbial inoculants with one or more PGPB strains. These strains can help a plant growth by several direct and indirect mechanisms (Figure 1). Employing various biopolymeric matrices as carriers for PGPB is a promising approach, as these matrices protect the microbes from harsh external conditions, and ensure the consistent and slow release of the bacteria near the target site (Pereira et al., 2023; Hazra et al., 2023). This chapter will explore the utility of some of biopolymeric matrices as nano-carriers for PGPB in agriculture, highlighting how this technology can reduce environmental harm. It will discuss recent advancements in encapsulating PGPB using materials like Biochar, Clay, alginate, starch, chitosan, and gelatin, and summarize progress made in this field.

Figure 1. Action mechanisms by which PGPB benefits plants. Image created with Biorender

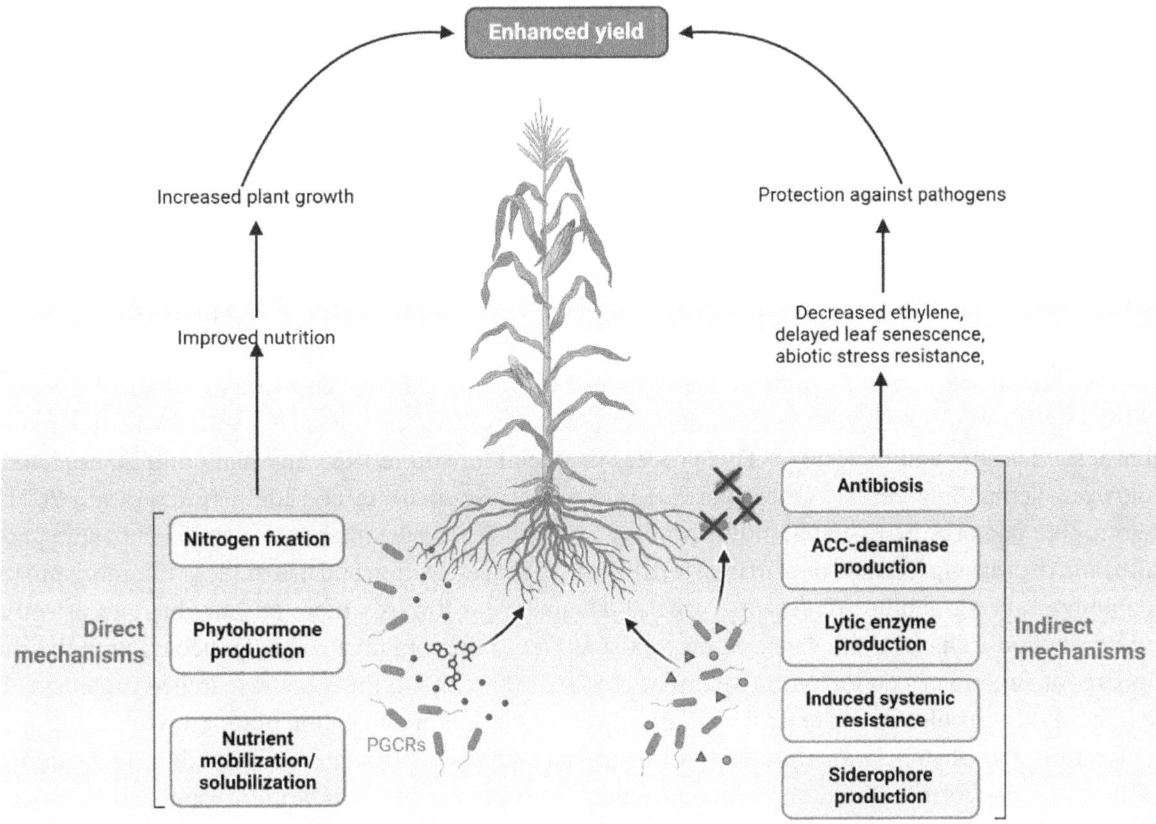

2. PLANT GROWTH PROMOTING MICROORGANISMS (PGPM)

Soil harbors a substantial quantity of bacteria, with the largest concentrations being located in the plant rhizosphere, namely surrounding the roots of plants (Upadhyay et al., 2022; Wei et al., 2023). These bacteria may promote plant development (i.e., plant growth-promoting bacteria), hinder plant growth (i.e., phytopathogens), or have no significant impact on plant growth (i.e., commensal bacteria) (Avinash et al., 2025). The fact that most plant roots typically emit a significant portion, i.e., between 5 and 30%, of all the carbon that the plant fixes during photosynthesis and provide this fixed carbon to soil microorganisms that utilize it as a food source is reflected in this typical bacterial localization (Ma et al., 2022). There is much evidence that various plants attract different kinds of soil microorganisms. This transpires due to the distinct composition of small molecules (predominantly amino acids, sugars, and organic acids) exuded by each plant's roots, which attract specific groups of the soil bacterial population.

The interest in PGPB arises from their capacity to enhance plant development and growth through: (i) augmentation of plant biomass, (ii) enhancement of nutrient content (including phosphorus, nitrogen, iron, and potassium), (iii) elongation of root and/or shoot length, (iv) acceleration of seed germination rates, (v) provision of protection against various pathogenic organisms (including nematodes, phytopathogenic bacteria, insects and fungi), and (vi) improvement of plant tolerance to diverse abiotic stresses (such as high temperatures, elevated salinity, root oxygen deficiency, flooding, and drought) (Khoso et al., 2024). This section provides a detailed overview of PGPR uses in field crops to enhance crop yield and quality.

Diazotrophic Bacteria's Significance in Biological Nitrogen Fixation (BNF)

Some diazotrophic microorganisms are responsible for the process known as biological nitrogen fixation (BNF), which involves the transformation of atmospheric nitrogen (N_2) into forms that plants can use, such as ammonium (NH_4^+). This process is crucial for soil fertility and plant output, especially in nitrogen-demanding cereals crops like rice and wheat (Soumare et al., 2020; Imran et al., 2021). Diazotrophic bacteria are essential in this process by providing a sustainable N_2 source, enhancing soil health, and decreasing reliance on artificial fertilizers. The investigation and use of these microorganisms in contemporary agriculture represent a crucial advancement towards more sustainable and effective farming methods. Diazotrophic bacteria may exist as free-living organisms, be connected with the rhizosphere, or function as endophytes (Mahmud et al., 2021). Among the most researched organisms in rice (Oryza sativa) cultivation are *Azospirillum* spp., recognised for enhancing plant growth via nitrogen (N_2) fixation and phytohormone production; Rhizobia, typically associated with nodule formation in legumes, can also form associations with non-leguminous plants; and *Burkholderia* spp. and *Herbaspirillum* spp., endophytic bacteria that inhabit the internal tissues of rice, aiding in nitrogen nutrition and promoting plant growth (Jaiswal et al., 2012).

Due to the fact that nitrogenase is very sensitive to oxygen, the process of nitrogen fixation is mediated by the enzyme nitrogenase. Nitrogenase is responsible for the conversion of nitrogen dioxide into NH_4^+ under anaerobic circumstances. This enzyme is protected from oxygen by diazotrophic bacteria via the use of a number of different techniques, including the creation of anaerobic microenvironments and the quick respiration of oxygen (Bennett et al., 2023). The resultant ammonium is either taken directly by plants or discharged into the soil, where it is assimilated and converted into amino acids and other vital macromolecules (Barbosa Santos et al., 2023; Zayed et al., 2023). By enriching the soil with readily accessible nitrogen, this method increases soil fertility and lessens the requirement for artificial fertilizers (Barbosa Santos et al., 2023). Plant production and growth are enhanced by the availability of fixed nitrogen (Zayed et al., 2023). Indole acetic acid (IAA) is one of the phytohormones produced by some diazotrophic bacteria that promote the growth of roots and vegetation (Egamberdieva et al., 2017). Utilizing diazotrophic bacteria improves soil health and lowers pollution while lessening the negative environmental effects of chemical fertilizers (Barbosa Santos et al., 2023; Zayed et al., 2023).

The ammonium produced is either discharged into the soil or taken directly by plants, which assimilate and convert it into amino acids and other important macromolecules (L'Espérance et al., 2024). This process provides the soil with accessible nitrogen, increasing fertility and minimizing the requirement for artificial fertilizers (Barbosa Santos et al., 2023; Zayed et al., 2023). The availability of fixed nitrogen promotes rice development and production (Di Benedetto et al., 2017; Kumar et al., 2021; Shahwar et al., 2023). Furthermore, certain diazotrophic bacteria create phytohormones such IAA, which promote root and vegetative growth (Kumar et al., 2021; Shahwar et al., 2023). The use of diazotrophic bacteria reduces the environmental effect of chemical fertilisers by enhancing soil health and lowering pollutants (Shahwar et al., 2023). Many companies make unfounded promises about their efficacy, although after decades of study, there is still no clear proof that the inoculation of non-legumes crops with diazotrophic bacteria routinely results in significant atmospheric N_2 fixation. Therefore, control is desperately needed to guarantee that inoculant products provide independent, verifiable proof of their ability to fix atmospheric N_2, safeguard nitrogenases, stay active all through the crop cycle, and show quantifiable impacts under field circumstances (Shahwar et al., 2023). By changing stress responses and increasing water usage efficiency, diazotrophic bacteria may eventually help crop plants tolerate environmental stress conditions like salt and drought (El-Saadony et al., 2024).

Mechanisms of Phosphate Solubilization and Their Effect on Plant Availability of Phosphorus

Photosynthesis, energy transmission, and nucleic acid synthesis all rely on phosphorus (P), an important plant nutrient (Khan et al., 2023; Bhat et al., 2024). But many soils contain phosphorus in insoluble forms that plants have a hard time absorbing (Bhat et al., 2024). To increase plant development and productivity, PGPMs, especially certain fungi and bacteria, play a important role in phosphate solubilisation. These microbes are multifunctional because they solubilize phosphate by a process that involves many processes, including the formation of organic acids, phosphatase enzymes, and siderophores (Pan & Cai, 2023; Silva et al., 2023). By reducing soil pH and dissolving insoluble inorganic phosphates (e.g., calcium, iron, and aluminium phosphates), the organic acids generated by PGPMs acidify the rhizosphere, allowing plants to absorb soluble phosphate ions (Silva et al., 2023). In order to make inorganic phosphate accessible to plants, phosphatase enzymes hydrolyse organic phosphate molecules (Pan & Cai, 2023; Silva et al., 2023). Siderophores also increase the availability of iron by chelating it from the environment and releasing the phosphate that was bound to iron complexes (Gomes et al., 2024).

A further intriguing method is used by some chemolithotrophic PGPMs, who liberate phosphate via the oxidation of mineral molecules. *Bacillus subtilis* is a good example since it produces organic acids that dissolve fluorapatite, the primary ingredient in rock phosphate (Ca5(PO4)3F) (Han et al., 2024; Bakki et al., 2024). Photosynthesis and biomolecule synthesis are both made more efficient by this process, which increases the availability of phosphorus and makes it easier to absorb other vital nutrients. Before applying PGPR to soil, it is crucial to study the varied microbial interactions between it and other soil occupants, as well as the right quantities of inoculants for various plant and soil conditions (Saeid & Prochownik, 2018; Han et al., 2024; Bakki et al., 2024). This will help ensure that biofertilizer usage is successful. A more sustainable agricultural system and healthier soil are the end results of these processes, which boost crop production and quality, make crops more resistant to environmental stress and diseases, and decrease the need for artificial phosphate fertilizers (Saeid & Prochownik, 2018; Bakki et al., 2024).

Phytohormone Synthesis: Gibberellins, Cytokinins, and Auxins and Their Effect on Plant Development

Small quantities of organic molecules generated by both plants and certain microbes, phytohormones are vital in controlling and organizing plant development and growth. Among the phytohormones most important are auxins, cytokinins, and gibberellins (Egamberdieva et al., 2017).

Auxins

Azospirillum spp., *Pseudomonas* spp., and *Bacillus* spp. synthesis auxins, fundamental molecules for plant development (Cruz-Hernández et al., 2022). Made via the tryptophan biosynthesis route, these plant hormones become IAA, the most often used auxin (Zhao, 2012). Their importance in plant growth is especially with regard to root development. Auxins improve water and nutrient absorption, therefore strengthening the plant, and increasing its resistance to stress by stimulating the elongation and branching of roots (Zhang et al., 2022). They also control the production and expansion of organs including leaves and flowers, therefore fostering a balanced and healthy growth of the plant (Liu et al., 2024). Auxins may be essential for plant adaptation under abiotic stress as they increase resistance to drought, salt, high temperatures (Liu et al., 2024).

Gibberellins

Gibberellins, essential phytohormones for growth and development, are produced by many microbes, including *Rhizobium* spp., *Gibberella fujikuroi*, and *Azospirillum* spp. (Salazar-Cerezo et al., 2018; Kildegaard et al., 2021; Nett et al., 2022). These diterpene molecules are synthesised via the mevalonic acid pathway (Kildegaard et al., 2021; Nett et al., 2022). Their primary role is to promote stem elongation, leading to enhanced plant height and improved sunshine exposure (Salazar-Cerezo et al., 2018; Kildegaard et al., 2021). Furthermore, gibberellins are essential in seed germination by alleviating dormancy and facilitating germination, so assuring expedited and uniform germination, which is critical for crop harvestmen (Kildegaard et al., 2021; Nett et al., 2022). Moreover, these hormones affect fruit development and growth, enhancing both size and quality, hence substantially augmenting crop production (Nett et al., 2022).

Cytokinins

Many different types of microbes, including Rhizobium, Pseudomonas, and Azotobacter, produce cytokinins, which are plant hormones that are crucial to a plant's growth and development (2022). Adenosine monophosphate (AMP) is the starting point for the production of modified adenines, which include zeatin (Kieber & Schaller, 2014). Their primary role is to promote stronger development by enhancing the creation of new tissues via stimulation of cell division in both shoots and roots (Wu et al., 2021). Delaying leaf senescence, which increases photosynthesis and plant productivity, is another important function of cytokinins (Hönig et al., 2018). Hormones like these work in tandem with auxins to control the production of both shoots and roots, balancing the plant's development from top to bottom (Mazzoni-Putman et al., 2021). Improvements in root structure and function brought about by the synergy of auxins, gibberellins, and cytokinins are crucial to plant development and productivity because they allow for more effective absorption of water and nutrients. The quality of agricultural goods is enhanced and crop productivity is significantly increased as a consequence of this hormonal interaction, which promotes strong and faster growth (Sosnowski et al., 2023). The phytohormones that PGPMs produce also help plants adapt and resist bad circumstances, so they can better handle biotic and abiotic stress and have more consistent and steady output (Voko et al., 2024).

Siderophores Production

Made by different microorganisms including bacteria and fungus in response to iron constraints in their environment, siderophores are low molecular weight organic molecules with a strong affinity for ferric ions (Fe^{3+}). Acquiring iron, a fundamental component for both microbes and plants, depends on this method. By improving the availability of iron in the soil and rhizosphere, siderophores helps to promote plant nutrition and therefore strengthen resistance to abiotic stress and pathogens (Truong et al., 2024). Moreover, their use in agriculture via PGPMs has demonstrated significant increase in plant development in salty soils with iron constraints (Truong et al., 2024).

Siderophoresis biosynthesis makes use of precursors including amino acids and carboxylic acids (Miethke & Marahiel, 2007). Depending on the generating microbe, among the most often occurring siderophores are catecholates, hydroxamates, and carboxylates (Xie et al., 2023). Microorganisms sense iron levels in their surroundings and, under iron shortage situations, stimulate siderophore production via regulators like Fur in *Escherichia coli* (Singh et al., 2022). As in *Pseudomonas* spp., where pyoverdine is delivered via membrane proteins carboxylates (Xie et al., 2023), siderophores chelate accessible iron and the siderophore–iron complexes are carried into microbial cells by specialized transport routes. This mechanism not only maximizes photosynthesis and other basic metabolic activities but also greatly increases root growth, hence boosting plant absorption of water and nutrients (Singh et al., 2022). In response to microbial interactions, plants can use many immunological defences including ferroptosis, a kind of iron-regulated cell death. In this regard, *Streptomyces hygroscopicus* OsiSh-2 is rather important in enhancing rice disease resistance by means of siderophoresis, hence controlling ferroptosis. This mechanism highlights its potential to improve crop yield and resilience as it balances mutualistic symbiosis with host defence (Singh et al., 2022).

In addition to reducing the availability of iron for pathogens, siderophores also limit the spread of plant maladies by contending for iron (Timofeeva et al., 2022). *Bacillus* spp. (bacillibactin and petrobactin), *Pseudomonas* spp. (pyoverdine and pyochelin), and *Rhizobium* spp. (rhizobactin) are notable examples of siderophore producers. These microorganisms are essential for the protection of legumes from pathogens, as well as the enhancement of processes such as nodulation and nitrogen fixation (Xie et al., 2023).

Biocontrol Control of Plant Pathogens

Biocontrol of pathogens by PGPMs is an important process that improves agricultural sustainability and productivity by inhibiting pathogen development and spread via competitive mechanisms and the synthesis of antimicrobial compounds (El-Saadony et al., 2022). In order to prevent illnesses, PGPMs compete with pathogens for vital resources like nitrogen and iron. They do this by colonising the same biological niches as the pathogens and sequestering nutrients via their siderophores (Dimkić et al., 2021). In addition, PGPMs produce enzymes and antibacterial compounds that break down pathogen cells. Trichoderma species, which create enzymes that fight fungal infections, Pseudomonas species, which make bacteriocins and siderophores, and Bacillus species, which make peptides that fight against microbes and enzymes that break them down, are all examples of PGPMs (Ayaz et al., 2022). By including these PGPMs into IPM plans, we may lessen our reliance on chemical pesticides, boost plant resilience, encourage sustainable farming methods, and ultimately improve plant health while decreasing disease risk (El-Saadony et al., 2022).

Resilience Against Abiotic Stress Conditions

Beneficial PGPMs are important for plants to survive and grow in harsh environments because they improve their ability to handle abiotic stress in many ways, such as by changing plant hormones, encouraging nutrient uptake, making exopolysaccharides that keep water in, activating phytohormones like abscisic acid to save water, creating metabolites and antioxidant enzymes to protect against oxidative stress, and dissolving nutrients in saline soils (L Morcillo & Manzanera, 2021; Moulick et al., 2024). Additionally, by strengthening internal plant systems, enhancing soil microbiota diversity and activity, and activating defence gene expression, these bacteria help plants become more resilient to harsh environmental factors including salt and drought (Chattaraj et al., 2025). When PGPMs are strategically used in agriculture, crop output is increased in areas with difficult circumstances, water and fertilizer dependence is decreased, and more environmentally friendly, sustainable agriculture is encouraged (De Andrade et al., 2023).

As previously stated, microbes may enhance plant development and alleviate abiotic and biotic stressors; however, Weinand et al., 2023 have documented divergent findings when assessing three Bacillus isolates in crop plants. Two isolates of *B. pumilus* suppressed fungal proliferation and decreased the occurrence of brown spot in certain types; however, they exhibited no conventional growth-promoting characteristics. Conversely, *B. megaterium* demonstrated auxin production capability, although it did not inhibit the illness. These results indicate that bacteria may enhance plant development without exhibiting conventional growth-promoting characteristics. Furthermore, despite the shown efficiency of several species, a considerable disparity persists between comprehending their mechanisms of action and their actual use as biofertilizers (Lee et al., 2024).

3. BIOCHAR AS NANO-CARRIERS

Biochar is created through the thermochemical conversion of biomass. Its application enhances carbon, nitrogen, and phosphorus levels in the soil. Biochar can be seen as an engineered nanocomposite at the molecular level, which allows for comparisons with the effects of nanoparticles (Bano et al., 2024). Adding biochar to soil has fewer negative impacts on both plants and/or the environment than traditional fertilizers. Additionally, it can help plants flourish in marginal or inappropriate soil types, like sandy, loamy, acidic, or polluted soils (Ajeng et al., 2020; Husna et al., 2019; Kari et al., 2021; Moulick et al., 2025). It has been demonstrated that biochar seeded with microbes improves plant growth, development, and nutrients quality because of its porosity, adsorbent capacity, and relatively high nutrient content (Ajeng et al., 2020; Husna et al., 2019; Kari et al., 2021; Qi et al., 2021).

Hydrochar, rich in nutrients, supports bacterial growth. This study looked at the effect of a special rhizobial inoculant made from hydrochar (using *Bradyrhizobium japonicum*) on how well soybeans grow. Experiments were conducted both in the field and in a greenhouse. The results showed significant improvements in root and shoot growth for soybeans treated with the inoculant, compared to those that were not, under both irrigated and drought conditions. Hydrochar-based inoculation mitigated drought stress effects, doubling nodule formation compared to a commercial inoculant (HISTICK). Nitrogen levels were 31% higher in plants inoculated with hydrochar-based *B. japonicum* compared to uninoculated ones, and 22% higher than those treated with HISTICK. Soybeans treated with hydrochar-based *B. japonicum* yielded 29% more grain under irrigation and 40% more under rainfed conditions. Thus, hydrochar-based *B. japonicum* significantly enhances soybean performance, especially under rainfed conditions (Egamberdieva, Ma, et al., 2020).

Ginseng, a valuable cash crop, faces issues due to long-term continuous cultivation, which disrupts soil balance and increases soil-borne diseases, hindering the industry. This study investigated how treating ginseng cultivation soil with a microbial inoculum containing the biocontrol strain *Frankia* F1 impacted soil health. Wheat, rice and corn straws were identified as the best carriers for the microbial inoculum. Among these, the inoculum made with corn straw biochar significantly improved soil quality. Specifically, soil pH (11.18%), organic matter (55.43%), total nitrogen (33.07%), available nitrogen (26.70%), available phosphorus (16.40%), and available potassium levels (9.10%) increased. Additionally, the activities of soil enzymes such as urease (52.73%), catalase (16.80%), and sucrase (43.80%) went up. The study concluded that using a microbial inoculum with corn straw biochar was the most effective way to enhance soil health in ginseng cultivation (Qi et al., 2021).

This study examined lupin growth and root disease incidence using different biochars. The tested biochar types included pyrolysis char from maize (MBC), pyrolysis char from wood (WBC), and hydrochar (HTC) from maize silage. The results showed no significant effects on lupin shoot and root growth with WBC at any concentration, while MBC had no impact except for a root dry weight increase at 2%. HTC at 2% significantly root dry weight increased by (54-75%) and increase in shoot dry weight by (21-25%). Plants grown with WBC and MBC at 2% and 3% concentrations exhibited disease. In contrast, plants grown without biochar or with HTC remained healthy, with no disease incidence. Furthermore, the endophytic bacteria *Pseudomonas putida* L2 and *Stenotrophomonas pavanii* L8 reduced disease symptoms in MBC- and WBC-amended soils compared to uninoculated plants. In summary, HTC significantly enhanced plant growth and prevented disease, while the use of specific endophytic bacteria helped control root rot in soils treated with other types of biochar (Egamberdieva, Shurigin, et al., 2020).

Thermal gasification of straw is an efficient method for producing bioenergy and gasification biochar (GB), which can crop productivity. A field study conducted in Denmark explored the impact of two different practices on soil and crops: traditional incorporation of straw into the soil versus removing the straw for thermal gasification and then applying the resulting GB back to the field. The researchers aimed to understand how these methods affect soil health and crop performance. Over the course of three years, two different rates of GB were applied to fields planted first with winter wheat, then with winter oilseed rape, and finally with winter wheat again. The findings revealed that GB application did not significantly increase soil organic carbon or crop yields. However, both straw and GB positively impacted bacteria and protist populations, though not earthworms. The high rate of GB increased soil potassium levels and pH, indicating its nutrient availability and liming effect (Hansen et al., 2017).

Incorporating biochar to soil can affect soil properties and plant growth, with outcomes varying based on the feedstocks and production methods (such as pyrolysis or gasification). This greenhouse research examined the impacts of two biochar types (pyrolysis and gasification) incorporated at a 5% volume rate into a peat-based growth medium, with rhizobia inoculation (with or without), on *Robinia pseudoacacia* (black locust) seedlings supplied with $15NH_4$ $15NO_3$. The results indicated that adding biochar without inoculation did not affect seedling growth. However, rhizobia inoculation elevated seedling growth, BNF, and N status. Inoculated seedlings exhibited lower δ15N values, suggesting that nitrogen from fertilization was supplemented by nitrogen from BNF. The combination of biochar type and inoculation had a significant effect on seedling growth Gasifier biochar combined with inoculation resulted in the best overall growth and nodule biomass. (Sun et al., 2020).

Another study by Husna and team evaluated four types of biochar—made from rice husk--(RH), coconut shell--(CS), corncob--(CC), and oil palm empty bunch--(OPEB)—to determine which was best for carrying. The biochars from OPEB, CC, and RH had similar bulk densities of 0.61, 0.66, and 0.64, respectively, while CS had the highest bulk density at 0.92. Due to its higher bulk density, CS had lower total pore space (65%) and water holding capacity (120.66%) compared to the other biochars. The chemical composition varied based on the raw materials used, with CS generally having higher nutrient concentrations indicated by its highest EC value. The C-organic content of CC, CS, RH, and OPEB biochars was over 50%, with values of 65.26%, 64.12%, 59.84%, and 51.24%, respectively, along with a high C/N ratio. The pH of OPEB, CS, and RH biochars was 6.26, 7.74, and 7.01, respectively, whereas CC had the highest pH at 9.95. CC also had the highest ash content (40.21%) and the lowest charcoal percentage (30%). To evaluate their suitability as PSM inoculants, these biochars were inoculated and stored for six months, with monthly checks on PSM population, pH, and moisture levels. The results indicated that different biochar types and storage times did not significantly affect PSM viability, pH, or moisture. PSM viability remained stable between log 7.07 cfu.g-1 and log 9.18 cfu.g-1 after six months, with CS showing the highest viability at log 9.18 cfu.g-1, a pH of 7.01, and 26.86% moisture. The study suggests that agricultural waste can be used to make biochar that maintains microbial viability for six months (Husna et al., 2019).

A study by Pavlicevic et al. (Pavlicevic et al., 2022) investigated use of Nanoclay, and Biochar as Carriers of PGDB in both watermelon and tomato plants. The experiments included multiple treatments: a control (C), treatment with bacteria (BAC), treatment with chitosan-coated mesoporous silica both alone (MSN) and with bacteria (MSN + B). Additional treatments involved nanoclay alone (NC) and with bacteria (NC + B), as well as three types of biochar treatments. The first biochar type, "Italian" biochar, was applied both alone (IB) and with bacteria (IB + B). The second type, Aries green biochar, was used alone (AB) and with bacteria (AB + B). Finally, Naked biochar was tested both alone (NB) and with bacteria (NB + B). The research explored the use of biochar and nanoparticles as carriers for beneficial bacteria to crops, assessing their effects on soil and plant health. The treatment led to an increase in nitrogen-fixing and phosphorus-solubilizing bacteria in the soil. Additionally, chlorophyll content increased by 1.2 to 1.3 times, cell viability improved by 1.1 to 1.5 times, and antioxidative properties enhanced by 1.1 to 1.4 times compared to the control plants. The treatments also enhanced the phosphorus (P) content by 1.1 to 1.6 times and nitrogen (N) content by 1.1 to 1.4 times in both tomato and watermelon plants, although these effects varied between species. In tomato plants, for example, the phosphorus content rose 1.2 times with chitosan-coated mesoporous silica nanoparticles containing adsorbed bacteria and 1.1 times with nanoclay containing adsorbed bacteria. Conversely, in watermelon plants, chitosan-coated mesoporous silica nanoparticles led to a 1.1-fold increase in phosphorus content, while nanoclay caused a 1.2-fold increase. Furthermore, tomatoes responded best to treatments with chitosan-coated mesoporous silica and "Italian" biochar combined with bacteria. This was consistent with the findings from antioxidant properties and viability tests, highlighting the benefit of these carriers. Differences in stem length and total plant mass could indicate higher nutrient transport in some treatments. In watermelon, significant increases in total mass and stem length were observed with all bacteria-adsorbed samples compared to controls, showing that watermelon was more sensitive to bacteria presence but less to carrier type. All treated plants exhibited increased stem length, with notable improvements from NC, AB, NC + B, and AB + B treatments. Tomatoes were more affected by MSN and IB carriers, while watermelon responded more to NC and AB carriers. In watermelon, a two-way ANOVA revealed an interaction between stem length and entire plant mass that was absent in tomatoes. This may be due to different growth cycles: tomatoes take about 80 days to fruit, while watermelons take around 60 days (Pavlicevic et al., 2022). Therefore, research is needed to understand why this interaction was observed in watermelon but not in tomatoes, and to better understand the responses of different plants to various treatments. The utilization of biochars and nanoparticles as carriers for PGPR positively influenced the growth of watermelon and tomato plants. The observed benefits included elevated levels of phosphorus, nitrogen, chlorophyll, antioxidative potential, cell viability, and total plant mass, along with an increase in N_2-fixing and phosphorus-solubilizing bacteria.

Gasification biochar has been less extensively researched as a delivery system compared to hydrochars or biochars produced via pyrolysis. Nonetheless, Sun et al. (Sun et al., 2020) discovered that gasification-produced biochar, when inoculated with rhizobia, had significantly greater positive effects on nitrogen content and the growth of black locust (*Robinia pseudoacacia*) seedlings than pyrolysis-produced biochar with the same rhizobia concentration. Likewise, Graziano et al. (Graziano et al., 2022) reported beneficial effects on soil and plants in their studies with wheat and maize.

Overall, the benefits observed varied depending on the plant species and their nutrient status. Additionally, the study highlights the potential to enhance the effects of plant growth-promoting rhizobacteria (PGPR) by using biochar and nanoparticles as carriers. These findings contribute to a better understanding of how to optimize plant health and productivity through innovative approaches in biochar and nanotechnology.

4. BIOPOLYMERIC MATRICES FOR DELIVERING PLANT-GROWTH-PROMOTING BACTERIA (PGPB)

Figure 2. Types of biopolymer matrix types as nanocarriers

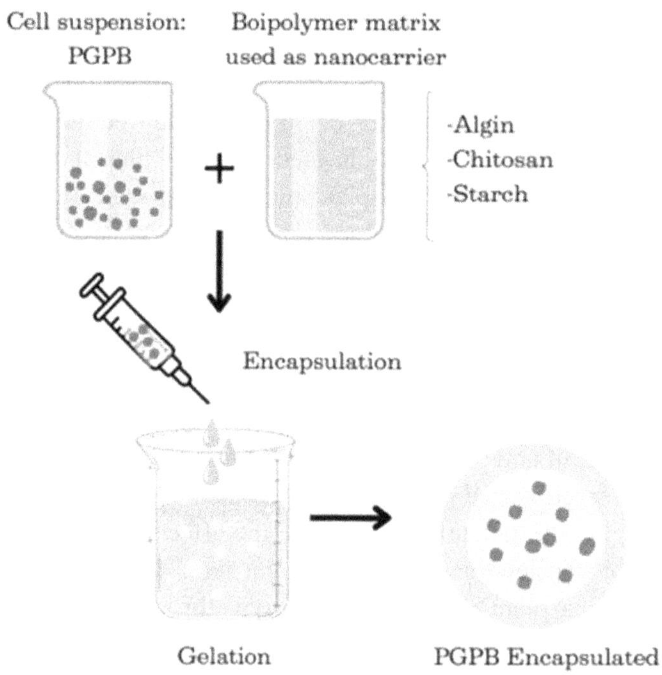

Bacterial cells are often immobilized and encapsulated in agriculture, pharmaceuticals, food, and other industries. This process involves placing the cells in a protective capsule. The capsule protects the cells from harmful environmental conditions, improves their stability, and enhances their survival during production, storage, handling, and rehydration. Biopolymers are derived from renewable resources, and many of them, such as alginate, starch, gelatin, and chitosan, are biodegradable (Figure 2). These biopolymers have been widely used as carriers for PGPB (Ashfaq et al., 2024; Pereira et al., 2023).

Rhizobacteria bioencapsulation protects bacteria in the soil and ensures they are released gradually. The encapsulation material's breakdown rate affects soil bacteria's activity. These capsules can be stored at room temperature for extended periods, offering a stable environment for the bacteria and minimizing the risks to their survival.

. Adding nutrients can turn capsules into bioreactors, boosting bacteria numbers in the soil. This method addresses issues with traditional peat inoculants, offering controlled release, protection against stresses, and less contamination during storage and transport.

The immobilization and encapsulation of PGPB cells can enhance the viability and stability of the bacteria during production, storage, handling, and rehydration (John et al., 2011; Schoebitz et al., 2013). In inoculant formulation, the carrier is a significant component. According to Bashan et al. (Yoav Bashan et al., 2014), Carrier materials offer a chemically stable and protective environment. They need to be non-toxic, easily accessible, affordable, free of preservatives, capable of maintaining humidity, and durable enough to be stored at room temperature for long periods. (John et al., 2011; Schoebitz et al., 2013). Bashan et al. emphasized that an ideal carrier should be user-friendly, enable controlled release of microorganisms, and be straightforward to produce and combine with other additives or nutrients. (Yoav Bashan et al., 2014). A method has been developed to inoculate both wet and dry seeds with PGPB) using alginate microbeads as a carrier. In this study, *Azospirillum brasilense* was used as the model PGPB. (Yoav Bashan et al., 2002). The microbeads were created by spraying a mixture of alginate solution and liquid bacterial culture at low pressure, resulting in small-diameter droplets. While some bacteria were lost during the process, the remaining bacteria within the microbeads were sufficient for seed inoculation, exceeding 10^{11} colony-forming units (CFU) per gram of inoculant. The bacterial population was further increased by allowing secondary multiplication. The bacteria were slowly released ranging from 10^4 to 10^6 CFU per gram. Tomato and wheat seedlings grown in infertile soil were effectively boosted in development by both wet and dry inoculants, and they biodegraded in 15 days in moist soil (Yoav Bashan et al., 2002). The primary techniques for encapsulating PGPB include emulsification, ionic gelation, and spray drying (Yoav Bashan et al., 2002). Incorporating various additives can reduce costs and enhance the stability, survival, encapsulation efficiency, mechanical strength, and swelling. Potential fillers include polymers, clays, osmoprotectants, and nutritional compounds (Yoav Bashan et al., 2002).

5. ALGINATE-BASED POLYMERIC CARRIERS

Sodium alginate is derived from brown algae, but seaweed is not its only source; certain bacteria, such as *Azotobacter vinelandii* (Yoav Bashan et al., 2002)and various *Pseudomonas* strains), also produce extracellular alginate. Alginates are linear macromolecules composed of two monomers, α-L-guluronic acid and β-D-mannuronic acid, linked by alpha 1–4 bonds, with molecular weights ranging from 20,000 to 200,000 Da. The properties of alginate, such as solubility and calcium complexation, vary depending on the seaweed source and manufacturing process. In order to generate an alginate solution, it is combined with cell culture and then injected into a $CaCl_2$solution with concentrations ranging from 0.05 to 0.1 M. The gelation process is completed in about 20–30 minutes while the mixture remains in this solution (Panichikkal, Puthiyattil, et al., 2021; Rodríguez-Rodríguez et al., 2020; Schoebitz et al., 2013).

Typically, polymeric matrices made from alginate are prepared in an aqueous solution, with concentrations ranging from 0.5% to 3.0%. Calcium chloride ($CaCl_2$) is commonly used in the process to help form and stabilize these matrices typically used at concentrations between 1.0% and 2.0%, is the most common crosslinking agent. Additional components, such as fillers and osmoprotectants, can enhance the mechanical and swelling properties of alginate materials, leading to more efficient encapsulation (Panichikkal, Puthiyattil, et al., 2021; Rodríguez-Rodríguez et al., 2020; Schoebitz et al., 2013). The main feature of alginate is its ability to form gels when in contact with calcium ions (Ca^{2+}), resulting in hydrogels with a three-dimensional matrix. These are usually made into beads, capsules, microcapsules, or nanocapsules. Alginate is biodegradable and has reactive carboxylic groups. It is also recognized as safe by the FDA, making it ideal for use as a carrier for plant-growth-promoting bacteria (PGPB). Alginate has been widely studied for this purpose, especially as biofertilizers or biopesticides.

Jung et al. (Jung et al., 1982)and Fages (Szopa et al., 2022) were pioneers in using alginates as carriers for *Azospirillum* and *Rhizobium*, respectively. Since their groundbreaking work, numerous studies have confirmed the benefits of alginate as a carrier PGPB. Bashan et al. (1999) demonstrated that encapsulating *A. brasilense* and *P. fluorescens* in alginate beads helps these bacteria survive for extended periods (Y

Bashan & Gonzalez, 1999). The porous structure of the alginate matrix protects the cells from mechanical stress, with bacteria settling into the pores. A study conducted by Schoebitz and colleagues (2012) found that after a year of storage, dried alginate beads containing *A. brasilense* were able to preserve 76% of their live cells. It was discovered by Zago et al. (2019) that *A. brasilense* could continue to grow for a period of ninety days when it was encased in alginate beads, which in turn ensured improved cell viability (Zago et al., 2019). However, Bashan et al. (2019) noted a minor drawback: the crosslinking of the alginate-calcium complex with the bacterial cell wall can kill many bacteria. Fortunately, this issue can be mitigated by incubating the alginate beads in fresh growth medium, where surviving bacteria will multiply and restore the original concentration.

Panichikkal et al. (Panichikkal, Prathap, et al., 2021) encapsulated *Pseudomonas* sp. in alginate beads together with zinc oxide nanoparticles and salicylic acid (SA). They found that the encapsulated *Pseudomonas* sp. had better biocontrol properties and plant growth-promoting in rice seedlings compared to the free-living microorganisms. Souza-Alonso et al. (Panichikkal, Prathap, et al., 2021) investigated the encapsulation of *P. libanensis* in alginate polymeric beads and observed that cell viability remained intact for up to ninety days, with the alginate beads gradually dissolving in the plant rhizosphere.

Calcium alginate beads have a loose structure that holds a lot of water. As a result, using alginate alone for bioencapsulating cells isn't enough to protect them during the drying process, often leading to somewhat deformed beads. To solve this issue, filler materials like starch can be added to increase the dry matter content in the beads. This not only enhances their mechanical strength but also allows for the gradual release of the cells into the soil. For instance, using just alginate beads resulted in a two-log reduction in *Azospirillum brasilense* cell numbers due to the beads containing 97-98% water. However, by incorporating starch into the bioencapsulation process, the water content was reduced to 65%, which significantly improved cell survival (Nussinovitch, 2010). Starch is not only inexpensive and abundant, but it also serves many purposes, including as an additive, adhesive, filler, and carbon source (Y. Bashan, 1986). Research on probiotic carriers has shown that starch provides protection because cells adhere to the granules.

When cells are stored for a lengthy period of time, the alginate matrix effectively shields them from mechanical stress and minimises the likelihood that they will die. The immobilisation of *A. lipoferum* in alginate beads resulted in a survival rate of 10^10 CFU/g after a year at RT. This survival rate is greater than the survival rate of traditional microbial inoculants and is noteworthy (Martínez-Cano et al., 2022). Additionally, *P. corrugata* and *Bacillus subtilis* immobilized in alginate beads were able to maintain a remarkable survival rate of 10^8 CFU/g after being stored for 3 years at 4°C. Even after this extended period, they retained their ability to promote plant growth effectively (Trivedi & Pandey, 2008). Furthermore, *A. brasilense* Cd and *P. fluorescens* 313 exhibited a survival rate of 10^5–10^6 CFU/g after 14 years in storage, retaining their capacity to stimulate wheat plant growth (Y Bashan & Gonzalez, 1999). These findings demonstrate that rhizobacteria can survive in alginate inoculants for long periods without losing their beneficial properties.

The two *P. fluorescens* strains were encapsulated in alginate-gelatin beads with different gelatin concentrations and applied as biopesticides to inoculate potato plants. This treatment led to decreased fungal growth and disease incidence, offered protection against harsh soil conditions, and promoted the establishment of PGPB in the rhizosphere. The positive effects of encapsulating microorganisms in alginate beads are attributed to mechanisms such as the production of antifungal agents (antibiotics and enzymes), siderophore production, phosphate solubilization, and ACC deaminase activity (Pour et al., 2019).

Saberi-Riseh and Moradi-Pour (Saberi-Rise & Moradi-Pour, 2020) discovered that encapsulation remarkably enhanced the survival rate of *Bacillus subtilis* compared to its free form. Similarly, Young et al. (Saberi-Rise & Moradi-Pour, 2020) found that encapsulating *B. subtilis* in alginate beads enriched with humic acid preserved high bacterial viability during a five-month storage period and effectively promoted lettuce plant growth. In addition, *B. megaterium* was encapsulated in alginate microcapsules in order to function as a biopesticide against *Rhizoctonia solani*, which is a prominent phytopathogen that affects rice plants. The cells that were encapsulated shown a higher resilience to ultraviolet radiation, and they continued to be effective in preventing the mycelial development of *R. solani* even after being exposed to ultraviolet light for a period of forty-eight hours (Saberi-Rise & Moradi-Pour, 2020)

Hernández-Montiel et al. (Hernández-Montiel et al., 2017) proposed that utilizing PGPB as biofertilizers that could be a viable alternative for agricultural management and sustainable tomato production. They emphasized that immobilizing *Pseudomonas putida* in alginate microcapsules offers protection and allows for a gradual release, which improves the bacteria's adhesion, permanence, and colonization on the roots, ultimately leading to increased productivity in tomato plants.

Starch, an affordable biopolymer, has been successfully integrated into alginate formulations as a filler. This addition enhances the beads' resistance to physical stress and UV radiation. Moreover, adding starch to the alginate beads can improve the survival rate of *Azospirillum brasilense* during encapsulation, especially for microorganisms that can utilize starch as a carbon source (Schoebitz et al., 2012). In addition, trehalose, used as an osmoprotectant additive, further boosts cell viability during the storage of *A. brasilense* in alginate beads (Zago et al., 2019). Moreover, Rohman et al., 2021 highlighted that although alginate beads are promising carriers for biofertilizer release, they have certain drawbacks, including low mechanical strength, subpar appearance, high porosity, and insufficient cell protection (Rohman et al., 2021).

6. STARCH-BASED POLYMERIC CAREER

Starches from various sources have been extensively studied both academically and industrially due to their renewable origins, biodegradability, affordability, and wide availability (Saberi Riseh et al., 2023a). As an important biopolymeric carrier for PGPB, starch offers significant advantages, including high solubility, controlled release, and non-toxicity (Saberi Riseh et al., 2023a). Additionally, starch is well-known for its properties in film formation and its capability to create micro- and nanoparticulate systems, as well as gels (Falcão et al., 2022). These qualities make starch an ideal material for use as a carrier for bioactive compounds and microorganisms.

Carbohydrates are commonly used in spray-dried encapsulations due to their low viscosity and good solubility (Reineccius, 1991). Starch is frequently used to encapsulate flavors and various other compounds (Reineccius, 1991; Goubet et al., 1998; López et al., 2012). Sodium alginate is frequently used to encapsulate microorganisms, and the resulting product can be used for various applications like immobilizing bacteria (Bashan, 1986; Bashan et al., 2002), fermentation, biological control agents (Bashan and Holguin, 1994), and plant growth stimulants (Bashan and Levanony, 1990; Schoebitz et al., 2012).

Starch can be sourced from plants and composed of D-glucose units and consists of two main fractions: amylose and amylopectin. Amylose features linear α-(1→4) linkages, whereas amylopectin includes both α-(1→4) linkages and approximately 5–6% α-(1→6) branch linkages. (Minakawa et al., 2019; Tester et al., 2004). The amylose-to-amylopectin ratio varies depending on the starch source, leading to starch granules with diverse physicochemical and functional properties that can influence their suitability for different industrial applications.

Starch is utilized to produce biodegradable polymeric matrices by exploiting the properties of amylose and amylopectin. These components enable the formation of gels and continuous networks stabilized by hydrogen bonds. The matrices can be created in aqueous solutions with starch concentrations between 1.5% and 3.0% or through other methods such as thermoplastic extrusion (Mali et al., 2005; Minakawa et al., 2019; Tester et al., 2004). Extrusion, which involves high temperatures for short periods, offers benefits such as high versatility, no effluent production, and ease of scalability for large-scale industrial applications (Marcelino et al., 2016).

Although the use of starch as a carrier for PGPB is not widely documented in the literature, its application for the encapsulation of bioactive compounds has been studied extensively in recent decades (Rohman et al., 2021). Starch has been utilized to create particles for encapsulating curcumin for pharmaceutical purposes films for encapsulating antioxidants in food packaging (Lu et al., 2019; Luo et al., 2020) and films to enhance the stability of phenolic compounds in food technology applications (Lu et al., 2019; Luo et al., 2020). Although pure starch is not his blend provides protection to the bacterial cells and allows the diffusion of essential compounds such as micronutrients and metabolites in inoculant formulations. The hydrophilic nature of starch molecules enables them to retain substantial amounts of water within their polymeric network, which supports bacterial survival and enhances the effectiveness of inoculation strategies. (Riseh et al., 2021; Rohman et al., 2021).

One of the main drawbacks of using starch matrices for microorganism encapsulation is their poor mechanical performance (Falcão et al., 2022; Saberi Riseh et al., 2023b; Tester et al., 2004). To overcome this limitation, adding plasticizers, fillers, and blending starch with other biodegradable polymers are viable strategies (Mali et al., 2005) found that plasticizers decrease intermolecular forces and increase the mobility between polymer chains, reducing discontinuities and brittle areas, thereby improving the mechanical properties of the materials.

Several investigators have documented the use of starch combined with other polymers to create effective carriers for PGPB. Marcelino et al. (2016) investigated the creation of solid formulations for PGPB inoculation using an extrusion process. In their study, starch was the primary component, of the formulation, along with sugarcane bagasse and other minor ingredients These solid polymeric matrices were able to sustain the viability of *Azospirillum brasilense* cells for up to 120 days at RT (Marcelino et al., 2016).When applied to soil 10 days before maize planting, these formulations supported plant growth by creating a favorable environment that maintained bacterial viability longer than liquid inoculant formulations.

Vercelheze et al. reported using a biodegradable polymeric matrix comprising cassava starch, polyvinyl alcohol, gelatin, and glycerol as a plasticizer. This combination was used as a coating for maize seeds and was able to maintain the viability of *Azospirillum brasilense* for up to 15 days after the bacteria were immobilized. The biodegradable coating proved to be a promising, cost-effective, and renewable material suitable for agricultural applications (Vercelheze et al., 2019).

Perez et al. (Perez et al., 2018) studied the application of starch-chitosan macrobeads for the controlled release of *Azospirillum brasilense* and *Pseudomonas fluorescens*. These macrobeads, made from a 20/80 chitosan-starch mixture with sodium tripolyphosphate were created using a simple dripping method. The study suggested that the hydrogel formation involved ionotropic-gelation and neutralization mechanisms. PGPB were either separately loaded or co-inoculated into the macrobeads. The bacteria-loaded macrobeads were then dried and stored at RT. The bacteria-maintained viability for at least 12 months, with recovery rates of around (10^9) CFU per gram for *A. brasilense* and (10^8) CFU per gram for *P. fluorescens*. Bacterial release in sterile saline solution exhibited a super Case-II transport mechanism. The polymeric matrix effectively released both types of PGPB in natural soils, indicating significant potential for the development of novel and improved biofertilizers (Perez et al., 2018)

A review by Saberi-Riseh et al. highlighted several advantages of using starch over other biopolymers. Starch is more economical and widely accessible compared to carboxymethyl cellulose and chitosan. Its extraction from various sources is simpler than that of alginate, which also suffers from high porosity and low mechanical strength, posing challenges for PGPB encapsulation. Additionally, unlike gelatin, starch does not require high concentrations to form a polymeric network and is not prone to melting at room temperature, as gelatin's melting point is approximately 35°C, which can impact storage stability (Saberi Riseh et al., 2023b).

7. CHITOSAN-BASED POLYMERIC MATRICES

Chitosan is a biodegradable, cationic polymer produced from chitin through an alkaline deacetylation process, transforming chitin into a useful material for various fields (Perez et al., 2018). Sourced from discarded shells of arthropods as well as fungal cell walls. Chitin is composed of N-acetyl-D-glucosamine and D-glucosamine subunits connected by β-(1→4) glycosidic bonds (Ferreira et al., 2020; Mali et al., 2005). Chitosan possesses antibacterial, antifungal, mucoadhesive, and gelling properties and used in various materials such as aerogels, microcapsules, hydrogels, nanoparticles, and electrospun fibers (Sharkawy et al., 2020).

Ramírez et al, 2010 reported that chitosan can safeguard plants against phythopathogen both pre- and post-harvest. It also enhances beneficial symbiotic interactions between plants and microorganisms, thereby regulating plant growth and development. Additionally, the chelating properties of chitosan facilitate the sequestration of nutrients and minerals, making them more available for plant uptake (Sharkawy et al., 2020).

Chitosan has received significant attention as a potential carrier for the encapsulation of PGPB due to its beneficial properties in agricultural systems. Chitosan exhibits several beneficial properties, including the ability to induce plant defense, enhancing antimicrobial properties, aiding in soil remediation, and activating defense mechanisms in field-grown plants. Additionally, it encourages the development of plants and improves the quality of the soil by limiting the leaching of anionic nutrient fertilisers such as phosphates and nitrates, increasing the activity of beneficial microbes, and enhancing the soil's ability to retain water (Sharkawy et al., 2020). Numerous studies have reported the successful use of chitosan as a carrier for PGPB to develop new biopesticides or biofertilizers (Ashfaq et al., 2024; Chanratana et al., 2018; Colin et al., 2024; Minakawa et al., 2019).

Methylobacterium oryzae CBMB20, known for its plant growth-promoting and biocontrol properties, was tested in various formulations to enhance tomato plant growth. These formulations included wet chitosan, wet alginate, dry chitosan, and dry alginate. The study found that the best results were achieved with wet chitosan (1.5% solution), prepared at a pH of 5.5–6.0, and with a contact time of 90 minutes. This method resulted in 80% bacterial survivability even after 90 days of storage at 4°C. Chitosan formulations demonstrated better entrapment efficiency, good resistance to degradation, and a slow release of cells during extended incubation. Under greenhouse conditions, both wet and dry chitosan formulations showed high bacteria survival in the spermosphere of treated soils even after 21 days. During the course of thirty days, the alginate formulation underwent total degradation, but the chitosan formulation only underwent partial degradation. Tomato plants treated with the wet chitosan formulation demonstrated increase in shoot and root length, as well as dry weight, compared to control .As a result, chitosan proves to be a more effective carrier than alginate for delivering bacteria to the plant rhizosphere, thus promoting enhanced plant growth (Chanratana et al., 2018).

Chanratana and the group explored that chitosan-immobilized *Methylobacterium oryzae* significantly help tomato plant in increasing dry weight, nutrient uptake (including nitrogen, phosphorus, potassium, and magnesium), and photosynthetic efficiency, while also limited electrolyte leakage under salt stress conditions. Using PGPB as bioinoculants to alleviate salt stress offers a sustainable and environmentally friendly approach to agriculture (Chanratana et al., 2018) However, maintaining bacterial populations in the soil for extended periods is challenging. This study demonstrated that chitosan-immobilized *Methylobacterium oryzae* CBMB20 can improve tomato plant growth under salt stress, enhancing dry weight, nutrient uptake, photosynthetic efficiency, and reducing electrolyte leakage. The bioinoculant also alleviated oxidative stress by increasing antioxidant enzyme activity and proline accumulation, and decreased excess Na+ influx into plant cells, improving the Na+/K+ ratio. Thus, chitosan-immobilized *Methylobacterium oryzae* CBMB20 shows potential as a bioinoculant to support plant growth in saline conditions (Chanratana et al., 2019).

This study examined the use of *Bacillus licheniformis* encapsulated in alginate-chitosan nanoparticle (CNP) beads with rice starch (RS) to enhance growth and reduce infection. The encapsulated bacteria were tested for their ability to produce IAA, fix nitrogen, produce 1-aminocyclopropane-1-carboxylic acid (ACC) deaminase, and exhibit antifungal activity against the pathogen *Sclerotium rolfsii*. In no way did the encapsulation have a detrimental effect on these advantageous characteristics. In addition, its PGP and biocontrol capabilities were evaluated by employing seedlings of *Capsicum annuum*, often known as chilli. Results showed that plants treated with the encapsulated bacteria supplemented with CNPs exhibited maximum growth enhancement, while those with CNPs and RS had enhanced disease suppression. The encapsulated *Bacillus licheniformis* significantly improved the growth of chili plants compared to free-living cells and effectively suppressed the disease caused by *Sclerotium rolfsii*. This indicates that encapsulated *B. licheniformis* with RS and CNPs is a promising delivery system for agricultural applications (Panichikkal, Puthiyattil, et al., 2021).

One study involving *Gaeumannomyces graminis* var. *tritici*, a major disease affecting wheat, explored methods to combat this issue. Applying bacterial inoculants to the soil can enhance plant nutrient absorption and boost probiotic bacteria efficiency. Encapsulation is a process that is used in agriculture with the purpose of further enhancing the efficiency of helpful microorganisms. Fourier transform infrared (FTIR) and X-ray diffraction (XRD) analyses confirmed that the prepared formulation was a mix of gellan and chitosan, while SEM revealed the spherical structure of the microcapsules. The bacterial release increased gradually, peaking at 10^9 CFU mL^-1 on the 50th day. Moreover, greenhouse experiments demonstrated that wheat plants treated with *Streptomyces fulvissimus* Uts22 microcapsules achieved 90% disease control, along with the highest root and shoot fresh and dry weights (Saberi-Riseh & Moradi-Pour, 2021). This study developed a new formulation to control wheat take-all disease using bacterial and nanoparticle-loaded chitosan-gellan gum microcapsules.

Table 1. The summary of the specific examples of nanocarriers

Nanocarriers	Source	Encapsulated Bacteria	Advantages	Disadvantages	References
Alginate	Brown algae, bacteria	*Azospirillum brasilense, Bacillus subtilis, Pseudomonas fluorescens*	Biodegradable, protects bacteria from environmental stress, long shelf-life	High porosity, low mechanical strength, requires crosslinking agents	(Bashan and Gonzalez, 1999)
Starch	Cereals, roots, tubers	*Azospirillum brasilense*	Biodegradable, low cost, abundant, controlled release	Poor mechanical performance, requires blending with other polymers	(Schoebitz, Simonin and Poncelet, 2012)
Chitosan	alkaline deacetylation of chitin	*Azospirillum brasilense, Pseudomonas fluorescens*	-Enhances plant-microbe interactions -Soil conditioning - Nutrient and mineral sequestration	Expensive, solubility issues	(Perez *et al.*, 2018)
Chitosan-Coated Mesoporous Silica Nanoparticles	Synthesized from tetraethylorthosilicate (TEOS) and chitosan	*Bacillus megaterium*	- High surface area and porosity - Controlled and slow release of bacteria - Enhances nutrient uptake	- May aggregate, reducing efficiency - Potential toxicity to beneficial soil microbes	(Pavlicevic *et al.*, 2022)
Nanoclay (Hydrophilic Bentonite)	Naturally occurring clay minerals	*Azospirillum brasilense, Pseudomonas putida*	- High bacterial loading capacity - Naturally abundant and inexpensive - Enhances soil structure and water retention	- Lower surface area compared to MSN - Slower bacterial release rate	(Pavlicevic *et al.*, 2022)
Naked Biochar (NB)	Pyrolyzed recycled wood	*Azotobacter vinelandii, Bacillus megaterium*	- High porosity for bacterial retention - Improves soil health and microbial diversity - Enhances nitrogen and phosphorus content	- Biochar properties vary with production conditions - Limited bacterial loading compared to MSN and nanoclay	(Pavlicevic *et al.*, 2022)
Aries Green Biochar (AB)	Gasified wood waste	*Azotobacter vinelandii, Bacillus megaterium*	- Higher surface area compared to NB - Improves soil microbial activity and nutrient bioavailability	- Can alter soil pH significantly - Limited bacterial retention compared to MSN	(Pavlicevic *et al.*, 2022)
"Italian" Biochar (IB)	Pyrogasified wood pellets	*Azotobacter vinelandii, Bacillus megaterium*	- High carbon content improves soil organic matter - Enhances plant stress resistance	- Lower surface area than AB and NB - Potential pH alterations affecting microbial activity	(Pavlicevic *et al.*, 2022)

8. POLYMERIC MATRICES BASED ON GELATIN

Gelatin (Gel) is derived from denatured proteins that are produced through the partial hydrolysis of collagen, using either acid (type A) or alkaline (type B) methods. The alkaline process specifically converts glutamine into glutamic acid and asparagine into aspartic acid, resulting in higher levels of these acids in type B gelatin. Gelatin is primarily sourced from pig skin, bovine hides, pig and cattle bones, and fish (Rodríguez-Rodríguez et al., 2020).

In spite of the fact that gelatin is used extensively in the food and pharmaceutical sectors, its utilization for the encapsulation of PGPB is much less prevalent in comparison to the utilisation of other biopolymers. Some studies have explored the use of gelatin in combination with alginate (Ferreira et al., 2020; Perez et al., 2018; Saberi Riseh et al., 2023b; Tu et al., 2015). Utilizing nanomaterials and biocompatible compounds for the nano-encapsulation of beneficial bacteria is essential for improving their effectiveness in challenging environmental conditions. This study investigated the encapsulation of two *P. fluorescens* strains (VUPF5 and T17-4) in alginate-gelatin nanocomposite beads with different gelatin concentrations. The swelling and moisture content of the beads increased with higher gelatin content. Encapsulation efficiency was highest at a 1.5% gelatin concentration, achieving 91.23% for VUPF5 and 87.23% for T17-4. Greenhouse experiments showed that inoculating potatoes with these bacterial strains and their nanocomposite beads reduced disease incidence (Pour et al., 2019). This encapsulation technique is effective in protecting plant probiotic bacteria from adverse soil conditions, facilitating their successful integration into the rhizosphere.

Researchers combined gelatin with sodium alginate (NaALG) to make a new microbial fungicide containing *Bacillus subtilis* SL-13 microspheres using an emulsification/internal gelation method (Pour et al., 2019). Under the microscope, the microcapsules appeared nearly spherical. Tests confirmed that gelatin and NaALG interacted when mixed. The highest encapsulation efficiency was 93.44% with 1.5% gelatin. The size, swelling, and breakdown of the beads increased as more gelatin was added. The encapsulated SL-13 bacteria remained viable at over 10^8 CFU/mL after being stored for 120 days at 25°C. The release of bacteria from the microcapsules increased quickly at first, then more gradually, reaching a peak of 10^10 CFU/mL on the 35th day. This process effectively creates rounded, uniform, and well-dispersed microcapsules of SL-13 using NaALG and gelatin, showing promise for future agricultural use. The figure below summarizes the overall effects of biopolymers carried PGPB on plants.

Figure 3. Summary of effect of encapsulation of PGPB on plants

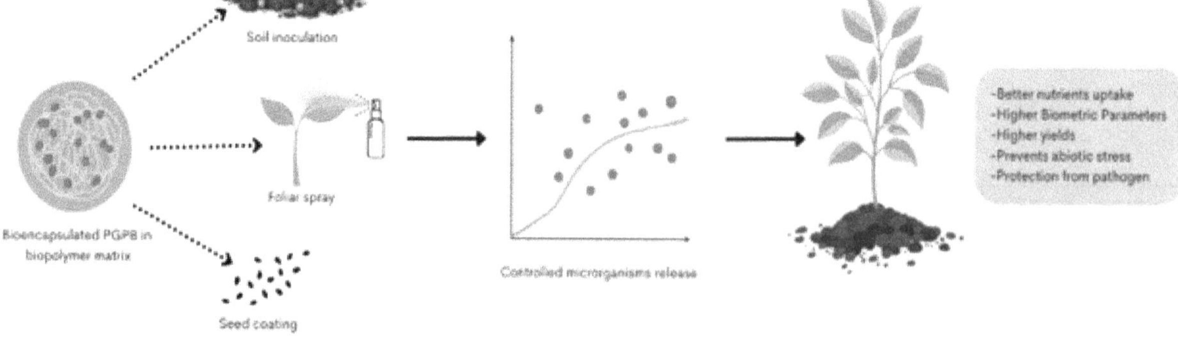

9. NANOPARTICLES AND PGPR: AN INTEGRATED STRATEGY TO MITIGATE ENVIRONMENTAL STRESS

PGPR and NPs have been widely discussed for their major effects on plants. However, in recent years, combining PGPR with NPs has shown to be more effective in plant development than employing either PGPR or NPs alone. Several studies have shown that PGPR and NPs have an important role in increasing agricultural production (Rajput et al., 2023; Khan et al., 2024). The production of sustainable agriculture was improved using *Enterobacter*, *Azospirillum*, *Pseudomonas*, *Agrobacterium*, and NPs. When PGPR and NPs are used together, they may increase crop production, dry-fresh biomass, plant height, and seed germination frequency more than when they are used separately (Barbosa Santos et al., 2023; Estefanía, 2023). There are several ways in which this combo may function. Better plant performance and productivity are a result of the direct processes, which include the PGPR's normal production of plant hormones like indole acetic acid, siderophores, etc., and increasing the availability of soil minerals via phosphate solubilization and N_2 fixation (Rehan et al., 2023). In addition to offering nutrient-rich substrate, the presence of NPs can help the PGPR tolerate greater numbers, which will increase the PGPR's efficiency and ultimately increase plant production (Gouda et al., 2017).

PGPR are generally acknowledged for mitigating various environmental challenges that hinder plant growth and development. Plants may enhance stress tolerance by using the capacity of PGPR to secrete exopolysaccharides under arid conditions (El-Saadony et al., 2024). In saline settings, it enhances water absorption, reduces stomatal conductance, elevates potassium accumulation at the expense of sodium, mitigates the adverse effects of soil salt, and amplifies antioxidant enzyme activities. All these modifications facilitate enhanced plant development in saline environment (Atta et al., 2023). It has been noted that PGPR enhances overall nutrient absorption while simultaneously immobilizing and diminishing the uptake of heavy metals by plants, thus mitigating the toxicity associated with heavy metals (Balasubramaniam et al., 2023). *Bacillus pumilus* associated with Ag-NPs shown superior efficacy in promoting onion plant development. The combined treatment demonstrated a decrease in salt-induced moisture retention in rhizospheric soil and an increase in bulb protein content, while reducing leaf flavonoids. Ag-NPs increased sucrose and proline concentrations. *Bacillus pumilus* shown greater significance under control circumstances for all agronomic growth responses, but *Pseudomonas moraviensis* exhibited potential resilience under saline conditions (Dobrzyński et al., 2020).

Furthermore, it has been shown that NPs enhance drought stress reduction, salt tolerance, and plant toxicity from both organic and inorganic soil pollutants (Chandrashekar et al., 2023). Higher water-holding capacity due to NPs' large surface area-to-volume ratio makes NP-amended soils more resistant to drought stress (Khan et al., 2024). In a similar vein, plants in NP-amended soil lessen saline conditions by reducing osmotic stress to higher soil water content and reducing Na^+ absorption owing to transitory Na^+ binding on NP sorption sites. The main way that NPs lessen the toxicity stress of heavy metal sorption, both inorganic and organic. All these NP uses against various environmental stressors have been examined by several research (Rajput et al., 2024). If both additions would work in concert, some recent research has investigated the potential of combining PGPR and NPs to improve the environmental conditions for plant growth. In these works, PGPR and NP synergies have been intensively investigated.

10. PGPR AND NANOPARTICLES IMPROVE SOIL QUALITY

The combination application of PGPR and NPs improves soil quality by performing numerous roles to reduce drought stress. When the soil water content was half the field capacity (50% of soil moisture level), combining NPs and PGPR significantly increased pH, EC, nitrate, phosphorus, extractable K, and organic matter when compared to using NPs and PGPR individually (Rajput et al., 2024). Sun et al. (2022) found that when FeO-NPs were treated with arsenic (As)-contaminated soil using the PGPR P. vermicola, the outcomes were significantly improved, producing more effective results. As a result, combining P. vermicola with FeO-NPs decreased As (arsenic) toxicity in Trachyspermum ammi seedlings, increasing plant growth and composition under metal stress as demonstrated by balanced exudation of organic acids (Sun et al., 2022). According to the findings of this study, improved water holding capacity and concomitant decrease in drought stress improve the survival and abundance of PGPR, allowing them to conduct their functions more effectively (Ahmad et al., 2022).

Salt affects the microbial community's composition and lowers microbial activity and biomass as indicators of soil quality. Additionally, in saline environments, Na^+ surpasses K^+ transport channels, which impedes and reduces plant growth (Al-Turki et al., 2023). Co-application of PGPR and NPs under salty environments has been shown to increase salt tolerance and plant growth, mostly via increasing the K^+/Na^+ ratio and decreasing Na^+ absorption. NPs and two endophytic bacteria, such as *Burkholderia phytofirmans* or *Enterobacter* sp., significantly decreased agricultural plant salt stress by reducing xylem Na^+ content absorption. Conversely, a combination of applications significantly increased the K^+ and K^+/Na^+ ratio, which decreased saline stress in plants (Giannelli et al., 2023). In the same experiment, the latter application to adsorption sites, as well as K^+ desorption via interaction combination, reduced the sodium adsorption ratio and Na^+ in soil. Significantly reducing the level of Na^+ in the soil solution following inoculation with NPs and *Paraburkholderia phytofirmans*—which may produce exopolysaccharides—relieved plant salinity stress (Govindasamy et al., 2020).

The combination of PGPRs with NPs yields synergistic advantages for soil quality by diminishing Na^+ levels and enhancing colonization efficiency. PGPR strains co-applied with nanoparticles in saline soil shown superior colonization effectiveness compared to PGPRs without nanoparticles in the soil (Rajput et al., 2023b). *Enterobacter* sp. with 5% nanoparticles exhibited superior colonization effectiveness in saline soil compared to *Burkholderia phytofirmans*, both with and without nanoparticles. In contrast to solo PGPR inoculation, the co-application of an endophytic PGPR with nanoparticles in saline soil resulted in a 150–250% increase in PGPR colonization inside the rhizosphere, root, and shoot interior bacterial populations (Zhang et al., 2024). In rhizospheric soil, nanoparticles exhibited a downregulation of the Na^+/K^+ ratio and enhanced the effectiveness of plant growth-promoting rhizobacteria root colonization, hence mitigating soil salinity stress. NPs and PGPRs preserve nutritional equilibrium by releasing mineral elements such as K^+, Ca^{2+}, and Mg^{2+} from the soil solution, hence reducing the concentration of Na^+ in the soil. Consequently, the soil Na^+/K^+ ratio progressively enhanced. Sodium ions in soil adhere to exopolysaccharides produced by plant growth-promoting rhizobacteria under stress conditions (Gupta et al., 2019). The genes that respond to stress in plants affected by nanoparticles (NPs) and plant growth-promoting rhizobacteria (PGPRs) might significantly vary based on the stress type, plant types or species, NP type, application method, and the strains of PGPRs used. This document outlines several gene categories potentially influenced by the presence of NPs and PGPRs in plants, as well as their overall roles in stress response (Fu & Yan, 2023).

11. DISADVANTAGES OF USING NPS AND PGPRS FOR SUSTAINABLE AGRICULTURE

Nanoparticles' unique characteristics, such their high reactivity and capacity to be tailored for applications, provide revolutionary possibilities for the agricultural industry. They do, however, pose health and environmental hazards. The toxicity of nanoparticles to non-target creatures, such as plants, animals, and soil microorganisms, is a concern. Research suggests that certain nanoparticles may disturb soil microbial populations and build up in the food chain. Since they are tiny, they are more mobile, which increases the possibility of environmental pollution and unidentified long-term consequences (Maharramov et al., 2019; Yonathan et al., 2022). Given their shown capacity to transcend biological boundaries, the possibility of nanoparticles entering the body via food intake also raises concerns about human health. Because the dangers connected with nanomaterials may not be sufficiently addressed by current frameworks, regulatory issues are substantial (Xuan et al., 2023). Despite these hazards, there are significant advantages to using nanoparticles in agriculture, which makes their safe and sustainable usage contingent upon thorough risk assessments, safe handling procedures, and well-defined regulatory rules (Gupta & Xie, 2018). Comprehensive risk evaluations, research into long-term consequences, and the development of biodegradable, environmentally and health-safe nanomaterials are all essential to ensuring a sustainable strategy. Additionally, reducing hazards requires creating precise regulatory rules and encouraging safe handling methods. By putting sustainability and safety first, the agriculture industry can use nanotechnology to solve the world's food security issues while preserving biodiversity and the environment (Prasad et al., 2017).

12. GLOBAL SAFETY REGULATION FOR NANOTECHNOLOGY

Humans, plants, animals, and whole ecosystems are among the many biotic elements that nanoparticles may negatively impact (Mishra & Sundaram, 2023). Understanding the intricate physical and chemical mechanisms that control the behavior and destiny of nanomaterials in the environment is crucial. Nanoparticles, for instance, could accumulate in soil, which may result in phytotoxicity and a reduction in soil fertility (Rajput et al, 2020). To ensure the safe and sustainable use of nanomaterials in agriculture, it is essential to thoroughly evaluate the toxicological impacts and environmental dangers. To minimize any environmental hazards while maximizing their advantages, biodegradable nanoparticle research must continue (Wahab et al., 2024). Additionally, investigating the potential of nanoparticles in soil remediation offers both exciting possibilities and formidable obstacles. Careful thought must be given to matters such nanoparticle toxicity, long-term environmental effects, and regulatory issues (Saifi et al., 2025). Effective use of nanoparticles in soil remediation requires a deep understanding of their fate and transport. To control the possible hazards that nanoparticles in agricultural applications may pose, regulatory agencies such as the Food and Drug Administration (FDA) and the Environmental Protection Agency (EPA) are essential. These organizations must address concerns about how nanomaterials affect the environment while making sure that advancements in nanotechnology support sustainable practices and food security (Atanda et al., 2025). To fully realize the transformational potential of nanoparticles in agriculture while addressing their potential hazards and environmental implications, further study, cooperation, and appropriate deployment are required. Successfully incorporating this cutting-edge technology into environmentally conscious and productively oriented sustainable practices may be crucial to the future of agriculture.

13. CONCLUSION

Reduced plant growth and crop mortality are common in important food and commercial cash crops when severe agroclimatic conditions are present. Since PGPR and NPs could significantly improve crop output and soil health in both favorable and unfavorable circumstances, our conclusion emphasizes the need of incorporating these technologies into environmentally responsible agricultural operations. Moreover, the synergy between PGPR and NPs not only provides a way to lessen reliance on synthetic fertilizers, but it also promises to make plants more resistant to the effects of climate change by enhancing plant performance, output, and the quality of the fruit or grain, in addition to the soil profile. According to the information presented in this chapter, however, there have only been a few field demonstrations carried out to evaluate the relevance of the interacting function that PGPR and NPs play in sustainable agriculture. It is necessary to do more study on the mechanistic aspects of the interaction between PGPR and NPs. Field demonstrations that last for an extended period might be an exceptionally effective method for evaluating the combined effects of PGPR and NPs. Independently conducted field experiments that lasted for a reasonable amount of time have been used to evaluate the potential of the PGPR and NPs to produce food in a sustainable manner. Moreover, the process of encapsulating PGPM has been shown to be an efficient approach for shielding them from potentially damaging environmental conditions. Out of all the many materials that have been investigated, the alginate-based matrices have been the subject of the most thorough investigation. On the other hand, starch derived from a wide variety of sources has a great deal of potential owing to its inexpensive cost and widespread availability. They provide a method to feed the expanding global population and the related growth in the need for food and energy over the future decades, while inflicting far less damage to the environment in comparison to traditional agricultural inputs. These cutting-edge products provide technical solutions that may be used to replace toxic chemical fertilizers and pesticides, maintain the fertility and stability of the soil, and improve agricultural yield and food security.

It is anticipated that future research will place a significant focus on conducting long-term field experiments, with the goal of delving more into the molecular foundations of PGPR and NP interactions. It is essential to conduct this research to get a knowledge of the dynamic interactions that occur in different kinds of soil, particularly those soils that are deficient in organic matter in tropical and subtropical locations. The goal is to create individualized application tactics that take use of the specific advantages offered by PGPR and NPs. This will result in a decreased dependence on synthetic fertilizers and an improvement in the environmental sustainability of agricultural systems. This in-depth investigation and use of PGPR and NPs has the potential to revolutionize agricultural operations to accommodate the growing demand for food on a global scale in a sustainable manner.

REFERENCES

Abdelaziz, A. M., Dacrory, S., Hashem, A. H., Attia, M. S., Hasanin, M., Fouda, H. M., Kamel, S., & ElSaied, H. (2021). Protective role of zinc oxide nanoparticles based hydrogel against wilt disease of pepper plant. *Biocatalysis and Agricultural Biotechnology*, *35*, 102083. https://doi.org/https://doi.org/10.1016/j.bcab.2021.102083. DOI: 10.1016/j.bcab.2021.102083

Ahmad, H. M., Fiaz, S., Hafeez, S., Zahra, S., Shah, A. N., Gul, B., Aziz, O., Mahmood-Ur-Rahman, , Fakhar, A., Rafique, M., Chen, Y., Yang, S. H., & Wang, X. (2022). Plant growth-promoting rhizobacteria eliminate the effect of drought stress in plants: A review. *Frontiers in Plant Science*, *13*, 875774. Advance online publication. DOI: 10.3389/fpls.2022.875774 PMID: 36035658

Ajeng, A. A., Abdullah, R., Ling, T. C., Ismail, S., Lau, B. F., Ong, H. C., Chew, K. W., Show, P. L., & Chang, J. S. (2020). Bioformulation of biochar as a potential inoculant carrier for sustainable agriculture. *Environmental Technology & Innovation*, *20*, 101168. DOI: 10.1016/j.eti.2020.101168

Ashfaq, M., Gupta, G., & Verma, N. (2024). Carbon-based nanocarriers for plant growth promotion: Fuelling when needed. *Nanoscale*, *17*(2), 616–634. DOI: 10.1039/D4NR03268C PMID: 39575969

Atanda, S. A., Shaibu, R. O., & Agunbiade, F. O. (2025). Nanoparticles in agriculture: Balancing food security and environmental sustainability. *Discov Agric*, *3*(1), 26. DOI: 10.1007/s44279-025-00159-x

Atta, K., Mondal, S., Gorai, S., Singh, A. P., Kumari, A., Ghosh, T., Roy, A., Hembram, S., Gaikwad, D. J., Mondal, S., Bhattacharya, S., Jha, U. C., & Jespersen, D. (2023). Impacts of salinity stress on crop plants: Improving salt tolerance through genetic and molecular dissection. *Frontiers in Plant Science*, *14*, 1241736. DOI: 10.3389/fpls.2023.1241736 PMID: 37780527

Avinash, S. H., Manasa, S. G., Nagaraju, Y., Rao, P. G., & Mandal, S. (2025). Molecular, physiological, and biochemical aspects of crop plants with PGPR under harsh environment. Husen, A. (Ed.). (2025). Agricultural Crop Improvement: Plant and Soil Relationships (1st ed.). CRC Press. 75–89. https://doi.org/DOI: 10.1201/9781032630366

Ayaz, M., Li, C., Ali, Q., Zhao, W., Chi, Y., Shafiq, M., Ali, F., Yu, X., Yu, Q., Zhao, J., Yu, J., Qi, R., & Huang, W. (2022). Bacterial and Fungal Biocontrol Agents for Plant Disease Protection: Journey from Lab to Field, Current Status, Challenges, and Global Perspectives. *Molecules (Basel, Switzerland)*, *28*(18), 6735. DOI: 10.3390/molecules28186735 PMID: 37764510

Bakki, M., Banane, B., Marhane, O., Esmaeel, Q., Hatimi, A., Barka, E. A., Azim, K., & Bouizgarne, B. (2024). Phosphate solubilizing *Pseudomonas* and *Bacillus* combined with rock phosphates promoting tomato growth and reducing bacterial canker disease. *Frontiers in Microbiology*, *15*, 1289466. DOI: 10.3389/fmicb.2024.1289466 PMID: 38765677

Balasubramaniam, T., Shen, G., Esmaeili, N., & Zhang, H. (2023). Plants' Response Mechanisms to Salinity Stress. *Plants*, *12*(12), 2253. DOI: 10.3390/plants12122253 PMID: 37375879

Bano, A., Aziz, M. K., Prasad, B., Ravi, R., Shah, M. P., Lins, P. V. D. S., Meili, L., & Prasad, K. S. (2024). The multifaceted power of biochar: A review on its role in pollution control, sustainable agriculture, and circular economy. *Environmental Chemistry and Ecotoxicology*, *7*, 286–304. DOI: 10.1016/j.enceco.2025.01.004

Barbosa Santos, C. H., Frezarin, E. T., Sales, L. R., & Rigobelo, E. C. (2023). Plant Growth-Promoting Rhizobacteria for Sustainable Agricultural Production. *Microorganisms*, *11*(4), 1088. DOI: 10.3390/microorganisms11041088 PMID: 37110511

Barbosa Santos, C. H., Frezarin, E. T., Sales, L. R., & Rigobelo, E. C. (2023). Plant Growth-Promoting Rhizobacteria for Sustainable Agricultural Production. *Microorganisms*, *11*(4), 1088. DOI: 10.3390/microorganisms11041088 PMID: 37110511

Bashan, Y. (1986). Alginate beads as synthetic inoculant carriers for slow release of bacteria that affect plant growth. *Applied and Environmental Microbiology*, *51*(5), 1089–1098. DOI: 10.1128/aem.51.5.1089-1098.1986 PMID: 16347055

Bashan, Y., de-Bashan, L. E., Prabhu, S. R., & Hernandez, J.-P. (2014). Advances in plant growth-promoting bacterial inoculant technology: Formulations and practical perspectives (1998-2013). *Plant and Soil*, *378*(1–2), 1–33. DOI: 10.1007/s11104-013-1956-x

Bashan, Y., & Gonzalez, L. E. (1999). Long-term survival of the plant-growth-promoting bacteria *Azospirillum brasilense* and *Pseudomonas fluorescens* in dry alginate inoculant. *Applied Microbiology and Biotechnology*, *51*(2), 262–266. DOI: 10.1007/s002530051391

Bashan, Y., Hernandez, J.-P., Leyva, L., & Bacilio, M. (2002). Alginate microbeads as inoculant carriers for plant growth-promoting bacteria. *Biology and Fertility of Soils*, *35*(5), 359–368. DOI: 10.1007/s00374-002-0481-5

Bennett, E. M., Murray, J. W., & Isalan, M. (2023). Engineering Nitrogenases for Synthetic Nitrogen Fixation: From Pathway Engineering to Directed Evolution. *Biodesign Research, 5*, 0005. https://doi.org/DOI: 10.34133/bdr.0005

Bhat, M. A., Mishra, A. K., Shah, S. N., Bhat, M. A., Jan, S., Rahman, S., Baek, K., & Jan, A. T. (2024). Soil and Mineral Nutrients in Plant Health: A Prospective Study of Iron and Phosphorus in the Growth and Development of Plants. *Current Issues in Molecular Biology*, *46*(6), 5194–5222. DOI: 10.3390/cimb46060312 PMID: 38920984

Castro, R. A., Dourado, M. N., de Almeida, J. R., Lacava, P. T., Nave, A., de Melo, I. S., de Azevedo, J. L., & Quecine, M. C. (2018). Mangrove endophyte promotes reforestation tree (*Acacia polyphylla*) growth. *Brazilian Journal of Microbiology : Publication of the Brazilian Society for Microbiology*, *49*(1), 59–66. DOI: 10.1016/j.bjm.2017.04.002 PMID: 28774638

Chandrashekar, H. K., Singh, G., Kaniyassery, A., Thorat, S. A., Nayak, R., Murali, T. S., & Muthusamy, A. (2023). Nanoparticle-mediated amelioration of drought stress in plants: A systematic review. *3 Biotech, 13*(10), 336. https://doi.org/DOI: 10.1007/s13205-023-03751-4

Chanratana, M., Han, G. H., Melvin Joe, M., Roy Choudhury, A., Sundaram, S., Halim, M. A., & Sa, T. (2018). Evaluation of chitosan and alginate immobilized Methylobacterium oryzae CBMB20 on tomato plant growth. *Archives of Agronomy and Soil Science*, *64*(11), 1489–1502. DOI: 10.1080/03650340.2018.1440390

Chanratana, M., Joe, M. M., Roy Choudhury, A., Anandham, R., Krishnamoorthy, R., Kim, K., Jeon, S., Choi, J., Choi, J., & Sa, T. (2019). Physiological response of tomato plant to chitosan-immobilized aggregated Methylobacterium oryzae CBMB20 inoculation under salinity stress. 3 Biotech, 9(11), 1–13. DOI: 10.1007/s13205-019-1923-1

Chattaraj, S., Samantaray, A., Ganguly, A., & Thatoi, H. (2025). Employing plant growth-promoting rhizobacteria for abiotic stress mitigation in plants: With a focus on drought stress. *Discover Applied Sciences*, *7*(1), 68. DOI: 10.1007/s42452-025-06468-6

Colin, C., Akpo, E., Perrin, A., Cornu, D., & Cambedouzou, J. (2024). Encapsulation in Alginates Hydrogels and Controlled Release: An Overview. *Molecules (Basel, Switzerland)*, *29*(11), 11–16. DOI: 10.3390/molecules29112515 PMID: 38893391

Cruz-Hernández, M. A., Mendoza-Herrera, A., Bocanegra-García, V., & Rivera, G. (2022). Azospirillum spp. From Plant Growth-Promoting Bacteria to Their Use in Bioremediation. *Microorganisms*, *10*(5), 1057. DOI: 10.3390/microorganisms10051057 PMID: 35630499

De Andrade, L. A., Santos, C. H., Frezarin, E. T., Sales, L. R., & Rigobelo, E. C. (2023). Plant Growth-Promoting Rhizobacteria for Sustainable Agricultural Production. *Microorganisms*, *11*(4), 1088. DOI: 10.3390/microorganisms11041088 PMID: 37110511

Di Benedetto, N. A., Corbo, M. R., Campaniello, D., Cataldi, M. P., Bevilacqua, A., Sinigaglia, M., & Flagella, Z. (2017). The role of Plant Growth Promoting Bacteria in improving nitrogen use efficiency for sustainable crop production: A focus on wheat. *AIMS Microbiology*, *3*(3), 413–434. DOI: 10.3934/microbiol.2017.3.413 PMID: 31294169

Dimkić, I., Janakiev, T., Petrović, M., Degrassi, G., & Fira, D. (2021). Plant-associated *Bacillus* and *Pseudomonas* antimicrobial activities in plant disease suppression via biological control mechanisms - A review. *Physiological and Molecular Plant Pathology*, *117*, 101754. DOI: 10.1016/j.pmpp.2021.101754

Dobrzyński, J., Jakubowska, Z., & Dybek, B. (2022). Potential of Bacillus pumilus to directly promote plant growth. *Frontiers in Microbiology*, *13*, 1069053. DOI: 10.3389/fmicb.2022.1069053 PMID: 36620067

Egamberdieva, D., Ma, H., Alimov, J., Reckling, M., Wirth, S., & Bellingrath-Kimura, S. D. (2020). Response of soybean to hydrochar-based rhizobium inoculation in loamy sandy soil. *Microorganisms*, *8*(11), 1–10. DOI: 10.3390/microorganisms8111674 PMID: 33126699

Egamberdieva, D., Shurigin, V., Alaylar, B., Ma, H., Müller, M. E. H., Wirth, S., Reckling, M., & Bellingrath-Kimura, S. D. (2020). The effect of biochars and endophytic bacteria on growth and root rot disease incidence of fusarium infested narrow-leafed lupin (Lupinus angustifolius L.). *Microorganisms*, *8*(4), 496. Advance online publication. DOI: 10.3390/microorganisms8040496 PMID: 32244470

Egamberdieva, D., Wirth, S. J., & Alqarawi, A. A., F., E., & Hashem, A. (. (2017). Phytohormones and Beneficial Microbes: Essential Components for Plants to Balance Stress and Fitness. *Frontiers in Microbiology*, *8*, 278255. DOI: 10.3389/fmicb.2017.02104 PMID: 29163398

El-Saadony, M. T., Saad, A. M., Mohammed, D. M., Fahmy, M. A., Elesawi, I. E., Ahmed, A. E., Algopishi, U. B., Elrys, A. S., Desoky, E. M., Mosa, W. F., Abd El-Mageed, T. A., Alhashmi, F. I., Mathew, B. T., AbuQamar, S. F., & El-Tarabily, K. A. (2024). Drought-tolerant plant growth-promoting rhizobacteria alleviate drought stress and enhance soil health for sustainable agriculture: A comprehensive review. *Plant Stress*, *14*, 100632. DOI: 10.1016/j.stress.2024.100632

El-Saadony, M. T., Saad, A. M., Mohammed, D. M., Fahmy, M. A., Elesawi, I. E., Ahmed, A. E., Algopishi, U. B., Elrys, A. S., Desoky, E. M., Mosa, W. F., Abd El-Mageed, T. A., Alhashmi, F. I., Mathew, B. T., AbuQamar, S. F., & El-Tarabily, K. A. (2024). Drought-tolerant plant growth-promoting rhizobacteria alleviate drought stress and enhance soil health for sustainable agriculture: A comprehensive review. *Plant Stress*, *14*, 100632. DOI: 10.1016/j.stress.2024.100632

El-Saadony, M. T., Saad, A. M., Soliman, S. M., Salem, H. M., Ahmed, A. I., Mahmood, M., El-Tahan, A. M., Ebrahim, M., Abd, A. A., El-Mageed, T. A., Negm, S. H., Selim, S., Babalghith, A. O., Elrys, A. S., El-Tarabily, K. A., & AbuQamar, S. F. (2022). Plant growth-promoting microorganisms as biocontrol agents of plant diseases: Mechanisms, challenges and future perspectives. *Frontiers in Plant Science*, *13*, 923880. DOI: 10.3389/fpls.2022.923880 PMID: 36275556

Estefanía, C. (2023). Plant-Growth-Promoting Rhizobacteria Improve Germination and Bioactive Compounds in Cucumber Seedlings. *Agronomy (Basel)*, *13*(2), 315. DOI: 10.3390/agronomy13020315

Falcão, L. de S., Coelho, D. B., Veggi, P. C., Campelo, P. H., Albuquerque, P. M., & de Moraes, M. A. (2022). Starch as a Matrix for Incorporation and Release of Bioactive Compounds: Fundamentals and Applications. *Polymers*, *14*(12), 1–16. DOI: 10.3390/polym14122361 PMID: 35745937

Ferreira, D. C. M., de Souza, A. L., da Silveira, J. V. W., Marim, B. M., Giraldo, G. A. G., Mantovan, J., Mali, S., & Pelissari, F. M. (2020). Chapter 17 - Chitosan nanocomposites for food packaging applications. In K. A. B. T.-M. H. N. for S. A.-F. and E. Abd-Elsalam (Ed.), Micro and Nano Technologies (pp. 393–435). Elsevier. https://doi.org/https://doi.org/10.1016/B978-0-12-821354-4.00017-0

Fu, B., & Yan, Q. (2023). Exopolysaccharide is required for motility, stress tolerance, and plant colonization by the endophytic bacterium *Paraburkholderia phytofirmans* PsJN. *Frontiers in Microbiology*, *14*, 1218653. Advance online publication. DOI: 10.3389/fmicb.2023.1218653 PMID: 37670984

Giannelli, G., Potestio, S., & Visioli, G. (2023). The Contribution of PGPR in Salt Stress Tolerance in Crops: Unravelling the Molecular Mechanisms of Cross-Talk between Plant and Bacteria. *Plants*, *12*(11), 2197. DOI: 10.3390/plants12112197 PMID: 37299176

Gomes, A. F., Almeida, M. C., Sousa, E., & Resende, D. I. (2024). Siderophores and metallophores: Metal complexation weapons to fight environmental pollution. *The Science of the Total Environment*, *932*, 173044. DOI: 10.1016/j.scitotenv.2024.173044 PMID: 38723971

González-Melendi, P., Fernández-Pacheco, R., Coronado, M. J., Corredor, E., Testillano, P. S., Risueño, M. C., Marquina, C., Ibarra, M. R., Rubiales, D., & Pérez-de-Luque, A. (2008). Nanoparticles as smart treatment-delivery systems in plants: Assessment of different techniques of microscopy for their visualization in plant tissues. *Annals of Botany*, *101*(1), 187–195. DOI: 10.1093/aob/mcm283 PMID: 17998213

Gouda, S., Kerry, R. G., Das, G., Paramithiotis, S., Shin, H., & Patra, J. K. (2017). Revitalization of plant growth promoting rhizobacteria for sustainable development in agriculture. *Microbiological Research*, *206*, 131–140. DOI: 10.1016/j.micres.2017.08.016 PMID: 29146250

Govindasamy, V., George, P., Kumar, M. *et al.* Multi-trait PGP rhizobacterial endophytes alleviate drought stress in a senescent genotype of sorghum [*Sorghum bicolor* (L.) Moench]. *3 Biotech* **10**, 13 (2020). https://doi.org/DOI: 10.1007/s13205-019-2001-4

Graziano, S., Caldara, M., Gullì, M., Bevivino, A., Maestri, E., & Marmiroli, N. (2022). A Metagenomic and Gene Expression Analysis in Wheat (T. durum) and Maize (Z. mays) Biofertilized with PGPM and Biochar. *International Journal of Molecular Sciences*, *23*(18), 10376. Advance online publication. DOI: 10.3390/ijms231810376 PMID: 36142289

Guo, H., White, J. C., Wang, Z., & Xing, B. (2018). Nano-enabled fertilizers to control the release and use efficiency of nutrients. *Current Opinion in Environmental Science & Health*, *6*, 77–83. https://doi.org/https://doi.org/10.1016/j.coesh.2018.07.009. DOI: 10.1016/j.coesh.2018.07.009

Gupta, A., Singh, S. K., Singh, M. K., Singh, V. K., Modi, A., Singh, P. K., & Kumar, A. (2019). Plant growth–promoting rhizobacteria and their functional role in salinity stress management. *Abatement of Environmental Pollutants*, 151-160. https://doi.org/DOI: 10.1016/B978-0-12-818095-2.00007-2

Gupta, R., & Xie, H. (2018). Nanoparticles in Daily Life: Applications, Toxicity and Regulations. *Journal of environmental pathology, toxicology and oncology: official organ of the International Society for Environmental Toxicology and Cancer, 37*(3), 209–230. https://doi.org/DOI: 10.1615/JEnvironPathol-ToxicolOncol.2018026009

Han, S. E., Kim, K. Y., & Htwe Maung, C. E. (2024). *Bacillus subtilis* PE7-Mediated Alleviation of Phosphate Starvation and Growth Promotion of Netted Melon (*Cucumis melo* L. Var. Reticulatus Naud.). *Microorganisms*, *12*(12), 2384. DOI: 10.3390/microorganisms12122384 PMID: 39770587

Hansen, V., Müller-Stöver, D., Imparato, V., Krogh, P. H., Jensen, L. S., Dolmer, A., & Hauggaard-Nielsen, H. (2017). The effects of straw or straw-derived gasification biochar applications on soil quality and crop productivity: A farm case study. *Journal of Environmental Management*, *186*, 88–95. DOI: 10.1016/j.jenvman.2016.10.041 PMID: 27815006

Hazra, S., Das, D., Moulick, D., & Hossain, A. (2023). Endophytes: The treasure house of bioactive compounds with potential applications in sustainable agriculture and other sectors. *Biocontrol Agents for Improved Agriculture*, 477-506. https://doi.org/DOI: 10.1016/B978-0-443-15199-6.00001-4

Hernández-Montiel, L. G., Chiquito-Contreras, C. J., Murillo-Amador, B., Vidal-Hernández, L., Quiñones-Aguilar, E. E., & Chiquito-Contreras, R. G. (2017). Efficiency of two inoculation methods of Pseudomonas putida on growth and yield of tomato plants. *Journal of Soil Science and Plant Nutrition*, *17*(4), 1003–1012. DOI: 10.4067/S0718-95162017000400012

Hönig, M., Plíhalová, L., Husičková, A., Nisler, J., & Doležal, K. (2018). Role of Cytokinins in Senescence, Antioxidant Defence and Photosynthesis. *International Journal of Molecular Sciences*, *19*(12), 4045. DOI: 10.3390/ijms19124045 PMID: 30558142

Husna, N., Budianta, D., Munandar, M., & Napoleon, A. (2019). Evaluation of several biochar types as inoculant carrier for indigenous Phosphate Solubilizing Microoorganism from acid sulphate soil. *Journal of Ecological Engineering*, *20*(6), 1–8. DOI: 10.12911/22998993/109078

Imran, A., Hakim, S., Tariq, M., Nawaz, M. S., Laraib, I., Gulzar, U., Hanif, M. K., Siddique, M. J., Hayat, M., Fraz, A., & Ahmad, M. (2021). Diazotrophs for Lowering Nitrogen Pollution Crises: Looking Deep Into the Roots. *Frontiers in Microbiology*, *12*, 637815. DOI: 10.3389/fmicb.2021.637815 PMID: 34108945

Jaiswal, S. K., Mohammed, M., Ibny, F. Y., & Dakora, F. D. (2021). Rhizobia as a Source of Plant Growth-Promoting Molecules: Potential Applications and Possible Operational Mechanisms. *Frontiers in Sustainable Food Systems*, *4*, 619676. DOI: 10.3389/fsufs.2020.619676

John, R. P., Tyagi, R. D., Brar, S. K., Surampalli, R. Y., & Prévost, D. (2011). Bio-encapsulation of microbial cells for targeted agricultural delivery. *Critical Reviews in Biotechnology*, *31*(3), 211–226. DOI: 10.3109/07388551.2010.513327 PMID: 20879835

Jung, G., Mugnier, J., Diem, H. G., & Dommergues, Y. R. (1982). Polymer-entrapped rhizobium as an inoculant for legumes. *Plant and Soil*, *65*(2), 219–231. DOI: 10.1007/BF02374652

Kari, A., Nagymáté, Z., Romsics, C., Vajna, B., Tóth, E., Lazanyi-Kovács, R., Rizó, B., Kutasi, J., Bernhardt, B., Farkas, É., & Márialigeti, K. (2021). Evaluating the combined effect of biochar and PGPR inoculants on the bacterial community in acidic sandy soil. Applied Soil Ecology, 160(March 2020). DOI: 10.1016/j.apsoil.2020.103856

Khan, A. R., Salam, A., Li, G., Iqbal, B., Ulhassan, Z., Liu, Q., Azhar, W., Liaquat, F., Shah, I. H., Ul Hassan, S. S., & Du, D. (2024). Nanoparticles and their crosstalk with stress mitigators: A novel approach towards abiotic stress tolerance in agricultural systems. *The Crop Journal*, *12*(5), 1280–1298. DOI: 10.1016/j.cj.2024.09.010

Khan, F., Siddique, A. B., Shabala, S., Zhou, M., & Zhao, C. (2023). Phosphorus Plays Key Roles in Regulating Plants' Physiological Responses to Abiotic Stresses. *Plants*, *12*(15), 2861. DOI: 10.3390/plants12152861 PMID: 37571014

Khoso, M. A., Wagan, S., Alam, I., Hussain, A., Ali, Q., Saha, S., Poudel, T. R., Manghwar, H., & Liu, F. (2024). Impact of plant growth-promoting rhizobacteria (PGPR) on plant nutrition and root characteristics: Current perspective. *Plant Stress*, *11*, 100341. DOI: 10.1016/j.stress.2023.100341

Kieber, J. J., & Schaller, G. E. (2014). Cytokinins. *The Arabidopsis Book / American Society of Plant Biologists*, *12*, e0168. DOI: 10.1199/tab.0168 PMID: 24465173

Kildegaard, K. R., Arnesen, J. A., Adiego-Pérez, B., Rago, D., Kristensen, M., Klitgaard, A. K., Hansen, E. H., Hansen, J., & Borodina, I. (2021). Tailored biosynthesis of gibberellin plant hormones in yeast. *Metabolic Engineering*, *66*, 1–11. DOI: 10.1016/j.ymben.2021.03.010 PMID: 33746070

Kumar, M., Ahmad, S., & Singh, R. (2022). Plant growth promoting microbes: Diverse roles for sustainable and ecofriendly agriculture. *Energy Nexus*, *7*, 100133. DOI: 10.1016/j.nexus.2022.100133

Kumar, S., Sindhu, S. S., & Kumar, R. (2021). Biofertilizers: An ecofriendly technology for nutrient recycling and environmental sustainability. *Current Research in Microbial Sciences*, *3*, 100094. DOI: 10.1016/j.crmicr.2021.100094 PMID: 35024641

L'Espérance, E., Bouyoucef, L. S., Dozois, J. A., & Yergeau, E. (2024). Tipping the plant-microbe competition for nitrogen in agricultural soils. *iScience*, *27*(10), 110973. DOI: 10.1016/j.isci.2024.110973 PMID: 39391734

Lee, S., Kim, A., Song, J., Choe, S., Jang, G., & Kim, Y. (2024). Plant growth-promoting rhizobacterium Bacillus megaterium modulates the expression of antioxidant-related and drought-responsive genes to protect rice (*Oryza sativa* L.) from drought. *Frontiers in Microbiology*, *15*, 1430546. DOI: 10.3389/fmicb.2024.1430546 PMID: 39234545

Liu, L., Yahaya, B. S., Li, J., & Wu, F. (2024). Enigmatic role of auxin response factors in plant growth and stress tolerance. *Frontiers in Plant Science*, *15*, 1398818. DOI: 10.3389/fpls.2024.1398818 PMID: 38903418

Lu, X., Li, C., & Huang, Q. (2019). Combining in vitro digestion model with cell culture model: Assessment of encapsulation and delivery of curcumin in milled starch particle stabilized Pickering emulsions. *International Journal of Biological Macromolecules*, *139*, 917–924. DOI: 10.1016/j.ijbiomac.2019.08.078 PMID: 31401275

Luo, K., Adra, H. J., & Kim, Y.-R. (2020). Preparation of starch-based drug delivery system through the self-assembly of short chain glucans and control of its release property. *Carbohydrate Polymers*, *243*, 116385. DOI: 10.1016/j.carbpol.2020.116385 PMID: 32532382

Ma, M., Jiang, X., Wang, Q., Guan, D., Li, L., Ongena, M., & Li, J. (2018). Isolation and identification of PGPR strain and its effect on soybean growth and soil bacterial community composition. *International Journal of Agriculture and Biology*, *20*(6), 1289–1297. DOI: 10.17957/IJAB/15.0627

Ma, W., Tang, S., Dengzeng, Z., Zhang, D., Zhang, T., & Ma, X. (2022). Root exudates contribute to belowground ecosystem hotspots: A review. *Frontiers in Microbiology*, *13*, 937940. DOI: 10.3389/fmicb.2022.937940 PMID: 36274740

Mahanta, S., Shree, J., Santra, S. C., Moulick, D., & Hossain, A. (2023). Deciphering of mycogenic nanoparticles by spectroscopic methods. In Kuddus, M., Ahmad, I. Z., & Hussain, C. M. (Eds.), *Myconanotechnology and application of nanoparticles in biology*. Academic Press., DOI: 10.1016/B978-0-443-15262-7.00003-6

Maharramov, A. M., Hasanova, U. A., Suleymanova, I. A., Osmanova, G. E., & Hajiyeva, N. E. (2019). The engineered nanoparticles in food chain: Potential toxicity and effects. *SN Applied Sciences*, *1*(11), 1362. DOI: 10.1007/s42452-019-1412-5

Mahmud, K., Missaoui, A., Lee, K., Ghimire, B., Presley, H. W., & Makaju, S. (2021). Rhizosphere microbiome manipulation for sustainable crop production. *Current Plant Biology*, *27*, 100210. DOI: 10.1016/j.cpb.2021.100210

Malgioglio, G., Rizzo, G. F., Nigro, S., du Prey, V. L., Herforth-Rahmé, J., Catara, V., & Branca, F. (2022). Plant-Microbe Interaction in Sustainable Agriculture: The Factors That May Influence the Efficacy of PGPM Application. *Sustainability (Basel), 14*(4), 1–28. DOI: 10.3390/su14042253

Mali, S., Sakanaka, L. S., Yamashita, F., & Grossmann, M. V. E. (2005). Water sorption and mechanical properties of cassava starch films and their relation to plasticizing effect. *Carbohydrate Polymers, 60*(3), 283–289. DOI: 10.1016/j.carbpol.2005.01.003

Marcelino, P. R. F., Milani, K. M. L., Mali, S., dos Santos, O. J. A. P., & de Oliveira, A. L. M. (2016). Formulations of polymeric biodegradable low-cost foam by melt extrusion to deliver plant growth-promoting bacteria in agricultural systems. *Applied Microbiology and Biotechnology, 100*(16), 7323–7338. DOI: 10.1007/s00253-016-7566-9 PMID: 27147530

Martínez-Cano, B., Mendoza-Meneses, C. J., García-Trejo, J. F., Macías-Bobadilla, G., Aguirre-Becerra, H., Soto-Zarazúa, G. M., & Feregrino-Pérez, A. A. (2022). Review and Perspectives of the Use of Alginate as a Polymer Matrix for Microorganisms Applied in Agro-Industry. *Molecules (Basel, Switzerland), 27*(13), 1–20. DOI: 10.3390/molecules27134248 PMID: 35807492

Mazzoni-Putman, S. M., Brumos, J., Zhao, C., Alonso, J. M., & Stepanova, A. N. (2021). Auxin Interactions with Other Hormones in Plant Development. *Cold Spring Harbor Perspectives in Biology, 13*(10), a039990. DOI: 10.1101/cshperspect.a039990 PMID: 33903155

Miethke, M., & Marahiel, M. A. (2007). Siderophore-Based Iron Acquisition and Pathogen Control. *Microbiology and Molecular Biology Reviews : MMBR, 71*(3), 413–451. DOI: 10.1128/MMBR.00012-07 PMID: 17804665

Minakawa, A. F. K., Faria-Tischer, P. C. S., & Mali, S. (2019). Simple ultrasound method to obtain starch micro- and nanoparticles from cassava, corn and yam starches. Food Chemistry, 283(May 2018), 11–18. DOI: 10.1016/j.foodchem.2019.01.015

Mishra, S., & Sundaram, B. (2023). Fate, transport, and toxicity of nanoparticles: An emerging pollutant on biotic factors. *Process Safety and Environmental Protection, 174*, 595–607. DOI: 10.1016/j.psep.2023.04.037

Morcillo, L., & Manzanera, M. (2021). The Effects of Plant-Associated Bacterial Exopolysaccharides on Plant Abiotic Stress Tolerance. *Metabolites, 11*(6), 337. DOI: 10.3390/metabo11060337 PMID: 34074032

Moulick, D., Khan, F. R., Siddiqui, N., Pradhan, J., Khamari, P., & Mahanta, S. (2024). *Applications of nanotechnology in abiotic stress management and biofortification. The Nanotechnology Driven Agriculture*. CRC Press.

Moulick, D., Mukherjee, A., Mahanta, S., Chowdhara, B., Pattnaik, B. K., Mukhopadhyay, A. K., Gujre, N., Shree, J., Santra, S. C., & Hazra, S. (2025). Nano-composites and nano-biochar: a novel solution to alleviate heavy metal stress in crops. In *Nanomaterials and nano-biochar in reducing soil stress* (pp. 183–224). Apple Academic Press.

Narang, C., Kant, R., & Mandal, S. (2024). Nanosensors for Soil and Nutrients Analytics. In Shahzad, R., Fiaz, S., Qayyum, A., Ul Islam, M., & Lee, I.-J. (Eds.), *Revolutionizing Agriculture: A Comprehensive Exploration of Agri-Nanotechnology* (pp. 177–200). Springer Nature Switzerland., DOI: 10.1007/978-3-031-76000-6_8

Nett, R. S., Bender, K. S., & Peters, R. J. (2022). Production of the plant hormone gibberellin by rhizobia increases host legume nodule size. *The ISME Journal, 16*(7), 1809–1817. DOI: 10.1038/s41396-022-01236-5 PMID: 35414717

Nussinovitch, A. (2010). Polymer macro- and micro-gel beads: Fundamentals and applications. Polymer Macro- and Micro-Gel Beads: Fundamentals and Applications, 1–303. DOI: 10.1007/978-1-4419-6618-6

Pan, L., & Cai, B. (2023). Phosphate-Solubilizing Bacteria: Advances in Their Physiology, Molecular Mechanisms and Microbial Community Effects. *Microorganisms, 11*(12), 2904. DOI: 10.3390/microorganisms11122904 PMID: 38138048

Panichikkal, J., Prathap, G., Nair, R. A., & Krishnankutty, R. E. (2021). Evaluation of plant probiotic performance of Pseudomonas sp. encapsulated in alginate supplemented with salicylic acid and zinc oxide nanoparticles. *International Journal of Biological Macromolecules, 166*, 138–143. DOI: 10.1016/j.ijbiomac.2020.10.110 PMID: 33096173

Panichikkal, J., Puthiyattil, N., Raveendran, A., Nair, R. A., & Krishnankutty, R. E. (2021). Application of Encapsulated Bacillus licheniformis Supplemented with Chitosan Nanoparticles and Rice Starch for the Control of Sclerotium rolfsii in Capsicum annuum (L.) Seedlings. *Current Microbiology, 78*(3), 911–919. DOI: 10.1007/s00284-021-02361-8 PMID: 33580333

Paramo, L. A., Feregrino-Pérez, A. A., Guevara, R., Mendoza, S., & Esquivel, K. (2020). Nanoparticles in agroindustry: Applications, toxicity, challenges, and trends. *Nanomaterials (Basel, Switzerland), 10*(9), 1–33. DOI: 10.3390/nano10091654 PMID: 32842495

Pattnaik, B. K., Sahu, C., & Moulick, D. (2024). Nanoparticles in soil–water–plant inference. In Roy, S., & Hossain, A. (Eds.), *Nanotechnol driven agric.* CRC Press. DOI: 10.1201/9781003376446-5

Pavlicevic, M., Abdelraheem, W., Zuverza-Mena, N., O'Keefe, T., Mukhtar, S., Ridge, G., Ranciato, J., Haynes, C., Elmer, W., Pignatello, J., Pagano, L., Caldara, M., Marmiroli, M., Maestri, E., Marmiroli, N., & White, J. C. (2022). Engineered Nanoparticles, Natural Nanoclay and Biochar, as Carriers of Plant-Growth Promoting Bacteria. *Nanomaterials (Basel, Switzerland), 12*(24), 4474. Advance online publication. DOI: 10.3390/nano12244474 PMID: 36558327

Pereira, J. F., Oliveira, A. L. M., Sartori, D., Yamashita, F., & Mali, S. (2023). Perspectives on the Use of Biopolymeric Matrices as Carriers for Plant-Growth Promoting Bacteria in Agricultural Systems. *Microorganisms, 11*(2), 1–17. DOI: 10.3390/microorganisms11020467 PMID: 36838432

Perez, J. J., Francois, N. J., Maroniche, G. A., Borrajo, M. P., Pereyra, M. A., & Creus, C. M. (2018). A novel, green, low-cost chitosan-starch hydrogel as potential delivery system for plant growth-promoting bacteria. *Carbohydrate Polymers, 202*, 409–417. https://doi.org/https://doi.org/10.1016/j.carbpol.2018.07.084. DOI: 10.1016/j.carbpol.2018.07.084 PMID: 30287017

Pour, M. M., Saberi-Riseh, R., Mohammadinejad, R., & Hosseini, A. (2019). Investigating the formulation of alginate- gelatin encapsulated Pseudomonas fluorescens (VUPF5 and T17-4 strains) for controlling Fusarium solani on potato. *International Journal of Biological Macromolecules*, *133*, 603–613. DOI: 10.1016/j.ijbiomac.2019.04.071 PMID: 31004642

Prasad, R., Bhattacharyya, A., & Nguyen, Q. D. (2017). Nanotechnology in Sustainable Agriculture: Recent Developments, Challenges, and Perspectives. *Frontiers in Microbiology*, *8*, 264976. DOI: 10.3389/fmicb.2017.01014 PMID: 28676790

Qi, Y., Liu, H., Wang, J., & Wang, Y. (2021). Effects of different straw biochar combined with microbial inoculants on soil environment in pot experiment. *Scientific Reports*, *11*(1), 1–13. DOI: 10.1038/s41598-021-94209-1 PMID: 34282227

Rajput, P., Singh, A., Agrawal, S., Ghazaryan, K., Rajput, V. D., Movsesyan, H., Mandzhieva, S., Minkina, T., & Alexiou, A. (2024). Effects of environmental metal and metalloid pollutants on plants and human health: Exploring nano-remediation approach. *Stress Biology*, *4*(1), 27. DOI: 10.1007/s44154-024-00156-y PMID: 38777953

Rajput, V., Minkina, T., Mazarji, M., Shende, S., Sushkova, S., Mandzhieva, S., Burachevskaya, M., Chaplygin, V., Singh, A., & Jatav, H. (2020). Accumulation of nanoparticles in the soil-plant systems and their effects on human health. *Annals of Agricultural Science*, *65*(2), 137–143. DOI: 10.1016/j.aoas.2020.08.001

Rajput, V. D., Singh, A., Minkina, T. M., Verma, K. K., & Singh, A. K. (Eds.). (2023). *Nanotechnology for Sustainable Agriculture: An Innovative and Eco-Friendly Approach* (1st ed.). Apple Academic Press., DOI: 10.1201/9781003333128

Rehan, M., Al-Turki, A. A., Abdelmageed, A. H., Abdelhameid, N. M., & Omar, A. F. (2023). Performance of Plant-Growth-Promoting Rhizobacteria (PGPR) Isolated from Sandy Soil on Growth of Tomato (*Solanum lycopersicum* L.). *Plants*, *12*(8), 1588. DOI: 10.3390/plants12081588 PMID: 37111812

Riseh, R. S., Ebrahimi-Zarandi, M., Gholizadeh Vazvani, M., & Skorik, Y. A. (2021). Reducing drought stress in plants by encapsulating plant growth-promoting bacteria with polysaccharides. *International Journal of Molecular Sciences*, *22*(23), 12979. Advance online publication. DOI: 10.3390/ijms222312979 PMID: 34884785

Rodríguez-Rodríguez, R., Espinosa-Andrews, H., Velasquillo-Martínez, C., & García-Carvajal, Z. Y. (2020). Composite hydrogels based on gelatin, chitosan and polyvinyl alcohol to biomedical applications: A review. *International Journal of Polymeric Materials*, *69*(1), 1–20. DOI: 10.1080/00914037.2019.1581780

Rohman, S., Kaewtatip, K., Kantachote, D., & Tantirungkij, M. (2021). Encapsulation of Rhodopseudomonas palustris KTSSR54 using beads from alginate/starch blends. *Journal of Applied Polymer Science*, *138*(12), 1–9. DOI: 10.1002/app.50084

Ryan, R. P., Germaine, K., Franks, A., Ryan, D. J., & Dowling, D. N. (2008). Bacterial endophytes: Recent developments and applications. *FEMS Microbiology Letters*, *278*(1), 1–9. DOI: 10.1111/j.1574-6968.2007.00918.x PMID: 18034833

Saberi-Rise, R., & Moradi-Pour, M. (2020). The effect of Bacillus subtilis Vru1 encapsulated in alginate – bentonite coating enriched with titanium nanoparticles against Rhizoctonia solani on bean. *International Journal of Biological Macromolecules, 152*, 1089–1097. DOI: 10.1016/j.ijbiomac.2019.10.197 PMID: 31751702

Saberi Riseh, R., Hassanisaadi, M., Vatankhah, M., & Kennedy, J. F. (2023a). Encapsulating biocontrol bacteria with starch as a safe and edible biopolymer to alleviate plant diseases: A review. Carbohydrate Polymers, 302(November 2022), 120384. DOI: 10.1016/j.carbpol.2022.120384

Saberi Riseh, R., Hassanisaadi, M., Vatankhah, M., & Kennedy, J. F. (2023b). Encapsulating biocontrol bacteria with starch as a safe and edible biopolymer to alleviate plant diseases: A review. *Carbohydrate Polymers, 302*, 120384. DOI: 10.1016/j.carbpol.2022.120384 PMID: 36604062

Saberi-Riseh, R., & Moradi-Pour, M. (2021). A novel encapsulation of *Streptomyces fulvissimus* Uts22 by spray drying and its biocontrol efficiency against *Gaeumannomyces graminis*, the causal agent of take-all disease in wheat. *Pest Management Science, 77*(10), 4357–4364. DOI: 10.1002/ps.6469 PMID: 33960105

Saeid, A., Prochownik, E., & Dobrowolska-Iwanek, J. (2018). Phosphorus Solubilization by Bacillus Species. *Molecules (Basel, Switzerland), 23*(11), 2897. DOI: 10.3390/molecules23112897 PMID: 30404208

Saifi, R.. (2025). Challenges and Opportunities of Nanomaterials in Sustainable Agriculture. In Rajan, R., Ahmad, F., Khoshariya, A., & Seth, C. S. (Eds.), *Sustainable Era of Nanomaterials. Sustainable Development and Biodiversity* (Vol. 40). Springer., DOI: 10.1007/978-981-96-4471-1_3

Salazar-Cerezo, S., Martínez-Montiel, N., García-Sánchez, J., Pérez-y-Terrón, R., & Martínez-Contreras, R. D. (2018). Gibberellin biosynthesis and metabolism: A convergent route for plants, fungi and bacteria. *Microbiological Research, 208*, 85–98. DOI: 10.1016/j.micres.2018.01.010 PMID: 29551215

Sampathkumar, K., Tan, K. X., & Loo, S. C. J. (2020). Developing Nano-Delivery Systems for Agriculture and Food Applications with Nature-Derived Polymers. *iScience, 23*(5), 101055. DOI: 10.1016/j.isci.2020.101055 PMID: 32339991

Schoebitz, M., López, M. D., & Roldán, A. (2013). Bioencapsulation of microbial inoculants for better soil-plant fertilization. A review. *Agronomy for Sustainable Development, 33*(4), 751–765. DOI: 10.1007/s13593-013-0142-0

Schoebitz, M., Simonin, H., & Poncelet, D. (2012). Starch filler and osmoprotectants improve the survival of rhizobacteria in dried alginate beads. *Journal of Microencapsulation, 29*(6), 532–538. DOI: 10.3109/02652048.2012.665090 PMID: 22372947

Shahwar, D., Mushtaq, Z., Mushtaq, H., Alqarawi, A. A., Park, Y., Alshahrani, T. S., & Faizan, S. (2023). Role of microbial inoculants as bio fertilizers for improving crop productivity: A review. *Heliyon, 9*(6), e16134. DOI: 10.1016/j.heliyon.2023.e16134 PMID: 37255980

Sharkawy, A., Barreiro, M. F., & Rodrigues, A. E. (2020). Chitosan-based Pickering emulsions and their applications: A review. *Carbohydrate Polymers, 250*(August), 116885. DOI: 10.1016/j.carbpol.2020.116885 PMID: 33049878

Silva, L. I., Pereira, M. C., Carvalho, A. M., Buttrós, V. H., Pasqual, M., & Dória, J. (2023). Phosphorus-Solubilizing Microorganisms: A Key to Sustainable Agriculture. *Agriculture*, *13*(2), 462. DOI: 10.3390/agriculture13020462

Singh, P., Chauhan, P. K., Upadhyay, S. K., Singh, R. K., Dwivedi, P., Wang, J., Jain, D., & Jiang, M. (2022). Mechanistic Insights and Potential Use of Siderophores Producing Microbes in Rhizosphere for Mitigation of Stress in Plants Grown in Degraded Land. *Frontiers in Microbiology*, *13*, 898979. DOI: 10.3389/fmicb.2022.898979 PMID: 35898908

Sosnowski, J., Truba, M., & Vasileva, V. (2023). The Impact of Auxin and Cytokinin on the Growth and Development of Selected Crops. *Agriculture*, *13*(3), 724. DOI: 10.3390/agriculture13030724

Soumare, A., Diedhiou, A. G., Thuita, M., Hafidi, M., Ouhdouch, Y., Gopalakrishnan, S., & Kouisni, L. (2020). Exploiting Biological Nitrogen Fixation: A Route Towards a Sustainable Agriculture. *Plants*, *9*(8), 1011. DOI: 10.3390/plants9081011 PMID: 32796519

Sun, Q., Liu, Y., Liu, H., & Dumroese, R. K. (2020). Interaction of biochar type and rhizobia inoculation increases the growth and biological nitrogen fixation of robinia pseudoacacia seedlings. *Forests*, *11*(6), 1–12. DOI: 10.3390/f11060711

Sun, Y., Ma, L., Ma, J., Li, B., Zhu, Y., & Chen, F. (2022). Combined application of plant growth-promoting bacteria and iron oxide nanoparticles ameliorates the toxic effects of arsenic in Ajwain (*Trachyspermum ammi* L.). *Frontiers in Plant Science*, *13*, 1098755. Advance online publication. DOI: 10.3389/fpls.2022.1098755 PMID: 36643291

Szopa, D., Mielczarek, M., Skrzypczak, D., Izydorczyk, G., Mikula, K., Chojnacka, K., & Witek-Krowiak, A. (2022). Encapsulation efficiency and survival of plant growth-promoting microorganisms in an alginate-based matrix – A systematic review and protocol for a practical approach. *Industrial Crops and Products*, *181*(March), 114846. Advance online publication. DOI: 10.1016/j.indcrop.2022.114846

Tester, R. F., Karkalas, J., & Qi, X. (2004). Starch - Composition, fine structure and architecture. *Journal of Cereal Science*, *39*(2), 151–165. DOI: 10.1016/j.jcs.2003.12.001

Tilman, D., Cassman, K. G., Matson, P. A., Naylor, R., & Polasky, S. (2002). Agricultural sustainability and intensive production practices. *Nature*, *418*(6898), 671–677. DOI: 10.1038/nature01014 PMID: 12167873

Timofeeva, A. M., Galyamova, M. R., & Sedykh, S. E. (2022). Bacterial Siderophores: Classification, Biosynthesis, Perspectives of Use in Agriculture. *Plants*, *11*(22), 3065. DOI: 10.3390/plants11223065 PMID: 36432794

Trivedi, P., & Pandey, A. (2008). Recovery of plant growth-promoting rhizobacteria from sodium alginate beads after 3 years following storage at 4°C. *Journal of Industrial Microbiology & Biotechnology*, *35*(3), 205–209. DOI: 10.1007/s10295-007-0284-7 PMID: 18087737

Truong, H., Fournier, C., Hichami, S., Dole, L., Leroy, F., Wendehenne, D., Nicolas-Francès, V., & Besson-Bard, A. (2024). Apo-siderophores promote growth of iron-deficient Arabidopsis plants by mobilizing iron from roots to shoots and reducing oxidative stress in roots. *Plant Stress*, *12*, 100488. DOI: 10.1016/j.stress.2024.100488

Tu, L., He, Y., Yang, H., Wu, Z., & Yi, L. (2015). Preparation and characterization of alginate-gelatin microencapsulated *Bacillus subtilis* SL-13 by emulsification/internal gelation. *Journal of Biomaterials Science. Polymer Edition*, *26*(12), 735–749. DOI: 10.1080/09205063.2015.1056075 PMID: 26159659

Upadhyay, S. K., Srivastava, A. K., Rajput, V. D., Chauhan, P. K., Bhojiya, A. A., Jain, D., Chaubey, G., Dwivedi, P., Sharma, B., & Minkina, T. (2022). Root Exudates: Mechanistic Insight of Plant Growth Promoting Rhizobacteria for Sustainable Crop Production. *Frontiers in Microbiology*, *13*, 916488. DOI: 10.3389/fmicb.2022.916488 PMID: 35910633

Vance, C. P. (2001). Symbiotic Nitrogen Fixation and Phosphorus Acquisition. Plant Nutrition in a World of Declining Renewable Resources. *Plant Physiology*, *127*(2), 390–397. DOI: 10.1104/pp.010331 PMID: 11598215

Vercelheze, A. E. S., Marim, B. M., Oliveira, A. L. M., & Mali, S. (2019). Development of biodegradable coatings for maize seeds and their application for Azospirillum brasilense immobilization. *Applied Microbiology and Biotechnology*, *103*(5), 2193–2203. DOI: 10.1007/s00253-019-09646-w PMID: 30685816

Vessey, J. K. (2003). Plant growth promoting rhizobacteria as biofertilizers. *Plant and Soil*, *255*(2), 571–586. DOI: 10.1023/A:1026037216893

Voko, M. P., Aremu, A. O., Makunga, N. P., Nisler, J., Doležal, K., & Masondo, N. A. (2024). The potential applications of cytokinins and cytokinin oxidase/dehydrogenase inhibitors for mitigating abiotic stresses in model and non-model plant species. *Current Plant Biology*, *40*, 100398. DOI: 10.1016/j.cpb.2024.100398

Wahab, A., Muhammad, M., Ullah, S., Abdi, G., Shah, G. M., Zaman, W., & Ayaz, A. (2024). Agriculture and environmental management through nanotechnology: Eco-friendly nanomaterial synthesis for soil-plant systems, food safety, and sustainability. *The Science of the Total Environment*, *926*, 171862. DOI: 10.1016/j.scitotenv.2024.171862 PMID: 38527538

Waller, F., Achatz, B., Baltruschat, H., Fodor, J., Becker, K., Fischer, M., Heier, T., Hückelhoven, R., Neumann, C., Von Wettstein, D., Franken, P., & Kogel, K. H. (2005). The endophytic fungus *Piriformospora indica* reprograms barley to salt-stress tolerance, disease resistance, and higher yield. *Proceedings of the National Academy of Sciences of the United States of America*, *102*(38), 13386–13391. DOI: 10.1073/pnas.0504423102 PMID: 16174735

Wei, X., Fu, T., He, G., Zhong, Z., Yang, M., Lou, F., & He, T. (2023). Characteristics of rhizosphere and bulk soil microbial community of Chinese cabbage (*Brassica campestris*) grown in Karst area. *Frontiers in Microbiology*, *14*, 1241436. DOI: 10.3389/fmicb.2023.1241436 PMID: 37789857

Weinand, T., El-Hasan, A., & Asch, F. (2023). Role of Bacillus spp. Plant Growth Promoting Properties in Mitigating Biotic and Abiotic Stresses in Lowland Rice (*Oryza sativa* L.). *Microorganisms*, *11*(9), 2327. DOI: 10.3390/microorganisms11092327 PMID: 37764171

Wu, W., Du, K., Kang, X., & Wei, H. (2021). The diverse roles of cytokinins in regulating leaf development. *Horticulture Research*, *8*(1), 118. DOI: 10.1038/s41438-021-00558-3 PMID: 34059666

Xie, B., Wei, X., Wan, C., Zhao, W., Song, R., Xin, S., & Song, K. (2023). Exploring the Biological Pathways of Siderophores and Their Multidisciplinary Applications: A Comprehensive Review. *Molecules (Basel, Switzerland)*, *29*(10), 2318. DOI: 10.3390/molecules29102318 PMID: 38792179

Xuan, L., Ju, Z., Skonieczna, M., Zhou, K., & Huang, R. (2023). Nanoparticles-induced potential toxicity on human health: Applications, toxicity mechanisms, and evaluation models. *MedComm*, *4*(4), e327. DOI: 10.1002/mco2.327 PMID: 37457660

Yonathan, K., Mann, R., Mahbub, K. R., & Gunawan, C. (2022). The impact of silver nanoparticles on microbial communities and antibiotic resistance determinants in the environment. *Environmental Pollution*, *293*, 118506. DOI: 10.1016/j.envpol.2021.118506 PMID: 34793904

Younas, A.. (2024). Applicability of Nanomaterials in Agriculture. In Shahzad, R., Fiaz, S., Qayyum, A., Ul Islam, M., & Lee, I. J. (Eds.), *Revolutionizing Agriculture: A Comprehensive Exploration of Agri-Nanotechnology. Nanotechnology in the Life Sciences*. Springer., DOI: 10.1007/978-3-031-76000-6_2

Zago, S. L., dos Santos, M. F., Konrad, D., Fiorini, A., Rosado, F. R., Missio, R. F., & Vendruscolo, E. C. G. (2019). Shelf Life of Azospirillum brasilense in Alginate Beads Enriched With Trehalose and Humic Acid. *Journal of Agricultural Science (Toronto)*, *11*(6), 269. DOI: 10.5539/jas.v11n6p269

Zayed, O., Hewedy, O. A., Abdelmoteleb, A., Ali, M., Youssef, M. S., Roumia, A. F., Seymour, D., & Yuan, C. (2023). Nitrogen Journey in Plants: From Uptake to Metabolism, Stress Response, and Microbe Interaction. *Biomolecules*, *13*(10), 1443. DOI: 10.3390/biom13101443 PMID: 37892125

Zhang, Q., Gong, M., Xu, X., Li, H., & Deng, W. (2022). Roles of Auxin in the Growth, Development, and Stress Tolerance of Horticultural Plants. *Cells*, *11*(17), 2761. DOI: 10.3390/cells11172761 PMID: 36078168

Zhang, Z., Zhou, Z., Feng, S., Guo, P., Wang, Y., Hao, B., Guo, W., & Li, F. Y. (2024). Synergistic effects of AMF and PGPR on improving saline-alkaline tolerance of *Leymus chinensis* by strengthening the link between rhizosphere metabolites and microbiomes. *Environmental Technology & Innovation*, *36*, 103900. DOI: 10.1016/j.eti.2024.103900

Zhao, Y. (2012). Auxin Biosynthesis: A Simple Two-Step Pathway Converts Tryptophan to Indole-3-Acetic Acid in Plants. *Molecular Plant*, *5*(2), 334–338. DOI: 10.1093/mp/ssr104 PMID: 22155950

Chapter 10
Nanocarriers for Delivering Pesticides

John Tsado Mathew
https://orcid.org/0000-0001-9514-5330
Department of Chemistry, Ibrahimm Badamasi Babangida University, Lapai, Nigeria

Jonathan Hussaini
https://orcid.org/0009-0004-9216-8691
Department of Chemistry, Ibrahimm Badamasi Babangida University, Lapai, Nigeria

Abel Inobeme
Department of Chemistry, Edo State University, Uzairue, Nigeria

Charles Oluwaseun Adetunji
Department of Microbiology, Edo State University, Uzairue, Nigeria

Musa Safiyanu Tanko
Department of Chemistry, Fedral University, Lafia, Nigeria

ABSTRACT

Nanocarriers have emerged as a promising solution for pesticide delivery, aiming to enhance efficiency, reduce environmental contamination, and improve targeted pest control. Traditional pesticide applications often result in excessive use, rapid degradation, and non-specific distribution, leading to environmental and health concerns. Nanocarriers, such as polymeric nanoparticles, lipid-based carriers, inorganic nanoparticles, and carbon-based materials, offer controlled release, enhanced bioavailability, and improved pesticide stability. These nanoformulations protect active ingredients from premature degradation, facilitate targeted delivery, and minimize off-target effects. Recent advancements in stimuli-responsive and biodegradable nanocarriers offer promising solutions for eco-friendly pest management. This review explores the types, synthesis, mechanisms, applications, challenges, and future perspectives of nanocarriers in pesticide delivery, emphasizing their potential in advancing precision agriculture.

DOI: 10.4018/979-8-3693-9869-2.ch010

1.0 INTRODUCTION

Agriculture plays a crucial role in global food security, yet it faces persistent challenges due to pests, diseases, and environmental factors. Pesticides have been a vital component of modern agriculture, aiding in pest control and crop protection. However, conventional pesticide formulations often suffer from inefficiencies such as rapid degradation, leaching, and non-target effects, leading to environmental pollution and health risks. The excessive use of pesticides contributes to soil and water contamination, bioaccumulation in food chains, and resistance development in pests. These challenges necessitate the development of more efficient and sustainable pesticide delivery systems (Tudi et al. 2021).

Nanotechnology has emerged as a revolutionary approach to overcoming these limitations. The application of nanocarriers for pesticide delivery offers several advantages, including enhanced solubility, improved bioavailability, targeted delivery, and controlled release. Nanocarriers can encapsulate active pesticide ingredients, protecting them from premature degradation due to environmental factors such as UV radiation, oxidation, and microbial degradation. This encapsulation ensures prolonged pesticide activity, reducing the frequency of application and minimizing wastage. Additionally, nanocarriers enable the controlled and site-specific release of pesticides, reducing the impact on non-target organisms and promoting environmental sustainability (Kapeleka & Mwema, 2024).

Polymeric Nanoparticles: Biodegradable and biocompatible carriers, such as polymers like chitosan, polylactic acid (PLA), and poly(lactic-co-glycolic acid) (PLGA), are widely used in pesticide delivery systems due to their stability and ability to provide controlled and sustained release of active ingredients. These polymers protect pesticides from premature degradation and allow for gradual release, enhancing efficacy and reducing environmental impact. Additionally, carriers like liposomes and nanoemulsions are effective in delivering hydrophobic pesticides. They improve solubility and absorption while minimizing pesticide volatilization and degradation, thereby increasing the efficiency of pest control measures (Begines et al. 2020).

Inorganic Nanocarriers: Silica nanoparticles, metal oxides like zinc oxide and titanium dioxide, and clay-based nanomaterials have been investigated for their stability and capacity to facilitate sustained pesticide release. These inorganic nanocarriers offer advantages such as high surface area and tunable properties, enhancing pesticide loading and controlled delivery. Carbon-based nanocarriers, including carbon nanotubes and graphene oxide, have also gained attention for their ability to improve pesticide loading and targeted delivery. However, the environmental impact of these materials requires further study. Concerns include potential toxicity to non-target organisms and persistence in ecosystems. Comprehensive environmental risk assessments are essential to ensure the safe and sustainable application of these nanomaterials in agriculture (Yin et al. 2023).

Stimuli-responsive nanocarriers represent a significant advancement in pesticide delivery systems, offering controlled release mechanisms that respond to specific environmental triggers such as pH, temperature, or enzymatic activity. This targeted approach ensures precise application, enhancing efficacy while minimizing excess pesticide use and environmental impact. For instance, pH-responsive nanocarriers are designed to release their payload in environments with specific pH levels. In agricultural applications, these carriers can be engineered to degrade in the alkaline conditions of certain insect digestive systems, ensuring that the pesticide is released precisely where it is most effective. Similarly, temperature-sensitive nanocarriers utilize polymers that undergo structural changes in response to temperature fluctuations, allowing for controlled pesticide release during periods of increased pest activity. Enzyme-responsive carriers release their contents upon interaction with specific enzymes present in target pests or pathogens, further enhancing the precision of pesticide application. The development of these smart delivery systems not only improves pest control efficiency but also contributes to sustainable agricultural practices by reducing the overall quantity of pesticides required and limiting their impact on non-target organisms and the environment (Camara et al. 2019).

Diffusion-Controlled Release: The pesticide slowly diffuses out of the nanocarrier over time, ensuring a sustained effect.pH-Responsive Release: Some nanocarriers are designed to release pesticides in response to changes in pH, making them ideal for targeting pests in specific environments. Enzyme-Triggered Release: Biodegradable polymeric carriers can release pesticides in response to enzymatic degradation, ensuring targeted action. External Stimuli-Triggered Release: Light, heat, or moisture can trigger the release of pesticides, allowing for precise application based on environmental conditions (Chaud et al. 2021).

Environmental and Health Implications: While nanocarriers offer numerous benefits in pesticide delivery, their environmental impact requires careful assessment. Concerns include the persistence of nanoparticles in soil and water, potential toxicity to non-target organisms, and bioaccumulation within ecosystems. Regulatory bodies are actively working to establish safety guidelines to ensure that nano-pesticides are both effective and environmentally friendly. Research into biodegradable and eco-friendly nanocarriers is crucial to addressing these concerns and promoting sustainable agricultural practices. Developing nanocarriers from biocompatible materials, such as chitosan or lignin, can enhance degradation and reduce ecological risks. Comprehensive environmental risk assessments are essential to understand the long-term effects of these advanced delivery systems and to guide the development of regulations that balance innovation with ecological safety (Xuan et al. 2023).

The integration of nanotechnology into pesticide delivery represents a significant advancement in agricultural pest management. By improving pesticide stability, bioavailability, and targeted delivery, nanocarriers can enhance efficiency while reducing environmental harm. However, further research is needed to optimize nanocarrier formulations, assess long-term environmental impacts, and establish regulatory frameworks. With continued advancements, nanocarrier-based pesticides have the potential to revolutionize agriculture by promoting sustainable and precision-based farming practices (Ali et al. 2023).

1.1 Types of Nanocarriers Used in Pesticide Delivery

The advent of nanocarriers has transformed modern agriculture by enhancing pesticide efficacy, stability, and controlled release. These advanced delivery systems address limitations of traditional pesticide formulations, such as rapid degradation, leaching, and environmental toxicity, by protecting active ingredients, reducing required dosages, and minimizing off-target effects. Key types of nanocarriers utilized in pesticide delivery include polymeric nanoparticles, which offer controlled and sustained release; inorganic nanoparticles like silica and metal oxides, valued for their stability and sustained release capabilities; lipid-based carriers such as liposomes and nanoemulsions, effective in delivering hydrophobic pesticides; carbon-based nanocarriers like carbon nanotubes and graphene oxide, noted for their high pesticide loading and targeted delivery; dendrimers, hyper-branched structures that provide precise control over size and functionality; and metal-organic frameworks (MOFs), porous crystalline materials with high loading capacity and tunable release properties. Each nanocarrier type presents unique advantages and challenges, with ongoing research focused on optimizing their design for effective and sustainable agricultural applications (Ali et al. 2024).

Polymeric Nanoparticles: Polymeric nanoparticles are extensively utilized in agriculture due to their biocompatibility, biodegradability, and capacity to provide controlled and sustained pesticide release. These nanoparticles are synthesized using natural or synthetic polymers, such as chitosan, alginate, poly(lactic-co-glycolic acid) (PLGA), and polycaprolactone (PCL). Common synthesis methods include emulsion polymerization, nanoprecipitation, and solvent evaporation techniques. The resulting nanoparticles can encapsulate active pesticide ingredients, protecting them from premature degradation and facilitating targeted delivery. This encapsulation enhances the pesticides' stability and bioavailability, allowing for lower dosages and reducing environmental contamination. Moreover, the controlled release properties of polymeric nanoparticles ensure a sustained release of pesticides over time, minimizing the need for frequent applications and thereby decreasing labor and operational costs. The use of biodegradable polymers ensures that the nanoparticles degrade into non-toxic byproducts, mitigating potential environmental impacts. However, challenges such as large-scale production, cost-effectiveness, and comprehensive environmental risk assessments remain to be addressed. Ongoing research focuses on optimizing synthesis methods, improving the efficiency of pesticide loading and release, and evaluating the long-term effects of polymeric nanoparticles on ecosystems to ensure sustainable agricultural practices (Chaud et al. 2021).

Chitosan: Chitosan, a biopolymer derived from chitin, is renowned for its antimicrobial properties, controlled release capabilities, and biodegradability. Its biocompatibility and non-toxic nature make it suitable for various applications, including agriculture and food packaging. Chitosan exhibits antibacterial activity against a broad spectrum of microorganisms, including Gram-negative and Gram-positive bacteria, yeast, and foodborne filamentous fungi. In agriculture, chitosan-based nanocomposites have been utilized to enhance plant growth and provide protection against pathogens. Its film-forming ability allows for the development of biodegradable packaging materials that can extend the shelf life of food products by inhibiting microbial growth. Additionally, chitosan's capacity for controlled release has been harnessed in drug delivery systems, where it facilitates the sustained release of therapeutic agents. The biodegradability of chitosan ensures that it breaks down into non-toxic byproducts, minimizing environmental impact. Ongoing research continues to explore and expand the potential applications of chitosan in various fields, highlighting its versatility and importance as a sustainable biopolymer (Inobeme et al. 2025).

Polylactic Acid (PLA): A biodegradable polymer that provides a slow-release mechanism, reducing pesticide application frequency. Poly(lactic-co-glycolic acid) (PLGA): A widely used polymer due to its ability to protect pesticides from environmental degradation while ensuring prolonged release. Polymeric nanoparticles encapsulate pesticides, shielding them from external environmental factors such as UV radiation and microbial degradation. Their controlled release properties help minimize pesticide wastage and prevent excessive residue accumulation in the environment.

Lipid-Based Nanocarriers Lipid-based nanocarriers, such as liposomes and nanoemulsions, have gained significant attention in pesticide delivery due to their biocompatibility, enhanced solubility, and targeted release mechanisms. Lipid-based carriers, such as liposomes and nanoemulsions, have significantly advanced pesticide delivery systems. Liposomes are spherical vesicles composed of lipid bilayers that encapsulate pesticides within their hydrophobic or hydrophilic compartments. They enhance pesticide stability and facilitate slow, controlled release, thereby reducing the risk of leaching. Nanoemulsions, stable mixtures of oil, water, and surfactants, improve the solubility and absorption of hydrophobic pesticides. By ensuring uniform pesticide dispersion, nanoemulsions lead to efficient pest control with minimal environmental impact. Overall, lipid-based carriers reduce pesticide volatilization and degradation, ensuring long-lasting effects with reduced application frequency (Lu et al. 2021).

Inorganic Nanocarriers: Inorganic nanocarriers, such as silica, clay, and metal oxides, provide excellent stability and controlled-release properties. These carriers can adsorb pesticides onto their surfaces or encapsulate them within their structures. Silica Nanoparticles: Porous silica nanocarriers can encapsulate pesticides and release them gradually in response to environmental stimuli such as pH or moisture. Clay-Based Nanocarriers: Clay minerals such as montmorillonite and kaolinite are used to improve pesticide retention in soil, reducing leaching and groundwater contamination. Metal Oxide Nanoparticles: Zinc oxide (ZnO) and titanium dioxide (TiO_2) nanoparticles exhibit antimicrobial and UV-protective properties, enhancing pesticide stability and efficacy. Inorganic nanocarriers are highly stable and resistant to environmental degradation, making them suitable for long-term pesticide delivery. However, their environmental persistence requires further study to assess potential risks (Yin et al. 2023).

Carbon-Based Nanocarriers: Carbon-based nanocarriers, such as carbon nanotubes (CNTs) and graphene oxide (GO), have been explored for their ability to enhance pesticide loading, targeted delivery, and controlled release. Carbon Nanotubes (CNTs): These cylindrical nanostructures possess high surface area and strong adsorption capacity, allowing efficient pesticide encapsulation and gradual release. Graphene Oxide (GO): GO sheets provide a large surface area for pesticide binding, enabling slow and targeted release while reducing pesticide losses. While carbon-based nanocarriers offer excellent pesticide delivery properties, concerns regarding their biodegradability and toxicity to non-target organisms require further research (Jampilek & Kralova, 2021).

Stimuli-Responsive Nanocarriers: Advanced nanocarriers capable of responding to specific environmental stimuli have been developed to enhance precision pesticide delivery. These include: pH-Responsive Nanocarriers: These release pesticides in response to soil or plant pH changes, ensuring targeted pest control. Temperature-Sensitive Nanocarriers: These deliver pesticides based on temperature fluctuations, releasing active ingredients when pest activity is highest. Light-Activated Nanocarriers: These release pesticides when exposed to specific wavelengths of light, reducing unnecessary pesticide dispersion. Stimuli-responsive nanocarriers offer enhanced precision and efficiency, reducing pesticide overuse and environmental contamination (Hu et al. 2023).

The use of nanocarriers in pesticide delivery has opened new avenues for sustainable agriculture by improving pesticide efficiency, reducing environmental risks, and ensuring controlled and targeted release. Polymeric nanoparticles, lipid-based nanocarriers, inorganic nanoparticles, carbon-based materials, and stimuli-responsive systems each offer unique advantages that enhance pesticide performance. However, further research is needed to address potential concerns related to toxicity, environmental persistence, and large-scale production. With continued advancements, nanocarrier-based pesticides hold great promise for revolutionizing pest management and promoting sustainable agricultural practices (Chaud et al. 2021).

1.2 Synthesis and Characterization of Nanocarriers

Table 1.: Nanocarriers have gained significant attention in pesticide delivery due to their ability to enhance stability, control release, and improve the bioavailability of active ingredients. The synthesis and characterization of nanocarriers are crucial steps in developing efficient and sustainable pesticide formulations. Various synthesis methods are employed depending on the type of nanocarrier, while multiple characterization techniques help determine their structural, chemical, and physical properties (Yin et al. 2023).

Synthesis of Nanocarriers: Polymeric nanoparticles are versatile carriers extensively utilized in pesticide delivery systems due to their customizable properties and efficient encapsulation capabilities. Common synthesis techniques include emulsion polymerization, nanoprecipitation, solvent evaporation, and ionic gelation. Nanoprecipitation, also known as the solvent displacement method, entails dissolving polymers in a water-miscible organic solvent, followed by rapid mixing with an aqueous solution, leading to immediate nanoparticle formation due to solvent diffusion These nanocarriers are synthesized using methods such as:

Emulsion Polymerization: This method involves the polymerization of monomers in an aqueous medium with surfactants to form nanoparticles. Solvent Evaporation: A polymer is dissolved in an organic solvent and emulsified in water, followed by solvent removal to form nanoparticles. Nanoprecipitation: This involves the rapid mixing of a polymer solution with a non-solvent to induce nanoparticle formation. Lipid-Based Nanocarriers: These nanocarriers, including liposomes and nanoemulsions, are synthesized using (Pulingam et al. 2022).

Thin-Film Hydration: A lipid film is hydrated with an aqueous phase to form vesicles. Microfluidization: High shear forces are applied to mix lipid and aqueous phases uniformly. High-Pressure Homogenization: Lipid dispersions are forced through a narrow channel to reduce droplet size. Inorganic Nanoparticles: These include silica, clay, and metal oxide nanoparticles, synthesized using: Sol-Gel Method: A precursor solution undergoes hydrolysis and condensation to form nanoparticles. Hydrothermal Synthesis: A high-temperature, high-pressure aqueous environment is used to grow nanoparticles. Co-Precipitation: Metal salts are precipitated in an alkaline medium to form nanoparticles. Carbon-Based Nanocarriers: These include carbon nanotubes and graphene oxide, synthesized through (Ashfaq et al. 2023).

Chemical Vapor Deposition (CVD): Hydrocarbon gases decompose at high temperatures, forming carbon nanostructures. Hydrothermal Method: Carbon precursors react in an autoclave to produce carbon nanoparticles. Stimuli-Responsive Nanocarriers: These are engineered to respond to environmental factors such as pH, temperature, or light, using functionalized polymers, hydrogels, or surface-modified nanoparticles (Manawi et al. 2018; Inobeme et al. 2024).

Characterization of Nanocarriers: Characterization is essential to ensure nanocarriers have the desired size, shape, surface properties, and stability. Common techniques include: Fourier Transform Infrared Spectroscopy (FTIR): Identifies functional groups and chemical interactions in nanocarriers. Scanning Electron Microscopy (SEM): Provides high-resolution images of nanocarrier morphology and surface structure. Transmission Electron Microscopy (TEM): Examines internal structures and nanoparticle size at the nanometer scale. Dynamic Light Scattering (DLS): Measures nanoparticle size distribution and zeta potential to assess stability. X-Ray Diffraction (XRD): Determines the crystalline nature and phase composition of nanoparticles (Altammar 2023).

The synthesis and characterization of nanocarriers are fundamental in developing efficient pesticide delivery systems. Proper synthesis techniques ensure nanoparticle stability and functionality, while advanced characterization methods help optimize their performance for sustainable agricultural applications. Nanocarriers, particularly polymeric nanoparticles, have garnered significant attention in agricultural applications due to their potential to enhance the delivery of agrochemicals such as pesticides. These carriers can improve efficiency through targeted delivery, controlled release, and reduced environmental impact. The synthesis methods for polymeric nanoparticles are broadly categorized into two approaches: bottom-up and top-down techniques. Bottom-up techniques involve in situ polymerization, where monomers undergo polymerization to form nanoparticles. This approach requires the use of solvents, initiators, and other agents, followed by separation and purification steps to obtain the desired nanoparticles. While effective, the need for these additional steps can increase costs and limit applicability in certain contexts, such as food and agriculture, where the presence of residual chemicals may be undesirable. Top-down techniques, on the other hand, utilize preformed natural or synthetic polymers to create nanoparticles through processes like mixing or emulsification, often with the application of external energy. These methods typically require fewer solvents and chemicals, making them more suitable for applications where safety and environmental considerations are paramount. In top-down approaches, the active components can be entrapped within the polymeric matrix or adsorbed on the surface, depending on the chemical nature of the polymer, surfactant, and active ingredient involved. Characterization of the synthesized nanocarriers is crucial to ensure their efficacy and safety in agricultural applications. Key parameters to assess include particle size, surface charge, morphology, encapsulation efficiency, and release profiles of the active ingredients. Advanced analytical techniques such as dynamic light scattering (DLS), scanning electron microscopy (SEM), transmission electron microscopy (TEM), and Fourier-transform infrared spectroscopy (FTIR) are commonly employed to evaluate these characteristics. Understanding these properties aids in optimizing the design and functionality of nanocarriers, ensuring that they meet the specific requirements for effective pesticide delivery. In conclusion, the careful synthesis and thorough characterization of nanocarriers are essential steps in developing efficient and sustainable pesticide delivery systems. By selecting appropriate synthesis methods and employing advanced characterization techniques, researchers can design nanocarriers that enhance pesticide efficacy, reduce environmental impact, and contribute to more sustainable agricultural practices.

Table 1. Synthesis and characterization of nanocarriers

Nanocarrier Type	Synthesis Methods	Characterization Techniques	References
Polymeric Nanoparticles	Emulsion polymerization, Solvent evaporation, Nanoprecipitation	FTIR, SEM, TEM, DLS, Zeta potential, TGA	Pulingam et al. 2022; Mathew et al. 2024a
Liquid Based Nanocarriers	Thin-film hydration, Microfluidization, High-pressure homogenization	TEM, DLS, FTIR, XRD, DSC	Ashfaq et al. 2023.
Inorganic Nanocarriers	Sol-gel method, Hydrothermal synthesis, Co-precipitation	XRD, SEM, TEM, FTIR, BET surface area analysis	Ghasempour et al. 2023.; Mathew et al. 2024b
Carbon Base Nanocarriers	Chemical vapor deposition (CVD), Hydrothermal, Arc discharge	Raman spectroscopy, XPS, TEM, BET, DLS	Mansuriya & Altintas, 2021; Mathew et al. 2024c
Stimuli-Responsive Nanocarriers	pH-sensitive polymers, Smart hydrogels, Functionalized nanoparticles	UV-Vis spectroscopy, TGA, FTIR, DLS, Zeta potential	Karimi et al. 2016.

1.3 Mechanisms of Pesticide Encapsulation and Release

Nanocarriers play a crucial role in enhancing pesticide delivery by encapsulating active ingredients and controlling their release. Encapsulation helps protect pesticides from environmental degradation, improves their solubility, and ensures targeted delivery, while controlled release mechanisms optimize efficiency and minimize ecological impact. The encapsulation and release of pesticides depend on the nature of the nanocarrier, the physicochemical properties of the pesticide, and external environmental factors (Ali et al. 2024).

Pesticide Encapsulation Mechanisms: Encapsulation refers to the process of enclosing pesticide molecules within a nanocarrier to enhance stability, prolong effectiveness, and control their release. Several techniques are used for encapsulating pesticides, depending on the type of nanocarrier (Kapeleka & Mwema, 2024).

Physical Encapsulation: This method involves trapping pesticide molecules within the structure of nanocarriers without altering their chemical composition. Examples include: Pesticides with low water solubility are embedded within the hydrophobic core of lipid-based carriers such as liposomes and nanoemulsions. Pesticide molecules adhere to the surface of porous nanocarriers like silica, clay, and carbon nanotubes via van der Waals forces or hydrogen bonding. Some nanocarriers, such as cyclodextrins, form host-guest inclusion complexes with pesticides, improving solubility and stability (Subramani & Ganapathyswamy, 2020; Mathew et al. 2025).

Chemical Encapsulation: Chemical interactions between nanocarriers and pesticides improve encapsulation efficiency and controlled release. These include: Functional groups on the nanocarrier surface form covalent bonds with pesticides, ensuring strong attachment and sustained release. Electrostatic attraction between charged nanocarriers and oppositely charged pesticide molecules helps in loading and stabilization. In polymeric nanoparticles and hydrogels, pesticides are chemically bound within the crosslinked polymer matrix, enhancing encapsulation (Fincheira et al. 2023).

Stimuli-Responsive Encapsulation: Advanced nanocarriers are designed to respond to external stimuli, ensuring on-demand pesticide release. These include: Nanocarriers release pesticides under acidic or alkaline conditions, useful for soil- or plant-specific targeting. Some polymeric nanocarriers swell or degrade at specific temperatures, triggering controlled release. UV- or visible-light-sensitive nanocarriers break down upon exposure to sunlight, releasing pesticides when needed. Mechanisms of Pesticide Release: The release of pesticides from nanocarriers follows different mechanisms, ensuring efficient and sustained action while minimizing environmental contamination (Fincheira et al. 2023).

Diffusion-Controlled Release: In polymeric nanocarriers, pesticide release primarily occurs through diffusion mechanisms. Initially, the pesticide molecules migrate from regions of higher concentration within the nanocarrier to lower concentration areas outside, driven by a concentration gradient. As the polymer matrix absorbs environmental moisture, it swells, creating pathways that facilitate the gradual diffusion of encapsulated pesticides. This controlled release ensures a sustained pesticide effect, minimizing the need for frequent applications and reducing environmental impact. The efficiency of this diffusion process is influenced by factors such as the physicochemical properties of the pesticide, the characteristics of the polymer matrix, and external environmental conditions. (Azeh et al. 2023).

pH-Triggered Release: Many nanocarriers are engineered to release pesticides in response to pH variations, enabling controlled delivery in specific agricultural environments. Polymeric and inorganic carriers can degrade in acidic conditions, such as those found in insect digestive systems or certain soil types, facilitating targeted pesticide release. For instance, pH-responsive copper-doped ZIF-8 metal–organic framework (MOF) nanoparticles have been developed to enhance pesticide delivery and translocation in plants. These nanocarriers exhibit acid-responsive release, with cumulative release rates of 79.5%, 69.9%, and 43.0% at pH values of 5, 7, and 9, respectively, after 48 hours. This pH-sensitive behavior ensures that pesticides are released primarily in acidic environments, such as insect guts, thereby increasing efficacy and reducing impact on non-target areas. Additionally, certain nanocarriers, like mesoporous silica, are designed to dissolve under alkaline conditions, allowing for controlled pesticide release in specific agricultural settings where such pH conditions prevail. The utilization of pH-responsive nanocarriers enhances the precision of pesticide application, contributing to more efficient and sustainable pest management practices (Adepu & Ramakrishna, 2021; Adetunji et al. 2022).

Enzyme-Responsive Release: This mechanism utilizes biological enzymes to break down the nanocarrier, releasing the pesticide at the target site. Biodegradable polymers release pesticides when exposed to soil microbes. Some nanocarriers degrade upon exposure to plant-secreted enzymes, ensuring localized pesticide release. Photo-responsive nanocarriers release pesticides upon exposure to sunlight or specific wavelengths of light. Some nanocarriers break down when exposed to UV light, releasing pesticides in response to sunlight. Gold or silver nanoparticles absorb light energy, generating heat that disrupts the nanocarrier and releases the pesticide (Pashirova et al. 2024)

Temperature-Responsive Release: Thermosensitive nanocarriers are engineered to release pesticides in response to temperature fluctuations, leveraging materials that undergo structural changes with varying temperatures. Polymers such as poly(N-isopropylacrylamide) (PNIPAAm) exhibit a phase transition at specific temperatures, known as the Critical Solution Temperature (CST). Below the CST, these polymers swell, while above it, they shrink, facilitating controlled pesticide release. This mechanism is particularly advantageous in agriculture, as it allows for the release of pesticides during warmer periods when pest activity is heightened. Additionally, inorganic nanocarriers can be designed to release their payload upon exposure to heat, making them suitable for pest control across various climatic conditions. The development of such temperature-responsive delivery systems enhances the precision and efficiency of pesticide application, contributing to more sustainable agricultural practices (Mathew et al. 2024b).

Controlled Burst Release: Nanocarriers have been engineered to release pesticides in response to specific environmental triggers, enabling rapid and targeted pest control. For instance, pH-responsive nanocarriers can release their payload upon encountering the alkaline conditions present in certain insect digestive systems, ensuring that the pesticide is activated precisely where needed. This targeted approach minimizes the impact on non-target organisms and reduces the overall quantity of pesticide required. Additionally, some nanocarriers are designed to encapsulate volatile pesticides, preventing premature evaporation. Upon release, these pesticides can effectively target airborne pests, enhancing their efficacy. The development of such stimuli-responsive nanocarriers represents a significant advancement in sustainable agriculture, offering precise pest control while mitigating environmental impact (Camara et al. 2019; Mathew et al. 2023a).

Nanocarrier-based pesticide delivery systems offer precise, controlled release mechanisms that enhance efficiency while reducing environmental harm. Various encapsulation techniques ensure pesticide stability, while different release mechanisms—such as diffusion, pH-sensitivity, enzyme response, and light activation—enable targeted and sustained action. Understanding these mechanisms is crucial for developing smart pesticide formulations that optimize agricultural productivity and sustainability (Fincheira et al. 2023).

1.4 Applications of Nanocarriers in Pesticide Delivery

The integration of nanocarriers into pesticide delivery systems has transformed modern agriculture by enhancing pesticide efficiency, stability, and targeted action. Traditional pesticide applications often face challenges such as rapid degradation, leaching, and environmental contamination. Nanocarriers address these issues by offering controlled release, improved bioavailability, and site-specific delivery of active ingredients. For instance, polymeric nanoparticles can encapsulate pesticides, protecting them from premature degradation and allowing for sustained release, thereby reducing the frequency of applications. Lipid-based carriers, like liposomes and nanoemulsions, improve the solubility and absorption of hydrophobic pesticides, ensuring uniform dispersion and effective pest control with minimal environmental impact. Additionally, inorganic nanocarriers, such as silica nanoparticles and metal oxides, provide stability and facilitate sustained pesticide release. The use of these nanocarriers not only enhances pest control efficacy but also contributes to sustainable agriculture by minimizing pesticide usage and reducing environmental contamination. Ongoing research focuses on developing biodegradable and eco-friendly nanocarriers to further promote sustainable agricultural practices (Ali et al. 2024).

Enhanced Pesticide Stability and Shelf Life: One of the significant advantages of nanocarriers is their ability to improve the stability of pesticides. Many conventional pesticides degrade quickly due to environmental factors such as UV radiation, temperature fluctuations, and microbial activity. Nanocarriers help: Protect against photodegradation: Encapsulation within nanoparticles, such as polymeric or lipid-based carriers, shields pesticides from harmful UV rays, prolonging their activity. Reduce volatilization: Some pesticides evaporate quickly when exposed to air, reducing their effectiveness. Nanocarriers, especially solid lipid nanoparticles, prevent rapid evaporation and ensure long-lasting effects. Minimize hydrolysis: Certain pesticides degrade when exposed to moisture. Nanoencapsulation, particularly using silica or chitosan nanoparticles, creates a protective barrier, preventing premature breakdown (Ali et al. 2024).

Controlled and Sustained Release of Pesticides: Nanocarriers allow the controlled release of pesticides, ensuring prolonged activity and reduced frequency of application. Different release mechanisms include: pH-sensitive release: Some nanocarriers release pesticides in response to pH variations in the soil or insect gut, ensuring targeted action. Temperature-responsive release: Polymers that expand or contract with temperature changes control pesticide release based on environmental conditions. Enzyme-triggered release: Some nanocarriers degrade in the presence of plant or microbial enzymes, ensuring site-specific pesticide delivery. Moisture-dependent release: Hydrogels and clay nanoparticles release pesticides gradually in response to soil moisture, optimizing pest control in dry and wet conditions. This controlled release minimizes the excessive use of pesticides, reducing environmental pollution and human health risks (Chaud et al. 2021; Adetunji et al. 2023a).

Targeted Pesticide Delivery for Improved Efficacy: Nanocarriers enable targeted pesticide delivery, reducing off-target effects and increasing efficiency. Several targeting mechanisms are employed: Foliage Targeting: Lipid-based nanoparticles, such as nanoemulsions, ensure better adhesion to plant surfaces, improving pesticide retention and absorption. Root-Specific Targeting: Nanocarriers like mesoporous silica nanoparticles transport pesticides directly to the root zone, preventing soil-borne pests and diseases. Insect-Specific Targeting: Some nanocarriers are designed to release pesticides only inside the insect gut, reducing harm to beneficial organisms. For example, chitosan nanoparticles interact with the insect's digestive enzymes, releasing active pesticides only upon ingestion. Pathogen-Specific Targeting: Functionalized nanoparticles can recognize and bind to specific fungal or bacterial pathogens, releasing fungicides or bactericides precisely at the infection site (Ali et al. 2024).

Reduction of Environmental Pollution and Eco-Toxicity: Traditional pesticide applications often lead to water and soil contamination due to leaching, runoff, and bioaccumulation. Nanocarriers help mitigate these issues by: Minimizing leaching: Encapsulated pesticides are released slowly, reducing the risk of groundwater contamination. Lowering pesticide residues: Controlled release reduces the overall amount of pesticide needed, decreasing residue accumulation in crops. Preventing non-target toxicity: Site-specific delivery ensures pesticides act only where needed, reducing harm to beneficial insects, pollinators, and aquatic life. Biodegradable nanocarriers: Many nanocarriers, such as chitosan and alginate, are biodegradable and do not persist in the environment, making them eco-friendly alternatives (Pathak et al. 2022).

Increased Solubility and Bioavailability of Pesticides: Many pesticides exhibit poor water solubility, which limits their absorption and effectiveness in agricultural applications. Nanocarriers have been developed to address this challenge by enhancing the solubility and bioavailability of hydrophobic pesticides. Nanoemulsions and liposomes are two types of nanocarriers that encapsulate hydrophobic pesticides, improving their dispersibility in water and increasing absorption. Cyclodextrin inclusion complexes are molecular carriers that enhance the solubility of poorly soluble pesticides, thereby increasing their bioavailability. Polymeric nanoparticles improve pesticide penetration into plant tissues, ensuring deeper and more effective pest control. By utilizing these nanocarrier systems, the limitations associated with poorly soluble pesticides can be overcome, leading to more efficient and sustainable agricultural practices (Ali et al. 2024).

Applications in Organic and Sustainable Farming: Nanocarriers are increasingly being used in organic and sustainable agriculture to improve the efficiency of bio-based pesticides. They help: Encapsulate botanical pesticides: Essential oils and plant extracts, such as neem oil, suffer from rapid degradation. Nanoencapsulation protects these bio-pesticides and enhances their longevity. Improve microbial pesticide delivery: Beneficial bacteria and fungi used as biopesticides can be encapsulated in nanoparticles, protecting them from environmental stress and ensuring controlled release. Reduce synthetic pesticide dependence: With improved efficiency and targeted delivery, nanocarriers lower the required pesticide dosage, promoting eco-friendly farming practices (Chaud et al. 2021; Musah et al. 2023).

The integration of nanocarriers in pesticide delivery has indeed enhanced efficiency, stability, and targeted application. However, concerns regarding their environmental and health safety persist, necessitating further research and regulatory approval. The large-scale production of nano-formulated pesticides remains costlier than conventional pesticides, posing economic challenges for widespread adoption. Additionally, farmers may require education and training to effectively implement nanotechnology-based pesticide solutions. Despite these challenges, ongoing research and development continue to refine nanocarrier-based pesticide delivery systems, aiming to make agriculture more efficient and sustainable (Wang et al. 2022).

Nanocarriers have transformed pesticide delivery in modern agriculture by enhancing stability, enabling controlled release, and allowing for targeted application, thereby improving environmental safety. By increasing the solubility of hydrophobic pesticides, these carriers enhance bioavailability and reduce the required dosage, leading to decreased environmental pollution. Furthermore, the development of smart, responsive pesticide systems utilizing nanotechnology offers precision in pest management, releasing active ingredients in response to specific environmental triggers. As research progresses, nanotechnology is poised to play an increasingly vital role in sustainable agriculture, ensuring higher crop yields while minimizing ecological harm (Fincheira et al. 2022).

1.5 Toxicity and Environmental Impact

Table 1.2 showed the incorporation of nanocarriers in pesticide delivery systems has led to notable improvements in pesticide efficiency, stability, and targeted application. However, concerns have emerged regarding their potential toxicity and environmental impact. The behavior of nanocarriers in the environment, their interactions with living organisms, and their long-term ecological consequences require thorough evaluation. Studies have indicated that certain nanopesticides may pose risks to non-target species, including beneficial insects, aquatic organisms, and soil microbes. The unique properties of nanomaterials, such as increased surface area and reactivity, can lead to unforeseen interactions within ecosystems. Additionally, the persistence and bioaccumulation potential of some nanocarriers raise concerns about their long-term environmental fate. Therefore, comprehensive risk assessments and the development of stringent regulatory frameworks are essential to ensure the safe and sustainable use of nanocarriers in agriculture (Ali et al. 2024).

Potential Toxicity to Non-Target Organisms: Nanocarriers in pesticide formulations offer enhanced efficiency and targeted delivery; however, they may pose risks to non-target organisms, including beneficial insects, aquatic life, and soil microbes. Metal-based nanoparticles, such as copper oxide (CuO), zinc oxide (ZnO), and titanium dioxide (TiO_2), can generate reactive oxygen species (ROS), leading to oxidative stress, DNA damage, and cytotoxicity in exposed organisms. For instance, studies have shown that ROS production by these nanoparticles can cause significant cellular damage, affecting the health of various non-target species. Additionally, the inhalation of nano-formulated pesticides by humans and animals may result in respiratory irritation and potential long-term health effects. Therefore, it is crucial to conduct comprehensive environmental risk assessments and establish regulatory guidelines to ensure the safe use of nanocarriers in agriculture (Adetunji et al. 2024; Mgadi et al. 2024).

Bioaccumulation and Food Chain Contamination: The application of nano-pesticides in agriculture has raised concerns about their potential to bioaccumulate in plant tissues, soil, and water bodies. When plants absorb these nano-formulated pesticides, there is a possibility that the nanoparticles may transfer through the food chain, potentially impacting human and animal health. Long-term exposure to certain nanoparticles has been associated with metabolic disruptions and immune system imbalances in living organisms. For instance, some studies have indicated that nanoparticles can induce toxic and genotoxic effects, leading to oxidative stress and DNA damage. Additionally, the persistence of these particles in the environment may result in their accumulation in various ecosystems, further contributing to potential health risks. Therefore, it is crucial to conduct comprehensive risk assessments and establish stringent regulatory frameworks to ensure the safe use of nano-pesticides in agriculture (Miteu et al. 2023).

Soil and Water Contamination: The application of nanocarriers in pesticide formulations has raised environmental concerns, particularly regarding their impact on soil health and aquatic ecosystems. Certain nanoparticles can disrupt essential soil microbial communities involved in nutrient cycling and plant growth. For instance, studies have shown that copper-based nanopesticides can alter soil microbial diversity and function, potentially affecting soil fertility. Moreover, some nanoparticles persist in the soil for extended periods, leading to long-term ecological consequences. The leaching of nano-formulated pesticides into water sources further exacerbates environmental risks. Nanoparticles can accumulate in aquatic organisms, such as fish and invertebrates, leading to toxicity and ecological imbalances. Research indicates that chronic exposure to these substances can result in sublethal effects, including altered reproduction and growth, which may have long-term ecological consequences. These findings underscore the importance of conducting comprehensive environmental risk assessments and developing regulatory frameworks to ensure the safe use of nanocarriers in agriculture (Etsuyankpa et al. 2024; Mgadi et al. 2024).

Airborne Exposure and Environmental Dispersion: The application of nano-pesticides introduces concerns regarding their potential to become airborne during spraying, leading to unintended inhalation by humans and wildlife. Wind dispersion can carry nanoparticles beyond the target area, resulting in environmental contamination and affecting ecosystems distant from the application site. This dispersion may cause unintended toxicity in non-target organisms, including beneficial insects, birds, and aquatic life. The small size and unique properties of nanoparticles can facilitate their entry into biological systems, potentially leading to bioaccumulation and unforeseen ecological impacts. Therefore, it is crucial to implement stringent application protocols and develop formulations that minimize drift to mitigate these risks (Kapeleka et al. 2024).

Degradability and Environmental Persistence: The environmental impact of nanocarriers in pesticide formulations is significantly influenced by their degradability. Biodegradable nanocarriers, such as those derived from chitosan and alginate, are designed to break down naturally in the environment, posing minimal ecological risks. In contrast, non-biodegradable nanoparticles, including certain metal oxides and synthetic polymers, can persist in soil and water systems for extended periods. This persistence may lead to the accumulation of these materials in the environment, potentially causing long-term ecological concerns. Therefore, the selection of biodegradable nanocarriers is crucial for minimizing environmental impact and promoting sustainable agricultural practices (Satchanska et al. 2024).

Regulatory and Safety Concerns: Given the potential risks associated with nano-formulated pesticides, it is imperative for regulatory agencies to establish comprehensive guidelines to ensure their safe application. Current regulatory frameworks may not fully address the unique properties and behaviors of nanomaterials, necessitating the development of specific evaluation protocols. Further research is essential to assess the long-term impacts of these substances on human health and the environment. Emphasis should be placed on developing eco-friendly, biodegradable nanocarriers, such as those derived from natural polymers like chitosan and alginate, to minimize toxicity while harnessing the benefits of nanotechnology in agriculture. Implementing stringent regulations and promoting sustainable nanocarrier designs will be crucial steps toward achieving safe and effective pest management solutions (Kumari et al. 2023).

Table 2. Toxicity and environmental impact of nanocarriers in pesticide delivery

Aspects	Potential Toxicity	Environmental Impact	References
Nanoparticles Accumulation	Some nanoparticles may accumulate in organisms, leading to cytotoxicity and oxidative stress.	Persistent nanoparticles can accumulate in soil and water, affecting microbial communities.	Xuan et al. 2023.
Ecotoxicity	Nanoparticles can be toxic to non-target organisms such as pollinators, fish, and beneficial insects.	Uncontrolled release may disrupt food chains and biodiversity.	Okagu et al. 2023.
Soil Health	High nanoparticle concentrations may alter soil microbial balance, affecting nutrient cycling.	Heavy metal-based nanoparticles (e.g., CuO, ZnO) may lead to soil contamination.	Alengebawy et al. 2021.
Water Contamination	Nanocarriers in pesticides may leach into groundwater, posing risks to aquatic organisms.	Nanoparticles can alter water chemistry and bioaccumulate in aquatic ecosystems.	Ali et al. 2024.
Airborne Risks	Inhalation of nano-formulated pesticides may pose respiratory health risks to humans and animals.	Wind dispersion of nanoparticles can lead to unintended environmental exposure.	Chaud et al. 2021; Akinhanmi et al. 2024
Bioaccumulation	Some nanoparticles, especially metal-based, can accumulate in plant tissues, entering the food chain.	Long-term accumulation in crops may pose risks to human and animal health.	Hlihor et al. 2022.

continued on following page

Table 2. Continued

Aspects	Potential Toxicity	Environmental Impact	References
Degradability	Biodegradable nanocarriers (e.g., chitosan, alginate) pose minimal toxicity risks.	Non-biodegradable nanoparticles may persist in the environment for years.	Mikušová & Mikuš, 2021;
Regulatory Concerns	Lack of standardized toxicity assessments and long-term studies	Need for strict environmental guidelines to regulate nano-based pesticides.	Kumari et al. 2023.

1.6 Regulatory and Safety Considerations

Table 1.3 revealed the integration of nanotechnology into pesticide formulations offers enhanced efficiency and targeted delivery. However, it also raises concerns regarding safety, environmental impact, and regulatory oversight. Nanopesticides may pose risks to non-target organisms, including beneficial insects, aquatic life, and soil microbes. Some nanoparticles can generate reactive oxygen species, leading to oxidative stress and toxicity in exposed organisms. Additionally, the persistence of certain nanomaterials in the environment can result in long-term ecological consequences. Given these potential risks, it is imperative for regulatory agencies to establish stringent guidelines for the safe use of nano-formulated pesticides. Comprehensive risk assessments are necessary to evaluate their long-term effects on human health and the environment. Developing eco-friendly, biodegradable nanocarriers is crucial to minimize toxicity while harnessing the benefits of nanotechnology in agriculture. Implementing robust regulations and safety measures will ensure that the advantages of nanotechnology are realized without compromising environmental integrity or public health (Adetunji et al. 2023b).

Many countries lack specific regulations for nano-formulated pesticides. Traditional pesticide regulations may not adequately address the unique properties of nanomaterials, such as their high reactivity and bioavailability. Regulatory agencies, including the Environmental Protection Agency (EPA) and the European Food Safety Authority (EFSA), are working toward developing specific frameworks for nano-based agricultural inputs. The absence of standardized toxicity studies and environmental impact assessments makes it difficult to evaluate the risks associated with nanocarriers. Long-term studies on bioaccumulation, ecotoxicity, and human health impacts are crucial. Regulatory bodies must implement standardized protocols for assessing nanocarrier-based pesticides, ensuring their safe application (Ali et al. 2024).

Workers handling nano-formulated pesticides may face risks of inhalation or dermal exposure. Proper protective measures, such as personal protective equipment (PPE) and safe handling protocols, are essential. Additionally, environmental monitoring programs should be established to track nanoparticle residues in soil, water, and crops to prevent contamination. There is currently a lack of clear labeling requirements for nano-based pesticides. Transparent labeling and risk-benefit communication can enhance public trust and ensure informed usage. Regulatory agencies should enforce mandatory labeling of products containing nanomaterials to improve transparency (Kapeleka et al. 2024).

Table 3. Regulatory and safety considerations for nanocarriers in pesticide delivery

Aspects	Regulatory Considerations	Safety Measures	References
Approval and Guidelines	Many countries lack specific regulations for nano-pesticides.	Establish clear safety guidelines for nano-based pesticides.	Kumari et al. 2023
Risk Assesment	Limited long-term toxicity studies hinder risk evaluation.	Conduct extensive ecotoxicological and human health studies.	Boretti 2024.
Standardization	No universal testing standards for nanocarrier-based pesticides.	Develop international protocols for nano-pesticide evaluation.	Kumari et al. 2023.
Environmental Monitoring	Challenges in tracking nano-pesticides in soil and water.	Implement regular monitoring programs for nano-pesticide residues.	Iavicoli et al. 2017; Inobeme et al. 2024
Worker and Consumer Safety	Potential inhalation and dermal exposure risks.	Use protective gear and proper handling procedures for nano-pesticides.	Lari et al. 2022.
Labeling and Transparency	Lack of clear labeling requirements for nano-formulated pesticides.	Mandate proper labeling to inform users of nano-ingredients.	Kumari et al. 2023.
Degradability and Disposal	Some nanocarriers persist in the environment, raising pollution concerns.	Promote the use of biodegradable and eco-friendly nanocarriers	Roy et al. 2021.
Public Perception and Acceptance	Concerns over unknown risks may limit adoption of nano-pesticides.	Increase public awareness through transparent risk-benefit communication	Lavicoli et al. 2017; Mathew et al. 2023b

1.7 Recent Advances and Future Prospects

Nanotechnology has significantly transformed the agricultural sector by enhancing the efficiency and precision of pesticide delivery. Recent advances in nanocarrier systems have improved pesticide stability, controlled release, and targeted application, reducing environmental contamination and improving crop protection. Despite these achievements, challenges such as toxicity concerns, regulatory gaps, and cost implications remain. This section explores the latest innovations in nanocarrier-based pesticide delivery and the future prospects of this technology (Ali et al. 2024).

One of the most promising innovations in nano-pesticides is the development of stimuli-responsive nanocarriers that release pesticides only when triggered by specific environmental conditions, such as pH changes, temperature fluctuations, or enzyme activity. For example, pH-sensitive nanocarriers can be designed to release pesticides in response to the alkaline conditions found in insect guts, ensuring selective toxicity. These smart carriers enhance efficiency while minimizing pesticide wastage and environmental pollution. Recent studies have demonstrated the potential of such systems in improving pesticide delivery and reducing environmental impact (Chaud et al. 2021; Olorunsola et al. 2024).

To address concerns over nanoparticle persistence in the environment, researchers are developing biodegradable nanocarriers made from natural polymers such as chitosan, starch, cellulose, and alginate. These carriers degrade into non-toxic byproducts after delivering the pesticide, reducing the risk of soil and water contamination. Additionally, lipid-based carriers, such as liposomes and solid lipid nanoparticles, are gaining popularity due to their biocompatibility and controlled release properties (Sampathkumar et al. 2020).

Nanoemulsions, composed of nano-sized oil droplets encapsulating active pesticides, have demonstrated significant improvements in pesticide solubility, bioavailability, and penetration into plant tissues. These formulations enhance the absorption of pesticides by insects and pathogens, thereby reducing the required dosage and minimizing environmental residues. Recent studies have highlighted the effectiveness of nanoemulsions in delivering natural pesticides, such as neem oil and essential oils, positioning them as viable alternatives to synthetic chemicals. The small droplet size of nanoemulsions, typically ranging from 20 to 200 nm, contributes to their high physical stability and efficient delivery of bioactive components. For instance, nanoemulsions incorporating essential oils have been shown to possess potential applications in treating pathogenic microbial contamination in food, indicating their broad-spectrum antimicrobial properties. Additionally, the use of nanoemulsions in botanical pesticide formulations has gained interest due to their ability to improve the delivery and efficacy of plant-derived antimicrobials. By leveraging nanoemulsion technology, it is possible to enhance the performance of natural pesticides, offering a sustainable and environmentally friendly approach to pest management in agriculture. (Kapeleka et al. 2024).

Recent advancements have led to the development of hybrid nanocarriers that combine different nanomaterials to achieve multifunctionality. For instance, silica-based nanocarriers modified with polymer coatings provide enhanced stability and sustained pesticide release. Similarly, nanocomposites that integrate metal nanoparticles with organic materials offer improved antimicrobial and pesticidal effects. These hybrid systems optimize pesticide efficiency while reducing application frequency (Shao et al. 2022; Oyewole et al. 2024).

The integration of nanocarriers with nano-sensors has enabled real-time monitoring of pesticide levels, soil conditions, and pest activity. Smart pesticide formulations equipped with nanosensors can detect pest infestations and release pesticides accordingly, ensuring precision-targeted applications. This approach, known as “precision agriculture,” minimizes pesticide overuse, reduces costs, and enhances environmental sustainability (Chaud et al. 2021).

Despite significant advancements, the commercial adoption of nano-formulated pesticides is limited by high production costs and complex manufacturing processes. Future research should focus on cost-effective production methods, such as green synthesis approaches, to make nanocarrier-based pesticides more affordable and accessible for large-scale agricultural use (Iavicoli et al. 2017; Utazi et al. 2024).

As nanotechnology-based pesticides become more prevalent, standardized regulatory frameworks are essential to ensure their safe application. Future efforts should focus on developing comprehensive guidelines for nanoparticle toxicity testing, risk assessment, and environmental impact studies. Regulatory agencies must collaborate with researchers to establish clear safety standards and public policies for nano-based pesticides. The next generation of nanocarriers will prioritize sustainability, biocompatibility, and environmental safety. Researchers are exploring bio-based nanoparticles, such as protein-based and polysaccharide-based carriers, that degrade naturally without leaving harmful residues. Additionally, advancements in AI and machine learning may further optimize nano-pesticide formulations, enabling smarter, data-driven pesticide applications (Kumari et al. 2023).

Future research may integrate CRISPR gene-editing technology with nanocarriers to enhance crop resistance to pests and diseases. Gene-edited plants, combined with controlled pesticide release from nanocarriers, could offer a revolutionary approach to pest management with minimal environmental impact. Widespread adoption of nano-pesticides depends on public awareness and acceptance. Transparency in safety studies, clear labeling, and educational campaigns on the benefits and risks of nanotechnology in agriculture will be crucial in gaining consumer trust (Komal et al. 2023).

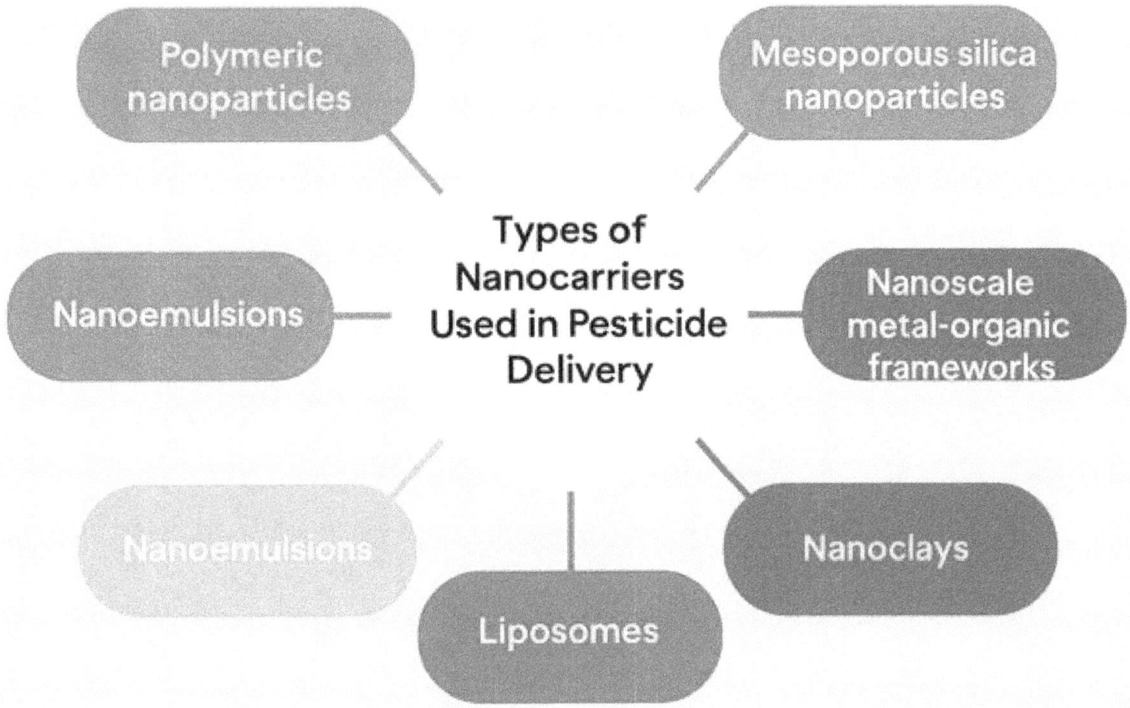

Nanocarrier-based pesticide delivery is a groundbreaking innovation that offers precise, efficient, and environmentally friendly pest control solutions. Recent advances in smart nanocarriers, biodegradable formulations, and nano-sensor integration have improved pesticide application, reducing chemical runoff and enhancing agricultural sustainability. However, challenges related to cost, safety, and regulations must be addressed to fully realize the potential of nano-pesticides. With continued research and innovation, nanotechnology is set to revolutionize pesticide delivery and contribute to a more sustainable future for global agriculture (Ali et al. 2024).

1.8 CONCLUSION

The application of nanocarriers in pesticide delivery represents a significant advancement in modern agriculture, offering enhanced efficacy, controlled release, and reduced environmental impact. Through the use of various nanocarrier systems such as polymeric nanoparticles, liposomes, nano-emulsions, and mesoporous silica nanoparticles, pesticide formulations have become more stable, bioavailable, and targeted. These innovations not only improve pest control efficiency but also minimize chemical runoff, thereby protecting soil and water ecosystems. Recent advancements, including stimuli-responsive nanocarriers, biodegradable materials, hybrid nano-formulations, and precision agriculture techniques, have further optimized pesticide delivery. These developments address key agricultural challenges such as pest resistance, environmental contamination, and the overuse of synthetic chemicals. Additionally, integrating nanosensors with pesticide formulations has introduced smart delivery systems that enhance monitoring and precision application

REFERENCES

Adepu, S., & Ramakrishna, S. (2021). Controlled Drug Delivery Systems: Current Status and Future Directions. *Molecules (Basel, Switzerland)*, *26*(19), 5905. DOI: 10.3390/molecules26195905 PMID: 34641447

Adetunji, C. O., Ogundolie F. A., Ajiboye, M. D., Mathew, J. T., Inobeme, A., Dauda, W. P., Ghazanfar, S., Olotu T., Olaniyan, O., Ijabadeniyi, O. A., Ajayi, O. O., and Adetunji, J. B. (2022). Nano-engineered Sensors for Food Processing. In a book: Bio- and Nano-sensing Technologies for Food Processing and Packaging, The Royal Society of Chemistry, 151-166. DOI: 10.1039/9781839167966-00151

Adetunji, C. O., Olotu, T., Mathew, J. T., Inobeme, A., Olaniyan, O. T., Adetunji, M. C., & Adebayo, O. O. (2024). Morphology, rheology, properties, and applications of nanostarch-filled polymer blends, Editor(s): Sabu Thomas, Soney C. George, Sharika T. Nair, In Micro and Nano Technologies, Nanofillers for Binary Polymer Blends, Elsevier, 443-464, DOI: 10.1016/B978-0-323-88655-0.00005-7

Akinhanmi, F. O., Aworunse, O. S., Oyewole, O. A., Adetunji, C. O., Mathew, J. T., Igiku, V., Eniola, K. I. T., & Yerima, M. B. (2024). Non-Target Effect, Environmental Impact, and Assessment of Nanobiofertilizer. In Book: Handbook of Agricultural Biotechnology. Wiley, 5, 253-270. Doi: DOI: 10.1002/9781394211548.ch10

Ali, S., Ahmad, N., Dar, M. A., Manan, S., Rani, A., Alghanem, S. M. S., Khan, K. A., Sethupathy, S., Elboughdiri, N., Mostafa, Y. S., Alamri, S. A., Hashem, M., Shahid, M., & Zhu, D. (2023). Nano-Agrochemicals as Substitutes for Pesticides: Prospects and Risks. *Plants*, *13*(1), 109. DOI: 10.3390/plants13010109 PMID: 38202417

Altammar, K. A. (2023). A review on nanoparticles: Characteristics, synthesis, applications, and challenges. *Frontiers in Microbiology*, *14*, 1155622. DOI: 10.3389/fmicb.2023.1155622 PMID: 37180257

Ashfaq, R., Rasul, A., Asghar, S., Kovács, A., Berkó, S., & Budai-Szűcs, M. (2023). Lipid Nanoparticles: An Effective Tool to Improve the Bioavailability of Nutraceuticals. *International Journal of Molecular Sciences*, *24*(21), 15764. DOI: 10.3390/ijms242115764 PMID: 37958750

Azeh, Y., Musah, M., Mathew, J. T., Badeggi, U. M., Estuyankpa, M. B., Muhammad, Z. D., Oluwaseyi, D. S., Umar, M. T., Abdulkarim, M. A., Muhammad, A. I., Ibrahim, A. A., Ismail, H., Muhammad, K. T., Umar, G., & Ahmed, T. (2023). In Situ Synthesis of Polyaniline Nanohybrid and Formulation of Polyaniline/Carboxymethyl cellulose/Ethylene Glycol Nanocomposite: Study of its Conducting and Antibacterial Properties. *Journal of Materials and Polymer Science*, *3*(4), 1–9. DOI: 10.47485/2832-9384.1041

Begines, B., Ortiz, T., Pérez-Aranda, M., Martínez, G., Merinero, M., Argüelles-Arias, F., & Alcudia, A. (2020). Polymeric Nanoparticles for Drug Delivery: Recent Developments and Future Prospects. *Nanomaterials (Basel, Switzerland)*, *10*(7), 1403. DOI: 10.3390/nano10071403 PMID: 32707641

Boretti, A. (2024). Comprehensive risk-benefit assessment of chemicals: A case study on glyphosate. *Toxicology Reports*, *13*, 101803. DOI: 10.1016/j.toxrep.2024.101803 PMID: 39606775

Camara, M. C., Campos, E. V. R., Monteiro, R. A., do Espirito Santo Pereira, A., de Freitas Proença, P. L., & Fraceto, L. F. (2019). Development of stimuli-responsive nano-based pesticides: Emerging opportunities for agriculture. *Journal of Nanobiotechnology*, *17*(1), 100. DOI: 10.1186/s12951-019-0533-8 PMID: 31542052

Chaud, M., Souto, E. B., Zielinska, A., Severino, P., Batain, F., Oliveira-Junior, J., & Alves, T. (2021). Nanopesticides in Agriculture: Benefits and Challenge in Agricultural Productivity, Toxicological Risks to Human Health and Environment. *Toxics*, *9*(6), 131. DOI: 10.3390/toxics9060131 PMID: 34199739

Etsuyankpa, M. B., Augustine, A. U., Musa, S. T., Mathew, J. T., Ismail, H., Salihu, A. M., & Mamman, A. (2024). An Overview of Wastewater Characteristics, Treatment and Disposal: A Review. *Journal of Applied Science & Environmental Management*, *28*(5), 1553–1572. DOI: 10.4314/jasem.v28i5.28

Fincheira, P., Hoffmann, N., Tortella, G., Ruiz, A., Cornejo, P., Diez, M. C., Seabra, A. B., Benavides-Mendoza, A., & Rubilar, O. (2023). Eco-Efficient Systems Based on Nanocarriers for the Controlled Release of Fertilizers and Pesticides: Toward Smart Agriculture. *Nanomaterials (Basel, Switzerland)*, *13*(13), 1978. DOI: 10.3390/nano13131978 PMID: 37446494

Ghasempour, A., Dehghan, H., Ataee, M., Chen, B., Zhao, Z., Sedighi, M., Guo, X., & Shahbazi, M. A. (2023). Cadmium Sulfide Nanoparticles: Preparation, Characterization, and Biomedical Applications. *Molecules (Basel, Switzerland)*, *28*(9), 3857. DOI: 10.3390/molecules28093857 PMID: 37175267

Hlihor, R. M., Ro ca, M., Hagiu-Zaleschi, L., Simion, I. M., Daraban, G. M., & Stoleru, V. (2022). Medicinal Plant Growth in Heavy Metals Contaminated Soils: Responses to Metal Stress and Induced Risks to Human Health. *Toxics*, *10*(9), 499. DOI: 10.3390/toxics10090499 PMID: 36136464

Hu, S., Yan, C., Fei, Q., Zhang, B., & Wu, W. (2023). MOF-based stimuli-responsive controlled release nanopesticide: Mini review. *Frontiers in Chemistry*, *11*, 1272725. DOI: 10.3389/fchem.2023.1272725 PMID: 37767340

Iavicoli, I., Leso, V., Beezhold, D. H., & Shvedova, A. A. (2017). Nanotechnology in agriculture: Opportunities, toxicological implications, and occupational risks. *Toxicology and Applied Pharmacology*, *329*, 96–111. DOI: 10.1016/j.taap.2017.05.025 PMID: 28554660

Inobeme, A., Adetunji, C. O., Maliki, M., Onyeachu, B. I., Kelani, T., Eziukwu, C. A., Olori, E., Mathew, J. T., and Bamigboye, M. O. (2023). Strategies to synthesize, advantages, and disadvantages of pharmaceutical nanoparticles. In book: Nanotechnology for Drug Delivery and Pharmaceuticals. Academic Press is an imprint of Elsevier, 371-402, .DOI: 10.1016/B978-0-323-95325-2.00006-7

Inobeme, A., Adetunji, C. O., Mathew, J. T., & Akram, M. (2025). *Extraction, properties, and modification of chitosan-based nanoparticles. Chitosan-Based Nanoparticles for Biomedical Applications*. Woodhead Publishing., DOI: 10.1016/B978-0-443-13997-0.00001-1

Jampilek, J., & Kralova, K. (2021). Advances in Drug Delivery Nanosystems Using Graphene-Based Materials and Carbon Nanotubes. *Materials (Basel)*, *14*(5), 1059. DOI: 10.3390/ma14051059 PMID: 33668271

Kapeleka, J. A., & Mwema, M. F. (2024). State of nano pesticides application in smallholder agriculture production systems: Human and environmental exposure risk perspectives. *Heliyon, 10*(20), e39225. DOI: 10.1016/j.heliyon.2024.e39225 PMID: 39492887

Kapeleka, J. A., & Mwema, M. F. (2024). State of nano pesticides application in smallholder agriculture production systems: Human and environmental exposure risk perspectives. *Heliyon, 10*(20), e39225. DOI: 10.1016/j.heliyon.2024.e39225 PMID: 39492887

Karimi, M., Eslami, M., Sahandi-Zangabad, P., Mirab, F., Farajisafiloo, N., Shafaei, Z., Ghosh, D., Bozorgomid, M., Dashkhaneh, F., & Hamblin, M. R. (2016). pH-Sensitive stimulus-responsive nanocarriers for targeted delivery of therapeutic agents. *Wiley Interdisciplinary Reviews. Nanomedicine and Nanobiotechnology, 8*(5), 696–716. DOI: 10.1002/wnan.1389 PMID: 26762467

Komal, J., Desai, H. R., Samal, I., Mastinu, A., Patel, R. D., Kumar, P. V. D., Majhi, P. K., Mahanta, D. K., & Bhoi, T. K. (2023). Unveiling the Genetic Symphony: Harnessing CRISPR-Cas Genome Editing for Effective Insect Pest Management. *Plants, 12*(23), 3961. DOI: 10.3390/plants12233961 PMID: 38068598

Kumari, R., Suman, K., Karmakar, S., Mishra, V., Lakra, S. G., Saurav, G. K., & Mahto, B. K. (2023). Regulation and safety measures for nanotechnology-based agri-products. *Frontiers in Genome Editing, 5*, 1200987. DOI: 10.3389/fgeed.2023.1200987 PMID: 37415849

Lari, S., Jonnalagadda, P. R., Yamagani, P., Medithi, S., Vanka, J., Pandiyan, A., Naidu, M., & Jee, B. (2022). Assessment of dermal exposure to pesticides among farmers using dosimeter and hand washing methods. *Frontiers in Public Health, 10*, 957774. DOI: 10.3389/fpubh.2022.957774 PMID: 36091511

Lu, H., Zhang, S., Wang, J., & Chen, Q. (2021). A Review on Polymer and Lipid-Based Nanocarriers and Its Application to Nano-Pharmaceutical and Food-Based Systems. *Frontiers in Nutrition, 8*, 783831. DOI: 10.3389/fnut.2021.783831 PMID: 34926557

Manawi, Y. M., Ihsanullah, , Samara, A., Al-Ansari, T., & Atieh, M. A. (2018). A Review of Carbon Nanomaterials' Synthesis via the Chemical Vapor Deposition (CVD) Method. *Materials (Basel), 11*(5), 822. DOI: 10.3390/ma11050822 PMID: 29772760

Mansuriya, B. D., & Altintas, Z. (2021). Carbon Dots: Classification, Properties, Synthesis, Characterization, and Applications in Health Care-An Updated Review (2018-2021). *Nanomaterials (Basel, Switzerland), 11*(10), 2525. DOI: 10.3390/nano11102525 PMID: 34684966

Mathew, J. T., Inobeme, A., Adetunji, C. O., & Oyewole, O. A. (2025). *Recent applications of chitosan-based nanoformulated metal and metal oxide nanoparticles. Chitosan-Based Nanoparticles for Biomedical Applications*. Elsevier., DOI: 10.1016/B978-0-443-13997-0.00029-1

Mathew, J. T., Inobeme, A., Azeh, Y., Musah, M., Abdulkadir, A., Shaba, E. Y., Etsuyankpa, M. B., Musa, T. S., Muhammad, A. I., Ismail, H., Kanwa, A. M., Mamman, A., & Hussaini, J. (2024). c. Carbon Capture and Storage Via Electrochemical and Bioelectrochemical Techniques: A Review. *Caliphate Journal of Science & Technology, 2*(2), 141–158. DOI: 10.4314/cajost.v6i2.3

Mathew, J. T., Inobeme, A., Musah, M., Azeh, Y., Abdullahi, A., Shaba, E. Y., Salihu, A. M., Muhammad, E. B., Josiah, J. G., Jibrin, N. A., Ismail, H., Muhammad, A. I., Maurice, J., Mamman, A., & Ndamitso, M. M. (2024). a. A Critical Review of Green Approach on Wastewater Treatment Strategies. *Journal of Applied Science & Environmental Management, 28*(2), 363–391. DOI: 10.4314/jasem.v28i2.9

Mathew, J. T., Musah, M., Azeh, Y., & Muhammed, M. (2023). a. Adsorptive Removal of Selected Toxic Metals from Pharmaceutical Wastewater using Fe_3O_4/ZnO Nanocomposite. *Dutse Journal of Pure and Applied Sciences, 9*(4a), 236–248. DOI: 10.4314/dujopas.v9i4a.22

Mathew, J. T., Musah, M., Azeh, Y., & Muhammed, M. (2023). b. Kinetic Study of Heavy Metals Removal from Pharmaceutical Wastewater Using Geopolymer/Fe_3O_4 Nanocomposite. *Bima Journal of Science and Technology, 7*(4), 152–163. DOI: 10.56892/bima.v7i4.539

Mathew, J. T., Musah, M., Azeh, Y., & Muhammed, M. (2024). b. Development of Fe3O4 Nanoparticles for the Removal of Some Toxic Metals from Pharmaceutical Wastewater. [CaJoST]. *Caliphate Journal of Science & Technology, 6*(1), 26–34. DOI: 10.4314/cajost.v6i1.4

Mgadi, K., Ndaba, B., Roopnarain, A., Rama, H., & Adeleke, R. (2024). Nanoparticle applications in agriculture: Overview and response of plant-associated microorganisms. *Frontiers in Microbiology, 15*, 1354440. DOI: 10.3389/fmicb.2024.1354440 PMID: 38511012

Mikušová, V., & Mikuš, P. (2021). Advances in Chitosan-Based Nanoparticles for Drug Delivery. *International Journal of Molecular Sciences, 22*(17), 9652. DOI: 10.3390/ijms22179652 PMID: 34502560

Miteu, G. D., Emmanuel, A. A., Addeh, I., Ojeokun, O., Olayinka, T., Godwin, J. S., Adeyemo, O. I., & Benneth, E. O. (2023). Nanoscience and technology as a pivot for sustainable agriculture and its One Health approach awareness. *Science in One Health, 2*, 100020. DOI: 10.1016/j.soh.2023.100020 PMID: 39077037

Musah, M., Mohammed, M., Mathew, J. T., & Umar, M. T. (2023). Evaluation of Seed Extracts of Azadirachta indica. Juss (Meliaceae) as Pesticide for the Treatment of Stored Bean Seeds. [DUJOPAS]. *Dutse Journal of Pure and Applied Sciences, 9*(2b), 98–108. DOI: 10.4314/dujopas.v9i2b.11

Okagu, I. U., Okeke, E. S., Ezeorba, W. C. F., Ndefo, J. C., & Ezeorba, T. P. C. (2023). Overhauling the ecotoxicological impact of synthetic pesticides using plants' natural products: A focus on Zanthoxylum metabolites. *Environmental Science and Pollution Research International, 30*(26), 67997–68021. DOI: 10.1007/s11356-023-27258-w PMID: 37148518

Olorunsola, A., Adetunji, C. O., Olaniyan, O., Adetunji, J. B., Akinbo, K., Adetuyi, B. O., Inobeme, A., Mathew, J. T., & Ajenifujah-Solebo, S. O. (2024). Application of Nanotechnology for the Production of Biopesticides, Bioinsecticides, Bioherbicides, Mosquitoe Repellants and Biofungicides. In book: Handbook of Agricultural Biotechnology, John Wiley & Sons, 293-309. Doi: DOI: 10.1002/9781119836483.ch11

Oyewole, O. A., Chimbekujwo, K. I., Oniha, M., Omoregie, I. P., Ayanda, O. I., Adetunji, C. O., & Mathew, J. T. (2024). Application of Nanobiofertilization for Bioremediation and Ecorestoration of Polluted Soil/Farmland. In Book: Handbook of Agricultural Biotechnology. Wiley, 5, 1-36. Doi:DOI: 10.1002/9781394211548.ch1

Pashirova, T., Salah-Tazdaït, R., Tazdaït, D., & Masson, P. (2024). Applications of Microbial Organophosphate-Degrading Enzymes to Detoxification of Organophosphorous Compounds for Medical Countermeasures against Poisoning and Environmental Remediation. *International Journal of Molecular Sciences*, *25*(14), 7822. DOI: 10.3390/ijms25147822 PMID: 39063063

Pathak, V. M., Verma, V. K., Rawat, B. S., Kaur, B., Babu, N., Sharma, A., Dewali, S., Yadav, M., Kumari, R., Singh, S., Mohapatra, A., Pandey, V., Rana, N., & Cunill, J. M. (2022). Current status of pesticide effects on environment, human health and it's eco-friendly management as bioremediation: A comprehensive review. *Frontiers in Microbiology*, *13*, 962619. DOI: 10.3389/fmicb.2022.962619 PMID: 36060785

Pulingam, T., Foroozandeh, P., Chuah, J. A., & Sudesh, K. (2022). Exploring Various Techniques for the Chemical and Biological Synthesis of Polymeric Nanoparticles. *Nanomaterials (Basel, Switzerland)*, *12*(3), 576. DOI: 10.3390/nano12030576 PMID: 35159921

Roy, A., Sharma, A., Yadav, S., Jule, L. T., & Krishnaraj, R. (2021). Nanomaterials for Remediation of Environmental Pollutants. *Bioinorganic Chemistry and Applications*, *1764647*, 1–16. Advance online publication. DOI: 10.1155/2021/1764647 PMID: 34992641

Sampathkumar, K., Tan, K. X., & Loo, S. C. J. (2020). Developing Nano-Delivery Systems for Agriculture and Food Applications with Nature-Derived Polymers. *iScience*, *23*(5), 101055. DOI: 10.1016/j.isci.2020.101055 PMID: 32339991

Satchanska, G., Davidova, S., & Petrov, P. D. (2024). Natural and Synthetic Polymers for Biomedical and Environmental Applications. *Polymers*, *16*(8), 1159. DOI: 10.3390/polym16081159 PMID: 38675078

Shao, C., Zhao, H., & Wang, P. (2022). Recent development in functional nanomaterials for sustainable and smart agricultural chemical technologies. *Nano Convergence*, *9*(1), 11. DOI: 10.1186/s40580-022-00302-0 PMID: 35235069

Shen, M., Liu, S., Jiang, C., Zhang, T., & Chen, W. (2023). Recent advances in stimuli-response mechanisms of nano-enabled controlled-release fertilizers and pesticides. *Eco-Environment & Health*, *2*(3), 161–175. DOI: 10.1016/j.eehl.2023.07.005 PMID: 38074996

Subramani, T., & Ganapathyswamy, H. (2020). An overview of liposomal nano-encapsulation techniques and its applications in food and nutraceutical. *Journal of Food Science and Technology*, *57*(10), 3545–3555. DOI: 10.1007/s13197-020-04360-2 PMID: 32903987

Tudi, M., Daniel Ruan, H., Wang, L., Lyu, J., Sadler, R., Connell, D., Chu, C., & Phung, D. T. (2021). Agriculture Development, Pesticide Application and Its Impact on the Environment. *International Journal of Environmental Research and Public Health*, *18*(3), 1112. DOI: 10.3390/ijerph18031112 PMID: 33513796

Utazi, E. B., Oyewole, O. A., Yakubu, J. G., Tsado, P. Y., Omoregie, I. P., Adetunji, C. O., Mathew, J. T., Igiku, V., Eniola, K. I. T., & Yerima, M. B. (2024). Toxicological Effects of Nanobiofertilizer on Water Body, Water Quality, Lower Plants, Zooplanktons, and Beneficial Microorganisms. In Book: Handbook of Agricultural Biotechnology. Wiley, 5, 201-232. Doi: DOI: 10.1002/9781394211548.ch8

Wang, D., Saleh, N. B., Byro, A., Zepp, R., Sahle-Demessie, E., Luxton, T. P., Ho, K. T., Burgess, R. M., Flury, M., White, J. C., & Su, C. (2022). Nano-enabled pesticides for sustainable agriculture and global food security. *Nature Nanotechnology*, *17*(4), 347–360. DOI: 10.1038/s41565-022-01082-8 PMID: 35332293

Xuan, L., Ju, Z., Skonieczna, M., Zhou, P. K., & Huang, R. (2023). Nanoparticles-induced potential toxicity on human health: Applications, toxicity mechanisms, and evaluation models. *MedComm*, *4*(4), e327. DOI: 10.1002/mco2.327 PMID: 37457660

Yin, J., Su, X., Yan, S., & Shen, J. (2023). Multifunctional Nanoparticles and Nanopesticides in Agricultural Application. *Nanomaterials (Basel, Switzerland)*, *13*(7), 1255. DOI: 10.3390/nano13071255 PMID: 37049348

Chapter 11
Nanocarrier-Enabled Agricultural Biotechnology and Crop Breeding:
Advances and Prospects

Shruti Awasthi
https://orcid.org/0000-0002-7648-1387
Garden City University, India

Chethana V. Chalapathy
Garden City University, India

Lahar Khandelwal
https://orcid.org/0009-0006-8218-3382
Garden City University, India

Preethi Rajesh
https://orcid.org/0009-0001-8077-2822
Garden City University, India

ABSTRACT

Nanocarrier-enabled agricultural biotechnology indicates an innovative approach to solving fundamental difficulties in modern agriculture, ranging from crop yield improvement to insect resistance and environmental stress reduction. Nanocarriers are nanosized delivery devices capable of precisely delivering agrochemicals, nutrients, or genetic material to plant cells. This method brings up new possibilities for crop breeding, genetic engineering, and pest management by providing unparalleled molecular control over biological processes. Nanocarriers are getting highly used in agriculture for gene delivery and pesticide applications. They are capable of transporting fertilizers, insecticides, and herbicides efficiently, decreasing environmental contamination and the need for repeated treatments. They have the potential to improve agricultural traits such as drought resistance and pest tolerance through gene transfer.

DOI: 10.4018/979-8-3693-9869-2.ch011

INTRODUCTION TO POLYMERIC NANOCARRIERS IN PLANT SCIENCE

Polymeric nanocarriers, made from biodegradable and biocompatible polymers, are gaining interest due to their potential to transform plant biotechnology and agriculture. These sub-microscopic particles can encapsulate and transport bioactive substances, reducing environmental impact and enhancing efficacy. This makes them an attractive option for enhancing crop yields and plant health in an era of sustainable agriculture.

Polymeric nanocarriers offer a sustainable and environmentally friendly approach to agriculture by encapsulating and releasing active ingredients gradually. This method reduces the need for agrochemicals, ensuring maximum effectiveness and minimal waste. Traditional methods, such as pesticides, herbicides, and fertilisers, are inefficient due to their quick degradation and non-specific dispersion.

By using polymeric nanocarriers, these chemicals can be encapsulated and released more precisely, minimizing toxicity, reducing dosage, and increasing efficacy (Mapossa, A., et al., 2023).

Polymeric nanocarriers are versatile, allowing them to carry genetic material and chemicals, making them useful for biotechnology and plant gene therapy. They can be used in agriculture for pest control and crop yield enhancement, reducing environmental impact by minimizing chemical runoff and contamination.

These nanocarriers can improve plant resilience to pests and diseases, maximize nutrient uptake, and reduce the need for broad-spectrum pesticides and fertilizers. They can also enhance crop yield and quality, leading to more sustainable agricultural practices.

SYNTHESIS METHODS OF POLYMERIC NANOCARRIERS

Several methods, such as chemical, physical, and biological ones, can be used to create polymeric nanocarriers. Each approach has its own benefits and drawbacks, enabling researchers to tailor its characteristics for particular agricultural uses.

Physical techniques control the size and form of particles, whereas chemical techniques use polymerisation reactions. Biological methods create nanocarriers with certain characteristics by using biomolecules or living things (Cao, Y., et al., 2023).

Figure 1. Plant-based polymeric nanocarriers

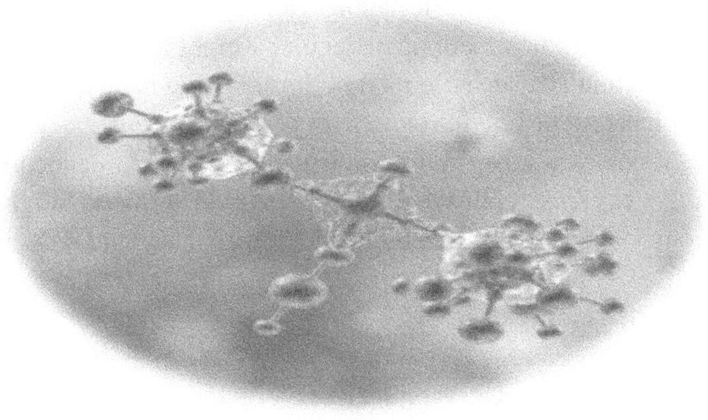

Chemical Synthesis Techniques

Chemical synthesis techniques are popular for creating polymeric nanocarriers, but they can be complex and require precise control. The toxicity of chemicals is crucial for agricultural product safety. Sustainability is essential, as some processes may generate harmful byproducts or waste.

Researchers should prioritize sustainability in chemical synthesis to minimize environmental impact and promote long-term viability (Cao, Y., et al., 2023). The following are the two main methods for chemical synthesis:

Emulsion Polymerization

In this process, monomers are polymerised in an emulsion system, which usually consists of an organic phase, surfactants, and water. Surfactant-formed micelles undergo polymerisation, which results in the creation of nanocarriers with precise size and shape.

Emulsion polymerisation is a cost-effective and customizable technique for creating nanocarriers with desired properties. It allows for the incorporation of functional groups and additives, enhancing their versatility and potential applications.

This process can be easily scaled up for industrial production, making it suitable for large-scale manufacturing. For example, targeting ligands can be used in cancer therapy (Singh, A., et al., 2020).

Polycondensation

Polycondensation reactions create polymeric nanocarriers by condensing two monomers and releasing a by-product. This allows precise adjustment of molecular weight and structure, enabling drug release characteristics. Polycondensation is a versatile method for tailoring nanocarrier properties to meet specific therapeutic needs, enhancing targeted drug delivery and disease efficacy.

Polycondensation reactions can attach tumour-targeting ligands to nanocarriers, enabling precise chemotherapy drug delivery to cancer cells while minimizing side effects (Singh, A., et al., 2020). This approach could revolutionize cancer treatment by improving efficacy and reducing toxicity.

Physical Synthesis Approaches

Physical methods of nanocarrier synthesis typically rely on mechanical forces and external stimuli to produce nanocarriers. These methods are advantageous for producing polymeric nanocarriers without the use of potentially harmful chemicals, making them ideal for applications where biocompatibility is essential (Chen, T., et al., 2023).

Figure 2. Physical synthesis approaches

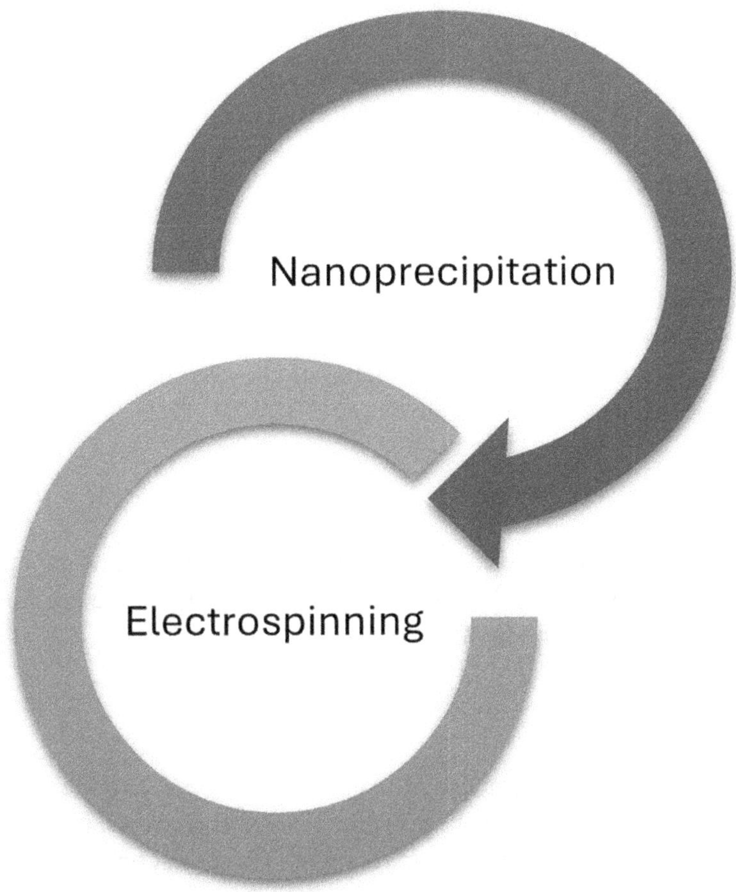

Nanoprecipitation

Nanoprecipitation is a straightforward and efficient method for preparing polymeric nanocarriers. It involves dissolving a polymer in an organic solvent and then adding this solution to a non-solvent, such as water.

The polymer precipitates into nanocarriers as the solvent diffuses into the non-solvent phase. This method is especially useful for encapsulating hydrophobic molecules and is characterized by its simplicity and rapid production time.

Electrospinning

In this technique, a polymer solution is subjected to a high voltage, causing it to form a jet that solidifies into nanofibers or nanoparticles as the solvent evaporates. Electrospinning allows for precise control over the morphology of the nanocarriers, making it suitable for creating nanofibers that can deliver nutrients or agrochemicals in a controlled manner over an extended period (Chen, T., et al., 2023).

Biological Methods for Nanocarrier Synthesis

Biological methods are gaining traction as eco-friendly alternatives for synthesizing polymeric nanocarriers. These techniques often rely on biological agents, such as microorganisms or plant extracts, to mediate the synthesis process.

Biological methods offer several advantages, including the elimination of toxic chemicals and the production of biocompatible nanocarriers (Malozyomov, B., et al., 2023).

Microbial Synthesis

Certain microorganisms, such as bacteria and fungi, can produce polymeric nanoparticles by reducing metal ions or synthesizing biopolymers. These biological agents can secrete enzymes that facilitate the polymerization of monomers, resulting in the formation of nanocarriers.

Green Synthesis

Green synthesis techniques involve the use of plant extracts to mediate the formation of polymeric nanocarriers. Plant-based compounds, such as polyphenols and flavonoids, can act as reducing agents, leading to the formation of nanoparticles in a process that is both biocompatible and environmentally benign.

CHARACTERIZATION TECHNIQUES FOR POLYMERIC NANOCARRIERS

Polymeric nanocarriers are crucial for agricultural applications due to their physical and chemical characteristics. Characterization methods like spectroscopic evaluation, zeta potential measurements, electron microscopy, and dynamic light scattering help scientists understand nanocarrier characteristics and determine the best application in agriculture (Kehayias, P., et al., 2017).

Structural Analysis

The stability and drug-release behavior of polymeric nanocarriers can be influenced by their molecular structure and chemical makeup, which can be determined through structural analysis. Techniques commonly used for structural analysis include:

Fourier Transform Infrared Spectroscopy (FTIR)

This technique is used to pinpoint particular functional groups in the nanocarriers' polymeric matrix. Researchers can identify any structural changes in the polymer and verify the existence of anticipated bonds by examining the infrared absorption spectra.

Nuclear Magnetic Resonance (NMR) Spectroscopy

This technique offers comprehensive information about the molecular makeup of the polymers that make up nanocarriers. NMR helps identify the arrangement of atoms in the polymer chain by examining the magnetic characteristics of atomic nuclei (Kehayias, P., et al., 2017).

Morphological Assessment

The way nanocarriers interact with plant cells is greatly influenced by their morphology, which includes their shape and surface properties. The following methods are used to evaluate the morphology of polymeric nanocarriers:

Techniques for examining the internal structure and surface morphology of nanocarriers include transmission electron microscopy (TEM) and scanning electron microscopy (SEM). Researchers can examine the core-shell structure of some polymeric nanocarriers and detect surface flaws thanks to SEM's high-resolution photographs of the exterior shape of the nanocarriers (Venditti, I., et al., 2017).

Figure 3. Structural analysis techniques

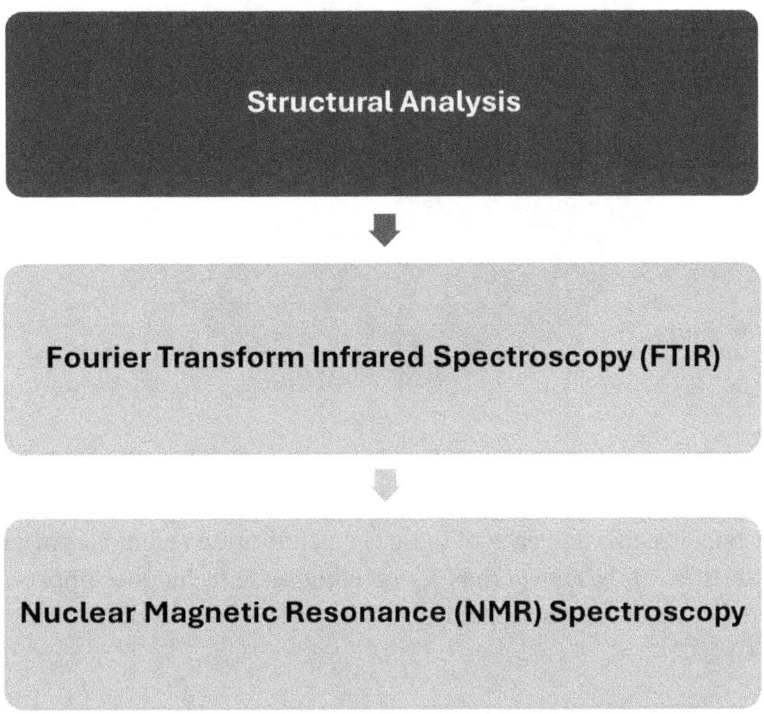

Size and Zeta Potential Measurement

Polymeric nanocarriers' stability, cellular uptake, and distribution within plant tissues are all influenced by their size and surface charge. Key techniques for measuring these parameters include:

Dynamic Light Scattering (DLS)

One popular technique for figuring out the size distribution of nanocarriers in solution is dynamic light scattering, or DLS. DLS offers a precise measurement of particle size by examining the light scattering brought on by the Brownian motion of particles.

Zeta Potential Analysis

Zeta potential quantifies the surface charge of nanocarriers, which influences how well they interact with plant cell membranes and how stable they are in suspension. Generally speaking, stable and less prone to aggregate nanocarriers have a high positive or negative zeta potential (Venditti, I., et al., 2017).

Figure 4. Key techniques for measuring size and zeta potential measurement

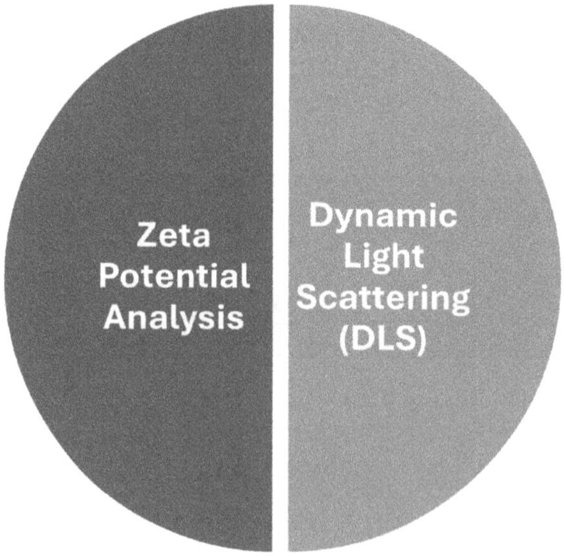

Researchers can improve the design of polymeric nanocarriers for particular agricultural applications by using the distinct insights that each of these characterisation techniques offers into the physical and chemical characteristics of these materials. Combining these techniques improves the efficacy and safety of nanocarriers for use in plant science by providing a thorough understanding of their behaviour (Venditti, I., et al., 2017).

DRUG LOADING MECHANISMS IN POLYMERIC NANOCARRIERS

Polymeric nanocarriers are crucial in plant science for medication loading, affecting their efficacy and release profile. Understanding drug loading mechanisms in nanocarriers can improve targeted distribution and regulated release of bioactive chemicals, thereby improving the stability and effectiveness of plant science applications.

Physical Encapsulation

Physical encapsulation is a widely used method for drug loading, enclosing active ingredients within polymeric nanocarriers. This technique ensures regulated release and delays premature destruction of bioactive substances, improving therapeutic efficacy in plant science applications. The hydrophobic core of amphiphilic polymeric nanocarriers can encapsulate hydrophobic materials, such as pesticides and herbicides, protecting them from sunlight and oxidation (Yang, Y., et al., 2022).

Chemical Conjugation

By joining active compounds to polymers or nanocarriers, a process known as chemical conjugation, therapeutic effects can be improved, and release can be precisely controlled. It improves stability and bioavailability, which makes it a versatile way to provide medication that has less side effects and more efficacy. In order to ensure accurate and controlled distribution, prodrug synthesis entails chemically changing the active ingredient and covalently binding it to the polymeric nanocarrier (Chavda, V. P., et al., 2023).

Surface Adsorption

Surface adsorption is one method of loading medications onto polymeric nanocarriers that permits effective release at the intended site. This technique improves drug delivery systems' therapeutic efficacy and design flexibility. Because electrostatic binding allows active molecules with opposing charges to adsorb to the surface of the nanocarrier, it is commonly used to load nutrients or genetic material. Weak bonding forces, such as hydrogen or Van der Waals bonds, can be used for rapid action and are less stable than covalent conjugation, despite enabling faster release profiles (Yang, Y., et al., 2022).

Strong chemical connections are created between the medication and nanocarrier during covalent conjugation, which improves stability and regulates release. However, in contrast to surface adsorption or electrostatic binding, this approach might restrict the flexibility of drug loading.

However, electrostatic binding is a common option for several applications because it provides a balance between stability and flexibility in drug loading. In the end, the particular needs of the drug delivery system under development should determine which approach is best.

Evaluation of Drug Loading Efficiency

DLE, or drug loading efficiency, is essential for polymeric nanocarriers to function at their best. To make sure the drug delivery system can successfully produce the intended therapeutic effect, it is crucial to thoroughly assess the DLE. Researchers can optimise the efficacy of the medication delivery system by identifying the most effective drug loading technique (Li, Ting, et al., 2022).

It calculates the percentage of active component that has been successfully loaded using fluorescence spectroscopy, HPLC, or UV-visible spectroscopy. Optimising DLE lowers waste and guarantees the best possible drug delivery. Improving patient outcomes and reducing possible side effects are two further benefits of achieving high drug loading efficiency.

Therefore, the creation of efficient drug delivery systems requires careful DLE evaluation and optimisation. Additionally, because less medication is required for each dose, greater drug loading efficiency can also result in cost savings. Researchers can guarantee continuous and dependable delivery of the active ingredient to the intended spot in the body by closely observing and modifying the drug loading procedure (Lv, Shixian et al., 2018).

Better patient compliance and more predictable therapeutic results may arise from this. All things considered, improving the effectiveness and safety of medication delivery systems depends heavily on high drug loading efficiency.

RELEASE KINETICS OF ACTIVE COMPOUNDS FROM NANOCARRIERS

The release rate of polymeric nanocarriers, which lessen waste and environmental impact, affects how useful they are in plant science (Caldeira, K., et al., 2003).

Mechanisms of Release

There are several ways that polymeric nanocarriers can release active chemicals, and these mechanisms can work alone or in concert:

Figure 5. Mechanisms of release

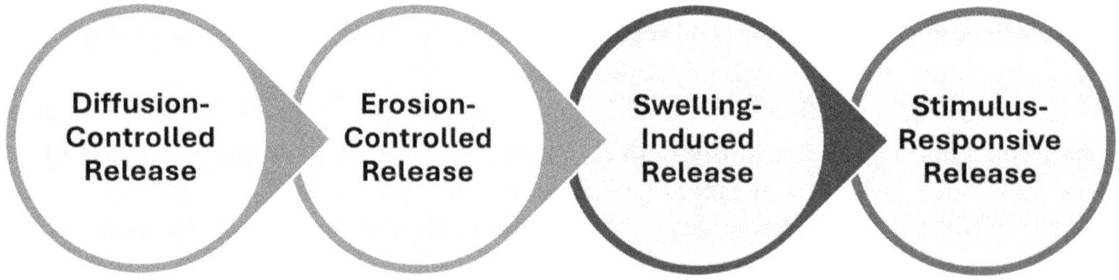

Active substances can be released from polymeric nanocarriers via a variety of methods. In diffusion-controlled release, the active component diffuses out of the nanocarrier through its matrix or polymeric shell under the influence of variables such as moisture, porosity, and size. Erosion-controlled release, which frequently uses biodegradable polymers like polylactic acid or polycaprolactone, happens as the polymer breaks down over time (Hyzinski-García, et al., 2014).

When specific polymers swell in reaction to external stimuli, active compounds are released by matrix rupture or diffusion. This phenomenon is known as swelling-induced release. Lastly, stimulus-responsive release lowers the possibility of off-target effects by enabling the active payload to be delivered in response to particular stimuli.

Factors Influencing Release Profiles

The kinetics of active drug release from polymeric nanocarriers are influenced by a number of factors, including:

The molecular weight, hydrophilicity, and crystallinity of the polymer, as well as its size and surface area, drug loading technique, and ambient circumstances, all affect the release rate of nanocarriers. Because they are exposed to the environment, smaller nanocarriers with larger surface areas exhibit faster release rates (Zhu, Tianzhen, et al., 2018).

The release profile is also influenced by the drug loading process, such as encapsulation, conjugation, or adsorption. The release behaviour of nanocarriers is also greatly influenced by environmental factors such as pH, temperature, and the presence of enzymes.

APPLICATIONS OF POLYMERIC NANOCARRIERS IN PLANT HEALTH

Polymeric nanocarriers offer innovative solutions in plant science and agriculture, improving plant health, resource utilization, and reducing environmental impact through efficient nutrient optimization, gene transfer, and targeted pesticide delivery.

Targeted Delivery of Agrochemicals

Polymeric nanocarriers are used for the efficient delivery of fungicides, herbicides, and pesticides. Traditional methods often cause environmental damage and high costs due to leaching, volatilization, and degradation. Polymeric nanocarriers offer targeted and regulated delivery methods, shielding active chemicals from environmental factors and UV light-induced degradation (Marinello, Francesco, et al., 2023).

They minimize non-target organism exposure and reduce administration frequency, ensuring steady chemical supply. Encapsulating fungicides in polymeric matrix increases their stability and solubility, making them effective against fungal infections. Additionally, nanocarriers can release fungicides in response to environmental cues like humidity or fungal enzyme presence.

Gene Transfer for Biotechnological Advancements

Polymeric nanocarriers are essential in plant genetic engineering, enabling the non-invasive transfer of genetic material like DNA, RNA, or CRISPR-Cas9. They protect genetic material from nuclease destruction and improve transgene expression and cellular absorption. These nanocarriers can also deliver small interfering RNA molecules to silence specific genes, supporting sustainable farming and reducing the need for artificial pesticides. This gene transfer offers a promising avenue for biotechnological advancements in agriculture, enhancing crop traits and increasing yield (Trivedi, Pankaj, et al., 2020).

Nutrient Delivery and Optimization

Polymeric nanocarriers are a new method of delivering nutrients to plants, allowing for delayed and targeted delivery of macronutrients and micronutrients. This method promotes environmental sustainability by reducing runoff and nutrient leaching and reduces the need for fertilizers (Thombare, Nandkishore, et al., 2021).

The use of nanocarriers also aids in micronutrient delivery, addressing deficiencies that hinder plant growth by selectively targeting specific tissues or cells. This approach reduces the amount of fertilizers needed, leading to cost savings for farmers and minimizing the risk of nutrient wastage and environmental harm. Overall, these advancements contribute to a more efficient and sustainable agricultural system.

Stress Management and Plant Immunity

Polymeric nanocarriers can deliver bioactive substances that enhance plant resistance to biotic and abiotic stressors. These substances help plants withstand harsh environmental conditions like drought and salinity, while also boosting their immune systems and resistance to pests and diseases (Trivedi, Pankaj, et al., 2020).

CHALLENGES AND LIMITATIONS OF POLYMERIC NANOCARRIERS IN AGRICULTURE

The use of polymeric nanocarriers in agriculture faces challenges due to technical, financial, environmental, and regulatory issues, which must be addressed to fully realize their potential in plant research.

Scalability and Cost of Production

Polymeric nanocarrier production faces challenges due to scalability, high material costs, and infrastructure requirements. High-quality polymers and sophisticated synthesis techniques may increase manufacturing costs, making them difficult for farmers, especially in poor countries. Additionally, many agricultural settings lack advanced machinery and controlled conditions for large-scale production (Yang, Y., et al., 2022).

Stability and Shelf-Life

Polymeric nanocarriers must remain stable during application and storage to be effective. Biodegradable polymers can degrade quickly, compromising ingredient release. Environmental factors like temperature, humidity, and ultraviolet light can affect their structural integrity, lowering their effectiveness (Mapossa, A., et al., 2023).

Environmental and Biosafety Concerns

Polymeric nanocarriers pose concerns about their potential impact on non-target creatures and the environment, as poorly made nanoparticles can accumulate in soil or water, potentially causing harm. Further ecotoxicological research is needed to assess potential hazards.

Regulatory Hurdles

The regulatory environment for polymeric nanocarriers in agriculture is evolving, causing confusion for users and developers. The lack of a recognized framework for safety evaluation and complicated approval procedures makes it challenging to obtain approvals, causing market delays (Lv, Shixian et al., 2018).

Farmer Awareness and Adoption

Farmers need education and training before adopting polymeric nanocarriers, as their advantages and uses are not widely known, especially in rural areas, and not all farmers have access to specialized tools or methods.

FUTURE DIRECTIONS AND EMERGING TRENDS IN POLYMERIC NANOCARRIER TECHNOLOGY

The field of polymeric nanocarrier technology in plant science is constantly evolving, with the next generation of agricultural applications likely to be shaped by current trends and future directions.

Development of Smart and Stimuli-Responsive Nanocarriers

Smart nanocarriers, which react to biological or environmental stimuli, offer a promising approach to accurate and regulated delivery systems. These nanocarriers can release their payload in response to acidic plant tissues or diseased areas, regulate nutrition or pesticide release under specific climatic conditions, and provide unmatched targeting and release precision (Venditti, I., et al., 2017).

Integration of Biodegradable and Eco-Friendly Materials

The sustainability trend is driving the development of nanocarriers using natural and biodegradable polymers like chitosan, cellulose, and alginate. Green synthesis methods, using non-toxic solvents and energy-efficient procedures, aim to minimize environmental impact.

Nanocarriers for Precision Agriculture

Precision agriculture utilizes data-driven methods to boost agricultural yields and optimize resource use. Polymeric nanocarriers, integrated with sensors, can release agrochemicals or nutrients, monitor soil health, and use AI algorithms and GPS-enabled spraying equipment for targeted delivery, reducing waste and increasing effectiveness.

Gene-Editing Applications

The development of CRISPR-Cas9 and other gene-editing technologies enables new applications for nanocarriers in plant biotechnology, including non-viral delivery systems and multiplex gene editing, which allows for complex agricultural trait alterations (Malozyomov, B., et al., 2023).

AI and Machine Learning in Nanocarrier Design

AI and machine learning are revolutionizing polymeric nanocarrier design and optimization, enabling predictive modeling and custom design for specific agricultural uses, thereby speeding up the development process.

Collaborative Research and Policy Development

The successful deployment of polymeric nanocarriers requires supportive policies, interdisciplinary cooperation, public-private partnerships, and uniform standards for testing, authorization, and application. These partnerships ensure affordability and innovation in international agriculture (Malozyomov, B., et al., 2023).

REFERENCES

Ahern, R. J., Hanrahan, J. P., Tobin, J. M., Ryan, K. B., & Crean, A. M. (2013). Comparison of fenofibrate-mesoporous silica drug-loading processes for enhanced drug delivery. *European Journal of Pharmaceutical Sciences*, *50*(3-4), 400–409. DOI: 10.1016/j.ejps.2013.08.026 PMID: 23981335

Caldeira, K., & Wickett, M. (2003). Oceanography: Anthropogenic carbon and ocean pH. *Nature*, *425*(6956), 365–365. DOI: 10.1038/425365a PMID: 14508477

Cao, Y., Si, J., Zheng, M., Zhou, Q., & Ge, Z. (2023). X-ray-responsive prodrugs and polymeric nanocarriers for multimodal cancer therapy. Chemical communications.

Chavda, V. P. (2023). *Advanced Phytochemical-Based Nanocarrier Systems for the Treatment of Breast Cancer* (Vol. 15). Cancers.

Chen, T., Peng, Y., Qiu, M., Yi, C., & Xu, Z. (2023). Recent advances in mixing-induced nanoprecipitation: From creating complex nanostructures to emerging applications beyond biomedicine. *Nanoscale*, *15*(8), 3594–3609. DOI: 10.1039/D3NR00280B PMID: 36727557

Hyzinski-García, M. C., Rudkouskaya, A., & Mongin, A. A. (2014). LRRC8A protein is indispensable for swelling-activated and ATP-induced release of excitatory amino acids in rat astrocytes. [n. pag.]. *The Journal of Physiology*, *592*(22), 592. DOI: 10.1113/jphysiol.2014.278887 PMID: 25172945

Kehayias, P., Jarmola, A., Mosavian, N., Fescenko, I., Benito, F., Laraoui, A., Smits, J., Bougas, L., Budker, D., Neumann, A., Brueck, S., & Acosta, V. (2017). Solution nuclear magnetic resonance spectroscopy on a nanostructured diamond chip. *Nature Communications*, *8*(1), 8. DOI: 10.1038/s41467-017-00266-4 PMID: 28775280

Li, T. (2022). An A-Mnse Nanorod as Anode for Superior Potassium-Ion Storage Via Synergistic Effects of Physical Encapsulation and Chemical Bonding. SSRN *Electronic Journal.*

Lv, S., Wu, Y., Cai, K., He, H., Li, Y., Lan, M., Chen, X., Cheng, J., & Yin, L. (2018). High Drug Loading and Sub-Quantitative Loading Efficiency of Polymeric Micelles Driven by Donor-Receptor Coordination Interactions. *Journal of the American Chemical Society*, *140*(4), 1235–1238. DOI: 10.1021/jacs.7b12776 PMID: 29332390

Malozyomov, B., Martyushev, N., Sorokova, S., Efremenkov, E. A., & Qi, M. (2023). Mathematical Modeling of Mechanical Forces and Power Balance in Electromechanical Energy Converter. *Mathematics*, *11*(10), 2394. DOI: 10.3390/math11102394

Mapossa, A., da Silva Júnior, A. H., de Oliveira, C. R. S., & Mhike, W. (2023). Thermal, Morphological and Mechanical Properties of Multifunctional Composites Based on Biodegradable Polymers/Bentonite Clay: A Review. *Polymers*, *15*(16), 15. DOI: 10.3390/polym15163443 PMID: 37631500

Mapossa, A., da Silva Júnior, A. H., de Oliveira, C. R. S., & Mhike, W. (2023). Thermal, Morphological and Mechanical Properties of Multifunctional Composites Based on Biodegradable Polymers/Bentonite Clay: A Review. *Polymers*, *15*(16), 15. DOI: 10.3390/polym15163443 PMID: 37631500

Karunathilake, E. M. B. M., Le, A. T., Heo, S., Chung, Y. S., & Mansoor, S. (2023). The path to smart farming: Innovations and opportunities in precision agriculture. *Agriculture*, *13*(8), 1593.

Singh, A., Hsu, M. H., Gupta, N., Khanra, P., Kumar, P., Verma, V. P., & Kapoor, M. (2020). Derivatized Carbon Nanotubes for Gene Therapy in Mammalian and Plant Cells. *ChemPlusChem*, *85*(3), 466–475. DOI: 10.1002/cplu.201900678 PMID: 32159284

Thombare, N.. (2021). *Guar gum based hydrogel as controlled micronutrient delivery system: Mechanism and kinetics of boron release for agricultural applications* (Vol. 112). Biopolymers.

Trivedi, P., Leach, J. E., Tringe, S. G., Sa, T., & Singh, B. K. (2020). Plant–microbiome interactions: From community assembly to plant health. *Nature Reviews. Microbiology*, *18*(11), 1–15. DOI: 10.1038/s41579-020-0412-1 PMID: 32788714

Venditti, I. (2017). Morphologies and functionalities of polymeric nanocarriers as chemical tools for drug delivery: A review. *Journal of King Saud University. Science*.

Yang, Y., Hou, X., Ma, S., Huang, S., Chen, J., Fang, Z., Nie, G., Xu, B., Serra, C., & Ding, S. (2022). Tetraphenylethylene-Based Nanogels by Physical Encapsulation Technology: An AIEgen Transparent Film Thermometers. *ACS Applied Polymer Materials*, *4*(3), 1974–1982. DOI: 10.1021/acsapm.1c01826

Zhang, X. (2019). Environmental DNA Shaping a New Era of Ecotoxicological Research. *Environmental Science & Technology*, *53*(10), 5605–5612. DOI: 10.1021/acs.est.8b06631 PMID: 31009204

Zhu, T., Wang, L., Sun, W., Wang, M., Yang, Z., Ji, T., Wang, S., Wang, Y., Xia, L., & Liu, G. (2018). Origin of Aragonite Scale Deposition on Carbon Steel at Ambient Circumstances. *Industrial & Engineering Chemistry Research*, *57*(1), 401–413. DOI: 10.1021/acs.iecr.7b03399

Chapter 12
The Role of Nanocarriers in Phytopathogen Management

Mohsen Mohamed Elsharkawy
https://orcid.org/0000-0003-4731-056X
Kafrelsheikh University, Egypt

ABSTRACT

The current approaches to controlling plant diseases rely primarily on agrochemicals that may harm both people and the environment. Nanotechnology can assist in minimizing the adverse effects of pesticides in a sustainable and environmentally friendly way by improving the solubility of pesticides, extending their shelf life, and lowering their toxicity. Similarly, biopesticides are great options for ecologically friendly agriculture. In agriculture, the application of NCs (nanocarriers), or nanoparticles that may encapsulate medications and release them selectively, is a developing field. The active components of biopesticides can be loaded into nanocarriers via encapsulation, entrapment, attachment, or absorption to create nano-biopesticides. Different types of nanocarriers, such as metal oxides, metals, non-metal oxides, carbon, polymers, lipids, etc., are used to make nano-biopesticides and RNAi pesticides. This chapter explores the use of nanocarriers to manage plant pests and deals with existing obstacles and future opportunities.

INTRODUCTION

Significant financial losses are caused by diseases attributed to phytopathogenic microorganisms, which impact crops like vineyards, olive groves, citrus orchards, and stone fruit, etc., as well as arable crops like vegetables, leguminous, and cereal plants. Controlling phytopathogens that cause diseases like canker and dieback on commercially and ecologically significant host plants is a significant issue in crops (Mondello et al., 2018). The pathogens' placement inside the trunk of grapevine hampers a straightforward delivery of antifungal medications, such as traditional spraying (Bertsch et al., 2013). Common therapies focus on preventative measures such as disinfecting pruning tools, protecting wounds, and applying fungicides in high dosages (Mondello et al., 2018). For advanced infestations, the entire

DOI: 10.4018/979-8-3693-9869-2.ch012

grapevine should be removed. To prevent further spread, contaminated items should be removed and burned (Gramaje et al., 2018).

Many artificial agricultural chemicals used to manage diseases are currently prohibited or under public pressure to decrease their use. Implementing integrated pest management (IPM) strategies, such as using biological pesticides or natural compounds, is a top priority (Broda, 2020). Nanocarriers contain lignin, which allows them to respond to external stimuli such as lignocellulolytic enzymes released by bacteria and fungi that cause wood diseases. This triggers the regulated release of curative medications. Lignin has already been studied as a renewable, plentiful, and cost-effective feedstock for creating complex nanostructures (Yiamsawas et al., 2017; Fischer et al., 2019). Lignin-derived chemicals serve as biodegradable components for nanomaterials, with recent research presenting lignin-based nanocarriers (Wurm & Weiss, 2014). Chitosan-g-poly (acrylamide) nanostructures were developed using a simple method (Pathania et al., 2016). Xylan nanocarriers were synthesized using toluene diisocyanate (Beckers et al., 2020). The manufacture of chitosan microcapsules was synthesized using maleic anhydride, glycidyl methacrylate terpolymer, and N-tert-butylacrylamide (Ciftci et al., 2020). Lignin–chitosan films were manufactured (Zou et al., 2019; Rosova et al., 2021).

Nanotechnology has been extensively employed in a number of industries during the last ten years, including electronics, aerospace, medicine, agriculture, and the life sciences (Bayda et al., 2019, An et al., 2022). Since they have been widely used to create nano-delivery applications and are thought to have the best chemical and physical characteristics of any material in the twenty-first century, nanomaterials are now an active area of research in the domains of contemporary medicine and agriculture (Wu & Li, 2022, Yan et al., 2022a). Nowadays, dsRNA may be delivered via a revolutionary technique called nanocarrier-mediated RNAi, which increases RNAi potency (Shen et al., 2014; Zheng et al., 2019; Li et al., 2019; Wang et al., 2023b). Hydrogen bonds, electrostatic interactions, and intermolecular forces allow nanocarriers to combine with nucleic acids. Nanocarriers improve RNAi efficacy by protecting dsRNA from nuclease destruction, increasing absorption through clathrin-activated endocytosis and facilitating promptly the endosomal escape (Ma et al., 2022). We thoroughly presented the evolution of RNAi methods in this chapter, outlined the benefits of RNA pesticides, described the developments, procedure, and function of nanocarrier-mediated RNAi, covered the use of RNA pesticides for managing diseases, and finally presented the present difficulties and potential future developments.

Pesticide treatment has shown to be the most successful technique in avoiding crop damage induced by pests. However, massive quantities of pesticides are released to the environment immediately following application due to several processes, including runoff, leaching, seepage, microbiological degradation, photodegradation, and others. Therefore, in order to achieve the necessary level of pest management within a certain time frame, pesticides are applied regularly and haphazardly. Considerably more pesticides are applied than essential, causing harm to growers and other workers in agriculture, as well as a continuing presence of residual pesticides in food items and the environment. So, the necessity to deal with pesticides or their adverse effects on organisms has strengthened the desire to discover alternatives to the conventional formulations of pesticides now accessible on the marketplace.

Many pest management experts are concentrating their efforts on creating non-toxic alternatives to synthetic pesticides in order to fight diseases and pests. Some of these alternative methods is the use of biopesticides. There are three different kinds of biopesticides: botanical, microbial, and plant-incorporated protectants (Kumar et al., 2021a). In this regard, biopesticides are very promising as they not only effectively control the pathogen but also lessen the harmful effects of conventional chemical pesticides (Kumar et al., 2021b). Biopesticides' potential for worldwide commercialization is increasing as a consequence of scientists and growers giving their application far more consideration due to their environmental safety. However, they have several disadvantages that keep them from being utilized and developing rapidly, even though they are environmentally sound. These include expensive production methods, inadequate storage space, environmental change sensitivity, the effectiveness of selective pest management, and inconsistent field application. In this regard, nano-biopesticides, which are a novel combination of nanomaterials and biopesticides, become a revolutionary way to overcome the disadvantages of biopesticides. Biopesticides can be delivered via synthetic nanomaterials, such as nanoparticles, which prevent degradation and provide regulated release (Abdollahdokht et al., 2022). As a particular category of biopesticides, nano-biopesticides are the most advanced class that combines biological agents and nanomaterials for improved efficacy. These novel formulations utilize the potential of nanomaterials to produce precise and effective delivery methods (Garg et al., 2023). Surface area, size, and reactivity are some of the distinct characteristics of nanomaterials that help make nano-biopesticides more efficient in killing pests while reducing their negative side effects. Nanomaterials may be classified into four primary groups according to their contents: organic, inorganic, organic-inorganic hybrids, and pure carbon nanostructures. Silver, titanium, copper, silica, zinc, gold, iron, aluminum, chitin, nano clay, chitosan, multi-walled carbon nanotubes, graphene, and biopolymers are the most often utilized NPs (nanoparticles) in agriculture. Cost-effectiveness, biodegradability, nontoxicity, and biocompatibility are important considerations when selecting a nanomaterial for applications in agriculture. Additionally, different kinds of nanocarriers are required based on their specific targets, which include different pest species, viruses, weeds, fungi, and crops. This implies that the nanocarriers must be compatible with the targets' physiological and physicochemical needs. Selecting nanostructured compounds for biopesticide application can take into account a number of factors, including crop condition and production, economic performance, nutrient loading capacity, environmental compatibility, nutrient usage efficiency, and nutrient release rate. The shape, size, and surface characteristics of penetrating nanocarriers that can be selected are strictly limited by the complex structure of plant cell walls (Oliveira & Ristroph, 2024). To maximize the combined impact in pest management and environmental risk mitigation, it is crucial to comprehend how nanomaterials and biopesticides interact with each other (Abdollahdokht et al., 2022). The kind of nanomaterial utilized, the active component of the biopesticide, and the desired application technique are some of the variables that must be taken into account when characterizing these formulations. Compared to traditional crop conservation efforts and large-scale operations, nano-biopesticides provide several benefits. They provide a number of advantages, including improved bioavailability, regulated release, and less environmental effect. These compositions reduce the environmental impact of pest management techniques by utilizing nanotechnology (Abdollahdokht et al., 2022). Microbial nano-biopesticides use microbes as the active component and add nanomaterials to increase stability and effectiveness. Utilizing the special characteristics of nanoparticles, plant-incorporated protectants in nano-biopesticides improve the transport and long-term release of biopesticides. Furthermore, naturally occurring compounds with pesticidal characteristics are used in nano-biopesticides, which enhance their distribution and overall effectiveness by employing nanoparticles. Nano-biopesticides provide advantages for the environment that go beyond their ability to manage certain pests. The use of traditional pesticides has decreased due to these formulations, which promote sustainable agriculture (Abdollahdokht et al., 2022). Additionally, using nano-biopesticides is consistent with precision agricultural concepts. Nanomaterials provide the precise and regulated distribution of biopesticides, enabling focused application and minimizing waste and undesirable impacts on species that are not the intended target.

Pest control and crop production can be enhanced by applying specific active substances in small amounts of nanoparticles at the targeted location. An important step toward more productive and ecologically friendly farming methods is the use of nanocarriers into pest management plans. In summary, a major development in pest control techniques is the transition from conventional pesticides to nano-biopesticides. The move to sustainable alternatives is required due to the environmental issues raised by conventional pesticides. The technique known as nano-formulation of pesticides was created and plays a significant part in plant conservation. Nano-biopesticides provide a potential alternative for targeted and effective pest management while reducing environmental impact. As the agriculture industry evolves, adopting these ecologically friendly techniques will be critical to guaranteeing food security and maintaining the optimal state of our ecosystems. A comprehensive understanding of nanotechnology is, in our opinion, crucial to achieving the objective of supporting global agriculture. This chapter focuses on the way nanocarriers in nano-biopesticide formulations might improve pesticide effectiveness while being ecologically responsible. It covers the complexities of synthesis, rigorous review, and strategic adaptation, providing clear insights for academics and policymakers working in the field of sustainable pest control.

NANOCARRIERS

Nanoparticles can provide plant protection through two processes: as separate nanoparticles providing plants with protection; alternatively, (b) fungicide nanoparticles that carry various active chemicals, such as fertilizers, which may be sprayed, or soaked into the roots, leaves, or seeds. In addition to acting as a carrier, nanoparticles can provide a number of benefits, including (i) increased solubility of limited water-soluble fungicides, (ii) longer shelf life, (iii) better site-specific absorption in the targeted microorganism, and (iv) less toxicity (Balaure et al., 2017). Additional benefit of the nanocarrier structure is the improvement of the stability and effectiveness of the utilized nanofungicides under various ecological conditions, which essentially lowers the amount and total number of uses thereby lowering the cost and harmfulness.

In order to create a potent formulation for use in the agroindustry, nanoparticles are also frequently employed as nanocarriers to protect, entrap, and adhere active particles. The common nanoparticles were used as fungicide nanocarriers or transporters. High delivery potential might be ensured by efficiently manufacturing silica nanoparticles with a regulated size, shape, and structure (Mody et al., 2014). Typically, they take the form of mesoporous nanoparticles of silica or PHSNs (porous-hollow silica nanoparticles), which are round particles with pore-like holes. PHSN and MSN often pack the fungicide in the interior core to provide a precise distribution by preventing the volatile active particles from leaving. The active components inside the nanoparticles are protected from UV radiation degradation by the shell construction of PHSNs. Research suggests that silica can enhance resistance of plants to abiotic and biotic stressors. Nanoparticles form of silica are a promising option for producing agro-industry items for fungal management (Worrall et al., 2018). Plant disease management was mostly achieved using chitosan and SNLs (solid-lipid nanoparticles). SLNs are emulsion-like particles composed of solid lipids at room temperature. The framework provided by SLNs allows for the entrapment of lipophilic-active particles without the need for organic solvents (Ekambaram et al., 2012). Surfactants are used to help settle and stabilize the SLN once it is submerged in water. Their main drawbacks are their limited loading activity and the possibility that the components will leak out of the composition while it is being stored (Tamjidi et al., 2013). SLNs have a number of potential benefits as nanocarriers, such as decreased leaching losses, altered bioactive component release patterns, and decreased environmental and human toxicity. These kinds of fungicide treatments provide a novel approach to controlling fungal infections in various plant species (Worrall et al., 2018). A biocompatible, non-toxic substance, and biodegradable chitosan nanoparticles have antimicrobial, antifungal, antiviral, antioxidant, bio-adhesive, anti-inflammatory, and adsorption-enhancing qualities (Maluin & Hussein 2020). Chitosan's antimicrobial actions against various microbes can be classified as either intracellular, extracellular, or both, depending on where the effects are targeted (Hosseinnejad & Jafari, 2016; Varlamov et al., 2018; Kravanja et al., 2019; Ke et al., 2021a). High-MW chitosan's possible antimicrobial actions involve minimizing nutrients absorption from the cells extracellularly, changing cell permeability, and functioning as a chelator of critical metals since it cannot pass though the cell wall (Kravanja et al., 2019, Ke et al., 2021b). In addition to its external antibacterial properties, low-MW chitosan also exhibits intracellular antimicrobial properties that impact RNA, mitochondrial activity, and protein synthesis (Kravanja et al., 2019; Seankamsorn et al., 2021). It has been observed that chitosan negatively affects pathogens (Seankamsorn et al., 2020). However, the degree of acetylation, microbiological characteristics, and bulk weight of chitosan all affect its antimicrobial activity. Chitosan nanoparticles are employed alone in the agriculture sector as potent antifungal substances and growth promoters to manage various fungal diseases. However, they have another application as agrochemical nanocarriers (Maluin et al., 2019). In order to provide a powerful nano delivery method for the formulation, this nanocarrier technology enables the encapsulation of active substances, either by ionic or covalent bonding or attached in chitosan polymeric matrix (Maluin & Hussein 2020). Chitosan is a beneficial pesticide that may be employed against harmful viruses, bacteria, and fungus with or without the binding of macronutrients. In order to combat *Fusarium* head-blight disease and wilt in wheat and chickpea plants, post-flowering stalk rot in maize, blast leaf in rice, blast disease in finger-millet, and leaf spot in maize, chitosan, either alone or in combination with other compounds, demonstrated considerable promise (Maluin et al., 2019). The hydrophobic properties of chitosan result in its poor water solubility. Therefore, to increase its dissolvability, chitosan is frequently mixed with copolymer, organic, and inorganic (Li et al., 2011). The properties of chitosan are enhanced by its reactive amine and hydroxyl groups, which allow for modification, graft reactions, and ionic contact. Chitosan adheres effectively to the leaves and stem epidermis of the plant, allowing bioactive particles to be absorbed and extending the duration of contact (Li et al., 2011).

Nanocarrier-Mediated RNAI Pesticides

To meet the demands of a rising worldwide population, conventional pesticides are widely used to reduce agricultural losses caused by plant diseases and pests. However, there is a serious risk to both farmers and the environment from its improper use (Kumar et al., 2019; Lykogianni et al., 2021). As a result, creating effective and environmentally responsible plant protection systems has emerged as a key global strategic need. RNAi (RNA interference) was initially discovered in *Caenorhabditis elegans* and is a highly stable, sequence-specific technique for preventing gene expression (Fire et al., 1998). A powerful reverse genetic technique for studying gene function, regulation, and interaction at the organismal and cellular levels, RNAi has rapidly gained popularity since its discovery. It also shows great promise for managing plant diseases (Nandety et al., 2015; Zhu & Palli, 2020). The fundamental idea behind RNA pesticides is that they cause developmental retardation or death in hazardous organisms by preventing the production of crucial functioning genes in resistant pests (Vogel et al., 2019; Christiaens et al., 2020; Singewar & Fladung, 2023; Zhao et al., 2023). The main obstacles are dsRNA accumulation in lysosomes, dsRNA instability in the organisms and environment, and low dsRNA delivery effectiveness across different physical and biological barriers (Wang et al., 2016; Song et al., 2019a; Zhu & Palli, 2020; Qiao et al., 2021; Hoang et al., 2022).

Nanotechnology allows the stabilization of genetic materials like dsRNA to increase their efficacy as a pest control agent, in addition to transferring genetic information to plants for improved gene editing. The development of cutting-edge genetic engineering methods like CRISPR-Cas9 has greatly expanded the potential for enhancing crop resilience and production. Grafting and controlled distribution of appropriate exogenous biomolecules for genome editing can be greatly facilitated by nanomaterials (Ulvila et al., 2006). Similarly, DNA plasmids encoding CRISPR may be delivered to plants in place of foreign DNA, allowing for the regulated production of Cas9 through a non-biolistic and non-pathogenic process (Xiao et al., 2015). The use of nanocarriers in agriculture has the potential to significantly alter crop development methods (Ye et al., 2021). However, the transport of nanoparticles across the plant cell wall is covered in a small number of studies (Avila et al., 2014). To comprehend the way synthetic nanoparticles alter the size, tensile strength, shape, aspect ratio, and physicochemical properties and features of these nanomaterials, as well as the way they impact their ability to internalize plant cells, deeper studies are required. Fludioxonil (FLU)-loaded glycine methyl ester-conjugated polysuccinimide (PGA) nanoparticles (FLU@PGA) may release fludioxonil in plant phloem in an alkaline environment with pH-sensitive regulated release (Winston et al., 2002). Analysis clearly showed that mesoporous silica nanoparticles can reduce 2,4-D sodium salt leakage from soil (Miyata et al., 2014). Additionally, using this nanoformulation had no detrimental impact on the nontarget species. The enzymatic degradation of the dsRNA in plants produces single-stranded small interfering RNA (siRNA), which silences high-sequence complementary RNA (Mahmoodi Chalbatani et al., 2019). Applying dsRNA topically to shield leaves from virus and insect assaults offers an alternative to transgenic RNAi since genetically engineered crops have acceptability and regulatory concerns (Kunte et al., 2020; Qiao et al., 2023). Due to its instability on the leaf surface, the effects of dsRNA-based pathogen defense are frequently transient; however, this issue may be resolved by using nanomaterials that attach the exogenous dsRNA to stabilize it, aid in the process of degradation prevention, and guarantee its successful release at the intended location (Saleh et al., 2006). It is commonly known that employing positively charged layered double-hydroxide (LDH) clay nanosheets as dsRNA nanocarriers works effectively. LDH nanosheets have the ability to bind substantial amounts of dsRNA and promote adhesion to the leaf surface. The carbonic acid generated by the humidity and CO_2 on the leaf surface breaks down LDH nanosheets, causing the bound dsRNA to be released gradually. Topical distribution of dsRNA with LDH assistance provides at least three times as much protection against the virus as does naked dsRNA (Zhou et al., 2013). Only a few taxa—coleopterans and nematodes, or beetles—benefit from dsRNA, leaving behind numerous noteworthy taxa like lepidopterans. This results from the lack of specific cellular machinery that enables dsRNA to be processed into siRNA by accessing the cytoplasm and avoiding the endosomes. Nanocarriers made of positively-charged polymers were used to address this problem; they bind the dsRNA and facilitate its passage across the cell membrane into the cytoplasm (Wang et al., 2014). The nanocarriers' ability to stimulate RNAi by applying dsRNA exogenously is very useful and widely applicable. Additionally, nanocarriers minimize the cost of dsRNA by lowering the quantity needed for effective therapy. The main contenders in this space are inexpensive biopolymers such as chitosan, which also have the advantage of being easily biodegradable. The issues of genetic engineering and the negative consequences of using pesticides with insufficient precision can be resolved by using this technology in the agricultural sector (Wang et al., 2021). Sustainable agricultural improvement can be achieved with nano-assisted genetic modification, although there are concerns about the long-term environmental impacts of using nanocarriers and their associated payloads. Protein, RNA, and DNA may be delivered using a variety of artificial nanocarriers which rarely biodegrade. On the other hand, insecticides, macronutrients, and non-actives are often delivered using nanocarriers that are biodegradable and biocompatible. Since their bioaccumulation may pose an environmental risk, biodegradable nanocarriers have a far greater degree of technological security than their non-biodegradable counterparts.

Nanocarriers can transmit plant hormones. Gibberellic acid, which is provided by chitosan and alginate NPs, tends to increase the amount of chlorophyll and enlarge the leaf area in common bean plants (Zhou et al., 2023). The administration of salicylic acid and abscisic acid using mesoporous NPs with corresponding controlled flow redox-responsive channels may also enhance plant stress responses (Das et al., 2015).

RNAI Approach for Improving Pesticides

A common post-transcriptional gene- silencing process in eukaryotes, RNAi is brought on by doublestranded RNA's effective and selective destruction of homologous mRNA (Zamore et al., 2000). There are now three main RNA interference (RNAi) pathways known to exist in organisms: Piwi-interacting RNA-pathway, microRNA pathway and siRNA (short/small interfering RNA) pathway (Zhu & Palli 2020). Researchers have used RNAi as a potent research technique to identify functional genes in several important insects on a huge scale. RNAi technique is a new and secure method of protecting plants in addition to being a potent research tool (Poreddy & Baldwin 2017). RNAi can be used to prevent the expression of key genes for pest development, leading to abnormalities in growth or even death (Zhang et al., 2017; Zotti et al., 2018). It is therefore appropriate for use in the manufacturing of RNA pesticides as it is precise, secure, and simple to use (Zhu & Palli 2020).

Currently, prolonged use of chemical pesticides causes pesticide resistance and contaminants, which pollute the environment, increase control expenses, and pose major biosecurity issues (Lykogianni et al., 2021). RNA pesticides are a novel approach to pest management that provide several benefits over conventional pesticides (Huang et al., 2019). RNAi is very specific, and its sequence mechanism of action has also drawn tremendous interest in pest management (Zamore et al., 2000, Wynant et al., 2014). Artificial microRNAs, dsRNAs, and hairpin-structured RNAs are examples of RNAi triggers that may be constructed specifically for targeting particular genes in dangerous species (Whyard et al., 2009).

Therefore, targeted control of a particular species can be achieved without endangering non-target species. Researchers found that the expression of dsRNA attacking *Blumeria graminis*' effector Avra10 in barley and wheat inhibits the development of the pathogen and reduces its pathogenicity (Nowara et al., 2010). MON87411, the first GM maize crop with dsRNA against the WCR, has been approved for cultivation in over 15 countries (Christiaens et al., 2020). However, the transformational RNAi-based strategy has high financial needs and public acceptability of GM crops. Compared to genetically modified crops, non-transformative RNA pesticides can be applied in the same practical and varied methods as synthetic pesticides, including sprays, petiole absorption, trunk injection, and root/seed soaking (Zotti et al., 2018, Dalakouras et al., 2020, Taning et al., 2020). For example, scientists have shown that the bacterially developed dsHvSnf7 applied on detached plant leaves resulted in mortality (98%) of *Henosepilachna vigintioctopunctata* (Lü et al., 2019). SIGS (Spray-induced gene-silencing) technology was used to treat agricultural diseases in a number of crops, including tomatoes and rice (Qiao et al., 2021,

Mosa & Youssef 2021, Ray et al., 2022). In conclusion, the performed studies have demonstrated the possibility of using non-transformative RNA-pesticides to manage plant diseases.

Nanocarrier-Mediated RNAI Pesticides Mode of Action

It was difficult to generate highly effective RNA pesticides because of the limited durability and delivery effectiveness of dsRNA. Recent research has indicated that the use of nanocarrier-loaded dsRNA may improve the durability of dsRNA (Taning et al., 2020, Yan et al., 2021). There are four primary phases in the action and function of nanocarrier-mediated dsRNA transport: endosomal escape, cellular absorption, nucleic acids binding, and the dispersion of nucleic acids or breakdown of nanocomplexes. In general, nanocarrier/dsRNA structures can be formed via electrostatic contact between positively-charged nanocarriers and negatively-charged dsRNA (Avila et al., 2014; Mahmoodi Chalbatani et al., 2019). For example, in mildly acidic environments, chitosan has a lot of positively charged amino groups that may electrostatically attach to negatively-charged dsRNA to create dsRNA/chitosan compounds (Qiao et al., 2023). Usually positively-charged, the dsRNA/nanocarrier compounds make it easier for them to attach to the negatively-charged cell membrane (Kunte et al., 2020). Nanocarrier/dsRNA compounds penetrate the cell primarily by receptor-mediated endocytosis after interacting with the cell membrane (Wang et al., 2024). Following cellular absorption, membrane-bound vesicles known as endosomes usually cover nanocarrier/dsRNA compounds. The final stage, compounds will be broken down in the lysosome alongside the endosome if they are unable to go away (Varkouhi et al., 2011). One of the most recognized theories on the process of endosomal escape is the "proton sponge" impact, however this is still debatable (Qiao et al., 2024). The dispersion of dsRNA/siRNA from a nanocarrier is necessary to initiate the RNAi cascade and carry out its biological impacts following endosomal escape. The method of dsRNA/siRNA releasing from a nanocarrier, which mainly comprises two releasing mechanisms, has not been well studied. One method is a delayed competitive displacing process, in which several highly charged polyanions, including chondroitin sulfate, heparin, and different analogues with a high selectivity for nanocarriers, replace and release dsRNA/siRNA (Kwon, 2011). An additional method relies on how nanocarriers react to intracellular stimuli. Intracellular stimuli like cytosolic reducers and acidic pH can change the chemical or physical characteristics of nanocarriers (Tzeng & Green 2013; Wang et al., 2014).

Pesticides Application of Nanocarrier-Mediated RNAI

Numerous nanocarriers were developed recently to distribute dsRNA and get through its delivery obstacles; they have been effectively used in pest control. Additionally, nanocarriers can increase the effectiveness of RNAi in plant pathogenic microbes. For example, *Phytophthora infestans*, *P. sojae*, and fungicide-resistant *P. capsici* exhibit outstanding control effects when exposed to dsRNA-CDs. Three distinct cationic lipid compositions were used to create artificial nanoparticles (AVs), which can increase dsRNA stability and provide longer-lasting RNAi-mediated defenses against *Botrytis cinerea* in plants before and after harvest.

The RNAi efficiency of various pest species differs substantially. A unique multicomponent nano-pesticide that co-delivers pesticide and dsRNA was developed to enhance the control impact of RNA pesticides (Qiao et al., 2024). High bioactivity and extended persistence can be attained by this co-delivery for long-term disease control. A plant elicitor and dsRNA were utilized to create a self-assembled multicomponent nano-bioprotectant against potato late blight, which shows a good protective effect when compared to the commonly used mancozeb fungicide (Wang et al., 2023a). RNA pesticides, which target plant diseases, may be sprayed directly onto crops. Sprayable RNA pesticides have been created by GreenLight Biosciences to combat several pests, including gray mold and powdery mildew (Yan et al., 2022b).

Nanocarriers Applications on Fungicides

Preliminary studies on nanofungicides were conducted by combining fungicides with solid wood (Liu et al., 2002b). Since that time, several investigations using a range of nanofungicides with antifungal properties have been conducted. FRAC (Nine fungicide-resistance action committee) groups were examined after being eliminated from the fungicide groups together with a number of essential oils, chitosan-dazomet, and chitosan-dazomet-hexaconazole. Silica, polymer mixtures, and chitosan were the most frequently researched nanoparticle carriers or transporters. The effectiveness of these nanofungicides was tested using a variety of fungal species. In order to improve tebuconazole's limited water solubility and capacity to stick to leaf surface, non-denatured nanosized empty cells that encase bacterial ghosts and gram-negative bacteria were used. *Pectobacterium cypripedii* was chosen as a bacterial ghost at the nanoscale because of its ability to stick to the plant. All plant types were evaluated using either the commercial tebuconazole or ghost-loaded tebuconazole procedures to suppress various fungal species (Hatfaludi et al., 2004). Maximum protection for the commercial products was demonstrated by the plants that were not affected by rainfall. A large number of groups were not responsive well to the fungal inoculation in plants subjected to intense rain one hour after the treatment, and the same was noted with commercial items. Pyraclostrobin, a limited water-soluble fungicide, was added to chitosan nanoparticle at varying quantities (Xu et al., 2014). After treatment, the nanofungicide was shown to be equally or less efficient than commercial pyraclostrobin in preventing the development of *C. gossypii*. On the seventh day following administration, there was a greater level of inhibition than with the active ingredient alone (Ilk et al., 2017). After two months of incubation on the culture plates of *F. oxysporum*, the kaempferol fungicide, which was likewise put into lecithin/chitosan, demonstrated an inhibition of around 67% effectiveness.

According to some research, chlorothalonil and tebuconazole were loaded onto various nanoparticles to improve the fungicides' poor solubility after being applied to solid wood (Liu et al., 2002 a, c). *Gloeophyllum trabeum*'s pine wood degradation was examined for two months in order to determine the mass loss. It has been noted that the hydrophobic chlorothalonil integrated onto various nanoparticles is inefficient due to formulation consistency and size, concentration rise, and reduced wood deterioration (Liu et al., 2002b). The surfactant-free approach of encapsulating tebuconazole and chlorothalonil resulted in a stable-aqueous solution with lower median particle sizes and more absorption in the wood (Liu et al., 2002b). In southern pine infected with *G. trabeum* and birchwood infected with *Trametes versicolor*, no significant decay was seen when a low concentration was used in comparison to the commercial fungicide (Liu et al., 2002c). Nanoparticles with higher stability and smaller size, devoid of surfactants, demonstrated greater resistance to fungal degradation (Liu et al., 2002c). Nanoencapsulation of hexaconazole, dazomet, and chitosan resulted in a stable liquid solution with tiny median particle sizes, leading to increased absorption in oil palm plants and improved treatment of BSR (basal stem rot) disease (Maluin et al., 2019). Encapsulating dazomet and hexaconazole in chitosan significantly improved absorption and reduced the impact of *G. boninense* on oil palms (Maluin et al., 2019). Although EOs (essential oils) have fungicidal properties, they tend to evaporate quickly when used on a broad scale.

Encapsulating five distinct EOs onto MSN had strong antifungal effects on the *A. niger* (Janatova et al., 2015). SLNs stabilized the EOs of *Zataria multiflora* and protected six different species of fungus.

Graphene Oxide Nanocarrier Applications to Control Plant Pathogens

Innovative nano-delivery methods for pesticides have garnered more interest in the last decade (Xin et al. 2020; Xu et al., 2021). The frequent deployment of pesticides might be reduced or eliminated with a more technical and logical design for the pesticide delivery method that uses rapid or site-specific releases with prolonged bioactivity (Singh et al., 2019; Dangi & Verma, 2021). Additionally, by lowering the amount and duration of pesticide applications, it can lessen the environmental danger of pesticides. Numerous nanomaterials were manufactured as vectors for the delivery of pesticides, including metal-inorganic compounds, mesoporous-silica nanoparticles, and graphene-based nanomaterials (Wang et al., 2020; Xu et al., 2020; Dong et al., 2021). Graphene oxide (GO) is regarded as a strong structure for pesticide delivery throughout several nanocarriers manufactured for pesticides because of its benefits, which include simple surface modification, a wide surface area, a high loading rate, and high soluble in water (Zhu et al., 2020; Chaudhary et al., 2021). For instance, using GO as a carrier for releasing emamectin benzoate, that can improve pesticides' distribution stability and long-term antipest effectiveness, was elucidated (Song et al., 2019b). Polydopamine-coated GO provides a greater capacity for loading hymexazol with pH-controlled discharge (Tong et al., 2018). In order to decrease the degradation of applied pesticides, it was demonstrated that GO can increase the leaf adherence of pesticides by acting as nanocarrier (Tong et al., 2018; Muda et al., 2020; Gao et al., 2021). The pesticide made using nanocarriers, however, should have straightforward ingredients, a straightforward manufacturing procedure, high bioactivity and multipurpose, as well as the right amount of pesticide, storage stability, and long-lasting effects. Pyraclostrobin (Pyr), a new strobilurin fungicide with a broad range, high effectiveness, and low toxicity, has curative and protective properties on plants (Dominguez et al., 2021). A novel GO-Pyr nanopesticide by combining the current Pyr pesticide with GO was created. The antifungal effectiveness against *Sclerotinia sclerotiorum* and *Fusarium graminearum* was investigated (Peng et al., 2022). A GO–Pyr nanostructure with strong antifungal activity was created when Pyr was loaded onto a GO nanocarrier. The crystal Pyr was distinguished by its smooth exterior, rectangular parallelepiped pattern, and range of sizes, whereas GO had a characteristic wrinkled shape (Peng et al., 2022). Large numbers of a rectangular-parallelepiped shape crystals appeared on the outer layer of GO as a result of pyr loading. The formation of these crystals is explained by Pyr's deposition over GO sheets, which could subsequently be confirmed by FT-IR. It is clear that the GO–Pyr spectra had all of the distinctive peaks GO and Pyr lacking the addition of a new peak, suggesting that Pyr was physically loaded onto GO without changing its chemical characteristics (Qin et al., 2017). The structural analysis of Pyr indicates that there was actually no loss of functions, but the GO–Pyr nanostructure lost a smaller weight than GO because the H-bond association between GO and Pyr occupied some functional groups that contain oxygen. The vast surface area and many functional groups on GO may be responsible for its better pesticide loading capacity. These features can offer a large number of sites of adsorption, resulting in effective pesticide administration and enhanced bioavailability (Tong et al., 2018, Song et al., 2021). A greater amount of pesticide loaded onto GO will increase the nanocomposite's antifungal activity. The prolonged release is in line with the ideal release of controlled treatment delivery systems. The pesticide's high activity and efficient concentration may be sustained for a considerable amount of time with a burst release followed by a steady release (Kumar et al., 2021). In order to continue providing strong antifungal activity, the GO-Pyr nanocomposite's early burst release following by gradual release may help maintain a suitable concentration for an extended period of time. Throughout storage, the GO-Pyr nanocomposites remained stable and did not precipitate or stratify (Peng et al., 2022). The GO-Pyr nanocomposite exhibits good adhesion and spreading capability on paraffin film surfaces, making it suitable for pesticide deposition and adsorption. Additionally, the GO–Pyr nanocomposites demonstrated greater antifungal efficacy than both GO and Pyr against the growth and development of the pathogens. GO has a lot of potential for pesticide administration and protecting plants due to its high efficiency, ease of preparation, versatility, and excellent biosafety. Consequently, GO could be a viable synergist for Pyr in the management of fungal diseases in plants. The pesticide delivery method using GO as a nanocarrier has potential for future crop protection and environmentally friendly farming due to the lack of hazardous organic additives and solvents and strong antimicrobial properties.

CHALLENGES

Growth enhancers' ability to effectively protect plants from different stressors was administered by nanocarriers (Zhao et al., 2020). These NMs access plant cells via means of ion channels, aquaporins (carrier proteins), endocytosis, and the creation of entirely new pores. NMs can also move apoplastically and symplastically in the xylem and phloem. Through sink activity, seeds, fruits, and flowers have a notable propensity to absorb ENPs from the phloem at comparatively higher concentrations. This raises concerns about the safety of both human and animal intake of these plant derivatives, in addition to the plant's toxicological effects. They might potentially infiltrate the food chain in any of these circumstances and have unforeseen repercussions. Furthermore, oxidative stress produced by too much Fe_3O_4 NPs might lower metabolic capacity in plant systems. A ZnO NPs may have a detrimental effect on chromosomal and cellular attributes and can be hazardous to health (Singh, 2019; Zheng et al., 2022). Additionally, it has been documented how the plant-associated microbiome is affected by short soil exposure following the application of varying concentrations of ZnO nanoparticles. Certain ENPs, including ZnO, SiO_2, and TiO_2, are photochemically active and may produce superoxide radicals by direct electron transfer when exposed to light. Research has demonstrated that metallic ENMs produce oxidative stress in plants, impede the detoxifying processes of ROS, and result in plant genotoxicity in cultivated plants (such as tomato and wheat) (Heikal et al., 2021). A different study compared the toxicity effects of three carbon nanomaterials (C60, rGO, and MWCNTs) on a rice mini-ecosystem. It was found that high exposure doses of CNMs caused negative effects on rice shoot height and root length, as well as alterations to root cortical cells and increased concentrations of phytohormones and antioxidant enzymes, while there were only minor changes in the predominant soil bacterial species. These results highlight the significance of carefully managing and regulating CNMs to avoid potential hazards to both organisms and environmental food safety (Hao et al., 2018). However, the research has also indicated that they can offer protection from oxidative damage. Therefore, it is necessary to do research on the mechanistic understanding of ENM metabolism in organisms and particular cells. Investigating potential adaptive mechanisms to the delayed effects of ENM exposure is also essential. The bioaccumulation of nanomaterials in food supply chains and their interactions with other environmental contaminants must be studied since they may have an adverse effect on important plant functions, which might compromise the sustainability of agriculture. The deployment of chemical or green ENMs into the field must be carefully monitored. NPs could be applied to agriculture if their non-toxic nature has been established in order to increase production and address important problems. A polymeric ENM coated with a plant-based insecticide is a novel agro-product that has been gaining interest since it is non-toxic. To maintain soil conditions and ecosystems, it is also critical to evaluate how NMs affect soil microbes in order to prevent unfavorable effects of microbial community. ENMs can be extremely detrimental to the functioning of an ecosystem since their buildup in treated soils may harm the soil microbiota and related trophic species (Zhang et al., 2020).

The long-term usage of standard pesticides can lead to pest resistance, pests can also acquire resistance to RNA insecticides. Unexpected consequences for non-targets are another issue related to the use of RNA pesticides, together with the resistance issue. Therefore, while developing RNA pesticides, researchers must reduce the dangers that dsRNAs could cause to non-target species in a sequence-specific way in order to mitigate the possible harms of RNAi products. Although using nanocarriers may increase the durability of RNA pesticides, it may also present additional risks to both humans and ecosystems, particularly due to their biotoxicity. For instance, at extremely high concentrations, SPc can harm the intestinal tissues, however it is not hazardous under working conditions. Therefore, when using it extensively, extremely strict application requirements need to be followed. Excessive applied levels of dsRNA show the potential to stimulate the immune system and overwhelm the RNAi core mechanism, both of which might be detrimental to organisms (De Schutter et al., 2022).

CONCLUSION

Nanotechnology has the ability to expand agricultural applications as well as improve the existing method of managing plant pathogens. The production of nanopesticides may lead to various benefits, including improved pesticide bioavailability and efficacy, decreased toxicity, and higher solubility of pesticides. It may also be possible to precisely release the active ingredients, target delivery, and extend the actives' shelf life. Pathogens on a range of crops have been successfully and safely managed with various nanoparticles and other nanomaterial forms, including agronanofungicides, nanobiopesticides and essential oil nanoformulations. There are two ways to apply nanoparticles for plant protection: (a) as individual nanoparticles that protect plants, or (b) as nanocarriers that carry various active ingredients which can be potentially applied to plant tissues. Nanoparticles are also commonly used as nanocarriers to encapsulate, entrap, and connect active particles to build a potent formulation to be applied in agroindustry. As a carrier, nanoparticles can provide a number of benefits, including (i) longer shelf life, (ii) greater site-specific absorption in the targeted microorganism, (iv) decreased toxicity, and (iii) increased solubility of the pesticides. Another benefit of the nanocarriers is that it increases the activity and stability of the nanopesticides under various ecological conditions.

Biologists should collaborate closely with specialists from many fields to have a deeper comprehension of the mechanisms and key interactions in a bio-nanotechnology system. Selecting a reliable and repeatable framework is also crucial for conducting effectiveness and biocompatibility studies at different cellular levels and pest host environment.

REFERENCES

Abdollahdokht, D., Gao, Y., Faramarz, S., Poustforoosh, A., Abbasi, M., Asadikaram, G., & Nematollahi, M. H. (2022). Conventional agrochemicals towards nano-biopesticides: An overview on recent advances. *Chemical and Biological Technologies in Agriculture*, *9*(1), 13. DOI: 10.1186/s40538-021-00281-0

Abdullahi, A., Ahmad, K., Ismail, I. S., Asib, N., Ahmed, O. H., Abubakar, A. I., Siddiqui, Y., & Ismail, M. R. (2020). Potential of using ginger essential oils-based nanotechnology to control tropical plant diseases. *The Plant Pathology Journal*, *36*(6), 515–535. DOI: 10.5423/PPJ.RW.05.2020.0077 PMID: 33312089

An, C., Sun, C., Li, N., Huang, B., Jiang, J., Shen, Y., Wang, C., Zhao, X., Cui, B., Wang, C., Li, X., Zhan, S., Gao, F., Zeng, Z., Cui, H., & Wang, Y. (2022). Nanomaterials and nanotechnology for the delivery of agrochemicals: Strategies towards sustainable agriculture. *Journal of Nanobiotechnology*, *20*(1), 11. DOI: 10.1186/s12951-021-01214-7 PMID: 34983545

Avila, L. A., Lee, S. Y., & Tomich, J. M. (2014). Synthetic in vitro delivery systems for plasmid DNA in eukaryotes. *Journal of Nanopharmaceutics and Drug Delivery*, *2*(1), 17–35. DOI: 10.1166/jnd.2014.1043

Balaure, P. C., Gudovan, D., & Gudovan, I. (2017). Nanopesticides: A new paradigm in crop protection. In *New Pesticides and Soil Sensors* (pp. 129–192). Academic Press. DOI: 10.1016/B978-0-12-804299-1.00005-9

Bayda, S., Adeel, M., Tuccinardi, T., Cordani, M., & Rizzolio, F. (2019). The history of nanoscience and nanotechnology: From chemical–physical applications to nanomedicine. *Molecules (Basel, Switzerland)*, *25*(1), 112. DOI: 10.3390/molecules25010112 PMID: 31892180

Beckers, S. J., Wetherbee, L., Fischer, J., & Wurm, F. R. (2020). Fungicide-loaded and biodegradable xylan-based nanocarriers. *Biopolymers*, *111*(12), e23413. DOI: 10.1002/bip.23413 PMID: 33306838

Bertsch, C., Ramírez-Suero, M., Magnin-Robert, M., Larignon, P., Chong, J., Abou-Mansour, E., Spagnolo, A., Clément, C., & Fontaine, F. (2013). Grapevine trunk diseases: Complex and still poorly understood. *Plant Pathology*, *62*(2), 243–265. DOI: 10.1111/j.1365-3059.2012.02674.x

Broda, M. (2020). Natural Compounds for Wood Protection against Fungi—A Review. *Molecules (Basel, Switzerland)*, *25*(15), 3538. DOI: 10.3390/molecules25153538 PMID: 32748877

Chaudhary, K., Kumar, K., Venkatesu, P., & Masram, D. T. (2021). Protein immobilization on graphene oxide or reduced graphene oxide surface and their applications: Influence over activity, structural and thermal stability of protein. *Advances in Colloid and Interface Science*, *289*, 102367. DOI: 10.1016/j.cis.2021.102367 PMID: 33545443

Christiaens, O., Niu, J., & Nji Tizi Taning, C. (2020). RNAi in insects: A revolution in fundamental research and pest control applications. *Insects*, *11*(7), 415. DOI: 10.3390/insects11070415 PMID: 32635402

Ciftci, N., Sargin, I., Arslan, G., Arslan, U., & Okudan, A. (2020). Ascorbic acid adsorption-release performance and antibacterial activity of chitosan-ter (GMA-MA-NTBA) polymer microcapsules. *Journal of Polymers and the Environment*, *28*(8), 2277–2288. DOI: 10.1007/s10924-020-01773-0

Dalakouras, A., Wassenegger, M., Dadami, E., Ganopoulos, I., Pappas, M. L., & Papadopoulou, K. (2020). Genetically modified organism-free RNA interference: Exogenous application of RNA molecules in plants. *Plant Physiology*, *182*(1), 38–50. DOI: 10.1104/pp.19.00570 PMID: 31285292

Dangi, K., & Verma, A. K. (2021). Efficient & eco-friendly smart nano-pesticides: Emerging prospects for agriculture. *Materials Today: Proceedings*, *45*, 3819–3824. DOI: 10.1016/j.matpr.2021.03.211

Das, S., Debnath, N., Cui, Y., Unrine, J., & Palli, S. R. (2015). Chitosan, carbon quantum dot, and silica nanoparticle mediated dsRNA delivery for gene silencing in *Aedes aegypti*: A Comparative Analysis. *ACS Applied Materials & Interfaces*, *7*(35), 19530–19535. DOI: 10.1021/acsami.5b05232 PMID: 26291176

De Schutter, K., Taning, C. N. T., Van Daele, L., Van Damme, E. J., Dubruel, P., & Smagghe, G. (2022). RNAi-based biocontrol products: Market status, regulatory aspects, and risk assessment. *Frontiers in Insect Science*, *1*, 818037. DOI: 10.3389/finsc.2021.818037 PMID: 38468882

Dominguez, A. N., Emmert, G. E., Gil, D. M., & Alvarez, R. M. S. (2021). Experimental and theoretical vibrational study of the fungicide pyraclostrobin. *Spectrochimica Acta. Part A: Molecular Spectroscopy*, *259*, 119888. DOI: 10.1016/j.saa.2021.119888 PMID: 34015601

Dong, J. T., Chen, W., Feng, J. G., Liu, X. Q., Xu, Y., Wang, C., Yang, W. C., & Du, X. Z. (2021). Facile, smart, and degradable metal-organic framework nanopesticides gated with FeIII-Tannic acid networks in response to seven biological and environmental stimuli. *ACS Applied Materials & Interfaces*, *13*(16), 19507–19520. DOI: 10.1021/acsami.1c04118 PMID: 33856772

Ekambaram, P., Sathali, A. A. H., & Priyanka, K. (2012). Solid lipid nanoparticles: A review. *Sci. Rev. Chem. Commun.*, *2*, 80–102.

Fire, A., Xu, S., Montgomery, M. K., Kostas, S. A., Driver, S. E., & Mello, C. C. (1998). Potent and specific genetic interference by double stranded RNA in *Caenorhabditis elegans.Nature*, *391*(6669), 806–881. DOI: 10.1038/35888 PMID: 9486653

Fischer, J., Beckers, S. J., Yiamsawas, D., Thines, E., Landfester, K., & Wurm, F. R. (2019). Targeted drug delivery in plants: Enzyme-responsive lignin nanocarriers for the curative treatment of the worldwide grapevine trunk disease. *Advanced Science (Weinheim, Baden-Wurttemberg, Germany)*, *6*(15), 1802315. DOI: 10.1002/advs.201802315 PMID: 31406660

Franco, P., & De Marco, I. (2020). The Use of Poly (N-vinyl pyrrolidone) in the Delivery of Drugs: A Review. *Polymers*, *12*(5), 1114. DOI: 10.3390/polym12051114 PMID: 32414187

Gao, X. D., Shi, F. Y., Peng, F., Shi, X. J., Cheng, C. H., Hou, W. L., Xie, H. C., Lin, X. H., & Wang, X. P. (2021). Formulation of nanopesticide with graphene oxide as the nanocarrier of pyrethroid pesticide and its application in spider mite control. *RSC Advances*, *11*(57), 36089–36097. DOI: 10.1039/D1RA06505J PMID: 35492771

Garg, D., Sridhar, K., Stephen Inbaraj, B., Chawla, P., Tripathi, M., & Sharma, M. (2023). Nano-Biofertilizer formulations for agriculture: A systematic review on recent advances and prospective applications. *Bioengineering (Basel, Switzerland)*, *10*(9), 1010. DOI: 10.3390/bioengineering10091010 PMID: 37760112

Gramaje, D., Úrbez-Torres, J. R., & Sosnowski, M. R. (2018). Managing grapevine trunk diseases with respect to etiology and epidemiology: Current strategies and future prospects. *Plant Disease*, *102*(1), 12–39. DOI: 10.1094/PDIS-04-17-0512-FE PMID: 30673457

Hao, Y., Ma, C., Zhang, Z., Song, Y., Cao, W., Guo, J., Zhou, G., Rui, Y., Liu, L., & Xing, B. (2018). Carbon nanomaterials alter plant physiology and soil bacterial community composition in a rice-soil-bacterial ecosystem. *Environmental Pollution*, *232*, 123–136. DOI: 10.1016/j.envpol.2017.09.024 PMID: 28947315

Hatfaludi, T., Liska, M., Zellinger, D., Ousman, J. P., Szostak, M., Jalava, K., & Lubitz, W. (2004). Bacterial ghost technology for pesticide delivery. *Journal of Agricultural and Food Chemistry*, *52*(18), 5627–5634. DOI: 10.1021/jf049489w PMID: 15373403

Heikal, Y. M., Sutan, N. A., Heikal, Y. M., & Sutan, N. A. (2021). Mechanisms of genotoxicity and oxidative stress induced by engineered nanoparticles in plants. *Induc. Genotoxicity Oxidative Stress Plants* 151–197, .DOI: 10.1007/978-981-16-2074-4_6

Hoang, B. T. L., Fletcher, S. J., Brosnan, C. A., Ghodke, A. B., Manzie, N., & Mitter, N. (2022). RNAi as a foliar spray: Efficiency and challenges to field applications. *International Journal of Molecular Sciences*, *23*(12), 6639. DOI: 10.3390/ijms23126639 PMID: 35743077

Hosseinnejad, M., & Jafari, S. M. (2016). Evaluation of different factors affecting antimicrobial properties of chitosan. *International Journal of Biological Macromolecules*, *85*, 467–475. DOI: 10.1016/j.ijbiomac.2016.01.022 PMID: 26780706

Huang, C.-Y., Wang, H., Hu, P., Hamby, R., & Jin, H. (2019). Small RNAs—Big players in plant-microbe interactions. *Cell Host & Microbe*, *26*(2), 173–182. DOI: 10.1016/j.chom.2019.07.021 PMID: 31415750

Ilk, S., Saglam, N., & Özgen, M. (2017). Kaempferol loaded lecithin/chitosan nanoparticles: Preparation, characterization, and their potential applications as a sustainable antifungal agent. *Artificial Cells, Nanomedicine, and Biotechnology*, *45*(5), 907–916. DOI: 10.1080/21691401.2016.1192040 PMID: 27265551

Janatova, A., Bernardos, A., Smid, J., Frankova, A., Lhotka, M., Kourimská, L., Pulkrabek, J., & Kloucek, P. (2015). Long-term antifungal activity of volatile essential oil components released from mesoporous silica materials. *Industrial Crops and Products*, *67*, 216–220. DOI: 10.1016/j.indcrop.2015.01.019

Kaushal, M. (2018). Role of microbes in plant protection using intersection of nanotechnology and biology. In Abd-Elsalam, K. A., & Prasad, R. (Eds.), *Nanobiotechnology Applications in Plant Protection* (pp. 111–135). Springer International Publishing.

Ke, C. L., Deng, F. S., Chuang, C. Y., & Lin, C. H. (2021a). Antimicrobial actions and applications of chitosan. *Polymers*, *13*, 904. DOI: 10.1007/978-3-319-91161-8_5

Ke, C. L., Liao, Y. T., & Lin, C. H. (2021b). MSS2 maintains mitochondrial function and is required for chitosan resistance, invasive growth, biofilm formation and virulence in *Candida albicans*. *Virulence*, *12*(1), 281–297. DOI: 10.1080/21505594.2020.1870082 PMID: 33427576

Kravanja, G., Primozic, M., Knez, Z., & Leitgeb, M. (2019). Chitosan-based (Nano) materials for novel biomedical applications. *Molecules (Basel, Switzerland)*, *24*(10), 1960. DOI: 10.3390/molecules24101960 PMID: 31117310

Kumar, A., Kanwar, R., & Mehta, S. K. (2021a). Eucalyptus oil-based nanoemulsion: A potent green nanowagon for controlled delivery of emamectin benzoate. *ACS Agricultural Science & Technology*, *1*(2), 76–88. DOI: 10.1021/acsagscitech.0c00061

Kumar, J., Ramlal, A., Mallick, D., & Mishra, V. (2021b). An overview of some biopesticides and their importance in plant protection for commercial acceptance. *Plants*, *10*(6), 1185. DOI: 10.3390/plants10061185 PMID: 34200860

Kumar, S., Nehra, M., Dilbaghi, N., Marrazza, G., Hassan, A. A., & Kim, K. H. (2019). Nano-based smart pesticide formulations: Emerging opportunities for agriculture. *Journal of Controlled Release*, *294*, 131–153. DOI: 10.1016/j.jconrel.2018.12.012 PMID: 30552953

Kunte, N., McGraw, E., Bell, S., Held, D., & Avila, L. A. (2020). Prospects, challenges and current status of RNAi through insect feeding. *Pest Management Science*, *76*(1), 26–41. DOI: 10.1002/ps.5588 PMID: 31419022

Kwon, Y. J. (2012). Before and after endosomal escape: Roles of stimuli-converting siRNA/polymer interactions in determining gene silencing efficiency. *Accounts of Chemical Research*, *45*(7), 1077–1088. DOI: 10.1021/ar200241v PMID: 22103667

Li, J., Qian, J., Xu, Y., Yan, S., Shen, J., & Yin, M. (2019). A facile-synthesized star polycation constructed as a highly efficient gene vector in pest management. *ACS Sustainable Chemistry & Engineering*, *7*(6), 6316–6322. DOI: 10.1021/acssuschemeng.9b00004

Li, M., Huang, Q., & Wu, Y. (2011). A novel chitosan-poly (lactide) copolymer and its submicron particles as imidacloprid carriers. *Pest Management Science*, *67*(7), 831–836. DOI: 10.1002/ps.2120 PMID: 21370387

Liu, Y., Laks, P., & Heiden, P. (2002a). Controlled release of biocides in solid wood. I. Efficacy against brown rot wood decay fungus (*Gloeophyllum trabeum*). *Journal of Applied Polymer Science*, *86*(3), 596–607. DOI: 10.1002/app.10896

Liu, Y., Laks, P., & Heiden, P. (2002b). Controlled release of biocides in solid wood. II. Efficacy against *Trametes versicolor* and *Gloeophyllum trabeum* wood decay fungi. *Journal of Applied Polymer Science*, *86*(3), 608–614. DOI: 10.1002/app.10897

Liu, Y., Laks, P., & Heiden, P. (2002c). Controlled release of biocides in solid wood. III. Preparation and characterization of surfactant-free nanoparticles. *Journal of Applied Polymer Science*, *86*(3), 615–621. DOI: 10.1002/app.10898

Lü, J., Liu, Z., Guo, W., Guo, M., Chen, S., Li, H., Yang, C., Zhang, Y., & Pan, H. (2019). Feeding delivery of dsHvSnf7 is a promising method for management of the pest *Henosepilachna vigintioctopunctata* (Coleoptera: Coccinellidae). *Insects*, *11*(1), 34. DOI: 10.3390/insects11010034 PMID: 31906124

Lykogianni, M., Bempelou, E., Karamaouna, F., & Aliferis, K. A. (2021). Do pesticides promote or hinder sustainability in agriculture? The challenge of sustainable use of pesticides in modern agriculture. *The Science of the Total Environment*, *795*, 148625. DOI: 10.1016/j.scitotenv.2021.148625 PMID: 34247073

Ma, Z., Zheng, Y., Chao, Z., Chen, H., Zhang, Y., Yin, M., Shen, J., & Yan, S. (2022). Visualization of the process of a nanocarrier-mediated gene delivery: Stabilization, endocytosis and endosomal escape of genes for intracellular spreading. *Journal of Nanobiotechnology*, *20*(1), 124. DOI: 10.1186/s12951-022-01336-6 PMID: 35264206

Mahmoodi Chalbatani, G., Dana, H., Gharagouzloo, E., Grijalvo, S., Eritja, R., Logsdon, C. D., Memari, F., Miri, S. R., Rad, M. R., & Marmari, V. (2019). Small interfering RNAs (siRNAs) in cancer therapy: A nano-based approach. *International Journal of Nanomedicine*, *14*, 3111–3128. DOI: 10.2147/IJN.S200253 PMID: 31118626

Maluin, F. N., & Hussein, M. Z. (2020). Chitosan-based agronanochemicals as a sustainable alternative in Crop Protection. *Molecules (Basel, Switzerland)*, *25*(7), 1611. DOI: 10.3390/molecules25071611 PMID: 32244664

Maluin, F. N., Mohd, Z. H., Nor Azah, Y., Sharida, F., Idris, A. S., Nur Hailini, Z. H., & Leona, D. J. D. (2019). Enhanced fungicidal efficacy on *Ganoderma boninense* by simultaneous co-delivery of hexaconazole and dazomet from their chitosan nanoparticles. *RSC Advances*, *9*(46), 27083–27095. DOI: 10.1039/C9RA05417K PMID: 35528577

Miyata, K., Ramaseshadri, P., Zhang, Y., Segers, G., Bolognesi, R., & Tomoyasu, Y. (2014). Establishing an in vivo assay system to identify components involved in environmental RNA interference in the western corn rootworm. *PLoS One*, *9*(7), e101661. DOI: 10.1371/journal.pone.0101661 PMID: 25003334

Mody, V. V., Cox, A., Shah, S., Singh, A., Bevins, W., & Parihar, H. (2014). Magnetic nanoparticle drug delivery systems for targeting tumor. *Applied Nanoscience*, *4*(4), 385–392. DOI: 10.1007/s13204-013-0216-y

Mondello, V., Songy, A., Battiston, E., Pinto, C., Coppin, C., Trotel-Aziz, P., Clément, C., Mugnai, L., & Fontaine, F. (2018). Grapevine trunk diseases: A review of fifteen years of trials for their control with chemicals and biocontrol agents. *Plant Disease*, *102*(7), 1189–1217. DOI: 10.1094/PDIS-08-17-1181-FE PMID: 30673583

Mosa, M. A., & Youssef, K. (2021). Topical delivery of host induced RNAi silencing by layered double hydroxide nanosheets: An efficient tool to decipher pathogenicity gene function of *Fusarium* crown and root rot in tomato. *Physiological and Molecular Plant Pathology*, *115*, 101684. DOI: 10.1016/j.pmpp.2021.101684

Muda, M. S., Kamari, A., Bakar, S. A., Yusoff, S. N. M., Fatimah, I., Phillip, E., & Din, S. M. (2020). Chitosan-graphene oxide nanocomposites as water-solubilising agents for rotenone pesticide. *Journal of Molecular Liquids*, *318*, 114066. DOI: 10.1016/j.molliq.2020.114066

Mydin, R. S. M. N., Zahidi, I. N. M., Ishak, N. N., Nik Ghazali, N. S. S., Moshawih, S., & Siddiquee, S. (2018). Potential of Calcium Carbonate Nanoparticles for Therapeutic Applications. *Malays. J. Med. Health Sci.*, *14*(Suppl. 1), 201–206.

Nandety, R. S., Kuo, Y. W., Nouri, S., & Falk, B. W. (2015). Emerging strategies for RNA interference (RNAi) applications in insects. *Bioengineered*, *6*(1), 8–19. DOI: 10.4161/21655979.2014.979701 PMID: 25424593

Nowara, D., Gay, A., Lacomme, C., Shaw, J., Ridout, C., Douchkov, D., Hensel, G., Kumlehn, J., & Schweizer, P. (2010). HIGS: Host-induced gene silencing in the obligate biotrophic fungal pathogen *Blumeria graminis.The Plant Cell*, *22*(9), 3130–3141. DOI: 10.1105/tpc.110.077040 PMID: 20884801

Oliveira, L. S. B. L., & Ristroph, D. K. (2024). Critical Review: Uptake and translocation of organic nanodelivery vehicles in plants. *Environmental Science & Technology*, *58*(13), 5646–5669. DOI: 10.1021/acs.est.3c09757 PMID: 38517744

Pathania, D., Gupta, D., Agarwal, S., Asif, M., & Gupta, V. K. (2016). Fabrication of chitosan-g-poly(acrylamide)/CuS nanocomposite for controlled drug delivery and antibacterial activity. *Materials Science and Engineering C*, *64*, 428–435. DOI: 10.1016/j.msec.2016.03.065 PMID: 27127073

Peng, Fei, Xiuping Wang, Wenjing Zhang, Xuejuan Shi, Caihong Cheng, Wenlong Hou, Xiaohu Lin, Xiaolu Xiao, & Jun Li. (2022). Nanopesticide formulation from pyraclostrobin and graphene oxide as a nanocarrier and application in controlling plant fungal pathogens. *Nanomaterials 12*, 7, 1112. .DOI: 10.3390/nano12071112

Poreddy, S., Li, J., & Baldwin, I. T. (2017). Plant-mediated RNAi silences midgut-expressed genes in congeneric lepidopteran insects in nature. *BMC Plant Biology*, *17*(1), 199. DOI: 10.1186/s12870-017-1149-5 PMID: 29132300

Qiao, H., Chen, J., Dong, M., Shen, J., & Yan, S. (2024). Nanocarrier-Based Eco-Friendly RNA Pesticides for Sustainable Management of Plant Pathogens and Pests. *Nanomaterials (Basel, Switzerland)*, *14*(23), 1874. DOI: 10.3390/nano14231874 PMID: 39683262

Qiao, H., Zhao, J., Wang, X., Xiao, L., Zhu-Salzman, K., Lei, J., Xu, D., Xu, G., Tan, Y., & Hao, D. (2023). An oral dsRNA delivery system based on chitosan induces G protein-coupled receptor kinase 2 gene silencing for *Apolygus lucorum* control. *Pesticide Biochemistry and Physiology*, *194*, 105481. DOI: 10.1016/j.pestbp.2023.105481 PMID: 37532313

Qiao, L., Lan, C., Capriotti, L., Ah-Fong, A., Nino Sanchez, J., Hamby, R., Heller, J., Zhao, H., Glass, N. L., Judelson, H. S., Mezzetti, B., Niu, D., & Jin, H. (2021). Spray-induced gene silencing for disease control is dependent on the efficiency of pathogen RNA uptake. *Plant Biotechnology Journal*, *19*(9), 1756–1768. DOI: 10.1111/pbi.13589 PMID: 33774895

Qin, H., Zhang, H., Li, L. X., Zhou, X. T., Li, J. P., & Kan, C. Y. (2017). Preparation and properties of lambda-cyhalothrin/polyurethane drug-loaded nanoemulsions. *RSC Advances*, *7*(83), 52684–52693. DOI: 10.1039/C7RA10640H

Ray, P., Sahu, D., Aminedi, R., & Chandran, D. (2022). Concepts and considerations for enhancing RNAi efficiency in phytopathogenic fungi for RNAi-based crop protection using nanocarrier-mediated dsRNA delivery systems. *Frontiers in Fungal Biology*, *3*, 977502. DOI: 10.3389/ffunb.2022.977502 PMID: 37746174

Rosova, E., Smirnova, N., Dresvyanina, E., Smirnova, V., Vlasova, E., Ivan'kova, E., Sokolova, M., Maslennikova, T., Malafeev, K., Kolbe, K., Kanerva, M., & Yudin, V. (2021). Biocomposite materials based on chitosan and lignin: Preparation and characterization. *Cosmetics*, *8*(1), 24. DOI: 10.3390/cosmetics8010024

Saleh, M. C., van Rij, R. P., Hekele, A., Gillis, A., Foley, E., O'Farrell, P. H., & Andino, R. (2006). The endocytic pathway mediates cell entry of dsRNA to induce RNAi Silencing. *Nature Cell Biology*, *8*(8), 793–802. DOI: 10.1038/ncb1439 PMID: 16862146

Seankamsorn, A., Cherdthong, A., So, S., & Wanapat, M. (2021). Influence of chitosan sources on intake, digestibility, rumen fermentation, and milk production in tropical lactating dairy cows. *Tropical Animal Health and Production*, *53*(2), 241. DOI: 10.1007/s11250-021-02697-0 PMID: 33811551

Seankamsorn, A., Cherdthong, A., & Wanapat, M. (2020). Combining crude glycerin with chitosan can manipulate in vitro ruminal efficiency and inhibit methane synthesis. *Animals (Basel)*, *10*(1), 37. DOI: 10.3390/ani10010037 PMID: 31878076

Shen, D., Zhou, F., Xu, Z., He, B., Li, M., Shen, J., Yin, M., & An, C. (2014). Systemically interfering with immune response by a fluorescent cationic dendrimer delivered gene suppression. *Journal of Materials Chemistry. B*, *2*(29), 4653–4659. DOI: 10.1039/C4TB00411F PMID: 32262277

Singewar, K., & Fladung, M. (2023). Double-stranded RNA (dsRNA) technology to control forest insect pests and fungal pathogens: Challenges and opportunities. *Functional & Integrative Genomics*, *23*(2), 185. DOI: 10.1007/s10142-023-01107-y PMID: 37243792

Singh, A., Dhiman, N., Kar, A. K., Singh, D., & Patnaik, S. (2019). Advances in controlled release pesticide formulations: Prospects to safer integrated pest management and sustainable agriculture. *Journal of Hazardous Materials*, *385*, 121525. DOI: 10.1016/j.jhazmat.2019.121525 PMID: 31740313

Singh, S. (2019). Zinc oxide nanoparticles impacts: Cytotoxicity, genotoxicity, developmental toxicity, and neurotoxicity. *Toxicology Mechanisms and Methods*, *29*(4), 300–311. DOI: 10.1080/15376516.2018.1553221 PMID: 30489211

Song, H., Fan, Y., Zhang, J., Cooper, A. M., Silver, K., Li, D., Li, T., Ma, E., Zhu, K. Y., & Zhang, J. (2019a). Contributions of dsRNases to differential RNAi efficiencies between the injection and oral delivery of dsRNA in *Locusta migratoria.Pest Management Science*, *75*(6), 1707–1717. DOI: 10.1002/ps.5291 PMID: 30525311

Song, S. J., Jiang, X. F., Shen, H., Wu, W. N., Shi, Q. Q., Wan, M. H., Zhang, J., Mo, H., & Shen, J. (2021). MXene (Ti_3C_2) Based pesticide delivery system for sustained release and enhanced pest control. *ACS Applied Bio Materials*, *4*(9), 6912–6923. DOI: 10.1021/acsabm.1c00607 PMID: 35006991

Song, S. J., Wang, Y. L., Xie, J., Sun, B. H., Zhou, N. L., Shen, H., & Shen, J. (2019b). Carboxymethyl chitosan modified carbon nanoparticle for controlled emamectin benzoate delivery: Improved solubility, pH-responsive release, and sustainable pest control. *ACS Applied Materials & Interfaces*, *11*(37), 34258–34267. DOI: 10.1021/acsami.9b12564 PMID: 31461267

Tamjidi, F., Shahedi, M., Varshosaz, J., & Nasirpour, A. (2013). Nanostructured lipid carriers (NLC): A potential delivery system for bioactive food molecules. *Innovative Food Science & Emerging Technologies*, *19*, 29–43. DOI: 10.1016/j.ifset.2013.03.002

Taning, C. N., Arpaia, S., Christiaens, O., Dietz-Pfeilstetter, A., Jones, H., Mezzetti, B., Sabbadini, S., Sorteberg, H. G., Sweet, J., Ventura, V., & Smagghe, G. (2020). RNA-based biocontrol compounds: Current status and perspectives to reach the market. *Pest Management Science*, *76*(3), 841–845. DOI: 10.1002/ps.5686 PMID: 31743573

Tong, Y. J., Shao, L. H., Li, X. L., Lu, J. Q., Sun, H. L., Xiang, S., Zhang, Z. H., Wu, Y., & Wu, X. M. (2018). Adhesive and stimulus sesponsive polydopamine-coated graphene oxide system for pesticide-loss control. *Journal of Agricultural and Food Chemistry*, *66*(11), 2616–2622. DOI: 10.1021/acs.jafc.7b05500 PMID: 29485869

Tzeng, S. Y., & Green, J. J. (2013). Subtle changes to polymer structure and degradation mechanism enable highly effective nanoparticles for siRNA and DNA delivery to human brain cancer. *Advanced Healthcare Materials*, *2*(3), 468–480. DOI: 10.1002/adhm.201200257 PMID: 23184674

Ulvila, J., Parikka, M., Kleino, A., Sormunen, R., Ezekowitz, R. A., Kocks, C., & Ramet, M. (2006). Double-stranded RNA is internalized by scavenger receptor-mediated endocytosis in Drosophila S2 Cells. *The Journal of Biological Chemistry*, *281*(20), 14370–14375. DOI: 10.1074/jbc.M513868200 PMID: 16531407

Varkouhi, A. K., Scholte, M., Storm, G., & Haisma, H. J. (2011). Endosomal escape pathways for delivery of biologicals. *Journal of Controlled Release*, *151*(3), 220–228. DOI: 10.1016/j.jconrel.2010.11.004 PMID: 21078351

Varlamov, V. P., & Mysyakina, I. S. (2018). Chitosan in biology, microbiology, medicine, and agriculture. *Microbiology*, *87*(5), 712–715. DOI: 10.1134/S0026261718050168

Vogel, E., Santos, D., Mingels, L., Verdonckt, T. W., & Broeck, J. V. (2019). RNA interference in insects: Protecting beneficials and controlling pests. *Frontiers in Physiology*, *9*, 1912. DOI: 10.3389/fphys.2018.01912 PMID: 30687124

Wang, K., Peng, Y., Pu, J., Fu, W., Wang, J., & Han, Z. (2016). Variation in RNAi efficacy among insect species is attributable to dsRNA degradation in vivo. *Insect Biochemistry and Molecular Biology*, *77*, 1–9. DOI: 10.1016/j.ibmb.2016.07.007 PMID: 27449967

Wang, X., Zheng, K., Cheng, W., Li, J., Liang, X., Shen, J., Dou, D., Yin, M., & Yan, S. (2021). Field application of star polymer-delivered chitosan to amplify plant defense against potato late blight. *Chemical Engineering Journal*, *417*, 129327. DOI: 10.1016/j.cej.2021.129327

Wang, Y., Li, C. N., Wang, T., Li, X. G., & Li, X. (2020). Polylactic Acid–Graphene Oxide-based Materials for loading and sustained release of poorly soluble pesticides. *Langmuir*, *36*(41), 12336–12345. DOI: 10.1021/acs.langmuir.0c02320 PMID: 32988210

Wang, Y., Li, M., Ying, J., Shen, J., Dou, D., Yin, M., Whisson, S. C., Birch, P. R., Yan, S., & Wang, X. (2023a). High-efficiency green management of potato late blight by a self-assembled multicomponent nano-bioprotectant. *Nature Communications*, *14*(1), 5622. DOI: 10.1038/s41467-023-41447-8 PMID: 37699893

Wang, Z., Li, Y., Zhang, B., Gao, X., Shi, M., Zhang, S., Zhong, S., Zheng, Y., & Liu, X. (2023b). Functionalized carbon dot-delivered RNA nano fungicides as superior tools to control Phytophthora pathogens through plant RdRP1 mediated spray-induced gene silencing. *Advanced Functional Materials*, *33*(22), 2213143. DOI: 10.1002/adfm.202213143

Wang, Z., Liu, G., Zheng, H., & Chen, X. (2014). Rigid nanoparticle-based delivery of anti-cancer siRNA: Challenges and opportunities. *Biotechnology Advances*, *32*(4), 831–843. DOI: 10.1016/j.biotechadv.2013.08.020 PMID: 24013011

Wanyika, H. (2013). Sustained release of fungicide metalaxyl by mesoporous silica nanospheres. *Journal of Nanoparticle Research*, *15*(8), 1831. DOI: 10.1007/s11051-013-1831-y

Whyard, S., Singh, A. D., & Wong, S. (2009). Ingested double-stranded RNAs can act as species-specific insecticides. *Insect Biochemistry and Molecular Biology*, *39*(11), 824–832. DOI: 10.1016/j.ibmb.2009.09.007 PMID: 19815067

Winston, W. M., Molodowitch, C., & Hunter, C. P. (2002). Systemic RNAi in *C. Elegans* requires the putative transmembrane protein SID-1. *Science*, *295*(5564), 2456–2459. DOI: 10.1126/science.1068836 PMID: 11834782

Worrall, E. A., Hamid, A., Mody, K. T., Mitter, N., & Pappu, H. R. (2018). Nanotechnology for plant disease management. *Agronomy (Basel)*, *8*(12), 285. DOI: 10.3390/agronomy8120285

Wu, H., & Li, Z. (2022). Recent advances in nano-enabled agriculture for improving plant performance. *The Crop Journal*, *10*(1), 1–12. DOI: 10.1016/j.cj.2021.06.002

Wurm, F. R., & Weiss, C. K. (2014). Nanoparticles from renewable polymers. *Frontiers in Chemistry*, *2*, 49. DOI: 10.3389/fchem.2014.00049 PMID: 25101259

Wynant, N., Santos, D., Van Wielendaele, P., & Vanden Broeck, J. (2014). Scavenger receptor-mediated endocytosis facilitates RNA interference in the desert locust, *Schistocerca gregaria.Insect Molecular Biology*, *23*(3), 320–329. DOI: 10.1111/imb.12083 PMID: 24528536

Xiao, D., Gao, X., Xu, J., Liang, X., Li, Q., Yao, J., & Zhu, K. Y. (2015). Clathrin-dependent endocytosis plays a predominant role in cellular uptake of double-stranded RNA in the red flour beetle. *Insect Biochemistry and Molecular Biology*, *60*, 68–77. DOI: 10.1016/j.ibmb.2015.03.009 PMID: 25863352

Xin, X. P., Judy, J. D., Sumerlin, B. B., & He, Z. L. (2020). Nano-enabled agriculture: From nanoparticles to smart nanodelivery systems. *Environmental Chemistry*, *17*(6), 413–425. DOI: 10.1071/EN19254

Xu, C. L., Cao, L. D., Bilal, M., Cao, C., Zhao, P. Y., Zhang, H. J., & Huang, Q. L. (2021). Multifunctional manganese-based carboxymethyl chitosan hydrogels for pH-Triggered pesticide release and enhanced fungicidal activity. *Carbohydrate Polymers*, *262*, 117933. DOI: 10.1016/j.carbpol.2021.117933 PMID: 33838810

Xu, C. L., Shan, Y. P., Bilal, M., Xu, B., Cao, L. D., & Huang, Q. L. (2020). Copper Ions Chelated mesoporous silica nanoparticles via dopamine chemistry for controlled pesticide release regulated by coordination bonding. *Chemical Engineering Journal*, *395*, 125093. DOI: 10.1016/j.cej.2020.125093

Xu, L., Cao, L. D., Li, F. M., Wang, X. J., & Huang, Q. L. (2014). Utilization of chitosan-lactide copolymer nanoparticles as controlled release pesticide carrier for pyraclostrobin against *Colletotrichum gossypii* Southw. *Journal of Dispersion Science and Technology*, *35*(4), 544–550. DOI: 10.1080/01932691.2013.800455

Yan, S., Ren, B. Y., & Shen, J. (2021). Nanoparticle-mediated double-stranded RNA delivery system: A promising approach for sustainable pest management. *Insect Science*, *28*(1), 21–34. DOI: 10.1111/1744-7917.12822 PMID: 32478473

Yan, S., Yin, M.-Z., & Shen, J. (2022a). Nanoparticle-based nontransformative RNA insecticides for sustainable pest control: Mechanisms, current status and challenges. *Entomologia Generalis*, *43*(1), 21–30. DOI: 10.1127/entomologia/2022/1618

Yan, Y., Zhu, X., Yu, Y., Li, C., Zhang, Z., & Wang, F. (2022b). Nanotechnology strategies for plant genetic engineering. *Advanced Materials*, *34*(7), e2106945. DOI: 10.1002/adma.202106945 PMID: 34699644

Ye, C., Hu, X. S., Wang, Z. W., Wei, D., Smagghe, G., Christiaens, O., Niu, J., & Wang, J. J. (2021). Involvement of clathrin-dependent endocytosis in cellular dsRNA uptake in aphids. *I. Insect Biochemistry and Molecular Biology*, *132*, 103557. DOI: 10.1016/j.ibmb.2021.103557 PMID: 33639241

Yiamsawas, D., Beckers, S. J., Lu, H., Landfester, K., & Wurm, F. R. (2017). Morphology-controlled synthesis of lignin nanocarriers for drug delivery and carbon materials. *ACS Biomaterials Science & Engineering*, *3*(10), 2375–2383. DOI: 10.1021/acsbiomaterials.7b00278 PMID: 33445295

Zamore, P. D., Tuschl, T., Sharp, P. A., & Bartel, D. P. (2000). RNAi: Double-stranded RNA directs the ATP-dependent cleavage of mRNA at 21 to 23 nucleotide intervals. *Cell*, *101*(1), 25–33. DOI: 10.1016/S0092-8674(00)80620-0 PMID: 10778853

Zhang, J., Khan, S. A., Heckel, D. G., & Bock, R. (2017). Next-generation insect-resistant plants: RNAi-mediated crop protection. *Trends in Biotechnology*, *35*(9), 871–882. DOI: 10.1016/j.tibtech.2017.04.009 PMID: 28822479

Zhang, P., Guo, Z., Zhang, Z., Fu, H., White, J. C., & Lynch, I. (2020). Nanomaterial transformation in the soil–plant system: Implications for food safety and application in agriculture. *Small*, *16*(21), 2000705. DOI: 10.1002/smll.202000705 PMID: 32462786

Zhao, J., Yan, S., Li, M., Sun, L., Dong, M., Yin, M., Shen, J., & Zhao, Z. (2023). NPFR regulates the synthesis and metabolism of lipids and glycogen via AMPK: Novel targets for efficient corn borer management. *International Journal of Biological Macromolecules*, *247*, 125816. DOI: 10.1016/j.ijbiomac.2023.125816 PMID: 37451386

Zhao, L., Lu, L., Wang, A., Zhang, H., Huang, M., Wu, H., Xing, B., Wang, Z., & Ji, R. (2020). Nano-biotechnology in agriculture: Use of nanomaterials to promote plant growth and stress tolerance. *Journal of Agricultural and Food Chemistry*, *68*(7), 1935–1947. DOI: 10.1021/acs.jafc.9b06615 PMID: 32003987

Zheng, Y., Chen, X., Joseph, N.D., Zhang, Y., Chen, H., & Gao, B. (2022). Occurrences and impacts of engineered nanoparticles in soils and groundwater. *Emerg. Contam. Soil Groundw. Syst.* 165–204, .DOI: 10.1016/B978-0-12-824088-5.00005-7

Zheng, Y., Hu, Y., Yan, S., Zhou, H., Song, D., Yin, M., & Shen, J. (2019). A polymer/detergent formulation improves dsRNA penetration through the body wall and RNAi-induced mortality in the soybean aphid *Aphis glycines.Pest Management Science*, *75*(7), 1993–1999. DOI: 10.1002/ps.5313 PMID: 30610748

Zhou, H., Wan, F., Jian, Y., Guo, F., Zhang, M., Shi, S., Yang, L., Li, S., Liu, Y., & Ding, W. (2023). Chitosan/dsRNA polyplex nanoparticles advance environmental RNA interference efficiency through activating clathrin-dependent endocytosis. *International Journal of Biological Macromolecules*, *253*, 127021. DOI: 10.1016/j.ijbiomac.2023.127021 PMID: 37741481

Zhou, J., Shum, K. T., Burnett, J. C., & Rossi, J. J. (2013). Nanoparticle-based delivery of RNAi therapeutics: Progress and challenges. *Pharmaceuticals (Basel, Switzerland)*, *6*(1), 85–107. DOI: 10.3390/ph6010085 PMID: 23667320

Zhu, K. Y., & Palli, S. R. (2020). Mechanisms, applications, and challenges of insect RNA interference. *Annual Review of Entomology*, *65*(1), 293–311. DOI: 10.1146/annurev-ento-011019-025224 PMID: 31610134

Zhu, X. Q., Zhang, J. Y., Xie, Q., & Hou, Z. L. (2020). High Sensitivity and Ultrafast Response Ethanol Sensors Based on Graphene Oxide. *ACS Applied Materials & Interfaces*, *12*(34), 38708–38713. DOI: 10.1021/acsami.0c12196 PMID: 32805966

Zotti, M., Dos Santos, E. A., Cagliari, D., Christiaens, O., Taning, C. N. T., & Smagghe, G. (2018). RNA interference technology in crop protection against arthropod pests, pathogens and nematodes. *Pest Management Science*, *74*(6), 1239–1250. DOI: 10.1002/ps.4813 PMID: 29194942

Zou, T., Sipponen, M. H., & Österberg, M. (2019). Natural shape-retaining microcapsules with shells made of chitosan-coated colloidal lignin particles. *Frontiers in Chemistry*, *7*, 370. DOI: 10.3389/fchem.2019.00370 PMID: 31192192

APPENDIX

Table 1. Utilization of nanoparticles as nanocarriers of fungicides

Nanoparticles	Pesticide	Pathogen	References
Nanoemulsion Chitosan	Chitosan	*Colletotrichum gloeosporioides, C. musae*	(Kaushal, 2018)
PVP copolymer	Tebuconazole, Chlorothalonil	*T. versicolor, G. trabeum*	(Liu et al., 2002a; Franco, & De Marco, 2020)
PVC	Tebuconazole, Chlorothalonil	*T. versicolor, G. trabeum*	(Liu et al., 2002b)
MSN	Metalaxyl	-	(Wanyika, 2013)
PLA– Chitosan	Pyraclostrobin	*C. gossypii*	(Xu et al., 2014]
Chitosan	Chitosan	*Sclerotium rolfsii, R. stolonifera, Thanatephorus cucumeris*	(Abdullahi et al., 2020)
Calcium carbonate	Validamycin	*R. solani*	(Mydin et al., 2018)

Chapter 13
Nanocarriers in Plant Disease Management

H. S. Supreeth
https://orcid.org/0009-0000-4412-6594
Christ University, India

Rashmi Ramakrishnan
Christ University, India

Jameel M. Al-Khayri
King Faisal University, Saudi Arabia

Mustafa I. Almaghasla
King Faisal University, Saudi Arabia

Praveen Nagella
https://orcid.org/0000-0001-6971-0527
Christ University, India

ABSTRACT

Almost 30-40% of the harvest is lost due to plant diseases worldwide. When traditional disease management strategies become less efficient, the advancement in management and incorporation of modern domains like nanotechnology can be the key factor to increase the efficiency of disease management. Nanocarriers are the particles in the nano range that are greatly diverse in makeup and transporting agents that target the specific site to induce biochemical interactions by releasing active compounds and defensive responses. The mechanism, despite being completely immunological, is a combination of chemical, physical, and biochemical interactions. A diverse range of nanocarriers is used to deliver bioactive molecules to control fungi, bacteria, nematodes, and other biotic limiting factors in agriculture. The usual makeup of nanocarriers varies from liposomes, chitosan, to inorganic metal carriers or carbon, silica-based materials. This chapter is a comprehensive analysis of morphology, loading capacity, biocompatibility of nanocarriers, and the host plant for effective disease control.

DOI: 10.4018/979-8-3693-9869-2.ch013

INTRODUCTION

Every living organism existing is associated with other creatures forming a strong symbiotic or antagonistic relationship, and the dependency among the creatures that form an ecosystem is unmatchable for various amenities (Chomicki et al., 2022). When this relationship has more antagonistic features, the host organism starts getting harmed because of the life processes of the other party involved, which makes it parasitic. This impacts the life activities and survival ability of the host organism negatively and affects the bioproduction of it too. In a global agricultural setup, the prevailing goal is to get the optimum yield from the crops with the minimum possible loss shared by various factors (Baloch, 2025). But the concept of microbial intervention challenges the production quality and quantity. The loss can be as huge as to destroy the yield of the whole country, which risks food security and global integrity (Pieter Johnson et al., 2024). The accidental intervention of *Phytophthora infestans* in Ireland caused severe late blight disease in potatoes, which resulted in great hunger and was titled as 'The Irish Potato Famine' (Bowbrick, 2024). Plant diseases are the diseases caused by various pathogens, biotic, abiotic, or physiological factors, which in turn result in loss in productivity. These diseases are widely studied under plant pathology (Ahmad et al., 2023). Plants, especially in agricultural fields, are more susceptible to the pathogenic diseases as they are monocultures and lack genetic diversity than the ones in their natural vegetation, where the complex ecosystem and beneficial micro relationships pose selective pressure on pathogens, resulting in mild infections. The usual plant pathogens are fungi, bacteria, viruses, nematodes, and some phytophagous insects and pests (figure 1), which can be broadly categorized under biotic stress factors (Michael Munster, 2022).

Figure 1. Various ways of plant infection

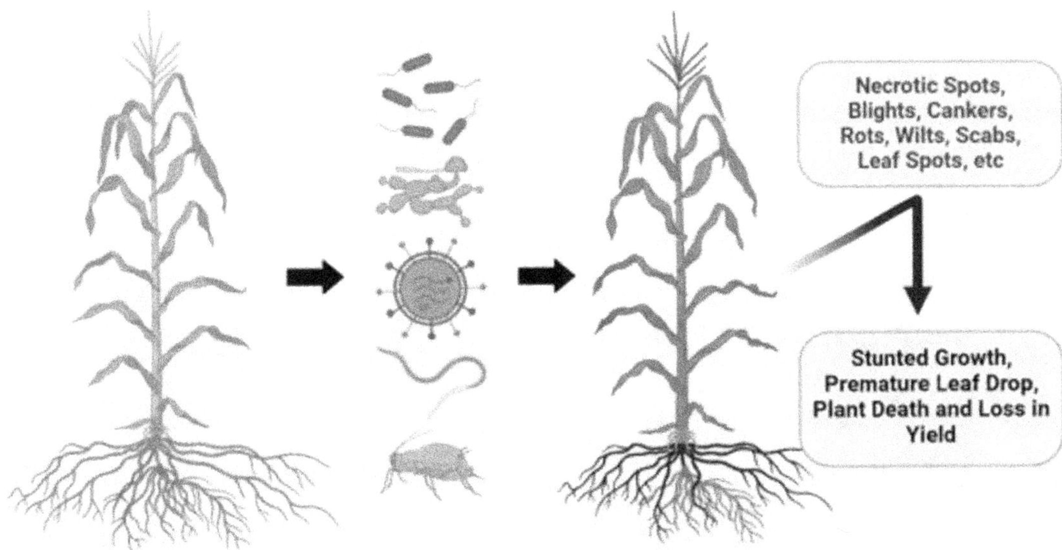

This chapter deals with the diseases that are affecting the agricultural crops, making them less efficient when magnified on the basis of production or yield quality and quantity. The disease caused by a few pathogens can be so strong as to damage the entire ecosystem but can be prevented or treated if the pathogenesis is studied properly to understand the morphological and anatomical complexities of them to enumerate an effective tackling mechanism that interrupts the disease-causing ability and the elimination of it from the system (Fu et al., 2020). Here, the focus is driven on the molecular level of understanding, where nanotechnology is combined with biotechnological applications to provide effective understanding and solutions to the plant diseases that are creating disastrous loss. The nanocarrier- based approach to deliver the magic biochemical that treats the disease through whatsoever mechanisms is not a recent discovery but an emerging subject that frames scope a great potential for research. So, the chapter discusses the management of diseases in agricultural crop in an effective way than the traditional version in all the ways using nanocarriers (Elmer et al., 2018).

Agricultural Management and Evolution of Nanocarriers

The concept of management of plants comes from the ancient to modern sophisticated advancements to tackle the challenges due to biotic factors such as pathogens, pests, and weeds, or abiotic factors like environmental stressors and nutrient deficiencies. This can adversely affect the productivity and quality of the yield. Plants face numerous problems that require effective management to ensure healthy growth and optimal yields (Kumari et al., 2023). These issues require an integrated approach of combining traditional practices with biological control methods, chemical treatments, and advanced technologies like nanocarriers to mitigate the effect in the best possible way that enhances agricultural productivity sustainably (Zhang et al., 2024). The loss in agriculture has varied targets on economic instability, food security issues, environmental degradation, and many more. With the pattern of rapid growth in the world population, it is estimated that the production efficiency of the agricultural practices has to be incremented by almost 60–70% to cater to the building necessities by 2050 (Cătălin Balaure et al., 2017). With the practice of chemical treatments such as pesticides, fungicides, herbicides and other biotic stress-limiting agents, there has to be a sequential increase in the percentage of usage of these chemical treatments as well. On a broad range, the ill effects of these chemical agents are well studied to cause intolerable damage to the ecosystem and the environment (Mittler, 2006).

Figure 2. Pie chart showing agricultural loss proportions

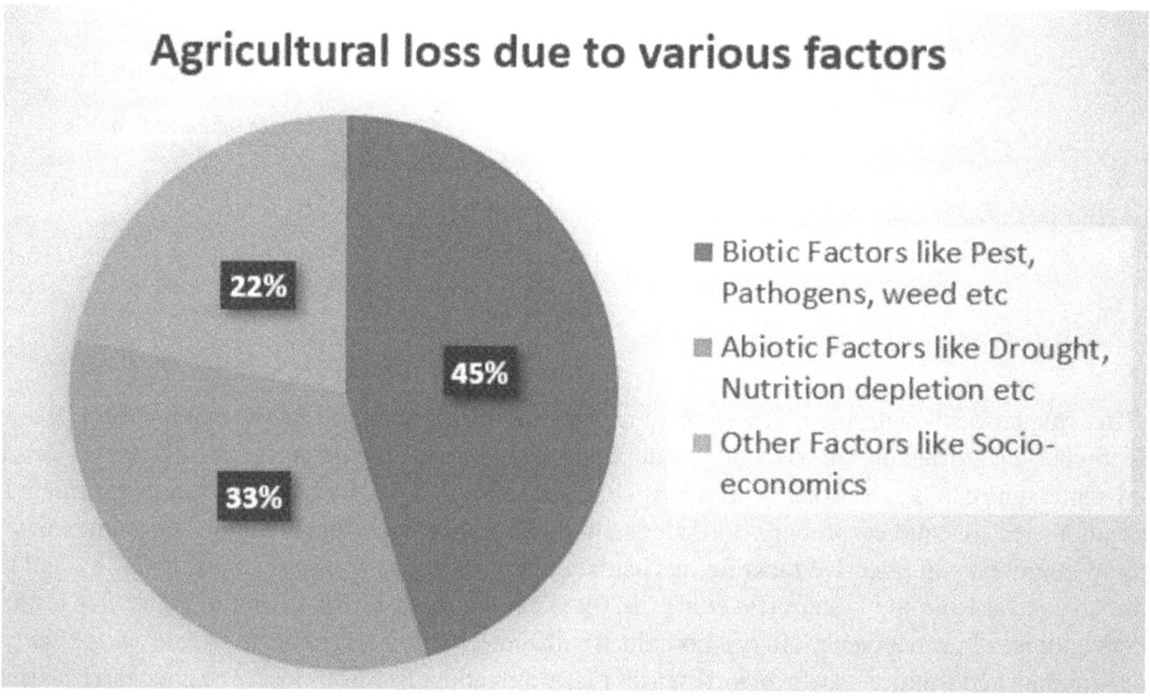

Agriculture has evolved over time in terms of management, and so does the pathogenesis of plants. Among various other factors like declining soil fertility, climate change, unscientific usage of chemicals and fertilizers, loss due to plant pathogens is considered major and complex too (figure 2). Approximately 40–45% of the loss in harvest annually is because of the various plant pathogens (Sinha et al., 2017). Traditionally, chemical methods such as the application of fungicides, insecticides, pesticides are used. But the major surveys suggest that long-term usage of these compounds can detain the water retention capacity of soil by abating the organic matter content, nutrient levels, thereby leading to loss. The analysis of the data shown in figure 3 clearly depicts the future demands and requirements (Cătălin Balaure et al., 2017).

Figure 3. Double bar graph showing use of chemical fertilizers over the time to maximise the production

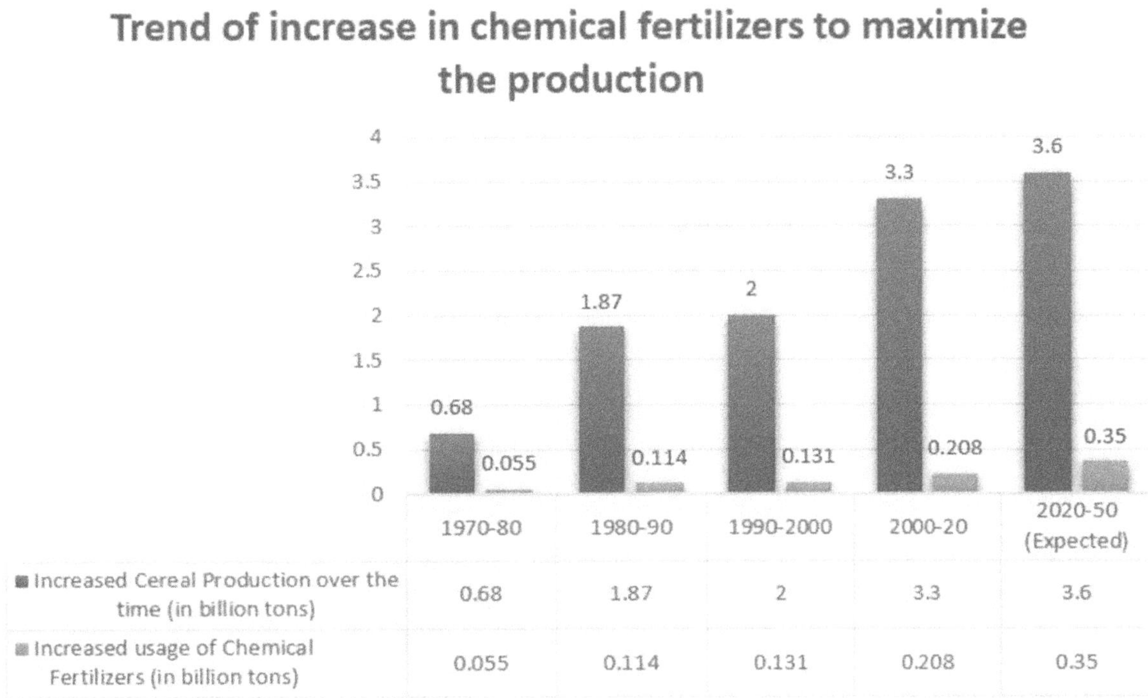

Nanocarriers provide numerous benefits over conventional disease management and farming methods. The incorporation of nanocarriers into agricultural practices permits us to address some of the most sensitive issues in food production and environmental conservation, setting the way for a more resilient and environmentally friendly agricultural network. By harnessing precision and versatility, the field of agriculture can be revolutionized by effectively managing the diseases and ultimately increasing productivity (Khoury & Makkouk, 2010).

The concept of nanotechnology evolved in the late 1950s with the foundational concepts led by the American physicist and Nobel Prize laureate Richard Feynman, who is considered to be the Father of Nanotechnology (Abu-Thabit & Makhlouf, 2018). His 1959 lecture, "There's Plenty of Room at the Bottom," is often credited with inspiring the field of nanotechnology (Durán-Lobato et al., 2021). Later on, the discovery of Scanning Tunnelling Microscopy, Atomic Force Microscopy allowed the scientific community to decode the particles which were in nanoscale, establishing the domain called Nanotechnology. As the field started expanding its applications, agriculture also became a part and plant science started harnessing benefits. Site targeted delivery of agrochemicals through nanocarriers in order to treat the diseases and other potential applications widened the field (Wasule et al., 2024). Major advances in nanomaterial synthesis and characterisation in the 2000s culminated in the development of more efficient and effective nanocarriers (figure 4) (Zhang et al., 2024).

Figure 4. Flowchart showing milestones of nanotechnology

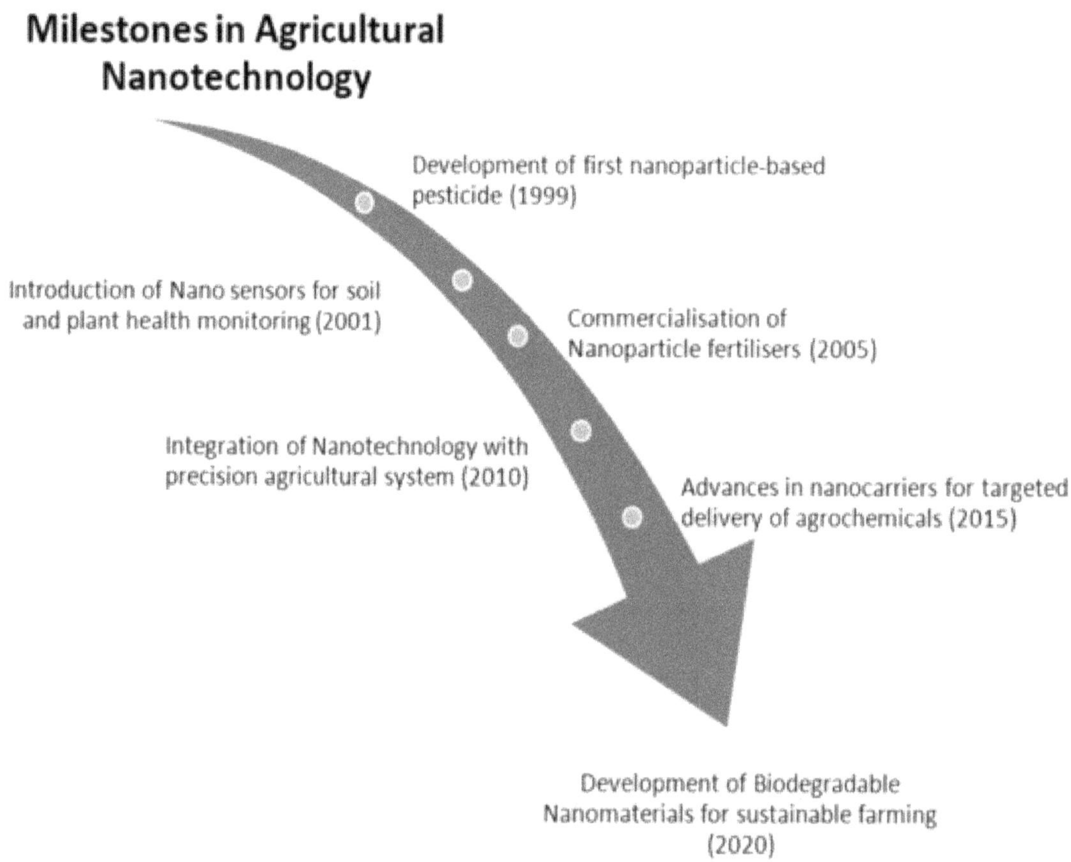

DIVERSITY OF NANOCARRIERS

Based on the compatibility and efficiency, there are many types of nanocarriers that are used to manage the plant diseases such as liposomes, nanoparticles, dendrimers, hybrid nanocarriers etc. Like shown in figure 5, these types are further classified based on structure and function (Sarvarian et al., 2022).

Liposomes as Nanocarriers

They are tiny spherical structures of phospholipid that are arranged in one or more layers in bilayer pattern which can encapsulate, protect and deliver both hydrophilic (water-soluble) and hydrophobic (fat-soluble) compounds to the target site (Meiwan Chen et al., 2012). Structurally it has an outer phospholipid bilayer, inner aqueous core which can store water soluble agrochemical and the hydrophobic interior region within the bilayer where fat soluble agrochemicals can be stored (Di Sotto et al., 2018). Liposomes are usually used to deliver fungicides such as Azoles (e.g., triadimefon) strobilurins (e.g., azoxystrobin), insecticides such as Pyrethroids (e.g., deltamethrin) neonicotinoids (e.g., imidacloprid), herbicides such as Glyphosate, atrazine, micronutrients such as Zn, Fe, or nitrogen fertilizers, beneficial microorganisms such as *Trichoderma, Bacillus thuringiensis,* PGRs and other agrochemicals (Bonechi et al., 2025). They are classified based on the structure, size, and the number of lipid bilayers into, unilamellar (single lipid bilayer), multilamellar (more than one concentric lipid layer) and charged liposomes. They enhance solubility, availability and targeted action of molecules of our interest to the plant which help to maximize the benefits of these chemicals while minimizing environmental impact by controlled release (Sarvarian et al., 2022).

Nanoparticles as Nanocarriers (Metallic, Polymeric, Silica and Carbon Based)

They are small solid particles made from various materials, including metals, polymers or carbon whose size ranges within 1 to 100 nm. Unlike other carriers, nanoparticles offer a large surface area-to-volume ratio which enhances bioavailability and solubility of beneficial agrochemicals. Similar to liposomes, this can also deliver fungicides, insecticides, nutrient solutions, PGRs and also nano emulsion mixtures with much more penetrance into the plant tissue (Alemzadeh et al., 2018). Metallic nanoparticles are the particles made up of Cu, Au, Ag, Zn etc. which are specially used for their antimicrobial property and high stability. They are chemically synthesised by reducing the metal salt solutions by strong reducing agents like hydrazine, sodium borohydride ($NaBH_4$) or citrate ions. This exothermic reaction yields in the formation of nanoparticles which are then stabilized using Polyvinyl alcohol (PVA), Sodium citrate or Chitosan to prevent agglomeration (Song & Cho, 2024). Silver nanoparticles (AgNPs) loaded with *azoxystrobin* enhances antifungal activity against *Botrytis cinerea* in grapes by triggering SAR and ROS accumulation (Alemzadeh et al., 2018). Polymeric nanoparticles are the biodegradable nanoparticles that are safe for environments and accumulate less toxicity. Usually, polymers such as polylactic acid (PLA), polyvinyl alcohol (PVA), and chitosan are used as they can be characterized to show controlled or triggered release of agrochemicals based on environmental or other stress factors. Chitosan nanoparticles loaded with *hexaconazole* suppresses *Fusarium oxysporum* in banana plants by inhibiting the biosynthesis of ergosterol in fungi. Chitosan itself acts as an elicitor, upregulating defence genes like PR1 and PAL. Silica nanoparticles are made up of silicon dioxide (SiO_2). They are usually mesoporous, having a pore size of 2-50 nm, or can be non-porous. They are resistant to heat, pH, and chemical interventions and offer a high surface area, which increases the adsorption rate (Saruchi et al., 2023). Carbon-based nanoparticles are unique carbon nanotubes, graphene, or fullerenes, known for their mechanical and electrical properties. In *Arabidopsis*, CNTs facilitate gene delivery without Agrobacterium, leading to stable expression of resistance traits and mild upregulation of oxidative stress defence genes. (Abbasi et al., 2019).

Dendrimers as Nanocarriers

Dendrimers are highly branched, tree-like macromolecules characterized by a central core that facilitates interaction with plant systems, multiple layers of branching units (Dendrons) that enhancing interactions with plant cells by increasing the surface area and the functional terminal groups that enhance solubility, reactivity, and targeting capabilities allowing for the effective delivery of agrochemicals and nutrients to plants (Bobbarala, 2012). They are synthesized layer by layer, starting from a core; this method is called the constructive (bottom-up) approach (Kumar et al., 2022). This allows for the incorporation of specific functional groups that can enhance plant uptake or target disease pathogens. Another version of dendrimer synthesis is the self-assembly technique, wherein the molecules organize based on interactions (Wang et al., 2023). This method can be useful for developing delivery systems that are responsive to plant environments. They offer some unique benefits, like mono disparity in terms of size, induces plant resistance, act as biosensors for pathogen detection, and an internal cavity for delivering nutrients or protective agents directly to the plant (Boas & Heegaard, 2004).

Hybrid Nanocarriers

Hybrid nanocarriers are the combination of two types of nanocarriers to leverage the advantages and improve the delivery of bioactive agents, including drugs, nutrients, and agrochemicals (Feng et al., 2024). Hybrid nanocarriers are gaining attention for their ability to enhance the effectiveness of plant disease management strategies because of their versatility, improved stability, and synergistic effects (Wen, 2019). A few combinations are inorganic-organic nanocomposites (silica-polymer nanocomposites), lipid-polymer hybrid nanocarriers (lipid-core polymeric shell nanocarriers, polymer-stabilized liposomes), metal-organic frameworks or MOFs (Zn-based MOFs, Fe-based MOFs), chitosan-based hybrid nanocarrier (chitosan-polymer blends, chitosan-silica nanocomposites) (figure 5) (Das, 2020).

Figure 5. Types of plant nanocarriers

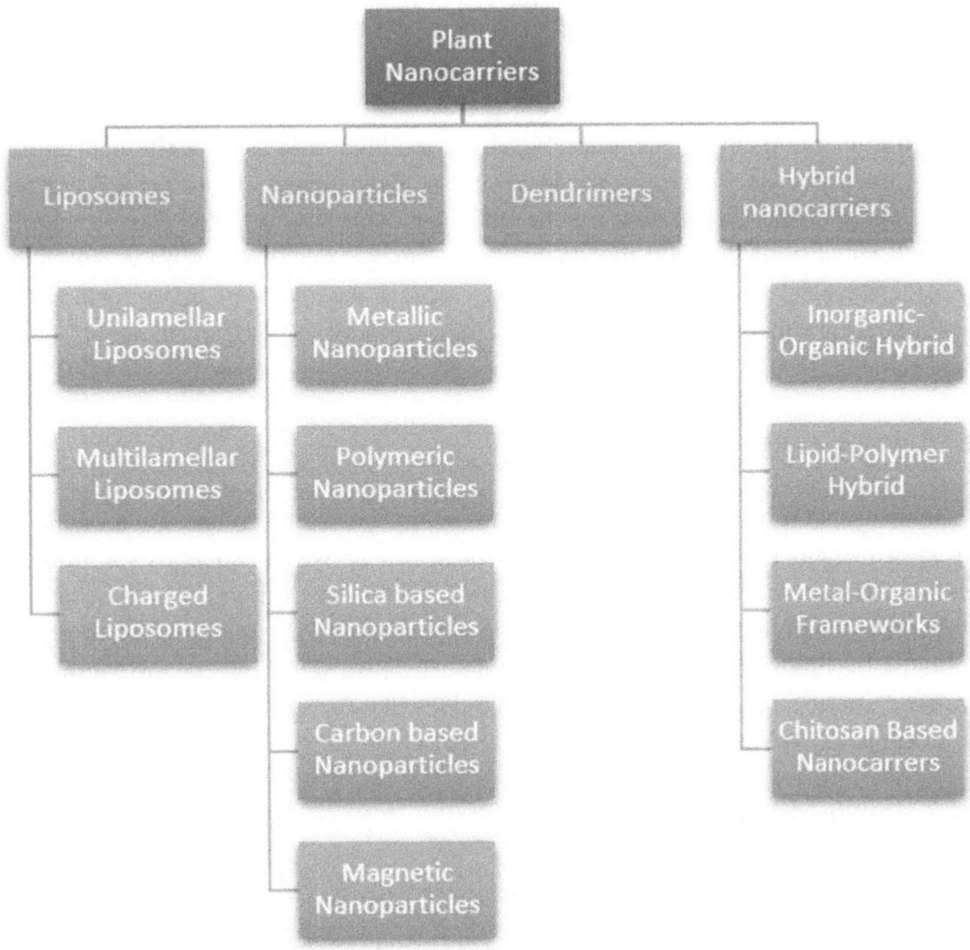

SYNTHESIS AND SELECTION OF NANOCARRIERS

The engineered nanoscale carriers that deliver the active compound play a pivotal role which requires precise synthesis design and mechanism (Kumari et al., 2023). The diverse variety caters efficient way to elicit response against all the complexities challenged by the pathogens but the selection procedure, as to which type of nanocarrier, the encapsulation of active compounds or agrochemical is done, decides the efficacy. While synthesizing the nanocarrier, factors like size, surface charge and loading capacity should be given priority (Ly et al., 2025). There are many strategies like the Top-Down approach, where bulk material is subjected to various physical exertions that makes it get reduced to nano size. This approach is typically used while developing metallic and crystalline nanoparticles. The opposite to this is another approach, the bottom-up approach where chemical reactions build up the nanocarriers (Nasrollahi & Alizadeh, 2024). This method fabricates the liposomes, polymeric and silica-based nanocarriers. Sometimes nanostructures are formed through scaffolds, using them as templates (Borane et al., 2024). Some of the techniques that fall under these three categories are:

Thin Hydration Method

Thin hydration is one of the most commonly used methods that follows the bottom-up approach, where phospholipids are dissolved in a volatile organic solvent such as chloroform. When the solvent is evaporated, a thin lipid film on the flask's walls is formed. This step is called film formation. The film is hydrated with an aqueous solution of our interested agrochemical solution by gentle shaking. The liposome suspension can be sonicated or extruded to achieve desired size and uniformity. This method is widely used to encapsulate fungicides or insecticides, RNA-based biocontrol agents, and hormones, which enhances their efficacy and reduces phytotoxicity (Mahesh Umbarkar et al., 2021).

Reverse Phase Evaporation

Reverse phase evaporation is a special technique to encapsulate larger volumes of the aqueous solutions of biocontrol agents, which helps treat macronutrient deficiencies. The aqueous solution of biocontrol agent is mixed with organic solvent and lipid mixture by continuous stirring, which forms a water-in-oil emulsion following the bottom-up approach. As the solvent evaporates under reduced pressure using a rotary evaporator, the lipids coalesce to form liposomes. Because of its high encapsulation efficiency and sustained release, it is used to treat hydrophilic fungicides like carbendazim and hydrophilic antibiotics like streptomycin, siRNAs molecules etc (Shi & Qi, 2021).

Layer-by-Layer (LbL) Assembly

LbL method helps synthesize polymeric nanocarriers with specific release profiles, through which targeted delivery of multiple agrochemicals is achieved, which can improve disease management strategies. A nanoparticle-based substrate is selected over which liposomes will be formed. Sequential deposition of charged lipids or polymers onto the substrate allows adsorption between each layer. During the assembly process, biocontrol agents, nutrients, agrochemicals or other active ingredients are incorporated into the multilayer structure, enabling controlled release profiles. After achieving the desired number of layers, the final structure can be detached from the substrate and characterized for further application (Ruano et al., 2021).

Physical methods like laser ablation, where a target metal is submerged in the liquid medium which is irradiated using a high-powered laser to generate nanoparticles, the evaporation-condensation method and the mechanical milling method are also used to produce metal nanoparticles that are free from chemical contaminants (Pechyen et al., 2024). These are further characterised by TEM, SEM, XRD etc (Angayarkanni & Philip, 2022).

TRANSPORTATION OF LOADED NANOCARRIERS AND GENE REGULATION IN HOST PLANT

Nanocarriers cross plant barriers using multiple pathways. Entry is often achieved through endocytosis, passive diffusion, or direct membrane interaction, including temporary disruption or charge-based adhesion. Once inside, they may dissolve, release ions, or provide controlled delivery of bioactive compounds. Their nanoscale size allows them to penetrate cell walls and reach target tissues. Facilitated transport mechanisms and phloem mobility further enhance their systemic reach. By combining these transport modes, nanocarriers improve the efficiency, stability, and targeting of plant protection agents making them an advanced alternative to conventional delivery systems in plant disease management (Sousa de Almeida et al., 2021).

The interaction of host plant species, pathogen involved, and the physicochemical nature of the nanoparticles used collectively influence the transcriptional reprogramming of defence-related genes and modulation of signalling pathways. Transcriptomic analysis reveals an upregulation of PR1, NPR1, and ICS1 genes in *Arabidopsis thaliana*, when AgNPs are integrated, which have been shown to activate the salicylic acid (SA)-dependent systemic acquired resistance (SAR) pathway and signal amplification (Baldassarre et al., 2020). Similarly, in tomato (*Solanum lycopersicum*) challenged particularly against necrotrophic pathogens like *Alternaria solani*, ZnONPs regulate genes like LOX2, AOS, and ERF1, triggering jasmonic acid (JA) and ethylene (ET) signalling cascades, which are associated with induced systemic resistance (ISR). The type and size of nanoparticles also influence the cellular pathways (Sousa de Almeida et al., 2021). Chitosan-based nanocarriers loaded with fungicides upregulate OsWRKY45, OsMPK6, and OsNPR3 which activate MAPK signalling cascades in *Oryza sativa*, especially in response to fungal pathogens. This modulates defence priming and transcriptional memory, enhancing the plant's ability to respond more effectively to subsequent infections (Wu et al., 2023). In *Glycine max* (soybean) and many leguminous crops, nano formulations of copper have shown to upregulate host R genes, including RPM1 and RPS2, suggesting a gene-for-gene resistance mechanism enhanced by nanoparticle-mediated delivery systems which downregulate *Phytophthora sojae* effectors concurrently (Angayarkanni & Philip, 2022).

The fact that the same nanoparticles can elicit divergent genetic responses in different plants is interesting. For example, carbon nanotubes (CNTs) activate oxidative stress-related genes such as CAT, SOD, and APX in maize, while in wheat, they induce a broader transcriptional response involving secondary metabolite biosynthesis genes like PAL and CHS, which are linked to phytoalexin production or phenylpropanoid biosynthesis in cereals in general (Wu et al., 2023).

APPLICATIONS OF NANOCARRIERS IN PLANT DISEASE MANAGEMENT

The incorporation of specific payloads to the above-mentioned various base nanocarriers is precisely delivered to the target site. The aid of nanocarriers benefits the mechanism of action of the payload through effective stabilization, solubilization, and controlled release (Mansour et al., 2025). The mechanism of action is very diverse to specific depending on the complexity and type of disease (figure 6). From specific antimicrobial compounds to boosting an overall plant's defence system, nanocarriers are unique (Kandhol et al., 2024).

Figure 6. General mechanism of action of nanocarriers inside the targeted plant host

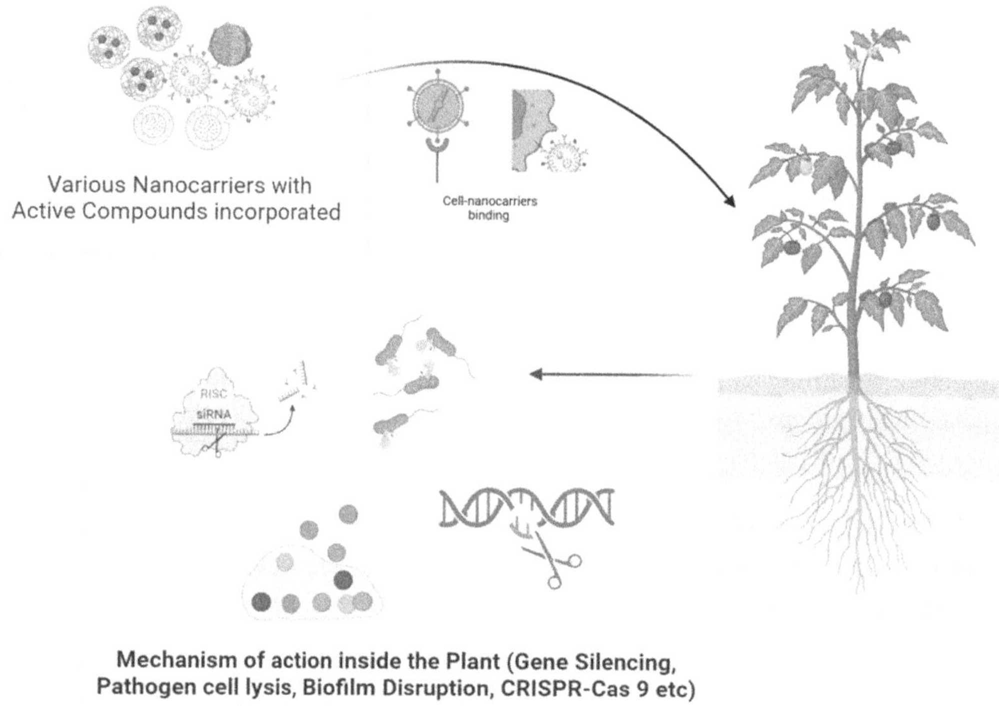

Treating Bacterial Diseases

Approximately 27% of the total loss is because of the bacterial diseases (Xiang et al., 2023). Bacterial pathogens such as *Xanthomonas oryzae, Pseudomonas syringae, Xanthomonas campestris, Erwinia amylovora, Ralstonia solanacearum* cause severe loss if the disease is left untreated in the initial stage (Xiang et al., 2023). Nanocarriers provide precision in targeting the disease-causing bacteria by delivering the antimicrobial agents directly to the infected area, which also prevents the off-target effects. Few bacterial strains have the ability to colonize, form aggregates, and build biofilms, which act as a protective layer for them against the antimicrobials by signal regulation through quorum sensing, EPS production, or c-di-GMP signalling (Xiang et al., 2023). Quorum sensing (QS) is a bacterial communication system that determines virulence. This can be inhibited by using quorum sensing inhibitors (QSIs) such as furanones and cinnamaldehyde derivatives, which suppress the QS regulatory gene. Nanocarriers have shown effective disruption of biofilms, thereby enhancing the activity of antimicrobial agents packed with nanocarriers. Varied types of nanocarrier makeups are used for treating bacterial infections based on their advantages and type of disease (Baldassarre et al., 2020). Polymeric nanocarriers are biodegradable and biocompatible which helps in controlled release of encapsulated agrochemical sustainably which is why, it is preferred more to deliver antimicrobials and bacteriophages to treat bacterial infection such as bacterial blight in rice (*Xanthomonas oryzae*), fire blight in apples (*Erwinia amylovora*), etc. Other than this, lipid based nanocarriers are used to deliver and increase the penetrance of hydrophobic agrochemicals or antimicrobial peptides. Inorganic nanoparticles like silver, zinc and copper nanoparticles directly work to disrupt cellular integrity of pathogens by producing ROS. Other than these, when the therapeutic molecule is poorly soluble in water, nano micelles are used to increase the bioavailability and dendrimers are used to increase the drug loading capacity (Pal et al., 2024).

Nanocarriers disrupt the cell wall of the bacteria, which causes structural damage that eventually causes cell death. While few nanoparticles like silver and zinc oxide nanoparticles produce ROS, which severely damage the cell wall, in others, nanocarrier payload can be antimicrobial peptides such as lysozymes, which can disrupt the peptidoglycan layer, essential oils, which degrade the lipid layer, and some other antibiotics against the pathogen (Avila-Quezada et al., 2022). Nanocarriers are loaded with dsRNA and siRNA for gene silencing. RNAi involves silencing of essential genes of the pathogen, which is essential for antibiotic resistance, virulence, etc. Introduction of PAMPs such as flagellin peptide can induce hypersensitivity reactions, which lead to the death of cells localized in the infected region. This prevents the further spread, and eventually the plant recovers. Salicylic acid, jasmonic acid, and β-glucans can aid in generating a systemic acquired immune response to tackle bacterial infections (Xiang et al., 2023).

Treating Viral Diseases

The viral infections can be more severe than the bacterial infections in many ways and so the pattern of treatment also changes significantly (Eid et al., 2025). Nanocarriers can deliver the antiviral compounds more efficiently to the target than the traditional way of application through spraying pesticides. The payload can contain dsRNA, siRNA for viral gene silencing that is essential for the formation of coat proteins, translation etc, any broad or narrow spectrum antiviral compounds such as ribavirin or quercetin. Encapsulation helps increase the efficiency of these antiviral compound coated nanoparticles. Silica nanoparticles disturb the association of viral components inside the host cell by delivering the peptides that selectively binds and blocks the coat protein forming sequence (Xiang et al., 2023). Helicase-primase inhibitors (HPIs) and nucleoside analogues are used to block the replication which stops the infection cycle. Small interfering RNAs can be introduced to the target site through polymeric nanoparticles as they are capable of controlled and sustained release which increases plant uptake. In *Cucumber mosaic virus* (CMV) and *Tobacco mosaic virus* (TMV) treatment, antiviral peptide sequence is introduced into the target site through lipid based nanocarriers as they can effortlessly deliver hydrophobic molecules into the system by increasing their solubility. In some cases, such as *Tomato yellow leaf curl virus* (TYLCV), there is a requirement of large loading of drugs for the efficient treatment. Hence carbon-based nanoparticles like carbon quantum dots, graphene oxide Nanoparticles are used as they can provide larger surface area. Metal-Organic Frameworks (MOFs) work in the similar way by increasing the absorption of formulation introduced by maximizing the porosity (Wang et al., 2022).

Viral infections can be brutal in the case of samplings or young soft plants. Introduction of elicitors and PAMPs can boost the immune response in these cases. The other promising technique to tackle viral infections is through CRISPR Cas method of gene manipulation (Islam et al., 2024). Usually, Cas 9 and 13 systems are used to cleave the viral genome which makes it non-functional leading to death of pathogens. It typically introduces plasmids or ribonucleoprotein complexes which are designed specifically. Other than these, silver nanoparticles produce ROS which damages the protein coat without any additional payload. Viral infection is majorly transmitted through insects like aphids, thrips and many more. Using insecticides as payload can repel these vectors which carry the virus and acts as prevention

(Wang et al., 2022). These multi directional and vast ways of using nanocarriers to treat viral diseases can greatly complement agriculture.

Treating Fungal Diseases

Fungal infections are one of the major reasons for agricultural loss in all the stages from germination to storage of yield. Fungal cellular structure is hard and rigid in nature, which makes it difficult for the agrochemicals to destroy it completely. The other difficulty is fungal spores, which get formed as a reproductive product. The fungal cell wall is primarily made of chitin and β-glucans, which can be used as targets while treating fungal infections (Mahra et al., 2025). Nanocarriers deliver the agents, which can disrupt their structures and weaken the pathogens. Azoles such as triazoles and imidazole are used in antifungal payloads that inhibit the biosynthesis of ergosterol, a key component of the outer membrane. Small molecules like farnesol-based inhibitors, tyrosol derivatives, oxylipin inhibitors, etc. can interfere with the mitogen-activated protein kinase (MAPK) pathways and quorum sensing communication in fungal strains, which avoids coordinated virulence, spore formation, and colonization (Ravinder Kumar et al., 2022). Pathogenicity is determined by the genes that a pathogen has. When using RNAi technology, these genes are silenced by the introduction of dsRNA or siRNA, and the disease can be modulated. Delivering antifungal peptides like defensins, thionine can disrupt the cell wall and make an easy target for the host defence system. Polymeric nanocarriers encapsulating azoxystrobin and tebuconazole are proven to produce prolonged protection against *Fusarium oxysporum* and *Botrytis cinerea* by controlled release and increasing the uptake by plant cells (Mahra et al., 2025). Other than this, to increase the adhesion of formulated nanoparticles on the plant surface, nano emulsions are used, which also provide essential oils and uniform distribution of applied agrochemicals. Few metallic nanocarriers like silver, zinc oxide and copper nanocarriers have an intrinsic antifungal nature, which destroys the fungal cells infecting plants by enhancing the efficiency of fungicide-coated and disrupting the cell wall structure (Machado et al., 2021).

Traditionally used fungicides like carbendazim, mancozeb, or metalaxyl can also be used as payloads, which can increase their efficacy by making them more soluble and stable. Other than these, ROS production using silver nanoparticles, usage of cellulase or protease enzymes to block the biofilm formation, and combining fungicides, RNAi molecules, and immune-priming agents to form a hybrid payload can create a greater impact. CRISPR Cas systems are also used to cleave the fungal genome at targeted sites (Zhou et al., 2024).

Treating Nematodes and Soil-Borne Pathogens

The microscopic roundworms, nematodes that target different parts of the plant to complete their life cycle, are one of the major reasons for the agricultural loss. Since they exhibit tissue-level organization, vitals like nerve function, reproduction, and circulation are targeted to succumb the organism (Mahra et al., 2025). Nanocarriers are made with neurotoxic substances like abamectin or carbamates and neurotoxin-producing peptides or analogs that disrupt the nervous system, making the nematode fail to perform life processes. The RNAi mechanism also targets the same vital processes. Nematodes are soil-borne pathogens that get attracted to plants through chemoreception. The nanocarriers loaded with phenolics or sulphur-containing molecules can inhibit this process and prevent the entry of nematodes (Sohlot et al., 2022). The other way of treating this is through the delivery of beneficial microbes like *Trichoderma spp.* and *Bacillus subtilis,* whose bioactive metabolites can in turn help treat the infection. Along with this, delivery of elicitors can activate PAMPs, systemic acquired resistance (SAR), or induced systemic resistance (ISR) pathways (Shikha Tiwari et al., 2023). The type of nanocarriers depends on the amount of payload required for the treatment; if the nervous system is targeted, not much of the loading capacity is required and can be loaded to polymeric or metallic nanocarriers, but if a high loading is required, for destruction of the digestive system or skin coat, carbon nanotubes are preferred. Additionally, clay-based nanocarriers are used to retain moisture so that long-term contact with the pathogen can be maintained, which increases the penetration and destruction (Su et al., 2023).

Nematodes lay eggs, which could be under incubation during the treatment with nematicides and hatch later. This further causes a recurring disease pattern and can be solved only if both nematodes and eggs are targeted with enzymes like chitinases and proteases. They also form associations like oomycetes and form biofilms. The same enzyme treatment can destroy these as well. Since the spectrum of soil-borne pathogens can be broad, targeted treatment is less preferred and a cocktail of payloads having multiple components is used by forming a hybrid nanocarrier (Rehab Y. Ghareeb et al., 2024).

Treating Phytophagous Insects and Pests

Unlike any other pathogens like bacteria, viruses, and fungi, the insects are highly mobile and cause direct and indirect infections at various sites. Hence the treatment strategy is designed carefully in consideration of the complex exoskeleton structure of insects, which is highly resistant to formulated nanocarrier spray (Proença et al., 2024). The target is made to damage the systematic rhythm of life processes of insects by targeting the nervous system, circulatory system, or even digestive system. This category also includes the vectors that bring in various infectants, which can be bacteria, viruses, or fungi. These vector insects, like aphids, are also made targets by this technique so that no further infection is achieved by those infectants. The arthropodal insects that feed on plants can destroy the crop within a very short time after infection (Chabbi et al., 2025). Primarily, nanocarriers with synthetic or botanical insecticides are used, but depending on the complexity of the infection, strategies are manipulated. Nanocarriers like Silica, metallic (ZnO, Ag, TiO_2), nano emulsions, and lipid nanoparticles with hormone analogs like methoprene or pyriproxyfen can interfere with the hormonal pathways and growth regulation pattern (Proença et al., 2024). Pheromone antagonists are used to avoid mating, and lectin inhibitors, Bt CRY genes are used to target gut enzymes and gut respectively, are applied through Electrostatic spray, seed coating, nano emulsion foliar sprays, and vapor treatment. Apart from this, unique approaches like insect repellent, growth disruptors, and larvicidal treatments are also employed (Chabbi et al., 2025).

AI SUPPORT FOR THE PRECISION OF NANOCARRIERS

Artificial intelligence is one fine technology that is adding benefits to all the conventional practices with much precision and advantages. The combination of artificial intelligence, nanotechnology, and conventional sciences can unlock so many doors that are now merely limitations. Starting from analysing the nanocarrier design, compatibility with the plant host, payload, loading capacity, real-time monitoring, efficiency, and many other revolutionary aspects can be monitored and recorded with much more ease, in fact, effortlessly (Rehab Y. Ghareeb et al., 2024). Although drawbacks of traditional method are effectively checked by the involvement of nanotechnology, the optimization of these nanocarriers in terms of their size, carrying and adhesion capacity, delivery time, and many other things are still yet to be optimized, which can be effectively checked by the incorporation of artificial intelligence into the mainstream (Nugroho et al., 2024).

The first and foremost challenge that occurs while designing any nanocarrier for formulation or encapsulation is its physical property characterization and making. The size, shape, loading capacity, composition, compatibility, charge, and many other characteristics of the nanocarrier decides its efficiency in terms of generating the expected response in the plants (Hu et al., 2024). It is also seen that even with modifications and optimizations, under various environmental conditions, the physiological status of plants changes, which again affects the functioning of nanocarriers. But with the incorporation of computational modelling and simulations using artificial intelligence, the behaviour of the nanocarrier under various conditions can be made AI-driven and can be controlled and optimized further for better interaction with plant tissues and pathogens (Pechyen et al., 2024). Along with that, AI can be used for the computation of all the previous data to obtain the best possible solution for particular infections through machine learning (ML) algorithms (Hu et al., 2024).

Artificial intelligence can be used to understand the structure-activity relationship modelling, which establishes the correlation between structural features of nanocarriers and the biological function of them. AI enables researchers to understand the effectiveness of various structures and modifications by reviewing computational studies, which can give better efficiency in eliciting a greater response inside the system by refining the nanocarrier formulation (Sahoo et al., 2024). Initially, through AI and ML algorithms, a compiled dataset of various nanocarriers is studied to understand correlation trend detection (Nugroho et al., 2024). In the study, AI detects the key structural aspect that determines the efficiency of the nanocarrier formulation, which is focused and prioritized for the design. AI-driven SAR technology in nanocarrier-based plant disease management can boost the ability to find more suitable nanocarrier formulations against any infection by reducing the wet lab work through minimized trial and error tests, as the AI rules out the least possible trails (Hu et al., 2024). Also, it enhances design precision and improves and promotes sustained usage of chemicals. Using biomarker identification techniques, AI can specifically target any protein, lipid, or any other biochemical entity that is specific to the pathogen, which makes the conjugated nanocarrier bind specifically to the target, increases the bioavailability, and reduces off-target effects (Nugroho et al., 2024).

POST HARVEST MANAGEMENT AND NANOCARRIERS

Post-harvest management refers to the set of procedures and measures that are followed in a broad view to safeguard the agricultural produce after harvesting in terms of its quality and shelf life (Neme et al., 2021). It can be seen that the significant amount of total agricultural harvest does not reach the purpose because of mechanical damage, microbial contamination, oxidation, pest infestations, and improper handling or storage (Milani et al., 2024). Although conventionally many procedures like cleaning, sorting, packaging, storage, transportation, and preservation are done, still a major chunk of harvest goes to waste. According to the Food and Agriculture Organization, 14% of the world's food is lost between harvest and retail, whereas fruits record 40% to 50% loss. In South Asia alone, major staple foods like rice and wheat record up to 20–25% loss after harvesting, even though these crops are grown, harvested, and are being stored for centuries together (Sánchez-Hernández et al., 2024).

The most important application of this technology is the development of nano-based antimicrobial coatings. Studies have shown that coatings developed through the use of silver nanoparticles (AgNPs), zinc oxide (ZnO NPs), and chitosan possess the potential to inhibit spoilage microorganisms, thereby improving shelf life (Kumar et al., 2022). Nanocarrier based coatings have prolonged antibacterial activity without leaving harmful residues, in contrast to traditional preservatives. An essential component of developing these food packaging alternatives is nanotechnology (Rahman et al., 2024). Conventional plastic films do not have adequate protection against microbial growth and infection followed by oxidative deterioration in comparison to nanocomposite films, which have nanoparticles that improve gas barrier properties and exhibit higher antimicrobial activity (Santiago-Aliste et al., 2023). One of the key post-harvest management issues relates to the control of ripening. Ethylene, a natural hormone linked to ripening, accelerates the breakdown of fruits and vegetables (Santiago-Aliste et al., 2023). Traditional ethylene absorbers, like potassium permanganate sachets, exhibit poor efficiency and require regular replacement. In contrast, nanocarriers that encapsulate 1-methylcyclopropene (1-MCP), a recognized ethylene inhibitor, allow for extended and controlled release, thus slowing ripening but not affecting the nutritional or sensory quality of the fruit. This technology has been used in commercial apple and banana storage with a considerable reduction in spoilage rates (González-Reza et al., 2023).

Nutrient degradation is one of the primary sources of post-harvest loss, particularly in perishable produce. Use of vitamins and antioxidants nanocarrier-encapsulated preserves the molecules from environmental factor-induced degradation, such as light, oxygen, and heat. For example, the addition of nano-encapsulated vitamin C to fruit juices prevents oxidation and provides nutritional stability for extended durations (Kumar et al., 2022). It is particularly effective in regions where refrigeration facilities are not available, and there is a need for preserving nutrients (Milani et al., 2024).

BIOSAFETY CHECK AND ENVIRONMENTAL IMPACT

A rich diversity of soil microorganisms can either be beneficial or sometimes disastrous. The sole purpose of using nanocarriers is for their precise and efficient delivery of treating mixtures, which help in managing the plant diseases efficiently. But sometimes, they can be toxic to the beneficial organisms too. For instance, the reactive oxygen species generated by silver nanoparticles not only disrupt the cell wall of pathogenic bacteria but also disrupt the beneficiary microbiota and plant cells themselves (Sharma et al., 2025). Hence, it is important to avoid overapplication of nanomaterials. Bioaccumulation is another serious concern where the chemicals used as nanocarriers remain in the system and start accumulating. When chitosan nanomaterials are used, as they are biodegradable, they leave minimal residues. Before using any material as nanocarriers, it is important to understand the pros and cons of that material to the environment, risk factors, and risk mitigation techniques like encapsulation to avoid bioleaching (Courtois et al., 2019).

LIMITATIONS OF NANOCARRIERS

Despite the nanocarrier-based disease management technique proven to be beneficial, the production cost of these nanocarriers, especially the carriers, which have to be engineered precisely is very high (Anjaneyulu et al., 2024). This can be a broken path for farmers who are less accessible to the resources. Nanocarriers that are made of metals and other non-biodegradable materials can accumulate to increase the toxicity, which can lead to bioaccumulation and intervention in the food chain adversely (Sharma et al., 2025). Other than all these, lack of on-field studies is the major concerning factor that makes people think it is less effective. The conception about nanocarriers can be built positively only if field studies are conducted and farmers start believing and understanding the incredibility (Anjaneyulu et al., 2024). There are a lot of factors involved in the complexity and compatibility of nanocarriers that were optimized in controlled conditions and agricultural fields. It is important to modulate the nanocarriers in order to achieve integrity and stability on the field over varied humidity, pH, temperature, and other parameters (Yonathan et al., 2022).

CONCLUSION AND FUTURE PROSPECTS

Nanocarriers potentially can overcome the drawbacks of traditional methods by enabling precise delivery of active ingredients, ensuring that pesticides, fungicides, or antimicrobial agents reach their targeted sites without unnecessary spreading into the surrounding ecosystem. This method not only reduces risks of toxicity but also increases the bioavailability of active ingredients, resulting in better overall efficacy while reducing the dosages of agrochemicals required. By minimizing the excessive use of chemical inputs, nanocarriers promote sustainable and organic farming practices, protecting soil health, water resources, and biodiversity. Although they have enormous potential, these challenges pose high costs, regulatory structures, and long-term environmental consequences and demand extensive field trials and refinements. Their mass adoption continues to be deterred by expensive production, limiting small-scale farmer access. Further, the regulatory ambiguity regarding nanoparticle safety and prolonged soil interaction necessitates in-depth research to ensure that such innovations do not present unexpected ecological or health consequences. Resolution of these issues by means of rigorous research, open risk assessment, and adaptive policy frameworks will be critical to ensuring the successful integration of nanocarrier technology into conventional farming.

In the future, it is assumed to be more localized and farmer-friendly by making it cost-efficient and more towards sustainability. The development of multi-functional nanocarriers (pathogen enzymes, pH shifts, ROS) is expected to improve soil quality along with treating the disease, which can serve a dual function. Integrated management of disease control, pest control, sustainability, and eco-friendly ways of using nanocarriers have to be optimized through field studies, which can unleash other potential challenges which could be complemented by CRISPR-Nanocarrier Synergy. Further, scaling up of nano-carrier production with advanced research can aid in reduction of loss of agricultural products, which in turn helps in food security and nutrition management and also phloem-mobile nanocarriers capable of systemic translocation to protect vascular tissues and manage phloem-residing pathogens through active field validation. One of the most essential future aspects in plant disease management with nanocarriers is the integration of artificial intelligence (AI) for improved accuracy and efficiency. Use of AI-driven predictive modelling and analytics allows precise use of nanocarriers through real-time evaluation of environmental factors, disease propagation, and plant health. Through use of machine learning models, it is possible to predict outbreaks of disease through assessment of climatic patterns, soil composition, and infection history, thus enabling proactive use of nanocarriers compared to reactive strategies. Use of intelligent nano sensors in fields allows real-time measurement of pathogen load, nutrient deficiency, and stress markers in plants, leading to selective release of encapsulated agrochemicals only where required. Such a method ensures crop protection with minimal use of resources, thus promoting the sustainability of nanocarrier-based agricultural technology.

REFERENCES

Abbasi, H., Antunes, M., & Velasco, J. I. (2019). Recent advances in carbon-based polymer nanocomposites for electromagnetic interference shielding. *Progress in Materials Science*, *103*, 319–373. DOI: 10.1016/j.pmatsci.2019.02.003

Abu-Thabit, N. Y., & Makhlouf, A. S. H. (2018). Historical development of drug delivery systems: From conventional macroscale to controlled, targeted, and responsive nanoscale systems. In Makhlouf, A. S. H., & Abu-Thabit, N. Y. (Eds.), *Stimuli Responsive Polymeric Nanocarriers for Drug Delivery Applications* (Vol. 1, pp. 3–41). Woodhead Publishing., DOI: 10.1016/B978-0-08-101997-9.00001-1

Ahmad, T. M., Haider, M. S., Randhir, T. O., Randhir, R., & Ahmad, S. R. (2023). Spatial analysis of factors influencing bacterial leaf blight in rice production. *Brazilian Journal of Biology*, *83*, e264249. DOI: 10.1590/1519-6984.264249 PMID: 36946787

Alemzadeh, E., Dehshahri, A., Izadpanah, K., & Ahmadi, F. (2018). Plant virus nanoparticles: Novel and robust nanocarriers for drug delivery and imaging. *Colloids and Surfaces. B, Biointerfaces*, *167*, 20–27. DOI: 10.1016/j.colsurfb.2018.03.026 PMID: 29625419

Alizadeh, M., Qarachal, J. F., & Sheidaee, E. (2025). Understanding the ecological impacts of nanoparticles: Risks, monitoring, and mitigation strategies. *Nanotechnology for Environmental Engineering*, *10*(1), 6. DOI: 10.1007/s41204-024-00403-7

Angayarkanni, A., & Philip, J. (2022). Synthesis of Nanoparticles and Nanofluids. *The Royal Society of Chemistry*, *1*, 1–40. DOI: 10.1039/9781839166457-00001

Anjaneyulu, B., Chauhan, V., Mittal, C., & Afshari, M. (2024). Innovative nanocarrier systems: A comprehensive exploration of recent developments in nano-biopesticide formulations. *Journal of Environmental Chemical Engineering*, *12*(5), 113693. DOI: 10.1016/j.jece.2024.113693

Avila-Quezada, G. D., Golinska, P., & Rai, M. (2022). Engineered nanomaterials in plant diseases: Can we combat phytopathogens? *Applied Microbiology and Biotechnology*, *106*(1), 117–129. DOI: 10.1007/s00253-021-11725-w PMID: 34913996

Baldassarre, F., Stradis, A. D., Altamura, G., Vergaro, V., Citti, C., Cannazza, G., Capodilupo, A. L., Dini, L., & Ciccarella, G. (2020). Application of calcium carbonate nanocarriers for controlled release of phytodrugs against *Xylella fastidiosa* pathogen. *Pure and Applied Chemistry*, *92*(3), 429–444. DOI: 10.1515/pac-2018-1223

Baloch, N. (2025). Microbial contributions to maize crop production: A comprehensive review of challenges and future perspectives. *Discover Agriculture*, *3*(1), 10. DOI: 10.1007/s44279-025-00164-0

Bernays, E. A. (2009). Phytophagous Insects. In Resh, V. H., & Cardé, R. T. (Eds.), *Encyclopedia of Insects* (2nd ed., pp. 798–800). Academic Press., DOI: 10.1016/B978-0-12-374144-8.00210-1

Boas, U., & Heegaard, H. (2004). Dendrimers in drug research. *Chemical Society Reviews*, *33*(1), 43–63. DOI: 10.1039/b309043b PMID: 14737508

Bobbarala, V., Sahu, S. C., Mishra, B. K., & Jena, B. K. (2012). A Search for Antibacterial Agents. In *A Search for Antibacterial Agents*. IntechOpen., DOI: 10.5772/1085

Bonechi, C., Tamasi, G., Donati, A., Bisozzi, F., Baglioni, M., Andreassi, M., Ietta, F., Leone, G., Magnani, A., & Rossi, C. (2025). Cationic liposomes as carriers of natural compounds from plant extract. *Biophysical Chemistry, 318*, 107381. DOI: 10.1016/j.bpc.2024.107381 PMID: 39733688

Borane, N., Aralekallu, S., Boddula, R., Singh, J., & Kurkuri, M. D. (2024). Recent trends in the "bottom-up" and "top down" techniques in the synthesis and fabrication of myriad carbonaceous nanomaterials. In Biswas, K., Mohanta, Y. K., Mohanta, T. K., & Saravanan, M. (Eds.), *Carbon-Based Nanomaterials in Biosystems* (pp. 91–120). Academic Press., DOI: 10.1016/B978-0-443-15508-6.00002-6

Bowbrick, P. (2024). Falsehoods and myths in famine research: The Bengal famine and Daoud. *Journal of International Development, 36*(6), 2513–2536. DOI: 10.1002/jid.3635

Carmen, B., Roberto, D., Carmen, B., & Roberto, D. (2011). Soil Bacteria Support and Protect Plants Against Abiotic Stresses. *In Abiotic Stress in Plants—Mechanisms and Adaptations*. IntechOpen.

Cătălin Balaure, P., Gudovan, D., & Gudovan, I. (2017). 4 - Nanopesticides: A new paradigm in crop protection. In Grumezescu, A. M. (Ed.), *New Pesticides and Soil Sensors* (pp. 129–192). Academic Press., DOI: 10.1016/B978-0-12-804299-1.00005-9

Chabbi, J., El Assimi, T., Blazic, R., El Haskouri, J., Kaddami, H., Youcef, H. B., Vidovic, E., El Kadib, A., & Lahcini, M. (2025). Eco-friendly catalysts: Chitosan-silver hybrid nanocarriers for efficient oxidative coupling of anilines to azobenzenes. *International Journal of Biological Macromolecules, 307*, 141996. DOI: 10.1016/j.ijbiomac.2025.141996 PMID: 40081687

Chen, M., Zhou, Y., Huang, J., Zhu, P., Peng, X., & Wang, Y. (2012). Liposome-Based Delivery Systems in Plant Polysaccharides. *Journal of Nanomaterials, 2012*(1), 682545. DOI: 10.1155/2012/682545

Chomicki, G., Beinart, R., Prada, C., Ritchie, K. B., & Weber, M. G. (2022). Editorial: Symbiotic relationships as shapers of biodiversity. *Frontiers in Ecology and Evolution, 10*, 850572. Advance online publication. DOI: 10.3389/fevo.2022.850572

Courtois, P., Rorat, A., Lemiere, S., Guyoneaud, R., Attard, E., Levard, C., & Vandenbulcke, F. (2019). Ecotoxicology of silver nanoparticles and their derivatives introduced in soil with or without sewage sludge: A review of effects on microorganisms, plants and animals. *Environmental Pollution, 253*, 578–598. DOI: 10.1016/j.envpol.2019.07.053 PMID: 31330350

Das, S., & Pattanayak, S. (2020). Nanotechnological approaches in sustainable agriculture and plant disease management. *Organic Agriculture, 2*, 13–32. DOI: 10.5772/intechopen.92463

Di Sotto, A., Paolicelli, P., Nardoni, M., Abete, L., Garzoli, S., Di Giacomo, S., Mazzanti, G., Casadei, M. A., & Petralito, S. (2018). SPC liposomes as possible delivery systems for improving bioavailability of the natural sesquiterpene? -caryophyllene: Lamellarity and drug-loading as key features for a rational drug delivery design. *Pharmaceutics, 10*(4), 4. Advance online publication. DOI: 10.3390/pharmaceutics10040274 PMID: 30551617

Durán-Lobato, M., López-Estévez, A. M., Cordeiro, A. S., Dacoba, T. G., Crecente-Campo, J., Torres, D., & Alonso, M. J. (2021). Nanotechnologies for the delivery of biologicals: Historical perspective and current landscape. *Advanced Drug Delivery Reviews*, *176*, 113899. DOI: 10.1016/j.addr.2021.113899 PMID: 34314784

Eid, N. A., Ibrahim, A. M. H., Elsharawy, A. A., & Salem, K. F. M. (2025). Nanofertilizers for Plant Viral Disease Management. In Kumar, P., & Dubey, R. C. (Eds.), *Nanofertilizers for Sustainable Agriculture: Assessing Impacts on Health, Environment, and Economy* (pp. 149–187). Springer Nature Switzerland. DOI: 10.1007/978-3-031-78649-5_7

Elmer, W., Ma, C., & White, J. (2018). Nanoparticles for plant disease management. *Current Opinion in Environmental Science & Health*, *6*, 66–70. DOI: 10.1016/j.coesh.2018.08.002

Fischer, J., Beckers, S. J., Yiamsawas, D., Thines, E., Landfester, K., & Wurm, F. R. (2019a). Targeted Drug Delivery in Plants: Enzyme-Responsive Lignin Nanocarriers for the Curative Treatment of the Worldwide Grapevine Trunk Disease Esca. *Advancement of Science*, *6*(15), 1802315. DOI: 10.1002/advs.201802315 PMID: 31406660

Fu, L., Wang, Z., Dhankher, O. P., & Xing, B. (2020). Nanotechnology as a new sustainable approach for controlling crop diseases and increasing agricultural production. *Journal of Experimental Botany*, *71*(2), 507–519. DOI: 10.1093/jxb/erz314 PMID: 31270541

Ghareeb, R. Y., Belal, E. B., El-Khateeb, N. M. M., & Shreef, B. A. (2024). Utilizing bio-synthesis of nanomaterials as biological agents for controlling soil-borne diseases in pepper plants: Root-knot nematodes and root rot fungus I. *BMC Plant Biology*, *24*, 110. https://link.springer.com/article/10.1186/s12870-024-04760-y. DOI: 10.1186/s12870-024-04760-y PMID: 38355449

González-Reza, R. M., Zambrano-Zaragoza, M. L., & Gutiérrez-Martínez, P. (2023). Introduction to Nanotechnological Applications in the Postharvest Processing of Fresh Horticultural Produce. In *Postharvest Nanotechnology for Fresh Horticultural Produce* (pp. 1–25). CRC Press., DOI: 10.1201/9781003142287-1

González-Reza, R. M., Zambrano-Zaragoza, M. L., & Gutiérrez-Martínez, P. (2023). Introduction to nanotechnological applications in the postharvest processing of fresh horticultural produce. In *Postharvest Nanotechnology for Fresh Horticultural Produce*. CRC Press. DOI: 10.1201/9781003142287-1

Hanna, E. A., Astete, C. E., Price, T., Tamez, C., Mendez, O. E., Garcia, A., Kewir, F. V., White, J. C., & Sabliov, C. M. (2024). Antifungal Efficacy of Nanodelivered Azoxystrobin against Rhizoctonia solani in Soybean (Glycine max). *ACS Agricultural Science & Technology*, *4*(3), 330–336. DOI: 10.1021/acsagscitech.3c00469

Hu, Y., Tang, J., & Yang, J. (2024). Introducing artificial intelligence technology to plant disease management for sustainable agriculture. *Crop Protection (Guildford, Surrey)*, *184*, 106764. DOI: 10.1016/j.cropro.2024.106764

Huston, D. C., Khudhir, M., & Hodda, M. (2022). Reliability and Utility of Standard Gene Sequence Barcodes for the Identification and Differentiation of Cyst Nematodes of the Genus Heterodera. *Journal of Nematology*, *54*(1), 20220024. DOI: 10.2478/jofnem-2022-0024 PMID: 35975224

Islam, M. R., Youngblood, M., Kim, H.-I., González-Gamboa, I., Monroy-Borrego, A. G., Caparco, A. A., Lowry, G. V., Steinmetz, N. F., & Giraldo, J. P. (2024). DNA delivery by virus-like nanocarriers in plant cells. *Nano Letters*, *24*(26), 7833–7842. DOI: 10.1021/acs.nanolett.3c04735 PMID: 38887996

Jiang, N., Yan, J., Liang, Y., Shi, Y., He, Z., Wu, Y., Zeng, Q., Liu, X., & Peng, J. (2020). Resistance genes and their interactions with bacterial blight/leaf streak pathogens (*Xanthomonas oryzae*) in Rice (*Oryza sativa* L.)—An updated review. *Rice (New York, N.Y.)*, *13*(1), 3. DOI: 10.1186/s12284-019-0358-y PMID: 31915945

Johnson, P. T. J., Dobson, A., Lafferty, K. D., Marcogliese, D. J., Memmott, J., Orlofske, S. A., Poulin, R., & Thieltges, D. W. (2010). When parasites become prey: Ecological and epidemiological significance of eating parasites. *Trends in Ecology & Evolution*, *25*(6), 362–371. DOI: 10.1016/j.tree.2010.01.005 PMID: 20185202

Kandhol, N., Singh, V. P., Pandey, S., Sharma, S., Zhao, L., Corpas, F. J., Chen, Z.-H., White, J. C., & Tripathi, D. K. (2024). Nanoscale materials and NO-ROS homeostasis in plants: Trilateral dynamics. *Trends in Plant Science*, *29*(12), 1310–1318. DOI: 10.1016/j.tplants.2024.06.009 PMID: 39379242

Karimi, N., Ghanbarzadeh, B., Hamishehkar, H., Keivani, F., Pezeshki, A., & Gholian, M. M. (2015). Phytosome and Liposome: The Beneficial Encapsulation Systems in Drug Delivery and Food Application. *Applied Food Biotechnology*, *2*(3), 17–27.

Khan, A., Ahmad, G., Haris, M., & Khan, A. A. (2023). Bio-organics Management: Novel Strategies to Manage Root-knot Nematode, Meloidogyne incognita Pest of Vegetable Crops. *Gesunde Pflanzen*, *75*(1), 193–209. DOI: 10.1007/s10343-022-00679-2

Khan, A., Khan, A., Ali, A., Fatima, S., & Siddiqui, M. A. (2023). Root-Knot Nematodes (Meloidogyne spp.): Biology, Plant-Nematode Interactions and Their Environmentally Benign Management Strategies. *Gesunde Pflanzen*, *75*(6), 2187–2205. DOI: 10.1007/s10343-023-00886-5

Khoury, W. E., & Makkouk, K. (2010). Integrated Plant Disease Management in Developing Countries. *Journal of Plant Pathology*, *92*(4), S35–S42.

Kumah, E. A., Fopa, R. D., Harati, S., Boadu, P., Zohoori, F. V., & Pak, T. (2023). Human and environmental impacts of nanoparticles: A scoping review of the current literature. *BMC Public Health*, *23*(1), 1059. DOI: 10.1186/s12889-023-15958-4 PMID: 37268899

Kumar, I., Bhattacharya, J., & Das, B. K. (2022). Efficacy of silver nanoparticles-based foliar spray application to control plant diseases, its effect on productivity, and risk assessment. *Arabian Journal of Geosciences*, *15*(5), 390. DOI: 10.1007/s12517-022-09652-9

Kumari, A., Rana, V., Yadav, S. K., & Kumar, V. (2023). Nanotechnology as a powerful tool in plant sciences: Recent developments, challenges and perspectives. *Plant Nano Biology*, *5*, 100046. DOI: 10.1016/j.plana.2023.100046

Liu, P., Yang, J., Liu, J., Duan, Y., & Wang, P. (2025). Multipurpose β-Cyclodextrin-Involved nanospherical Aggregates: Self-Assembly Process, foliar Adhesion, and promising applications for kiwifruit and rice Protections. *Chemical Engineering Journal*, *507*, 160385. DOI: 10.1016/j.cej.2025.160385

Ly, H. Q., Chen, Y.-J., Nguyen, V. T., & Tseng, C.-L. (2025). Optimization of the poloxamer 407-conjugated gelatin to synthesize pH-sensitive nanocarriers for controlled paclitaxel delivery. *Journal of Polymer Research*, *32*(2), 38. DOI: 10.1007/s10965-024-04248-1

Machado, T. O., Beckers, S. J., Fischer, J., Müller, B., Sayer, C., de Araújo, P. H. H., Landfester, K., & Wurm, F. R. (2020). Bio-based lignin nanocarriers loaded with fungicides as a versatile platform for drug delivery in plants. *Biomacromolecules*, *21*(7), 2755–2763. DOI: 10.1021/acs.biomac.0c00487 PMID: 32543851

Machado, T. O., Beckers, S. J., Fischer, J., Sayer, C., de Araújo, P. H. H., Landfester, K., & Wurm, F. R. (2021). Cellulose nanocarriers via miniemulsion allow Pathogen-Specific agrochemical delivery. *Journal of Colloid and Interface Science*, *601*, 678–688. DOI: 10.1016/j.jcis.2021.05.030 PMID: 34091315

Mahra, S., Tripathi, S., Tiwari, K., Sharma, S., Mathew, S., Kumar, V., & Sharma, S. (2025). Harnessing nanotechnology for sustainable agriculture: From seed priming to encapsulation. *Plant Nano Biology*, *11*, 100124. DOI: 10.1016/j.plana.2024.100124

Mansour, R. S. H., Tekade, R. K., & Ali, A. A. (2025). PEGylated nanocarriers for solubilization. In Tekade, R. K., & Jain, N. K. (Eds.), *PEGylated Nanocarriers in Medicine and Pharmacy* (pp. 137–169). Springer Nature., DOI: 10.1007/978-981-97-7316-9_5

Milani, M. H., Gohari, G., Manganaris, G. A., & Fotopoulos, V. (2024). 10—Impacts of nanocomposites on the postharvest physiology and shelf life of agricultural crops. In Thamaraiselvan, C., Lau, W. J., Maldonado, A. J., & Othman, N. H. (Eds.), *Nanocomposites for Environmental, Energy, and Agricultural Applications* (pp. 315–347). Woodhead Publishing., DOI: 10.1016/B978-0-443-13935-2.00010-3

Mittler, R. (2006). Abiotic stress, the field environment and stress combination. *Trends in Plant Science*, *11*(1), 15–19. DOI: 10.1016/j.tplants.2005.11.002 PMID: 16359910

Munster, M. (2022). Diseases and disorders. NC State Extension Publications | *Browse Popular Publications.*https://content.ces.ncsu.edu/extension-gardener-handbook/5-diseases-and-disorders

Nancarrow, N., Aftab, M., Hollaway, G., Rodoni, B., & Trębicki, P. (2021). Yield Losses Caused by Barley Yellow Dwarf Virus-PAV Infection in Wheat and Barley: A Three-Year Field Study in South-Eastern Australia. *Microorganisms*, *9*(3), 645. DOI: 10.3390/microorganisms9030645 PMID: 33808907

Nasrollahi, S., & Alizadeh, N. (2024). Synthesis of functionalized superparamagnetic iron oxide nanoparticles for ibuprofen and naproxen hydrophobic drugs delivery. *Materials Chemistry and Physics*, *326*, 129775. DOI: 10.1016/j.matchemphys.2024.129775

Neme, K., Nafady, A., Uddin, S., & Tola, Y. B. (2021). Application of nanotechnology in agriculture, postharvest loss reduction and food processing: Food security implication and challenges. *Heliyon*, *7*(12), e08539. DOI: 10.1016/j.heliyon.2021.e08539 PMID: 34934845

Noman, M., Ahmed, T., Wang, J., Ijaz, M., Shahid, M., Islam, M. S., Azizullah, , Manzoor, I., Li, D., & Song, F. (2023). Nano-enabled crop resilience against pathogens: Potential, mechanisms and strategies. *Crop Health*, *1*(1), 15. DOI: 10.1007/s44297-023-00015-8

Nugroho, H., Chew, J. X., Eswaran, S., & Tay, F. S. (2024a). Resource-optimized cnns for real-time rice disease detection with ARM cortex-M microprocessors. *Plant Methods*, *20*(1), 159. DOI: 10.1186/s13007-024-01280-6 PMID: 39415246

Nugroho, H., Chew, J. X., Eswaran, S., & Tay, F. S. (2024b). Resource-optimized cnns for real-time rice disease detection with ARM cortex-M microprocessors. *Plant Methods*, *20*(1), 159. DOI: 10.1186/s13007-024-01280-6 PMID: 39415246

Pal, G., Ingole, K. D., Yavvari, P. S., Verma, P., Kumari, A., Chauhan, C., Chaudhary, D., Srivastava, A., Bajaj, A., & Vemanna, R. S. (2024). Exogenous application of nanocarrier-mediated double-stranded RNA manipulates physiological traits and defence response against bacterial diseases. *Molecular Plant Pathology*, *25*(1), e13417. DOI: 10.1111/mpp.13417 PMID: 38279851

Pechyen, C., Tangnorawich, B., Toommee, S., Marks, R., & Parcharoen, Y. (2024). Green synthesis of metal nanoparticles, characterization, and biosensing applications. *Sensors International*, *5*, 100287. DOI: 10.1016/j.sintl.2024.100287

Peng, F., Wang, X., Zhang, W., Shi, X., Cheng, C., Hou, W., Lin, X., Xiao, X., & Li, J. (2022). Nanopesticide formulation from pyraclostrobin and graphene oxide as a nanocarrier and application in controlling plant fungal pathogens. *Nanomaterials (Basel, Switzerland)*, *12*(7), 7. Advance online publication. DOI: 10.3390/nano12071112 PMID: 35407232

Proença, P. L. de F., Campos, E. V. R., Costa, T. G., de Lima, R., Preisler, A. C., de Oliveira, H. C., da Rocha, C. M., de Andrade, D. J., Goncalves, K. C., Polanczyk, R. A., & Fraceto, L. F. (2024). Curcumin and carvacrol co-loaded zein nanoparticles: Comprehensive preparation and assessment of biological activities in pest control. *Plant Nano Biology*, *8*, 100067. DOI: 10.1016/j.plana.2024.100067

Rahman, F. U., Zhu, Q., Wu, Z., Li, X., Chen, W., Xiong, T., & Zhu, X. (2024). Current insights into the biocontrol and biotechnological approaches for postharvest disease management of *Botrytis cinerea*. *Postharvest Biology and Technology*, *216*, 113055. DOI: 10.1016/j.postharvbio.2024.113055

Ruano, M., Mateos-Maroto, A., Ortega, F., Ritacco, H., Rubio, J. E. F., Guzman, E., & Rubio, R. G. (2021). Fabrication of robust capsules by sequential assembly of polyelectrolytes onto charged liposomes. *Langmuir*, *37*(20), 6189–6200. DOI: 10.1021/acs.langmuir.1c00341 PMID: 33945690

Sahoo, L., Mohapatra, D., Raghuvanshi, H. R., Kumar, S., Kaur, R., Anshika, , Sapna, , Chawla, R., & Afreen, N. (2024). Transforming agriculture through Artificial Intelligence: Advancements in plant disease detection, Applications, and challenges. *Journal of Advances in Biology & Biotechnology*, *27*(5), 381–388. DOI: 10.9734/jabb/2024/v27i5796

Sánchez-Hernández, E., Santiago-Aliste, A., Correa-Guimarães, A., Martín-Gil, J., Gavara-Clemente, R. J., & Martín-Ramos, P. (2024). Carvacrol encapsulation in chitosan–carboxymethylcellulose–alginate nanocarriers for postharvest tomato protection. *International Journal of Molecular Sciences*, *25*(2), 2. Advance online publication. DOI: 10.3390/ijms25021104 PMID: 38256176

Sanghi, A. K., Kumar, R., Chaubey, K. K., Pal, A. K., Sharma, S., Dayal, D., Kumar, A., & Verma, D. K. (2025). Application of Chitin/Chitosan Based Agro Nano Chemicals in Crop Protection. In Bachheti, R. K., Bachheti, A., & Husen, A. (Eds.), *Chitin-Based Nanoparticles for the Agriculture Sectors* (pp. 155–177). Springer Nature., DOI: 10.1007/978-981-96-0920-8_9

Santiago-Aliste, A., Sánchez-Hernández, E., Buzón-Durán, L., Marcos-Robles, J. L., Martín-Gil, J., & Martín-Ramos, P. (2023). Uncaria tomentosa-Loaded Chitosan Oligomers–Hydroxyapatite–Carbon Nitride Nanocarriers for Postharvest Fruit Protection. *Agronomy (Basel)*, *13*(9), 9. Advance online publication. DOI: 10.3390/agronomy13092189

Saruchi, K., Kumar, V., Ghfar, A. A., & Pandey, S. (2023). Controlled release of harmful pesticide dichlorvos through synthesized biodegradable aloe vera–acrylic acid-based hydrogel and its utilization in soil water management. *Journal of Analytical Science and Technology*, *14*(1), 12. DOI: 10.1186/s40543-023-00374-5

Sarvarian, P., Samadi, P., Gholipour, E., Shams Asenjan, K., Hojjat-Farsangi, M., Motavalli, R., Motavalli Khiavi, F., & Yousefi, M. (2022). Application of Emerging Plant-Derived Nanoparticles as a Novel Approach for Nano-Drug Delivery Systems. *Immunological Investigations*, *51*(4), 1039–1059. DOI: 10.1080/08820139.2021.1891094 PMID: 33627016

Sharma, M. M. M., Kapoor, D., Loyal, A., Kumar, R., Sharma, P., & Husen, A. (2025). Plant response to silver nanoparticles in terms of growth, development, production, and protection: An overview. In Husen, A. (Ed.), *Plant Response to Silver Nanoparticles: Plant Growth, Development, Production, and Protection* (pp. 1–22). Springer Nature., DOI: 10.1007/978-981-97-7352-7_1

Sharma, P., Sharma, M. M. M., Kapoor, D., Loyal, A., & Husen, A. (2025). Beneficial and Adverse Effects of Silver Nanoparticles on Rhizosphere Biology. In Husen, A. (Ed.), *Plant Response to Silver Nanoparticles: Plant Growth, Development, Production, and Protection* (Vol. 2, pp. 43–57). Springer Nature., DOI: 10.1007/978-981-97-7352-7_3

Sinha, K., Ghosh, J., & Sil, P. C. (2017). New pesticides: A cutting-edge view of contributions from nanotechnology for the development of sustainable agricultural pest control. In Grumezescu, A. M. (Ed.), *New Pesticides and Soil Sensors* (pp. 47–79). Academic Press., DOI: 10.1016/B978-0-12-804299-1.00003-5

Sohlot, M., Das, S., & Debnath, N. (2022). Recent developments in silica nanoparticle-based drug delivery system. In Hameed, S., & Rehman, S. (Eds.), *Nanotechnology for Infectious Diseases* (pp. 237–261). Springer., DOI: 10.1007/978-981-16-9190-4_11

Song, B., & Cho, C.-W. (2024). Applying polyvinyl alcohol to the preparation of various nanoparticles. *Journal of Pharmaceutical Investigation*, *54*(3), 249–266. DOI: 10.1007/s40005-023-00649-4

Sousa de Almeida, M., Susnik, E., Drasler, B., Taladriz-Blanco, P., Petri-Fink, A., & Rothen-Rutishauser, B. (2021). Understanding nanoparticle endocytosis to improve targeting strategies in nanomedicine. *Chemical Society Reviews*, *50*(9), 5397–5434. DOI: 10.1039/D0CS01127D PMID: 33666625

Su, C., Li, Y., Liu, S., Feng, H., Wang, J., & Yan, S. (2023). Star polymer soil delivery nanoplatform for applying biological agents in the field to control plant rhizosphere diseases. *Journal of Controlled Release*, *364*, 406–419. DOI: 10.1016/j.jconrel.2023.10.053 PMID: 37924956

Thanh, V. M., Bui, L. M., Bach, L. G., Nguyen, N. T., Thi, H. L., & Hoang Thi, T. T. (2019). Origanum majorana L. Essential Oil-Associated Polymeric Nano Dendrimer for Antifungal Activity against *Phytophthora infestans.Materials (Basel)*, *12*(9), 1446. DOI: 10.3390/ma12091446 PMID: 31060208

Tiwari, S., Dubey, N. K., & Kumar, C. (2023). Phytochemicals as an Eco-Friendly Source for Sustainable Management of Soil-Borne Plant Pathogens in Soil Ecosystem—Agroecological Approaches for Sustainable Soil Management—*Wiley Online Library*. https://onlinelibrary.wiley.com/doi/abs/10.1002/9781119911999.ch13

Umbarkar, M., Thakare, S., Surushe, T., Giri, A., & Chopade, V. (2021). Formulation and Evaluation of Liposome by Thin Film Hydration Method: Formulation of Liposome. *Journal of Drug Delivery and Therapeutics*, *11*(1), 72–76. DOI: 10.22270/jddt.v11i1.4677

Wang, J., Hao, K., Yu, F., Shen, L., Wang, F., Yang, J., & Su, C. (2022). Field application of nanoliposomes delivered quercetin by inhibiting specific hsp70 gene expression against plant virus disease. *Journal of Nanobiotechnology*, *20*(1), 16. DOI: 10.1186/s12951-021-01223-6 PMID: 34983536

Warghane, A., Saini, R., Shri, M., Andankar, I., Ghosh, D. K., & Chopade, B. A. (2024). Application of nanoparticles for management of plant viral pathogen: Current status and future prospects. *Virology*, *592*, 109998. DOI: 10.1016/j.virol.2024.109998 PMID: 38301447

Wen, X. (2019). One-pot route to graft long-chain polymer onto silica nanoparticles and its application for high-performance poly (L -lactide) nanocomposites. *RSC Advances*, *9*(24), 13908–13915. DOI: 10.1039/C9RA01360A PMID: 35519579

Wu, T., Bi, Y., Yu, Y., Zhou, Z., Yuan, B., Ding, X., Zhang, Q., Chen, X., Yang, H., Liu, H., & Chu, Z. (2023). Activated Expression of Rice DMR6-like Gene OsS3H Partially Explores the Susceptibility to Bacterial Leaf Streak Mediated by Knock-Out OsF3H04g. *International Journal of Molecular Sciences*, *24*(17), 13263. DOI: 10.3390/ijms241713263 PMID: 37686066

Xiang, H., Meng, J., Shao, W., Zeng, D., Ji, J., Wang, P., Zhou, X., Qi, P., Liu, L., & Yang, S. (2023). Plant protein-based self-assembling core–shell nanocarrier for effectively controlling plant viruses: Evidence for nanoparticle delivery behavior, plant growth promotion, and plant resistance induction. *Chemical Engineering Journal*, *464*, 142432. DOI: 10.1016/j.cej.2023.142432

Yonathan, K., Mann, R., Mahbub, K. R., & Gunawan, C. (2022). The impact of silver nanoparticles on microbial communities and antibiotic resistance determinants in the environment. *Environmental Pollution*, *293*, 118506. DOI: 10.1016/j.envpol.2021.118506 PMID: 34793904

Zhang, M., Ma, W., Tao, R., Fan, Q., Zhang, M., Qin, D., Cao, X., Li, J., Xiong, R., & Huang, C. (2024). Nanomaterials: Recent advances in plant disease diagnosis and treatment. *Nano Today*, *57*, 102326. DOI: 10.1016/j.nantod.2024.102326

Zhao, Z., Lacombe, J., Simon, L., Sanchez-Ballester, N. M., Khanishayan, A., Shaik, N., Case, K., Dugas, P. Y., Repellin, M., Lollo, G., Soulairol, I., Harris, A. F., Gordon, M., Begu, S., & Zenhausern, F. (2024). Physical, biochemical, and biological characterization of olive-derived lipid nanovesicles for drug delivery applications. *Journal of Nanobiotechnology*, *22*(1), 453–550. DOI: 10.1186/s12951-024-02964-w PMID: 39558361

Chapter 14
Potential Impact of Nanocarriers in Prevention and Management of Plant Fungal Diseases

Maryam Abdul Ajees
https://orcid.org/0009-0005-5456-4316
Manipal School of Life Sciences, Manipal Academy of Higher Education, India

Vibha Kamath
https://orcid.org/0009-0006-9439-6625
Manipal School of Life Sciences, Manipal Academy of Higher Education, India

Vyasraj G. Bhat
https://orcid.org/0009-0002-6418-8869
Manipal School of Life Sciences, Manipal Academy of Higher Education, India

Mukirala Vasanth Kumar
https://orcid.org/0009-0004-1916-3237
Manipal School of Life Sciences, Manipal Academy of Higher Education, India

Krupa Bhat
https://orcid.org/0009-0006-2540-7968
Manipal School of Life Sciences, Manipal Academy of Higher Education, India

Pawas Kumar
https://orcid.org/0009-0008-0559-6049
Manipal School of Life Sciences, Manipal Academy of Higher Education, India

Srinivasamurthy Vanishree
Manipal School of Life Sciences, Manipal Academy of Higher Education, India

Arya Kaniyassery
Manipal School of Life Sciences, Manipal Academy of Higher Education, India

Annamalai Muthusamy
https://orcid.org/0009-0008-0342-1691
Manipal School of Life Sciences, Manipal Academy of Higher Education, India

ABSTRACT

The application of nanocarriers in agriculture has become a groundbreaking method to increase nutrient delivery, boost plant health, and successfully fight fungal infections. These nanoscale structures increase agrochemical stability, solubility, and bioavailability, guaranteeing long-term efficacy while minimizing negative environmental effects. The synthesis and characterization of nanocarriers involve sophisticated techniques such as sol-gel processes, self-assembly, and in situ polymerization, ensuring optimal performance. Despite their vast potential, challenges such as production scalability, regulatory

DOI: 10.4018/979-8-3693-9869-2.ch014

compliance, and long-term environmental safety must be addressed to facilitate widespread adoption. This chapter provides an in-depth analysis of nanocarrier technology in agriculture, highlighting its role in improving plant nutrient uptake, disease resistance, and environmental sustainability. With continued research and development, nanocarriers are poised to revolutionize modern agriculture, paving the way for sustainable and high-yield farming practices.

1. INTRODUCTION TO NANOCARRIERS IN AGRICULTURE

Nanocarriers in agriculture are an innovative approach to enhance the efficiency of pesticide delivery, nutrient release, and soil remediation (An et al., 2022). Targeted distribution of active chemicals, enhanced stability, and controlled release are just a few benefits of these nanoscale systems, which include liposomes, nanoparticles, and nanotubes. Nanocarriers can reduce waste, lessen their negative effects on the environment, and support sustainable farming methods by increasing the bioavailability of pesticides and fertilizers (Singh et al., 2021). Nanocarriers are a promising technology for next agricultural developments because of their ability to maximize crop protection, improve soil health, and increase overall agricultural production (Fu et al., 2020). Nanotechnology in agriculture is the use of nanoscience and nanomaterials to improve a variety of agricultural processes, such as crop production, insect control, and soil management (Pramanik et al., 2020). To increase the effectiveness, sustainability, and productivity of agricultural techniques, nanoparticles, nanosensors, nanocarriers, and nanomaterials are used. Nanoscale material manipulation makes it possible to create sophisticated instruments that precisely target certain agricultural requirements, resulting in more efficient solutions (Chhipa & Joshi 2016).

The promise of nanotechnology to address the increasing global issues of environmental sustainability, climate change, and food security is therefore significant in agriculture. Nanotechnology can lower the quantity of chemicals needed, decrease environmental damage, and improve agricultural production and quality by enabling the targeted and controlled delivery of nutrients, fertilizers, and pesticides (Munir et al., 2024). Furthermore, nanotechnology provides novel approaches to early disease detection, plant health monitoring, and better water and fertilizer management, all of which support more resource-efficient and sustainable agricultural methods. With these developments, nanotechnology has the potential to make agriculture a more environmentally responsible and productive sector (Wahab et al., 2024).

1.1 Evolution of Nanocarrier Technology for Plant Health

The use of fungicides dates to the early 19th century when sulfur-based compounds were first employed for fungal disease management. A breakthrough occurred in 1882 with the introduction of Bordeaux mixture, a combination of copper sulfate and lime, which became one of the most effective and widely used fungicides (Morton & Staub., 2008). Throughout the 20th century, the development of synthetic fungicides, such as dithiocarbamates in the 1940s and systemic fungicides like benzimidazoles and triazoles in the 1950s-1970s, revolutionized plant protection (Mohamed et al., 2009). However, the widespread application of these chemicals raised concerns regarding resistance development and environmental toxicity, particularly in the late 20th century.

The 21st century saw the emergence of nanocarriers as an alternative approach to traditional fungicides (Asmawi et al., 2024). Early research in the 2000s focused on nanoencapsulation, a technique designed to protect active ingredients from degradation and enable controlled release. By the 2010s, studies on metal nanoparticles, such as silver and zinc oxide, demonstrated their antifungal properties, offering promising alternatives to chemical fungicides (Djearamane et al., 2022). More recently, advancements in polymer-based nanocarriers, liposomal formulations, and nanogels have further improved the efficacy and sustainability of fungicide applications (Mondéjar-López et al., 2024).

Theoretically, nanocarrier-based fungicide delivery relies on several key mechanisms. Nanoencapsulation prevents premature degradation of active ingredients, ensuring prolonged efficacy. Targeted delivery is achieved through functionalized nanoparticles, which enhance site-specific action and minimize environmental impact. Additionally, nanoparticles improve penetration into plant surfaces, increasing the solubility and transport of antifungal agents (Sun et al., 2018). These properties collectively make nanocarrier-based fungicides a more efficient and environmentally friendly alternative to traditional chemical fungicides. The evolution of nanocarrier technology for plant health has progressed significantly, driven by the need for more efficient and sustainable methods of delivering agrochemicals, nutrients, and genetic material to plants (Mujtaba et al., 2021). Initially, the use of conventional pesticide and fertilizer delivery systems often led to inefficiencies, such as poor bioavailability, rapid degradation, or environmental contamination (An et al., 2022). In response to these challenges, researchers began exploring the potential of nanocarriers—nanoscale materials capable of encapsulating active compounds and delivering them precisely to targeted sites in plants. Early developments in nanocarrier technology focused on simple nanoparticles, which were designed to improve the solubility and stability of agrochemicals (Vurro et al., 2019). These nanoparticles were often based on metals, lipids, or polymers that could encapsulate or adsorb pesticides, ensuring controlled and sustained release over time. As the technology advanced, more sophisticated nanocarriers, such as liposomes, dendrimers, and carbon-based nanoparticles, emerged. These carriers offered enhanced targeting capabilities, allowing for the precise delivery of agrochemicals to specific plant tissues, minimizing waste, and reducing off-target effects (Ganesan et al., 2024).

In parallel, nanocarriers have also been adapted to improve the delivery of essential nutrients, fertilizers, and even genetic material for plant growth regulation (Mujtaba et al., 2021). This evolution has facilitated better nutrient absorption, reduced leaching, and enhanced plant resistance to environmental stresses, such as drought or disease. More recent advancements include the use of responsive nanocarriers that release their payload in response to specific environmental triggers, such as pH or temperature changes, enabling even greater precision in managing plant health (Shishir et al., 2021). Overall, the evolution of nanocarrier technology for plant health reflects a growing trend toward precision agriculture, where the focus is on optimizing resource use, enhancing crop yield, and reducing the environmental impact of agricultural practices. As the technology continues to advance, the potential for nanocarriers to revolutionize plant health management and contribute to sustainable agriculture grows exponentially. Table 1 illustrates the difference between traditional fungicides and nanocarriers. It highlights key aspects such as mode of action, efficiency, target specificity, resistance development, etc. This comparison provides insights into how nanocarriers enhance the use of traditional fungicides by improving targeted delivery and reduces side effects.

Table 1. Comparison of conventional chemical fungicides and nanocarriers

Parameter	Traditional Fungicides	Nanocarriers	References
Historical Development	Widely used since 19th century with the introduction of copper-based fungicides like Bordeaux mixture	Emerged in early 21st century as a nanotechnology driven approach to improve plant health	Morton & Staub., 2008 ; Asmawi et al., 2024
Mode of Action	Directly kills or inhibits fungal growth through chemical toxicity	Enhances bioavailability, controlled release, and targetted delivery	Gupta., 2018 ; Ganesan et al., 2024
Efficiency	Rapid action, but may require frequent application due to degradation	Highly efficient due to sustained release and improved penetration	An et al., 2022
Target Specificity	Often non-specific, affecting both harmful and beneficial microbes.	High specificity, reducing off-target effects	Vega et al., 2022 ;
Resistance Development	Continuous use leads to fungal resistance	Delays resistance due to enhanced penetration and multiple mode of action	Lucas et al., 2015 ; Jabr-Milane etal., 2008
Application Methods	Spraying, dusting, or soil incorporation	Nanoemulsions, nanocapsules, or nanogels	Morton & Staub., 2008 ; Osman et al., 2024

1.2 Scope of Nanocarriers in Sustainable Agriculture

The scope of nanocarriers in sustainable agriculture is vast, offering innovative solutions to address critical challenges in food production, environmental preservation, and resource management. Nanocarriers have the potential to transform various aspects of agriculture, from crop protection and nutrient delivery to soil health management, all while promoting sustainability and reducing environmental impact (Younis et al., 2021). One of the key applications of nanocarriers in sustainable agriculture is in the efficient delivery of agrochemicals, such as pesticides and herbicides. Traditional chemical applications often result in significant wastage, contamination of surrounding ecosystems, and resistance development in pests (Ali et al., 2023). Nanocarriers, on the other hand, enable controlled and targeted release of active ingredients, ensuring that they are delivered directly to the intended site with minimal leakage or evaporation. This reduces the frequency and quantity of pesticide use, minimizing environmental pollution and harm to non-target organisms (Vega et al., 2020). In the realm of fertilization, nanocarriers offer the potential to enhance nutrient uptake by plants. Conventional fertilizers are often inefficient, with a large portion of nutrients being lost to leaching or volatilization. Nanocarriers can encapsulate nutrients in a way that ensures slow, controlled release, improving their availability to plants and reducing nutrient runoff. This contributes to healthier soils, better crop yields, and less reliance on synthetic fertilizers, which are energy-intensive to produce and can contribute to soil degradation (Osman et al., 2024).

Nanocarriers are also showing promise in promoting plant health and resistance to diseases, pests, and environmental stresses. For example, nanocarriers can deliver bioactive compounds or plant growth regulators, enhancing plant immunity, stress tolerance, and growth in adverse conditions such as drought, salinity, or extreme temperatures. This is particularly important in the context of climate change, as crops need to adapt to shifting environmental conditions to maintain food security (Jogaiah et al., 2021). Additionally, nanocarriers offer the potential to improve soil health through the targeted delivery of soil amendments, such as beneficial microbes or organic matter. This can help restore soil fertility, promote biodiversity, and reduce the need for chemical soil treatments (Behl et al., 2024). Furthermore, the use of nanocarriers in precision agriculture, which focuses on delivering resources exactly where and when they are needed, aligns with sustainable practices by reducing resource waste and improving overall farm efficiency. Overall, the scope of nanocarriers in sustainable agriculture extends far beyond simple pesticide and fertilizer delivery. By enhancing resource efficiency, reducing environmental impact, and promoting plant health, nanocarriers hold the potential to contribute significantly to the future of sustainable farming, offering a pathway toward more resilient, productive, and eco-friendly agricultural systems.

1.3 Application and Potential of Nanocarriers in Suitable Agricultural Practices

Nanocarriers have transformed agriculture by offering innovative solutions to some of the most difficult problems facing contemporary farming. These nanoscale delivery technologies provide accurate, effective, and environmentally friendly ways to safeguard crops, manage resources, and improve plant health (Fracis et al., 2023). Nanocarriers open the door to a more productive and sustainable future by incorporating nanotechnology into agricultural methods. This helps to solve problems like resource waste, environmental pollution, and the rising need for food security (Khan et al., 2022). This section examines the many uses of nanocarriers, emphasizing how they have revolutionized several agricultural fields.

2. NANOCARRIERS AS MODE OF DELIVERY

Nanocarrier interact with fungal pathogens through multiple mechanisms to enhance drug delivery and therapeutic efficacy. Functionalized nanocarriers can specifically target fungal cells (Mahapatra et al., 2025) by recognizing pathogen-associated molecular patterns, facilitating selective uptake. Once at the infection site, nanocarriers improve drug penetration (Nene et al., 2021) across fungal cells walls and membranes through surface modifications, increasing intracellular drug concentration. Controlled drug release ensure sustained antifungal activity, reduced resistance development and minimizing toxicity (Mejia et al., 2021). Additionally, nanocarriers can enhance immune response modulation, promoting fungal clearance while reducing host cell damage (Tarunnum et al., 2024).

2.1 Pesticide and Herbicide Delivery

Nanocarriers are widely employed to enhance the delivery and efficacy of pesticides and herbicides (Banerjee et al., 2022). By encapsulating these agrochemicals, nanocarriers enable controlled and sustained release, thereby reducing the frequency of applications and minimizing the loss of active ingredients. This targeted approach not only curtails environmental contamination but also mitigates the development of resistance in pests and weeds. Moreover, nanocarriers improve the solubility and stability of agrochemicals, ensuring their prolonged effectiveness in the field (Kumar et al., 2019; Oliveira et al., 2020)

2.2 Nutrient and Fertilizer Delivery

The application of nanocarriers in nutrient and fertilizer delivery significantly enhances plant nutrient uptake. Conventional fertilizers often face challenges such as leaching, volatilization, and poor absorption (Dutta et al., 2022). Nanocarriers address these issues by encapsulating essential nutrients and releasing them gradually, leading to improved plant absorption and reduced waste. This not only minimizes environmental pollution but also promotes soil fertility and increases crop productivity while lowering fertilizer input requirements (Rahman et al., 2021).

2.3 Plant Growth Regulation and Biostimulants

Nanocarriers facilitate the delivery of plant growth regulators (PGRs) and biostimulants, which are critical for enhancing plant growth, stress tolerance, and overall health (Yadav et al., 2024). The targeted delivery provided by nanocarriers optimizes the effects of these compounds, ensuring maximum efficiency. This application is particularly beneficial for improving crop yields and aiding plants in coping with environmental stresses and diseases (Rajasaheb et al., 2024).

2.4 Disease and Pest Control

Nanocarriers play a pivotal role in disease and pest management by enabling the precise delivery of antimicrobial agents, fungicides, and insecticides (Qiao et al., 2024). This targeted delivery reduces the number of chemical inputs required, minimizes off-target effects, and lessens environmental impact. Furthermore, nanocarriers can be designed to carry natural biocides or biocontrol agents, offering an eco-friendly alternative to traditional chemical pesticides (Kumar et al., 2019).

2.5 Soil Health and Remediation

The role of nanocarriers extends to soil health improvement by facilitating the delivery of soil amendments, including beneficial microbes, enzymes, and organic compounds (Xin et al., 2020). These amendments enhance soil fertility, microbial diversity, and structure, which collectively support healthier plant growth (Bose and pal 2023). Additionally, nanocarriers are employed in soil remediation processes to address pollutants such as heavy metals and pesticides, thereby improving soil quality and sustainability (Singh et al., 2021).

2.6 Gene Delivery and Genetic Engineering

Nanocarriers have potential applications in genetic engineering, particularly for the delivery of genetic material such as DNA or RNA into plant cells (Mujtaba et al., 2021). This approach facilitates the development of crops with enhanced traits, including disease resistance, drought tolerance, and improved nutrient uptake. Nanocarrier-based gene delivery offers higher precision and efficiency compared to traditional methods, advancing the capabilities of agricultural biotechnology (Squire et al., 2023; Raza et al., 2023).

Figure 1. Application of nanocarrier in agriculture (Adopted and modified from Shang et al., 2019)

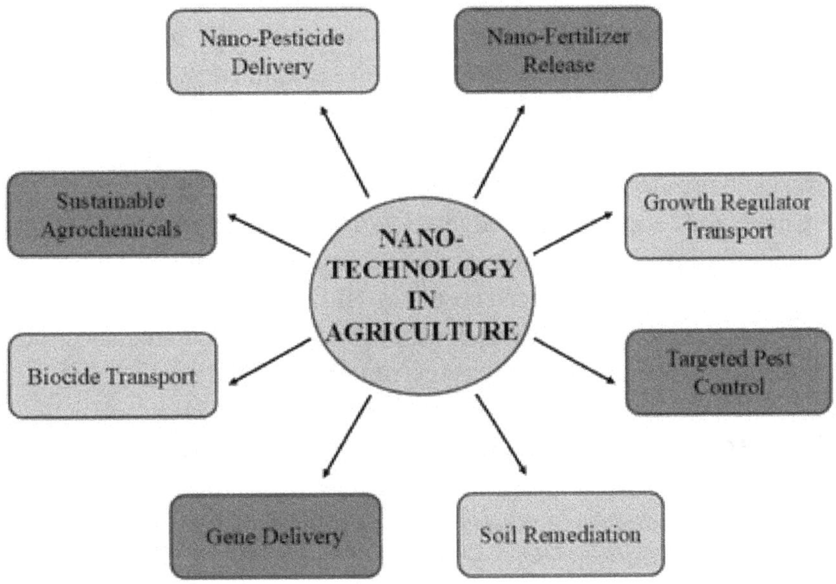

Overall, nanocarriers represent a transformative advancement in agricultural science, offering innovative solutions to enhance sustainability, efficiency, and productivity. By enabling precise delivery of agrochemicals, nutrients, biostimulants, and genetic material, these nanoscale systems address critical challenges such as resource wastage, environmental degradation, and pest resistance. Their applications in soil health improvement, disease management, and diagnostics further highlight their versatility and potential to revolutionize farming practices. While significant progress has been made, continued research is essential to optimize nanocarrier technologies, understand their long-term environmental interactions, and address existing gaps. As the agricultural sector evolves to meet the growing global food demand, nanocarriers will undoubtedly play a pivotal role in shaping a more resilient and sustainable future. Therefore, the review is aimed to provide a comprehensive overview of how nanocarriers can address key agricultural challenges, including the efficient delivery of nutrients, protection against pests and diseases, and sustainable farming practices. It covers the various types and classes of nanocarriers, exploring their unique properties and applications in agriculture. In addition, it delves into the synthesis, characterization, and optimization methods used to develop these nanocarriers, ensuring their effectiveness and safety. Furthermore, it highlights the role of nanocarriers in enhancing plant nutrient delivery and protection, as well as their environmental and biological interactions. Lastly, the chapter identifies existing research gaps and outlines future directions for the development and application of nanocarrier technology in agriculture (Fig. 1).

3. OVERVIEW OF KEY AGRICULTURAL CHALLENGES ADDRESSED BY NANOCARRIERS

Nanotechnology plays a vital role in agriculture by improving food quality safety, enhancing nutrient absorption, and reducing the use of agricultural inputs (Prasad et al., 2017). Recent developments in nanotechnology applications in sustainable agriculture are examined, with particular attention paid to nanopesticides, nanofertilizers, nanobiosensors, and nano-enabled soil remediation, through biotic and abiotic remediation techniques (Usman et al., 2020). The adaptation of nano-delivery systems, initially created for precision medicine, to agricultural applications is examined; by protecting, transporting, and releasing bioactive molecules at designated target areas (Vega-Vásquez et al., 2020). Nanotechnology provides a wide range of solutions for problems in forestry, agriculture and the environment such as sustainable resource management and targeted pesticide delivery (Iavicoli et al., 2017). Building on nanotechnology potential for sustainable agriculture, nanopesticides, which are nanostructures that deliver agrochemical ingredients (Acl) more precisely and efficiently (Chaud et al., 2021).

3.1 Targeted Nutrient Delivery for Optimized Plant Growth

Nanotechnology makes it possible to create novel nanoscale materials with a wide range of uses, especially in agriculture. Nanocarriers play vital role for providing plants with micronutrients like iron, zinc, copper, manganese, and cobalt as well as macronutrients like potassium, phosphorous, and nitrogen (Jampílek et al., 2017). By enhancing nutrient delivery and lowering losses in comparison to conventional fertilisers, nano- fertilisers present a viable way to increase crop yields and nutrient usage efficiency (Iqbal et al., 2019). Phyto- nanotechnology provides considerable promise to improve plant production by targeted delivery of nutrients and controlled release of agrochemicals, boosting yields and disease resistance while minimizing environmental effect (Sharma et al., 2019). Compared to traditional urea, rosin adduct-coated urea fertiliser offers a more effective and sustainable fertilization technique by lowering loss and increasing maize growth and chlorophyll content through a regulated release of urea (Mumtaz et al., 2019). To enhance phosphorous (P) availability for plants, biodegradable polymer (PNCs) was created utilizing polyhydroxyalkanoate (PHA) and calcium phosphate nanoparticles (Ca-P-NPs), building on developments in nutrient delivery, P-driven eutrophication was addressed by these PNCs, which decreased soil phosphorous loss by 80% (Sigmon et al., 2021). To rectify nutrient imbalances in aquaponic systems, scientists analysed customized fish meals to increase potassium availability which frequently lower nitrogen and phosphorous, higher potassium levels in these systems have been shown to promote important plant functions like nutrient intake and photosynthesis this increases growth (Duarte et al., 2024). Nanotechnology provides novel ways to maximize nutrient delivery and boost soil health, such as nano fertilizers and nano enhanced soil amendments, these developments open the door to more sustainable agricultural methods (Tomar et al., 2024). Expanding the role of nanotechnology in agriculture, it enhances crop growth, yield, and quality by improving nutrient use efficiency and reducing waste, they reduce environmental effect, shield plants from stress and sustainable option for efficient farming (Qureshi et al., 2018).

3.2 Precision Farming and Agrochemical Delivery

Agrochemicals are paying more attention to nanomaterials because of their special qualities, such as their small size and excellent solubility (An et al., 2022). Ineffective targeting, timing, and dosing of agrichemicals have a negative impact on the environment and public health, to improve efficiency and decrease waste, this study created sensitive, biodegradable, and sustainable nanostructures for accurate agrochemical distribution (Xu et al., 2022). Controlled release systems have the potential to improve the sustainability and efficacy of agricultural inputs growth hormones, nutrients and pesticides (Campos et al., 2014).

3.2.1 Herbicides

The agronomic use of nanotechnology in plants enables the controlled release of herbicides, for targeted delivery. Engineered nanoparticles (ENP) act as magic bullets delivering herbicides in a controlled and precise manner, such advancements in in precision delivery have the potential to improve crop production and reduced environmental impacts (Banerjee et al., 2022). The smart sprayer integrates a real-time machine vision system with individually controlled nozzles to detect weed infestations, ensuring precise herbicide application (Tian et al., 2002).

3.2.2 Pesticides

Although pesticides are essential for crop protection, excessive use of them damages the environment and increases insect resistance, using materials like chitosan, cellulose, and alginate, tailored pesticide administration is made possible by nanotechnology, particularly biopolymeric nanoformulations which provide precision crop protection. These nanoformulations are economical and environmentally beneficial since they improve pesticide stability, lower toxicity, and promote microbial activity (Kumar et al., 2022).

3.2.3 Fungicides

Fungi have evolved over a billion years, with some species posing significant challenges to agriculture, while fungicides have boosted productivity, their overuse has caused health and environmental concerns (Tleuovaet et al., 2020). By providing targeted pesticide delivery, controlled- release fertilizers, enhanced biosensing, and quick disease diagnosis, nanotechnology presents intriguing answers for precision agriculture, mitigating the shortcomings of traditional farming and possibly transforming the industry (Dulhan et al., 2017).

3.3 Enhancing Water Use Efficiency through Controlled Irrigation Systems

To satisfy rising needs and reduce environmental consequences through improved management, decreased losses, and smart water allocation, irrigated agriculture is under increasing pressure to increase water use efficiency (Howell et al., 2001). The need to increase water use efficiency in irrigated agriculture has been brought to light by growing demand for freshwater brought on by population growth and drought, the best irrigation schedule may be possible with smart irrigation systems that use wireless technologies and sophisticated control techniques. The superiority of closed-loop systems and the advantages of combining soil, plant, and weather-based monitoring and control techniques (Bwambale et al., 2022). To solve freshwater scarcity, agriculture must move towardss maximising water production through systems- based optimisation of irrigation, crop selection, and precision technology. This calls for collaboration between water users in agricultural and non- agricultural sectors (Evans et al., 2008). By autonomously controlling water distribution based on real- time field data and resource availability, an intelligent IoT- multiagent system automates irrigation and increases water use efficiency (Jiménez et al., 2022).

A key strategy to increase agricultural water use efficiency, especially for cereal crops in water-stressed areas like South Asia, and guarantee future food security is to automate drip irrigation using sensor networks and Internet of Things technology (Sidhu et al., 2021). By automatically modifying irrigation based on real- time data and encouraging sustainable habits, an AI- powered smart irrigation system maximises water efficiency in agriculture by analysing soil temperature and moisture using deep learning (Alex et al., 2024). Most of the agriculture in the United States is irrigated by sprinklers, which are more efficient than conventional techniques yet frequently ignore field variations due to uniform applications, by utilizing sensors, wireless technologies, and site-specific variable irrigation management (SS-VRIM) takes advantage to improve water use efficiency (O' Shaughnessy et al., 2015). Recycled stormwater provides a sustainable substitute for the increasingly scarce water used for agricultural irrigation, in addition to showing the necessity of controlling nutrient leaching to avoid eutrophication in natural streams (Nnadi et al., 2015). By alternating between frying and irrigating different areas of the root system, controlled alternative partial root-zone irrigation (CAPRI) increases water consumption efficiency without appreciably lowering yield, by modifying Stomatal behavior and encouraging carbon redistribution improve crop yield (Kang et al., 2004).

3.4. Post-Harvest Applications of Nanocarriers for Crop Protection and Quality Retention

Nanotechnology presents intriguing possibilities for raising agricultural output and lowering post-harvest losses through improved delivery systems, precision sensing, and sustainable substitutes for traditional fertilisers and storage techniques (Babu et al., 2022). From seed to post- harvest, nanotechnology could completely transform agriculture through increased crop output, more efficient use of resources, and the creation of cutting- edge packaging (Shukla et al., 2019). To reduce food loss, increase shelf life, and enhance food security while striking a balance between sustainability and preservation, integrated post- harvest management solutions that include physical, chemical, and innovative packaging techniques are essential (Lalpekhlua et al., 2024). In horticulture, nanotechnology has a wide range of uses, from improving crop quality and production with nanopesticides and fertilisers to lowering post- harvest losses with antimicrobial nanomaterials and facilitating accurate monitoring with nanosensors, all of which contribute to future food security (Rana et al., 2021). Using nanoformulations and nanopackaging, nanotechnology improves the stability and effectiveness of natural products, providing longer shelf life for a variety of crops, better post- harvest disease management, and decreased food spoilage (Alghuthaymi et al., 2020). By employing nanomaterials in coatings, packaging, and diagnostics to shield agriproducts from pathogens and environmental influences while improving shelf life, aesthetic appeal, and nutritional content, nanotechnology provides creative ways to lower post- harvest losses (Bhardwaj et al., 2023) (Fig. 2).

Figure 2. Applications of nanocarriers in plant growth and sustainable agriculture (Adopted and modified from Rana et al., 2021)

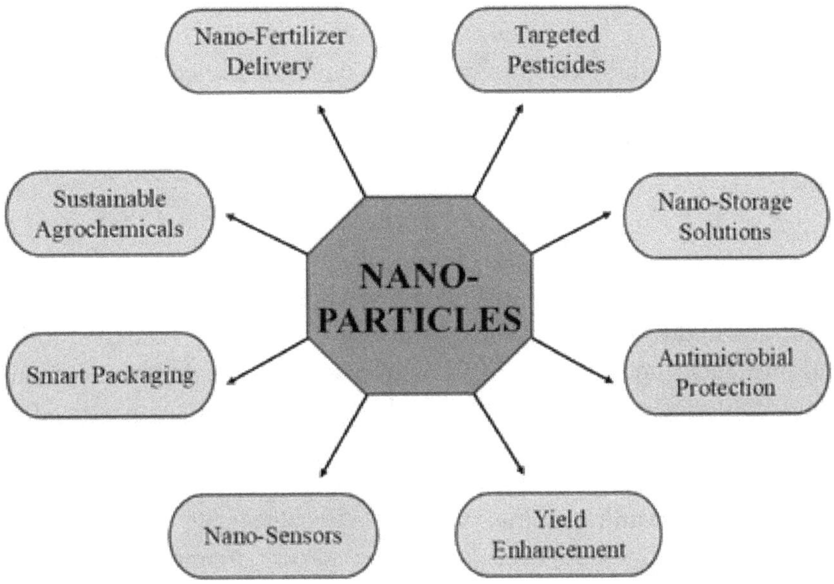

Overall, nanotechnology has transformed agriculture by improving post-harvest management, nutrition delivery, pesticide precision, and water use efficiency, it permits the creation of targeted delivery nanopesticides, and fertilizers, which improve crop growth and yield while lowering waste and environmental effect. To alleviate freshwater scarcity, smart irrigation systems and internet of Things-driven technologies optimize water utilization, post-harvest uses such as coatings and nanopackaging reduce food waste and increase shelf life. Agricultural production, resource efficiency, and environmental protection can all be increased by incorporating nanotechnology into sustainable practices, this opens the door for creative solutions to address the problem facing global food security and guarantee agricultural sustainability over the long term.

4. TYPES AND CLASSES OF NANOCARRIERS

Nanocarriers in agriculture refer to nanoscale materials designed to deliver active agents such as nutrients, pesticides, fungicides, and biostimulants to plants in a controlled and efficient manner (Chariou et al., 2020). These nanocarriers can be broadly classified into three main types based on their composition and structure: organic nanocarriers, inorganic nanocarriers, and hybrid nanocarriers. Organic nanocarriers include liposomes, polymeric nanoparticles, and protein-based carriers, which are often biodegradable and biocompatible. Inorganic nanocarriers, such as metal and metal oxide nanoparticles, are known for their stability and enhanced delivery efficiency. Hybrid nanocarriers combine both organic and inorganic materials to harness the benefits of each, providing improved functionality and versatility. The development and use of these nanocarriers offer significant potential for improving plant health management, enhancing nutrient uptake, and reducing the environmental impact of conventional agricultural practices.

4.1 Organic Nanocarriers

Organic nanocarriers are synthesized from natural or synthetic organic materials and are known for their biocompatibility, biodegradability, and versatility (Romero et al., 2012; Basak et al., 2024). Major types of organic nanocarriers include liposomes, protein-based nanocarriers, and polymeric nanoparticles (Lu et al., 2021).

Liposomes

Liposomes are spherical vesicles composed of lipid bilayers, with a hydrophilic core and hydrophobic outer layers (Guimarães et al., 2021). They are widely used in agriculture due to their ability to encapsulate both hydrophilic and hydrophobic substances, making them highly versatile for delivering various agrochemicals and nutrients. Liposomes can fuse with plant cell membranes, allowing the controlled release of the encapsulated substance directly into the plant cells (Alghuthaymi et al., 2021). This targeted delivery method minimizes the use of chemical pesticides and fertilizers, reducing environmental impact while enhancing crop protection and growth. Liposomes are used extensively in delivering bio-pesticides, fungicides, herbicides, and growth regulators, where the encapsulated substances are protected from degradation during transport and release over an extended period (Banerjee et al., 2022).

Protein-Based Nanocarriers

These are derived from proteins such as albumin, zein, or casein. Protein-based nanocarriers are highly biocompatible and biodegradable, making them environmentally friendly. They can be engineered to deliver various active ingredients, such as plant growth regulators, pesticides, and nutrients (Zhang et al., 2022). The biopolymers used to create protein-based nanocarriers can be cross-linked or modified to enhance their stability, loading capacity, and release profiles. For example, protein nanocarriers can be used to deliver micronutrients like zinc or iron to crops suffering from nutrient deficiencies (Sharma et al., 2022). Additionally, protein-based carriers can protect sensitive bioactive compounds like biostimulants and natural pesticides, enhancing their shelf-life and effectiveness (Chadha., 2021).

Polymeric Nanocarriers

These nanocarriers are made from synthetic or natural polymers such as polylactic acid (PLA), poly (lactic-co-glycolic acid) (PLGA), and chitosan (Ahuja et al., 2020). Polymeric nanoparticles offer several advantages, including controlled release, increased solubility of poorly water-soluble compounds, and protection of active ingredients from degradation. These carriers are widely used in delivering both nutrients and agrochemicals (Shakiba et al., 2020). Their surface can be easily modified to increase bioactivity or target specific plant tissues. For example, the use of chitosan-based nanoparticles can enhance the delivery of nitrogen and phosphorus to plants (Mujtaba et al., 2020), promoting healthier growth, or they can serve as delivery systems for herbicides or fungicides, ensuring their slow and sustained release, and reducing the need for frequent applications.

4.2 Inorganic Nanocarriers

Inorganic nanocarriers are typically synthesized from inorganic materials like metals, metal oxides, or silica (Liu et al., 2021). These nanocarriers offer advantages in terms of stability, high loading capacity, and the potential for targeted delivery of agrochemicals, fertilizers, and biocontrol agents. The two main types of inorganic nanocarriers are metal-based nanoparticles and metal-oxide nanoparticles (Kumar et al., 2021).

Metal Nanoparticles

Metal nanoparticles, such as silver, gold, and copper, have been used in agricultural applications due to their unique properties (Jiang et al., 2022). Silver nanoparticles are particularly well-known for their antimicrobial properties and are often used in delivering biocides or fungicides to plants, providing prolonged protection against pathogens (Sánchez-López et al., 2020). Gold nanoparticles are used for their ability to encapsulate and release a wide range of active compounds, including nutrients and pesticides, in a controlled manner. They also exhibit excellent stability and can be functionalized to increase their specificity for certain plant species or tissues (Fratoddi et al., 2014). Copper nanoparticles are another example, providing a controlled release of copper-based fungicides and bactericides, thus preventing the spread of plant diseases while minimizing toxicity to non-target organisms (Kasana et al., 2017).

Metal Oxide Nanoparticles

Metal oxide nanoparticles, such as zinc oxide (ZnO), titanium dioxide (TiO_2), and iron oxide (Fe_2O_3), are widely used in plant health management due to their unique physical and chemical properties (Meena et al., 2020). ZnO nanoparticles are particularly useful for delivering micronutrients like zinc to plants, which is crucial for various physiological functions, including chlorophyll synthesis and enzyme activation (Adil et al., 2022). TiO_2 nanoparticles, on the other hand, are commonly used for their photocatalytic properties, which can help in controlling fungal infections under sunlight (Serov et al., 2024). These nanoparticles can also be used for improving soil health, as they help in breaking down organic matter, thereby enriching the soil with nutrients that are essential for plant growth.

Silica Nanoparticles

Silica-based nanocarriers are another promising class of inorganic materials for agricultural applications. They are typically used as carriers for nutrients, agrochemicals, and biocontrol agents (Wang et al., 2024). Silica nanoparticles offer high surface area, ease of modification, and the ability to control the release rate of encapsulated substances (Kwon et al., 2013). They can also be engineered to provide enhanced nutrient delivery to plants or to serve as carriers for fungicides, ensuring the controlled release of these agents over an extended period, which can reduce the frequency of applications and minimize environmental contamination.

4.3 Hybrid Nanocarriers

Hybrid nanocarriers combine the advantages of both organic and inorganic materials to create systems that leverage the unique properties of each component. These nanocarriers are designed to offer superior loading capacity, stability, and controlled release profiles (Park et al., 2020). By combining the biocompatibility and biodegradability of organic materials with the high stability and functionality of inorganic components, hybrid nanocarriers provide an ideal solution for a range of agricultural applications. Hybrid nanocarriers are typically developed by integrating organic materials such as lipids or polymers with inorganic materials like metal nanoparticles or silica (Kashapov et al., 2021). For instance, the combination of polymeric materials with metal nanoparticles can result in nanocarriers with both the controlled release properties of the polymer and the antimicrobial or catalytic properties of the metal. These hybrid systems are used for controlled nutrient delivery, enhanced protection against diseases, and the targeted delivery of biocontrol agents. They are particularly useful for applications where the synergy between the organic and inorganic components enhances the performance of the nanocarrier in complex environments like soil or plant tissues.

4.4 Ingredients Delivered by Nanocarriers

The versatility of nanocarriers extends beyond their structural types; they are also effective in delivering a wide variety of ingredients in agricultural practices. These ingredients include nutrients, agrochemicals, and biostimulants, each with specific uses and benefits for plant health.

Nutrients

One of the primary applications of nanocarriers in agriculture is the delivery of essential plant nutrients. Nanocarriers allow for the controlled and sustained release of macronutrients such as nitrogen, phosphorus, and potassium (NPK) as well as micronutrients like zinc, iron, and manganese (Vega-Vásquez et al., 2020). By using nanocarriers, nutrients can be delivered more efficiently to plants, reducing nutrient loss through leaching and improving plant uptake. For instance, polymeric nanoparticles can release nitrogen gradually, ensuring that plants have access to it over an extended period, which improves nitrogen use efficiency and reduces environmental pollution (Beig et al., 2023).

Agrochemicals

Agrochemicals such as pesticides, herbicides, and fungicides are essential for crop protection, but their widespread use has raised concerns regarding environmental contamination and resistance (Boudh & Singh., 2019). Nanocarriers offer a solution to these challenges by providing a more controlled and targeted delivery of agrochemicals. Liposomes and metal nanoparticles, for example, can encapsulate and deliver pesticides in a way that minimizes chemical degradation, reduces the number of agrochemicals needed, and targets specific pests or pathogens (Vurro et al., 2019). This controlled release not only enhances the effectiveness of the agrochemical but also reduces the environmental impact associated with excessive pesticide use.

Biostimulants

Biostimulants are natural or synthetic substances that enhance plant growth and development by stimulating physiological processes (Du Jardin et al., 2015). These include compounds like plant hormones, amino acids, and seaweed extracts. Nanocarriers play a significant role in delivering these biostimulants to plants, where they can improve seed germination, root growth, and stress tolerance (Sahu et al., 2023). For example, protein-based nanocarriers can be used to deliver growth regulators like auxins and cytokinins, promoting faster growth and increased resilience to environmental stressors such as drought or salinity.

Overall, the development and application of nanocarriers in agriculture represent a significant advancement in plant health management. Whether in the form of organic, inorganic, or hybrid nanocarriers, these delivery systems are revolutionizing the way nutrients, agrochemicals, and biostimulants are administered to plants. By enabling controlled release, enhancing bioavailability, and reducing environmental impacts, nanocarriers have the potential to make agricultural practices more sustainable, efficient, and environmentally friendly. As research in this field continues to evolve, it is expected that nanocarriers will play an increasingly important role in addressing global challenges related to food security, crop protection, and sustainable farming practices (Table 2).

Table 2. Overview of different types of nanocarriers in agriculture

Type of Nanocarrier	Composition	Delivery Mechanism	Advantages	Limitations	Applications	References
Liposomes	Lipid bilayers encapsulating active agents	Diffusion and fusion with plant cell membranes for controlled release.	Biocompatible, non-toxic, and suitable for hydrophilic and hydrophobic molecules.	High production cost and limited stability under field conditions.	Delivery of pesticides, fertilizers, and plant hormones	Zhang et al., 2024 Biondo et al., 2022
Protein-Based Nanocarriers	Proteins like zein and casein	Release through enzymatic degradation in soil or plant systems	Biodegradable, eco-friendly, and enhances nutrient bioavailability	Sensitivity to environmental conditions like pH, temperature etc	Delivery of micronutrients (Zn, Fe) and natural pesticides.	Gomes & Sobral 2021
Polymeric Nanoparticles	Polymers like chitosan, PLGA, and alginate	Biodegradable release via diffusion or polymer degradation	Slow-release mechanisms, enhanced stability, and protection of active ingredients.	Regulatory issues and complex synthesis processes	Sustained nutrient delivery, encapsulation of biocontrol agents.	Shakiba et al., 2020 Vinzant et al., 2023
Metal Nanoparticles	Metals like silver, gold, and zinc oxide	Direct uptake by plants or controlled release through oxidation/reduction	Antimicrobial properties, dual functionality as delivery and protective agents.	Potential toxicity to soil and non-target organisms	Delivery of micronutrients, pest control, and pathogen inhibition.	Gehlout et al.,2022 Hyder et al., 2023
Metal Oxide Nanoparticles	ZnO, TiO_2, Fe_2O_3	Release via controlled dissolution in soil or plant systems.	Enhanced solubility, improved plant uptake, and stress tolerance.	Environmental persistence and regulatory constraints	ZnO nanoparticles for micronutrient delivery and UV protection.	Hyder et al., 2023 Cartwright et al., 2020
Silica Nanocarriers	Mesoporous silica nanoparticles	Controlled release via pore size manipulation and surface functionalization.	High loading capacity and compatibility with diverse molecules.	High synthesis cost and potential soil accumulation.	Delivery of agrochemicals, biostimulants, and genetic materials	Rastogi et al.,2019 Wang et al., 2023
Hybrid Nanocarriers	Combination of organic and inorganic components	Multi-modal release mechanisms combining properties of both materials.	Improved efficiency, tailored release profiles, and versatility.	Complex synthesis and high production costs.	Delivery of fertilizers and pesticides with dual benefits (e.g., nutrient + pest control)	Takeshita et al., 2023
Nanoclays	Layered silicate minerals	Adsorption and desorption of active compounds for slow release.	Low cost, high stability, and soil-friendly properties.	Limited compatibility with some active agents	Delivery of herbicides, fungicides, and soil conditioners.	Merino et al., 2020
Nanoemulsions	Oil-water emulsions stabilized by surfactants	Slow diffusion of active agents through plant surfaces.	High bioavailability and enhanced penetration of active ingredients.	Stability issues during storage and field application.	Delivery of hydrophobic pesticides and plant growth stimulants.	Nguyen et al., 2022

5. SYNTHESIS, CHARACTERIZATION, AND OPTIMIZATION METHODS OF NANOCARRIERS

The field of nanocarrier synthesis and characterization has evolved significantly, offering promising applications across various sectors, particularly in agriculture. This chapter examines the current state of nanocarrier development, focusing on synthesis techniques, characterization methods, and the challenges in scaling up for agricultural applications. Sustainable nanotechnology for agriculture encourages the development of nanomaterials that will reduce the reliance on traditional chemicals, such as pesticides for crops or antibiotics for livestock, or the environmental impact of agrochemicals. While the ecological nanotechnology community has developed significant knowledge on the applications and effects of inorganic nanoparticles, an opportunity exists to expand upon the applications of polymeric nanocarriers that can improve the efficiency of delivery of traditional agrochemicals while also providing advantages of biocompatibility and biodegradability.

5.1 Synthesis Approaches

5.1.1 Bottom-Up Techniques

Bottom-up synthesis approaches are a cornerstone in the fabrication of nanocarriers, leveraging the assembly of materials at the molecular or atomic level to create nanoscale structures with high precision. This method emulates natural biological processes such as self-assembly and crystallization, enabling the formation of nanocarriers with tailored physicochemical properties, including size, shape, surface functionality, and drug-loading capacity. Bottom-up techniques such as sol-gel processes, co-precipitation, and molecular self-assembly are widely employed to produce nanostructures with uniform morphology and controlled functionality, crucial for applications in drug delivery, diagnostics, and theranostics. These methods are advantageous over top-down approaches in terms of scalability, cost-efficiency, and the ability to achieve nanoscale resolution without requiring extensive post-processing. For instance, surfactant-templated self-assembly has been extensively utilized to fabricate mesoporous silica nanoparticles with well-defined pore architectures, facilitating high drug loading and controlled release profiles (Slowing et al., 2008). Similarly, the co-precipitation of lipids and polymers enables the production of liposomes and polymeric micelles, respectively, with surface functionalities tailored to target-specific cellular receptors (Peer et al., 2007). By exploiting the intrinsic properties of molecular interactions, bottom-up approaches provide a versatile platform for designing nanocarriers with enhanced biocompatibility and functionality, advancing their potential in precision medicine. The steps of synthesis have been summarized in Figure 3.

Figure 3. Bottom-up technique of nanoparticle synthesis (Adopted and modified from Habiba et al., 2014, Created using Biorender.com)

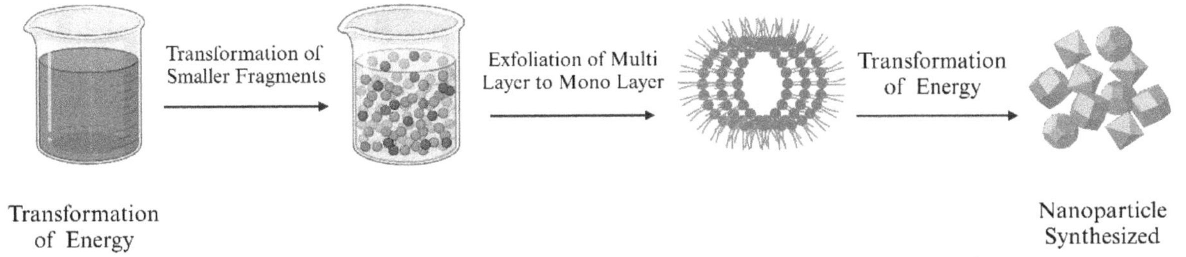

5.1.2. In Situ Polymerization

In situ polymerization is an innovative technique for synthesizing polymeric nanoparticles directly within a reaction medium, providing a robust platform for the development of antifungal agents. This approach involves the polymerization of monomers in the presence of a stabilizer or template, allowing the formation of nanoparticles with controlled size, morphology, and functionalization. In the context of fungal disease management, in situ polymerization has garnered significant attention due to its ability to encapsulate and deliver antifungal agents efficiently, improving their stability, bioavailability, and targeted delivery. Polymers such as polylactic acid (PLA), poly(lactic-co-glycolic acid) (PLGA), chitosan, and polyethylene glycol (PEG) are frequently employed in this process to fabricate nanoparticles. These materials are biocompatible and can be functionalized with antifungal agents, such as amphotericin B, itraconazole, or fluconazole, enhancing their therapeutic efficacy (Patil et al., 2019). The controlled release properties of polymeric nanoparticles reduce the systemic toxicity associated with conventional antifungal treatments while ensuring prolonged drug release at the site of infection (Prabhu et al., 2015). Recent studies have explored the use of in situ polymerization to encapsulate phytochemicals with antifungal properties, such as curcumin or eugenol, demonstrating synergistic effects against fungal pathogens like *Candida albicans* and *Aspergillus fumigatus* (Agrawal et al., 2020; Rajasekar et al., 2018). Furthermore, surface-functionalized nanoparticles synthesized through in situ polymerization can be engineered to target specific fungal cell components, such as β-glucans or chitin, offering enhanced specificity and reducing off-target effects (Li et al., 2021). The technique is versatile and can be adapted to various delivery systems, including hydrogels and micelles, enabling localized antifungal drug delivery (Mohanraj & Chen, 2006). Additionally, nanoparticles formed by in situ polymerization exhibit stability under physiological conditions and protect the encapsulated drug from degradation, addressing key challenges in fungal disease management (Sharma et al., 2021). Innovative applications, such as stimuli-responsive nanoparticles that release antifungal agents in response to pH changes or fungal enzymatic activity, are also being developed using in situ polymerization techniques (Mishra et al., 2022). These smart delivery systems can enhance therapeutic outcomes, particularly in resistant fungal infections. Overall, in situ polymerization offers a promising strategy for the next generation of antifungal therapeutics, addressing the limitations of traditional treatments and paving the way for targeted, efficient, and patient-friendly antifungal therapies.

Advantages of In Situ Polymerization:

1. Controlled Molecular Assembly lies at the heart of in situ polymerization, offering a high degree of precision in nanocarrier synthesis. This technique involves carefully orchestrating the reaction conditions to guide the self-assembly of monomers into polymeric chains, ultimately forming the desired nanocarrier structure. By manipulating factors such as temperature, pH, and the presence of surfactants or templates, researchers can exert fine control over the resulting nanoparticles' size, shape, and morphology. This control is crucial for tailoring nanocarriers to specific applications, as their performance is often intimately linked to their physical characteristics. The controlled assembly process allows for the creation of uniform and well-defined nanocarriers, ensuring consistent properties and predictable behavior in delivery systems. Furthermore, the versatility of in situ polymerization enables the incorporation of various functional groups and targeting moieties during the synthesis process, further enhancing the specificity and efficacy of the nanocarriers (Vauthier, C et al., 2009).
2. **Precise size distribution control** is paramount in in situ polymerization, ensuring the uniformity and consistency of the synthesized nanocarriers. This meticulous control over size distribution is achieved through the precise manipulation of reaction parameters, including temperature, pH, monomer concentration, and the presence of stabilizing agents. By carefully adjusting these parameters, researchers can fine-tune the nucleation and growth kinetics of the nanoparticles, leading to a population of nanocarriers with a narrow size distribution. This uniformity is critical for achieving predictable and reproducible behavior in various applications, such as drug delivery, where consistent size ensures uniform diffusion and cellular uptake. Moreover, precise size control minimizes the risk of aggregation and sedimentation, enhancing the stability and shelf-life of nanocarrier formulations. The ability to tailor the size distribution to specific requirements allows for the optimization of nanocarrier performance in diverse applications, from targeted drug delivery to controlled release of agrochemicals (Vauthier et al., 2009).

5.1.3 Self-Assembly Method

Self-assembly is a powerful strategy in nanotechnology that enables the spontaneous organization of molecules into well-defined, stable nanostructures under specific conditions. In the context of fungal disease management, self-assembled nanoparticles offer a promising avenue for the development of targeted, effective, and minimally invasive treatments. These nanoparticles can be engineered to carry antifungal agents, improving their solubility, stability, and bioavailability. Self-assembly occurs due to non-covalent interactions such as van der Waals forces, hydrogen bonding, hydrophobic effects, and electrostatic interactions, allowing for the formation of nanocarriers like micelles, liposomes, and dendrimers.

Advantages of Self-Assembly Method:

- One of the primary advantages of self-assembly is its ability to provide enhanced control over nanoparticle properties, including size, shape, and surface functionality. This level of precision is crucial for ensuring effective delivery to fungal infection sites while minimizing off-target effects. For instance, self-assembled polymeric micelles have shown the ability to encapsulate hydrophobic antifungal drugs like amphotericin B, improving their therapeutic index and reducing nephrotoxicity (Mouton et al., 2016). Additionally, liposomes self-assembled from phospholipids have been widely studied for their potential to deliver antifungal agents directly to fungal cells while sparing host tissues (Sharma et al., 2020).

- Another significant advantage is the cost-effectiveness of self-assembly methods. The spontaneous nature of the process eliminates the need for complex machinery or high-energy input, making the technique suitable for large-scale production. Moreover, the versatility of self-assembly enables the incorporation of functional molecules, such as targeting ligands and imaging agents, without requiring extensive chemical modifications. For fungal diseases, this translates into the development of smart delivery systems that can actively target fungal pathogens while monitoring treatment efficacy (Wang et al., 2018).
- Overall, the application of self-assembly in nanoparticle fabrication represents a transformative approach in fungal disease management. By harnessing the simplicity and efficiency of self-assembly, researchers can create advanced therapeutic platforms that address the limitations of traditional antifungal therapies and improve patient outcomes.

5.1.4 Solution Methods for Nanoparticle Synthesis in Fungal Plant Management

Nanoparticle (NP) synthesis for fungal plant management involves various solution-based methods that enable the production of stable and effective nanoparticles. The **solution methods** primarily include **chemical, physical, and biological (green) synthesis** approaches, each offering different advantages and limitations. These methods are widely explored for their ability to produce nanoparticles with desirable properties such as high antifungal efficacy, controlled release, and biocompatibility.

Chemical Methods

Chemical solution-based synthesis is the most commonly used approach for nanoparticle production. This includes methods like **sol-gel synthesis, chemical reduction, co-precipitation, and hydrothermal synthesis**. In the **sol-gel method**, metal precursors undergo hydrolysis and condensation reactions in a solution, forming a gel that, upon drying, yields nanoparticles with high uniformity. **Chemical reduction**, another widely used method, involves the reduction of metal salts using reducing agents such as sodium borohydride or citrate, leading to metallic nanoparticles. **Co-precipitation** relies on the precipitation of metal hydroxides from aqueous solutions, followed by heat treatment to obtain nanoparticles. **Hydrothermal synthesis** involves the reaction of precursors in a high-temperature aqueous medium under controlled pressure, allowing the formation of metal oxide nanoparticles with defined crystalline structures (Sulak., 2025). While chemical methods ensure precise control over nanoparticle properties, they often involve toxic reagents and may lead to environmental hazards.

Physical Methods

Physical methods utilize mechanical, thermal, or radiation-based techniques to generate nanoparticles from bulk materials. One common **solution-based physical method** is **laser ablation**, where a high-energy laser beam irradiates a metal target submerged in a liquid medium, leading to the formation of nanoparticles. Another approach, **electrochemical synthesis**, involves the dissolution of a metal anode in an electrolyte solution, where electrochemical reactions lead to the deposition of nanoparticles. **Microwave-assisted synthesis** is another promising technique that uses microwave radiation to heat solutions containing metal precursors, leading to rapid and controlled nanoparticle formation. These methods typically produce high-purity nanoparticles without chemical contaminants; however, they often require high energy input and specialized equipment, making them less suitable for large-scale applications.

Biological (Green) Solution Methods

Biological or green synthesis methods employ microorganisms (bacteria, fungi, and algae) or plant extracts to reduce metal ions into nanoparticles in an aqueous solution. **Fungal-mediated synthesis** is particularly promising, as fungi produce extracellular enzymes and metabolites that facilitate NP formation while also acting as natural stabilizers. **Plant extract-based synthesis** involves using phytochemicals such as flavonoids, polyphenols, and proteins to reduce and cap nanoparticles, making the process non-toxic and environmentally friendly. Green synthesis methods offer sustainable and biocompatible solutions for NP production, but their scalability and reproducibility remain challenges (Table 3).

Table 3. Solution-based nanoparticle synthesis methods

Method	Solution Used	Advantages	Disadvantages	Reference
Sol-Gel Synthesis	Metal alkoxides, alcohol, acid/base catalyst	Produces uniform and stable nanoparticles, easy to control	Requires long processing times, solvent toxicity	Ahlawat,et al., 2014
Chemical Reduction	Metal salts, reducing agents ($NaBH_4$, citrate)	Simple and cost-effective, widely used	Uses toxic chemicals, risk of environmental contamination	Wang, H et al., 2005
Co-precipitation	Metal salt solutions, precipitating agents	Fast and scalable, minimal byproduct formation	Poor size control, agglomeration risk	Rao, Srinivasa, et al., 2011
Hydrothermal Synthesis	Aqueous metal precursor solution under high pressure	Produces high-purity crystalline nanoparticles	Requires high temperature and pressure	Madathil, et al., 2007
Laser Ablation	Metal targets in a liquid medium	Chemical-free, produces high-purity nanoparticles	Requires high energy, expensive equipment	Semaltian (2010).
Electrochemical Synthesis	Metal electrodes in electrolyte solution	High-purity, controlled nanoparticle deposition	Requires controlled environment, costly setup	KhaydarovRashid, et al., 2009
Microwave-Assisted Synthesis	Metal precursors in aqueous solution	Rapid heating, energy-efficient	Requires specialized microwave reactors	Palchik, et al., 2000
Fungal-Mediated Synthesis	Fungal biomass or culture filtrate in aqueous solution	Eco-friendly, cost-effective, produces stable nanoparticles	Slow synthesis rate, scalability issues	Anjum, et al., 2023
Plant Extract-Based Synthesis	Plant extracts with metal precursors	Non-toxic, biocompatible, renewable	Difficult to standardize, variable composition	Hano, & Abbasi, (2021).

5.1.5 Sol-Gel Process

The sol-gel method is a versatile wet chemical technique for synthesizing nanomaterials, particularly metal oxide nanoparticles. This process involves transitioning a solution, or "sol," into a solid "gel" phase through hydrolysis and polycondensation reactions. Typically, metal alkoxides serve as precursors, which, when dissolved in water or alcohol, undergo hydrolysis to form a gel. The resulting wet gel is then dried and calcined to produce nanoparticles. This method offers advantages such as cost-effectiveness, low reaction temperatures, and precise control over the chemical composition of the products. The drying process of the gel is crucial, as it significantly influences the properties of the final product. Depending on the desired application, different drying techniques are employed. For instance, in alcoholic solutions, drying may involve burning off the alcohol. The choice of drying method affects the porosity, surface area, and structural integrity of the nanoparticles (Bokov, Dmitry, et al., 2021).

The sol-gel process plays a crucial role across various fields, including optics and electronics, where it is instrumental in fabricating thin films and coatings for optical devices, sensors, and electronic components, with nanoscale control over thickness and composition enhancing performance in antireflective coatings and dielectric layers. In energy-related applications, sol-gel-derived materials contribute to the development of batteries, fuel cells, and photovoltaic devices by enabling the production of electrode materials with high surface area and controlled porosity, improving energy storage and conversion efficiency. The technique is also widely used in surface engineering, facilitating the creation of protective coatings that offer wear and corrosion resistance, making them valuable in industries such as automotive and aerospace. In the biomedical field, sol-gel-derived nanomaterials are employed in biosensors for detecting biological molecules and in drug delivery systems, where their biocompatibility and tunable properties allow for controlled therapeutic release. Additionally, sol-gel technology finds use in separation technologies like chromatography, as the porous nature of the materials enhances the effective separation of chemical species, improving both analytical and preparative separation techniques. In agriculture, sol-gel materials are utilized for the controlled release of fertilizers, pesticides, and other agrochemicals, enhancing crop protection and nutrient delivery while minimizing environmental impact and chemical waste (Bokov, Dmitry, et al., 2025). Overall, the sol-gel method is a conventional and industrially viable approach for synthesizing nanoparticles with diverse chemical compositions. Its versatility and ability to produce materials with tailored properties make it a valuable technique in advancing nanotechnology and its applications (Bokov, Dmitry, et al., 2025) (Fig. 4).

Figure 4. Step by step preparation of nanoparticles by sol-gel process (Adopted and modified from Bokov et al., 2020, Created using Biorender.com)

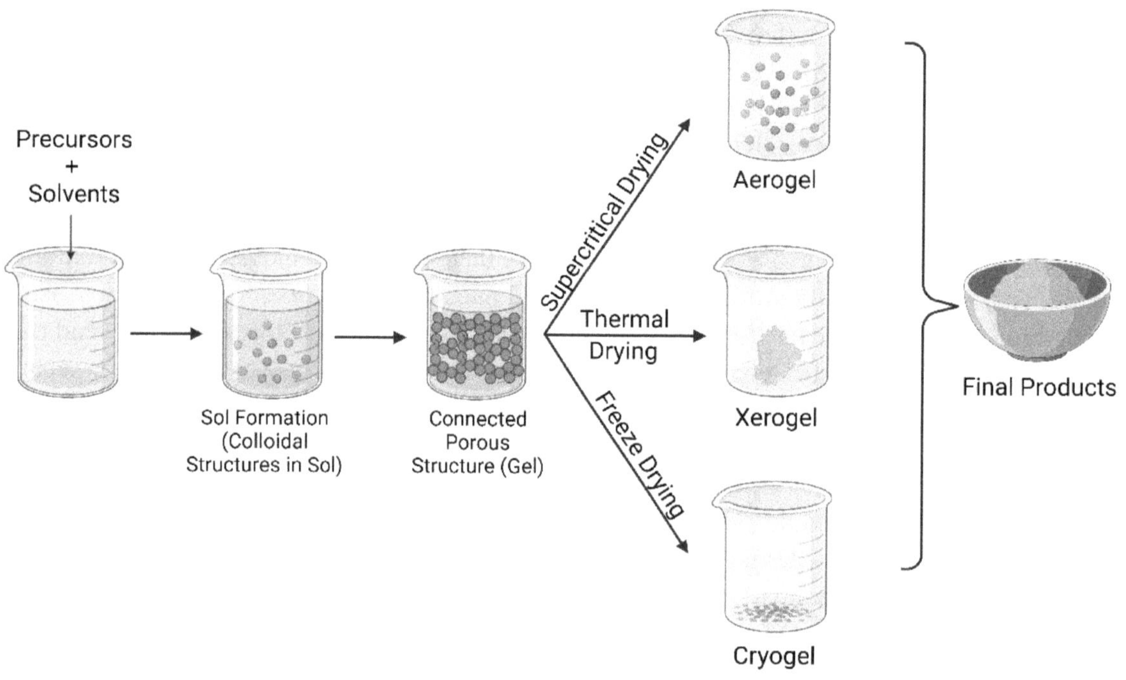

Advantages of Sol gel Process:

- One of the significant benefits of the sol-gel method is its ability to produce materials with high purity. Since the process allows for molecular-level mixing of precursors, it ensures a homogeneous composition, leading to products with superior purity compared to other synthesis methods (Esposito, S. 2019).
- Another advantage is the precise control over the particle size and distribution. The sol-gel process enables the synthesis of nanoparticles with a narrow size distribution, which is crucial for applications requiring uniformity in particle size (Esposito, S. 2019).
- The sol-gel method also allows for the incorporation of various dopants into the matrix, enabling the tailoring of material properties to suit specific applications. This versatility makes it a powerful tool for designing materials with desired optical, electronic, or mechanical properties (Esposito, S. 2019).
- Furthermore, the sol-gel process is cost-effective and operates at relatively low temperatures compared to traditional solid-state methods. This reduces energy consumption and allows for the synthesis of materials that might decompose at higher temperatures (Hench, L. L., & West, J. K. 1990).

In summary, the sol-gel method offers several advantages, including high product purity, precise control over particle size, versatility in doping, and cost-effectiveness. These benefits make it a preferred choice for synthesizing nanomaterials for various applications.

5.2 Top-Down Technique

The top-down method includes the breakdown of huge matter into nano-sized particles. It is a sustractive method that provides for breaking of superstructure bulk matter and confining it to smaller structures by chemical or mechanical means. Some techniques involving top-down involve approaches are mechanical grinding, attrition, and milling (Aarthye, P. *et al.*, 2021). Top-down approaches are simple, yet they produce irregular shaped tiny particles and since there is a difficulty face to maintain a proper size and shape if the particle is the major downfall of this approach (Abid, Namra, et al., 2022). The following are few methods, which are employed under top-down approach:

5.2.1. Mechanical Milling or Ball Process

Ball milling is a fundamental and effective top-down method that produces nanomaterials by transferring kinetic energy from grinding media to the material that is being reduced. This method is useful to produce various nanoparticles and metal alloys using different materials. The interaction of balls, and between balls and vessel walls, produces drastic phase transformation at high temperature and pressure. The ball miling process has repercussions on various factors such as crystal size, particle size and surface, mechanical dislocations and modification of meta-stable phases. Miling process can be used for the growth of these techniques which includes high capacity, safety, simplicity, reliability, long term maintainance and universality (Neikov, O. D. 2009; Janot, R., & Guérard, D. 2002). With merits come certain limitations like energy wastage, destructive/destroying/disordering crystals structure, noise and contamination while working, sturdy weights (Li, Lu Hua, et al., 2011). The steps of synthesis have been summarized in Figure 5 and the steps of synthesis has been summarized in Figure 3.

Figure 5. Top down technique for nanoparticle synthesis (Adopted and modified from Habiba et al., 2014, Created using Biorender.com)

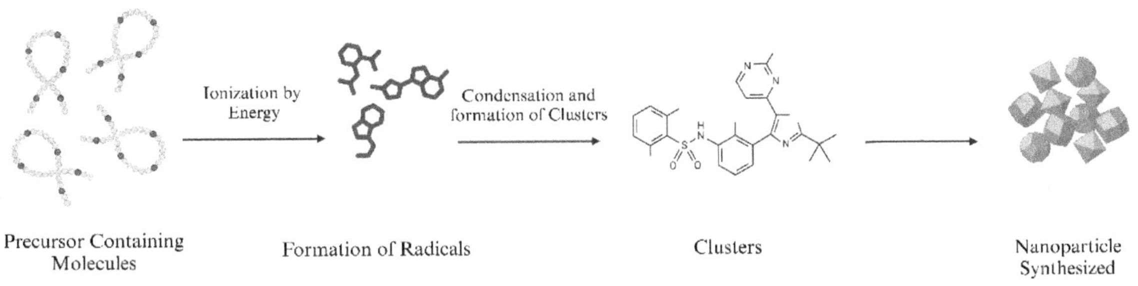

5.2.2. Lithography

This method of synthesis of nanoparticles requires expensive equipments and instruments and are also dreadful with respect to energy requirement but are capable to make micron-sized articles. Since ages lithography techniques were used for producing computers and printed circuit boards, but nowadays we used nanoimprint lithography method, which is different from conventional lithographic, initial step involves formation off template followed by stamping of soft polymeric method to form patterns. The formation of template matrix requires a latex sphere through nanospheres lithographic method. Various types of lithographic printing are used nowaday out of which some are photo-lithography, electron beam lithography, soft lithography, nanoimprint lithography, focused ion lithography, and dip pin lithography (Haynes, C. L., et al., 2001).

5.2.3. Nanocrystal Formation

The synthesis of uniform nanocrystals is essential due to the direct correlation between their size uniformity and the consistency of their chemical and physical properties. The hot-injection method involves the rapid injection of precursors into a hot solvent, causing a sudden increase in monomer concentration that induces burst nucleation, where numerous nuclei form simultaneously, followed by controlled growth as the monomer concentration decreases. This approach effectively results in nanocrystals with narrow size distributions. On the other hand, the heat-up method involves gradually heating a mixture of precursors and solvents, leading to a controlled increase in monomer concentration. This gradual process allows simultaneous nucleation and growth, yielding uniform nanocrystals and is particularly advantageous for large-scale production since it avoids the challenges of rapid precursor injection. Both methods rely on balancing nucleation and growth to control size distribution, with the hot-injection method emphasizing rapid nucleation followed by diffusion-controlled growth, while the heat-up method focuses on controlled heating to ensure uniform crystal development. Understanding these formation mechanisms is crucial for the tailored synthesis of nanocrystals with desired properties, facilitating advancements in diverse fields such as electronics, catalysis, and biomedical research (Kwon, S. G et al., 2011).

5.3 Characterization of Nanoparticles

Modern nanocarrier characterization employs various sophisticated analytical techniques, as outlined in recent research.

5.3.1 Analysis Techniques

Dynamic Light Scattering (DLS)

Dynamic Light Scattering (DLS) is a widely used technique for determining the size distribution of nanoparticles in suspension. This method measures the Brownian motion of particles and relates it to their hydrodynamic diameter (Smith et al., 2020). Studies have demonstrated that DLS can provide accurate measurements of particle size distribution with high reproducibility, making it essential for quality control in nanocarrier production (Hoo, Christopher M., et al., 2008).

Particle Size Distribution Analysis

Particle size distribution analysis is crucial for assessing the homogeneity and stability of nanocarriers. Consistent particle size ensures uniform drug delivery and bioavailability. Techniques such as DLS and nanoparticle tracking analysis (NTA) have been utilized in numerous studies to measure particle size distributions accurately (Hassellöv, Martin, et al., 2008).

Zeta Potential Measurements

Zeta potential measurements assess the surface charge of nanoparticles, which influences stability and aggregation behavior. A high absolute zeta potential typically indicates good colloidal stability (Garcia et al., 2018). Recent publications have highlighted zeta potential analysis as a predictor of nanocarrier stability in various biological media (Clogston, J. D., et al., 2011).

Solution Behavior Studies

Solution behavior studies investigate how nanocarriers interact with solvents and biological fluids. Techniques like DLS and fluorescence correlation spectroscopy (FCS) help evaluate aggregation tendencies and stability under physiological conditions. Research has confirmed the importance of these studies for predicting in vivo performance (Deng, Z., et al., 2008).

5.3.2 Microscopy Techniques

Transmission Electron Microscopy (TEM)

TEM provides high-resolution images of nanocarriers, allowing direct observation of particle morphology and size. TEM has been extensively used in studies to verify particle uniformity and core-shell structures in liposomal and polymeric nanocarriers (Falsafi, Seid Reza, et al., 2020).

Scanning Electron Microscopy (SEM)

SEM offers detailed surface morphology images, which are critical for assessing the surface structure and porosity of nanocarriers. Recent research articles have demonstrated SEM's utility in evaluating surface modifications and functionalization of nanoparticles (Falsafi, Seid Reza, et al., 2020).

Atomic Force Microscopy (AFM)

AFM provides three-dimensional surface topography and mechanical property measurements at the nanoscale. AFM has been utilized in studies for characterizing the roughness and mechanical stiffness of polymeric nanoparticles (Falsafi, Seid Reza, et al., 2020).

5.3.3 Spectroscopic Methods

FTIR Analysis

Fourier Transform Infrared Spectroscopy (FTIR) is employed for surface chemistry characterization, allowing the identification of functional groups and molecular interactions on the nanocarrier surface. FTIR studies have been widely cited for confirming the presence of targeting ligands and surface modifications in functionalized nanocarriers (Mourdikoudis, S., et al., 2018).

Surface Chemistry Characterization

Surface chemistry characterization involves techniques such as X-ray Photoelectron Spectroscopy (XPS) and FTIR to analyze surface functionalization and elemental composition. XPS studies have confirmed successful grafting of functional groups on polymeric nanocarriers, enhancing their targeting capabilities (Fernandez-Garcia, M., et al., 2004).

Structural Confirmation

Structural confirmation techniques, including Nuclear Magnetic Resonance (NMR) and Raman spectroscopy, provide molecular-level insights into nanocarrier composition and structure. Recent research emphasizes the role of NMR in confirming the successful encapsulation of therapeutic agents within the nanocarrier core (Fendler, J. H. (Ed.). 2008). These characterization methods collectively offer a comprehensive understanding of nanocarrier properties, essential for their successful application in drug delivery and other biomedical fields.

5.4 Challenges in Up- Scaling for Agricultural Use

Scaling up the production of nanoparticles and nanocarriers for agricultural applications presents several challenges across production, environmental considerations, and regulatory compliance.

5.4.1. Production Challenges

Achieving batch-to-batch consistency, cost-effective manufacturing, and quality control maintenance is crucial for the reliable application of nanocarriers in agriculture. Ensuring uniformity between production batches is essential to maintain consistent efficacy and safety profiles, as variations can undermine their performance. Consistent physicochemical properties, such as size, charge, and encapsulation efficiency, are necessary to guarantee predictable behavior. However, the scalability of nanocarrier production is often hindered by high costs associated with sophisticated equipment, raw materials, and stringent quality control measures. Developing economically viable manufacturing processes, including innovations in synthesis methods and the use of abundant natural materials, could help reduce costs. Moreover, quality control requires rigorous characterization techniques to monitor parameters like particle size distribution, surface charge, and stability. Scaling up production demands the development of robust protocols using advanced analytical tools and standardized procedures to maintain product integrity and performance (Ditta, A., et al., 2015).

5.4.2. Environmental Considerations

Biodegradability, eco-toxicity assessment, and sustainable production methods are crucial factors in determining the environmental impact of nanocarriers. Biodegradability plays a key role, as non-biodegradable nanoparticles can accumulate in the environment, posing potential ecological risks (Corsi, Ilaria, et al., 2023). Designing nanocarriers from biodegradable materials ensures they break down into non-toxic byproducts, minimizing long-term environmental persistence. To safeguard ecosystems, comprehensive eco-toxicity assessments are essential, evaluating the effects of nanocarriers on non-target organisms such as soil microbes, plants, and animals. Understanding nanoparticle interactions with environmental components requires standardized testing protocols and long-term impact studies to prevent ecological harm (Boros, B. V., et al., 2020). Additionally, the synthesis of nanocarriers should prioritize sustainable production methods that align with green chemistry principles, reducing hazardous waste and energy consumption. Environmentally friendly approaches like the use of renewable resources and solvent-free synthesis not only lower ecological impact but also promote public acceptance of nanotechnology in agriculture.

5.4.3 Economic Barriers

Economic obstacles like high development and manufacturing costs, market competition, consumer awareness and acceptance, infrastructure requirements, and long-term return on investment (ROI) are impeding the adoption of nanocarriers, which hold promise for the management of fungal diseases. Advanced research and development (R&D) capabilities, such as specialized machinery, knowledgeable staff, and complex manufacturing procedures, are necessary for the synthesis of nanocarriers. For developing nations, where R&D budgets are frequently constrained, these requirements present serious financial obstacles (Salamanca-Buentello et al., 2005). Intellectual property (IP) rights are frequently linked to the synthesis of nanocarriers, which can obstruct access to and transfer of technology in developing nations (Sampat, 2009). Due to their narrow profit margins, many smallholder farmers might not have the funds to invest in new technologies (Lowder et al., 2016).

5.4.4 Economic Feasibility

By increasing crop output, decreasing losses, and cutting input costs through improved active component efficiency, nanocarriers in fungal disease management provide economic viability. Thier precise delivery reduces the need for fungicides, which lowers application costs and their negative effects on the environment. Reducing chemical residues can improve the marketability and health of the soil. Optimizing nano formulation efficiency and reducing production costs are necessary for widespread adoption, though, to guarantee affordability and optimize financial gains inn sustainable agriculture (Su et., 2022). Stimulus-responsive nanocarriers optimize pesticide use through precise, controlled release, increasing the economic feasibility of managing fungal diseases. This increases crop marketability, lowers application costs, and lessens the impact on the environment. By increasing efficiency with lower pesticide quantities, nanocarriers maximize revenue while promoting sustainability, however, further advancements in cost-effective formulations are essential for widescale adoption (Ma et al., 2024). Through increasing productivity, cutting waste, and fostering sustainability, nanotechnology improves the economic feasibility of both cosmetics and agriculture. Nanocarriers optimize the use of pesticides in agriculture, reducing expenses and their impact on the environment. In cosmetics, they improve bioavailability, turning waste materials into valuable products. This collaboration reduces manufacturing costs and satisfies the markets increasing need for environmentally friendly products. However, for widespread adoption and to optimize the economic and environmentally benefits, it is imperative tpo overcome legislative and technical obstacles (Nhan et al., 2024).

Through resource optimization, waste reduction, and efficiency enhancement, nanotechnology provides an economically feasible and sustainable solution for a variety of businesses, in cosmetics, nanocarriers increase bioavailability, turning waste materials into valuable products, while in agriculture, they improve pesticide delivery, reducing expenses and environmental effect.

6. NANOCARRIERS FOR PLANT NUTRIENT DELIVERY

The use of biodegradable nanocarriers for advanced nutrient delivery holds the potential to transform crop biofortification and satisfy the world's food needs (Dutta et al., 2022). As evidenced by urea-doped hydroxyapatite nanoparticles, nanocarriers present a viable way to enhance nutrient delivery to plants, boosting growth and reducing environmental effect in comparison to traditional fertilisers (Pradhan et al., 2021). In comparison to conventional fertilisation techniques, nanocarriers enable improved delivery of micro- and macronutrients to promote crop growth, yield, and representing a promising new frontier in plant nutrition (Kaur et al., 2024). With carefully thought-out, short-range stable formulations, liposome-based nanoparticles provide a very efficient way to supply nutrients to plants, improving nutrient uptake and translocation while combating environmental issues (Karny et al., 2018). In contrast to traditional fertilisers, nanocarriers- like chitosan- coated nano-zeolites (Takkar et al., 2024), morphology-tailored nanohydroxyapatite (Tang et al., 2023), graphene-based nanocarriers (Bhattacharya et al., 2022) - offer a revolutionary way to improve phosphorus delivery to plants by overcoming soil fixation and enhancing nutrient uptake.

6.1 Macronutrient Delivery: Efficient Uptakes and Transport of Nitrogen, Phosphorus, and Potassium (NPK)

Nano fertilizers have the potential to improve the uptake and delivery of vital macronutrients such as nitrogen, phosphorus, and potassium (NPK) in plants (Alrbaihat et al., 2022). The importance of nitrogen and sulphur fertilisation for the best delivery of macronutrients and crop productivity was demonstrated by the increase in spring wheat grain yield (Klikocka et al., 2018). Agriculture has become more intense due to rising food demand, and to compensate for soil depletion and guarantee the best possible plant development and output, vital macronutrients like NPK must be applied externally (Sinha et al., 2020). As a promising macronutrient nano-fertilizer, calcium phosphate nanoparticles (CaP) offer better nutrient use efficiency and controlled release of nitrogen and phosphorus than traditional fertilisers, which may help promote more sustainable farming methods (Carmona et al., 2022). The methods of macronutrient uptake, transport, and distribution from the soil to the plant vasculature are revealed by single-cell transcriptomics, which shows varied expression patterns of NPK transporters in various root cell types (Lhamo et al., 2021). To effectively absorb and use vital macronutrients like NPK especially in situations where nutrients are scarce, plants have developed a variety of strategies, such as specialised transporters and channels (Chen et al., 2008). Compared to traditional fertilisation techniques, foliar application of chitosan-NPK nanoparticles has demonstrated potential in improving the transport and uptake of macronutrients in wheat, resulting in higher yield and shorter life cycle (Abdel-Aziz et al., 2016).

6.2 Micronutrient Delivery: Enhancing Bioavailability of Zinc, Iron, Manganese, and Other Trace Elements

To improve micronutrient bioavailability and meet global nutritional difficulties, biotechnological treatments such as genetic modification and genome editing are required to address micronutrient deficits in staple crops that are made worse by pollution, soil conditions, and climate change (Banerjee et al., 2023). A complicated interaction between soil characteristics, root exudates, rhizosphere microbial activity, and plant physiology determines micronutrient bioavailability in plants, which in turn affects nutrient uptake, transport, and ultimately, the nutritional value of edible plant portions (Kumar et al., 2016). The successful delivery of composite micronutrient nanoparticles comprising Ni, Cu, Zn, and Fe to plant tissues can improve the uptake of these vital components and perhaps solve several nutritional shortages (Tombuloglu et al., 2020). Even while eight trace elements- Fe, Zn, Mn, Cu, Ni, B, Mo, and

Cl- are now known to be necessary for plant growth, research on the physiology, absorption processes, and functions of micronutrients in plant development is still ongoing (Welch et al., 1995).

6.3 Mode of Action: Controlled Release, Slow-Release Mechanisms, and Root Uptake

Multiple, interacting release mechanisms are frequently used in controlled release systems, which are intended to maximise medication delivery by regulating the release's timing and location. These mechanisms must be carefully considered during the design and implementation process (Siegel et al., 2011). Compared to traditional fertilisers, slow-release fertilisers (SRFs) provide a controlled nutrient release, increasing nutrient use efficiency and reducing environmental losses. Compound- type SRFs have special potential for further research (Fu et al., 2018). Slow-release and controlled-release nitrogen fertilisers are crucial for increasing nutrient usage efficiency and reducing nitrogen losses in agricultural systems (Varadachari et al., 2010). Stabilised and slow-release fertilisers provide a practical way to improve agricultural nutrient usage efficiency by controlling nutrient release and reducing losses (Trenkel et al., 2021). With release mechanisms that vary depending on the type of material and are governed by variables like soil, temperature, and moisture, slow- and controlled- release fertilisers provide better nutrient use efficiency and less leaching than standard fertilisers (Morgan et al., 2009). The release kinetics of controlled- and slow-release fertilisers and agrochemicals are characterised using mathematical models, such as zero-order, first-order, Higuchi, and Korsmeyer-Peppas, which help to optimise nutrient management techniques and provide insights into their mode of action (Lakshani et al., 2023). By reducing nutrient loss and offering a customised release profile that responds to environmental cues, slow- and controlled-release nanofertilizers- which are frequently made using nanocomposites or coating techniques- offer increased nutrient use efficiency and support more environmentally friendly farming methods (Haydar et al., 2024).

6.4 Impact on Plant Physiology: Enhancing Photosynthesis, Root Development, and Stress Tolerance

Focusing on important physiological functions like photosynthesis, carbon absorption, assimilate translocation, and stress response systems, genetic engineering presents exciting opportunities to increase plant productivity and stress tolerance (Nowicka et al., 2018). Crops can be made more salinity tolerant by taking use of the physiological, biochemical, and molecular responses that plants have to salt stress. These responses include antioxidant activity, osmotic adjustment, and polyamine control (Shahid et al., 2020). To increase plant growth and stress tolerance in the face of climate change, it is essential to comprehend how encironmental conditions such as temperature, nutrient levels, and water availability affect photosynthetic CO_2 uptake (Morales et al., 2020). By improving antioxidant activity, proline and sugar content, leaf area, and seed quality, foliar application of zinc and iron nanoparticles can lessen the detrimental effects of water shortage stress on pinto beans (Fatollahpour et al., 2020). Basil's morphological characteristics, photosynthetic pigments, and antioxidant capacity can all be enhanced by foliar application of biosynthesis copper and zinc nanoparticles, which will increase yield and quality (Abbasifar et al., 2020). Although more research is required to maximise targeted release and reduce environmental impact, nanoparticles present a promising way to improve plant growth, physiology, and stress tolerance by boosting nutrient delivery and possibly increasing photosynthetic efficiency (Nafees et al., 2020). Depending on the characteristics of the nanoparticle, the type of plant, and the surrounding environment, nanoparticles can have a wide range of effects on plant physiology, from improving photosynthesis, root development, and stress tolerance to causing toxicity. This emphasises the necessity of carefully assessing and optimising their agricultural applications (Ahmad et al., 2022). Rewatering cotton after a brief period of water stress at growth stages might boost photosynthesis and root activity, increasing water-use efficiency without appreciably affecting output (Luo et al., 2016).

Beyond the constraints of conventional fertilisers, nanocarriers provide accurate and effective nutrient delivery, revolutionising plant nutrition. Through the encapsulation of nutrients in biodegradable nanoscale structures, they maximize nutrient utilization efficiency, reduce environmental impact, and improve macro and micronutrient intake. The timely availability of nutrients is guaranteed by controlled-release systems, and beneficial effects on plant physiology, such as improved photosynthesis and stress tolerance, lead to higher crop yields and better plant health. Although further research is necessary, to fully understand their potential and guarantee safe use, nanocarriers offer a revolutionary approach to sustainable agriculture, tackling issues of global food security and encouraging ecologically friendly farming methods.

7. NANOCARRIERS FOR PLANT PROTECTION

Nanocarriers are increasingly being recognized as a groundbreaking solution in plant protection, offering a sustainable alternative to conventional chemical pesticides. As the agricultural sector faces growing challenges, such as increased pest resistance and environmental concerns, nanocarriers provide a promising pathway forward. This chapter explores the current state and future potential of nanocarriers in plant protection, highlighting their benefits, applications, and the obstacles that must be addressed. Nanocarriers are diminutive delivery systems designed to transport active agents directly to target sites within plants. These carriers, typically ranging from 1 to 100 nanometers in size, can penetrate plant tissues more effectively than traditional methods. This capability enhances the efficacy of active agents and reduces the quantity required, thereby minimizing environmental impact. By encapsulating biological plant protection agents (BPPAs), nanocarriers improve stability and provide controlled release, ensuring sustained effectiveness (Pinto, T. V., et al., 2022). One of the most promising applications of nanocarriers is in the delivery of genetic materials such as miRNA, RNA interference (RNAi), and CRISPR-Cas biomolecules. These technologies enable the modification of gene expressions in plants, enhancing their resistance to pests and diseases. For example, RNAi can silence specific genes in pests, effectively disabling their ability to harm crops. CRISPR-Cas systems allow for precise genome editing, offering the potential to create crops that are inherently more resistant to environmental stresses and pathogens (Mujtaba, M., et al., 2021). Nanocarriers also facilitate the efficient delivery of traditional pesticides. Encapsulated nanopesticides can release their active ingredients in a controlled manner, ensuring that they reach target pests at the optimal time and concentration. This targeted delivery not only enhances pesticide effectiveness but also minimizes off-target effects, reducing harm to beneficial insects and the surrounding ecosystem (Victoria, J., et al., 2024)

7.1 Recent Advancements in Nanocarriers Technology in Agriculture Sector

Recent advancements in nanotechnology have led to the development of bioactive antifeedants using essential oils and plant extracts. These formulations are designed to deter pests from feeding on crops without causing harm to non-target organisms. The use of natural compounds aligns with the growing demand for organic and environmentally friendly agricultural practices. Researchers have enhanced the stability and bioavailability of these natural compounds using nanocarriers, making them more effective as plant protection agents (Melanie, M., et al., 2022). Furthermore, studies have explored the use of lignin-based nanocarriers for pesticide delivery. Lignin, a natural polymer found in plant cell walls, offers a biodegradable and renewable material for nanocarrier construction. Lignin-based nanocarriers have been shown to improve the dispersion and effectiveness of pesticides, offering a sustainable approach to plant protection (Beckers, S., et al., 2020). Another exciting development is the use of polymeric nanocarriers that can respond to environmental stimuli, such as changes in pH or temperature. These smart carriers can adjust their release rates based on environmental conditions, ensuring that active agents are dispensed precisely when and where they are needed. Such innovations not only enhance the precision of plant protection strategies but also reduce the need for multiple applications, thereby decreasing the overall use of chemicals in agriculture (Beckers, S., et al., 2020).

7.2 Targeted Delivery of Pesticides, Insecticides, and Fungicides

Nanoparticles have emerged as promising tools for the targeted of agrochemicals, including pesticides, insecticides, and fungicides. Their small size and tunable surface properties allow for effective transport and controlled release at specific plant tissues or pests. Various types of nanoparticles such as polymeric nanoparticles, liposomes, and inorganic nanoparticles (e.g., silica and gold nanoparticles) have been developed for this purpose. Targeted delivery offers several advantages, including enhanced bioavailability of active compounds, reduction in chemical wastage, and improved efficacy against pests and pathogens. For instance, polymeric nanoparticles loaded with imidacloprid have demonstrated increased efficiency in pest control compared to conventional formulations (Zhao, Jingyu, et al. 2022). Liposome-based delivery systems for fungicides like azoxystrobin have also shown enhanced retention on plant surfaces, resulting in prolonged protection (Chen, Peizhou, et al., 2020)

7.3 Encapsulation of Biocontrol Agents and Natural Pesticides

Encapsulation techniques using nanoparticles are being increasingly explored for the delivery of biocontrol agents and natural pesticides derived from botanical extracts and microbial metabolites. Encapsulation protects the bioactive agents from environmental degradation, ensuring their stability and prolonged activity. Commonly used materials for encapsulation include chitosan, alginate, and silica nanoparticles. Chitosan nanoparticles loaded with neem oil have shown potent insecticidal activity with extended shelf life and slow-release properties (Perlatti, B., et al., 2013). Similarly, silica nanoparticles encapsulating Bacillus thuringiensis formulations have been effective against caterpillars while maintaining the viability of the bacterial spores (Sehaole, T. P. (2012). Encapsulation not only improves the shelf life but also reduces the volatilization and degradation of natural pesticides, thereby enhancing their field applicability and reducing the frequency of applications.

7.4 Mode of Action: Controlled Release for Prolonged Effectiveness and Reduced Resistance

Nanoparticle-based delivery systems are designed to provide controlled and sustained release of agrochemicals, leading to prolonged effectiveness and reduced resistance development in pests and pathogens (Kashyap, P. L., et al., 2015). Controlled release mechanisms often rely on polymer matrices or porous structures that allow gradual diffusion of the active ingredient. For instance, poly(lactic-co-glycolic acid) (PLGA) nanoparticles have demonstrated sustained release of herbicides, reducing the need for repeated applications (Schnoor, Brian, et al., 2018). The slow release minimizes pesticide exposure peaks, which is crucial in delaying the onset of resistance. Furthermore, nanoparticles can be functionalized with ligands that respond to environmental triggers such as pH or enzymatic activity, ensuring the release occurs only when pests are present (Zhao, Ming, et al., 2022). This precision reduces the selection pressure on pest populations, further preventing resistance development.

7.5 Advantages: Minimization of Environmental Contamination, Reduced Pesticide Leaching, and Runoff

The application of nanoparticle-based delivery systems offers several environmental benefits, including minimized contamination, reduced pesticide leaching, and runoff. Conventional pesticide application methods often result in excessive chemical use, with a significant portion being washed off into water bodies, causing environmental damage (Kashyap, et al., 2015). Nanoparticles reduce this impact through enhanced adhesion to plant surfaces and targeted delivery. For example, mesoporous silica nanoparticles modified with hydrophobic coatings have shown increased retention on leaf surfaces, reducing pesticide runoff by 40% (Lin, Hanchen, et al., 2024). Additionally, the reduced frequency of pesticide application due to controlled release also lowers the overall chemical burden on the environment. Studies have -reported a 50% reduction in pesticide use with nanoparticle formulations without compromising pest control efficiency (Yadav, A., & Yadav, K. 2018). In conclusion, nanoparticle-based delivery systems for plant protection offer a promising alternative to conventional agrochemical applications. By enabling targeted delivery, controlled release, and reduced environmental contamination, these systems contribute to sustainable agriculture while maintaining crop productivity. Further research focusing on large-scale application methods and long-term safety assessments will be crucial for their widespread adoption (Fig. 6).

Figure 6. Application of nanocarriers in plant protection (Adopted and modified from Worrall et al., 2018, Created using Biorender.com)

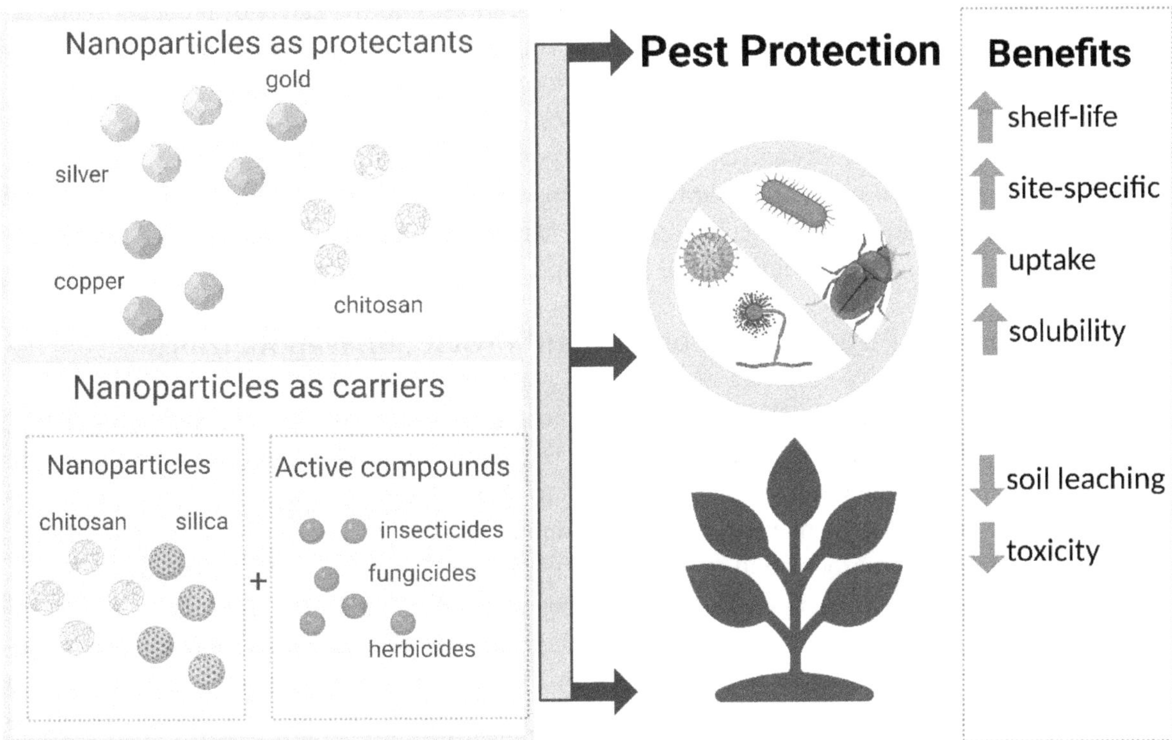

7.6 Application and Benefits of Nanocarriers in Plant Protection

Nanocarriers hold great promise for transforming plant protection by offering a sustainable and efficient alternative to traditional chemical pesticides (Kumar et al., 2019). With their ability to enhance the delivery and efficacy of active agents, nanocarriers can play a pivotal role in addressing the growing challenges of pest resistance and environmental sustainability in agriculture. As research continues to advance and regulatory frameworks evolve, nanocarriers are poised to become a cornerstone of modern plant protection strategies, contributing to a more resilient and sustainable agricultural future. Through continued innovation and collaboration, the agricultural industry can harness the power of nanotechnology to meet the demands of a growing global population while preserving the health of our planet.

The integration of nanocarriers into plant protection strategies offers numerous benefits like, nanocarriers enhance the efficacy of agricultural practices by improving the delivery and stability of active agents, ensuring they reach target sites in optimal condition, which can lead to higher crop yields and reduced losses from pest and disease damage. They contribute to environmental sustainability by reducing the quantity of chemicals needed and minimizing off-target effects, aligning with global efforts to decrease agriculture's environmental footprint. Although the initial development and implementation of nanocarrier technology may involve higher costs, the long-term benefits, such as reduced pesticide use and increased crop productivity, can result in significant cost savings for farmers. Their versatility allows them to be tailored for delivering a wide range of active agents, including synthetic pesticides, biological molecules, and natural compounds, facilitating the development of customized plant protection solutions for specific agricultural needs. Furthermore, nanocarriers improve safety by reducing chemical applications and precisely targeting pests, thereby lowering the risk of exposure to harmful substances for both farmers and consumers, ultimately enhancing food safety. (Beckers, S., et al., 2020).

In conclusion, the future of agriculture relies heavily on innovative solutions like nanocarriers that can bridge the gap between productivity and environmental stewardship. By embracing these advancements, we can ensure the long-term viability of our food systems and the well-being of our ecosystems. As we move forward, the integration of nanotechnology into plant protection will be instrumental in creating a more sustainable and prosperous agricultural landscape.

7.7. Challenges and Future Direction of Plant Protection Nanocarriers

The introduction of nanotechnology in agriculture presents several challenges, including regulatory hurdles, as existing frameworks may not fully address the unique properties and risks associated with nanocarriers, highlighting the need for clear guidelines to ensure their safe use. Public perception also plays a critical role in acceptance, making transparent communication and education essential to address concerns about safety and environmental impact. Scalability remains a significant challenge, as the successful performance of nanocarriers in laboratory studies must be consistently replicated in real-world agricultural settings. Economic factors, particularly the high costs associated with developing and implementing nanocarrier technology, can be a barrier for small-scale farmers, emphasizing the importance of reducing production costs and improving accessibility. Continued research and development are crucial to optimize nanocarrier formulations and delivery mechanisms, requiring collaboration among researchers, industry stakeholders, and regulatory bodies to accelerate innovation and address existing barriers. (Ahmad, Razi, et al. 2021)

8. ENVIRONMENTAL AND BIOLOGICAL INTERACTION OF NANOCARRIERS

Through tailored release, biodegradability, and increased pesticide delivery efficiency, polymeric nanoparticles present a possible means of enhancing agricultural sustainability (Shakiba et al., 2020). To predict behaviour and guarantee safe application in biomedical and environmental fields, it is necessary to thoroughly characterise the physiochemical properties of nanomaterials and their intricate interactions with biomolecules using both experimental and computational methods (Chen et al., 2017). Despite the enormous technical potential of two-dimensional materials, their biological interactions must be carefully examined, considering chemical, electrical, and physical aspects, to assure their safe and effective development for a range of applications (Wang et al., 2016). By increasing crop productivity, improving food quality, and reducing agricultural inputs, nanotechnology has the potential to completely transform the agro-food industry. However, to address potential concerns and enable widespread implementation, more research, clearer regulations, and findings are required (Acharya et al., 2020). The usage of nanoparticles in agriculture is growing because of their special qualities, which allow for the targeted administration of agrochemicals and nutrients for increased crop yield and growth, they have potential for sustainable agriculture, building on developments in sustainable nanotechnology and NanoBio-interactions, and the goal is to optimize nanomaterials to maximize agricultural output (Singh et al., 2015). Concerns regarding the possible toxicity of manufactured nanoparticles to the environment and human health are raised by their extensive use in a variety of industries, calling for more study to completely comprehend and lessen their negative impacts (Prajitha et al., 2019). Though the possible environmental effects of nanoparticle release still need careful study, green synthesis of nanomaterials using plants offers a safer and more ecologically friendly alternative to standard chemical processes, even though nanotechnology offers a wide range of uses (Rai et al., 2018).

8.1 Nanocarrier Fate in Soil: Biodegradation, Leaching, and Accumulation

The fate, shape, and bioavailability of artificial nanomaterials in natural soil environments are greatly influenced by soil properties such as salinity, texture, pH, organic matter, and saturation (Cornelis et al., 2014). Although nanotechnology has advantages in agriculture, further study and standardised risk assessment frameworks are needed to guarantee the safe and sustainable use of nanomaterials due to their possible environmental influence on soil life. (Singh et al., 2022). Leaching of nanoparticles in soil requires modified assessment methods to account for the unique properties of nanocarriers, which may affect their mobility. These modifications are crucial for accurately predicting the potential environmental impact of nanopesticides (López-Cabeza et al., 2020). The accumulation of nanoparticles in plants through the soil concerns about their transfer into edible tissues and potential human health risks, understanding the uptake mechanisms and environmental behaviour of nanoparticles is essential for evaluating their impact (Rajput et al., 2020) (Fig. 7).

Figure 7. Fate of nanomaterials in soil and plants (Adopted and modified from Usman et al., 2020, Created using Biorender.com)

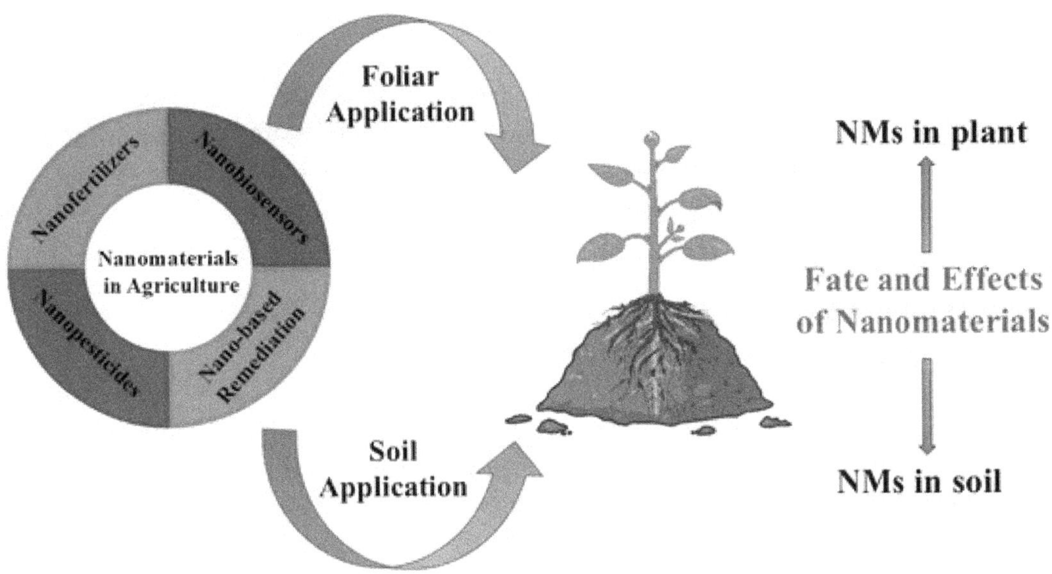

8.2 Environmental Impacts of the Nanocarrier

However, a thorough assessment of the environmental impact of nanocarriers in the treatment of fungal diseases is necessary. Nanocarriers provide tailored delivery and less chemical use, but questions about their permanence, bioaccumulation, and interactions with water and soil ecosystems still need to be answered. To reduce unexpected environmental effects and maximize their advantages in treatment of fungal diseases, sustainable development of nanocarriers should prioritize biodegradable materials, environmentally friendly synthesis techniques (Campos et al., 2022). Possible hazards include the buildup of nanoparticles in water and soil, unforeseen impacts on helpful microorganisms, and long-term ecological repercussions. To address these issues and guarantee sustainable agricultural uses, eco-friendly polymer selection, regulated degradation studies, and exacting toxicity evaluations are needed (Pascoli et al., 2018). When it comes to managing fungal diseases, polymeric nanocarriers minimize environmental pollution, improve targeted distribution, and decrease chemical discharge, although its biodegradability reduces long-term dangers, formulations that are environmentally friendly must be optimized for sustainable development (Shakiba et al., 2020)

8.3 Interaction with Soil Microbiota and Plant Rhizosphere

Plant development, nutrient cycling, and soil health are all significantly impacted by the relationship between the rhizosphere and soil microorganisms. Bacteria and fungi are examples soil microorganisms that affect disease resistance, nutrient availability and root development (Rajput et al., 2018). Interaction between soil microbiota and the rhizosphere involve intricate systems of communication, such as symbiosis like arbuscular mycorrhizal associations (Lambers et al., 2009). The interaction of nanoparticles with soil microbiota is crucial for understanding their influence on soil health and aquatic ecosystems. Their effect on microbes is greatly influenced by variables such nanoparticle concentration, size, and shape (Shende et al., 2021). Food production and biogeochemical cycles depends on soil microbiota, yet to overuse of metallic nanoparticles in agriculture compromises microbial diversity and structure, endangering ecosystems (da Silva Junior et al., 2022). Through interactions between roots and biotic and abiotic elements of the soil in the rhizosphere where energy, nutrition, and signal exchanges generate circumstances different from bulk soil plant survival and crop productivity are dependent (Pinton et al., 2000). Plant health is impacted by the rhizosphere, a centre of microbial interactions where root exudates affect both diseases and beneficial microbes, this interaction investigates the ways in which these interactions and farming methods can support the health and growth of plants (Raaijmakers et al.,2009).

8.4 Toxicological Assessments: Nanocarriers' Impact on Non-Target Organisms and Biodiversity

Although nanotechnology has made significant strides in many areas, it is important to carefully evaluate the risks associated with nanopesticides, especially regarding their possible toxicity and genotoxicity to humans and non-target animals like fruit flies, which provide a useful model for examining these impacts (Demir et al., 2020). *Caenorhabditis elegans* a model organism chosen for its genetic significance and susceptibility to environmental pollutants, demonstrated variable toxicity to pesticide-loaded nanocarriers (Jacques et al., 2017). Nanopesticides impact non-target organisms such as soil microorganisms, plants, worms, insects, algae, daphids, and fishes, necessitating advanced methods to analyse their interactions with biological and environmental systems (Côa et al., 2020). High sensitivity in terrestrial and aquatic invertebrates, vertebrates, microbes and plants is revealed by the toxicity of nanoencapsulated atrazine (nATZ) to non-target species. *D. dimilis* is the most susceptible species, according to a species sensitivity distribution (SSD) analysis (Hennig et al., 2023). Climate change-driven increases in pesticide use provide hazards to aquatic ecosystems and human health, underscoring the need for mitigating measures (Punniyakotti et al., 2024). Although nanocarrier has potential, its effects on biodiversity must be carefully assessed. Stakeholder education and public perception are also essential for reducing environmental dangers (Babatunde et al., 2019). Innovative nanoagrochemicals made possible by nanotechnology effectively control pests while causing least amount of environmental damage but concerns about nanopesticides and nanocarriers possible detrimental effects on biodiversity (Shekhar et al., 2021).

8.5 Long-Term Safety

Evaluating the environmental impact, bioaccumulation, and potential toxocity of nanocarriers is necessary to guarantee their long-term safety in the treatment of fungal diseases, although biodegradable nanocarriers based on biopolymers provide a sustainable, more research to fully understand their impact on microbial diversity and persistence in soil. It is necessary to address the risks of fungal infections developing resistance and accumulation of nanoparticles in ecosystems. Concerns about human health, particularly those resulting from exposure to the food chain, emphasize the necessity of stringent regulatory control (Machado et al.,2022). Thorough analyses of the stability, biodegradability, and potential bioaccumulation of nanocarriers in soil and plant systems are necessary to guarantee their long-term safety in the treatment of fungal diseases. Although undesired environmental persistence, toxicity to non-target organisms, microbial resistance are issues that need to be addressed, nanocarriers improve efficacy and targeted delivery (Huang et al., 2024). Evaluating the environmental impact, bioaccumulation, and impacts on non-target organisms of nanocarriers is necessary to ensure their long-term safety in the treatment of fungal diseases. Although using Generally Recognized As Safe (GRAS) and biodegradable products can reduce risks, their safe and sustainable use use requires thorough research (Bilia et al., 2019).

8.6 Case Studies: Safe Application of Nanocarriers in Field Trials

Innovative agricultural tools like slow-release fertilizers and nano-pesticides are made possible by nanotechnology, but careful testing is necessary to ensure safe field use (Acharya et al., 2020). Case studies show safer nanoparticles like alginate/chitosan and chitosan/tripolyphosphate nanocarriers loaded with GA3 significantly enhance agricultural productivity, field trails on *Solanum lycopersicum* for sustainable farming was proved by field testing that showed increase in fruit yield (Pereira et al., 2019). Zein nanoparticles loaded with neem oil showed stability and decreased toxicity when compared to neem oil alone, they also showed no negative effects on Caenorhabditis elegans, soil microbiota or *Allium cepa.* The soil nitrogen cycle enzyme gene counts were maintained, and DNA damage was reduced by these nanoparticles (Pascoli et al., 2019). Mesoporous silica-based nanomaterials (MSNs) enable regulated delivery, pesticide removal, degradation, and detection, hence increasing the safety and efficacy of pesticides. They are perfect for environmentally friendly farming practices because of their low toxicity, great stability, and adjustable size (Kong et al., 2021) . A field investigation on wheat plants exposed to nanoceria (nCeO2) revealed that high concentrations changed the microstructures of the root and leaf cells and decreased the amount of chlorophyll. The grain protein rose by as much as 32.6%, although neither biomass nor yield were significantly impacted (Du et al., 2015). By attaching to the pathogen surface, entering the fungal cells, and damaging the internal organelles and membranes, Zinc oxide nanoparticles (ZnO NPs) made from *Trychyspermum ammi* seeds efficiently inhibited *Cercospora canescens,* the cause of leaf spot in mung beans (Rafiq et al., 2022). Through the disruptions of DNA replication, enzyme activity, and biosynthetic pathways, silver nanoparticles (Ag NPs) produced from Moringa oleifera leaves shown high antifungal action against *Aspergillus flavus,* a key generator of aflatoxins in rice (Mohamed et al., 2022). By causing cellular membrane leakage and structural deformation, copper oxide nanoparticles (CuO NPs) made from *Cassia fistula* leaves effectively inhibited Fusarium oxysporum, which causes tomato plant withering (Ashraf et., 2021)

8.7 Regulatory Concerns

The regulation of nanocarriers in fungal disease management is essential to ensure their safe and effectiveness integration into agricultural practices. Global regulatory are changing to evaluate the toxicity, effectivenss, and environmental impact of formulations based on nanoparticles. Standardized criteria for the approval, labelling, and monitoring of pesticides and fungicides based on nanotechnology are being developed by organizations including the EPA, EFSA, and FAO/WHO. These rules place a strong Empasis on risk assessment, policies also sustainable development by restriciting the use of hazardous materials and encouraging biodegradable nanocarriers (Nizamani et al.,2024). The main regulatory issues surrounding nanocarriers in treatment of fungal diseases are their long-term sustainability, hazards to human health, and environmental safety, for their safe and successful implementation in sustainable agriculture, future regulatory initiatives should support openness and stimulate environmentally appropriate nanocarriers compositions, and unify international regulations (Alghuthaymi et al.,2021). Potential hazards to the environment and human health, such as toxicity and bioaccumulation, are the main regulatory issues for nanocarriers used in the treatment of fungal diseases, to guarantee their safe and sustainable use in agriculture, standardized safety evalutions, responsible application standards, and international legislation necessary (Sonawane et al., 2022). Environment impact, toxicity, and bioaccumulation are regulatory considerations for nanocarriers used in the treatment of fungal diseases, to stop contamination and the emergence of resistance, standardized protocols for the synthesis, application, and disposal are required (Rameez et al., 2024).

8.8 Regulatory Compliance

Nanocarriers in agriculture offer promising advancements, yet their deployment necessitates strict adherence to safety standards, environmental regulations, and application guidelines to mitigate potential risks. Ensuring human and environmental safety involves comprehensive testing for toxicity, allergenicity, and other hazards. Regulatory bodies worldwide have established frameworks to oversee the safe use of nanotechnology in agricultural products. For instance, in India, the Environment (Protection) Act of 1986 empowers the Ministry of Environment, Forest and Climate Change to regulate hazardous substances, including nanomaterials. Additionally, the Food Safety and Standards Authority of India (FSSAI) has issued guidelines for the use of nanotechnology in food products, ensuring consumer safety (Kumari, Ritika, et al., 2023). Environmental regulations are crucial, particularly regarding the release of nanoparticles into ecosystems. In the European Union, the European Chemicals Agency (ECHA) and the European Food Safety Authority (EFSA) govern nanotechnology in foods and pesticides, specifying permissible levels, disposal methods, and monitoring requirements to prevent environmental contamination (Iavicoli, Ivo et al., 2017). Establishing clear application guidelines is essential for the safe and effective use of nanocarriers in agriculture. This includes recommendations on dosages, application methods, and timing to maximize benefits while minimizing risks. Developing these guidelines requires collaboration among researchers, industry stakeholders, and regulatory bodies. For example, the Indian Council of Agricultural Research (ICAR) has provided guidelines on nanotechnology research in agriculture, promoting responsible development and application of nanotechnologies in the sector (Chávez-Hernández, Jorge Antonio, et al., 2024). Addressing these challenges is essential for the successful integration of nanotechnology into sustainable agricultural practices, while nanotechnology holds significant potential for enhancing agricultural productivity and sustainability, it is imperative to navigate its application within the established safety standards, environmental regulations, and application guidelines to ensure it does not pose risks to human health or the environment.

Antimicrobial nanoparticles, intelligent food packaging, and nano-encapsulated agrochemicals are just a few of the notable advancements in the agri-feed –food industries made possible by nanotechnology, in EU and non-EU nations to give a general overview of the safety evaluation frameworks and regulatory actions for nanomaterials in the food and agribusiness sectors (Amenta et al., 2015) .

Nanopesticides have advantages like increased effectiveness and lower ingredient utilization, their environmental risk assessment necessitates modification, with emphasis on topics like exposure and ecotoxicity, this examines the risk assessment methodologies (Kookana et al., 2014). Although there is a promise for sustainable food production using nanotechnology in agriculture, especially in the development of nanoagrochemicals, there are no concerns about changed risk profiles. Environmental destiny does not necessarily lessen environmental impact, however a review of 78 studies revealed that nanoagrochemicals were 20-30% more effective than traditional products (Kah et al., 2018). Effective risk assessments are necessary for the commercialization of nanotechnologies to address safety, health, and environmental concerns. Current models concentrate on screening-level evaluations, while higher-tier, more quantitative models that account for uncertainly are becoming increasingly popular (Hristozov et al., 2016). Important concerns regarding the regulation, hazards, and advantages of nanopesticides and nanofertilizers have been brought up by their growing use in agriculture. To encourage sustainable agrochemicals improvements, analysis addresses the research and regulatory development (Kah et al., 2015)

8.9 Safety Evaluation Frameworks

The regulatory frameworks for nanocarriers in fungal disease management vary globally, with nations actively assessing their policies to ensure the safe use of nano-based products in agriculture. Approval of nanocarrier -based fungicides requires demonstrating their safety for consumers and the environment. Several countries have established regulations for nano-agrochemicals, biofertilizers, and pest control agents, incorporating risk assessments, labelling requirements and toxicity evaluations. Regulatory approaches differ, with some nations integrating nanotechnology into existing agricultural policies, while others have developed specific guidelines for nano-based agro-products, standardized safety assessments and international collaborations are essential to facilitate the widespread adoption of nanocarriers in fungal disease management while ensuring environmental and human health protection (Kumari et al., 2023). Safety, environmental effect, and sustainable use are the main focuses of regulatory frameworks for nanocarriers in the treatment of fungal diseases. Nations are creating specialized regulations for nano pesticides and fertilizers or incorporating nanotechnology into their current agricultural policy. To ensure responsible adoption and maximize efficacy in the control of fungal diseases, standardized safety studies, toxicity evaluations, and international collaboration are essential (Nizamani et al., 2024). Environmental protection, sustainability, and safety are guaranteed by efficient regulatory frameworks for nanocarriers in the treatment of fungal diseases. To promote sustainable agricultural practices, responsible adoption requires standardized evaluations, precise norms, and international cooperation that strikes a balance between innovation and risk management.

8.10 Commercially Approved Nanocarriers for Agricultural Purposes

Commercially available nanoparticles are extensively utilized in a variety of fields, such as manufacturing, agriculture, and medicine. Because of their improved effectiveness, controlled release characteristics, and lower environemental impact as compared to traditional formulations, nanoparticles such as insecticides, fertilizers and soil amendments are becoming more and more popular in agriculture. By enhancing agrochemical efficiency through controlled release, engineered metal Organic Frameworks (MOFs) provide a sustainable option that lessens the demand for traditional pesticides amd fertilizers. They are perfect nanocarriers for altering the loading and release characteristics of agrochemicals in agriculture because of their biocompatibility and biodegradability (Basak et al., 2024). Polymeric nanoparticles are proposed as efficient nanocarriers for delivering agrochemicals like pesticides, offering benefits such as improved targeting, slow release, and biodegradability. The creation, synthesis, and characterization of customized polymeric nanocarriers are maximized their usage in agriculture (Shakiba et al., 2020). Since biopolymer nanocarriers are biodegradable, they provide an eco-friendly way to distribute agrochemicals over an extended period, increasing productivity and minimizing environmental effect, it is necessary to comprehend and reduce the hazards associated with nanotechnology (Machado et al., 2022). Casein nanoparticles coated with iron oxide have several advantages for agriculture, such as accurate pesticide and nutrition delivery, because of their core-shell structure, these nanoparticles allow for the controlled and focused delivery of bioactive compounds. They are useful for sustainable agricultural methods since their strucuture and composition may be determined through a variety of analytical methods (Burvey et al,. 2023)

Overall, by increasing agricultural output, reducing environmental effects, and optimizing pesticide delivery nanotechnology has enormous potential to transform agriculture. (MOFs), polymeric nanoparticles, and biopolymer-based carriers are examples of engineered nanocarriers that provide targeted delivery, controlled release, and biodegradability, all of which promote more environmentally friendly farming methods. Their extensive use, however, prompts questions regarding possible toxicity, effects on environment and consequences on biodiversity, necessitating more study and regulatory frameworks, it is important to evaluate how nanocarriers interact with non-target species, plant rhizospheres and soil microbiota. Notwithstanding obstacles, the agro-food industry sustainable expansion by efficient nanomaterials.

8.11 Scalability

Cost-effective and sustainable large-scale production is essential for the scalability of nanocarriers in the treatment of fungal diseases, as it guarantees formulation consistency and efficacy. Precision farming combined with nanocarriers can improve targeted disease management while reducing environmental effects. To enable broad adoption and useful implementation in agricultural systems including regulatory approval and safety evelations (Vega-Vásquez et al., 2020). For the large-scale synthesis of nanocarriers, the utilization of plant-based expression systems is a viable strategy, as shown in the manufacture of ferritin. Scalable nanocarriers can be created for use in agriculture by refining formulation procedures, purification methods, and production factors (Knödler et al., 2023). To effectively control fungal diseases, biopolymer based nanocarriers must be scalable, which requires retaining bioactivity retaining bioactivity during mass production, optimizimg large-scale synthesis, and guaranteeing cost-effectiveness (Xiang et al., 2023). Polymeric nanocarriers must be made economically without sacrificing stability or bioactivity to be used on a large scale in the treatment of fungal diseases, field uses will require standardized delivery and formulation procedures to minimize regulatory barriers and enable broad use in a variety of crops (Zhang et al.,2024)

8.12 Global Adoptation

The global adoptation of nanocarriers in fungal disease management depends on their ability to enhance the sustained release of antifungal agents while minimizing environmental impacts. Biopolymer-based nanocarriers reduce microplastic pollution and eco-toxicological concerns because they are renewable and biodegradable, making them sustainable substitute for traditional agrochemicals. Regulatory frameworks, economic viability, and public opinion of nanotechnology all affect their broad adoptation. Transparent policies, a clear risk assessment and proven advantages in enhancing soil health and agricultural productivity are necessary for successful commercialization. Globally, the use of nanocarriers in the management of fungal diseases can spur a change to more efficient and sustainable farming methods as climate changes exacerbates agricultural problems (Machado et al., 2022). The improvements in plant transport systems for nanoparticles further boosts the use of nanocarriers in the management of fungal diseases worldwide. Effective delivery techniques, such as direct injection, and foliar spray, improve nanoparticles absorption and guarantee the targeted and continuous release of antifungal drugs. It is essential to comprehend vascular interactions in plants, especially xylem and phloem transport, to maximize the delivery of agrochemicals using nanocarriers. Even though foliar spray has a higher uptake efficiency than soil soaking, nanoparticles formulations need to be modified for better absorption and systemic distribution in future studies. Combining these developments with biopolymer-based nanocarriers can increase adoptation world-wide and provide economically and environmentally sound solutions for the management of fungal disease in agriculture (Su et al., 2019).

Effectiveness sustainability, and regulatory approval are key factors in the worldwide use of nanocarriers in the treatment of fungal diseases. Effective fungal control ensured by optimizing delivery techniques such as direct injection and foliar treatment, which improve nanoparticles uptake and transport. Nanocarriers based on biopolymers further promote environmentally friendly solutions by lowering reliance on agrochemicals.

9. RESEARCH GAPS

The need for a more thorough understanding of the long-term environmental impact, biodegradability and interactions within the soil-plant ecosystem of nanocarriers is one of the research gaps in fungal disease management. For evaluating their toxicity, stability, and effectiveness in actual agricultural settings, there are no established methods. Furthermore, optimising targeted delivery mechanisms to increase accuracy and reduce non-target impacts presents difficulties. To enhance large-scale, economical biogenic synthesis techniques and guarantee regulatory compliance for a secure and long-lasting application in plant disease management, more research is needed (Omran et al., 2022). The necessity for more thorough field testing to confirm the practical effectiveness and scalability of nano-based fungicides is underscored by their restricted commercial availability. Furthermore, little research has been done on tailoring nanocarriers formulations for various crop-pathogen systems to guarantee reliable performance across a range of agricultural circumstances (Kutawa et al.,2021). There are still important research gaps in myconanotechnology optimisation for targeted and controlled administration of antifungal medicines, despite the tremendous promise of nanotechnology in the management of fungal diseases. Although fungi have been widely studied for the creation of nanoparticles, little is known about the molecular interactions of fungal- derived nanocarriers with plant-pathogen complexes. To improve effectiveness and reduce unforeseen environmental effects, more research is also needed to understand the mechanisms controlling nanoparticle uptake, translocation and degradation in plant and soil ecosystems. Additionally unexplored is the use of nano-based gene transfer in the creation of disease-resistant crops (Rai et al.,2015). More study is required to improve the best application techniques for various crops and environmental circumstances to close the gap between lab research and practical use of nanocarriers in the management of fungal diseases. A thorough investigation of formulation stability and transport effectiveness is necessary due to the heterogeneity in nanoparticle interactions with plant species, which raises questions regarding consistency and repeatability in field settings. Furthermore, there is still a lack of study on the long-term effects of nanocarriers on plant health and soil microbiota, which calls for thorough risk evaluations (Pérez-de-Luque et al.,2017). Understanding their dual involvement in pathogen control and plant nutrition is another research gap in nanocarrier- based fungal disease management. Examining the effects of these multipurpose nanomaterials on microbial populations, plant immunity and overall crop resilience may lead to novel approaches to integrated nutrition and disease control. Furthermore, to maximise effectiveness while lowering environmental dangers, it is crucial to ascertain the ideal concentration and synergistic effects of nano-formulations in actual agricultural contexts. Filling in these gaps can help control plant health in a more comprehensive and long-term manner (Kaur et al.,2024).

10. FUTURE DIRECTIONS

Enhanced formulation efficacy, increasing targeted delivery, and reducing environmental impact are the main goals of future developments for nanocarriers in the treatment of fungal diseases. While lessening reliance on agrochemicals, developments in nanocomposites can aid in the fight against resistant fungus strains. To maintain sustainability and regulatory compliance, biodegradable and biopolymer –based nanocarriers should be given priority in research, to further enhance disease management, nanosensors can be integrated for early pathogen detection and precise application of nano-based fungicides. Researchers, legislators, and industry stakeholders must work together to improve nanocarrier technology, resolve safety issues and enable broad adoption for long-term and efficient fungal disease management in agriculture (Alghuthaymi et al., 2021). Multifunctional nanoparticles that combine antifungal qualities with effects that promote plant growth should be investigated as future directions for nanocarriers in the therapy of fungal diseases. Synergistic pairings of metal nanoparticles with biopolymers or chemicals produced fron plants can improve efficacy and decrease toxicity in nanobioformulations, for better fungal targeting and absorption, research should also concentrate on surface changes and nanoparticle shape customization. Creating uniform safety evaluations and legal requirements would help ensure that nanocarriers are used responsibly in agricultural disease control (Cruz-Luna et al., 2021). The integration of nanotechnology with cutting-edge diagnostic technologies for early and accurate pathogen detection should be the main emphasis of future developments for nanocarriers in the managements of fungal diseases. Rapid on-site illness identification, tailored treatment, and a reduction in needless pesticide use can all acheived by developing portable nanosensor-based systems. Widespread implementation will require uniform safety standards, as well as increased commercial availability and scalability of nanoparticle-based solutions, plant dieases managment can be made more sustainable and effective by taking a comprehensive approach that combines precision agriculture and nanotechnology (Ray et al., 2023).

Future studies should concentrate on creating biodegradable nanocarriers that enhance soil health and nutrient uptake in addition to controlling fungal infections. Optimizing their efficacy while reducing environmental hazards requires an understanding of how they interact plant roots and soil bacteria, field research must be extended outside of lab settings to evaluate their scalability and long-term effects, precision combined with nanotechnology can also improve targeted fungicide application, lowering chemical runoff and encouraging sustainable crop disease management (Maharshi et al., 2025)

AUTHOR CONTRIBUTIONS

AM and AK conceived the idea; MAA, VK, VGB, MVK, KB, PK, SV, AK and AM carried out the literature survey, prepared tables and figures, wrote the draft and edited the article, and all authors approved the article.

ACKNOWLEDGEMENTS

We thank Manipal Academy of Higher Education (MAHE), Manipal, Karnataka, India, TIFAC-CORE and FIST, DST New Delhi, DBT BUILDER grant (BT/INF/22/SP43065/2021), DBT, New Delhi, and K-FIST, VGST, Govt. of Karnataka for the facilities. The authors (VS and AK) are thankful to MAHE for the T.M.A. Pai Ph.D. scholarship. AK is thankful to the Indian Council of Medical Research (ICMR), Govt. of India for financial assistance as Project Associate-I through a research grant (No. 59/08/2022-TRM/BMS). We are grateful to Dr. B.S. Satish Rao, Professor and Director, Manipal School of Life Sciences, MAHE for encouragement and support.

REFERENCES

Aarthye, P., & Sureshkumar, M. (2021). Green synthesis of nanomaterials: An overview. *Materials Today: Proceedings*, *47*, 907–913. DOI: 10.1016/j.matpr.2021.04.564

Abbasifar, A., Shahrabadi, F., & ValizadehKaji, B. (2020). Effects of green synthesized zinc and copper nano-fertilizers on the morphological and biochemical attributes of basil plant. *Journal of Plant Nutrition*, *43*(8), 1104–1118. DOI: 10.1080/01904167.2020.1724305

Abdel-Aziz, H. M., Hasaneen, M. N., & Omer, A. M. (2016). Nano chitosan-NPK fertilizer enhances the growth and productivity of wheat plants grown in sandy soil. *Spanish Journal of Agricultural Research*, *14*(1), e0902–e0902. DOI: 10.5424/sjar/2016141-8205

Abid, N., Khan, A. M., Shujait, S., Chaudhary, K., Ikram, M., Imran, M., & Maqbool, M. (2022). Synthesis of nanomaterials using various top-down and bottom-up approaches, influencing factors, advantages, and disadvantages: A review. *Advances in Colloid and Interface Science*, *300*, 102597. DOI: 10.1016/j.cis.2021.102597 PMID: 34979471

Acharya, A., & Pal, P. K. (2020). Agriculture nanotechnology: Translating research outcome to field applications by influencing environmental sustainability. *NanoImpact*, *19*, 100232. DOI: 10.1016/j.impact.2020.100232

Adil, M., Bashir, S., Bashir, S., Aslam, Z., Ahmad, N., Younas, T., Asghar, R. M. A., Alkahtani, J., Dwiningsih, Y., & Elshikh, M. S. (2022). Zinc oxide nanoparticles improved chlorophyll contents, physical parameters, and wheat yield under salt stress. *Frontiers in Plant Science*, *13*, 932861. DOI: 10.3389/fpls.2022.932861 PMID: 35991444

Agrawal, U., Sharma, R., & Gupta, M. (2020). Antifungal efficacy of curcumin-loaded nanoparticles. *Drug Delivery*, *27*(1), 192–204.

Ahlawat, D. S., Kumari, R., Rachna, , & Yadav, I. (2014). Synthesis and characterization of sol–gel prepared silver nanoparticles. *International Journal of Nanoscience*, *13*(01), 1450004. DOI: 10.1142/S0219581X14500045

Ahmad, A., Hashmi, S. S., Palma, J. M., & Corpas, F. J. (2022). Influence of metallic, metallic oxide, and organic nanoparticles on plant physiology. *Chemosphere*, *290*, 133329. DOI: 10.1016/j.chemosphere.2021.133329 PMID: 34922969

Ahmad, R., Srivastava, S., Ghosh, S., & Khare, S. K. (2021). Phytochemical delivery through nanocarriers: A review. *Colloids and Surfaces. B, Biointerfaces*, *197*, 111389. DOI: 10.1016/j.colsurfb.2020.111389 PMID: 33075659

Ahuja, R., Panwar, N., Meena, J., Singh, M., Sarkar, D. P., & Panda, A. K. (2020). Natural products and polymeric nanocarriers for cancer treatment: A review. *Environmental Chemistry Letters*, *18*(6), 2021–2030. DOI: 10.1007/s10311-020-01056-z

Alex, B., Jignasa, G., Madhubabu, K., & Gopi, A. (2024, August). AI-Driven Smart Irrigation: Enhancing Agricultural Water Efficiency Through Intelligent Valve Regulation in Piped and Micro Irrigation Networks. In *2024 First International Conference on Pioneering Developments in Computer Science & Digital Technologies (IC2SDT)* (pp. 76-81). IEEE. https://doi.org/DOI: 10.1109/IC2SDT62152.2024.10696274

Alghuthaymi, M., Abd-Elsalam, K. A., Paraliker, P., & Rai, M. (2020). Mono and hybrid nanomaterials: Novel strategies to manage postharvest diseases. *Multifunctional Hybrid Nanomaterials for Sustainable Agri-Food and Ecosystems*, 287-317. https://do i.org/10.1016/B978-0-12-821354-4.00013-3

Alghuthaymi, M. A., Ahmad, A., Khan, Z., Khan, S. H., Ahmed, F. K., Faiz, S., Nepovimova, E., Kuča, K., & Abd-Elsalam, K. A. (2021). Exosome/liposome-like nanoparticles: New carriers for CRISPR genome editing in plants. *International Journal of Molecular Sciences*, *22*(14), 7456. DOI: 10.3390/ijms22147456 PMID: 34299081

Alghuthaymi, M. A., C, R., P, R., Kalia, A., Bhardwaj, K., Bhardwaj, P., Abd-Elsalam, K. A., Valis, M., & Kuca, K. (2021). Nanohybrid antifungals for control of plant diseases: Current status and future perspectives. *Journal of Fungi (Basel, Switzerland)*, *7*(1), 48. DOI: 10.3390/jof7010048 PMID: 33450851

Ali, S., Ahmad, N., Dar, M. A., Manan, S., Rani, A., Alghanem, S. M. S., Khan, K. A., Sethupathy, S., Elboughdiri, N., Mostafa, Y. S., Alamri, S. A., Hashem, M., Shahid, M., & Zhu, D. (2023). Nano-agrochemicals as substitutes for pesticides: Prospects and risks. *Plants*, *13*(1), 109. DOI: 10.3390/plants13010109 PMID: 38202417

Alrbaihat, M. R. (2022, May). Agricultural nano fertilizers: macronutrient types and applications review. In *International Conference on Geotechnical Engineering-IRAQ* (pp. 306-316). Singapore: Springer Nature Singapore. https://doi.org/DOI: 10.1007/978-981-19-7358-1_26

Amenta, V., Aschberger, K., Arena, M., Bouwmeester, H., Moniz, F. B., Brandhoff, P., & Peters, R. J. (2015). Regulatory aspects of nanotechnology in the agri/feed/food sector in EU and non-EU countries. *Regulatory Toxicology and Pharmacology*, *73*(1), 463–476. DOI: 10.1016/j.yrtph.2015.06.016 PMID: 26169479

An, C., Sun, C., Li, N., Huang, B., Jiang, J., Shen, Y., Wang, C., Zhao, X., Cui, B., Wang, C., Li, X., Zhan, S., Gao, F., Zeng, Z., Cui, H., & Wang, Y. (2022). Nanomaterials and nanotechnology for the delivery of agrochemicals: Strategies towards sustainable agriculture. *Journal of Nanobiotechnology*, *20*(1), 11. DOI: 10.1186/s12951-021-01214-7 PMID: 34983545

Anjum, S., Vyas, A., & Sofi, T. (2023). Fungi-mediated synthesis of nanoparticles: Characterization process and agricultural applications. *Journal of the Science of Food and Agriculture*, *103*(10), 4727–4741. DOI: 10.1002/jsfa.12496 PMID: 36781932

Ashraf, H., Anjum, T., Riaz, S., Ahmad, I. S., Irudayaraj, J., Javed, S., Qaiser, U., Naseem, S., & Naseem, S. (2021). Inhibition mechanism of green-synthesized copper oxide nanoparticles from Cassia fistula towards Fusarium oxysporum by boosting growth and defense response in tomatoes. *Environmental Science. Nano*, *8*(6), 1729–1748. DOI: 10.1039/D0EN01281E

Asmawi, A. A., Adam, F., Azman, N. A. M., & Rahman, M. B. A. (2024). Advancements in the nanodelivery of azole-based fungicides to control oil palm pathogenic fungi. *Heliyon*, *10*(18), e37132. DOI: 10.1016/j.heliyon.2024.e37132 PMID: 39309766

Babatunde, D. E., Denwigwe, I. H., Babatunde, O. M., Gbadamosi, S. L., Babalola, I. P., & Agboola, O. (2019). Environmental and societal impact of nanotechnology. *IEEE Access : Practical Innovations, Open Solutions*, *8*, 4640–4667. DOI: 10.1109/ACCESS.2019.2961513

Babu, P. J., Saranya, S., Longchar, B., & Rajasekhar, A. (2022). Nanobiotechnology-mediated sustainable agriculture and post-harvest management. *Current Research in Biotechnology*, *4*, 326–336. DOI: 10.1016/j.crbiot.2022.07.004

Banerjee, S., Mazumder, S., Chatterjee, D., Bose, S., & Majee, S. B. (2022). Nanotechnology for cargo delivery with a special emphasis on pesticide, herbicide, and fertilizer. In *Nano-enabled Agrochemicals in Agriculture* (pp. 105–144). Academic Press. DOI: 10.1016/B978-0-323-91009-5.00002-1

Banerjee, S., Roy, P., Nandi, S., & Roy, S. (2023). Advanced biotechnological strategies towards the development of crops with enhanced micronutrient content. *Plant Growth Regulation*, *100*(2), 355–371. DOI: 10.1007/s10725-023-00968-4 PMID: 36686885

Basak, S., Bhattacharyya, P., Pudake, R. N., Lokhande, P. E., Rednam, U., & Chakrabarti, S. (2024). Metal-organic framework as nanocarriers for agricultural applications: A review. *Frontiers in Nanotechnology*, *6*, 1385981. DOI: 10.3389/fnano.2024.1385981

Beckers, S., Peil, S., & Wurm, F. R. (2020). Pesticide-loaded nanocarriers from lignin sulfonates—A promising tool for sustainable plant protection. *ACS Sustainable Chemistry & Engineering*, *8*(50), 18468–18475. DOI: 10.1021/acssuschemeng.0c05897 PMID: 33381356

Beig, B., Niazi, M. B. K., Jahan, Z., Zia, M., Shah, G. A., & Iqbal, Z. (2023). Biodegradable polymer encapsulated nickel nanoparticles for slow release urea promotes Rhode grass yield and nitrogen recovery. *Journal of Polymers and the Environment*, *31*(5), 1866–1883. DOI: 10.1007/s10924-022-02729-2

Bhardwaj, S., Lata, S., & Garg, R. (2023). Application of nanotechnology for preventing postharvest losses of agriproducts. *The Journal of Horticultural Science & Biotechnology*, *98*(1), 31–44. DOI: 10.1080/14620316.2022.2091488

Bhattacharya, N., Cahill, D. M., Yang, W., & Kochar, M. (2022). Graphene as a nano-delivery vehicle in agriculture – current knowledge and future prospects. *Critical Reviews in Biotechnology*, *43*(6), 851–869. DOI: 10.1080/07388551.2022.2090315 PMID: 35815813

Bilia, A. R., Piazzini, V., Risaliti, L., Vanti, G., Casamonti, M., Wang, M., & Bergonzi, M. C. (2019). Nanocarriers: A successful tool to increase solubility, stability and optimise bioefficacy of natural constituents. *Current Medicinal Chemistry*, *26*(24), 4631–4656. DOI: 10.2174/0929867325666181101110050 PMID: 30381065

Biondo, F., Baldassarre, F., Vergaro, V., & Ciccarella, G. (2022). Controlled biocide release from smart delivery systems: Materials engineering to tune release rate, biointeractions, and responsiveness. In *Nanotechnology-Based Sustainable Alternatives for the Management of Plant Diseases* (pp. 31–147). Elsevier. DOI: 10.1016/B978-0-12-823394-8.00010-X

Bokov, D., Jalil, A. T., Chupradit, S., Suksatan, W., Ansari, M. J., Shewael, I. H., Valiev, G. H., & Kianfar, E. (2020). Nanomaterial by Sol-Gel Method: Synthesis and Application. *Advances in Materials Science and Engineering*, *2021*(1), 5102014. DOI: 10.1155/2021/5102014

Boros, B. V., & Ostafe, V. (2020). Evaluation of ecotoxicology assessment methods of nanomaterials and their effects. *Nanomaterials (Basel, Switzerland)*, *10*(4), 610. DOI: 10.3390/nano10040610 PMID: 32224954

Bose, B., & Pal, H. (2023). Application of microbial nanotechnology in sustainable agriculture through soil remediation. In *Environmental Applications of Microbial Nanotechnology* (pp. 253–274). Elsevier. DOI: 10.1016/B978-0-323-91744-5.00018-7

Boudh, S., & Singh, J. S. (2019). Pesticide contamination: environmental problems and remediation strategies. *Emerging and eco-friendly approaches for waste management*, 245-269.

Burvey, P., Jain, P., & Singh, A. (2023). Strategic role of casein coated iron oxide nano-particles for delivering nano nutrients for enhance soil quality and their potential. *Materials Today: Proceedings*, *89*, 96–105. DOI: 10.1016/j.matpr.2023.05.739

Bwambale, E., Abagale, F. K., & Anornu, G. K. (2022). Smart irrigation monitoring and control strategies for improving water use efficiency in precision agriculture: A review. *Agricultural Water Management*, *260*, 107324. DOI: 10.1016/j.agwat.2021.107324

Campos, E. V. R., de Oliveira, J. L., & Fraceto, L. F. (2014). Applications of controlled release systems for fungicides, herbicides, acaricides, nutrients, and plant growth hormones: A review. *Advanced Science, Engineering and Medicine*, *6*(4), 373–387. DOI: 10.1166/asem.2014.1538

Campos, E. V. R., do Espirito Santo Pereira, A., de Oliveira, J. L., Villarreal, G. P. U., & Fraceto, L. F. (2022). Nature-based nanocarrier system: An eco-friendly alternative for improving crop resilience to climate changes. *Anthropocene Science*, *1*(3), 396–403. DOI: 10.1007/s44177-022-00029-x

Carmona, F. J., Guagliardi, A., & Masciocchi, N. (2022). Nanosized calcium phosphates as novel macronutrient nano-fertilizers. *Nanomaterials (Basel, Switzerland)*, *12*(15), 2709. DOI: 10.3390/nano12152709 PMID: 35957141

Cartwright, A., Jackson, K., Morgan, C., Anderson, A., & Britt, D. W. (2020). A review of metal and metal-oxide nanoparticle coating technologies to inhibit agglomeration and increase bioactivity for agricultural applications. *Agronomy (Basel)*, *10*(7), 1018. DOI: 10.3390/agronomy10071018

Chadha, S. (2021). Recent advances in nano-encapsulation technologies for controlled release of biostimulants and antimicrobial agents. *Advances in nano-fertilizers and nano-pesticides in agriculture*, 29-55.

Chariou, P. L., Ortega-Rivera, O. A., & Steinmetz, N. F. (2020). Nanocarriers for the delivery of medical, veterinary, and agricultural active ingredients. *ACS Nano*, *14*(3), 2678–2701. DOI: 10.1021/acsnano.0c00173 PMID: 32125825

Chaud, M., Souto, E. B., Zielinska, A., Severino, P., Batain, F., Oliveira-Junior, J., & Alves, T. (2021). Nanopesticides in Agriculture: Benefits and Challenge in Agricultural Productivity, Toxicological Risks to Human Health and Environment. *Toxics*, *9*(6), 131. DOI: 10.3390/toxics9060131 PMID: 34199739

Chávez-Hernández, J. A., Velarde-Salcedo, A. J., Navarro-Tovar, G., & Gonzalez, C. (2024). Safe nanomaterials: From their use, application and disposal to regulations. Nanoscale advances.Chen, H., Zhang, M., Zhai, G. (2018). Lipid-based nanoparticles in the delivery of amphotericin B for fungal infections. *International Journal of Nanomedicine*, *13*, 867–883.

Chen, P., Ference, C., Sun, X., Lin, Y., Tan, L., & Zhong, T. (2020). Antimicrobial efficacy of liposome-encapsulated citral and its effect on the shelf life of shatangju Mandarin. *Journal of Food Protection*, *83*(8), 1315–1322. DOI: 10.4315/JFP-20-115 PMID: 32294203

Chen, R., & Riviere, J. E. (2017). Biological and environmental surface interactions of nanomaterials: Characterization, modeling, and prediction. *Wiley Interdisciplinary Reviews. Nanomedicine and Nanobiotechnology*, *9*(3), e1440. DOI: 10.1002/wnan.1440 PMID: 27863136

Chen, Y. F., Wang, Y., & Wu, W. H. (2008). Membrane transporters for nitrogen, phosphate and potassium uptake in plants. *Journal of Integrative Plant Biology*, *50*(7), 835–848. DOI: 10.1111/j.1744-7909.2008.00707.x PMID: 18713394

Chhipa, H., & Joshi, P. (2016). Nanofertilisers, nanopesticides and nanosensors in agriculture. *Nanoscience in food and agriculture 1*, 247-282.

Clogston, J. D., & Patri, A. K. (2011). Zeta potential measurement. Characterization of nanoparticles intended for drug delivery, 63-70.

Côa, F., Bortolozzo, L. S., Petry, R., Da Silva, G. H., Martins, C. H., de Medeiros, A. M., ... & Martinez, D. S. T. (2020). Environmental toxicity of nanopesticides against non-target organisms: the state of the art. *Nanopesticides: From research and development to mechanisms of action and sustainable use in agriculture*, 227-279.

Cornelis, G., Hund-Rinke, K., Kuhlbusch, T., Van den Brink, N., & Nickel, C. (2014). Fate and bioavailability of engineered nanoparticles in soils: A review. *Critical Reviews in Environmental Science and Technology*, *44*(24), 2720–2764. DOI: 10.1080/10643389.2013.829767

Corsi, I., Venditti, I., Trotta, F., & Punta, C. (2023). Environmental safety of nanotechnologies: The eco-design of manufactured nanomaterials for environmental remediation. *The Science of the Total Environment*, *864*, 161181. DOI: 10.1016/j.scitotenv.2022.161181 PMID: 36581299

Cruz-Luna, A. R., Cruz-Martínez, H., Vásquez-López, A., & Medina, D. I. (2021). Metal nanoparticles as novel antifungal agents for sustainable agriculture: Current advances and future directions. *Journal of Fungi (Basel, Switzerland)*, *7*(12), 1033. DOI: 10.3390/jof7121033 PMID: 34947015

da Silva Júnior, A. H., Mulinari, J., de Oliveira, P. V., de Oliveira, C. R. S., & Júnior, F. W. R. (2022). Impacts of metallic nanoparticles application on the agricultural soils microbiota. *Journal of Hazardous Materials Advances*, *7*, 100103. DOI: 10.1016/j.hazadv.2022.100103

de Oliveira, C., Mulinari, J., Reichert, F., & Júnior, A. (2020, September). Nano-delivery systems of pesticides active agents for agriculture applications-An overview. In *Ciência, Tecnologia e Inovação: Do Campo à Mesa,Proceeding of International Agribusiness Congress,Recife, Brazil* (pp. 25-27). DOI: 10.31692/ICIAGRO.2020.0051

Demir, E. (2020). Drosophila as a model for assessing nanopesticide toxicity. *Nanotoxicology*, *14*(9), 1271–1279. DOI: 10.1080/17435390.2020.1815886 PMID: 32969292

Deng, Z., Xu, S., & Li, S. (2008). Understanding a relaxation behavior in a nanoparticle suspension for drug delivery applications. *International Journal of Pharmaceutics*, *351*(1-2), 236–243. DOI: 10.1016/j.ijpharm.2007.10.001 PMID: 18093763

Ditta, A., Arshad, M., & Ibrahim, M. (2015). Nanoparticles in sustainable agricultural crop production: applications and perspectives. Nanotechnology and plant sciences: nanoparticles and their impact on plants, 55-75.

Djearamane, S., Xiu, L. J., Wong, L. S., Rajamani, R., Bharathi, D., Kayarohanam, S., De Cruz, A. E., Tey, L.-H., Janakiraman, A. K., Aminuzzaman, M., & Selvaraj, S. (2022). Antifungal properties of zinc oxide nanoparticles on Candida albicans. *Coatings*, *12*(12), 1864. DOI: 10.3390/coatings12121864

Du, W., Gardea-Torresdey, J. L., Ji, R., Yin, Y., Zhu, J., Peralta-Videa, J. R., & Guo, H. (2015). Physiological and biochemical changes imposed by CeO2 nanoparticles on wheat: A life cycle field study. *Environmental Science & Technology*, *49*(19), 11884–11893. DOI: 10.1021/acs.est.5b03055 PMID: 26368651

Du Jardin, P. (2015). Plant biostimulants: Definition, concept, main categories and regulation. *Scientia Horticulturae*, *196*, 3–14. DOI: 10.1016/j.scienta.2015.09.021

Duarte, S. F. P., & Cerozi, B. S. (2024). Enhancing plant growth in aquaponic systems via potassium manipulation in fish feeds: A pilot study of tailored feeds bridging nutritional gaps in aquaponics. *Agricultural Systems*, *218*, 104001. DOI: 10.1016/j.agsy.2024.104001

Duhan, J. S., Kumar, R., Kumar, N., Kaur, P., Nehra, K., & Duhan, S. (2017). Nanotechnology: The new perspective in precision agriculture. *Biotechnology Reports (Amsterdam, Netherlands)*, *15*, 11–23. DOI: 10.1016/j.btre.2017.03.002 PMID: 28603692

Dutta, S., Pal, S., Panwar, P., Sharma, R. K., & Bhutia, P. L. (2022). Biopolymeric nanocarriers for nutrient delivery and crop biofortification. *ACS Omega*, *7*(30), 25909–25920. DOI: 10.1021/acsomega.2c02494 PMID: 35936412

Dutta, S., Pal, S., Panwar, P., Sharma, R. K., & Bhutia, P. L. (2022). Biopolymeric nanocarriers for nutrient delivery and crop biofortification. *ACS Omega*, *7*(30), 25909–25920. DOI: 10.1021/acsomega.2c02494 PMID: 35936412

Esposito, S. (2019). "Traditional" Sol-Gel Chemistry as a Powerful Tool for the Preparation of Supported Metal and Metal Oxide Catalysts. *Materials (Basel)*, *12*(4), 668. DOI: 10.3390/ma12040668 PMID: 30813441

Evans, R. G., & Sadler, E. J. (2008). Methods and technologies to improve efficiency of water use. *Water Resources Research*, *44*(7), 2007WR006200. Advance online publication. DOI: 10.1029/2007WR006200

Falsafi, S. R., Rostamabadi, H., Assadpour, E., & Jafari, S. M. (2020). Morphology and microstructural analysis of bioactive-loaded micro/nanocarriers via microscopy techniques; CLSM/SEM/TEM/AFM. *Advances in Colloid and Interface Science*, *280*, 102166. DOI: 10.1016/j.cis.2020.102166 PMID: 32387755

Fatollahpour Grangah, M., Rashidi, V., Mirshekari, B., Khalilvand Behrouzyar, E., & Farahvash, F. (2020). Effects of nano-fertilizers on physiological and yield characteristics of pinto bean cultivars under water deficit stress. *Journal of Plant Nutrition*, *43*(19), 2898–2910. DOI: 10.1080/01904167.2020.1799000

Fendler, J. H. (Ed.). (2008). *Nanoparticles and nanostructured films: preparation, characterization, and applications*. John Wiley & Sons.

Fernandez-Garcia, M., Martinez-Arias, A., Hanson, J. C., & Rodriguez, J. A. (2004). Nanostructured oxides in chemistry: Characterization and properties. *Chemical Reviews*, *104*(9), 4063–4104. DOI: 10.1021/cr030032f PMID: 15352786

Francis, S. P., Rene Christena, L., Francis, M. P., & Abdul, M. H. (2023). Nanotechnology: A potential approach for nutraceuticals. *Current Nutrition and Food Science*, *19*(7), 673–681. DOI: 10.2174/1573401319666221024162943

Fratoddi, I., Venditti, I., Cametti, C., & Russo, M. V. (2014). Gold nanoparticles and gold nanoparticle-conjugates for delivery of therapeutic molecules. Progress and challenges. *Journal of Materials Chemistry. B*, *2*(27), 4204–4220. DOI: 10.1039/C4TB00383G PMID: 32261559

Fu, J., Wang, C., Chen, X., Huang, Z., & Chen, D. (2018). Classification research and types of slow controlled release fertilizers (SRFs) used-a review. *Communications in Soil Science and Plant Analysis*, *49*(17), 2219–2230. DOI: 10.1080/00103624.2018.1499757

Fu, L., Wang, Z., Dhankher, O. P., & Xing, B. (2020). Nanotechnology as a new sustainable approach for controlling crop diseases and increasing agricultural production. *Journal of Experimental Botany*, *71*(2), 507–519. DOI: 10.1093/jxb/erz314 PMID: 31270541

Ganesan, M., Tenchov, R., Iyer, K. A., Ralhan, K., Diaz, L. L., Bird, R. E., ... & Zhou, Q. (2024). Nanoscale materials at work: Mapping emerging applications in energy, medicine, and beyond.

Gao, W., Thamphiwatana, S., Angsantikul, P., & Zhang, L. (2014). Nanoparticle-based local antimicrobial drug delivery. *Advanced Drug Delivery Reviews*, *78*, 1–14. PMID: 28939377

Gomes, A., & Sobral, P. J. D. A. (2021). Plant protein-based delivery systems: An emerging approach for increasing the efficacy of lipophilic bioactive compounds. *Molecules (Basel, Switzerland)*, *27*(1), 60. DOI: 10.3390/molecules27010060 PMID: 35011292

Guimarães, D., Cavaco-Paulo, A., & Nogueira, E. (2021). Design of liposomes as drug delivery system for therapeutic applications. *International Journal of Pharmaceutics*, *601*, 120571. DOI: 10.1016/j.ijpharm.2021.120571 PMID: 33812967

Gupta, P. K. (2018). Toxicity of fungicides. In *Veterinary toxicology* (pp. 569–580). Academic Press.

Habiba, K., Makarov, V. I., Weiner, B. R., & Morell, G. (2014). Fabrication of nanomaterials by pulsed laser synthesis. *Manufacturing nanostructures, 10*, 263-292.

Hamidi, M., Azadi, A., & Rafiei, P. (2008). Hydrogel nanoparticles in drug delivery. *Advanced Drug Delivery Reviews*, *60*(15), 1638–1649. DOI: 10.1016/j.addr.2008.08.002 PMID: 18840488

Hano, C., & Abbasi, B. H. (2021). Plant-based green synthesis of nanoparticles: Production, characterization and applications. *Biomolecules*, *12*(1), 31. DOI: 10.3390/biom12010031 PMID: 35053179

Hassellöv, M., Readman, J. W., Ranville, J. F., & Tiede, K. (2008). Nanoparticle analysis and characterization methodologies in environmental risk assessment of engineered nanoparticles. *Ecotoxicology (London, England)*, *17*(5), 344–361. DOI: 10.1007/s10646-008-0225-x PMID: 18483764

Haydar, M. S., Ghosh, D., & Roy, S. (2024). Slow and controlled release nanofertilizers as an efficient tool for sustainable agriculture: Recent understanding and concerns. *Plant Nano Biology*, *100058*, 100058. Advance online publication. DOI: 10.1016/j.plana.2024.100058

Haynes, C. L., & Van Duyne, R. P. (2001). Nanosphere lithography: A versatile nanofabrication tool for studies of size-dependent nanoparticle optics. *The Journal of Physical Chemistry B*, *105*(24), 5599–5611. DOI: 10.1021/jp010657m

Hench, L. L., & West, J. K. (1990). The sol-gel process. *Chemical Reviews*, *90*(1), 33–72. DOI: 10.1021/cr00099a003

Hennig, T. B., Bandeira, F. O., Puerari, R. C., Fraceto, L. F., & Matias, W. G. (2023). A systematic review of the toxic effects of a nanopesticide on non-target organisms: Estimation of protective concentrations using a species sensitivity distribution (SSD) approach–The case of atrazine. *The Science of the Total Environment*, *871*, 162094. DOI: 10.1016/j.scitotenv.2023.162094 PMID: 36764548

Hoo, C. M., Starostin, N., West, P., & Mecartney, M. L. (2008). A comparison of atomic force microscopy (AFM) and dynamic light scattering (DLS) methods to characterize nanoparticle size distributions. *Journal of Nanoparticle Research*, *10*(S1), 89–96. DOI: 10.1007/s11051-008-9435-7

Howell, T. A. (2001). Enhancing water use efficiency in irrigated agriculture. *Agronomy Journal*, *93*(2), 281–289. DOI: 10.2134/agronj2001.932281x

Hristozov, D., Gottardo, S., Semenzin, E., Oomen, A., Bos, P., Peijnenburg, W., van Tongeren, M., Nowack, B., Hunt, N., Brunelli, A., Scott-Fordsmand, J. J., Tran, L., & Marcomini, A. (2016). Frameworks and tools for risk assessment of manufactured nanomaterials. *Environment International*, *95*, 36–53. DOI: 10.1016/j.envint.2016.07.016 PMID: 27523267

Huang, L., Luo, S., Tong, S., Lv, Z., & Wu, J. (2024). The development of nanocarriers for natural products. *Wiley Interdisciplinary Reviews. Nanomedicine and Nanobiotechnology*, *16*(3), e1967. DOI: 10.1002/wnan.1967 PMID: 38757428

Hyder, S., Ul-Nisa, M., Shahid, H., Gohar, F., Gondal, A. S., Riaz, N., & Iqbal, R. (2023). Recent trends and perspectives in the application of metal and metal oxide nanomaterials for sustainable agriculture. *Plant Physiology and Biochemistry*, *202*, 107960. DOI: 10.1016/j.plaphy.2023.107960 PMID: 37591032

Iavicoli, I., Leso, V., Beezhold, D. H., & Shvedova, A. A. (2017). Nanotechnology in agriculture: Opportunities, toxicological implications, and occupational risks. *Toxicology and Applied Pharmacology*, *329*, 96–111. DOI: 10.1016/j.taap.2017.05.025 PMID: 28554660

Iavicoli, I., Leso, V., Beezhold, D. H., & Shvedova, A. A. (2017). Nanotechnology in agriculture: Opportunities, toxicological implications, and occupational risks. *Toxicology and Applied Pharmacology*, *329*, 96–111. DOI: 10.1016/j.taap.2017.05.025 PMID: 28554660

Ijaz, I., Gilani, E., Nazir, A., & Bukhari, A. (2020). Detail review on chemical, physical and green synthesis, classification, characterizations and applications of nanoparticles. *Green Chemistry Letters and Reviews*, *13*(3), 223–245. DOI: 10.1080/17518253.2020.1802517

Iqbal, M., Umar, S., & Mahmooduzzafar, F. (2019). Nano-fertilization to enhance nutrient use efficiency and productivity of crop plants. *Nanomaterials and plant potential*, 473-505.

Jabr-Milane, L. S., van Vlerken, L. E., Yadav, S., & Amiji, M. M. (2008). Multi-functional nanocarriers to overcome tumor drug resistance. *Cancer Treatment Reviews*, *34*(7), 592–602. DOI: 10.1016/j.ctrv.2008.04.003 PMID: 18538481

Jacques, M. T., Oliveira, J. L., Campos, E. V., Fraceto, L. F., & Ávila, D. S. (2017). Safety assessment of nanopesticides using the roundworm Caenorhabditis elegans. *Ecotoxicology and Environmental Safety*, *139*, 245–253. DOI: 10.1016/j.ecoenv.2017.01.045 PMID: 28160702

Jain, K. K. (2005). Advances in the delivery of systemic antifungal agents. *Advanced Drug Delivery Reviews*, *57*(8), 1063–1085.

Jampílek, J., & Kráľová, K. (2017). Nanomaterials for delivery of nutrients and growth-promoting compounds to plants. *Nanotechnology: An agricultural paradigm*, 177-226. https://doi.org/DOI: 10.1007/978-981-10-4573-8_9

Janot, R., & Guérard, D. (2002). Ball-milling: The behavior of graphite as a function of the dispersal media. *Carbon*, *40*(15), 2887–2896. DOI: 10.1016/S0008-6223(02)00223-3

Jiang, Y., Zhou, P., Zhang, P., Adeel, M., Shakoor, N., Li, Y., Li, M., Guo, M., Zhao, W., Lou, B., Wang, L., Lynch, I., & Rui, Y. (2022). Green synthesis of metal-based nanoparticles for sustainable agriculture. *Environmental Pollution*, *309*, 119755. DOI: 10.1016/j.envpol.2022.119755 PMID: 35839973

Jiménez, A. F., Cárdenas, P. F., & Jiménez, F. (2022). Intelligent IoT-multiagent precision irrigation approach for improving water use efficiency in irrigation systems at farm and district scales. *Computers and Electronics in Agriculture*, *192*, 106635. DOI: 10.1016/j.compag.2021.106635

Jogaiah, S., Singh, H. B., Fraceto, L. F., & De Lima, R. (Eds.). (2020). *Advances in nano-fertilizers and nano-pesticides in agriculture: A smart delivery system for crop improvement*. Woodhead Publishing.

Kah, M. (2015). Nanopesticides and nanofertilizers: Emerging contaminants or opportunities for risk mitigation? *Frontiers in Chemistry*, *3*, 64. DOI: 10.3389/fchem.2015.00064 PMID: 26636068

Kah, M., Kookana, R. S., Gogos, A., & Bucheli, T. D. (2018). A critical evaluation of nanopesticides and nanofertilizers against their conventional analogues. *Nature Nanotechnology*, *13*(8), 677–684. DOI: 10.1038/s41565-018-0131-1 PMID: 29736032

Kang, S., & Zhang, J. (2004). Controlled alternate partial root-zone irrigation: Its physiological consequences and impact on water use efficiency. *Journal of Experimental Botany*, *55*(407), 2437–2446. DOI: 10.1093/jxb/erh249 PMID: 15361526

Karny, A., Zinger, A., Kajal, A., Shainsky-Roitman, J., & Schroeder, A. (2018). Therapeutic nanoparticles penetrate leaves and deliver nutrients to agricultural crops. *Scientific Reports*, *8*(1), 1–10. DOI: 10.1038/s41598-018-25197-y PMID: 29773873

Kasana, R. C., Panwar, N. R., Kaul, R. K., & Kumar, P. (2017). Biosynthesis and effects of copper nanoparticles on plants. *Environmental Chemistry Letters*, *15*(2), 233–240. DOI: 10.1007/s10311-017-0615-5

Kashapov, R., Ibragimova, A., Pavlov, R., Gabdrakhmanov, D., Kashapova, N., Burilova, E., Zakharova, L., & Sinyashin, O. (2021). Nanocarriers for biomedicine: From lipid formulations to inorganic and hybrid nanoparticles. *International Journal of Molecular Sciences*, *22*(13), 7055. DOI: 10.3390/ijms22137055 PMID: 34209023

Kashyap, P. L., Xiang, X., & Heiden, P. (2015). Chitosan nanoparticle based delivery systems for sustainable agriculture. *International Journal of Biological Macromolecules*, *77*, 36–51. DOI: 10.1016/j.ijbiomac.2015.02.039 PMID: 25748851

Kaur Takkar, M., & Gumber, K. (2024). Agro-nanocarriers: An emerging trend for better plant nutrition. *Journal of Plant Nutrition*, *47*(2), 332–350. DOI: 10.1080/01904167.2023.2275073

Khan, F., Pandey, P., & Upadhyay, T. K. (2022). Applications of nanotechnology-based agrochemicals in food security and sustainable agriculture: An overview. *Agriculture*, *12*(10), 1672. DOI: 10.3390/agriculture12101672

Khaydarov, R. A., Khaydarov, R. R., Gapurova, O., Estrin, Y., & Scheper, T. (2009). Electrochemical method for the synthesis of silver nanoparticles. *Journal of Nanoparticle Research*, *11*(5), 1193–1200. DOI: 10.1007/s11051-008-9513-x

Klikocka, H., Marks, M., Barczak, B., Szostak, B., Podleśna, A., & Podleśny, J. (2018). Response of spring wheat to NPK and S fertilization. The content and uptake of macronutrients and the value of ionic ratios. *Open Chemistry*, *16*(1), 1059–1065. DOI: 10.1515/chem-2018-0116

Knödler, M., Opdensteinen, P., Sankaranarayanan, R. A., Morgenroth, A., Buhl, E. M., Mottaghy, F. M., & Buyel, J. F. (2023). Simple plant-based production and purification of the assembled human ferritin heavy chain as a nanocarrier for tumor-targeted drug delivery and bioimaging in cancer therapy. *Biotechnology and Bioengineering*, *120*(4), 1038–1054. DOI: 10.1002/bit.28312 PMID: 36539373

Kong, X. P., Zhang, B. H., & Wang, J. (2021). Multiple roles of mesoporous silica in safe pesticide application by nanotechnology: A review. *Journal of Agricultural and Food Chemistry*, *69*(24), 6735–6754. DOI: 10.1021/acs.jafc.1c01091 PMID: 34110151

Kookana, R. S., Boxall, A. B., Reeves, P. T., Ashauer, R., Beulke, S., Chaudhry, Q., Cornelis, G., Fernandes, T. F., Gan, J., Kah, M., Lynch, I., Ranville, J., Sinclair, C., Spurgeon, D., Tiede, K., & Van den Brink, P. J. (2014). Nanopesticides: Guiding principles for regulatory evaluation of environmental risks. *Journal of Agricultural and Food Chemistry*, *62*(19), 4227–4240. DOI: 10.1021/jf500232f PMID: 24754346

Kumar, J. A., Krithiga, T., Manigandan, S., Sathish, S., Renita, A. A., Prakash, P., Prasad, B. S. N., Kumar, T. R. P., Rajasimman, M., Hosseini-Bandegharaei, A., Prabu, D., & Crispin, S. (2021). A focus to green synthesis of metal/metal based oxide nanoparticles: Various mechanisms and applications towards ecological approach. *Journal of Cleaner Production*, *324*, 129198. DOI: 10.1016/j.jclepro.2021.129198

Kumar, R., Kumar, N., Rajput, V. D., Mandzhieva, S., Minkina, T., Saharan, B. S., Kumar, D., Sadh, P. K., & Duhan, J. S. (2022). Advances in biopolymeric nanopesticides: A new eco-friendly/eco-protective perspective in precision agriculture. *Nanomaterials (Basel, Switzerland)*, *12*(22), 3964. DOI: 10.3390/nano12223964 PMID: 36432250

Kumar, S., Nehra, M., Dilbaghi, N., Marrazza, G., Hassan, A. A., & Kim, K. H. (2019). Nano-based smart pesticide formulations: Emerging opportunities for agriculture. *Journal of Controlled Release*, *294*, 131–153. DOI: 10.1016/j.jconrel.2018.12.012 PMID: 30552953

Kumar, V., Kumar, M., Shrivastava, N., Bisht, S., Sharma, S., & Varma, A. (2016). *Interaction among rhizospheric microbes, soil, and plant roots: Influence on micronutrient uptake and bioavailability. Plant, Soil and Microbes* (Vol. 2). Mechanisms and Molecular Interactions., DOI: 10.1007/978-3-319-29573-2_8

Kumari, R., Suman, K., Karmakar, S., Mishra, V., Lakra, S. G., Saurav, G. K., & Mahto, B. K. (2023). Regulation and safety measures for nanotechnology-based agri-products. *Frontiers in Genome Editing*, *5*, 1200987. DOI: 10.3389/fgeed.2023.1200987 PMID: 37415849

Kutawa, A. B., Ahmad, K., Ali, A., Hussein, M. Z., Abdul Wahab, M. A., Adamu, A., Ismaila, A. A., Gunasena, M. T., Rahman, M. Z., & Hossain, M. I. (2021). Trends in nanotechnology and its potentialities to control plant pathogenic fungi: A review. *Biology (Basel)*, *10*(9), 881. DOI: 10.3390/biology10090881 PMID: 34571758

Kwon, S., Singh, R. K., Perez, R. A., Abou Neel, E. A., Kim, H. W., & Chrzanowski, W. (2013). Silica-based mesoporous nanoparticles for controlled drug delivery. *Journal of Tissue Engineering*, *4*, 2041731413503357. DOI: 10.1177/2041731413503357 PMID: 24020012

Kwon, S. G., & Hyeon, T. (2011). Formation mechanisms of uniform nanocrystals via hot-injection and heat-up methods. *Small*, *7*(19), 2685–2702. DOI: 10.1002/smll.201002022 PMID: 21805623

Lakshani, N., Wijerathne, H. S., Sandaruwan, C., Kottegoda, N., & Karunarathne, V. (2023). Release Kinetic Models and Release Mechanisms of Controlled-Release and Slow-Release Fertilizers. *ACS Agricultural Science & Technology*, *3*(11), 939–956. DOI: 10.1021/acsagscitech.3c00152

Lalpekhlua, K., Tirkey, A., Saranya, S., & Babu, P. J. (2024). Post-harvest Management Strategies for Quality Preservation in Crops. *International Journal of Vegetable Science*, *30*(5), 587–635. DOI: 10.1080/19315260.2024.2392551

Lam, S. J., O'Brien-Simpson, N. M., Pantarat, N., Sulistio, A., Wong, E. H. H., Chen, Y.-Y., Lenzo, J. C., Holden, J. A., Blencowe, A., Reynolds, E. C., & Qiao, G. G. (2016). Combating multidrug-resistant Gram-negative bacteria with structurally nanoengineered antimicrobial peptide polymers. *Nature Microbiology*, *1*(11), 16162. DOI: 10.1038/nmicrobiol.2016.162 PMID: 27617798

Lambers, H., Mougel, C., Jaillard, B., & Hinsinger, P. (2009). Plant-microbe-soil interactions in the rhizosphere: an evolutionary perspective.

Lhamo, D., & Luan, S. (2021). Potential networks of nitrogen-phosphorus-potassium channels and transporters in Arabidopsis roots at a single cell resolution. *Frontiers in Plant Science*, *12*, 689545. DOI: 10.3389/fpls.2021.689545 PMID: 34220911

Li, C., Mei, H., & Zhou, Z. (2021). Surface-modified nanoparticles in fungal disease management. *Nanomedicine (London)*, *16*(7), 583–598.

Li, L. H., Chen, Y., Behan, G., Zhang, H., Petravic, M., & Glushenkov, A. M. (2011). Large-scale mechanical peeling of boron nitride nanosheets by low-energy ball milling. *Journal of Materials Chemistry*, *21*(32), 11862–11866. DOI: 10.1039/c1jm11192b

Lin, H., Ma, N., He, L., Xu, P., Wang, F., & You, C. (2024). High deposition and precise stimulus-response release performance of lignin-coated dendritic mesoporous organosilica nanoparticles for efficient pesticide utilization. *International Journal of Biological Macromolecules*, *259*, 129163. DOI: 10.1016/j.ijbiomac.2023.129163 PMID: 38181906

Liu, Q., Kim, Y. J., Im, G. B., Zhu, J., Wu, Y., Liu, Y., & Bhang, S. H. (2021). Inorganic nanoparticles applied as functional therapeutics. *Advanced Functional Materials*, *31*(12), 2008171. DOI: 10.1002/adfm.202008171

López-Cabeza, R. (2020). Methods for understanding the fate of nanopesticides in soil and water. *Nanopesticides: From Research and Development to Mechanisms of Action and Sustainable Use in Agriculture*, 111-136.

Lowder, S. K., Skoet, J., & Raney, T. (2016). The number, size, and distribution of farms, smallholder farms, and family farms worldwide. *World Development*, *87*, 16–29. DOI: 10.1016/j.worlddev.2015.10.041

Lu, H., Zhang, S., Wang, J., & Chen, Q. (2021). A review on polymer and lipid-based nanocarriers and its application to nano-pharmaceutical and food-based systems. *Frontiers in Nutrition*, *8*, 783831. DOI: 10.3389/fnut.2021.783831 PMID: 34926557

Lucas, J. A., Hawkins, N. J., & Fraaije, B. A. (2015). The evolution of fungicide resistance. *Advances in Applied Microbiology*, *90*, 29–92. DOI: 10.1016/bs.aambs.2014.09.001 PMID: 25596029

Luo, H. H., Zhang, Y. L., & Zhang, W. F. (2016). Effects of water stress and rewatering on photosynthesis, root activity, and yield of cotton with drip irrigation under mulch. *Photosynthetica*, *54*(1), 65–73. DOI: 10.1007/s11099-015-0165-7

Ma, C., Li, G., Xu, W., Qu, H., Zhang, H., Bahojb Noruzi, E., & Li, H. (2024). Recent advances in stimulus-responsive nanocarriers for pesticide delivery. *Journal of Agricultural and Food Chemistry*, *72*(16), 8906–8927. DOI: 10.1021/acs.jafc.4c00997 PMID: 38602422

Machado, T. O., Grabow, J., Sayer, C., de Araújo, P. H., Ehrenhard, M. L., & Wurm, F. R. (2022). Biopolymer-based nanocarriers for sustained release of agrochemicals: A review on materials and social science perspectives for a sustainable future of agri-and horticulture. *Advances in Colloid and Interface Science*, *303*, 102645. DOI: 10.1016/j.cis.2022.102645 PMID: 35358807

Madathil, A. N. P., Vanaja, K. A., & Jayaraj, M. K. (2007). Synthesis of ZnO nanoparticles by hydrothermal method. In *Nanophotonic materials IV* (Vol. 6639, pp. 47–55). SPIE.

Mahapatra, M., Mohapatra, S., & Jena, G. K. (2025). Nanostructured Lipid Carriers: A Revolutionary Approach in Antifungal Drug Delivery. *Biomedical Materials & Devices*. Advance online publication. DOI: 10.1007/s44174-025-00277-6

Maharshi, A., Rachana, D., Pavantheja, V., Geethikalal, V., Harikrishna, K., Pranitha, C., & Dunna, M. (2025). Challenges and Future Directions of Nanotechnology in Plant Sciences. In *Plant-Based Nanoparticle Synthesis for Sustainable Agriculture* (pp. 307–320). CRC Press. DOI: 10.1201/9781003477730-22

Martinez, L. R., Casadevall, A., & Fries, B. C. (2018). Nanotechnology in antifungal therapies. *Fungal Biology Reviews*, *32*(2), 80–90.

Meena, J., Gupta, A., Ahuja, R., Singh, M., Bhaskar, S., & Panda, A. K. (2020). Inorganic nanoparticles for natural product delivery: A review. *Environmental Chemistry Letters*, *18*(6), 2107–2118. DOI: 10.1007/s10311-020-01061-2

Mejía, S. P., Sánchez, A., Vásquez, V., & Orozco, J. (2021). Functional nanocarriers for delivering itraconazole against fungal intracellular infections. *Frontiers in Pharmacology*, *12*, 685391. DOI: 10.3389/fphar.2021.685391 PMID: 34262456

Melanie, M., Miranti, M., Kasmara, H., Malini, D. M., Husodo, T., Panatarani, C., Joni, I. M., & Hermawan, W. (2022). Nanotechnology-Based Bioactive Antifeedant for Plant Protection. *Nanomaterials (Basel, Switzerland)*, *12*(4), 630. DOI: 10.3390/nano12040630 PMID: 35214959

Merino, D., Tomadoni, B., Salcedo, M. F., Mansilla, A. Y., Casalongué, C. A., & Alvarez, V. A. (2020). Nanoclay as carriers of bioactive molecules applied to agriculture. *Handbook of nanomaterials and nanocomposites for energy and environmental applications*, 1-22.

Mishra, R., Rana, S., & Gupta, S. (2022). Stimuli-responsive nanoparticles for fungal therapy. *Advanced Drug Delivery Reviews*, *182*, 114120.

Mohamed, G. G., Ibrahim, N. A., & Attia, H. A. (2009). Synthesis and anti-fungicidal activity of some transition metal complexes with benzimidazole dithiocarbamate ligand. *Spectrochimica Acta. Part A: Molecular and Biomolecular Spectroscopy*, *72*(3), 610–615. DOI: 10.1016/j.saa.2008.10.051 PMID: 19119060

Mohamed, Y. M. A., & Elshahawy, I. E. (2022). Antifungal activity of photo-biosynthesized silver nanoparticles (AgNPs) from organic constituents in orange peel extract against phytopathogenic Macrophomina phaseolina. *European Journal of Plant Pathology*, *162*(3), 725–738. DOI: 10.1007/s10658-021-02434-1

Mohanraj, V. J., & Chen, Y. (2006). Nanoparticles – A review. *Tropical Journal of Pharmaceutical Research*, *5*(1), 561–573.

Mondéjar-López, M., García-Simarro, M. P., Navarro-Simarro, P., Gómez-Gómez, L., Ahrazem, O., & Niza, E. (2024). A review on the encapsulation of "eco-friendly" compounds in natural polymer-based nanoparticles as next generation nano-agrochemicals for sustainable agriculture and crop management. *International Journal of Biological Macromolecules*, *280*, 136030. DOI: 10.1016/j.ijbiomac.2024.136030 PMID: 39332563

Morales, F., Ancín, M., Fakhet, D., González-Torralba, J., Gámez, A. L., Seminario, A., Soba, D., Ben Mariem, S., Garriga, M., & Aranjuelo, I. (2020). Photosynthetic metabolism under stressful growth conditions as a bases for crop breeding and yield improvement. *Plants*, *9*(1), 88. DOI: 10.3390/plants9010088 PMID: 31936732

Morgan, K. T., Cushman, K. E., & Sato, S. (2009). Release mechanisms for slow-and controlled-release fertilizers and strategies for their use in vegetable production. *HortTechnology*, *19*(1), 10–12. DOI: 10.21273/HORTTECH.19.1.10

Morton, V., & Staub, T. (2008). A short history of fungicides. *APSnet Features, 308*(1).

Mourdikoudis, S., Pallares, R. M., & Thanh, N. T. (2018). Characterization techniques for nanoparticles: Comparison and complementarity upon studying nanoparticle properties. *Nanoscale*, *10*(27), 12871–12934. DOI: 10.1039/C8NR02278J PMID: 29926865

Mouton, J. W., Theuretzbacher, U., & Craig, W. A.. (2016). Tissue concentrations: Their relevance for antimicrobial efficacy. *Clinical Microbiology and Infection*, *14*(2), 109–118.

Mujtaba, M., Khawar, K. M., Camara, M. C., Carvalho, L. B., Fraceto, L. F., Morsi, R. E., Elsabee, M. Z., Kaya, M., Labidi, J., Ullah, H., & Wang, D. (2020). Chitosan-based delivery systems for plants: A brief overview of recent advances and future directions. *International Journal of Biological Macromolecules*, *154*, 683–697. DOI: 10.1016/j.ijbiomac.2020.03.128 PMID: 32194112

Mujtaba, M., Wang, D., Carvalho, L. B., Oliveira, J. L., Espirito Santo Pereira, A. D., Sharif, R., Jogaiah, S., Paidi, M. K., Wang, L., Ali, Q., & Fraceto, L. F. (2021). Nanocarrier-mediated delivery of miRNA, RNAi, and CRISPR-Cas for plant protection: Current trends and future directions. *ACS Agricultural Science & Technology*, *1*(5), 417–435. DOI: 10.1021/acsagscitech.1c00146

Mumtaz, I., Majeed, Z., Ajab, Z., Ahmad, B., Khurshid, K., & Mubashir, M. (2019). Optimized tuning of rosin adduct with maleic anhydride for smart applications in controlled and targeted delivery of urea for higher plant's uptake and growth efficiency. *Industrial Crops and Products*, *133*, 395–408. DOI: 10.1016/j.indcrop.2019.02.036

Munir, A., Salah, M. A., Ali, M., Ali, B., Saleem, M. H., Samarasinghe, K. G. B. A., De Silva, S. I. S., Ercisli, S., Iqbal, N., & Anas, M. (2024). Advancing Agriculture: Harnessing Smart Nanoparticles for Precision Fertilization. *BioNanoScience*, *14*(4), 3846–3863. DOI: 10.1007/s12668-024-01597-5

Nafees, M., Ali, S., Rizwan, M., Aziz, A., Adrees, M., Hussain, S. M., ... & Junaid, M. (2020). Effect of nanoparticles on plant growth and physiology and on soil microbes. *Nanomaterials and Environmental Biotechnology*, 65-85. https://doi.org/DOI: 10.1007/978-3-030-34544-0_5

Neikov, O. D. (2009). Mechanical crushing and grinding. In *Handbook of non-ferrous metal powders* (pp. 45-62). DOI: 10.1016/B978-1-85617-422-0.00002-1

Nene, S., Shah, S., Rangaraj, N., Mehra, N. K., Singh, P. K., & Srivastava, S. (2021). Lipid based nanocarriers: A novel paradigm for topical antifungal therapy. *Journal of Drug Delivery Science and Technology*, *62*, 102397. DOI: 10.1016/j.jddst.2021.102397

Nguyen, M. H., Tran, T. N. M., & Vu, N. B. D. (2022). Antifungal activity of essential oil-encapsulated lipid nanoemulsions formulations against leaf spot disease on tomato caused by Alternaria alternata. *Archiv für Phytopathologie und Pflanzenschutz*, *55*(2), 235–257. DOI: 10.1080/03235408.2021.2015887

Nhani, G. B. B., Di Filippo, L. D., de Paula, G. A., Mantovanelli, V. R., da Fonseca, P. P., Tashiro, F. M., Monteiro, D. C., Fonseca-Santos, B., Duarte, J. L., & Chorilli, M. (2024). High-tech sustainable beauty: Exploring nanotechnology for the development of cosmetics using plant and animal by-products. *Cosmetics*, *11*(4), 112. DOI: 10.3390/cosmetics11040112

Nizamani, M. M., Hughes, A. C., Zhang, H. L., & Wang, Y. (2024). Revolutionizing agriculture with nanotechnology: Innovative approaches in fungal disease management and plant health monitoring. *The Science of the Total Environment*, *172473*, 172473. Advance online publication. DOI: 10.1016/j.scitotenv.2024.172473 PMID: 38615773

Nnadi, E. O., Newman, A. P., Coupe, S. J., & Mbanaso, F. U. (2015). Stormwater harvesting for irrigation purposes: An investigation of chemical quality of water recycled in pervious pavement system. *Journal of Environmental Management*, *147*, 246–256. DOI: 10.1016/j.jenvman.2014.08.020 PMID: 25256837

Nowicka, B., Ciura, J., Szymańska, R., & Kruk, J. (2018). Improving photosynthesis, plant productivity and abiotic stress tolerance–current trends and future perspectives. *Journal of Plant Physiology*, *231*, 415–433. DOI: 10.1016/j.jplph.2018.10.022 PMID: 30412849

O'Shaughnessy, S. A., Evett, S. R., Andrade, A., Workneh, F., Price, J. A., & Rush, C. M. (2015). Site-specific variable rate irrigation as a means to enhance water use efficiency. In *2015 ASABE/IA Irrigation Symposium: Emerging Technologies for Sustainable Irrigation-A Tribute to the Career of Terry Howell, Sr. Conference Proceedings* (pp. 1-21). American Society of Agricultural and Biological Engineers. https://doi.org/DOI: 10.13031/irrig.20152140448

Omran, B. A., & Baek, K. H. (2022). Control of phytopathogens using sustainable biogenic nanomaterials: Recent perspectives, ecological safety, and challenging gaps. *Journal of Cleaner Production*, *372*, 133729. DOI: 10.1016/j.jclepro.2022.133729

Osman, A. I., Zhang, Y., Farghali, M., Rashwan, A. K., Eltaweil, A. S., Abd El-Monaem, E. M., Mohamed, I. M. A., Badr, M. M., Ihara, I., Rooney, D. W., & Yap, P. S. (2024). Synthesis of green nanoparticles for energy, biomedical, environmental, agricultural, and food applications: A review. *Environmental Chemistry Letters*, *22*(2), 841–887. DOI: 10.1007/s10311-023-01682-3

Palchik, O., Zhu, J., & Gedanken, A. (2000). Microwave assisted preparation of binary oxide nanoparticles. *Journal of Materials Chemistry*, *10*(5), 1251–1254. DOI: 10.1039/a908795h

Park, W., Shin, H., Choi, B., Rhim, W. K., Na, K., & Han, D. K. (2020). Advanced hybrid nanomaterials for biomedical applications. *Progress in Materials Science*, *114*, 100686. DOI: 10.1016/j.pmatsci.2020.100686

Parveen, S., Misra, R., & Sahoo, S. K. (2012). Nanoparticles: A boon to drug delivery, therapeutics, diagnostics, and imaging. *Nanomedicine; Nanotechnology, Biology, and Medicine*, *8*(2), 147–166. DOI: 10.1016/j.nano.2011.05.016 PMID: 21703993

Pascoli, M., Jacques, M. T., Agarrayua, D. A., Avila, D. S., Lima, R., & Fraceto, L. F. (2019). Neem oil based nanopesticide as an environmentally-friendly formulation for applications in sustainable agriculture: An ecotoxicological perspective. *The Science of the Total Environment*, *677*, 57–67. DOI: 10.1016/j.scitotenv.2019.04.345 PMID: 31051383

Pascoli, M., Lopes-Oliveira, P. J., Fraceto, L. F., Seabra, A. B., & Oliveira, H. C. (2018). State of the art of polymeric nanoparticles as carrier systems with agricultural applications: A minireview. *Energy, Ecology & Environment*, *3*(3), 137–148. DOI: 10.1007/s40974-018-0090-2

Patil, S., Das, S., & Dhavale, D. (2019). Nanoparticles for antifungal drug delivery: An emerging paradigm. *Pharmaceutics*, *11*(6), 305.

Peer, D., Karp, J. M., Hong, S., Farokhzad, O. C., Margalit, R., & Langer, R. (2007). Nanocarriers as an emerging platform for cancer therapy. *Nature Nanotechnology*, *2*(12), 751–760. DOI: 10.1038/nnano.2007.387 PMID: 18654426

Peer, D., Karp, J. M., Hong, S., Farokhzad, O. C., Margalit, R., & Langer, R. (2007). Nanocarriers as an emerging platform for cancer therapy. *Nature Nanotechnology*, *2*(12), 751–760. DOI: 10.1038/nnano.2007.387 PMID: 18654426

Pereira, A. D. E. S., Oliveira, H. C., & Fraceto, L. F. (2019). Polymeric nanoparticles as an alternative for application of gibberellic acid in sustainable agriculture: A field study. *Scientific Reports*, *9*(1), 7135. DOI: 10.1038/s41598-019-43494-y PMID: 31073210

Pérez-de-Luque, A. (2017). Interaction of nanomaterials with plants: What do we need for real applications in agriculture? *Frontiers in Environmental Science*, *5*, 12. DOI: 10.3389/fenvs.2017.00012

Perlatti, B., de Souza Bergo, P. L., Fernandes, J. B., & Forim, M. R. (2013). Polymeric nanoparticle-based insecticides: a controlled release purpose for agrochemicals. In *Insecticides-Development of safer and more effective technologies*. IntechOpen. DOI: 10.5772/53355

Pinto, T. V., Silva, C. A., Siquenique, S., & Learmonth, D. A. (2022). Micro-and nanocarriers for encapsulation of biological plant protection agents: A systematic literature review. *ACS Agricultural Science & Technology*, *2*(5), 838–857. DOI: 10.1021/acsagscitech.2c00113

Pinton, R., Varanini, Z., & Nannipieri, P. (2000). The rhizosphere as a site of biochemical interactions among soil components, plants, and microorganisms. In *The rhizosphere* (pp. 17–34). CRC press.

Prabhu, P., & Patravale, V. (2015). Polymeric nanoparticles for targeted treatment in oncology: Current insights. *International Journal of Nanomedicine*, *10*, 1001–1018. PMID: 25678788

Pradhan, S., Durgam, M., & Mailapalli, D. R. (2021). Urea loaded hydroxyapatite nanocarrier for efficient delivery of plant nutrients in rice. *Archives of Agronomy and Soil Science*, *67*(3), 371–382. DOI: 10.1080/03650340.2020.1732940

Prajitha, N., Athira, S. S., & Mohanan, P. V. (2019). Bio-interactions and risks of engineered nanoparticles. *Environmental Research*, *172*, 98–108. DOI: 10.1016/j.envres.2019.02.003 PMID: 30782540

Pramanik, P., Krishnan, P., Maity, A., Mridha, N., Mukherjee, A., & Rai, V. (2020). Application of nanotechnology in agriculture. *Environmental Nanotechnology*, *4*, 317–348.

Prasad, R., Bhattacharyya, A., & Nguyen, Q. D. (2017). Nanotechnology in sustainable agriculture: Recent developments, challenges, and perspectives. *Frontiers in Microbiology*, *8*, 1014. DOI: 10.3389/fmicb.2017.01014 PMID: 28676790

Punniyakotti, P., Vinayagam, S., Rajamohan, R., Priya, S. D., Moovendhan, M., & Sundaram, T. (2024). Environmental Fate and Ecotoxicological Behaviour of Pesticides and Insecticides in Non-Target Environments: Nanotechnology-Based Mitigation Strategies. *Journal of Environmental Chemical Engineering*, *113349*(5), 113349. Advance online publication. DOI: 10.1016/j.jece.2024.113349

Qiao, H., Chen, J., Dong, M., Shen, J., & Yan, S. (2024). Nanocarrier-Based Eco-Friendly RNA Pesticides for Sustainable Management of Plant Pathogens and Pests. *Nanomaterials (Basel, Switzerland)*, *14*(23), 1874. DOI: 10.3390/nano14231874 PMID: 39683262

Qureshi, A., Singh, D. K., & Dwivedi, S. (2018). Nano-fertilizers: A novel way for enhancing nutrient use efficiency and crop productivity. *International Journal of Current Microbiology and Applied Sciences*, *7*(2), 3325–3335. DOI: 10.20546/ijcmas.2018.702.398

Raaijmakers, J. M., Paulitz, T. C., Steinberg, C., Alabouvette, C., & Moënne-Loccoz, Y. (2009). The rhizosphere: a playground and battlefield for soilborne pathogens and beneficial microorganisms. https://doi.org/DOI: 10.1007/s11104-008-9568-6

Rafiq, H., Aftab, Z. E. H., Anjum, T., Ali, B., Akram, W., Bashir, U., Mirza, F. S., Aftab, M., Ali, M. D., Li, G., & Li, G. (2022). Bio-fabrication of zinc oxide nanoparticles to rescue mung bean against cercospora leaf spot disease. *Frontiers in Plant Science*, *13*, 1052984. DOI: 10.3389/fpls.2022.1052984 PMID: 36523618

Rafiq, H., Aftab, Z. E. H., Anjum, T., Ali, B., Akram, W., Bashir, U., Mirza, F. S., Aftab, M., Ali, M. D., Li, G., & Li, G. (2022). Bio-fabrication of zinc oxide nanoparticles to rescue mung bean against cercospora leaf spot disease. *Frontiers in Plant Science*, *13*, 1052984. DOI: 10.3389/fpls.2022.1052984 PMID: 36523618

Rahman, M. H., Haque, K. S., & Khan, M. Z. H. (2021). A review on application of controlled released fertilizers influencing the sustainable agricultural production: A Cleaner production process. *Environmental Technology & Innovation*, *23*, 101697. DOI: 10.1016/j.eti.2021.101697

Rai, M., & Kratosova, G. (2015). Management of phytopathogens by application of green nanobiotechnology: Emerging trends and challenges. *Acta Agraria Debreceniensis*, (66), 15–22. DOI: 10.34101/actaagrar/66/1884

Rai, P. K., Kumar, V., Lee, S., Raza, N., Kim, K. H., Ok, Y. S., & Tsang, D. C. (2018). Nanoparticle-plant interaction: Implications in energy, environment, and agriculture. *Environment International*, *119*, 1–19. DOI: 10.1016/j.envint.2018.06.012 PMID: 29909166

Rajasekar, S., Akhil, V., & Pattanayak, S. P. (2018). Eugenol-loaded nanoparticles for antifungal activity. *Journal of Biomedical Materials Research*, *106*(12), 3046–3054.

Rajesaheb, K. S., Subramanian, S., Boominathan, P., Thenmozhi, S., & Gnanachitra, M. (2024). Biostimulant in improving crop yield and soil health. *Communications in Soil Science and Plant Analysis*, •••, 1–36.

Rajput, V., Minkina, T., Mazarji, M., Shende, S., Sushkova, S., Mandzhieva, S., Burachevskaya, M., Chaplygin, V., Singh, A., & Jatav, H. (2020). Accumulation of nanoparticles in the soil-plant systems and their effects on human health. *Annals of Agricultural Science*, *65*(2), 137–143. DOI: 10.1016/j.aoas.2020.08.001

Rajput, V. D., Minkina, T., Sushkova, S., Tsitsuashvili, V., Mandzhieva, S., Gorovtsov, A., Nevidomskyaya, D., & Gromakova, N. (2018). Effect of nanoparticles on crops and soil microbial communities. *Journal of Soils and Sediments*, *18*(6), 2179–2187. DOI: 10.1007/s11368-017-1793-2

Rameez, M., Khan, N., Ahmad, S., & Ahmad, M. M. (2024). Bionanocomposites: A new approach for fungal disease management. *Biocatalysis and Agricultural Biotechnology*, *103115*, 103115. Advance online publication. DOI: 10.1016/j.bcab.2024.103115

Rana, R. A., Siddiqui, M. N., Skalicky, M., Brestic, M., Hossain, A., Kayesh, E., Popov, M., Hejnak, V., Gupta, D., Mahmud, N., & Islam, T. (2021). Prospects of nanotechnology in improving the productivity and quality of horticultural crops. *Horticulturae*, *7*(10), 332. DOI: 10.3390/horticulturae7100332

Rao, B. S., Kumar, B. R., Reddy, V. R., Rao, T. S., & Chalapathi, G. J. C. L. (2011). Preparation and characterization of CdS nanoparticles by chemical co-precipitation technique. *Chalcogenide Letters*, *8*(3), 177–185.

Rastogi, A., Tripathi, D. K., Yadav, S., Chauhan, D. K., Živčák, M., Ghorbanpour, M., ... & Brestic, M. (2019). Application of silicon nanoparticles in agriculture. *3 Biotech, 9*, 1-11.

Ray, M. K., Mishra, A. K., Mohanta, Y. K., Mahanta, S., Chakrabartty, I., Kungwani, N. A., Avula, S. K., Panda, J., & Pudake, R. N. (2023). Nanotechnology as a promising tool against phytopathogens: A futuristic approach to agriculture. *Agriculture*, *13*(9), 1856. DOI: 10.3390/agriculture13091856

Raza, A., Charagh, S., Salehi, H., Abbas, S., Saeed, F., Poinern, G. E., Siddique, K. H. M., & Varshney, R. K. (2023). Nano-enabled stress-smart agriculture: Can nanotechnology deliver drought and salinity-smart crops? *Journal of Sustainable Agriculture and Environment*, *2*(3), 189–214. DOI: 10.1002/sae2.12061

Romero, G., & Moya, S. E. (2012). Synthesis of organic nanoparticles. In *Frontiers of nanoscience* (Vol. 4, pp. 115-141). DOI: 10.1016/B978-0-12-415769-9.00004-2

Sahu, S., Gautam, S., Singh, A., Lohani, P., Sharma, C., Pathak, P., Kumar, A., & Singh, H. (2023). Synthesis and characterization of chitosan-zinc-salicylic acid nanoparticles: A plant biostimulant. *International Journal of Biological Macromolecules*, *253*, 127602. DOI: 10.1016/j.ijbiomac.2023.127602 PMID: 37875188

Salamanca-Buentello, F., Persad, D. L., Court, E. B., Martin, D. K., Daar, A. S., & Singer, P. A. (2005). Nanotechnology and the developing world. *PLoS Medicine*, *2*(5), e97. DOI: 10.1371/journal.pmed.0020097 PMID: 15807631

Sampat, B. N. (2010). When do applicants search for prior art? *The Journal of Law & Economics*, *53*(2), 399–416. DOI: 10.1086/651959

Sánchez-López, E., Gomes, D., Esteruelas, G., Bonilla, L., Lopez-Machado, A. L., Galindo, R., Cano, A., Espina, M., Ettcheto, M., Camins, A., Silva, A. M., Durazzo, A., Santini, A., Garcia, M. L., & Souto, E. B. (2020). Metal-based nanoparticles as antimicrobial agents: An overview. *Nanomaterials (Basel, Switzerland)*, *10*(2), 292. DOI: 10.3390/nano10020292 PMID: 32050443

Schnoor, B., Elhendawy, A., Joseph, S., Putman, M., Chacón-Cerdas, R., Flores-Mora, D., Bravo-Moraga, F., Gonzalez-Nilo, F., & Salvador-Morales, C. (2018). Engineering atrazine loaded poly (lactic-co-glycolic acid) nanoparticles to ameliorate environmental challenges. *Journal of Agricultural and Food Chemistry*, *66*(30), 7889–7898. DOI: 10.1021/acs.jafc.8b01911 PMID: 30039704

Sehaole, T. P. (2012). *Evaluation of different media formulations on spore production and toxicity of Bacillus thuringiensis subsp. aizawai* (Doctoral dissertation, North-West University).

Semaltianos, N. G. (2010). Nanoparticles by laser ablation. *Critical Reviews in Solid State and Material Sciences*, *35*(2), 105–124. DOI: 10.1080/10408431003788233

Serov, D. A., Gritsaeva, A. V., Yanbaev, F. M., Simakin, A. V., & Gudkov, S. V. (2024). Review of antimicrobial properties of titanium dioxide nanoparticles. *International Journal of Molecular Sciences*, *25*(19), 10519. DOI: 10.3390/ijms251910519 PMID: 39408848

Shahid, M. A., Sarkhosh, A., Khan, N., Balal, R. M., Ali, S., Rossi, L., Gómez, C., Mattson, N., Nasim, W., & Garcia-Sanchez, F. (2020). Insights into the physiological and biochemical impacts of salt stress on plant growth and development. *Agronomy (Basel)*, *10*(7), 938. DOI: 10.3390/agronomy10070938

Shakiba, S., Astete, C. E., Paudel, S., Sabliov, C. M., Rodrigues, D. F., & Louie, S. M. (2020). Emerging investigator series: polymeric nanocarriers for agricultural applications: synthesis, characterization, and environmental and biological interactions. *Environmental Science. Nano*, *7*(1), 37–67. DOI: 10.1039/C9EN01127G

Shang, Y., Hasan, M. K., Ahammed, G. J., Li, M., Yin, H., & Zhou, J. (2019). Applications of Nanotechnology in Plant Growth and Crop Protection: A Review. *Molecules (Basel, Switzerland)*, *24*(14), 2558. DOI: 10.3390/molecules24142558 PMID: 31337070

Sharma, A., Jain, N., & Reddy, A. D.. (2020). Advances in the development of nanomedicines for antifungal therapeutics. *Expert Opinion on Drug Delivery*, *17*(7), 913–928.

Sharma, D., Dhuriya, Y. K., Sharma, J., & Gupta, M. (2019). Nanoelements: An agricultural paradigm for targeted plant nutrition therapeutic approach. *Nanotechnology for Agriculture: Crop Production & Protection*, 73-83.

Sharma, P., Kumar, S., & Mittal, A. (2021). In situ polymerization of nanoparticles for antifungal applications. *Materials Today. Communications*, *26*, 102182.

Sharma, S., Singh, N., & Rastogi, M. (2022). Nanotechnology to Address Micronutrient and Macronutrient Deficiency. In *Nanotechnology* (pp. 35–59). CRC Press. DOI: 10.1201/9781003220350-3

Shekhar, S., Sharma, S., Kumar, A., Taneja, A., & Sharma, B. (2021). The framework of nanopesticides: A paradigm in biodiversity. *Materials Advances*, *2*(20), 6569–6588. DOI: 10.1039/D1MA00329A

Shende, S. S., Rajput, V. D., Gorovtsov, A. V., Harish, Saxena, P., Minkina, T. M., ... & Kizilkaya, R. (2021). Interaction of nanoparticles with microbes. *Plant-Microbes-Engineered Nano-particles (PM-ENPs) Nexus in Agro-Ecosystems: Understanding the Interaction of Plant, Microbes and Engineered Nano-particles (ENPS)*, 175-188.

Shishir, M. R. I., Gowd, V., Suo, H., Wang, M., Wang, Q., Chen, F., & Cheng, K. W. (2021). Advances in smart delivery of food bioactive compounds using stimuli-responsive carriers: Responsive mechanism, contemporary challenges, and prospects. *Comprehensive Reviews in Food Science and Food Safety*, *20*(6), 5449–5488. DOI: 10.1111/1541-4337.12851 PMID: 34668321

Shukla, P., Chaurasia, P., Younis, K., Qadri, O. S., Faridi, S. A., & Srivastava, G. (2019). Nanotechnology in sustainable agriculture: Studies from seed priming to post-harvest management. *Nanotechnology for Environmental Engineering*, *4*(1), 1–15. DOI: 10.1007/s41204-019-0058-2

Sidhu, R. K., Kumar, R., Rana, P. S., & Jat, M. L. (2021). Automation in drip irrigation for enhancing water use efficiency in cereal systems of South Asia: Status and prospects. *Advances in Agronomy*, *167*, 247–300. DOI: 10.1016/bs.agron.2021.01.002

Siegel, R. A., & Rathbone, M. J. (2011). Overview of controlled release mechanisms. In *Fundamentals and applications of controlled release drug delivery* (pp. 19–43). Springer US., DOI: 10.1007/978-1-4614-0881-9_2

Sigmon, L. R., Adisa, I. O., Liu, B., Elmer, W. H., White, J. C., Dimkpa, C. O., & Fairbrother, D. H. (2021). Biodegradable polymer nanocomposites provide effective delivery and reduce phosphorus loss during plant growth. *ACS Agricultural Science & Technology*, *1*(5), 529–539. DOI: 10.1021/acsagscitech.1c00149

Singh, A., Singh, N. B., Hussain, I., Singh, H., & Singh, S. C. (2015). Plant-nanoparticle interaction: An approach to improve agricultural practices and plant productivity. *International Journal of Pharmaceutical Science Invention*, *4*(8), 25–40.

Singh, D., & Gurjar, B. R. (2022). Nanotechnology for agricultural applications: Facts, issues, knowledge gaps, and challenges in environmental risk assessment. *Journal of Environmental Management*, *322*, 116033. DOI: 10.1016/j.jenvman.2022.116033

Singh, H., Sharma, A., Bhardwaj, S. K., Arya, S. K., Bhardwaj, N., & Khatri, M. (2021). Recent advances in the applications of nano-agrochemicals for sustainable agricultural development. *Environmental Science. Processes & Impacts*, *23*(2), 213–239. DOI: 10.1039/D0EM00404A PMID: 33447834

Singh, R. P., Handa, R., & Manchanda, G. (2021). Nanoparticles in sustainable agriculture: An emerging opportunity. *Journal of Controlled Release*, *329*, 1234–1248. DOI: 10.1016/j.jconrel.2020.10.051 PMID: 33122001

Sinha, D., & Tandon, P. K. (2020). An overview of nitrogen, phosphorus and potassium: Key players of nutrition process in plants. *Sustainable solutions for elemental deficiency and excess in crop plants*, 85-117. https://doi.org/DOI: 10.1007/978-981-15-8636-1_5

Slowing, I. I., Vivero-Escoto, J. L., Wu, C. W., & Lin, V. (2008). Mesoporous silica nanoparticles as controlled release drug delivery and gene transfection carriers. *Advanced Drug Delivery Reviews*, *60*(11), 1278–1288. DOI: 10.1016/j.addr.2008.03.012 PMID: 18514969

Slowing, I. I., Vivero-Escoto, J. L., Wu, C. W., & Lin, V. S. (2008). Mesoporous silica nanoparticles as controlled release drug delivery and gene transfection carriers. *Advanced Drug Delivery Reviews*, *60*(11), 1278–1288. DOI: 10.1016/j.addr.2008.03.012 PMID: 18514969

Sonawane, H., Shelke, D., Chambhare, M., Dixit, N., Math, S., Sen, S., Borah, S. N., Islam, N. F., Joshi, S. J., Yousaf, B., Rinklebe, J., & Sarma, H. (2022). Fungi-derived agriculturally important nanoparticles and their application in crop stress management–Prospects and environmental risks. *Environmental Research*, *212*, 113543. DOI: 10.1016/j.envres.2022.113543 PMID: 35613631

Squire, H. J., Tomatz, S., Voke, E., González-Grandío, E., & Landry, M. (2023). The emerging role of nanotechnology in plant genetic engineering. *Nature Reviews Bioengineering*, *1*(5), 314–328. DOI: 10.1038/s44222-023-00037-5

Su, Y., Ashworth, V., Kim, C., Adeleye, A. S., Rolshausen, P., Roper, C., White, J., & Jassby, D. (2019). Delivery, uptake, fate, and transport of engineered nanoparticles in plants: A critical review and data analysis. *Environmental Science. Nano*, *6*(8), 2311–2331. DOI: 10.1039/C9EN00461K

Su, Y., Zhou, X., Meng, H., Xia, T., Liu, H., Rolshausen, P., Roper, C., McLean, J. E., Zhang, Y., Keller, A. A., & Jassby, D. (2022). Cost–benefit analysis of nanofertilizers and nanopesticides emphasizes the need to improve the efficiency of nanoformulations for widescale adoption. *Nature Food*, *3*(12), 1020–1030. DOI: 10.1038/s43016-022-00647-z PMID: 37118298

Sulak, M., & Yardımcı, B. K. (2025). *Green synthesis of tin and titanium nanoparticles using edible plant extracts: exploring their anticancer and antifungal activities*. Biomass Conv. Bioref., DOI: 10.1007/s13399-025-06745-9

Sun, Q., Li, J., & Le, T. (2018). Zinc oxide nanoparticle as a novel class of antifungal agents: Current advances and future perspectives. *Journal of Agricultural and Food Chemistry*, *66*(43), 11209–11220. DOI: 10.1021/acs.jafc.8b03210 PMID: 30299956

Takeshita, V., Campos, E. V., Rodrigues, J. S., & Fraceto, L. F. (2023). Opinion: Hybrid nanoparticle systems–Two-way delivery approach for agriculture. *Plant Nano Biology*, *6*, 100053. DOI: 10.1016/j.plana.2023.100053

Takkar, M. K., & Gumber, K. (2024). Engineered Nano-Carriers for Stimuli Controlled Release of Micronutrient in Plants. https://doi.org/DOI: 10.21203/rs.3.rs-4677814/v1

Tamayo, L. A., Zapata, P. A., & Vejar, N. D. (2014). Nanoparticle-loaded hydrogels for antifungal treatments. *Journal of Biomedical Nanotechnology*, *10*(7), 1419–1430.

Tang, S., Liang, J., Li, O., Shao, N., Jin, Y., Ni, J., Fei, X., & Li, Z. (2023). Morphology-Tailored Hydroxyapatite Nanocarrier for Rhizosphere-Targeted Phosphorus Delivery. *Small*, *19*(14), 2206954. DOI: 10.1002/smll.202206954 PMID: 36599675

Tarannum, N., Pooja, K., Jakhar, S., & Mavi, A. (2024). Nanoparticles assisted intra and transdermic delivery of antifungal ointment: An updated review. *Discover Nano*, *19*(1), 11. DOI: 10.1186/s11671-023-03932-3 PMID: 38195832

Tian, L. (2002). Development of a sensor-based precision herbicide application system. *Computers and Electronics in Agriculture*, *36*(2-3), 133–149. DOI: 10.1016/S0168-1699(02)00097-2

Tleuova, A. B., Wielogorska, E., Talluri, V. P., Štěpánek, F., Elliott, C. T., & Grigoriev, D. O. (2020). Recent advances and remaining barriers to producing novel formulations of fungicides for safe and sustainable agriculture. *Journal of Controlled Release*, *326*, 468–481. DOI: 10.1016/j.jconrel.2020.07.035 PMID: 32721524

Tomar, B., Patle, T., Parihar, S. S., Singh, P., & Tomar, S. S. (2024). Enhancing Nutrient Uptake With Nano Fertilizers and Soil Amendments. In *Harnessing NanoOmics and Nanozymes for Sustainable Agriculture* (pp. 43–59). IGI Global., DOI: 10.4018/979-8-3693-1890-4.ch003

Tombuloglu, H., Ercan, I., Alshammari, T., Tombuloglu, G., Slimani, Y., Almessiere, M., & Baykal, A. (2020). Incorporation of micro-nutrients (nickel, copper, zinc, and iron) into plant body through nanoparticles. *Journal of Soil Science and Plant Nutrition*, *20*(4), 1872–1881. DOI: 10.1007/s42729-020-00258-2

ME Trenkel, T. (2021). *Slow-and controlled-release and Stabilized Fertilizers: an option for enhancing nutrient use effiiency in agriculture*. International Fertilizer Industry Association (IFA).

Usman, M., Farooq, M., Wakeel, A., Nawaz, A., Cheema, S. A., Rehman, H., Ashraf, I., & Sanaullah, M. (2020). Nanotechnology in agriculture: Current status, challenges and future opportunities. *The Science of the Total Environment*, *721*, 137778. DOI: 10.1016/j.scitotenv.2020.137778 PMID: 32179352

Varadachari, C., & Goertz, H. M. (2010). *Slow-release and controlled-release nitrogen fertilizers*. Indian Nitrogen Group, Society.

Vauthier, C., & Bouchemal, K. (2009). Methods for the preparation and manufacture of polymeric nanoparticles. *Pharmaceutical Research*, *26*(5), 1025–1058. DOI: 10.1007/s11095-008-9800-3 PMID: 19107579

Vega-Vásquez, P., Mosier, N. S., & Irudayaraj, J. (2020). Nanoscale drug delivery systems: From medicine to agriculture. *Frontiers in Bioengineering and Biotechnology*, *8*, 79. DOI: 10.3389/fbioe.2020.00079 PMID: 32133353

Victoria, J., Tripathi, S., Prakash, V., Tiwari, K., Mahra, S., Sharma, A., Rana, S., Kandhol, N., Sahi, S., Tripathi, D. K., & Sharma, S. (2024). Encapsulated nanopesticides application in plant protection: Quo vadis? *Plant Physiology and Biochemistry*, *206*, 108225. DOI: 10.1016/j.plaphy.2023.108225 PMID: 38147708

Vinzant, K., Rashid, M., & Khodakovskaya, M. V. (2023). Advanced applications of sustainable and biological nano-polymers in agricultural production. *Frontiers in Plant Science*, *13*, 1081165. DOI: 10.3389/fpls.2022.1081165 PMID: 36684740

Vurro, M., Miguel-Rojas, C., & Pérez-de-Luque, A. (2019). Safe nanotechnologies for increasing the effectiveness of environmentally friendly natural agrochemicals. *Pest Management Science*, *75*(9), 2403–2412. DOI: 10.1002/ps.5348 PMID: 30672106

Wahab, A., Muhammad, M., Ullah, S., Abdi, G., Shah, G. M., Zaman, W., & Ayaz, A. (2024). Agriculture and environmental management through nanotechnology: Eco-friendly nanomaterial synthesis for soil-plant systems, food safety, and sustainability. *The Science of the Total Environment*, *926*, 171862. DOI: 10.1016/j.scitotenv.2024.171862 PMID: 38527538

Wang, H., Qiao, X., Chen, J., & Ding, S. (2005). Preparation of silver nanoparticles by chemical reduction method. *Colloids and Surfaces. A, Physicochemical and Engineering Aspects*, *256*(2-3), 111–115. DOI: 10.1016/j.colsurfa.2004.12.058

Wang, L., Hu, C., & Shao, L. (2018). The antimicrobial activity of nanoparticles: Present situation and prospects for the future. *International Journal of Nanomedicine*, *13*, 1227–1249. PMID: 28243086

Wang, M., Hu, Z., Yang, T., Pei, H., & Zhang, F. (2023). A dual pesticide–fertilizer silicon-base nanocomposite to synergistically control fungal disease and provide nutrition. *Environmental Science. Nano*, *10*(12), 3462–3475. DOI: 10.1039/D3EN00500C

Wang, Z., Thavarajah, N., & Fernando, X. (2024). Exploring Silica Nanoparticles: A Sustainable Solution for Pest Control in Sri Lankan Rice Farming. *Technologies*, *12*(11), 210. DOI: 10.3390/technologies12110210

Wang, Z., Zhu, W., Qiu, Y., Yi, X., von dem Bussche, A., Kane, A., Gao, H., Koski, K., & Hurt, R. (2016). Biological and environmental interactions of emerging two-dimensional nanomaterials. *Chemical Society Reviews*, *45*(6), 1750–1780. DOI: 10.1039/C5CS00914F PMID: 26923057

Welch, R. M., & Shuman, L. (1995). Micronutrient nutrition of plants. *Critical Reviews in Plant Sciences*, *14*(1), 49–82. DOI: 10.1080/07352689509701922

Worrall, E. A., Hamid, A., Mody, K. T., Mitter, N., & Pappu, H. R. (2018). Nanotechnology for Plant Disease Management. *Agronomy (Basel)*, *8*(12), 285. DOI: 10.3390/agronomy8120285

Xiang, H., Meng, J., Shao, W., Zeng, D., Ji, J., Wang, P., Zhou, X., Qi, P., Liu, L., & Yang, S. (2023). Plant protein-based self-assembling core–shell nanocarrier for effectively controlling plant viruses: Evidence for nanoparticle delivery behavior, plant growth promotion, and plant resistance induction. *Chemical Engineering Journal*, *464*, 142432. DOI: 10.1016/j.cej.2023.142432

Xu, T., Wang, Y., Aytac, Z., Zuverza-Mena, N., Zhao, Z., Hu, X., Ng, K. W., White, J. C., & Demokritou, P. (2022). Enhancing agrichemical delivery and plant development with biopolymer-based stimuli responsive core–shell nanostructures. *ACS Nano*, *16*(4), 6034–6048. DOI: 10.1021/acsnano.1c11490 PMID: 35404588

Yadav, A., Alhan, S., & Dhania, G. (2024). Nanobiostimulants for Enhancing Plant Growth and Yield. *Nanobiostimulants: Emerging Strategies for Agricultural Sustainability*, 111-134.

Yadav, A., & Yadav, K. (2018). Nanoparticle-based plant disease management: tools for sustainable agriculture. Nanobiotechnology applications in plant protection, 29-61.

Yang, W., Gan, L., & Han, S. (2019). Advances in polymeric nanoparticles for antifungal treatments. *Current Drug Targets*, *20*(5), 451–460.

Younis, S. A., Kim, K. H., Shaheen, S. M., Antoniadis, V., Tsang, Y. F., Rinklebe, J., Deep, A., & Brown, R. J. (2021). Advancements of nanotechnologies in crop promotion and soil fertility: Benefits, life cycle assessment, and legislation policies. *Renewable & Sustainable Energy Reviews*, *152*, 111686. DOI: 10.1016/j.rser.2021.111686

Zhang, R., Han, Y., Xie, W., Liu, F., & Chen, S. (2022). Advances in protein-based nanocarriers of bioactive compounds: From microscopic molecular principles to macroscopical structural and functional attributes. *Journal of Agricultural and Food Chemistry*, *70*(21), 6354–6367. DOI: 10.1021/acs.jafc.2c01936 PMID: 35603429

Zhang, W., Zhao, J., & Fang, J.. (2020). Nanotechnology-based strategies for treatment of fungal infections. *Journal of Materials Chemistry. B*, *8*(5), 994–1007.

Zhang, Y., Cao, Y., Jiang, W., Ma, Q., Shin, J., Sun, H., Cui, J., Chen, Y., Giraldo, J. P., Strano, M. S., Lowry, G. V., Sheen, J., & Marelli, B. (2024). Polymeric Nanocarriers Autonomously Cross the Plant Cell Wall and Enable Protein Delivery for Stress Sensing. *Advanced Materials*, *36*(41), 2409356. DOI: 10.1002/adma.202409356 PMID: 39149770

Zhang, Y., Huang, C., & Xiong, R. (2024). Advanced materials for intracellular delivery of plant cells: Strategies, mechanisms and applications. *Materials Science and Engineering R Reports*, *160*, 100821. DOI: 10.1016/j.mser.2024.100821

Zhang, Y., & Zhang, S. (2017). Development of nanoparticles for antifungal delivery. *Nanomedicine; Nanotechnology, Biology, and Medicine*, *13*(7), 2303–2312. PMID: 28712917

Zhao, J., Song, R., Li, H., Zheng, Q., Li, S., Liu, L., Li, X., Bai, L., & Liu, K. (2022). New formulation to accelerate the degradation of pesticide residues: Composite nanoparticles of imidacloprid and 24-epibrassinolide. *ACS Omega*, *7*(33), 29027–29037. DOI: 10.1021/acsomega.2c02820 PMID: 36033692

Zhao, M., Li, P., Zhou, H., Hao, L., Chen, H., & Zhou, X. (2022). pH/redox dual responsive from natural polymer-based nanoparticles for on-demand delivery of pesticides. *Chemical Engineering Journal*, *435*, 134861. DOI: 10.1016/j.cej.2022.134861

Chapter 15
Nanotechnological Green Synthesis and Its Impact on Plant Science

Elif Esra Altuner
https://orcid.org/0000-0001-7663-6898
University of Kocaeli Health and Tecnology, Turkey

Pınar Bozbeyoğlu
University of Gumushane, Turkey

Yener Tekeli
University of Adiyaman, Turkey

Fatih Sen
University of Kutahya Dumlupinar, Turkey

ABSTRACT

Nanotechnology is a multidisciplinary field that manipulates materials at the nanoscale, typically between 1 and 100 nanometers. It bridges various disciplines, including physics, chemistry, biology, engineering, and materials science, creating materials with unique properties that classical theories cannot predict. Nanotechnology is central to advancements in energy, medical therapies, environmental solutions, and electronics. Nanotechnology's ability to create more efficient energy storage systems, such as high-capacity batteries, and its potential to revolutionize healthcare by enabling personalized medicine further demonstrate its vast impact. These innovations offer solutions to pressing issues like clean energy, health care, and ecological preservation. This chapter focuses on green synthesis methods in plant science, highlighting environmentally friendly techniques for producing nanomaterials. These sustainable methods not only contribute to eco-friendly agricultural practices but also promote environmental conservation and sustainability in the field of nanotechnology.

DOI: 10.4018/979-8-3693-9869-2.ch015

1. INTRODUCTION

Nanotechnology can be defined as a multidisciplinary field that involves the design, production and application of substances and materials at the atomic and molecular level, i.e. between 1 and 100 nanometers (Altuner, Bekmezci, et al., 2022; Altuner, Gur, et al., 2021; Bekmezci et al., 2022). Since this technology inherently involves the interaction of many branches of science, it requires an in-depth understanding of fields such as physics (Timuralp et al., 2024), chemistry (Altuner, 2023; Karimi, Altuner, Aygun, et al., 2023), biology (Kocak et al., 2024; Ozdemir et al., 2022), engineering and materials science (Doner et al., 2023; Timuralp et al., 2024). The basis of nanotechnology is the emergence of new physical and chemical properties that classical physics and chemistry can not predict, thanks to this extreme reduction in the size of materials. For example, nanomaterials can exhibit higher surface area and better reactivity than their larger counterparts (Timuralp et al., 2024). These properties offer revolutionary innovations in various applications such as energy storage (Al Janabi et al., 2024; Darabi et al., 2023), water purification (Altuner, Gulbagca, et al., 2023; Darabi et al., 2023), targeted drug delivery systems in medicine (Alizadeh et al., 2022) and innovative electronic components (Cui et al., 2001; Hong et al., 2007; Westermayr et al., 2021; M. Zhang et al., 2021). Therefore, nanotechnology plays a critical role in today's technological development, providing significant opportunities in both basic science research and industrial applications (Dauthal et al., 2015) As can be understood from here, nanotechnology works together with multidisciplinary branches of science.

Its roots can be traced back to Richard Feynman's iconic "There is Plenty of Room at the Bottom" speech in 1959, in which Feynman opened the door to nanotechnology by demonstrating the possibilities of atomic-level structure (Stylios, 2013). By the mid-1980s, the invention of advanced imaging techniques such as scanning tunnelling microscopy (STM) and atomic force microscopy (AFM) provided the ability to observe and manipulate individual atoms and molecules. The 1990s and 2000s marked a significant turning point with the synthesis of nanomaterials (e.g., nanowires, graphene, and nanoparticles) and the discovery of their physical, chemical, and biological properties (Y. Liu et al., 2003). Today, nanotechnology has revolutionary applications in many fields, including electronics, medicine, energy, the environment, food, and materials science (Madou, 2011). Nanobiotechnology, in particular, is gaining significant importance thanks to its innovative approaches to critical health issues, such as early diagnosis of diseases and the development of treatment methods in agriculture and plant science (Gökdemir et al., 2023).

Using plants with nanotechnological methods is essential in increasing productivity and ensuring environmental sustainability and food security (K. K. Verma et al., 2022). One of the most important contributions of nanotechnology to plants is increasing agricultural productivity. Bioavailability can be increased by using nanomaterials in various ways in plants' growth and development processes. For example, unlike traditional fertilizers, nano fertilizers help plants absorb nutrients more efficiently, allowing them to be used more effectively by the plant. Thus, they allow plants to produce more products with less fertilizer and reduce agricultural input costs (Maity et al., 2022). This is vital, especially for communities suffering from food insufficiency in developing countries. Improving nutritional quality through more efficient nutrient intake can positively affect public health. In addition, nanotechnology can increase plants' resistance to diseases. For example, nanomaterials developed using nanotechnological methods can be applied to plants and protect against pathogens (Polat, 2013). The health status of plants and environmental conditions can be monitored using nanosensors so that disease symptoms can be detected at early stages (Pérez-Moreno et al., 2021). This makes disease management in agriculture more effective while reducing chemical pesticide use (A Malysheva, 2015; D Kühnel, 2014; Lombi, 2013; N Sozer, 2009). Since excessive use of chemical pesticides negatively affects ecosystems, developing more environmentally friendly solutions against diseases with nanotechnical methods is essential (A Malysheva, 2015; Gilbertson et al., 2020; Gillings, 2013). One of the most striking applications of nanotechnology in plant use is observed in the field of plant biotechnology. Using nanotechnological methods such as genetic engineering and CRISPR technology, the genetic structures of plants can be changed to obtain more resistant and high-yield varieties (C. Sun et al., 2008). Such applications increase the capacity to cope with challenges such as climate change and environmental pollution while providing a sustainable food production approach. Another critical aspect of plant use of nanotechnology is its contribution to increasing environmental sustainability (Destoumieux-Garzón et al., 2018; Hofmann et al., 2020; McEwen et al., 2018; R Ganesh Pillai, 2017). It is possible to develop nanotechnological solutions in a wide range of areas, from managing water resources to protecting soil health (Ainiwaer et al., 2024; Berkner et al., 2022; Du et al., 2020; Liao et al., 2023). For example, by using nano-filters, it is possible to remove pollutants from water used in agriculture and improve water quality (C. Sun et al., 2008). In addition, cleaning heavy metals in the soil with nanomaterials increases the productivity of agricultural lands and is an essential step for environmentally sound agricultural practices.

Nanoparticles play a crucial role in various fields, from medicine to electronics (Karimi et al., 2024; Thapa et al., 2010; Timuralp et al., 2024). Understanding their classification is essential for harnessing their potential and driving innovation. By organizing nanoparticles into distinct categories based on size, composition, and shape, we can unlock new applications and improve existing technologies. Embracing this knowledge will lead to advancements that benefit society as a whole (Darabi et al., 2024). Nanoparticles have an essential place in nanotechnology and are defined as particles with a size of 1 to 100 nanometers (Altuner, Akin, et al., 2022; Altuner, Arıkan, et al., 2021; Altuner, Bekmezci, et al., 2021). These miniature structures, nanotubes, nanosheets, nanospheres and nanorods, exhibit various physical and chemical properties and offer significant innovations in many industries and branches of science (Maity et al., 2022). Nanotubes are lightweight and highly durable materials formed by the combination of carbon atoms arranged in single or multilayer structures in a cylindrical form (Choi et al., 2010). They have gained an important place, mainly in electronics (Hong et al., 2007), energy (Doner et al., 2023; Murshed et al., 2014), and nanomedicine (Alizadeh et al., 2022). For example, they attract attention in materials science and electronics due to their high strength and lightness. Nanolayers are materials in the form of thin layers and are usually single or a few atoms thick (Correa et al., 2016; Rashtbari et al., 2021). These structures can increase catalysis and energy storage efficiency by increasing the surface area in various applications (Rashtbari et al., 2021). In this way, they are used in surface engineering and coating applications. Nanospheres are particles with a spherical shape and nanometer-sized dimensions; these structures can be effectively used in targeting and drug delivery systems (Kröger et al., 1999; Umegaki et al., 2009). Nanorods, usually cylindrical and manufactured from various materials, offer significant advantages in optical and electronic applications, especially in light absorption (F. Li et al., 2021; X. Liu, 2005). Finally, quantum dots are semiconductor crystals at the nanometer scale and have unique optical properties (Pan et al., 2018; S. Zhu et al., 2015). These properties can be used in many technologies, especially light-emitting diodes (LEDs) (B. G. Kumar et al., 2018), lasers (Calabro et al., 2019; S. C. Singh et al., 2010), and solar cells (Nozik, 2002; Pan et al., 2018). All these nanostructures inspire research in various areas, from basic science to engineering, while laying the foundations for future innovations

2. NANOPARTICLES AND THEIR CLASSIFICATION

When we look at the recent history of nanoparticles, in the 20th century, the famous physicist Richard Feynman, in this speech that laid the foundations of nanotechnology, suggested that structures at the atomic level could be designed (Langton, 2019). Thanks to the electron microscope developed by Ernst Ruska, scientists were able to directly observe nanoscale structures for the first time (Rajender Singh, 2024). Today, nanoparticles are used in many areas such as biomedicine, electronics, energy, the environment, and the defense industry (P Wang, 2016). The future of nanotechnology will go even further by merging with areas such as quantum materials and bio-nanotechnology. Nanoparticles (NPs) are diverse in nature and can be classified into different types based on their morphology, size, and shape. Among these, organic nanoparticles are prominent and encompass various structures, such as ferritin, micelles, dendrimers, and liposomes. Organic nanoparticles are characterized by their non-toxic and biodegradable properties, which make them suitable for a range of biomedical applications. Notably, certain types of organic nanoparticles, like micelles and liposomes, possess a spherical shape that allows them to encapsulate drugs and other compounds effectively (Khalil et al., 2020). These nanoparticles are also recognized for their sensitivity to heat and light, which can be strategically utilized in drug delivery systems. Due to their unique properties and functionalities, organic nanoparticles have gained significant attention in the field of targeted drug delivery, where they can enhance the effectiveness and reduce the side effects of therapeutic agents (figure 1).

Figure 1. Organic and inorganic nanoparticles (Khalil et al., 2020)

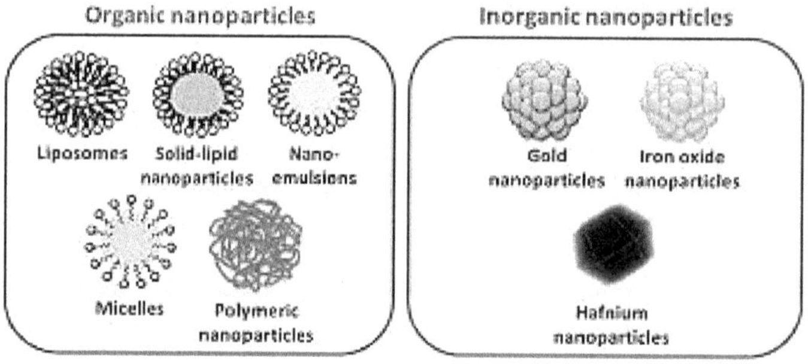

"Reprinted (adapted) with permission from (Khalil et al., 2020) Copyright (2024) IntechOpen- Current and Future Aspects of Nanomedicine"

The most prevalent form of organic nanoparticles is the nanosphere or nanocapsule, which serves as a carrier for various bioactive substances (Guterres et al., 2007). Their ability to deliver drugs directly to specific sites within the body improves treatment outcomes and minimizes systemic exposure. In contrast, inorganic nanoparticles do not contain carbon in their structure (Altuner, Bekmezci, et al., 2021). These nanoparticles are similarly non-toxic and possess unique properties such as biocompatibility and hydrophilicity, making them stable in various environments. Inorganic nanoparticles are generally more robust than their organic counterparts, which allows for their use in a broader range of applications (Boulaiz et al., 2011; Chiozzi et al., 2020). They can be classified into two primary categories: metal nanoparticles, which include gold (Youssef et al., 2014) and silver nanoparticles (Chen et al., 2015), and metal oxide nanoparticles (Reddy, 2017), such as titanium dioxide (TÜRKASLAN et al., 2020) and zinc oxide nanoparticles (Ameen et al., 2023). Each type of inorganic nanoparticle exhibits distinct chemical and physical properties, expanding their potential uses in areas like drug delivery, imaging, and therapeutic applications in medicine.

2.1. Metal-Based NPs

Metal nanoparticles: Metals synthesize pure metal nanoparticles using destructive or constructive methods. Nanoparticles of all metals can be synthesized (Panigrahi et al., 2004). Aluminium (Ash et al., 2002), gold (Qian et al., 2006), iron (N. Kocak et al., 2012; S-J Park, 2000), lead (N. Kocak et al., 2012), silver (Melaiye et al., 2005), cobalt (Darabi et al., 2024; Zare et al., 2024), zinc (Kaur et al., 2021), cadmium (X. Liu, 2005)and copper (Banik et al., 2018)are well-known metal nanoparticles. Nanoparticles exhibit distinctive properties due to their small size (10-100 nm), surface area/volume ratio, surface charge, pore size, surface charge density, structure (crystalline and amorphous), shape (spherical, rod, hexagonal, tetragonal, cylindrical and irregular), colour and environmental factors (sunlight, humidity, air and heat) (Chang et al., 2010; J. Wang et al., 2010).

Metal oxide nanoparticles: Synthesizing metal oxide nanoparticles changes the properties of the relevant metal nanoparticles, such as oxidizing iron nanoparticles to iron oxide nanoparticles (Alizadeh et al., 2022; Ameen et al., 2023). Metal oxide nanoparticles include zinc oxide (Altuner, Gulbagca, et al., 2023; Karimi, Altuner, Gulbagca, et al., 2023; Kolodziejczak-Radzimska et al., 2014), silicon dioxide (Altuner, Ozalp, et al., 2022; Sprenger, 2013), iron (Baker et al., 2004), aluminium (Ash et al., 2002), cerium (Xu et al., 2014), titanium (ALTUNER, 2023), and magnetite (Sheng et al., 2014).

Ceramic nanoparticles: Ceramic nanoparticles also known as non-metallic solids, are synthesized by heating or sequential cooling (C. Gu et al., 2019). Researchers are focusing on these nanoparticles due to their wide applications, such as photodegradation of dyes, photocatalysis (Hojjati-Najafabadi et al., 2021), and visual applications .(Feng et al., 2018)

Biological nanoparticles or bio-nanoparticles: Biological or bio-nanoparticles are aggregations of atoms or molecules prepared in a biological system with at least one dimension in the 1-100 nm range (Altuner, Erduran, et al., 2023; Erduran et al., 2023). All nanoparticles are naturally occurring nanoparticles. These nanoparticles are divided into two categories intracellular and extracellular (Chou et al., 2010; Shivaji et al., 2011). Magnetosomes are examples of intracellular structures, while lipoproteins and viruses are examples of extracellular structures. Magnetosomes (Bazylinski et al., 2004), exosomes (Bagheri Hashkavayi et al., 2020), ferritin (Zhantong Wang et al., 2017), lipoproteins (Ng et al., 2011), and viruses (Destito et al., 2009) are examples of bionanoparticles.

2.2. Carbon-Based NPs

A nanomaterial is a material that has a structural component or unit smaller than a micrometre in at least one dimension. The building blocks of matter (~0.2 nm) are considered nanomaterials (Smith et al., 2005).

Carbon-based nanomaterials: Carbon-based nanomaterials are composed entirely of carbon. Fullerenes, carbon nanotubes (CNTs) (C. Li et al., 2003), graphene (Ozalp et al., 2019), carbon black (Athreya et al., 2010), and carbon nanofibers (B. K. Balan et al., 2012) are classes of carbon-based nanoparticles.

Fullerenes: Fullerenes consist of nanomaterials in the form of hollow cages (Goldshleger, 2007). They have become a significant commercial interest due to their electrical conductivity, electronics, structure, strength, and versatility (Goldshleger, 2007; Shibata et al., 1997).

Carbon nanotubes (CNTs): Carbon Nanotubes are long tubular structures with a 1-2 nm diameter (Göksu et al., 2016). Depending on their diameter, carbon nanotubes can be estimated as semiconducting or metallic. The structure of the CNT resembles a graphite sheet rolled onto itself. Based on rounding, CNTs are classified into single-walled (SWCNTs) (Bavandpour et al., 2021), double-walled (DWCNTs) (Pu et al., 2009) and multi-walled (MWCNTs) (Motsaathebe et al., 2021) (figure 2).

Figure 2. Types of CNTs (Chatzichristos et al., 2022)

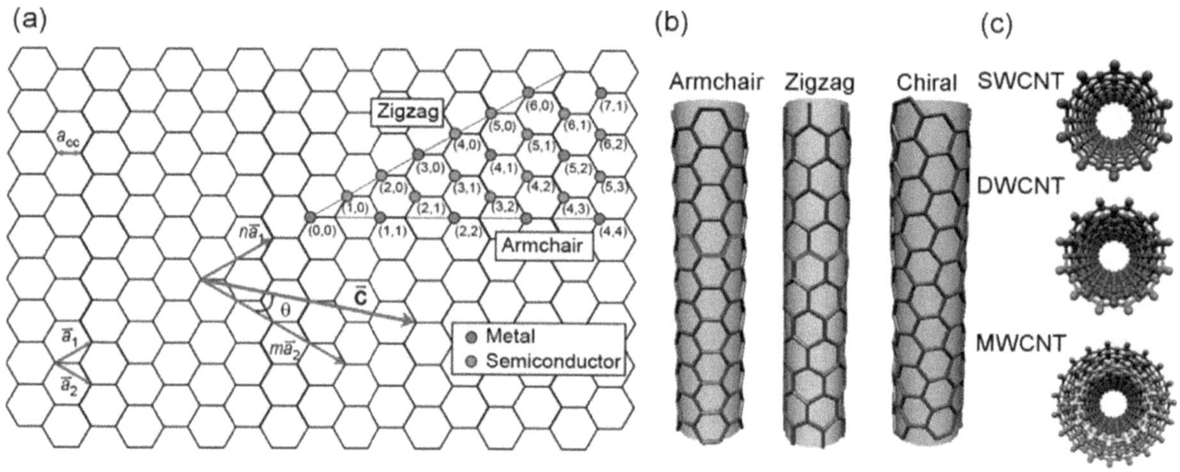

"Reprinted (adapted) with permission from (Chatzichristos et al., 2022) Copyright (2024) MDPI- Nanomaterials"

3. NPs DEPENDING ON SIZE

0-D: These NPs have a structure that meets all three dimensions (x, y, z). Their dimensions are between 1-100 nm (Bolsenga, 1983). They are primarily spherical. Most of the NPs and quantum dots are in 0-D NPs class (Ateia et al., 2024).

1-D: These NPs meet only 2 dimensions (x, y). Their dimensions are larger than 100 nm. They can be crystalline, amorphous or metallic. All nanotubes, nanorods, nanofibers and nanowires fall into this category (Yang et al., 2020).

2-D: These NPs meet one dimension at the nanoscale, while the other two do not fall below the nanoscale. They can be crystalline, amorphous, metallic, polymeric and ceramic. Nanosheets, nanocoatings, graphene and nanowalls are examples of 2-D nanoparticles (Novoselov et al., 2016; Z. Sun et al., 2016).

3-D: These NPs have three arbitrary dimensions beyond the nanoscale and are more significant than 100 nm. They are nanocrystalline in structure and have the properties of NPs at the nanoscale. Graphene nano gels and curved balls are a few examples of 3-D NPs (Ijaz et al., 2020; Saravanan et al., 2022). All dimesions of the shematic illustration is given in figure 3.

Figure 3. The shematic illustration of all dimensions carbon nps (Nikolova et al., 2020)

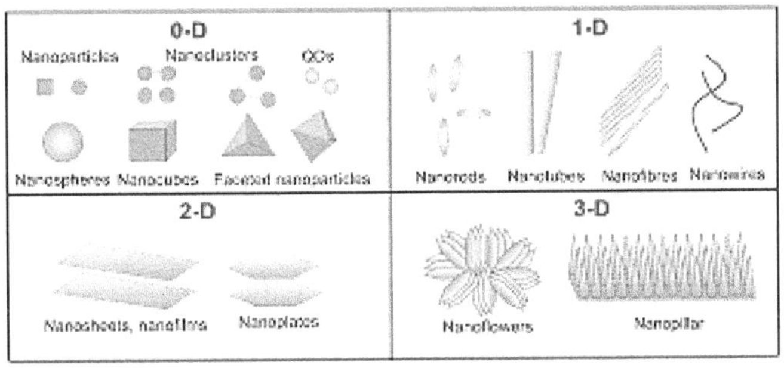

"Reprinted (adapted) with permission from (Nikolova et al., 2020) Copyright (2024) MDPI- Biomimetics"

4. SYNTHESIS OF NANOPARTICLES

Numerous innovative techniques for synthesizing nanoparticles are continually emerging, reflecting the dynamic nature of materials science and nanotechnology (Ijaz et al., 2020). The synthesis of nanoparticles serves as a fundamental cornerstone in these fields, underpinning a diverse array of industrial and scientific applications ranging from drug delivery systems and catalysis to electronics and environmental remediation. Generally, nanoparticle synthesis can be categorized into three primary methods: physical, chemical and biological methods. Physical, chemical, and biological, each with distinct advantages and challenges (Ijaz et al., 2020; Ying et al., 2022a).

The physical methods of synthesis encompass several techniques, including aerosol generation, ultraviolet radiation exposure, and thermal decomposition. These methods rely on physical processes to produce nanoparticles, often requiring extreme conditions. For instance, in the aerosol process, a flame temperature of approximately 2400 °K is necessary to create atomized aerosol droplets, which subsequently form metallic nanoparticles as they cool and solidify. Similarly, the production of palladium oxide nanoparticles (PdO NPs) through plasma-assisted physical vapor deposition demands three thermal cycles, ranging from 250 °C to 800 °C, resulting in substantial energy consumption (Hano et al., 2021)

Chemical methods for nanoparticle synthesis involve a variety of reactions conducted under controlled reactor conditions. Common chemical processes include sol-gel techniques, coacervation, and chemical vapor deposition. While these methods can offer high levels of precision in particle size and morphology, they often utilize multiple chemical species or reactive molecules. This complexity can inadvertently increase the reactivity and toxicity of the resulting nanoparticles, raising potential concerns for human health and environmental safety due to unpredictable compositional outcomes and byproducts (Hussain et al., 2016).

Biological methods present an alternative approach, utilizing microorganisms such as bacteria or fungi, as well as plant extracts, to facilitate the synthesis of nanoparticles. These biological systems can offer a more environmentally friendly option, often occurring under mild conditions and with less toxic byproducts. The ability of living organisms to produce nanoparticles through natural processes adds an innovative dimension to nanomaterials synthesis and highlights the potential for sustainable practices in the field. In summary, while various methods for nanoparticle synthesis exist, each with its own set of characteristics, the ongoing research and development in this area continue to advance our understanding and applications of nanoparticles across multiple disciplines. The choice of synthesis method often depends on the desired properties of the nanoparticles and the specific application in question.

5. THE ROLE OF PLANTS IN NANOPARTICLE SYNTHESIS

Concerns regarding the extensive consumption of energy, the release of hazardous chemicals into the environment, and the reliance on complex equipment and challenging synthesis conditions in traditional methods for producing synthetic nanoparticles are prompting a significant transition toward greener synthesis strategies. These green synthesis methods represent a promising fusion of nanotechnology and green chemistry practices (Archives et al., 2023; F. Khan et al., 2022). One notable advantage of green synthesis is its use of bioactive components that can be extracted from various plant sources. These components not only provide a cost-effective solution for nanoparticle production but are also more environmentally friendly compared to conventional methods. As a result, biological methods have emerged as the most favored approach for synthesizing nanoparticles (Ijaz et al., 2020).

There is a discernible shift in the scientific community toward developing environmentally sustainable synthesis techniques. These methods often incorporate biological agents such as plants, microorganisms, enzymes, and biomolecules, which serve as reducing and stabilizing agents during the nanoparticle synthesis process. This trend reflects a growing recognition of the need for sustainable and eco-friendly practices in nanotechnology (Archives et al., 2023). By harnessing the capabilities of these biological agents, researchers are not only improving the efficacy and efficiency of nanoparticle production but also minimizing the ecological impact of these processes (Prajapati et al., 2024). The fact that plant extracts act as reducers and stabilizers in nanomaterial synthesis increases the effectiveness of this method and appeals to a wide range of applications in various fields, especially in the medical (irem Kaya et al., 2020), electronics (Cui et al., 2001), environmental engineering (Naseer et al., 2020) and agricultural sectors (K. K. Verma et al., 2022). The remarkable biological diversity found in plants, along with the unique properties of their natural substances, enables researchers to manipulate various characteristics of nanoparticles (Hofmann et al., 2020). These characteristics include size, shape, and functionality, which are essential for a wide range of applications in fields such as medicine, electronics, and environmental remediation. Notably, the environmental effects, recycling potential, and biodegradability of nanoparticles derived from plant sources are vital considerations in the quest for sustainable development. By employing green synthesis methods to create these plant-derived nanoparticles, we can effectively safeguard natural resources, offer innovative solutions to pressing challenges, and open new avenues for scientific research and industrial advancements (F. Khan et al., 2022). In the broader context of clean and green technological development, green nanotechnology emerges as a significant subfield within environmental science (Adeyemi et al., 2022). This discipline focuses on developing technologies that are not only efficient but also environmentally friendly. Specifically, environmentally friendly technology is defined as practices that prioritize the protection of the environment and the prevention of hazardous waste generation. This includes concepts drawn from green chemistry and green engineering principles. Green chemistry, in particular, provides an eco-conscious framework for synthesizing products by emphasizing the use of safer, biodegradable materials while minimizing the creation of hazardous byproducts (Dahl et al., 2007). The foundational principles of green chemistry advocate for the utilization of non-toxic, environmentally benign resources that require less energy and are cost-effective (Voukkali et al., 2024). By adhering to these principles, researchers and industries can contribute significantly to reducing environmental impact while fostering innovation and sustainability in various applications (Soni et al., 2022). The concept of environmentally friendly "green chemistry" in the biosynthesis of nanoparticles has been applied to develop clean and environmentally friendly nanoparticles using organisms such as bacteria, fungi, plants, actinomycetes, etc., and this process is called "green synthesis" (Hano et al., 2021) In this context, green synthesis is based on principles such as efficient use of natural resources, integration of renewable energy sources and minimizing the use of harmful chemicals. In addition, using natural components such as biomolecules, herbal extracts, and microorganisms aims to reduce costs and environmental impact (Karpuz et al., 2024). Green synthesis techniques are increasingly used in producing nanomaterials, drugs, catalysts and other chemical products, thus ensuring that more ecological and human health-friendly products are obtained. In this context, green synthesis is considered an essential strategy in chemistry in terms of cost-effectiveness and environmental sustainability (Jadoun et al., 2020; Karpuz et al., 2024).

Indigenous medicinal plants are found all over the world. The World Health Organization (WHO) has approved the use of these medicinal plants in the treatment of various diseases such as Parkinson's disease, Alzheimer's disease, HIV infection and AIDS, COVID-19, diabetes, etc. These plants' place is significant in producing environmentally sustainable, safe and cost-effective medicines. Chemical-reducing agents are used to produce nanoparticles that can be toxic, but green nanoparticle synthesis using non-toxic phytochemical components is a frequently used method in nanomedicine science (Raveendran et al., 2003). Phytomedicine, first practised by the ancient Chinese and Indians (Ayurveda), has become a unique field of science for treating recurrent and conventionally incurable diseases. Phytomedicine development and global marketing began to grow in the twentieth century and are expected to reach five trillion dollars by 2020 (Pal et al., 2015).

Plants play a crucial role in the field of green synthesis by harnessing the power of various microorganisms (Bahrulolum et al., 2021), including fungi (Molnár et al., 2018), bacteria (Gopalakrishnan et al., 2021), and algae (Dahoumane et al., 2017) as well as utilizing extracts from their leaves (Lebaschi et al., 2017), flowers (H. Kumar et al., 2020), roots (Behravan et al., 2019), bark (Pindiga et al., 2022), fruits (Dubey et al., 2010), and seeds (Usmani et al., 2019). These green materials are rich in bioactive compounds such as polyphenols and proteins, which can effectively replace traditional chemical reagents in the synthesis process. This substitution allows for the reduction of metal ions to a lower valence state without the harmful environmental implications often associated with chemical methods. The synthesis of metal nanoparticles using green materials occurs under specific conditions, which include optimal temperature, concentration of reactants, and ambient air quality (Duan et al., 2015). When these conditions are met, metal nanoparticles can be produced with remarkable precision. One notable advantage of green synthesis is that it can yield nanoparticles of superior quality compared to those produced through conventional chemical routes. For example, Fe_3O_4 nanoparticles synthesized through green methods exhibit particle sizes ranging from 2 to 80 nanometers, which is considerably smaller than the 87 to 400 nanometers typically seen with particles synthesized via wet chemical methods (Almasoud et al., 2021; Ying et al., 2022b). This smaller size is particularly important for applications that require high surface area and reactivity. One of the key benefits of green synthesis is its ability to utilize natural resources, such as biological agents, herbal extracts, and microorganisms. The presence of various beneficial bioactive components in these natural materials plays a significant role in controlling not only the size but also the shape of the nanoparticles being produced (Martínez-Ballesta et al., 2018). This level of control is vital, as it determines the suitability of nanoparticles for diverse applications, ranging from medical uses and environmental remediation to advancements in nanotechnology. In this context, plants are often referred to as the "chemical factories" of nature. They provide effective reducing and capping agents, which are essential for stabilizing nanoparticles during synthesis. The notable bioactive compounds involved in this process include phenolic acids, flavonoids, terpenoids, alkaloids, and proteins. These components not only assist in the synthesis but also enhance the functionality of the resulting nanoparticles, making them valuable for various applications across multiple scientific disciplines and industries.

6. APPLICATIONS OF PLANT-DERIVED NANOPARTICLES

Green synthesis is of critical importance for the integration of environmentally friendly and sustainable methods into nanotechnology. Green synthesis of nanoparticles reduces the negative impacts on the environment by minimizing the use of chemical solvents and harmful substances. In this context, nanotechnology applications based on environmentally friendly methods are compatible with the sustainability goals in modern science and contribute to the development of green technologies of the future (A. Sharma et al., 2019). This important trend highlighted in the chapter has potential applications not only at the laboratory scale but also at the industrial scale. The use of nanoparticles in medical, agricultural, and environmental fields has the potential to create a great impact in combating global challenges such as food safety, health, and environmental sustainability (Yaqoob et al., 2020). In particular, the use of plant-derived nanoparticles in food safety offers significant opportunities for increasing productivity in agriculture and providing environmentally friendly production processes. Similarly, biological applications of nanoparticles in the medical field can create revolutionary innovations in areas such as cancer treatment, drug delivery systems, and biosensors. In the environmental field, these nanoparticles can fulfill important tasks such as water purification, pollution prevention, and environmental monitoring, thus contributing to sustainable development. There has been a significant increase in scientific publications in nanotechnology in the last decade. Green synthesized nanomaterials play an essential role in applying nanotechnology to various fields. Nanoparticles (NPs) are in high demand across various sectors due to their diverse applications in electronics (M. Das et al., 2019), environmental science (S. Kalia et al., n.d.), energy (Basri et al., 2010), and biomedicine (A. Sharma et al., 2019). Among them, silver (Ag) and gold (Au) nanoparticles have attracted significant attention for their biological potential (Yaqoob et al., 2020). Plant-derived green NPs are a safer alternative to chemically synthesized ones, generally resulting in fewer side effects for humans. In nanomedicine, these green NPs offer benefits such as antimicrobial, antiparasitic, and anti-inflammatory properties, tailored to specific needs. In agriculture, they enhance precision farming through controlled release of agrochemicals, targeted delivery of biomolecules, and improved plant disease management. Additionally, in food science, green NPs improve processing, storage, and packaging methods, ensuring greater safety (Peralta-Videa et al., 2016). In bioengineering, they act as biocatalysts and biosensors, fostering innovation. Finally, in cosmetics, they are key ingredients in sunscreens, anti-aging products, and nanoemulsions that deliver vital bioactive compounds (Hano et al., 2021). Embracing green NPs is essential for creating safer, effective solutions across multiple industries. NPs synthesized by green methods show perfect antibacterial, antifungal and antiparasitic activities. Gusseme et al. reported that biogenic silver produced using Lactobacillus fermentum was used to remove viruses from drinking water (Soundararajan et al., 2022). Aquapure and QSI-Nano are commercially available for household water treatment systems containing silver as a disinfectant. Magnetotactic bacteria (MTB) NPs have been used for gene delivery, and PEI-associated MTB-NPs have been used to deliver b-galactosidase plasmids both in vitro and in vivo. Co-transfection of siRNA with quantum dots generates fluorescent NPs that monitor nucleic acid delivery and improve transfection rates. NPs, ranging in size from 1 to 100 nm, bind to the HIV-1 virus, preventing it from infecting host cells (Özdemir et al., 2023; Schnell et al., 1997). These nanoparticles can damage bacterial cells and are effective in remediating hazardous waste. Green nanomaterials efficiently treat contaminated water and minimize the need for cleaning chemicals. Iron nanoparticles (FeNPs) play a crucial role in water disinfection and soil remediation, while also serving as environmentally friendly fertilizers that enhance crop production (Zia-ur-Rehman et al., 2023). Additionally, NPs are utilized in solar energy coatings and battery technologies, improving the efficiency of chemical reactions.

6.1. Therapeutic and Medical Applications

Nanomedicine has emerged as a cutting-edge field that offers innovative strategies for understanding the complexities of biological systems, as well as enhancing traditional methods of disease diagnosis and treatment (Maciej Serda et al., 2024). By leveraging advanced technologies, nanomedicine provides valuable tools for exploring medical science at the cellular level, enabling researchers to investigate cellular behaviors and interactions with unprecedented precision (Wong et al., 2013). One significant advancement in nanomedicine is the use of gold nanoparticles (AuNPs), which have garnered considerable attention due to their remarkable properties (M. S. Khan et al., 2013). These nanoparticles are biocompatible and non-toxic, making them ideal candidates for medical applications (A. Gautam et al., 2013). Furthermore, gold nanoparticles exhibit excellent optical absorbance characteristics, allowing for their use in various diagnostic imaging techniques and targeted drug delivery systems. Researchers believe that the unique properties of AuNPs can revolutionize the way diseases are diagnosed and treated, potentially leading to more effective therapies and improved patient outcomes (A. Gautam et al., 2013). Figure 8 shows an illustration of the use of nanoparticles synthesized from natural sources in therapeutic and medical applications.

6.1.1. Anticancer Properties

Nanomedicine is a groundbreaking field that harnesses the power of nanotechnology to revolutionize the treatment, screening, and diagnosis of numerous diseases, particularly cancer (Shastri et al., 2024). This cutting-edge approach seeks to deliver proactive and empowering solutions that encompass a comprehensive array of procedures (van Droffelaar et al., 2018). These include the early prediction of disease onset, precise diagnosis of cancer, and the creation of effective prevention strategies. Furthermore, nanomedicine focuses on the development of targeted medications that can attack cancerous cells while minimizing damage to healthy tissue, enhancing the effectiveness of treatments and improving patient outcomes (Hejmady et al., 2020). Through its innovative techniques and applications, nanomedicine is poised to significantly transform healthcare and improve the way we combat various diseases. One of the most groundbreaking aspects of nanomedicine is its potential in cancer prevention, which is an area of active research (Pei et al., 2022). Numerous studies have demonstrated that plant-derived NPs exhibit notable anti-cancer properties. For instance, zinc oxide nanoparticles (ZnONPs) synthesized from the leaf extract of Cassia auriculata have shown promise in therapeutic applications (Asif et al., 2023). These nanoparticles were found to have no adverse effects on normal human breast cells (MCF-12A), suggesting their safety in non-cancerous tissues. In contrast, they exhibit significant tumoricidal activity against

MCF-7 breast cancer cells, illustrating their potential as a selective treatment for malignant growths (J. Lee et al., 2014). Research has demonstrated that green AuNPs, synthesized from the seed extract of the plant Trachyspermum ammi, exhibit significant anti-cancer properties (Sargazi et al., 2022). These nanoparticles have been found to effectively inhibit cellular growth in HepG2 liver cancer cell lines, suggesting their potential as a therapeutic agent in the treatment of liver cancer. The mechanisms by which these nanoparticles exert their inhibitory effects may involve the induction of apoptosis or disruption of cellular metabolic processes, warranting further investigation into their efficacy and mechanisms of action in cancer therapy (Gupta, 2013). This inhibition occurs in a concentration-dependent manner, meaning that higher concentrations of the nanoparticles lead to more significant reductions in cell growth (Soenen et al., 2009). The mechanism behind this anticancer effect is associated with apoptosis, a form of programmed cell death, which is instigated by reactive oxygen species (ROS) produced by the nanoparticles (Mohammadinejad et al., 2019). The role of ROS is critical, as it may influence mitochondrial functions within the cells. Specifically, the induction of apoptosis may be linked to ROS-driven gene expression changes, particularly involving the activation of the Caspase-3 pathway, which is vital for the apoptotic process. Furthermore, the disruption of mitochondrial membrane potential caused by these plant-based nanoparticles exacerbates this apoptotic response, paving the way for novel approaches in cancer treatment and prevention. Research in this area continues to evolve, highlighting the significant potential of utilizing plant-derived NPs as safe and effective agents for cancer prevention, thereby opening new avenues for therapeutic interventions (Hano et al., 2021). Gurunatahan et al. explored the therapeutic potential of AgNPs in breast cancer treatment (Gurunathan et al., 2013). They developed these nanoparticles using the culture supernatant from Bacillus funiculus and investigated their cytotoxic effects on MDA-MB-231 breast cancer cells, recognized for their aggressive nature. This research paves the way for innovative cancer therapies that could significantly improve patient outcomes. The researchers treated the breast cancer cells with varying concentrations of the synthesized AgNPs, specifically ranging from 5 to 25 µg/mL, over a 24-hour period (Mohanta et al., 2022). Their findings revealed that the AgNPs effectively inhibited cell growth in a dose-dependent manner. This inhibition was associated with several biochemical changes in the cells, including the activation of lactate dehydrogenase (LDH) and caspase-3, both of which are markers indicative of cell damage and apoptosis. Additionally, they observed an increase in the production of ROS, which plays a significant role in inducing oxidative stress and triggering cell death. In a related study, Firdhouse et al. examined the cytotoxic effects of nanosilver on prostate cancer cells (Firdhouse et al., 2013). The researchers synthesized AgNPs using an aqueous extract of Alternanthera sessilis, which is recognized for its medicinal properties (Firdhouse et al., 2013; Nadaf et al., 2022). Their findings showed that the biosynthesized AgNPs enhanced antiproliferative activity and significantly decreased the viability of prostate cancer cells (Nadaf et al., 2022). Notably, morphological assessments revealed several changes in the cancer cells, including shrinkage, oxidative stress responses, and curling, all of which are indicative of impending cell death. Apoptosis was remarkably high, with a reported 95% occurrence at a concentration of 25 µg/mL for prostate cancer cells (Firdhouse et al.) In contrast, the study found that exposure to AgNPs resulted in complete growth inhibition of breast cancer cells in the examined conditions. Beyond the scope of cancer treatment, the medicinal properties of Origanum vulgare (commonly known as oregano) are well-documented. According to some researchers, oregano (Origanum vulgare) has long been recognized for its medicinal properties and is used in the treatment of a variety of health conditions (Ghosh et al., 2022). These include respiratory disorders such as asthma and bronchitis, where oregano's antimicrobial and anti-inflammatory properties can provide relief. Additionally, it is effective in managing indigestion and has been shown to alleviate symptoms associated with gastrointestinal discomfort. Oregano is also beneficial for individuals suffering from rheumatoid arthritis due to its potential to reduce inflammation and pain in affected joints (Greff et al., 2023a). In terms of urinary tract issues, the herb's antibacterial properties can aid in combating urinary tract infections. Furthermore, oregano has been utilized in treating various dental problems, including gingivitis and oral infections. These insights emphasize the significance of not only investigating novel treatments, such as AgNPs, but also revisiting and incorporating traditional remedies like oregano into healthcare practices to develop comprehensive and effective solutions for improving health and wellbeing.

6.1.2. Anti-Diabetic Treatment

Kannan et al. conducted a study in which they biosynthesized AgNPs using the aqueous leaf extract of Lonicera japonica (K. Balan et al., 2016). The primary aim of this research was to explore the anti-diabetic properties of these nanoparticles. The findings showed that the AgNPs exhibited strong inhibitory effects on two critical carbohydrate digestive enzymes: α-amylase and α-glucosidase (Wahab et al., 2023). The inhibitory potency was quantified using half-maximal inhibitory concentration (IC50) values, which were calculated to be 54.56 μg mL-1 for α-amylase and 37.86 μg mL-1 for α-glucosidase. These results suggest that AgNPs have significant potential in managing diabetes by effectively reducing carbohydrate digestion, indicating their promise as a form of nanomedicine for diabetic treatment. In a compelling study by Prabhu et al. (2018), AgNPs derived from the aqueous leaf extract of Pouteria sapota were rigorously evaluated for their anti-diabetic properties, both in vitro and in vivo (Greff et al., 2023b). The in vivo tests employed a rat model with diabetes induced by streptozotocin, showcasing the potential of these nanoparticles in combating diabetes effectively. The results were notable, as a significant decrease in blood sugar levels was observed among the treated rats, indicating that the AgNPs could play a role in glucose regulation and management of diabetes (Prabhu et al., 2018). Additionally, AgNPs derived from the leaves of Punica granatum were examined for their effects not only on diabetes-related enzymes but also on cancer cell viability (Greff et al., 2023b; (Prabhu et al., 2018). This study highlighted the dual function of these nanoparticles, showcasing their potent inhibition of α-amylase and α-glucosidase while demonstrating high effectiveness in promoting cell death in cancer cells. Ying Guo et al. investigated the anti-diabetic effects of AuNPs derived from the herb Fritillaria cirrhosa (Y. Guo et al., 2020). When tested on diabetic rats induced by streptozotocin, the AuNPs demonstrated a significant activation of pancreatic islet cells, highlighting their potential to enhance insulin production and secretion. Moreover, Venkatachalam et al. focused on isolating AuNPs from the plant Cassia auriculata to explore their anti-diabetic effects in experimental rats (Venkatachalam et al., 2013). In this study, the AuNPs were administered orally, and the results revealed a decrease in glucose levels in rats that had been induced with diabetes via alloxan. Specifically, in Group IV, where the AuNPs were injected directly into diabetic rats, researchers speculated that the observed reduction in glucose levels could be due to several mechanisms: the nanoparticles may interact with peripheral tissues to inhibit hepatic gluconeogenesis, enhance glucose uptake in muscle and fat tissues, or stimulate the regeneration of remaining beta cells in the pancreas (Y. Guo et al., 2020). This multifaceted approach suggests a promising avenue for utilizing nanoparticles in diabetic management and treatment strategies.

6.1.3. Neuroregenerative Treatment

Neurodegenerative diseases, such as stroke and Parkinson's disease, have a significant impact on neuronal cells that are essential for maintaining proper neurological function. Understanding and addressing these conditions is vital for promoting brain health and enhancing overall well-being (Bhattacharya et al., 2022). Neurodegenerative diseases, such as stroke, Parkinson's disease, and injuries to the spinal cord or brain, have a profound impact on neuronal cells, which are critical for proper neurological function (Jacob et al., 2018; Langer et al., 2003; Numata et al., 2010). These conditions lead to significant disruptions in the communication pathways within the nervous system, ultimately resulting in a decline in cognitive and motor abilities. For instance, a stroke can cause sudden damage to brain tissue due to interrupted blood flow, which may result in difficulties with speech, movement, and coordination (Pollak et al., 2002; Tysnes et al., 2017). Parkinson's disease, characterized by the progressive loss of dopamine-producing neurons, often leads to tremors, stiffness, and balance problems, affecting a person's ability to carry out everyday tasks (Freed et al., 2001; Nellore et al., 2013; Pezzella et al., 1997). Additionally, spinal cord injuries can impede the transmission of signals between the brain and the rest of the body, resulting in paralysis or severe motor impairment (Kishima et al., 2004). The consequences of these diseases extend beyond physical limitations; they can also impact a person's emotional well-being and quality of life, making it challenging to engage in social interactions, work, or maintain independence. Overall, these neurodegenerative conditions impose a heavy burden on individuals and their families, highlighting the urgent need for research and effective treatments (Olasehinde et al., 2017; Tiwari et al., 2018). To address these challenges, regenerative therapy is essential, as it focuses on slowing the progression of neuronal degeneration. This approach aims to preserve neuronal activity and functionality, which encompasses not only the survival of existing neurons but also the stimulation of growth and repair processes (Agatonovic-Kustrin et al., 2018; Goldsmith et al., 2014; D. Y. Lee et al., 2017; Nischwitz et al., 2008; Page et al., 2021; Silva, 2006). By enhancing the growth and maintenance of neuronal cells, regenerative therapies can help restore lost connections, foster neuroprotection, and ultimately improve the quality of life for individuals affected by these debilitating diseases (Bensadoun et al., 2003; Cheng et al., 2010). AgNPs with a size of 30 nm were biosynthesized from bacterial E.coli to treat human neuroblastoma cells (SH-SY5Y). It has been reported that AgNPs increase neurite length and improve different neuronal markers such as β-tubulin III, neurogenin-1, synaptophysin, Drd-2, Gap-43 and Map2 expression (Cheng et al., 2010).

6.1.4. Antimicrobial Activity

Antibiotic resistance has become one of the most pressing public health challenges of the last few years and is anticipated to escalate further if not addressed (Charbgoo et al., 2017; Raula et al., 2007). This phenomenon occurs when bacteria adapt and develop the ability to withstand the effects of antimicrobial agents that were previously effective in treating infections (Y. Gu et al., 2012; Na et al., 2021; Q. Zhang et al., 2014). The rapid evolution of bacterial genomes plays a pivotal role in this resistance, allowing for quick adaptation to various environmental pressures, including exposure to antibiotics.

In light of this growing crisis, researchers are exploring innovative approaches to develop new treatments. One promising avenue is the use of biogenic NPs, which are engineered from natural biological materials (Fayaz et al., 2010; Y. Zhu, 2017). These nanoparticles have demonstrated encouraging results in targeting and inhibiting multidrug-resistant bacteria. Their unique properties, such as improved ability to penetrate bacterial cells and their potential to disrupt bacterial mechanisms, make them a compelling option in combatting resistant pathogens (Francino, 2016; Jacobs et al., 2013). Continued research into biogenic NPs could lead to more effective therapies that provide new hope in the fight against antibiotic resistance and enhance our ability to manage infectious diseases (El-Belely et al., 2021; Göksu et al., 2020; Mahalakshmi et al., 2020). Many antibiotics used in clinical practice to combat microbial infections caused by various pathogens have limitations in their efficacy and application. These limitations often arise from the development of antibiotic resistance and the potential for adverse side effects, which constrain the dosages that can be safely administered to patients. These new treatments aim to minimize the risks associated with traditional antibiotics while effectively targeting pathogens (Raula et al., 2007). In this regard, AgNPs have demonstrated remarkable effectiveness, consistently highlighted by numerous researchers (Sathishkumar et al., 2016). NPs enhance antimicrobial activity when combined with organic and inorganic compounds. Silver is well-known for its antibacterial properties, with biogenic silver nanoparticles being 20 times more effective than chemically produced ones. Plant extracts are selected for this purpose based on their biological value [125]. For example, green silver nanoparticles from Carissa carandas leaf extract effectively inhibit Shigella flexneri, while reduced graphene oxide-coated palladium-silver nanostructures from stevia leaf extract target Escherichia coli (Reetika Singh et al., 2021). Copper oxide nanoparticles from Caesalpinia bonducella extract also display significant antibacterial activity due to reactive oxygen species (ROS) production and their ability to disrupt bacterial membranes (Letchumanan et al., 2021). Iron nanoparticles were tested against various bacteria, showing strong antibacterial effects. Combining these nanoparticles with plant extracts enhances their ability to damage microbial cell walls. Additionally, AgNPs from Sisymbrium irio in Saudi Arabia effectively inhibit multidrug-resistant Pseudomonas aeruginosa and Acinetobacter baumannii (Parashar et al., 2021). Moreover, the antifungal activity of nano xylan obtained from corn cob xylan was identified against Candida albicans, Candida parapsilosis and Cryptococcus neoformans, whereas AgNPs obtained from Clerodendrum inerme leaf extract showed dual antibacterial and antifungal activity against a wide range of human pathogenic strains (Kar et al., 2021). Interestingly, AuNPs produced from the same C. inerme extract also showed similar inhibitory capacity. The authors concluded that these NPs may have improved antimicrobial Activity due to the synergistic effect of bioactive absorbed phytochemicals from this plant. The anti-biofilm effect of AuNPs produced from T. and seed extract was also observed against Listeria monocytogenes and Serratia marcescens, most probably due to intracellular ROS production (Tungare et al., 2024). ZnONPs also showed potential antimicrobial activity as evidenced by the Activity of ZnONPs derived from Cinnamomum verum bark extract against E. coli and Staphylococcus aureus (P. Das et al., 2021). ZnO nanoparticles (ZnONPs) from C. auriculata effectively disrupt bacterial cell integrity (Kaushal et al., 2024). Copper oxide nanoparticles (CuONPs) from Cymbopogon citratus have strong antimicrobial and antibiofilm properties, with variations linked to bacterial cell wall differences. Manganese oxide nanoparticles (MnONPs) from Abutilon indicum target both Gram-negative and Gram-positive bacteria, while nickel oxide nanoparticles (NiONPs) from stevia are more effective against Gram-negative bacteria (Kaushal et al., 2024). The antimicrobial effects depend on the nanoparticle type and the plant extract used.

6.2. Wound Healing Applications

NPs have gained significant attention in the medical field for their therapeutic properties, especially due to their effective antibacterial and antifungal activities (Tian et al., 2007). Among these, silver nanoparticles (AgNPs) stand out for their notable effectiveness in promoting wound healing (Bardania et al., 2020). In a detailed study conducted by Shankar in 2015, the researchers explored the wound-healing potential of AgNPs specifically in the context of burn injuries (Shankar et al., 2015). This investigation was particularly innovative as it involved the utilization of the peel from the Lansium domesticum fruit, a natural source that may enhance the healing effects of the nanoparticles (Abdallah et al., 2022). The findings of this study contribute to a deeper understanding of how AgNPs can be applied in clinical settings to improve the healing process for burn-related wounds, highlighting the importance of both the nanoparticles and the natural compounds derived from the fruit peel. To assess the effectiveness of healing, Ag NPs were successfully integrated into Pluronic F127 gels. The findings revealed that a 0.1% w/w AgNP injection significantly enhanced the healing activity and demonstrated even greater efficacy when combined with hydroxyproline (Zhoufeng Wang et al., 2012). Furthermore, it has been reported that the tensile strength of wounds treated in this manner, as well as the time required for closure, were significantly higher than those in other treatment groups. The observed improvement in wound healing was closely associated with an increase in collagen production, a protein that plays a crucial role in the wound-healing process by providing structure and support to the healing tissue. In their study, Shankar et al. presented compelling evidence supporting this correlation (Kaushal et al., 2024) They demonstrated that with the application of the treatment, the diameter of the wounds reduced progressively over time, indicating accelerated healing. Furthermore, research conducted by Arunachalam in 2013 highlighted the significant effectiveness of AgNPs synthesized from the Indigofera aspalathoides plant (Narayanan et al., 2021). This study detailed the bioactivity of these nanoparticles, which not only contributed to faster wound closure but also exhibited antimicrobial properties, enhancing the overall healing process and preventing infection (Kaushal et al., 2024). The findings suggest that the combination of increased collagen production and the unique properties of silver nanoparticles could play a vital role in future wound management strategies (Narayanan et al., 2021). These nanoparticles were formulated into a preservative-free hydrogel dressing, which proved to be beneficial for a variety of wounds, including cuts, fresh burns, and dry wounds, showcasing their versatility and potential application in wound care. NPs made from Ag that have been synthesized using various plant materials, including Orchidantha chinensis, Euphorbia milii leaf extract, Cratoxylum formosum, Phoebe lanceolate, Scurrula parasitic, Ceratostigma minus, Mucuna birdwoodiana, Myrsine africana, and Lindera strychnifolia, have demonstrated significant effectiveness in promoting wound healing (Kaushal et al., 2024). Additionally, the herbal tea derived from Stachys lavandulifolia has shown promise in this regard. Beyond plant sources, Ag and Au nanoparticles produced from actinobacterial metabolites have also exhibited notable benefits in facilitating the wound healing process. Research indicates that ointments containing silver nanoparticles (AgNPs) and combinations of silver and gold nanoparticles (Ag/AuNPs) at a concentration of 10% can achieve remarkable outcomes, healing nearly 100% of excision wounds within a period of 19 to 21 days (Pormohammad et al., 2021). This performance stands in stark contrast to other commonly used standard ointments, highlighting the potential of these green nanoparticles. Their application in clinical settings could lead to scarless wound healing, representing a significant advancement in wound care.

6.3. Biosensor Applications

Biosensors are sophisticated devices designed to sense and measure various substances by leveraging biological interactions (Ahovan et al., 2020; Kavita V, 2017). These devices operate by detecting specific biological responses, such as enzyme activity or antibody-antigen binding, which occur when a target analyte is present (Asal et al., 2018; Songa et al., 2016). The biosensor then converts these biological interactions into a quantitative or qualitative format through processes known as transduction and electromechanical interpretation (S. Liu et al., 2020; Sahin et al., 2013). The ability to analyze and study these small biological interactions with high precision is crucial in several fields (figure 4).

Figure 4. Synthesized NPs form natural resources and applications for biosensors (Durmaz et al., 2023)

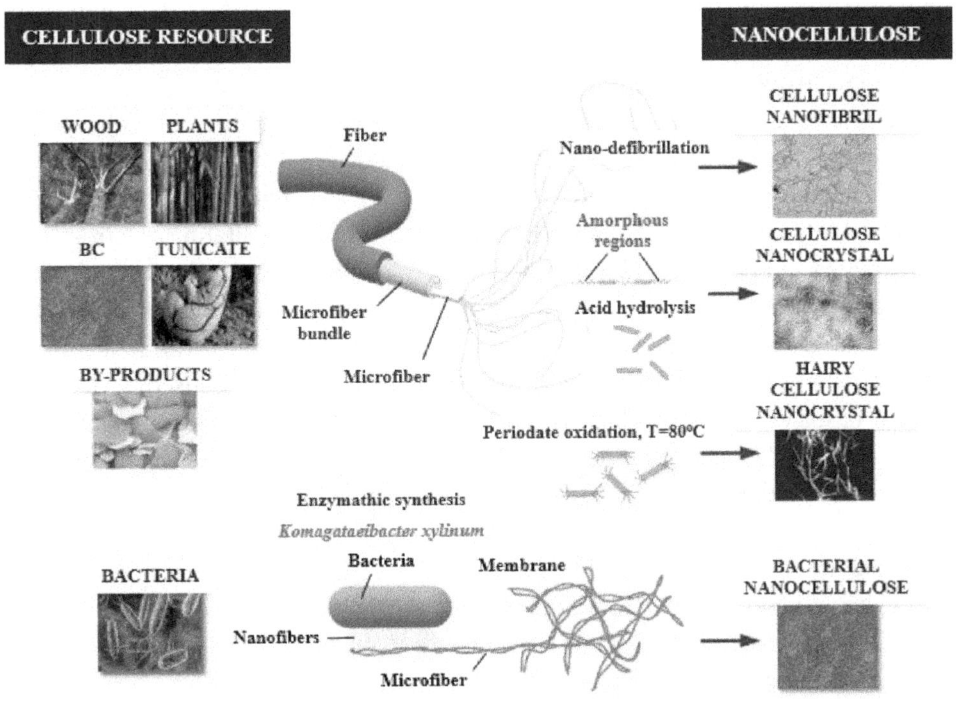

In medical diagnostics, for instance, biosensors play a vital role in identifying biomarkers for diseases, enabling early detection and monitoring of health conditions (Hernandez et al., 2012; Ozalp et al., 2019). In the food industry, they are employed to ensure food safety and quality by detecting contaminants, pathogens, or spoilage indicators (Zhou et al., 2020). Additionally, biosensors are essential tools in environmental monitoring, where they help assess pollution levels and the health of ecosystems by detecting toxins or harmful microorganisms (Ahovan et al., 2020). Overall, the development and implementation of biosensors significantly enhance our ability to monitor biological processes, contributing to advancements in healthcare, food safety, and environmental conservation. Pandey et al. developed room-temperature AgNPs from guar gum, leading to a new optical ammonia sensor (Pandey et al., 2013). This innovative sensor is capable of diagnosing ammonia levels in various biological fluids, including plasma, sweat, and saliva, paving the way for crucial biomedical applications in humans (Antony, 2024). Furthermore, the researchers also synthesized AuNPs from guar gum to enhance the same optical ammonia sensor (L. Sharma et al., 2022). This advancement showcases remarkable reproducibility and sensitivity, achieving an impressive detection limit of just 1 part per billion, making it an invaluable tool for medical diagnostics. Chandrakant et al. developed an optical fibre-based sensor to monitor peroxide concentration as low as 0.01 mM using AgNPs synthesized from locust bean gum (Tagad et al., 2013). In a study conducted by Gayda et al. extracellular metabolites derived from the yeast Ogataea polymorpha were employed to produce a variety of environmentally friendly NPs (Gayda et al., 2019). Among the different types synthesized, the palladium-based green nanoparticles (Pd-based GNPs) emerged as the most effective candidates for the development of amperometric biosensors (Raj et al., 2019). These innovative biosensors have significant potential applications as carriers for enzymes, enhancing their functionality in a range of biotechnological contexts, including biosensors and biofuel cells. Furthermore, Ghodake et al. provided valuable insights into the use of AgNPs coated with gallic acid (Y. K. Gautam et al., 2021). They showed that gallic acid-coated silver nanoparticles could serve as a nanoprobe for the colorimetric detection of aminoglycosides [249], (Ebralidze et al., 2019). This method highlights the versatility of NPs in biological sensing applications, showcasing their ability to produce a color change that indicates the presence of specific compounds, thereby offering a promising tool for the detection of important pharmacological agents (Qin et al., 2022; Wu et al., 2023). The amine groups found in antimicrobial growth promoters (AMGs) play a crucial role as molecular bridges, facilitating electrostatic binding interactions between adjacent particles. This unique interaction contributes to the formation of aggregates of AgNPs, which are increasingly significant in various applications. In a notable advancement in water purification, Ghodake et al. introduced a revolutionary method designed to eliminate bacterial contaminants from water sources (V. N. Ghodake et al., 2018). Their research focused on developing highly effective cellulose-based water filters. These filters were constructed using a sophisticated layer-by-layer (LBL) assembly technique, which combined the filtering fabric of polypropylene and polyethylene (PP/PE) with Ag NPs (G. Ghodake et al., 2020; V. N. Ghodake et al., 2018). This ingenious design aimed not only to enhance the structural integrity of the filters but also to improve their efficiency in removing bacteria during water treatment processes. The findings of this study were impressive, demonstrating that PP/PE fabrics, when embedded with a higher density of AgNPs, exhibited extraordinary antimicrobial properties (G. Ghodake et al., 2020). Specifically, these enhanced materials significantly inhibited the growth of pathogenic bacteria, achieving a remarkable reduction of 96.7% for Escherichia coli and an astounding 97.9% for Staphylococcus aureus (Ghodake et al., 2018). Such results underscore the potential of these innovative water filters in providing safer drinking water by effectively combating harmful bacterial contaminants. These significant reductions highlight the potential of using engineered materials to combat bacterial contamination in water sources. Furthermore, the researchers produced paper filters through a similar LBL assembly of cellulose filters combined with AgNP-PP/PE fabrics. These paper filters demonstrated impressive inactivation rates for the same bacterial strains, achieving efficiencies of 99.4% against E. coli and 98.7% against S. Aureus (Haider et al., 2018). The promising results of this method suggest that it could play a vital role in preventing waterborne diseases, offering a practical solution for improving water quality in various settings, including developing regions where access to clean water is a persistent challenge (Haider et al., 2018; Hano et al., 2021).

6.4. Bioimaging Applications

Nanotechnology plays a crucial role in enhancing the imaging techniques used to study biological systems at the cellular and molecular levels (Bieberstein et al., 2013). In a notable study by Mukherjee et al., colloidal AgNPs were synthesized using a natural extract from the leaves of the Olax scandens plant (Mukherjee et al., 2014). This innovative approach highlights the potential of green chemistry in the production of nanoparticles (Dahl et al., 2007). The researchers conducted a study that revealed green-synthesized AgNPs exhibit a unique red fluorescence when introduced to B16F10 melanoma cells (Gao et al., 2020). This property makes them particularly promising for applications in bioimaging, as it allows for the visualization of cancer cells in a non-invasive manner. The underlying mechanism for the observed fluorescence is linked to the presence of specific phytochemicals and proteins that are extracted from the leaves of the Olax scandens plant using a methanolic solvent. These fluorescent components interact with the AgNPs, enhancing their visibility under appropriate imaging conditions (Mukherjee et al., 2014). This innovative approach not only highlights the potential of using natural extracts to create functional nanoparticles but also opens new avenues for improving cancer detection and monitoring. These compounds interact with the silver nanoparticles, resulting in a significant enhancement of their fluorescence within cellular environments. This interaction allows for precise visualization of cellular processes and structures, which is invaluable for researchers studying cell biology (Gao et al., 2020). Additionally, another advancement in bioimaging has come from the synthesis of silver sulfide NPs from the mycelium of the edible mushroom Pleurotus ostreatus (A. Kalia et al., 2018). These nanoparticles possess an average size ranging from 10 to 15 nm and have been reported to exhibit intense luminescent properties. The unique luminescence of these silver sulfide NPs suggests they could serve as promising new fluorescent probes in various cell biology studies. Researchers are exploring the potential applications of these nanoparticles in imaging techniques, which could significantly improve our understanding of cellular dynamics and interactions. Collectively, these findings from the researches underscore the exciting prospects of using nanotechnology to develop advanced imaging tools for biological studies, paving the way for innovative methodologies in scientific research (Gokdemir et al., 2022).

6.5. Water Treatment Applications

Access to clean and safe drinking water remains a pressing challenge, particularly in developing countries where infrastructure and resources are often limited (Oturan et al., 2014). In these regions, conventional water treatment methods frequently rely on the use of harsh chemicals, which can have detrimental effects on both human health and the environment (Presentato et al., 2020; Tijing et al., 2020). These techniques not only pose risks of pollution but also may lead to the formation of additional harmful byproducts in the water supply [264]. In light of these concerns, researchers are increasingly exploring nanomaterial-based (NT-based) adsorbents and catalysts as effective and environmentally friendly alternatives to traditional chemical treatments (Afshan et al., 2023). These innovative materials offer enhanced performance due to their large surface area and unique properties, making them suitable for various applications in water purification. One particularly promising application involves the use of AgNPs, which has gained recognition from the World Health Organization (WHO) for its efficacy in water treatment (Afshan et al., 2023; Zahoor et al., 2021). AgNPs exhibit antibacterial properties and can effectively eliminate pathogens from drinking water, making them a valuable tool in improving water quality. A notable study conducted by Luo et al. presented a novel method for synthesizing AgNPs using the biopolymer xylan (Luo et al., 2014). This natural polymer presents a sustainable approach to nanoparticle production, minimizing the reliance on toxic chemicals typically used in synthesis processes. The researchers reported that the xylan-AgNPs demonstrated a remarkable capability to detect mercury ions (Hg^{2+}) in water samples [266], (Luo et al., 2014). Their findings revealed a detection limit of just 4.6 nanometers for Hg^{2+}, which is significantly lower than the maximum allowable concentration for drinking water set by the WHO, which is 30 nm (Afshan et al., 2023). This advancement not only provides a reliable means of detecting harmful heavy metals such as mercury in water supplies but also highlights the potential of biopolymer-based materials in the creation of safer and more sustainable water treatment technologies. As the global demand for clean drinking water continues to rise, the application of NT-based solutions like xylan-Ag NPs represents a vital step toward improving water safety and public health in vulnerable communities (Luo et al., 2014). By fabrication method, some researchers produced Au NPs with an average size of 16 nm using pineapple (Rajam et al., 2019). They reported its antibacterial activity against some gram-positive and gram-negative pathogens present in water, thereby inhibiting the growth of bacteria (Tavares et al., 2020). Various green nanocomposites have emerged as a new solution for water purification. Khodadadi et al. biosynthesized AgNPs from Achillea millefolium L. extract and waste peach seed shell for the reduction of substances such as 4-nitrophenol (4-NP), Methyl Orange (MO) and Methylene Blue (MB) in water (Rocha et al., 2024). AgNPs synthesized from Vaccinium macrocarpon fruit extract and Ag NP/clinoptilolite were also used for similar purposes (Khodadadi et al., 2017). In addition, copper/graphene composites and manganese dioxide/oxide materials, derived from the leaf extract of Cuscuta reflexa, were also investigated for their ability to facilitate similar reduction processes. These materials demonstrate potential as effective catalysts due to their favorable structural characteristics and high reactivity. Moreover, a thorough study was conducted on the catalytic reduction of nitroarenes and various organic dyes in aqueous solutions using copper nanoparticles (CuNPs) supported on sodium borosilicate glass, which was obtained from Acalypha indica L. leaf extract (Kaiser et al., 2015). The results indicated that this particular catalyst not only proved efficient in reducing nitroarenes and dyes but also exhibited remarkable stability. Importantly, the catalyst could be recycled several times without experiencing a significant decline in its catalytic activity, making it a promising option for sustainable and environmentally friendly remediation processes. NPs play a crucial role in combating harmful microorganisms by effectively modifying or coating surfaces with antimicrobial agents. It is essential for antimicrobial coatings to exhibit strong antibacterial properties while ensuring low toxicity to humans and the environment. This compelling combination is why green nanoparticles are increasingly being chosen for water treatment solutions (Ameta et al., 2021).

6.6. Agricultural Applications

Agriculture is an integral pillar of developing economies, providing livelihoods for more than 50% of the population (Rosegrant et al., 2005). Its importance cannot be overstated, as it not only supports the economic framework but also contributes to food security and community well-being. However, the agricultural sector is currently grappling with a myriad of pressing challenges. One critical issue is the sustainable use of natural resources, which are becoming increasingly strained under the pressures of overexploitation and degradation (Amekawa, 2011). Climate change presents a significant threat, altering weather patterns and affecting crop yields, while urbanization is consuming fertile land, reducing the area available for agriculture (Pawlak et al., 2020). Moreover, the excessive flow and accumulation of pesticides and fertilizers are harming ecosystems and compromising food safety. As the global population continues to grow, the demand for food is rising sharply, placing additional strain on already limited agricultural resources (Pawlak et al., 2020). To tackle these challenges, it is imperative that we implement innovative practices and policies that prioritize sustainability, maximize productivity, and ensure equitable access to resources. Addressing these multifaceted issues is essential for the resilience and future prosperity of the agricultural sector (Ditta, 2012). Nanofertilizers mainly include nitrogen fertilizers, potassium fertilizers, nanoporous zeolite, zinc nanofertilizers, nanoherbicides and nanopesticides. In some studies, ZnONPs have shown broad agricultural interest due to the anti-phytopathogenic activity of ZnONPs obtained from lemon fruit against soft rot bacterial pathogen Dickeya dadantii (Kanakari et al., 2023). In addition, it has been reported that ZnONPs produced using Eucalyptus globulus extract have fungicidal activity against significant apple pathogens orchards and have an effect against fungi. Titanium dioxide nanoparticles (TiO_2NPs) from lemon fruit exhibit antibacterial activity comparable to zinc oxide nanoparticles (ZnONPs) against Dickeya dadantii (Kanakari et al., 2023). AgNPs from wheat effectively reduce salinity stress in wheat by modifying abscisic acid levels and enhancing ion homeostasis and antioxidant defenses. Additionally, ZnONPs demonstrate low toxicity while stimulating the antioxidant response in flax shoots (Ditta, 2012).

6.7. Environmental Protection Applications

Phytogenic nanoparticles play a crucial role in environmental protection. A clean and pollution-free environment is essential for sustaining both human and animal life. As a result, phytogenic nanoparticles are now utilized in nearly every environmental protection action plan. Their applications range from simple bioremediation to the degradation of dyes, removal of metals, and elimination of organic pollutants from wastewater (P. Verma et al., 2020).

6.7.1. Bioremediation of Metal Ions

Bioremediation of metal ions plays a crucial role in environmental clean-up efforts, primarily due to the remarkable ability of certain plants to absorb and accumulate metal ions from the soil and water (Malik, 2004). This process harnesses the natural mechanisms of plants to remove toxic metals, thereby reducing their concentration in contaminated environments (Chaudhary et al., 2017). In recent years, researchers have developed various phytogenic nanoparticles derived from metals. These nanoparticles, which are created using plant extracts, offer innovative solutions for detoxifying polluted sites since they often exhibit enhanced stability and efficacy compared to traditional methods (Hassen et al., 1998). Moreover, bioremediation is not limited to plants; bacterial populations also play a significant role. Certain bacteria can produce nanoparticles capable of binding and removing metal particles from the environment. This approach not only aids in cleaning up metal contaminants but can also convert harmful ions into less toxic forms, facilitating safer disposal and reducing environmental hazards (Veglió et al., 1997). Together, these biotechnological innovations highlight the potential of using biological systems, both plant and microbial, to address metal pollution effectively. Arsenic is the most common contaminant in groundwater and is highly toxic, potentially causing dysfunction in the human immune system and leading to various infections. The World Health Organization (WHO) has classified arsenic as a Group I carcinogen, meaning there is sufficient evidence to conclude that it can cause cancer in humans (Mandal et al., 2002; Rachana Singh et al., 2015). To safeguard public health, the WHO has established a permissible limit for arsenic in drinking water at less than 10 μg/mL (Mandal et al., 2002; Rachana Singh et al., 2015). Recent advancements in nanotechnology have led to the development of phytogenic magnetic nanoparticles from plant materials. These nanoparticles effectively absorb and remove contaminants from water and soil, providing a sustainable solution for arsenic pollution and enhancing environmental health (Singh et al., 2015). Additionally, research studies on bioremediation have successfully used metal nanoparticles to combat E. coli, proving their effectiveness in improving public health. Recent research has explored the production of nanoparticles using Eucalyptus beads. Additionally, the extremophilic organism Deinococcus radiodurans R1 strain has demonstrated significant potential for bioremediation. Some researhers developed a practical and rapid method for arsenic removal by producing magnetic iron nanoparticles that immobilize the D. radiodurans R1 strain (K. Guo et al., 2022). They utilized living cells for the in vivo synthesis of nanoparticles, which were then characterized using techniques such as X-ray Diffraction (XRD) and Scanning Electron Microscopy (SEM) (K. Guo et al., 2021). The removal procedure focused on calculating the removal efficiency percentage, which showed that the magnetic iron nanoparticles had an adsorption capacity of approximately 95 mg/g (Hui et al., 2018).

A significant number of research studies have focused on the removal of heavy toxic metals, including chromium (Cr III), lead (Pb II), arsenic (As V), cobalt (Co II), nickel, and calcium, by utilizing plant extracts as coating materials. These studies highlight the potential of biogenic approaches to address environmental contamination (Bozbeyoglu et al., 2023). Phytogenic metallic nanoparticles, which are derived from these plant extracts, are believed to function primarily through an electrostatic adsorption mechanism. This process allows them to effectively interact with and capture specific metal ions, leading to their removal from contaminated environments. The use of these natural extracts not only offers a sustainable alternative to traditional methods of metal removal but also leverages the inherent properties of plants to mitigate pollution levels.

6.7.2. Dye Degradation

Toxic dyes found in industrial effluents are a major environmental concern due to their potential to contaminate natural water bodies. Many of these dyes are classified as carcinogenic, which means they can contribute to cancer development in living organisms (Tichapondwa et al., 2020). The presence of toxic dyes in aquatic ecosystems can severely disrupt the photosynthetic processes of aquatic plants, leading to decreased oxygen levels and rendering these environments unsuitable for marine and freshwater life (Miri et al., 2018). One commonly studied toxic dye is methylene blue. Researchers have explored various methods for its remediation, with one promising approach involving the use of iron oxide nanoparticles extracted from Mallow leaves (Goudarzi et al., 2016). This method capitalizes on the natural properties of the nanoparticles to effectively degrade the dye. The effectiveness of this remediation process is evidenced by a significant decrease in the peak absorbance intensity measured at a wavelength of 668 nm, indicating a reduction in dye concentration (Razzaghi et al., 2017; P. Verma et al., 2020). Beyond iron oxide nanoparticles, a range of metal oxides has also been investigated for their photocatalytic ability to degrade synthetic dyes. These include zinc oxide, titanium oxide, selenium oxide, and copper oxide, each offering unique advantages in the degradation process. The use of these materials in photocatalysis is an active area of research, aiming to develop sustainable and efficient solutions to combat water pollution caused by toxic dyes.

6.7.3. Remediation of Nitrophenol

Pollutants significantly harm the health of the environment. One notable organic pollutant found in water is 4-nitrophenol, which can lead to loss of consciousness in both humans and animals due to its ability to form methemoglobin. In the study conducted by Yardily et al., researchers investigated the reduction of 4-nitrophenol, a common organic pollutant. They confirmed the effectiveness of the reduction process by measuring a significant decrease in the peak intensity at 400 nm, which is indicative of the presence of 4-nitrophenol. This observation suggests that the compound was effectively transformed during the process. The study found that the complete transformation required a duration of 25 minutes, as documented in the research conducted by Verma et al. (P. Verma et al., 2020). This finding highlights the efficiency of the reduction method employed in the study.

6.7.4. Metal Detection Properties

Phytogenic nanoparticles can detect toxic metals, which is one important application of nanoparticles in nano biosensors. In the study conducted by Singh et al., silver nanoparticles were used in sun-dried and fresh mango leaves to detect mercury and lead ions (Rajat Singh et al., 2023). Other green nanoparticles have also been developed to detect lead, mercury, and zinc ions (Hano et al., 2021).

7. EFFECTS OF NANOPARTICLES ON PLANTS

As nanotechnology applications have grown, the use of nanomaterials with a high surface-to-volume ratio has also increased. The functions and applications of nanomaterials vary based on the size and composition of the nanoparticles (Tunca, 2016). When used as fertilizers, nanomaterials release nutrients to plants slowly, requiring only very small amounts compared to conventional chemical fertilizers. This results in a positive impact on both the plants and the environment by reducing the ecological risks associated with chemical fertilizers (Cinisli et al., 2019). The excessive and unregulated use of chemical pesticides and fertilizers in agriculture can adversely affect both the environment and human health, primarily in efforts to boost plant yield. Some studies have shown that nanoparticles applied to plants promote plant growth and development. For example, in one study, nanoparticles such as Cu_2O (0-160 ppm) and TiO_2 (0.05-0.2 g L^{-1}) were shown to promote tomato growth by increasing germination, root/shoot elongation and transpiration (Mony et al., 2022). It has been observed that nanoparticles positively affect many physiological activities of plants, including photosynthesis. Nanoparticles of TiO_2, CeO_2 and ZnO are important in this category. For example, TiO_2 nanoparticles significantly improved spinach growth by (i) improving light absorption, (ii) increasing the production of the RUBISCO enzyme, and (iii) reducing oxidative stress mediated by UV radiation in the chloroplast (Mony et al., 2022). In addition, nanoparticle-mediated targeted delivery of nucleotides, proteins, and other phytoactive molecules has the potential for genetic modification and regulation of plant metabolism. Using nanotechnology to control and release agrochemicals can reduce the damage of plant protection products, reduce nutrient losses in fertilizers, and increase crop yields through optimized nutrient management. Various nanomaterials are being designed to release agrochemicals in a controlled manner. For example, mesoporous silica nanoparticles protect plants from damage caused by agrochemicals. They load pesticides such as avermectin into the nucleus. This process protects the pesticide from photodegradation while allowing its release. Similarly, to protect plants against insects, they can enter the cuticular lipids of insects through photoabsorption, causing mortality. Nanoparticle-based herbicides are also used to control parasitic weeds at low doses. In addition, the use of nanoparticle-containing fertilizers increases the nutrient utilization efficiency of plants. For example, the use of fertilizers that provide low phosphate ($PO4^{-3}$) release in the plant contributes to the protection of the environment and the productivity of agricultural areas by reducing the risk of eutrophication in water (Gokdemir et al., 2022).

The potential environmental and health effects of plant-derived nanoparticles produced by nanotechnological green synthesis are one of the most critical areas of today's research. In particular, more research is needed on the long-term effects of nanoparticles on ecosystems, their bioaccumulation potential, and their interactions with organisms (Tunca, 2016). Environmental toxicity can threaten biodiversity due to the accumulation of nanoparticles in water and soil ecosystems, which can harm ecosystem functioning. Potential risks to human health are also a major concern. Inhalation, skin contact, or ingestion of nanoparticles can lead to health problems. In addition, the potential for bioaccumulation and toxic effects in the human body of these nanoparticles can lead to serious health threats. In this context, comprehensive studies on the safety and toxicology of nanoparticles are required (Gokdemir et al., 2022). On the other hand, the establishment of international regulatory frameworks for the production and use of nanoparticles is critical for safe and sustainable use. Such regulations are necessary to minimize both environmental and human health risks.

The potential toxic effects of nanoparticles on plants, soil, and ecosystems are a growing concern, especially in agricultural systems and environmental settings. Plant-derived nanoparticles can accumulate in soil and have adverse effects on plant growth and soil microorganisms (Hano et al., 2021). The bioaccumulation capacity of nanoparticles can pose long-term toxicological risks to soil and aquatic ecosystems. Nanoparticles used in agriculture can disrupt ecosystem balance by interacting with non-target species, which can lead to a decrease in biodiversity. In particular, the leaching of nanoparticles into water resources can have toxic effects on aquatic organisms (Tunca, 2016). In addition, the accumulation of these nanoparticles in agricultural products can pose threats to human health through their entry into the food chain. The long-term sustainability of the use of nanotechnology in agriculture requires meticulous examination of such environmental and health risks. Therefore, further research on the potential long-term effects of the release and accumulation of nanoparticles in the environment is important.

8. CONCLUSION

In conclusion, nanotechnological green synthesis offers a transformative approach to advancing plant science by utilizing environmentally friendly methods to produce nanomaterials. This innovative technique minimizes the reliance on harmful chemicals and toxic reagents typically used in traditional synthesis processes, thereby enhancing the efficiency and sustainability of agricultural practices. The application of green-synthesized nanoparticles can significantly improve plant health and productivity by facilitating better nutrient uptake, enhancing disease resistance, and promoting sustainable methods that mitigate environmental degradation. Moreover, integrating nanotechnology into plant science holds great promise for revolutionary advancements in crop yield, food security, and environmental sustainability. By using biological entities and processes for synthesizing nanoscale materials, researchers can develop eco-friendly fertilizers, pesticides, and growth enhancers that align with sustainable agriculture principles. As challenges posed by climate change, resource scarcity, and population growth continue to escalate, nanotechnological green synthesis emerges as a vital tool for cultivating resilient and high-yielding plant varieties. Ultimately, this approach enhances the scientific understanding of how nanomaterials interact with plants and paves the way for innovative solutions that support both agricultural productivity and ecological balance in the years ahead.

REFERENCES

Abdallah, H. M., Mohamed, G. A., & Ibrahim, S. R. M. (2022). Lansium domesticum—A Fruit with Multi-Benefits: Traditional Uses, Phytochemicals, Nutritional Value, and Bioactivities. *Nutrients 2022, Vol. 14, Page 1531, 14*(7), 1531. DOI: 10.3390/nu14071531

Adeyemi, J. O., Oriola, A. O., Onwudiwe, D. C., & Oyedeji, A. O. (2022). Plant Extracts Mediated Metal-Based Nanoparticles: Synthesis and Biological Applications. *Biomolecules 2022, Vol. 12, Page 627, 12*(5), 627. DOI: 10.3390/biom12050627

Afshan, N., Bari, A., Khan, A., & Shah, Y. (2023). Advanced Bionanomaterials Used to Recover Pollutants From Wastewater. *Sustainable Nanomaterials for Biosystems Engineering*, 271–296. DOI: 10.1201/9781003333517-13

Agatonovic-Kustrin, S., Kettle, C., & Morton, D. W. (2018). A molecular approach in drug development for Alzheimer's disease. *Biomedicine and Pharmacotherapy*, *106*, 553–565. DOI: 10.1016/j.biopha.2018.06.147 PMID: 29990843

Ahovan, Z. A., Hashemi, A., De Plano, L. M., Gholipourmalekabadi, M., & Seifalian, A. (2020). Bacteriophage Based Biosensors: Trends, Outcomes and Challenges. *Nanomaterials 2020, Vol. 10, Page 501, 10*(3), 501. DOI: 10.3390/nano10030501

Ainiwaer, M., Jia, H., Zhang, T., Huang, J., Zhang, N., Yin, X., Peng, L., Li, H., & Zeng, X. (2024). Effective co-immobilization of arsenic and cadmium in contaminated soil by sepiolite-modified nano-zero-valent iron and its impact on the soil bacterial community. *Scientific Reports*, *14*(1), 26178. DOI: 10.1038/s41598-024-77066-6 PMID: 39478164

Al Janabi, M. A. Y., El Houda Tiri, R. N., Cherif, A., Altuner, E. E., Lee, C. J., Sen, F., Dragoi, E. N., Karimi, F., & Kalikeri, S. (2024). Hydrogen Generation by Methanolysis of NaBH4 via Efficient CuFe2O4 Nanoparticle Catalyst: A Kinetic Study and DNN Model. *Topics in Catalysis*, *67*(9–12), 843–852. DOI: 10.1007/s11244-024-01904-0

Alizadeh, M., Asrami, P. N., Altuner, E. E., Gulbagca, F., Tiri, R. N. E., Aygun, A., Kaynak, İ., Sen, F., & Cheraghi, S. (2022). An ultra-sensitive rifampicin electrochemical sensor based on Fe3O4 nanoparticles anchored Multiwalled Carbon nanotube modified glassy carbon electrode. *Chemosphere*, *309*, 136566. DOI: 10.1016/j.chemosphere.2022.136566 PMID: 36152837

Almasoud, N., Alhaik, H., Almutairi, M., Houjak, A., Hazazi, K., Alhayek, F., Aljanoubi, S., Alkhaibari, A., Alghamdi, A., Soliman, D. A., Alomar, T. S., & Awad, M. A. (2021). Green nanotechnology synthesized silver nanoparticles: Characterization and testing its antibacterial activity. *Green Processing and Synthesis*, *10*(1), 510–528. DOI: 10.1515/gps-2021-0048

Altuner, E. E., Akin, M., Bayat, R., Bekmezci, M., Burhan, H., & Sen, F. (2022). Challenges in commercialization of carbon nanomaterial-based sensors. *Carbon Nanomaterials-Based Sensors: Emerging Research Trends in Devices and Applications*, 381–392. DOI: 10.1016/B978-0-323-91174-0.00020-2

Altuner, E. E.ALTUNER. (2023). Electrochemical sensitivity, selectivity, and sensory properties of a natural polymer-supported titanium nanocomposite electrode towards lysine amino acid. *Muş Alparslan Üniversitesi Fen Bilimleri Dergisi*, *11*(1), 28–32. DOI: 10.18586/msufbd.1268440

Altuner, E. E., Arıkan, K., Burhan, H., Ozdemir, S., & Şen, F. (2021). Commercial aspects of direct alcohol fuel cells. *Nanomaterials for Direct Alcohol Fuel Cells*, 511–524. DOI: 10.1016/B978-0-12-821713-9.00012-3

Altuner, E. E., Bekmezci, M., Bayat, R., Akin, M., Isik, I., & Şen, F. (2021). Dendrimer-based nanocomposites for alcohol fuel cells. *Nanomaterials for Direct Alcohol Fuel Cells*, 337–352. DOI: 10.1016/B978-0-12-821713-9.00007-X

Altuner, E. E., Bekmezci, M., & Sen, F. (2022). Manufacturing Techniques of Magnetic Polymer Nanocomposites. *Handbook of Magnetic Hybrid Nanoalloys and Their Nanocomposites*, 303–318. DOI: 10.1007/978-3-030-90948-2_12

Altuner, E. E., Erduran, V., & Sen, F. (2023). Green synthesized nanomaterials for bioimaging. In *Synthesis of Bionanomaterials for Biomedical Applications Micro and Nano Technologies* (pp. 265–286). Elsevier., DOI: 10.1016/B978-0-323-91195-5.00023-4

Altuner, E. E., Gulbagca, F., Tiri, R. N. E., Aygun, A., & Sen, F. (2023). Highly efficient palladium-zinc oxide nanoparticles synthesized by biogenic methods: Characterization, hydrogen production and photocatalytic activities. *Chemical Engineering Journal Advances*, *14*, 100465. DOI: 10.1016/j.ceja.2023.100465

Altuner, E. E., Gur, T., & Şen, F. (2021). Ternary/quaternary nanomaterials for direct alcohol fuel cells. *Nanomaterials for Direct Alcohol Fuel Cells*, 157–172. DOI: 10.1016/B978-0-12-821713-9.00001-9

Altuner, E. E., Ozalp, V. C., Yilmaz, M. D., Sudagidan, M., Aygun, A., Acar, E. E., Tasbasi, B. B., & Sen, F. (2022). Development of electrochemical aptasensors detecting phosphate ions on TMB substrate with epoxy-based mesoporous silica nanoparticles. *Chemosphere*, *297*, 134077. DOI: 10.1016/j.chemosphere.2022.134077 PMID: 35218784

Ameen, F., Altuner, E. E., Elhouda Tiri, R. N., Gulbagca, F., Aygun, A., Sen, F., Majrashi, N., Orfali, R., & Dragoi, E. N. (2023). Highly active iron (II) oxide-zinc oxide nanocomposite synthesized Thymus vulgaris plant as bioreduction catalyst: Characterization, hydrogen evolution and photocatalytic degradation. *International Journal of Hydrogen Energy*, *48*(55), 21139–21151. DOI: 10.1016/j.ijhydene.2022.11.229

Amekawa, Y. (2011). Agroecology and Sustainable Livelihoods: Towards an Integrated Approach to Rural Development. *Journal of Sustainable Agriculture*, *35*(2), 118–162. DOI: 10.1080/10440046.2011.539124

Ameta, R., Rai, A. K., Vyas, S., Bhatt, J. P., & Ameta, S. C. (2021). *Green synthesis of nanocomposites using plant extracts and their applications. Handbook of Greener Synthesis of Nanomaterials and Compounds* (Vol. 1). Fundamental Principles and Methods., DOI: 10.1016/B978-0-12-821938-6.00020-7

Antony, A. (2024). Flexible and Wearable Biosensors: Revolutionizing. *Health Monitor*, •••, 237–258. DOI: 10.1007/978-981-97-3048-3_12

Pattoo, T. A. (2023). *Flora to Nano: Sustainable Synthesis of Nanoparticles via Plant-Mediated Green Chemistry*. Plant Science Archives.

Asal, M., Özen, Ö., Şahinler, M., & Polatoğlu, İ. (2018). Recent Developments in Enzyme, DNA and Immuno-Based Biosensors. *Sensors 2018, Vol. 18, Page 1924, 18*(6), 1924. DOI: 10.3390/s18061924

Ash, B. J., Schadler, L. S., & Siegel, R. W. (2002). Glass transition behavior of alumina/polymethylmethacrylate nanocomposites. *Materials Letters, 55*(1–2), 83–87. DOI: 10.1016/S0167-577X(01)00626-7

Asif, N., Amir, M., & Fatma, T. (2023). Recent advances in the synthesis, characterization and biomedical applications of zinc oxide nanoparticles. *Bioprocess and Biosystems Engineering 2023 46:10, 46*(10), 1377–1398. DOI: 10.1007/s00449-023-02886-1

Ateia, E. E., Rabie, O., & Mohamed, A. T. (2024). Assessment of the correlation between optical properties and CQD preparation approaches. *The European Physical Journal Plus, 139*(1), 24. DOI: 10.1140/epjp/s13360-023-04811-7

Athreya, S. R., Kalaitzidou, K., & Das, S. (2010). Processing and characterization of a carbon black-filled electrically conductive Nylon-12 nanocomposite produced by selective laser sintering. *Materials Science and Engineering A, 527*(10–11), 2637–2642. DOI: 10.1016/j.msea.2009.12.028

Bagheri Hashkavayi, A., Cha, B. S., Lee, E. S., Kim, S., & Park, K. S. (2020). Advances in Exosome Analysis Methods with an Emphasis on Electrochemistry. *Analytical Chemistry, 92*(19), 12733–12740. DOI: 10.1021/acs.analchem.0c02745 PMID: 32902258

Bahrulolum, H., Nooraei, S., Javanshir, N., Tarrahimofrad, H., Mirbagheri, V. S., Easton, A. J., & Ahmadian, G. (2021). Green synthesis of metal nanoparticles using microorganisms and their application in the agrifood sector. *Journal of Nanobiotechnology 2021 19:1, 19*(1), 1–26. DOI: 10.1186/s12951-021-00834-3

Baker, C., Shah, S. I., & Hasanain, S. K. (2004). Magnetic behavior of iron and iron-oxide nanoparticle/polymer composites. *Journal of Magnetism and Magnetic Materials, 280*(2–3), 412–418. DOI: 10.1016/j.jmmm.2004.03.037

Balan, B. K., & Kurungot, S. (2012). Tuning the functionality of a carbon nanofiber-Pt-RuO2 system from charge storage to electrocatalysis. *Inorganic Chemistry, 51*(18), 9766–9774. DOI: 10.1021/ic301147s PMID: 22946658

Balan, K., Qing, W., Wang, Y., Liu, X., Palvannan, T., Wang, Y., Ma, F., & Zhang, Y. (2016). Antidiabetic activity of silver nanoparticles from green synthesis using Lonicera japonica leaf extract. *RSC Advances, 6*(46), 40162–40168. DOI: 10.1039/C5RA24391B

Banik, M., Patra, M., Dutta, D., Mukherjee, R., & Basu, T. (2018). A simple robust method of synthesis of copper-silver core-shell nano-particle: Evaluation of its structural and chemical properties with anticancer potency. *Nanotechnology, 29*(32), 325102. DOI: 10.1088/1361-6528/aac372 PMID: 29742069

Bardania, H., Mahmoudi, R., Bagheri, H., Salehpour, Z., Fouani, M. H., Darabian, B., Khoramrooz, S. S., Mousavizadeh, A., Kowsari, M., Moosavifard, S. E., Christiansen, G., Javeshghani, D., Alipour, M., & Akrami, M. (2020). Facile preparation of a novel biogenic silver-loaded Nanofilm with intrinsic anti-bacterial and oxidant scavenging activities for wound healing. *Scientific Reports 2020 10:1, 10*(1), 1–14. DOI: 10.1038/s41598-020-63032-5

Basri, S., Kamarudin, S. K., Daud, W. R. W., & Yaakub, Z. (2010). Nanocatalyst for direct methanol fuel cell (DMFC). In International Journal of Hydrogen Energy (Vol. 35, Issue 15, pp. 7957–7970). Elsevier Ltd. DOI: 10.1016/j.ijhydene.2010.05.111

Bavandpour, R., Rajabi, M., Karimi-Maleh, H., & Asghari, A. (2021). Application of deep eutectic solvent and SWCNT-ZrO2 nanocomposite as conductive mediators for the fabrication of simple and rapid electrochemical sensor for determination of trace anti-migration drugs. *Microchemical Journal, 165*, 106141. DOI: 10.1016/j.microc.2021.106141

Bazylinski, D. A., & Frankel, R. B. (2004). Magnetosome formation in prokaryotes. *Nature Reviews Microbiology 2004 2:3, 2*(3), 217–230. DOI: 10.1038/nrmicro842

Behravan, M., Hossein Panahi, A., Naghizadeh, A., Ziaee, M., Mahdavi, R., & Mirzapour, A. (2019). Facile green synthesis of silver nanoparticles using Berberis vulgaris leaf and root aqueous extract and its antibacterial activity. *International Journal of Biological Macromolecules, 124*, 148–154. DOI: 10.1016/j.ijbiomac.2018.11.101 PMID: 30447360

Bekmezci, M., Altuner, E. E., & Sen, F. (2022). Biological Characterization of Magnetic Hybrid Nanoalloys. *Handbook of Magnetic Hybrid Nanoalloys and Their Nanocomposites*, 1–18. DOI: 10.1007/978-3-030-34007-0_28-1

Bensadoun, J. C., De Almeida, L. P., Fine, E. G., Tseng, J. L., Déglon, N., & Aebischer, P. (2003). Comparative study of GDNF delivery systems for the CNS: Polymer rods, encapsulated cells, and lentiviral vectors. *Journal of Controlled Release, 87*(1–3), 107–115. DOI: 10.1016/S0168-3659(02)00353-X PMID: 12618027

Berkner, S., Schwirn, K., & Voelker, D. (2022). Too advanced for assessment? Advanced materials, nanomedicine and the environment. *Environmental Sciences Europe, 34*(1), 71. Advance online publication. DOI: 10.1186/s12302-022-00647-7 PMID: 35990895

Bhattacharya, T., Soares, G. A. B. E., Chopra, H., Rahman, M. M., Hasan, Z., Swain, S. S., & Cavalu, S. (2022). Applications of Phyto-Nanotechnology for the Treatment of Neurodegenerative Disorders. *Materials 2022, Vol. 15, Page 804, 15*(3), 804. DOI: 10.3390/ma15030804

Bieberstein, A., Roosen, J., Marette, S., Blanchemanche, S., & Vandermoere, F. (2013). Consumer choices for nano-food and nano-packaging in France and Germany. In European Review of Agricultural Economics (Vol. 40, Issue 1, pp. 73–94). DOI: 10.1093/erae/jbr069

Bolsenga, S. J. (1983). Spectral Reflectances of Snow and Fresh-Water Ice from 340 Through 1 100 nm. *Journal of Glaciology, 29*(102), 296–305. DOI: 10.3189/S0022143000008352

Boulaiz, H., Alvarez, P. J., Ramirez, A., Marchal, J. A., Prados, J., Rodríguez-Serrano, F., Perán, M., Melguizo, C., & Aranega, A. (2011). Nanomedicine: Application Areas and Development Prospects. *International Journal of Molecular Sciences*, *12*(5), 3303–3321. DOI: 10.3390/ijms12053303 PMID: 21686186

Bozbeyoglu, P., & Gündoğdu, A. (2023). Adsorption of hexavalent chromium from aqueous solution onto corn cobs – activated carbon. *Turkish Journal of Analytical Chemistry*, *5*(2), 107–117. DOI: 10.51435/turkjac.1394304

Calabro, R. L., Yang, D. S., & Kim, D. Y. (2019). Controlled Nitrogen Doping of Graphene Quantum Dots through Laser Ablation in Aqueous Solutions for Photoluminescence and Electrocatalytic Applications. *ACS Applied Nano Materials*, *2*(11), 6948–6959. DOI: 10.1021/acsanm.9b01433

Chang, P. R., Jian, R., Yu, J., & Ma, X. (2010). Fabrication and characterisation of chitosan nanoparticles/plasticised-starch composites. *Food Chemistry*, *120*(3), 736–740. DOI: 10.1016/j.foodchem.2009.11.002

Charbgoo, F., Ramezani, M., & Darroudi, M. (2017). Bio-sensing applications of cerium oxide nanoparticles: Advantages and disadvantages. *Biosensors & Bioelectronics*, *96*, 33–43. DOI: 10.1016/j.bios.2017.04.037 PMID: 28458132

Chatzichristos, A., & Hassan, J. (2022). Current Understanding of Water Properties inside Carbon Nanotubes. *Nanomaterials 2022, Vol. 12, Page 174*, *12*(1), 174. DOI: 10.3390/nano12010174

Chaudhary, A., Shirodkar, S., & Sharma, A. (2017). Characterization of Nickel Tolerant Bacteria Isolated from Heavy Metal Polluted Glass Industry for its Potential Role in Bioremediation. *Soil & Sediment Contamination*, *26*(2), 184–194. DOI: 10.1080/15320383.2017.1267110

Chen, L. X., Zheng, J. N., Wang, A. J., Wu, L. J., Chen, J. R., & Feng, J. J. (2015). Facile synthesis of porous bimetallic alloyed PdAg nanoflowers supported on reduced graphene oxide for simultaneous detection of ascorbic acid, dopamine, and uric acid. *Analyst*, *140*(9), 3183–3192. DOI: 10.1039/C4AN02200A PMID: 25756085

Cheng, Z., Zhang, J., Liu, H., Li, Y., Zhao, Y., & Yang, E. (2010). Central nervous system penetration for small molecule therapeutic agents does not increase in multiple sclerosis- and Alzheimer's disease-related animal models despite reported blood-brain barrier disruption. *Drug Metabolism and Disposition: the Biological Fate of Chemicals*, *38*(8), 1355–1361. DOI: 10.1124/dmd.110.033324 PMID: 20427691

Chiozzi, V., & Rossi, F. (2020). Inorganic-organic core/shell nanoparticles: Progress and applications. In Nanoscale Advances (Vol. 2, Issue 11, pp. 5090–5105). Royal Society of Chemistry. DOI: 10.1039/D0NA00411A

Choi, W., Abrahamson, J. T., Strano, J. M., & Strano, M. S. (2010). Carbon nanotube-guided thermopower waves. *Materials Today*, *13*(10), 22–33. DOI: 10.1016/S1369-7021(10)70183-8

Chou, L. Y. T., Ming, K., & Chan, W. C. W. (2010). Strategies for the intracellular delivery of nanoparticles. *Chemical Society Reviews*, *40*(1), 233–245. DOI: 10.1039/C0CS00003E PMID: 20886124

Cinisli, K. T., Uçar, S., & Dikbaş, N. (2019). Use of nanomaterials in agriculture. *Yuzuncu Yil University Journal of Agricultural Sciences*, *29*(4), 817–831. DOI: 10.29133/yyutbd.595658

Correa, S., Dreaden, E. C., Gu, L., & Hammond, P. T. (2016). Engineering nanolayered particles for modular drug delivery. *Journal of Controlled Release*, *240*, 364–386. DOI: 10.1016/j.jconrel.2016.01.040 PMID: 26809005

Cui, Y., & Lieber, C. M. (2001). Functional Nanoscale Electronic Devices Assembled Using Silicon Nanowire Building Blocks. *Science*, *291*(5505), 851–853. DOI: 10.1126/science.291.5505.851 PMID: 11157160

Dahl, J. A., Maddux, B. L. S., & Hutchison, J. E. (2007). Toward greener nanosynthesis. In Chemical Reviews (Vol. 107, Issue 6, pp. 2228–2269). DOI: 10.1021/cr050943k

Dahoumane, S. A., Mechouet, M., Wijesekera, K., Filipe, C. D. M., Sicard, C., Bazylinski, D. A., & Jeffryes, C. (2017). Algae-mediated biosynthesis of inorganic nanomaterials as a promising route in nanobiotechnology-a review. In Green Chemistry (Vol. 19, Issue 3, pp. 552–587). Royal Society of Chemistry. DOI: 10.1039/C6GC02346K

Darabi, R., Karimi-Maleh, H., Altuner, E. E., Gulbagca, F., Tiri, R. N. E., Chinnathambi, A., Brindhadevi, K., Halvaci, E., & Sen, F. (2024). Cobalt Nanoparticles Supported Active Carbon from Chitosan Biopolymer Using Thermal Method: Synthesis, Characterization, and Hydrogen Production. *Topics in Catalysis*, •••, 1–10. DOI: 10.1007/s11244-024-01998-6

Darabi, R., Karimi-Maleh, H., Shahidi, S. A., Gulbagca, F., Nour Elhouda Tiri, R., Monadi Sefidan, A., Dehdashtian, N., Alhrasishawi, W., Aygun, A., Esra Altuner, E., & Sen, F. (2023). Hexamethylenetetramine-based nanomaterial synthesis, characterization, investigation of photocatalytic activity against Rhodamine B dye and kinetic study for hydrogen fuel production. *Fuel*, *352*, 128841. DOI: 10.1016/j.fuel.2023.128841

Das, M., Sarmah, S., & Sarkar, D. (2019). Photo sensing property of nanostructured CdS-porous silicon (PS):p-Si based MSM hetero-structure. *Journal of Materials Science Materials in Electronics*, *30*(12), 11239–11249. DOI: 10.1007/s10854-019-01470-2

Das, P., Ghosh, S., & Nayak, B. (2021). Phyto-fabricated Nanoparticles and Their Anti-biofilm Activity: Progress and Current Status. *Frontiers in Nanotechnology*, *3*, 739286. DOI: 10.3389/fnano.2021.739286

Dauthal, P., & Mukhopadhyay, M. (2015). Biofabrication, characterization, and possible bio-reduction mechanism of platinum nanoparticles mediated by agro-industrial waste and their catalytic activity. *Journal of Industrial and Engineering Chemistry*, *22*, 185–191. DOI: 10.1016/j.jiec.2014.07.009

Destito, G., Schneemann, A., & Manchester, M. (2009). Biomedical nanotechnology using virus-based nanoparticles. In *Current Topics in Microbiology and Immunology* (Vol. 327, pp. 95–122). Springer., DOI: 10.1007/978-3-540-69379-6_5

Destoumieux-Garzón, D., Mavingui, P., Boetsch, G., Boissier, J., Darriet, F., Duboz, P., Fritsch, C., Giraudoux, P., Le Roux, F., Morand, S., Paillard, C., Pontier, D., Sueur, C., & Voituron, Y. (2018). The one health concept: 10 years old and a long road ahead. *Frontiers in Veterinary Science*, *5*(FEB), 14. Advance online publication. DOI: 10.3389/fvets.2018.00014 PMID: 29484301

Ditta, A. (2012). How helpful is nanotechnology in agriculture? *Advances in Natural Sciences: Nanoscience and Nanotechnology*, *3*(3), 033002. DOI: 10.1088/2043-6262/3/3/033002

Doner, N., Burhan, H., Bayat, R., Esra Altuner, E., & Sen, F. (2023). Energy Generation through Thermopower Waves using Multi-walled Carbon Nanotube Yarn. *Fuel, 331*, 125906. DOI: 10.1016/j.fuel.2022.125906

Du, J., Tang, J., Xu, S., Ge, J., Dong, Y., Li, H., & Jin, M. (2020). ZnO nanoparticles: Recent advances in ecotoxicity and risk assessment. *Drug and Chemical Toxicology, 43*(3), 322–333. DOI: 10.1080/01480545.2018.1508218 PMID: 30264597

Duan, H., Wang, D., & Li, Y. (2015). Green chemistry for nanoparticle synthesis. *Chemical Society Reviews, 44*(16), 5778–5792. DOI: 10.1039/C4CS00363B PMID: 25615873

Dubey, S. P., Lahtinen, M., & Sillanpää, M. (2010). Tansy fruit mediated greener synthesis of silver and gold nanoparticles. *Process Biochemistry (Barking, London, England), 45*(7), 1065–1071. DOI: 10.1016/j.procbio.2010.03.024

Durmaz, E., Sertkaya, S., Yilmaz, H., Olgun, C., Ozcelik, O., Tozluoglu, A., & Candan, Z. (2023). Lignocellulosic Bionanomaterials for Biosensor Applications. *Micromachines 2023, Vol. 14, Page 1450, 14*(7), 1450. DOI: 10.3390/mi14071450

Ebralidze, I. I., Laschuk, N. O., Poisson, J., & Zenkina, O. V. (2019). Colorimetric Sensors and Sensor Arrays. *Nanomaterials Design for Sensing Applications*, 1–39. DOI: 10.1016/B978-0-12-814505-0.00001-1

El-Belely, E. F., Farag, M. M. S., Said, H. A., Amin, A. S., Azab, E., Gobouri, A. A., & Fouda, A. (2021). Green Synthesis of Zinc Oxide Nanoparticles (ZnO-NPs) Using Arthrospira platensis (Class: Cyanophyceae) and Evaluation of their Biomedical Activities. *Nanomaterials 2021, Vol. 11, Page 95, 11*(1), 95. DOI: 10.3390/nano11010095

Erduran, V., Altuner, E. E., & Şen, F. (2023). Stability and validation of bionanomaterials. *Synthesis of Bionanomaterials for Biomedical Applications*, 251–263. DOI: 10.1016/B978-0-323-91195-5.00021-0

Fayaz, A. M., Balaji, K., Girilal, M., Yadav, R., Kalaichelvan, P. T., & Venketesan, R. (2010). Biogenic synthesis of silver nanoparticles and their synergistic effect with antibiotics: A study against gram-positive and gram-negative bacteria. *Nanomedicine; Nanotechnology, Biology, and Medicine, 6*(1), 103–109. DOI: 10.1016/j.nano.2009.04.006 PMID: 19447203

Feng, Y., Lai, S., Yang, L., Riedel, R., & Yu, Z. (2018). Polymer-derived porous Bi2WO6/SiC(O) ceramic nanocomposites with high photodegradation efficiency towards Rhodamine B. *Ceramics International, 44*(7), 8562–8569. DOI: 10.1016/j.ceramint.2018.02.061

Firdhouse, M. J., & Lalitha, P. (2013). Biosynthesis of silver nanoparticles using the extract of Alternanthera sessilis-antiproliferative effect against prostate cancer cells. *Cancer Nanotechnology, 4*(6), 137–143. DOI: 10.1007/s12645-013-0045-4 PMID: 26069509

Francino, M. P. (2016). Antibiotics and the human gut microbiome: Dysbioses and accumulation of resistances. *Frontiers in Microbiology, 6*(JAN), 164577. DOI: 10.3389/fmicb.2015.01543 PMID: 26793178

Freed, C. R., Greene, P. E., Breeze, R. E., Tsai, W.-Y., DuMouchel, W., Kao, R., Dillon, S., Winfield, H., Culver, S., Trojanowski, J. Q., Eidelberg, D., & Fahn, S. (2001). Transplantation of Embryonic Dopamine Neurons for Severe Parkinson's Disease. *The New England Journal of Medicine*, *344*(10), 710–719. DOI: 10.1056/NEJM200103083441002 PMID: 11236774

Ganesh, R., & Pillai, A. B. (2017). Perceptions and attitude effects on nanotechnology acceptance: An exploratory framework. *Journal of Nanoparticle Research*, *19*(2), 19. DOI: 10.1007/s11051-016-3733-2

Gao, H., Fan, P., Xu, Q., Li, Y., Wang, J., Weng, J., Zhou, J., & Feng, B. (2020). In Vitro and in Vivo Antitumor Activity of Silver Nanoparticles on B16 Melanoma. *Nano*, *15*(12), 2050163. Advance online publication. DOI: 10.1142/S1793292020501635

Gautam, A., & Van Veggel, F. C. J. M. (2013). Synthesis of nanoparticles, their biocompatibility, and toxicity behavior for biomedical applications. *Journal of Materials Chemistry. B*, *1*(39), 5186–5200. DOI: 10.1039/c3tb20738b PMID: 32263325

Gautam, Y. K., Sharma, K., Tyagi, S., Ambedkar, A. K., Chaudhary, M., & Pal Singh, B. (2021). Nanostructured metal oxide semiconductor-based sensors for greenhouse gas detection: Progress and challenges. *Royal Society Open Science*, *8*(3), 201324. Advance online publication. DOI: 10.1098/rsos.201324 PMID: 33959316

Gayda, G. Z., Demkiv, O. M., Stasyuk, N. Y., Serkiz, R. Y., Lootsik, M. D., Errachid, A., Gonchar, M. V., & Nisnevitch, M. (2019). Metallic Nanoparticles Obtained via "Green" Synthesis as a Platform for Biosensor Construction. *Applied Sciences 2019, Vol. 9, Page 720, 9*(4), 720. DOI: 10.3390/app9040720

Ghodake, G., Shinde, S., Saratale, G. D., Kadam, A., Saratale, R. G., & Kim, D. Y. (2020). Water Purification Filter Prepared by Layer-by-layer Assembly of Paper Filter and Polypropylene-polyethylene Woven Fabrics Decorated with Silver Nanoparticles. *Fibers and Polymers*, *21*(4), 751–761. DOI: 10.1007/s12221-020-9624-2

Ghodake, V. N., Naik, S. V., Bhukhanwala, K. N., Kande, K. V., Bhor, N. J., & Patravale, V. B. (2018). Nanoengineered Systems for Biopesticides. *Handbook of Nanomaterials for Industrial Applications*, 243–259. DOI: 10.1016/B978-0-12-813351-4.00015-8

Ghosh, S., Nandi, S., & Basu, T. (2022). Nano-Antibacterials Using Medicinal Plant Components: An Overview. *Frontiers in Microbiology*, *12*, 768739. DOI: 10.3389/fmicb.2021.768739 PMID: 35273578

Gilbertson, L. M., Pourzahedi, L., Laughton, S., Gao, X., Zimmerman, J. B., Theis, T. L., Westerhoff, P., & Lowry, G. V. (2020). Guiding the design space for nanotechnology to advance sustainable crop production. *Nature Nanotechnology*, *15*(9), 801–810. DOI: 10.1038/s41565-020-0706-5 PMID: 32572231

Gillings, M. R. (2013). Evolutionary consequences of antibiotic use for the resistome, mobilome, and microbial pangenome. *Frontiers in Microbiology*, *4*(JAN). Advance online publication. DOI: 10.3389/fmicb.2013.00004 PMID: 23386843

Gökdemir, F. Ş., Gündoğdu, M., Muftareviç, S., Sunar, A., & Eyidoğan, F. (2023). *Nanopartiküllerin tarımsal bilimlerdeki önemi ve kullanım alanları. 49*(1), 11–17.

Gokdemir, G., Doner, N., Sert, Z., Sen, F., & Ciddi, K. (2022). Effects of CuO, TiO2 and graphite microparticles on the heat transfer properties of greases. *Engineering Science and Technology, an International Journal, 30*, 101044. DOI: 10.1016/j.jestch.2021.08.001

Göksu, H., Çelik, B., Yıldız, Y., Şen, F., & Kılbaş, B. (2016). Superior Monodisperse CNT-Supported CoPd (CoPd@CNT) Nanoparticles for Selective Reduction of Nitro Compounds to Primary Amines with NaBH4 in Aqueous Medium. *ChemistrySelect, 1*(10), 2366–2372. DOI: 10.1002/slct.201600509

Göksu, H., Zengin, N., Burhan, H., Cellat, K., & Şen, F. (2020). A Novel Hydrogenation of Nitroarene Compounds with Multi Wall Carbon Nanotube Supported Palladium/Copper Nanoparticles (PdCu@MWCNT NPs) in Aqueous Medium. *Scientific Reports 2020 10:1, 10*(1), 1–8. DOI: 10.1038/s41598-020-64988-0

Goldshleger, N. F. (2007). Fullerenes and Fullerene-based materials in catalysis. *Fullerene Science and Technology, 9*(3), 255–280. DOI: 10.1081/FST-100104493

Goldsmith, M., Abramovitz, L., & Peer, D. (2014). Precision nanomedicine in neurodegenerative diseases. *ACS Nano, 8*(3), 1958–1965. DOI: 10.1021/nn501292z PMID: 24660817

Gopalakrishnan, V., & Muniraj, S. (2021). Neem flower extract assisted green synthesis of copper nanoparticles – Optimisation, characterisation and anti-bacterial study. *Materials Today: Proceedings, 36*, 832–836. DOI: 10.1016/j.matpr.2020.07.013

Goudarzi, M., & Salavati-Niasari, M. (2016). Controllable synthesis of new Tl2S2O3 nanostructures via hydrothermal process; Characterization and investigation photocatalytic activity for degradation of some anionic dyes. *Journal of Molecular Liquids, 219*, 851–857. DOI: 10.1016/j.molliq.2016.04.056

Greff, B., Sáhó, A., Lakatos, E., & Varga, L. (2023a). Biocontrol Activity of Aromatic and Medicinal Plants and Their Bioactive Components against Soil-Borne Pathogens. *Plants 2023, Vol. 12, Page 706, 12*(4), 706. DOI: 10.3390/plants12040706

Greff, B., Sáhó, A., Lakatos, E., & Varga, L. (2023b). Biocontrol Activity of Aromatic and Medicinal Plants and Their Bioactive Components against Soil-Borne Pathogens. *Plants 2023, Vol. 12, Page 706, 12*(4), 706. DOI: 10.3390/plants12040706

Gu, C., Guo, C., Dong, X., Hu, Z., Wu, P., Su, Z., Lu, Y., Xu, B., Yu, Z., & Liu, A. (2019). Core-shell structured iron-containing ceramic nanoparticles: Facile fabrication and excellent electromagnetic absorption properties. *Journal of the American Ceramic Society, 102*(12), 7098–7107. DOI: 10.1111/jace.16619

Gu, Y., Chen, X., Lee, J. H., Monteiro, D. A., Wang, H., & Lee, W. Y. (2012). Inkjet printed antibiotic- and calcium-eluting bioresorbable nanocomposite micropatterns for orthopedic implants. *Acta Biomaterialia, 8*(1), 424–431. DOI: 10.1016/j.actbio.2011.08.006 PMID: 21864730

Guo, K., Cheng, C., Chen, L., Xie, J., Li, S., He, S., & Xiao, F. (2021). Uranium enrichment performence and uranium stress mechanism of Deinococcus radiodurans. *Journal of Radioanalytical and Nuclear Chemistry, 330*(3), 1191–1206. DOI: 10.1007/s10967-021-08018-x

Guo, K., Cheng, C., Chen, L., Xie, J., Li, S., He, S., & Xiao, F. (2022). Application of Deinococcus radiodurans in the treatment of environmental pollution by heavy metals and radionuclides. *Journal of Radioanalytical and Nuclear Chemistry*, *331*(2), 655–664. DOI: 10.1007/s10967-021-08141-9

Guo, Y., Jiang, N., Zhang, L., & Yin, M. (2020). Green synthesis of gold nanoparticles from Fritillaria cirrhosa and its anti-diabetic activity on Streptozotocin induced rats. *Arabian Journal of Chemistry*, *13*(4), 5096–5106. DOI: 10.1016/j.arabjc.2020.02.009

Gupta, N. (2013). Novel Applications of Functionally Graded Nano, Optoelectronic and Thermoelectric Materials Article in. *International Journal of Materials. Mechanics and Manufacturing*, *1*(3), 221–224. Advance online publication. DOI: 10.7763/IJMMM.2013.V1.47

Gurunathan, S., Raman, J., Abd Malek, S. N., John, P. A., & Vikineswary, S. (2013). Green synthesis of silver nanoparticles using Ganoderma neo-japonicum Imazeki: A potential cytotoxic agent against breast cancer cells. *International Journal of Nanomedicine*, *8*, 4399–4413. DOI: 10.2147/IJN.S51881 PMID: 24265551

Guterres, S. S., Alves, M. P., & Pohlmann, A. R. (2007). Polymeric Nanoparticles, Nanospheres and Nanocapsules, for Cutaneous Applications. Https://Doi.Org/10.1177/117739280700200002*, 2*, 117739280700200. DOI: 10.1177/117739280700200002

Haider, A., Haider, S., Kang, I. K., Kumar, A., Kummara, M. R., Kamal, T., & Han, S. S. (2018). A novel use of cellulose based filter paper containing silver nanoparticles for its potential application as wound dressing agent. *International Journal of Biological Macromolecules*, *108*, 455–461. DOI: 10.1016/j.ijbiomac.2017.12.022 PMID: 29222019

Hano, C., & Abbasi, B. H. (2021). Plant-Based Green Synthesis of Nanoparticles: Production, Characterization and Applications. *Biomolecules 2022, Vol. 12, Page 31, 12*(1), 31. DOI: 10.3390/biom12010031

Hassen, A., Saidi, N., Cherif, M., & Boudabous, A. (1998). Effects of heavy metals on Pseudomonas aeruginosa and Bacillus thuringiensis. *Bioresource Technology*, *65*(1–2), 73–82. DOI: 10.1016/S0960-8524(98)00011-X

Hejmady, S., Pradhan, R., Alexander, A., Agrawal, M., Singhvi, G., Gorain, B., Tiwari, S., Kesharwani, P., & Dubey, S. K. (2020). Recent advances in targeted nanomedicine as promising antitumor therapeutics. *Drug Discovery Today*, *25*(12), 2227–2244. DOI: 10.1016/j.drudis.2020.09.031 PMID: 33011342

Hernandez, F. J., & Ozalp, V. C. (2012). Graphene and Other Nanomaterial-Based Electrochemical Aptasensors. *Biosensors 2012, Vol. 2, Pages 1-14, 2*(1), 1–14. DOI: 10.3390/bios2010001

Hofmann, T., Lowry, G. V., Ghoshal, S., Tufenkji, N., Brambilla, D., Dutcher, J. R., Gilbertson, L. M., Giraldo, J. P., Kinsella, J. M., Landry, M. P., Lovell, W., Naccache, R., Paret, M., Pedersen, J. A., Unrine, J. M., White, J. C., & Wilkinson, K. J. (2020). Technology readiness and overcoming barriers to sustainably implement nanotechnology-enabled plant agriculture. *Nature Food*, *1*(7), 416–425. DOI: 10.1038/s43016-020-0110-1

Hojjati-Najafabadi, A., Davar, F., Enteshari, Z., & Hosseini-Koupaei, M. (2021). Antibacterial and photocatalytic behaviour of green synthesis of Zn0.95Ag0.05O nanoparticles using herbal medicine extract. *Ceramics International*, *47*(22), 31617–31624. DOI: 10.1016/j.ceramint.2021.08.042

Hong, S., & Myung, S. (2007). Nanotube electronics: A flexible approach to mobility. In Nature Nanotechnology (Vol. 2, Issue 4, pp. 207–208). Nature Publishing Group. DOI: 10.1038/nnano.2007.89

Hui, M., Shengyan, P., Yaqi, H., Rongxin, Z., Anatoly, Z., & Wei, C. (2018). A highly efficient magnetic chitosan "fluid" adsorbent with a high capacity and fast adsorption kinetics for dyeing wastewater purification. *Chemical Engineering Journal*, *345*, 556–565. DOI: 10.1016/j.cej.2018.03.115

Hussain, I., Singh, N. B., Singh, A., Singh, H., & Singh, S. C. (2016). Green synthesis of nanoparticles and its potential application. *Biotechnology Letters*, *38*(4), 545–560. DOI: 10.1007/s10529-015-2026-7 PMID: 26721237

Ijaz, I., Gilani, E., Nazir, A., & Bukhari, A. (2020). Detail review on chemical, physical and green synthesis, classification, characterizations and applications of nanoparticles. *Green Chemistry Letters and Reviews*, *13*(3), 59–81. DOI: 10.1080/17518253.2020.1802517

Kaya, S. I., Kurbanoglu, S., Yavuz, E., Demiroglu Mustafov, S., Sen, F., & Ozkan, S. A. (2020). Carbon-based ruthenium nanomaterial-based electroanalytical sensors for the detection of anticancer drug Idarubicin. *Scientific Reports*, *10*(1), 11057.

Jacob, J., Haponiuk, J. T., Thomas, S., & Gopi, S. (2018). Biopolymer based nanomaterials in drug delivery systems: A review. *Materials Today. Chemistry*, *9*, 43–55. DOI: 10.1016/j.mtchem.2018.05.002

Jacobs, M., Nagaraj, V. J., Mertz, T., Selvam, A. P., Ngo, T., & Prasad, S. (2013). An electrochemical sensor for the detection of antibiotic contaminants in water. *Analytical Methods*, *5*(17), 4325–4329. DOI: 10.1039/c3ay40994e

Jadoun, S., Arif, R., Jangid, N. K., & Meena, R. K. (2020). Green synthesis of nanoparticles using plant extracts: a review. In Environmental Chemistry Letters (Vol. 19, Issue 1, pp. 355–374). Springer. DOI: 10.1007/s10311-020-01074-x

Kaiser, B., Vogg, G., Fürst, U. B., & Albert, M. (2015). Parasitic plants of the genus Cuscuta and their interaction with susceptible and resistant host plants. *Frontiers in Plant Science*, *6*(FEB), 118928. DOI: 10.3389/fpls.2015.00045 PMID: 25699071

Kalia, A., & Kaur, G. (2018). *Biosynthesis of Nanoparticles Using Mushrooms*. 351–360. DOI: 10.1007/978-3-030-02622-6_17

Kalia, S., Kango, S., Kumar, A., Haldorai, Y., Kumari, B., & Kumar, R. (2014). Magnetic polymer nanocomposites for environmental and biomedical applications. *Colloid & Polymer Science*, *292*, 2025–2052.

Kanakari, E., & Dendrinou-Samara, C. (2023). Fighting Phytopathogens with Engineered Inorganic-Based Nanoparticles. *Materials 2023, Vol. 16, Page 2388*, *16*(6), 2388. DOI: 10.3390/ma16062388

Kar, P., Banerjee, S., Chhetri, A., & Sen, A. (2021). Synthesis, physicochemical characterization and biological activity of synthesized Silver and Rajat Bhasma nanoparticles using Clerodendrum inerme. *Journal of Phytology*, *13*, 64–71. DOI: 10.25081/jp.2021.v13.7026

Karimi, F., Altuner, E. E., Aygun, A., Bayat, R., Rajendran, S., & Sen, F. (2023). Synthesis of Silver Nanoparticles by Biogenic Methods: Characterization and Development of a Sensor Sensible to Pharmaceutical Medicine Paracetamol. *Topics in Catalysis*, *1*, 1–9. DOI: 10.1007/S11244-023-01887-4/TABLES/1

Karimi, F., Altuner, E. E., Aygun, A., Bayat, R., Rajendran, S., & Sen, F. (2024). Synthesis of Silver Nanoparticles by Biogenic Methods: Characterization and Development of a Sensor Sensible to Pharmaceutical Medicine Paracetamol. *Topics in Catalysis*, *67*(9–12), 585–593. DOI: 10.1007/s11244-023-01887-4

Karimi, F., Altuner, E. E., Gulbagca, F., Tiri, R. N. E., Sen, F., Javadi, A., & Dragoi, E. N. (2023). Facile bio-fabrication of ZnO@AC nanoparticles from chitosan: Characterization, hydrogen generation, and photocatalytic properties. *Environmental Research*, *216*, 114668. DOI: 10.1016/j.envres.2022.114668 PMID: 36397611

Karpuz, Ö., Baltacı, C., Gül, A., Gülen, J., Bozbeyoğlu, P., & Aydoğan, N. (2024). Green synthesis of iron and silver nanoparticles from aqueous extract of buckwheat husk waste: Antibacterial, cytotoxic, and dye decolorization properties. *Biomass Conversion and Biorefinery*, *2024*, 1–33. DOI: 10.1007/s13399-024-06287-6

Kaur, P., & Choudhury, D. (2021). Functionality of receptor targeted zinc-insulin quantum clusters in skin tissue augmentation and bioimaging. *Journal of Drug Targeting*, *29*(5), 541–550. DOI: 10.1080/1061186X.2020.1864740 PMID: 33307859

Kaushal, P., Maity, D., & Awasthi, R. (2024). Nano-green: Harnessing the potential of plant extracts for sustainable antimicrobial metallic nanoparticles. *Journal of Drug Delivery Science and Technology*, *94*, 105488. DOI: 10.1016/j.jddst.2024.105488

Kavita, V. (2017). DNA Biosensors-A Review. *Journal of Bioengineering & Biomedical Science*, *7*(2). Advance online publication. DOI: 10.4172/2155-9538.1000222

Khalil, I. A. H., Arida, I. A., Ahmed, M., Khalil, I. A. H., Arida, I. A., & Ahmed, M. (2020). *Introductory Chapter: Overview on Nanomedicine Market*. Current and Future Aspects of Nanomedicine., DOI: 10.5772/intechopen.91890

Khan, F., Shariq, M., Asif, M., Siddiqui, M. A., Malan, P., & Ahmad, F. (2022). Green Nanotechnology: Plant-Mediated Nanoparticle Synthesis and Application. *Nanomaterials 2022, Vol. 12, Page 673, 12*(4), 673. DOI: 10.3390/nano12040673

Khan, M. S., Vishakante, G. D., & Siddaramaiah, H. (2013). Gold nanoparticles: A paradigm shift in biomedical applications. *Advances in Colloid and Interface Science*, *199–200*, 44–58. DOI: 10.1016/j.cis.2013.06.003 PMID: 23871224

Khodadadi, B., Bordbar, M., Yeganeh-Faal, A., & Nasrollahzadeh, M. (2017). Green synthesis of Ag nanoparticles/clinoptilolite using Vaccinium macrocarpon fruit extract and its excellent catalytic activity for reduction of organic dyes. *Journal of Alloys and Compounds*, *719*, 82–88. DOI: 10.1016/j.jallcom.2017.05.135

Kishima, H., Poyot, T., Bloch, J., Dauguet, J., Condé, F., Dollé, F., Hinnen, F., Pralong, W., Palfi, S., Déglon, N., Aebischer, P., & Hantraye, P. (2004). Encapsulated GDNF-producing C2C12 cells for Parkinson's disease: A pre-clinical study in chronic MPTP-treated baboons. *Neurobiology of Disease*, *16*(2), 428–439. DOI: 10.1016/j.nbd.2004.03.012 PMID: 15193299

Kocak, N., Sahin, M., Kücükkolbasi, S., & Erdogan, Z. O. (2012). Synthesis and characterization of novel nano-chitosan Schiff base and use of lead (II) sensor. *International Journal of Biological Macromolecules*, *51*(5), 1159–1166. DOI: 10.1016/j.ijbiomac.2012.09.003 PMID: 22982811

Kocak, Y., Aygun, A., Altuner, E. E., Gonca, S., Berikten, D., Tiri, R. N. E., & Sen, F. (2024). *Eco-friendly production of platinum nanoparticles: physicochemical properties, evaluation of biological and catalytic activities. 21*, 51–62. doi: DOI: 10.1007/S13762-023-05232-W/FIGURES/7

Kolodziejczak-Radzimska, A., & Jesionowski, T. (2014). Zinc Oxide—From Synthesis to Application: A Review. *Materials 2014, Vol. 7, Pages 2833-2881, 7*(4), 2833–2881. DOI: 10.3390/ma7042833

Kröger, N., Deutzmann, R., & Sumper, M. (1999). Polycationic peptides from diatom biosilica that direct silica nanosphere formation. *Science*, *286*(5442), 1129–1132. DOI: 10.1126/science.286.5442.1129 PMID: 10550045

Kühnel, D., & Nickel, C. (2014). The OECD expert meeting on ecotoxicology and environmental fate: Towards the development of improved OECD guidelines for the testing of nanomaterials. *The Science of the Total Environment*, *472*, 347–353. DOI: 10.1016/j.scitotenv.2013.11.055 PMID: 24461369

Kumar, B. G., Sadeghi, S., Melikov, R., Aria, M. M., Jalali, H. B., Ow-Yang, C. W., & Nizamoglu, S. (2018). Structural control of InP/ZnS core/shell quantum dots enables high-quality white LEDs. *Nanotechnology*, *29*(34), 345605. DOI: 10.1088/1361-6528/aac8c9 PMID: 29846177

Kumar, H., Bhardwaj, K., Kuča, K., Kalia, A., Nepovimova, E., Verma, R., & Kumar, D. (2020). Flower-Based Green Synthesis of Metallic Nanoparticles: Applications beyond Fragrance. *Nanomaterials 2020, Vol. 10, Page 766, 10*(4), 766. DOI: 10.3390/nano10040766

Langer, K., Balthasar, S., Vogel, V., Dinauer, N., Von Briesen, H., & Schubert, D. (2003). Optimization of the preparation process for human serum albumin (HSA) nanoparticles. *International Journal of Pharmaceutics*, *257*(1–2), 169–180. DOI: 10.1016/S0378-5173(03)00134-0 PMID: 12711172

Langton, C. G. (2019). Nano Technology With Feynman Machines: Scanning Tunneling Engineering and Artificial Life. *Artificial Life*, *14*(6), 443–500. DOI: 10.4324/9780429032769-21

Lebaschi, S., Hekmati, M., & Veisi, H. (2017). Green synthesis of palladium nanoparticles mediated by black tea leaves (Camellia sinensis) extract: Catalytic activity in the reduction of 4-nitrophenol and Suzuki-Miyaura coupling reaction under ligand-free conditions. *Journal of Colloid and Interface Science*, *485*, 223–231. DOI: 10.1016/j.jcis.2016.09.027 PMID: 27665075

Lee, D. Y., Kim, H. W., Yang, H., & Sung, S. H. (2017). Hydrolyzable tannins from the fruits of Terminalia chebula Retz and their α-glucosidase inhibitory activities. *Phytochemistry*, *137*, 109–116. DOI: 10.1016/j.phytochem.2017.02.006 PMID: 28213992

Lee, J., Chatterjee, D. K., Lee, M. H., & Krishnan, S. (2014). Gold nanoparticles in breast cancer treatment: Promise and potential pitfalls. *Cancer Letters*, *347*(1), 46–53. DOI: 10.1016/j.canlet.2014.02.006 PMID: 24556077

Letchumanan, D., Sok, S. P. M., Ibrahim, S., Nagoor, N. H., & Arshad, N. M. (2021). Plant-Based Biosynthesis of Copper/Copper Oxide Nanoparticles: An Update on Their Applications in Biomedicine, Mechanisms, and Toxicity. *Biomolecules 2021, Vol. 11, Page 564, 11*(4), 564. DOI: 10.3390/biom11040564

Li, C., Wang, D., Liang, T., Li, G., Wang, X., Cao, M., & Liang, J. (2003). Oxidation behavior of CNTs and the electric double layer capacitor made of the CNT electrodes. *Science in China Series E: Technological Sciences 2003 46:4, 46*(4), 349–354. DOI: 10.1360/03ye0067

Li, F., Ni, B., Zheng, Y., Huang, Y., & Li, G. (2021). A simple and efficient voltammetric sensor for dopamine determination based on ZnO nanorods/electro-reduced graphene oxide composite. *Surfaces and Interfaces*, *26*, 101375. DOI: 10.1016/j.surfin.2021.101375

Liao, Y., Xu, D., Cao, Y., & Zhu, Y. G. (2023). Advancing sustainable agriculture: Enhancing crop nutrition with next-generation nanotech-based fertilizers. *Nano Research*, *16*(12), 13205–13225. DOI: 10.1007/s12274-023-6284-8

Liu, S., Zhao, Y., Hao, W., Zhang, X. D., & Ming, D. (2020). Micro- and nanotechnology for neural electrode-tissue interfaces. *Biosensors & Bioelectronics*, *170*, 112645. Advance online publication. DOI: 10.1016/j.bios.2020.112645 PMID: 33010703

Liu, X. (2005). A facile route to preparation of sea-urchinlike cadmium sulfide nanorod-based materials. *Materials Chemistry and Physics*, *91*(1), 212–216. DOI: 10.1016/j.matchemphys.2004.11.015

Liu, Y., & Simon, J. D. (2003). *Isolation and Biophysical Studies of Natural Eumelanins: Applications of Imaging Technologies and Ultrafast Spectroscopy. Pigment Cell Research.* Wiley.

Lombi, E., Donner, E., Taheri, S., Tavakkoli, E., Jämting, Å. K., McClure, S., Naidu, R., Miller, B. W., Scheckel, K. G., & Vasilev, K. (2013). Transformation of four silver/silver chloride nanoparticles during anaerobic treatment of wastewater and post-processing of sewage sludge. *Environmental Pollution*, *176*, 193–197. DOI: 10.1016/j.envpol.2013.01.029 PMID: 23434771

Luo, Y., Shen, S., Luo, J., Wang, X., & Sun, R. (2014). Green synthesis of silver nanoparticles in xylan solution via Tollens reaction and their detection for Hg2+. *Nanoscale*, *7*(2), 690–700. DOI: 10.1039/C4NR05999A PMID: 25429650

Madou, M. J. (Ed.). (2011). *Fundamentals of Microfabrication and Nanotechnology, Three-Volume Set.* CRC press.

Mahalakshmi, S., Hema, N., & Vijaya, P. P. (2020). In Vitro Biocompatibility and Antimicrobial activities of Zinc Oxide Nanoparticles (ZnO NPs) Prepared by Chemical and Green Synthetic Route— A Comparative Study. *BioNanoScience*, *10*(1), 112–121. DOI: 10.1007/s12668-019-00698-w

Maity, D., Gupta, I., & Saha, S. (2022). Biosynthesized metal oxide nanoparticles for sustainable agriculture: Next-generation nanotechnology for crop production, protection and management. *Nanoscale*, *14*(38), 38. DOI: 10.1039/D2NR03944C PMID: 36124943

Malik, A. (2004). Metal bioremediation through growing cells. *Environment International*, *30*(2), 261–278. DOI: 10.1016/j.envint.2003.08.001 PMID: 14749114

Malysheva, A., Lombi, E., & Voelcker, N. H. (2015). Bridging the divide between human and environmental nanotoxicology. *Nature Nanotechnology*, *10*(10), 835–844. DOI: 10.1038/nnano.2015.224 PMID: 26440721

Mandal, B. K., & Suzuki, K. T. (2002). Arsenic round the world: A review. *Talanta*, *58*(1), 201–235. DOI: 10.1016/S0039-9140(02)00268-0 PMID: 18968746

Martínez-Ballesta, Mc., Gil-Izquierdo, Á., García-Viguera, C., & Domínguez-Perles, R. (2018). Nanoparticles and Controlled Delivery for Bioactive Compounds: Outlining Challenges for New "Smart-Foods" for Health. *Foods 2018, Vol. 7, Page 72, 7*(5), 72. DOI: 10.3390/foods7050072

McEwen, S. A., & Collignon, P. J. (2018). Antimicrobial resistance: A One Health perspective. *Microbiology Spectrum*, *6*(2), 6.2.10. Advance online publication. DOI: 10.1128/microbiolspec.ARBA-0009-2017 PMID: 29600770

Melaiye, A., Sun, Z., Hindi, K., Milsted, A., Ely, D., Reneker, D. H., Tessier, C. A., & Youngs, W. J. (2005). Silver(I)-imidazole cyclophane gem-diol complexes encapsulated by electrospun tecophilic nanofibers: Formation of nanosilver particles and antimicrobial activity. *Journal of the American Chemical Society*, *127*(7), 2285–2291. DOI: 10.1021/ja040226s PMID: 15713108

Miri, A., Shahraki Vahed, H. O., & Sarani, M. (2018). Biosynthesis of silver nanoparticles and their role in photocatalytic degradation of methylene blue dye. *Research on Chemical Intermediates*, *44*(11), 6907–6915. DOI: 10.1007/s11164-018-3529-3

Mohammadinejad, R., Moosavi, M. A., Tavakol, S., Vardar, D. Ö., Hosseini, A., Rahmati, M., Dini, L., Hussain, S., Mandegary, A., & Klionsky, D. J. (2019). Necrotic, apoptotic and autophagic cell fates triggered by nanoparticles. *Autophagy*, *15*(1), 4–33. DOI: 10.1080/15548627.2018.1509171 PMID: 30160607

Mohanta, Y. K., Mishra, A. K., Nayak, D., Patra, B., Bratovcic, A., Avula, S. K., Mohanta, T. K., Murugan, K., & Saravanan, M. (2022). Exploring Dose-Dependent Cytotoxicity Profile of Gracilaria edulis-Mediated Green Synthesized Silver Nanoparticles against MDA-MB-231 Breast Carcinoma. *Oxidative Medicine and Cellular Longevity*, *2022*(1), 3863138. DOI: 10.1155/2022/3863138 PMID: 35251470

Molnár, Z., Bódai, V., Szakacs, G., Erdélyi, B., Fogarassy, Z., Sáfrán, G., Varga, T., Kónya, Z., Tóth-Szeles, E., Szucs, R., & Lagzi, I. (2018). Green synthesis of gold nanoparticles by thermophilic filamentous fungi. *Scientific Reports*, *8*(1), 3943. Advance online publication. DOI: 10.1038/s41598-018-22112-3 PMID: 29500365

Mony, C., Kaur, P., Rookes, J. E., Callahan, D. L., Eswaran, S. V., Yang, W., & Manna, P. K. (2022). Nanomaterials for enhancing photosynthesis: Interaction with plant photosystems and scope of nanobionics in agriculture. *Environmental Science. Nano*, *9*(10), 3659–3683. DOI: 10.1039/D2EN00451H

Motsaathebe, P. C., & Fayemi, O. E. (2021). Serotonin electrochemical detection in tomatoes at MWCNT-AONP nanocomposite modified electrode. *Materials Research Express*, *8*(11), 115004. DOI: 10.1088/2053-1591/ac31fd

Mukherjee, S., Chowdhury, D., Kotcherlakota, R., Patra, S., Vinothkumar, B., Bhadra, M. P., Sreedhar, B., & Patra, C. R. (2014). Potential Theranostics Application of Bio-Synthesized Silver Nanoparticles (4-in-1 System). *Theranostics*, *4*(3), 316–335. DOI: 10.7150/thno.7819 PMID: 24505239

Murshed, S. M. S., & Nieto De Castro, C. A. (2014). Superior thermal features of carbon nanotubes-based nanofluids – A review. *Renewable & Sustainable Energy Reviews*, *37*, 155–167. DOI: 10.1016/j.rser.2014.05.017

Na, G., Zhang, W., Gao, H., Wang, C., Li, R., Zhao, F., Zhang, K., & Hou, C. (2021). Occurrence and antibacterial resistance of culturable antibiotic-resistant bacteria in the Fildes Peninsula, Antarctica. *Marine Pollution Bulletin*, *162*, 111829. DOI: 10.1016/j.marpolbul.2020.111829 PMID: 33243441

Nadaf, S. J., Jadhav, N. R., Naikwadi, H. S., Savekar, P. L., Sapkal, I. D., Kambli, M. M., & Desai, I. A. (2022). Green synthesis of gold and silver nanoparticles: Updates on research, patents, and future prospects. *OpenNano*, *8*, 100076. DOI: 10.1016/j.onano.2022.100076

Narayanan, M., Divya, S., Natarajan, D., Senthil-Nathan, S., Kandasamy, S., Chinnathambi, A., Alahmadi, T. A., & Pugazhendhi, A. (2021). Green synthesis of silver nanoparticles from aqueous extract of Ctenolepis garcini L. and assess their possible biological applications. *Process Biochemistry (Barking, London, England)*, *107*, 91–99. DOI: 10.1016/j.procbio.2021.05.008

Naseer, A., Ali, A., Ali, S., Mahmood, A., Kusuma, H. S., Nazir, A., Yaseen, M., Khan, M. I., Ghaffar, A., Abbas, M., & Iqbal, M. (2020). Biogenic and eco-benign synthesis of platinum nanoparticles (Pt NPs) using plants aqueous extracts and biological derivatives: Environmental, biological and catalytic applications. *Journal of Materials Research and Technology*, *9*(4), 9093–9107. DOI: 10.1016/j.jmrt.2020.06.013

Nellore, J., Pauline, C., & Amarnath, K. (2013). Bacopa monnieri Phytochemicals Mediated Synthesis of Platinum Nanoparticles and Its Neurorescue Effect on 1-Methyl 4-Phenyl 1,2,3,6 Tetrahydropyridine-Induced Experimental Parkinsonism in Zebrafish. *Journal of Neurodegenerative Diseases*, *2013*, 1–8. DOI: 10.1155/2013/972391 PMID: 26317003

Ng, K. K., Lovell, J. F., & Zheng, G. (2011). Lipoprotein-inspired nanoparticles for cancer theranostics. *Accounts of Chemical Research*, *44*(10), 1105–1113. DOI: 10.1021/ar200017e PMID: 21557543

Nikolova, M. P., & Chavali, M. S. (2020). Metal Oxide Nanoparticles as Biomedical Materials. *Biomimetics 2020, Vol. 5, Page 27*, *5*(2), 27. DOI: 10.3390/biomimetics5020027

Nischwitz, V., Berthele, A., & Michalke, B. (2008). Speciation analysis of selected metals and determination of their total contents in paired serum and cerebrospinal fluid samples: An approach to investigate the permeability of the human blood-cerebrospinal fluid-barrier. *Analytica Chimica Acta*, *627*(2), 258–269. DOI: 10.1016/j.aca.2008.08.018 PMID: 18809082

Novoselov, K. S., Mishchenko, A., Carvalho, A., & Castro Neto, A. H. (2016). 2D materials and van der Waals heterostructures. In Science (Vol. 353, Issue 6298). American Association for the Advancement of Science. DOI: 10.1126/science.aac9439

Nozik, A. J. (2002). Quantum dot solar cells. *Physica E, Low-Dimensional Systems and Nanostructures*, *14*(1–2), 115–120. DOI: 10.1016/S1386-9477(02)00374-0

Numata, K., & Kaplan, D. L. (2010). Silk-based delivery systems of bioactive molecules. *Advanced Drug Delivery Reviews*, *62*(15), 1497–1508. DOI: 10.1016/j.addr.2010.03.009 PMID: 20298729

Olasehinde, T. A., Olaniran, A. O., Okoh, A. I., & Koulen, P. (2017). Therapeutic potentials of microalgae in the treatment of Alzheimer's disease. *Molecules (Basel, Switzerland)*, *22*(3), 480. Advance online publication. DOI: 10.3390/molecules22030480 PMID: 28335462

Oturan, M. A., Aaron, J.-J., Oturan, M. A., & Aaron, J.-J. (2014). Advanced Oxidation Processes in Water/Wastewater Treatment: Principles and Applications. *Critical Reviews in Environmental Science and Technology*, *44*(23), 2577–2641. Https://Doi.Org/10.1080/10643389.2013.829765. DOI: 10.1080/10643389.2013.829765

Ozalp, V. C., Karabiyik, G., Bayrac, A. T., & Guvenc Tuna, B. (2019). Graphene-Based Electrochemical Aptasensors. In B. Palys (Ed.), Handbook of Graphene, Biosensors and Advanced Sensor: Vol. vol.6 (University of Warsaw, p. 465). Warsaw: Scrivener Publishing. DOI: 10.1002/9781119468455.ch101

Ozdemir, S., Turkan, Z., Kilinc, E., Altuner, E. E., & Sen, F. (2022). Anoxybacillus flavithermus loaded ɣ-Fe2O3 magnetic nanoparticles as an efficient magnetic sorbent for the preconcentrations of Cu(II) and Mn(II). *Food and Chemical Toxicology*, *168*, 113334. DOI: 10.1016/j.fct.2022.113334 PMID: 35952822

Özdemir, S., Turkan, Z., Kılınc, E., Altuner, E. E., & Sen, F. (2023). Bioaccumulation, resistance, and remediation of Mn(II) and Cu(II) and their impacts on antioxidant enzymes of Anoxybacillus flavithermus. *International Journal of Environmental Science and Technology*, *20*(10), 10823–10834. DOI: 10.1007/s13762-023-05133-y

Page, M. J., McKenzie, J. E., Bossuyt, P. M., Boutron, I., Hoffmann, T. C., Mulrow, C. D., Shamseer, L., Tetzlaff, J. M., Akl, E. A., Brennan, S. E., Chou, R., Glanville, J., Grimshaw, J. M., Hróbjartsson, A., Lalu, M. M., Li, T., Loder, E. W., Mayo-Wilson, E., McDonald, S., & Moher, D. (2021). The PRISMA 2020 statement: An updated guideline for reporting systematic reviews. *Journal of Clinical Epidemiology*, *134*, 178–189. DOI: 10.1016/j.jclinepi.2021.03.001 PMID: 33789819

Pal, G. D., Shaikh, M., Forsyth, C. B., Ouyang, B., Keshavarzian, A., & Shannon, K. M. (2015). Abnormal lipopolysaccharide binding protein as marker of gastrointestinal inflammation in Parkinson disease. *Frontiers in Neuroscience*, *9*(SEP), 152896. DOI: 10.3389/fnins.2015.00306 PMID: 26388718

Pan, Z., Rao, H., Mora-Seró, I., Bisquert, J., & Zhong, X. (2018). Quantum dot-sensitized solar cells. *Chemical Society Reviews*, *47*(20), 7659–7702. DOI: 10.1039/C8CS00431E PMID: 30209490

Pandey, S., Goswami, G. K., & Nanda, K. K. (2013). Green synthesis of polysaccharide/gold nanoparticle nanocomposite: An efficient ammonia sensor. *Carbohydrate Polymers*, *94*(1), 229–234. DOI: 10.1016/j.carbpol.2013.01.009 PMID: 23544533

Panigrahi, S., Kundu, S., Ghosh, S. K., Nath, S., & Pal, T. (2004). General method of synthesis for metal nanoparticles. *Journal of Nanoparticle Research*, *6*(4), 411–414. DOI: 10.1007/s11051-004-6575-2

Parashar, S., Sharma, M. K., Garg, C., & Garg, M. (2021). Green Synthesized Silver Nanoparticles as Silver Lining in Antimicrobial Resistance: A Review. *Current Drug Delivery*, *19*(2), 170–181. DOI: 10.2174/1567201818666210331123022 PMID: 33797368

Park, S.-J., Kim, S., Lee, S., Khim, Z. G., Char, K., & Hyeon, T. (2000). Synthesis and magnetic studies of uniform iron nanorods and nanospheres. *Journal of the American Chemical Society*, *122*(35), 8581–8582. DOI: 10.1021/ja001628c

Pawlak, K., & Kołodziejczak, M. (2020). The Role of Agriculture in Ensuring Food Security in Developing Countries: Considerations in the Context of the Problem of Sustainable Food Production. *Sustainability 2020, Vol. 12, Page 5488*, *12*(13), 5488. DOI: 10.3390/su12135488

Pei, Z., Chen, S., Ding, L., Liu, J., Cui, X., Li, F., & Qiu, F. (2022). Current perspectives and trend of nanomedicine in cancer: A review and bibliometric analysis. *Journal of Controlled Release*, *352*, 211–241. DOI: 10.1016/j.jconrel.2022.10.023 PMID: 36270513

Peralta-Videa, J. R., Huang, Y., Parsons, J. G., Zhao, L., Lopez-Moreno, L., Hernandez-Viezcas, J. A., & Gardea-Torresdey, J. L. (2016). Plant-based green synthesis of metallic nanoparticles: scientific curiosity or a realistic alternative to chemical synthesis? *Nanotechnology for Environmental Engineering 2016 1:1*, *1*(1), 1–29. DOI: 10.1007/s41204-016-0004-5

Pérez-Moreno, A., Fabián, F. L., Hermes, P. H., Edgar, V. N., Ileana, V. R., & Ali, A. (2021). Nanoscience and nanotechnology regarding food packaging and nanomaterials to extending the postharvest life and the shelf life of foods. *Food Losses, Sustainable Postharvest and Food Technologies*, 313–384. DOI: 10.1016/B978-0-12-821912-6.00001-8

Pezzella, A., D'Ischia, M., Napolitano, A., Misuraca, G., & Prota, G. (1997). Iron-mediated generation of the neurotoxin 6-hydroxydopamine quinone by reaction of fatty acid hydroperoxides with dopamine: A possible contributory mechanism for neuronal degeneration in parkinson's disease. *Journal of Medicinal Chemistry*, *40*(14), 2211–2216. DOI: 10.1021/jm970099t PMID: 9216840

Pindiga, N. Y., Ya'u, K. A., Abubakar, A., & Madugu, Y. D. (2022). Green Synthesis and Characterization of Cobalt, Iron and Copper Nanoparticles Derived from the Stem-Bark Extract of Khaya Senegalensis (Mahogany) and Its Antimirobial Activity. *Online Journal of Chemistry*, *2*(1), 1–14. DOI: 10.31586/ojc.2022.158

Polat, V. (2013). *Polikarbazol/Kitosan iletken kompozitlerinin elektrokimyasal sentezi ve karakterizasyonu*. Fen Bilimleri Enstitüsü.

Pollak, P., Fraix, V., Krack, P., Moro, E., Mendes, A., Chabardes, S., Koudsie, A., & Benabid, A. L. (2002). Treatment results: Parkinson's disease. *Movement Disorders*, *17*(S3, SUPPL. 3), S75–S83. Advance online publication. DOI: 10.1002/mds.10146 PMID: 11948759

Pormohammad, A., Monych, N. K., Ghosh, S., Turner, D. L., & Turner, R. J. (2021). Nanomaterials in Wound Healing and Infection Control. *Antibiotics (Basel, Switzerland)*, *10*(5), 473. DOI: 10.3390/antibiotics10050473 PMID: 33919072

Prabhu, S., Vinodhini, S., Elanchezhiyan, C., & Rajeswari, D. (2018). Evaluation of antidiabetic activity of biologically synthesized silver nanoparticles using Pouteria sapota in streptozotocin-induced diabetic rats. *Journal of Diabetes*, *10*(1), 28–42. DOI: 10.1111/1753-0407.12554 PMID: 28323393

Prajapati, D., Jabborova, D., Saharan, B. S., Singh, N., Patani, A., Singh, S., & Joshi, C. (2024). Bionanotechnology: A Paradigm for Advancing Environmental Sustainability. *Indian Journal of Microbiology, 2024*, 1–27. DOI: 10.1007/s12088-024-01389-1 PMID: 40371027

Presentato, A., Piacenza, E., Scurria, A., Albanese, L., Zabini, F., Meneguzzo, F., Nuzzo, D., Pagliaro, M., Martino, D. C., Alduina, R., & Ciriminna, R. (2020). A new water-soluble bactericidal agent for the treatment of polymicrobial infections. *Preprints, 9*(9), 2020080106. DOI: 10.20944/preprints202008.0106.v1

Pu, S. N., Yin, W. Y., Mao, J. F., & Liu, Q. H. (2009). Crosstalk prediction of single- and double-walled carbon-nanotube (SWCNT/ DWCNT) bundle interconnects. *IEEE Transactions on Electron Devices, 56*(4), 560–568. DOI: 10.1109/TED.2009.2014429

Qian, L., & Yang, X. (2006). Polyamidoamine dendrimers-assisted electrodeposition of gold-platinum bimetallic nanoflowers. *The Journal of Physical Chemistry B, 110*(33), 16672–16678. DOI: 10.1021/jp063302h PMID: 16913805

Qin, M., Li, J., & Song, Y. (2022). Toward High Sensitivity: Perspective on Colorimetric Photonic Crystal Sensors. *Analytical Chemistry, 94*(27), 9497–9507. DOI: 10.1021/acs.analchem.2c01804 PMID: 35759455

Raj, M., Moon, J. M., Goyal, R. N., Park, D. S., & Shim, Y. B. (2019). Simultaneous detection of ATP metabolites in human plasma and urine based on palladium nanoparticle and poly(bromocresol green) composite sensor. *Biosensors & Bioelectronics, 126*, 758–766. DOI: 10.1016/j.bios.2018.11.056 PMID: 30554097

Rajam, B. M., Ramasamy, P., & Mahalingam, U. (2019). Antimicrobial activity of green synthesized plasmonic nanoparticles. *Materials for Biomedical Engineering: Bioactive Materials for Antimicrobial, Anticancer, and Gene Therapy*, 117–151. DOI: 10.1016/B978-0-12-818435-6.00005-0

Rashtbari, S., Dehghan, G., Khataee, S., Amini, M., & Khataee, A. (2021). Dual enzymes-mimic activity of nanolayered manganese-calcium oxide for fluorometric determination of metformin. *Chemosphere, 133063*. Advance online publication. DOI: 10.1016/j.chemosphere.2021.133063 PMID: 34848228

Raula, J., Kuivanen, A., Lähde, A., Jiang, H., Antopolsky, M., Kansikas, J., & Kauppinen, E. I. (2007). Synthesis of L-leucine nanoparticles via physical vapor deposition at varying saturation conditions. *Journal of Aerosol Science, 38*(12), 1172–1184. DOI: 10.1016/j.jaerosci.2007.08.009

Raveendran, P., Fu, J., & Wallen, S. L. (2003). Completely "Green" Synthesis and Stabilization of Metal Nanoparticles. *Journal of the American Chemical Society, 125*(46), 13940–13941. DOI: 10.1021/ja029267j PMID: 14611213

Razzaghi, M., Karimi, A., Aghdasinia, H., & Joghataei, M. T. (2017). Oxidase-Peroxidase sequential polymerization for removal of a dye from contaminated water by horseradish peroxidase (HRP)/glucose oxidase (GOx)/polyurethane hybrid catalyst. *Korean Journal of Chemical Engineering, 34*(11), 2870–2878. DOI: 10.1007/s11814-017-0183-1

Reddy, K. R. (2017). Green synthesis, morphological and optical studies of CuO nanoparticles. *Journal of Molecular Structure, 1150*, 553–557. DOI: 10.1016/j.molstruc.2017.09.005

Rocha, V., Lago, A., Silva, B., Barros, Ó., Neves, I. C., & Tavares, T. (2024). Immobilization of biogenic metal nanoparticles on sustainable materials – green approach applied to wastewater treatment: A systematic review. *Environmental Science. Nano*, *11*(1), 36–60. DOI: 10.1039/D3EN00623A

Rosegrant, M. W., Ringler, C., Benson, T., Diao, X., Resnick, D., Thurlow, J., Torero, M., & Orden, D. (2005). *Agriculture and achieving the Millennium Development Goals*. World Bank. Retrieved from https://hdl.handle.net/10568/158139

Sahin, O. G., Gulce, H., & Gulce, A. (2013). Polyvinylferrocenium based platinum electrodeposited amperometric biosensors for lysine detection. *Journal of Electroanalytical Chemistry (Lausanne, Switzerland)*, *690*, 1–7. DOI: 10.1016/j.jelechem.2012.11.021

Saravanan, A., Kumar, P. S., Hemavathy, R. V., Jeevanantham, S., Jawahar, M. J., Neshaanthini, J. P., & Saravanan, R. (2022). A review on synthesis methods and recent applications of nanomaterial in wastewater treatment: Challenges and future perspectives. *Chemosphere*, *307*, 135713. DOI: 10.1016/j.chemosphere.2022.135713 PMID: 35843436

Sargazi, S., Laraib, U., Er, S., Rahdar, A., Hassanisaadi, M., Zafar, M. N., Díez-Pascual, A. M., & Bilal, M. (2022). Application of Green Gold Nanoparticles in Cancer Therapy and Diagnosis. *Nanomaterials 2022, Vol. 12, Page 1102*, *12*(7), 1102. DOI: 10.3390/nano12071102

Sathishkumar, P., Preethi, J., Vijayan, R., Mohd Yusoff, A. R., Ameen, F., Suresh, S., Balagurunathan, R., & Palvannan, T. (2016). Anti-acne, anti-dandruff and anti-breast cancer efficacy of green synthesised silver nanoparticles using Coriandrum sativum leaf extract. *Journal of Photochemistry and Photobiology. B, Biology*, *163*, 69–76. DOI: 10.1016/j.jphotobiol.2016.08.005 PMID: 27541567

Schnell, M. J., Johnson, J. E., Buonocore, L., & Rose, J. K. (1997). Construction of a novel virus that targets HIV-1-infected cells and controls HIV-1 infection. *Cell*, *90*(5), 849–857. DOI: 10.1016/S0092-8674(00)80350-5 PMID: 9298897

Shankar, S., & Rhim, J. W. (2015). Amino acid mediated synthesis of silver nanoparticles and preparation of antimicrobial agar/silver nanoparticles composite films. *Carbohydrate Polymers*, *130*, 353–363. DOI: 10.1016/j.carbpol.2015.05.018 PMID: 26076636

Sharma, A., & Das, J. (2019). Small molecules derived carbon dots: Synthesis and applications in sensing, catalysis, imaging, and biomedicine. In Journal of Nanobiotechnology (Vol. 17, Issue 1, p. 92). BioMed Central Ltd. DOI: 10.1186/s12951-019-0525-8

Sharma, L., & Tagad, C. (2022). Biosynthesis of polysaccharides-stabilized metal nanoparticles for chemical and biosensing applications. *Polysaccharide Nanoparticles: Preparation and Biomedical Applications*, 553–583. DOI: 10.1016/B978-0-12-822351-2.00021-8

Shastri, D. H., & Gandhi, S. (2024). Nanorevolution Unleashing the Power of Nanotechnology. *Current Nanomedicine*, *14*(3), 227–246. DOI: 10.2174/0124681873279660231226070118

Sheng, W., Liu, J., Liu, S., Lu, Q., Kaplan, D. L., & Zhu, H. (2014). One-step synthesis of biocompatible magnetite/silk fibroin core-shell nanoparticles. *Journal of Materials Chemistry. B*, *2*(42), 7394–7402. DOI: 10.1039/C4TB01125B PMID: 32261964

Shibata, T., Ishii, T., Nozawa, H., & Tamamura, T. (1997). High-aspect-ratio nanometer-pattern fabrication using fullerene-incorporated nanocomposite resists for dry-etching application. *Japanese Journal of Applied Physics*, *36*(12, SUPPL. B), 7642–7645. DOI: 10.1143/JJAP.36.7642

Shivaji, S., Madhu, S., & Singh, S. (2011). Extracellular synthesis of antibacterial silver nanoparticles using psychrophilic bacteria. *Process Biochemistry (Barking, London, England)*, *46*(9), 1800–1807. DOI: 10.1016/j.procbio.2011.06.008

Silva, G. A. (2006). Neuroscience nanotechnology: Progress, opportunities and challenges. *Nature Reviews. Neuroscience*, *7*(1), 65–74. DOI: 10.1038/nrn1827 PMID: 16371951

Singh, Rajender. (2024). Introduction to Transmission Electron Microscopy (TEM). *Transmission Electron Microscopy Sample Preparation*, 1–20. DOI: 10.1007/978-3-031-82967-3_1

Singh, R., Mehra, R., Walia, A., Gupta, S., Chawla, P., kumar, H., Thakur, A., Kaushik, R., & Kumar, N. (2023). Colorimetric sensing approaches based on silver nanoparticles aggregation for determination of toxic metal ions in water sample: A review. *International Journal of Environmental Analytical Chemistry*, *103*(6), 1361–1376. DOI: 10.1080/03067319.2021.1873315

Singh, R., Singh, S., Parihar, P., Singh, V. P., & Prasad, S. M. (2015). Arsenic contamination, consequences and remediation techniques: A review. *Ecotoxicology and Environmental Safety*, *112*, 247–270. DOI: 10.1016/j.ecoenv.2014.10.009 PMID: 25463877

Singh, S. C., Mishra, S. K., Srivastava, R. K., & Gopal, R. (2010). Optical properties of selenium quantum dots produced with laser irradiation of water suspended Se nanoparticles. *The Journal of Physical Chemistry. C, Nanomaterials and Interfaces*, *114*(41), 17374–17384. DOI: 10.1021/jp105037w

Smith, D. K., Hirst, A. R., Love, C. S., Hardy, J. G., Brignell, S. V., & Huang, B. (2005). Self-assembly using dendritic building blocks—Towards controllable nanomaterials. *Progress in Polymer Science*, *30*(3–4), 220–293. DOI: 10.1016/j.progpolymsci.2005.01.006

Soenen, S. J. H., Illyes, E., Vercauteren, D., Braeckmans, K., Majer, Z., De Smedt, S. C., & De Cuyper, M. (2009). The role of nanoparticle concentration-dependent induction of cellular stress in the internalization of non-toxic cationic magnetoliposomes. *Biomaterials*, *30*(36), 6803–6813. DOI: 10.1016/j.biomaterials.2009.08.050 PMID: 19765821

Songa, E. A., & Okonkwo, J. O. (2016). Recent approaches to improving selectivity and sensitivity of enzyme-based biosensors for organophosphorus pesticides: A review. *Talanta*, *155*, 289–304. DOI: 10.1016/j.talanta.2016.04.046 PMID: 27216686

Soni, R. A., Rizwan, M. A., & Singh, S. (2022). Opportunities and potential of green chemistry in nanotechnology. *Nanotechnology for Environmental Engineering*, *7*(3), 661–673. DOI: 10.1007/s41204-022-00233-5

Soundararajan, D., Ramana, L. N., Shankaran, P., & Krishnan, U. M. (2022). Nanoparticle-based strategies to target HIV-infected cells. *Colloids and Surfaces. B, Biointerfaces*, *213*, 112405. DOI: 10.1016/j.colsurfb.2022.112405 PMID: 35255375

Sozer, N., & Kokini, J. L. (2009). Nanotechnology and its applications in the food sector. *Trends in Biotechnology*, *27*(2), 82–89. DOI: 10.1016/j.tibtech.2008.10.010 PMID: 19135747

Sprenger, S. (2013). Epoxy resin composites with surface-modified silicon dioxide nanoparticles: A review. *Journal of Applied Polymer Science*, *130*(3), 1421–1428. DOI: 10.1002/app.39208

Stylios, G. K. (2013). There is plenty of room at the bottom, R.P. Feynman. *International Journal of Clothing Science and Technology*, *25*(5). Advance online publication. DOI: 10.1108/IJCST-06-2013-0067

Sun, C., & Gallaghar, R. G., E. caspersun@comcast. ne. (2008). *Nanomaterials: new challenges in environmental health and safety*.

Sun, Z., Martinez, A., & Wang, F. (2016). Optical modulators with 2D layered materials. In Nature Photonics (Vol. 10, Issue 4, pp. 227–238). Nature Publishing Group. DOI: 10.1038/nphoton.2016.15

Tagad, C. K., Dugasani, S. R., Aiyer, R., Park, S., Kulkarni, A., & Sabharwal, S. (2013). Green synthesis of silver nanoparticles and their application for the development of optical fiber based hydrogen peroxide sensor. *Sensors and Actuators. B, Chemical*, *183*, 144–149. DOI: 10.1016/j.snb.2013.03.106

Tavares, T. D., Antunes, J. C., Padrão, J., Ribeiro, A. I., Zille, A., Amorim, M. T. P., Ferreira, F., & Felgueiras, H. P. (2020). Activity of Specialized Biomolecules against Gram-Positive and Gram-Negative Bacteria. *Antibiotics 2020, Vol. 9, Page 314*, *9*(6), 314. DOI: 10.3390/antibiotics9060314

Thapa, D., Kulkarni, N., Mishra, S. N., Paulose, P. L., & Ayyub, P. (2010). Enhanced magnetization in cubic ferrimagnetic CuFe2O4 nanoparticles synthesized from a citrate precursor: The role of Fe2+. *Journal of Physics. D, Applied Physics*, *43*(19), 195004. DOI: 10.1088/0022-3727/43/19/195004

Tian, J., Wong, K. K. Y., Ho, C. M., Lok, C. N., Yu, W. Y., Che, C. M., Chiu, J. F., & Tam, P. K. H. (2007). Topical delivery of silver nanoparticles promotes wound healing. *ChemMedChem*, *2*(1), 129–136. DOI: 10.1002/cmdc.200600171 PMID: 17075952

Tichapondwa, S. M., Newman, J. P., & Kubheka, O. (2020). Effect of TiO2 phase on the photocatalytic degradation of methylene blue dye. *Physics and Chemistry of the Earth Parts A/B/C*, *118–119*, 102900. DOI: 10.1016/j.pce.2020.102900

Tijing, L. D., Dizon, J. R. C., Ibrahim, I., Nisay, A. R. N., Shon, H. K., & Advincula, R. C. (2020). 3D printing for membrane separation, desalination and water treatment. In *Applied Materials Today* (Vol. 18). Elsevier Ltd., DOI: 10.1016/j.apmt.2019.100486

Timuralp, C., Ercan, E., Ercetin, U., Aygun, A., Esra Altuner, E., & Sen, F. (2024). Nanoencapsulation of phase change material with CuO nanoparticles: Development of thermal properties for energy storage. *Inorganic Chemistry Communications*, *170*, 113258. DOI: 10.1016/j.inoche.2024.113258

Tiwari, P., Nayak, P., Prusty, S. K., & Sahu, P. K. (2018). Phytochemistry and pharmacology of tinospora cordifolia: A review. *Systematic Reviews in Pharmacy*, *9*(1), 70–78. DOI: 10.5530/srp.2018.1.14

Tunca, E. Ü. (2016). Nanoteknolojinin Temeli Nanopartiküller Ve Nanopartiküllerin Fitoremediasyon U. *Ordu University Journal of Science and Tecnology, 5*(2), 23–34. Retrieved from https://dergipark.org .tr/en/pub/ordubtd/issue/11070/132200

Tungare, K., Gupta, J., Bhori, M., Garse, S., Kadam, A., Jha, P., Jobby, R., Amanullah, M., & Vijayakumar, S. (2024). Nanomaterial in controlling biofilms and virulence of microbial pathogens. *Microbial Pathogenesis, 192*, 106722. DOI: 10.1016/j.micpath.2024.106722 PMID: 38815775

Türkaslan, B., Aktaş, M., & Akar, S.TÜRKASLAN. (2020). B., AKTAŞ, M., & AKAR, S. (2020). Fonksiyonelleştirilmiş Titanyum Dioksit Nanotüp Sentezi. *Süleyman Demirel Üniversitesi Fen Edebiyat Fakültesi Fen Dergisi, 15*(1), 167–176. DOI: 10.29233/sdufeffd.666020

Tysnes, O. B., & Storstein, A. (2017). Epidemiology of Parkinson's disease. *Journal of Neural Transmission (Vienna, Austria), 124*(8), 901–905. DOI: 10.1007/s00702-017-1686-y PMID: 28150045

Umegaki, T., Yan, J. M., Zhang, X. B., Shioyama, H., Kuriyama, N., & Xu, Q. (2009). Hollow Ni-SiO2 nanosphere-catalyzed hydrolytic dehydrogenation of ammonia borane for chemical hydrogen storage. *Journal of Power Sources, 191*(2), 209–216. DOI: 10.1016/j.jpowsour.2009.02.054

Usmani, A., Mishra, A., Jafri, A., Arshad, M., & Siddiqui, M. A. (2019). Green Synthesis of Silver Nanocomposites of Nigella sativa Seeds Extract for Hepatocellular Carcinoma. *Current Nanomaterials, 4*(3), 191–200. DOI: 10.2174/2468187309666190906130115

van Droffelaar, B., & Jacobs, M. (2018). Nature-Based Training Program Fosters Authentic Leadership. *Journal of Leadership Studies, 12*(3), 7–18. DOI: 10.1002/jls.21569

Vegliό, F., Beolchini, F., & Gasbarro, A. (1997). Biosorption of toxic metals: An equilibrium study using free cells of Arthrobacter sp. *Process Biochemistry (Barking, London, England), 32*(2), 99–105. DOI: 10.1016/S0032-9592(96)00047-7

Venkatachalam, M., Singaravelu, G., Govindaraju, K., & Ahn, J. S. (2013). *PTP 1B inhibitory action of a phytochemical propanoic acid, 2-(3-acetoxy-4,4,14-trimethylandrost-8-en-17-yl)*. Current Science, JSTOR. Retrieved from https://www.jstor.org/stable/24097522

Verma, K. K., Song, X. P., Joshi, A., Tian, D. D., Rajput, V. D., Singh, M., Arora, J., Minkina, T., & Li, Y. R. (2022). Recent Trends in Nano-Fertilizers for Sustainable Agriculture under Climate Change for Global Food Security. *Nanomaterials 2022, Vol. 12, Page 173, 12*(1), 173. DOI: 10.3390/nano12010173

Verma, P., & Ratan, J. K. (2020). Assessment of the negative effects of various inorganic water pollutants on the biosphere—an overview. *Inorganic Pollutants in Water*, 73–96. DOI: 10.1016/B978-0-12-818965-8.00005-6

Voukkali, I., Papamichael, I., Loizia, P., Economou, F., Stylianou, M., Naddeo, V., & Zorpas, A. A. (2024). Fashioning the Future: Green chemistry and engineering innovations in biofashion. *Chemical Engineering Journal, 497*, 155039. DOI: 10.1016/j.cej.2024.155039

Wahab, M., & Janaswamy, S. (2023). A review on biogenic silver nanoparticles as efficient and effective antidiabetic agents. *Functional Food Science - Online ISSN: 2767-3146, 3*(7), 93–107. DOI: 10.31989/ffs.v3i7.1119

Wang, J., Zhang, S., & Zhang, Y. (2010). Fabrication of chronocoulometric DNA sensor based on gold nanoparticles/poly(l-lysine) modified glassy carbon electrode. *Analytical Biochemistry, 396*(2), 304–309. DOI: 10.1016/j.ab.2009.10.004 PMID: 19818728

Wang, P., Lombi, E., Zhao, F.-J., & Kopittke, P. M. (2016). Nanotechnology: A new opportunity in plant sciences. *Trends in Plant Science*, *21*(8), 699–712. DOI: 10.1016/j.tplants.2016.04.005 PMID: 27130471

Wang, Zhantong, Gao, H., Zhang, Y., Liu, G., Niu, G., & Chen, X. (2017). Functional ferritin nanoparticles for biomedical applications. *Frontiers of Chemical Science and Engineering 2017 11:4, 11*(4), 633–646. DOI: 10.1007/s11705-017-1620-8

Wang, Z., Liao, F., Guo, T., Yang, S., & Zeng, C. (2012). Synthesis of crystalline silver nanoplates and their application for detection of nitrite in foods. *Journal of Electroanalytical Chemistry (Lausanne, Switzerland)*, *664*, 135–138. DOI: 10.1016/j.jelechem.2011.11.006

Westermayr, J., & Marquetand, P. (2021). Machine Learning for Electronically Excited States of Molecules. *Chemical Reviews*, *121*(16), 9873–9926. DOI: 10.1021/acs.chemrev.0c00749 PMID: 33211478

Wong, I. Y., Bhatia, S. N., & Toner, M. (2013). Nanotechnology: Emerging tools for biology and medicine. *Genes & Development*, *27*(22), 2397–2408. DOI: 10.1101/gad.226837.113 PMID: 24240230

Wu, Y., Feng, J., Hu, G., Zhang, E., & Yu, H. H. (2023). Colorimetric Sensors for Chemical and Biological Sensing Applications. *Sensors 2023, Vol. 23, Page 2749, 23*(5), 2749. DOI: 10.3390/s23052749

Xu, C., & Qu, X. (2014). Cerium oxide nanoparticle: a remarkably versatile rare earth nanomaterial for biological applications. *NPG Asia Materials 2014 6:3, 6*(3), e90–e90. DOI: 10.1038/am.2013.88

Yang, B., Wang, L., Zhang, M., Luo, J., Lu, Z., & Ding, X. (2020). Fabrication, Applications, and Prospects of Aramid Nanofiber. *Advanced Functional Materials*, *30*(22), 2000186. DOI: 10.1002/adfm.202000186

Yaqoob, S. B., Adnan, R., Rameez Khan, R. M., & Rashid, M. (2020). Gold, Silver, and Palladium Nanoparticles: A Chemical Tool for Biomedical Applications. *Frontiers in Chemistry*, *8*, 520643. DOI: 10.3389/fchem.2020.00376 PMID: 32582621

Ying, S., Guan, Z., Ofoegbu, P. C., Clubb, P., Rico, C., He, F., & Hong, J. (2022a). Green synthesis of nanoparticles: Current developments and limitations. *Environmental Technology & Innovation*, *26*, 102336. DOI: 10.1016/j.eti.2022.102336

Ying, S., Guan, Z., Ofoegbu, P. C., Clubb, P., Rico, C., He, F., & Hong, J. (2022b). Green synthesis of nanoparticles: Current developments and limitations. *Environmental Technology & Innovation*, *26*, 102336. DOI: 10.1016/j.eti.2022.102336

Youssef, A. M., Abdel-Aziz, M. S., & El-Sayed, S. M. (2014). Chitosan nanocomposite films based on Ag-NP and Au-NP biosynthesis by Bacillus Subtilis as packaging materials. *International Journal of Biological Macromolecules*, *69*, 185–191. DOI: 10.1016/j.ijbiomac.2014.05.047 PMID: 24875320

Zahoor, M., Nazir, N., Iftikhar, M., Naz, S., Zekker, I., Burlakovs, J., Uddin, F., Kamran, A. W., Kallistova, A., Pimenov, N., & Khan, F. A. (2021). A Review on Silver Nanoparticles: Classification, Various Methods of Synthesis, and Their Potential Roles in Biomedical Applications and Water Treatment. *Water 2021, Vol. 13, Page 2216, 13*(16), 2216. DOI: 10.3390/w13162216

Zare, N., Karimi, F., Altuner, E. E., Mashkoor, H., Tiri, R. N. E., Gulbagca, F., Bijad, M., Cherif, A., & Sen, F. (2024). Synthesis of Bimetallic Nickel-Cobalt Nanoparticles Supported by Functional Multi-Walled Carbon Nanotubes by Chemical Methods: Characterization and Hydrogen Energy Generation. *Topics in Catalysis*, *67*(9–12), 606–614. DOI: 10.1007/s11244-024-01907-x

Zhang, M., Li, G., Huang, L., Ran, P., Huang, J., Yu, M., Yuqian, H., Guo, J., Liu, Z., & Ma, X. (2021). Versatile fabrication of liquid metal nano-ink based flexible electronic devices. *Applied Materials Today*, *22*, 100903. DOI: 10.1016/j.apmt.2020.100903

Zhang, Q., Jia, A., Wan, Y., Liu, H., Wang, K., Peng, H., Dong, Z., & Hu, J. (2014). Occurrences of three classes of antibiotics in a natural river basin: Association with antibiotic-resistant Escherichia coli. *Environmental Science & Technology*, *48*(24), 14317–14325. DOI: 10.1021/es503700j PMID: 25427067

Zhou, Q., & Tang, D. (2020). Recent advances in photoelectrochemical biosensors for analysis of mycotoxins in food. *Trends in Analytical Chemistry*, *124*, 115814. DOI: 10.1016/j.trac.2020.115814

Zhu, S., Song, Y., Zhao, X., Shao, J., Zhang, J., & Yang, B. (2015). The photoluminescence mechanism in carbon dots (graphene quantum dots, carbon nanodots, and polymer dots): current state and future perspective. *Nano Research 2014 8:2*, *8*(2), 355–381. DOI: 10.1007/s12274-014-0644-3

Zhu, Y., Zhao, Y., Li, B., Huang, C.-L., Zhang, S.-Y., Yu, S., Chen, Y.-S., Zhang, T., Gillings, M. R., & Su, J.-Q. (2017). Continental-scale pollution of estuaries with antibiotic resistance genes. *Nature Microbiology*, *2*(4), 16270. DOI: 10.1038/nmicrobiol.2016.270 PMID: 28134918

Zia-ur-Rehman, M., Mfarrej, M. F. B., Usman, M., Anayatullah, S., Rizwan, M., Alharby, H. F., Abu Zeid, I. M., Alabdallah, N. M., & Ali, S. (2023). Effect of iron nanoparticles and conventional sources of Fe on growth, physiology and nutrient accumulation in wheat plants grown on normal and salt-affected soils. *Journal of Hazardous Materials*, *458*, 131861. DOI: 10.1016/j.jhazmat.2023.131861 PMID: 37336110

Compilation of References

Aacharya, R., & Chhipa, H. (2020). Nanocarbon fertilizers: Implications of carbon nanomaterials in sustainable agriculture production. In Abd-Elsalam, K. A. (Ed.), *Carbon Nanomaterials for Agri-Food and Environmental Applications* (pp. 297–321). Elsevier., DOI: 10.1016/B978-0-12-819786-8.00015-3

Aarthye, P., & Sureshkumar, M. (2021). Green synthesis of nanomaterials: An overview. *Materials Today: Proceedings*, *47*, 907–913. DOI: 10.1016/j.matpr.2021.04.564

Abasi, F., Raja, N. I., Mashwani, Z. U. R., Amjad, M. S., Ehsan, M., Mustafa, N., Haroon, M., & Proćków, J. (2022). Biogenic silver nanoparticles as a stress alleviator in plants: A mechanistic overview. *Molecules (Basel, Switzerland)*, *27*(11), 3378. DOI: 10.3390/molecules27113378 PMID: 35684312

Abbasifar, A., Shahrabadi, F., & ValizadehKaji, B. (2020). Effects of green synthesized zinc and copper nano-fertilizers on the morphological and biochemical attributes of basil plant. *Journal of Plant Nutrition*, *43*(8), 1104–1118. DOI: 10.1080/01904167.2020.1724305

Abbasi, H., Antunes, M., & Velasco, J. I. (2019). Recent advances in carbon-based polymer nanocomposites for electromagnetic interference shielding. *Progress in Materials Science*, *103*, 319–373. DOI: 10.1016/j.pmatsci.2019.02.003

Abbas, Q., Yousaf, B., Ullah, H., Ali, M. U., Ok, Y. S., & Rinklebe, J. (2020). Environmental transformation and nano-toxicity of engineered nano-particles (ENPs) in aquatic and terrestrial organisms. *Critical Reviews in Environmental Science and Technology*, *50*(23), 2523–2581. DOI: 10.1080/10643389.2019.1705721

Abdallah, H. M., Mohamed, G. A., & Ibrahim, S. R. M. (2022). Lansium domesticum—A Fruit with Multi-Benefits: Traditional Uses, Phytochemicals, Nutritional Value, and Bioactivities. *Nutrients 2022, Vol. 14, Page 1531*, *14*(7), 1531. DOI: 10.3390/nu14071531

Abdallah, N. A., Prakash, C. S., & McHughen, A. G. (2015). Genome editing for crop improvement: Challenges and opportunities. *GM Crops and Food: Biotechnology in Agriculture and the Food Chain*, *6*(4), 183–205. DOI: 10.1080/21645698.2015.1129937 PMID: 26930114

Abdelaziz, A. M., Dacrory, S., Hashem, A. H., Attia, M. S., Hasanin, M., Fouda, H. M., Kamel, S., & ElSaied, H. (2021). Protective role of zinc oxide nanoparticles based hydrogel against wilt disease of pepper plant. *Biocatalysis and Agricultural Biotechnology*, *35*, 102083. https://doi.org/https://doi.org/10.1016/j.bcab.2021.102083. DOI: 10.1016/j.bcab.2021.102083

Abdel-Aziz, H. M., Hasaneen, M. N., & Omer, A. M. (2016). Nano chitosan-NPK fertilizer enhances the growth and productivity of wheat plants grown in sandy soil. *Spanish Journal of Agricultural Research*, *14*(1), e0902–e0902. DOI: 10.5424/sjar/2016141-8205

Abdollahdokht, D., Gao, Y., Faramarz, S., Poustforoosh, A., Abbasi, M., Asadikaram, G., & Nematollahi, M. H. (2022). Conventional agrochemicals towards nano-biopesticides: An overview on recent advances. *Chemical and Biological Technologies in Agriculture*, *9*(1), 13. DOI: 10.1186/s40538-021-00281-0

Abdulhussein, A. T., Kannarpady, G. K., Wright, A. B., Ghosh, A., & Biris, A. S. (2016). Current trend in fabrication of complex morphologically tunable superhydrophobic nano scale surfaces. *Applied Surface Science*, *384*, 311–332. DOI: 10.1016/j.apsusc.2016.04.186

Abdullahi, A., Ahmad, K., Ismail, I. S., Asib, N., Ahmed, O. H., Abubakar, A. I., Siddiqui, Y., & Ismail, M. R. (2020). Potential of using ginger essential oils-based nanotechnology to control tropical plant diseases. *The Plant Pathology Journal*, *36*(6), 515–535. DOI: 10.5423/PPJ.RW.05.2020.0077 PMID: 33312089

Abel, S., Jule, L. T., Belay, F., Shanmugam, R., Dwarampudi, L. P., Nagaprasad, N., & Krishnaraj, R. (2021). Application of Titanium Dioxide Nanoparticles Synthesized by Sol-Gel Methods in Wastewater Treatment. *Journal of Nanomaterials*, *2021*(1), 3039761. DOI: 10.1155/2021/3039761

Abid, N., Khan, A. M., Shujait, S., Chaudhary, K., Ikram, M., Imran, M., & Maqbool, M. (2022). Synthesis of nanomaterials using various top-down and bottom-up approaches, influencing factors, advantages, and disadvantages: A review. *Advances in Colloid and Interface Science*, *300*, 102597. DOI: 10.1016/j.cis.2021.102597 PMID: 34979471

Abu-Thabit, N. Y., & Makhlouf, A. S. H. (2018). Historical development of drug delivery systems: From conventional macroscale to controlled, targeted, and responsive nanoscale systems. In Makhlouf, A. S. H., & Abu-Thabit, N. Y. (Eds.), *Stimuli Responsive Polymeric Nanocarriers for Drug Delivery Applications* (Vol. 1, pp. 3–41). Woodhead Publishing., DOI: 10.1016/B978-0-08-101997-9.00001-1

Acharya, A., & Pal, P. K. (2020). Agriculture nanotechnology: Translating research outcome to field applications by influencing environmental sustainability. *NanoImpact*, *19*, 100232. DOI: 10.1016/j.impact.2020.100232

Adepu, S., & Ramakrishna, S. (2021). Controlled Drug Delivery Systems: Current Status and Future Directions. *Molecules (Basel, Switzerland)*, *26*(19), 5905. DOI: 10.3390/molecules26195905 PMID: 34641447

Adetunji, C. O., Ogundolie F. A., Ajiboye, M. D., Mathew, J. T., Inobeme, A., Dauda, W. P., Ghazanfar, S., Olotu T., Olaniyan, O., Ijabadeniyi, O. A., Ajayi, O. O., and Adetunji, J. B. (2022). Nano-engineered Sensors for Food Processing. In a book: Bio- and Nano-sensing Technologies for Food Processing and Packaging, The Royal Society of Chemistry, 151-166. DOI: 10.1039/9781839167966-00151

Adetunji, C. O., Olotu, T., Mathew, J. T., Inobeme, A., Olaniyan, O. T., Adetunji, M. C., & Adebayo, O. O. (2024). Morphology, rheology, properties, and applications of nanostarch-filled polymer blends, Editor(s): Sabu Thomas, Soney C. George, Sharika T. Nair, In Micro and Nano Technologies, Nanofillers for Binary Polymer Blends, Elsevier, 443-464, DOI: 10.1016/B978-0-323-88655-0.00005-7

Adeyemi, J. O., Oriola, A. O., Onwudiwe, D. C., & Oyedeji, A. O. (2022). Plant Extracts Mediated Metal-Based Nanoparticles: Synthesis and Biological Applications. *Biomolecules 2022, Vol. 12, Page 627, 12*(5), 627. DOI: 10.3390/biom12050627

Adhikari, T., Kundu, S., Biswas, A. K., Tarafdar, J. C., & Rao, A. S. (2015). Characterization of Zinc Oxide Nano Particles and Their Effect on Growth of Maize (*Zea mays* L.) Plant. *Journal of Plant Nutrition, 38*(10), 1505–1515. DOI: 10.1080/01904167.2014.992536

Adil, M., Bashir, S., Bashir, S., Aslam, Z., Ahmad, N., Younas, T., Asghar, R. M. A., Alkahtani, J., Dwiningsih, Y., & Elshikh, M. S. (2022). Zinc oxide nanoparticles improved chlorophyll contents, physical parameters, and wheat yield under salt stress. *Frontiers in Plant Science, 13*, 932861. DOI: 10.3389/fpls.2022.932861 PMID: 35991444

Aesaert, S., Impens, L., Coussens, G., Van Lerberge, E., Vanderhaeghen, R., Desmet, L., Vanhevel, Y., Bossuyt, S., Wambua, A. N., Van Lijsebettens, M., Inzé, D., De Keyser, E., Jacobs, T. B., Karimi, M., & Pauwels, L. (2022). Optimized Transformation and Gene Editing of the B104 Public Maize Inbred by Improved Tissue Culture and Use of Morphogenic Regulators. *Frontiers in Plant Science, 13*, 883847. DOI: 10.3389/fpls.2022.883847 PMID: 35528934

Afshan, N., Bari, A., Khan, A., & Shah, Y. (2023). Advanced Bionanomaterials Used to Recover Pollutants From Wastewater. *Sustainable Nanomaterials for Biosystems Engineering*, 271–296. DOI: 10.1201/9781003333517-13

Agatonovic-Kustrin, S., Kettle, C., & Morton, D. W. (2018). A molecular approach in drug development for Alzheimer's disease. *Biomedicine and Pharmacotherapy, 106*, 553–565. DOI: 10.1016/j.biopha.2018.06.147 PMID: 29990843

Agrawal, U., Sharma, R., & Gupta, M. (2020). Antifungal efficacy of curcumin-loaded nanoparticles. *Drug Delivery, 27*(1), 192–204.

Ahanger, M. A., & Agarwal, R. M. (2017). Salinity stress induced alterations in antioxidant metabolism and nitrogen assimilation in wheat (Triticum aestivum L) as influenced by potassium supplementation. *Plant Physiology and Biochemistry, 115*, 449–460. DOI: 10.1016/j.plaphy.2017.04.017 PMID: 28478373

Ahern, R. J., Hanrahan, J. P., Tobin, J. M., Ryan, K. B., & Crean, A. M. (2013). Comparison of fenofibrate-mesoporous silica drug-loading processes for enhanced drug delivery. *European Journal of Pharmaceutical Sciences, 50*(3-4), 400–409. DOI: 10.1016/j.ejps.2013.08.026 PMID: 23981335

Ahlawat, D. S., Kumari, R., Rachna, , & Yadav, I. (2014). Synthesis and characterization of sol–gel prepared silver nanoparticles. *International Journal of Nanoscience, 13*(01), 1450004. DOI: 10.1142/S0219581X14500045

Ahmad, A., Hashmi, S. S., Palma, J. M., & Corpas, F. J. (2022). Influence of metallic, metallic oxide, and organic nanoparticles on plant physiology. *Chemosphere, 290*, 133329. DOI: 10.1016/j.chemosphere.2021.133329 PMID: 34922969

Ahmad, H. M., Fiaz, S., Hafeez, S., Zahra, S., Shah, A. N., Gul, B., Aziz, O., Mahmood-Ur-Rahman, , Fakhar, A., Rafique, M., Chen, Y., Yang, S. H., & Wang, X. (2022). Plant growth-promoting rhizobacteria eliminate the effect of drought stress in plants: A review. *Frontiers in Plant Science, 13*, 875774. Advance online publication. DOI: 10.3389/fpls.2022.875774 PMID: 36035658

Ahmadi-Majd, M., Mousavi-Fard, S., Rezaei Nejad, A., & Fanourakis, D. (2022). Carbon nanotubes in the holding solution stimulate flower opening and prolong vase life in carnation. *Chemical and Biological Technologies in Agriculture*, *9*(1), 15. DOI: 10.1186/s40538-021-00264-1

Ahmadi-Majd, M., Rezaei Nejad, A., Mousavi-Fard, S., & Fanourakis, D. (2022). Postharvest application of single, multi-walled carbon nanotubes and nanographene oxide improves rose keeping quality. *The Journal of Horticultural Science & Biotechnology*, *97*(3), 346–360. DOI: 10.1080/14620316.2021.1993755

Ahmad, R., Srivastava, S., Ghosh, S., & Khare, S. K. (2021). Phytochemical delivery through nanocarriers: A review. *Colloids and Surfaces. B, Biointerfaces*, *197*, 111389. DOI: 10.1016/j.colsurfb.2020.111389 PMID: 33075659

Ahmad, S., Belwal, V., Punia, S. S., Ram, M., Dalip, , Rajput, S. S., Kunwar, R., Meena, M. K., Gupta, D., Kumawat, G. L., Hussain, T., & Mohamed, H. I. (2022). Role of plant secondary metabolites and phytohormones in drought tolerance: A review. *Gesunde Pflanzen*, *75*(4), 729–746. DOI: 10.1007/s10343-022-00795-z

Ahmad, T. M., Haider, M. S., Randhir, T. O., Randhir, R., & Ahmad, S. R. (2023). Spatial analysis of factors influencing bacterial leaf blight in rice production. *Brazilian Journal of Biology*, *83*, e264249. DOI: 10.1590/1519-6984.264249 PMID: 36946787

Ahmad, Z., Tahseen, S., Wasi, A., Ganie, I. B., Shahzad, A., Emamverdian, A., Ramakrishnan, M., & Ding, Y. (2022). Nanotechnological Interventions in Agriculture. *Nanomaterials (Basel, Switzerland)*, *2022*(12), 2667. DOI: 10.3390/nano12152667 PMID: 35957097

Ahmar, S., Mahmood, T., Fiaz, S., Mora-Poblete, F., Shafique, M. S., Chattha, M. S., & Jung, K. H. (2021). Advantage of Nanotechnology-Based Genome Editing System and Its Application in Crop Improvement. *Frontiers in Plant Science*, *12*, 663849. DOI: 10.3389/fpls.2021.663849 PMID: 34122485

Ahmed, T., Noman, M., Qi, Y., White, J. C., & Qi, X. (2024). Protein nanocarrier-enabled plant genetic engineering systems. *Trends in Biotechnology*, *0*(0). Advance online publication. DOI: 10.1016/j.tibtech.2024.08.009 PMID: 39277495

Ahovan, Z. A., Hashemi, A., De Plano, L. M., Gholipourmalekabadi, M., & Seifalian, A. (2020). Bacteriophage Based Biosensors: Trends, Outcomes and Challenges. *Nanomaterials 2020, Vol. 10, Page 501, 10*(3), 501. DOI: 10.3390/nano10030501

Ahuja, R., Panwar, N., Meena, J., Singh, M., Sarkar, D. P., & Panda, A. K. (2020). Natural products and polymeric nanocarriers for cancer treatment: A review. *Environmental Chemistry Letters*, *18*(6), 2021–2030. DOI: 10.1007/s10311-020-01056-z

Ainiwaer, M., Jia, H., Zhang, T., Huang, J., Zhang, N., Yin, X., Peng, L., Li, H., & Zeng, X. (2024). Effective co-immobilization of arsenic and cadmium in contaminated soil by sepiolite-modified nano-zero-valent iron and its impact on the soil bacterial community. *Scientific Reports*, *14*(1), 26178. DOI: 10.1038/s41598-024-77066-6 PMID: 39478164

Ajayan, P. M., & Lijima, S. (1993). Capillarity-induced filling of carbon nanotubes. *Nature*, *361*(6410), 333–334. DOI: 10.1038/361333a0

Ajeng, A. A., Abdullah, R., Ling, T. C., Ismail, S., Lau, B. F., Ong, H. C., Chew, K. W., Show, P. L., & Chang, J. S. (2020). Bioformulation of biochar as a potential inoculant carrier for sustainable agriculture. *Environmental Technology & Innovation*, *20*, 101168. DOI: 10.1016/j.eti.2020.101168

Akinhanmi, F. O., Aworunse, O. S., Oyewole, O. A., Adetunji, C. O., Mathew, J. T., Igiku, V., Eniola, K. I. T., & Yerima, M. B. (2024). Non-Target Effect, Environmental Impact, and Assessment of Nanobiofertilizer. In Book: Handbook of Agricultural Biotechnology. Wiley, 5, 253-270. Doi: DOI: 10.1002/9781394211548.ch10

Al Janabi, M. A. Y., El Houda Tiri, R. N., Cherif, A., Altuner, E. E., Lee, C. J., Sen, F., Dragoi, E. N., Karimi, F., & Kalikeri, S. (2024). Hydrogen Generation by Methanolysis of NaBH4 via Efficient CuFe2O4 Nanoparticle Catalyst: A Kinetic Study and DNN Model. *Topics in Catalysis*, *67*(9–12), 843–852. DOI: 10.1007/s11244-024-01904-0

Albanese, A., Tang, P. S., & Chan, W. C. (2012). The effect of nanoparticle size, shape, and surface chemistry on biological systems. *Annual Review of Biomedical Engineering*, *14*(1), 1–16. DOI: 10.1146/annurev-bioeng-071811-150124 PMID: 22524388

Alemzadeh, E., Dehshahri, A., Izadpanah, K., & Ahmadi, F. (2018). Plant virus nanoparticles: Novel and robust nanocarriers for drug delivery and imaging. *Colloids and Surfaces. B, Biointerfaces*, *167*, 20–27. DOI: 10.1016/j.colsurfb.2018.03.026 PMID: 29625419

Alex, B., Jignasa, G., Madhubabu, K., & Gopi, A. (2024, August). AI-Driven Smart Irrigation: Enhancing Agricultural Water Efficiency Through Intelligent Valve Regulation in Piped and Micro Irrigation Networks. In *2024 First International Conference on Pioneering Developments in Computer Science & Digital Technologies (IC2SDT)* (pp. 76-81). IEEE. https://doi.org/DOI: 10.1109/IC2SDT62152.2024.10696274

Alghuthaymi, M., Abd-Elsalam, K. A., Paraliker, P., & Rai, M. (2020). Mono and hybrid nanomaterials: Novel strategies to manage postharvest diseases. *Multifunctional Hybrid Nanomaterials for Sustainable Agri-Food and Ecosystems*, 287-317. https://do i.org/10.1016/B978-0-12-821354-4.00013-3

Alghuthaymi, M. A., Ahmad, A., Khan, Z., Khan, S. H., Ahmed, F. K., Faiz, S., Nepovimova, E., Kuča, K., & Abd-Elsalam, K. A. (2021). Exosome/liposome-like nanoparticles: New carriers for CRISPR genome editing in plants. *International Journal of Molecular Sciences*, *22*(14), 7456. DOI: 10.3390/ijms22147456 PMID: 34299081

Alghuthaymi, M. A., C, R., P, R., Kalia, A., Bhardwaj, K., Bhardwaj, P., Abd-Elsalam, K. A., Valis, M., & Kuca, K. (2021). Nanohybrid antifungals for control of plant diseases: Current status and future perspectives. *Journal of Fungi (Basel, Switzerland)*, *7*(1), 48. DOI: 10.3390/jof7010048 PMID: 33450851

Ali, A., Bang, S. W., Chung, S. M., & Staub, J. E. (2014). Plant transformation via pollen tube-mediated gene transfer. Plant. *Plant Molecular Biology Reporter*, *33*(3), 742–747. DOI: 10.1007/s11105-014-0839-5

Ali, A., Shahbaz, M., Ölmez, F., Fatima, N., Umar, U. U. D., Ali, M., Akram, M., Seelan, J. S. S., & Baloch, F. S. (2024). RNA interference: A promising biotechnological approach to combat plant pathogens, mechanism and future prospects. *World Journal of Microbiology & Biotechnology*, *40*(11), 339. DOI: 10.1007/s11274-024-04143-3 PMID: 39358476

Ali, N., Bilal, M., Khan, A., Ali, F., & Iqbal, H. M. N. (2020). Effective exploitation of anionic, nonionic, and nanoparticle-stabilized surfactant foams for petroleum hydrocarbon contaminated soil remediation. *The Science of the Total Environment*, *704*, 135371. DOI: 10.1016/j.scitotenv.2019.135391 PMID: 31806317

Ali, O., Ramsubhag, A., & Jayaraman, J. (2021). Biostimulant properties of seaweed extracts in plants: Implications towards sustainable crop production. *Plants*, *10*(3), 531. DOI: 10.3390/plants10030531 PMID: 33808954

Ali, S., Ahmad, N., Dar, M. A., Manan, S., Rani, A., Alghanem, S. M. S., Khan, K. A., Sethupathy, S., Elboughdiri, N., Mostafa, Y. S., Alamri, S. A., Hashem, M., Shahid, M., & Zhu, D. (2023). Nano-Agrochemicals as Substitutes for Pesticides: Prospects and Risks. *Plants*, *13*(1), 109. DOI: 10.3390/plants13010109 PMID: 38202417

Ali, S., Mehmood, A., & Khan, N. (2021). Uptake, translocation, and consequences of nanomaterials on plant growth and stress adaptation. *Journal of Nanomaterials*, *2021*(1), 6677616. DOI: 10.1155/2021/6677616

Alizadeh, M., Asrami, P. N., Altuner, E. E., Gulbagca, F., Tiri, R. N. E., Aygun, A., Kaynak, İ., Sen, F., & Cheraghi, S. (2022). An ultra-sensitive rifampicin electrochemical sensor based on Fe3O4 nanoparticles anchored Multiwalled Carbon nanotube modified glassy carbon electrode. *Chemosphere*, *309*, 136566. DOI: 10.1016/j.chemosphere.2022.136566 PMID: 36152837

Alizadeh, M., Qarachal, J. F., & Sheidaee, E. (2025). Understanding the ecological impacts of nanoparticles: Risks, monitoring, and mitigation strategies. *Nanotechnology for Environmental Engineering*, *10*(1), 6. DOI: 10.1007/s41204-024-00403-7

Almasoud, N., Alhaik, H., Almutairi, M., Houjak, A., Hazazi, K., Alhayek, F., Aljanoubi, S., Alkhaibari, A., Alghamdi, A., Soliman, D. A., Alomar, T. S., & Awad, M. A. (2021). Green nanotechnology synthesized silver nanoparticles: Characterization and testing its antibacterial activity. *Green Processing and Synthesis*, *10*(1), 510–528. DOI: 10.1515/gps-2021-0048

Aloo, B. N., Tripathi, V., Makumba, B. A., & Mbega, E. R. (2022). Plant growth-promoting rhizobacterial biofertilizers for crop production: The past, present, and future. *Frontiers in Plant Science*, *13*, 1002448. DOI: 10.3389/fpls.2022.1002448 PMID: 36186083

Alphandéry, E. (2020). Natural metallic nanoparticles for application in nano oncology. *International Journal of Molecular Sciences*, *21*(12), 4412. DOI: 10.3390/ijms21124412 PMID: 32575884

Alrbaihat, M. R. (2022, May). Agricultural nano fertilizers: macronutrient types and applications review. In *International Conference on Geotechnical Engineering-IRAQ* (pp. 306-316). Singapore: Springer Nature Singapore. https://doi.org/DOI: 10.1007/978-981-19-7358-1_26

Altammar, K. A. (2023). A review on nanoparticles: Characteristics, synthesis, applications, and challenges. *Frontiers in Microbiology*, *14*, 1155622. DOI: 10.3389/fmicb.2023.1155622 PMID: 37180257

Altpeter, F., & James, V. A. (2005). Genetic transformation of turf-type bahiagrass (Paspalum notatum Flugge) by biolistic gene transfer. *International Turfgrass Society Research Journal*, *10*, 485–489.

Altpeter, F., Springer, N. M., Bartley, L. E., Blechl, A., Brutnell, T. P., Citovsky, V., & Voytas, D. F. (2016). Advancing crop transformation in the era of genome editing. *The Plant Cell, 28*(7), 1510–1520. DOI: 10.1105/tpc.16.00196 PMID: 27335450

Altuner, E. E., Akin, M., Bayat, R., Bekmezci, M., Burhan, H., & Sen, F. (2022). Challenges in commercialization of carbon nanomaterial-based sensors. *Carbon Nanomaterials-Based Sensors: Emerging Research Trends in Devices and Applications*, 381–392. DOI: 10.1016/B978-0-323-91174-0.00020-2

Altuner, E. E., Arıkan, K., Burhan, H., Ozdemir, S., & Şen, F. (2021). Commercial aspects of direct alcohol fuel cells. *Nanomaterials for Direct Alcohol Fuel Cells*, 511–524. DOI: 10.1016/B978-0-12-821713-9.00012-3

Altuner, E. E., Bekmezci, M., & Sen, F. (2022). Manufacturing Techniques of Magnetic Polymer Nanocomposites. *Handbook of Magnetic Hybrid Nanoalloys and Their Nanocomposites*, 303–318. DOI: 10.1007/978-3-030-90948-2_12

Altuner, E. E., Bekmezci, M., Bayat, R., Akin, M., Isik, I., & Şen, F. (2021). Dendrimer-based nanocomposites for alcohol fuel cells. *Nanomaterials for Direct Alcohol Fuel Cells*, 337–352. DOI: 10.1016/B978-0-12-821713-9.00007-X

Altuner, E. E., Gur, T., & Şen, F. (2021). Ternary/quaternary nanomaterials for direct alcohol fuel cells. *Nanomaterials for Direct Alcohol Fuel Cells*, 157–172. DOI: 10.1016/B978-0-12-821713-9.00001-9

Altuner, E. E.ALTUNER. (2023). Electrochemical sensitivity, selectivity, and sensory properties of a natural polymer-supported titanium nanocomposite electrode towards lysine amino acid. *Muş Alparslan Üniversitesi Fen Bilimleri Dergisi, 11*(1), 28–32. DOI: 10.18586/msufbd.1268440

Altuner, E. E., Erduran, V., & Sen, F. (2023). Green synthesized nanomaterials for bioimaging. In *Synthesis of Bionanomaterials for Biomedical Applications Micro and Nano Technologies* (pp. 265–286). Elsevier., DOI: 10.1016/B978-0-323-91195-5.00023-4

Altuner, E. E., Gulbagca, F., Tiri, R. N. E., Aygun, A., & Sen, F. (2023). Highly efficient palladium-zinc oxide nanoparticles synthesized by biogenic methods: Characterization, hydrogen production and photocatalytic activities. *Chemical Engineering Journal Advances, 14*, 100465. DOI: 10.1016/j.ceja.2023.100465

Altuner, E. E., Ozalp, V. C., Yilmaz, M. D., Sudagidan, M., Aygun, A., Acar, E. E., Tasbasi, B. B., & Sen, F. (2022). Development of electrochemical aptasensors detecting phosphate ions on TMB substrate with epoxy-based mesoporous silica nanoparticles. *Chemosphere, 297*, 134077. DOI: 10.1016/j.chemosphere.2022.134077 PMID: 35218784

Ameen, F., Altuner, E. E., Elhouda Tiri, R. N., Gulbagca, F., Aygun, A., Sen, F., Majrashi, N., Orfali, R., & Dragoi, E. N. (2023). Highly active iron (II) oxide-zinc oxide nanocomposite synthesized Thymus vulgaris plant as bioreduction catalyst: Characterization, hydrogen evolution and photocatalytic degradation. *International Journal of Hydrogen Energy, 48*(55), 21139–21151. DOI: 10.1016/j.ijhydene.2022.11.229

Amekawa, Y. (2011). Agroecology and Sustainable Livelihoods: Towards an Integrated Approach to Rural Development. *Journal of Sustainable Agriculture, 35*(2), 118–162. DOI: 10.1080/10440046.2011.539124

Amenta, V., Aschberger, K., Arena, M., Bouwmeester, H., Moniz, F. B., Brandhoff, P., & Peters, R. J. (2015). Regulatory aspects of nanotechnology in the agri/feed/food sector in EU and non-EU countries. *Regulatory Toxicology and Pharmacology*, *73*(1), 463–476. DOI: 10.1016/j.yrtph.2015.06.016 PMID: 26169479

Ameta, R., Rai, A. K., Vyas, S., Bhatt, J. P., & Ameta, S. C. (2021). *Green synthesis of nanocomposites using plant extracts and their applications. Handbook of Greener Synthesis of Nanomaterials and Compounds* (Vol. 1). Fundamental Principles and Methods., DOI: 10.1016/B978-0-12-821938-6.00020-7

Ananthakumar, B., Nanjundeshwara, N., Lakshmaiah, A., & Raghavan, R. (2024). Synthetic Gene Circuits as a Promising Approach in Cancer Immunotherapy. *Critical Developments in Cancer Immunotherapy*, 391–418. DOI: 10.4018/979-8-3693-3976-3.ch011

An, C., Sun, C., Li, N., Huang, B., Jiang, J., Shen, Y., Wang, Y., Zhao, X., Cui, B., Wang, C., Li, X., Zhan, S., Gao, F., Zeng, Z., Cui, H., & Wang, Y. (2022). Nanomaterials and nanotechnology for the delivery of agrochemicals: Strategies towards sustainable agriculture. *Journal of Nanobiotechnology*, *20*(1), 11. DOI: 10.1186/s12951-021-01214-7 PMID: 34983545

Angayarkanni, A., & Philip, J. (2022). Synthesis of Nanoparticles and Nanofluids. *The Royal Society of Chemistry*, *1*, 1–40. DOI: 10.1039/9781839166457-00001

Anjaneyulu, B., Chauhan, V., Mittal, C., & Afshari, M. (2024). Innovative nanocarrier systems: A comprehensive exploration of recent developments in nano-biopesticide formulations. *Journal of Environmental Chemical Engineering*, *12*(5), 113693. DOI: 10.1016/j.jece.2024.113693

Anjum, N. A., Rodrigo, M. A. M., Moulick, A., Heger, Z., Kopel, P., Zítka, O., Adam, V., Lukatkin, A. S., Duarte, A. C., Pereira, E., & Kizek, R. (2016). Transport phenomena of nanoparticles in plants and animals/humans. *Environmental Research*, *151*, 233–243. DOI: 10.1016/j.envres.2016.07.018 PMID: 27504871

Anjum, S., Vyas, A., & Sofi, T. (2023). Fungi-mediated synthesis of nanoparticles: Characterization process and agricultural applications. *Journal of the Science of Food and Agriculture*, *103*(10), 4727–4741. DOI: 10.1002/jsfa.12496 PMID: 36781932

Antony, A. (2024). Flexible and Wearable Biosensors: Revolutionizing. *Health Monitor*, •••, 237–258. DOI: 10.1007/978-981-97-3048-3_12

Arora, S., Sharma, P., Kumar, S., Nayan, R., Khanna, P. K., & Zaidi, M. G. H. (2012). Gold-nanoparticle induced enhancement in growth and seed yield of *Brassica juncea.Plant Growth Regulation*, *66*(3), 303–310. DOI: 10.1007/s10725-011-9649-z

Artursson, P., Edman, P., Laakso, T., & Sjoholm, I. (1984). Characterization of polyacryl starch microparticles as carriers for proteins and drugs. *Journal of Pharmaceutical Sciences*, *73*(11), 1507–1513. DOI: 10.1002/jps.2600731103 PMID: 6440978

Arunkumar, T., Karthikeyan, R., Ram Subramani, R., Viswanathan, K., & Anish, M. (2020). Synthesis and characterisation of multi-walled carbon nanotubes (MWCNTs). *International Journal of Ambient Energy*, *41*(4), 452–456. DOI: 10.1080/01430750.2018.1472657

Asal, M., Özen, Ö., Şahinler, M., & Polatoğlu, İ. (2018). Recent Developments in Enzyme, DNA and Immuno-Based Biosensors. *Sensors 2018, Vol. 18, Page 1924, 18*(6), 1924. DOI: 10.3390/s18061924

Ash, B. J., Schadler, L. S., & Siegel, R. W. (2002). Glass transition behavior of alumina/polymethylmethacrylate nanocomposites. *Materials Letters, 55*(1–2), 83–87. DOI: 10.1016/S0167-577X(01)00626-7

Ashfaq, M., Gupta, G., & Verma, N. (2025). Carbon-based nanocarriers for plant growth promotion: Fuelling when needed. *Nanoscale, 17*(2), 616–634. DOI: 10.1039/D4NR03268C PMID: 39575969

Ashfaq, R., Rasul, A., Asghar, S., Kovács, A., Berkó, S., & Budai-Szűcs, M. (2023). Lipid Nanoparticles: An Effective Tool to Improve the Bioavailability of Nutraceuticals. *International Journal of Molecular Sciences, 24*(21), 15764. DOI: 10.3390/ijms242115764 PMID: 37958750

Ashraf, H., Anjum, T., Riaz, S., Ahmad, I. S., Irudayaraj, J., Javed, S., Qaiser, U., Naseem, S., & Naseem, S. (2021). Inhibition mechanism of green-synthesized copper oxide nanoparticles from Cassia fistula towards Fusarium oxysporum by boosting growth and defense response in tomatoes. *Environmental Science. Nano, 8*(6), 1729–1748. DOI: 10.1039/D0EN01281E

Asif, N., Amir, M., & Fatma, T. (2023). Recent advances in the synthesis, characterization and biomedical applications of zinc oxide nanoparticles. *Bioprocess and Biosystems Engineering 2023 46:10, 46*(10), 1377–1398. DOI: 10.1007/s00449-023-02886-1

Askary, M., Talebi, S. M., Amini, F., & Bangan, A. D. B. (2017). Effects of iron nanoparticles on *Mentha piperita* L. under salinity stress. *Biologija (Vilnius, Lithuania), 63*(1). Advance online publication. DOI: 10.6001/biologija.v63i1.3476

Aslani, F., Bagheri, S., Muhd Julkapli, N., Juraimi, A. S., Hashemi, F. S. G., & Baghdadi, A. (2014). Effects of engineered nanomaterials on plants growth: An overview. *TheScientificWorldJournal, 2014*(1), 641759. DOI: 10.1155/2014/641759 PMID: 25202734

Asmawi, A. A., Adam, F., Azman, N. A. M., & Rahman, M. B. A. (2024). Advancements in the nanodelivery of azole-based fungicides to control oil palm pathogenic fungi. *Heliyon, 10*(18), e37132. DOI: 10.1016/j.heliyon.2024.e37132 PMID: 39309766

Atanda, S. A., Shaibu, R. O., & Agunbiade, F. O. (2025). Nanoparticles in agriculture: Balancing food security and environmental sustainability. *Discov Agric, 3*(1), 26. DOI: 10.1007/s44279-025-00159-x

Ateia, E. E., Rabie, O., & Mohamed, A. T. (2024). Assessment of the correlation between optical properties and CQD preparation approaches. *The European Physical Journal Plus, 139*(1), 24. DOI: 10.1140/epjp/s13360-023-04811-7

Athreya, S. R., Kalaitzidou, K., & Das, S. (2010). Processing and characterization of a carbon black-filled electrically conductive Nylon-12 nanocomposite produced by selective laser sintering. *Materials Science and Engineering A, 527*(10–11), 2637–2642. DOI: 10.1016/j.msea.2009.12.028

Atta, K., Mondal, S., Gorai, S., Singh, A. P., Kumari, A., Ghosh, T., Roy, A., Hembram, S., Gaikwad, D. J., Mondal, S., Bhattacharya, S., Jha, U. C., & Jespersen, D. (2023). Impacts of salinity stress on crop plants: Improving salt tolerance through genetic and molecular dissection. *Frontiers in Plant Science, 14*, 1241736. DOI: 10.3389/fpls.2023.1241736 PMID: 37780527

Avellan, A., Yun, J., Zhang, Y., Spielman-Sun, E., Unrine, J. M., Thieme, J., Li, J., Lombi, E., Bland, G., & Lowry, G. V. (2019). Nanoparticle Size and Coating Chemistry Control Foliar Uptake Pathways, Translocation, and Leaf-to-Rhizosphere Transport in Wheat. *ACS Nano*, *13*(5), 5291–5305. DOI: 10.1021/acsnano.8b09781 PMID: 31074967

Avila, L. A., Lee, S. Y., & Tomich, J. M. (2014). Synthetic in vitro delivery systems for plasmid DNA in eukaryotes. *Journal of Nanopharmaceutics and Drug Delivery*, *2*(1), 17–35. DOI: 10.1166/jnd.2014.1043

Avila-Quezada, G. D., Golinska, P., & Rai, M. (2022). Engineered nanomaterials in plant diseases: Can we combat phytopathogens? *Applied Microbiology and Biotechnology*, *106*(1), 117–129. DOI: 10.1007/s00253-021-11725-w PMID: 34913996

Avinash, S. H., Manasa, S. G., Nagaraju, Y., Rao, P. G., & Mandal, S. (2025). Molecular, physiological, and biochemical aspects of crop plants with PGPR under harsh environment. Husen, A. (Ed.). (2025). Agricultural Crop Improvement: Plant and Soil Relationships (1st ed.). CRC Press. 75–89. https://doi.org/DOI: 10.1201/9781032630366

Ayaz, M., Li, C., Ali, Q., Zhao, W., Chi, Y., Shafiq, M., Ali, F., Yu, X., Yu, Q., Zhao, J., Yu, J., Qi, R., & Huang, W. (2022). Bacterial and Fungal Biocontrol Agents for Plant Disease Protection: Journey from Lab to Field, Current Status, Challenges, and Global Perspectives. *Molecules (Basel, Switzerland)*, *28*(18), 6735. DOI: 10.3390/molecules28186735 PMID: 37764510

Azameti, M. K., & Imoro, A. W. M. (2023). Nanotechnology: A promising field in enhancing abiotic stress tolerance in plants. *Crop Design*, 100037.

Azeh, Y., Musah, M., Mathew, J. T., Badeggi, U. M., Estuyankpa, M. B., Muhammad, Z. D., Oluwaseyi, D. S., Umar, M. T., Abdulkarim, M. A., Muhammad, A. I., Ibrahim, A. A., Ismail, H., Muhammad, K. T., Umar, G., & Ahmed, T. (2023). In Situ Synthesis of Polyaniline Nanohybrid and Formulation of Polyaniline/Carboxymethyl cellulose/Ethylene Glycol Nanocomposite: Study of its Conducting and Antibacterial Properties. *Journal of Materials and Polymer Science*, *3*(4), 1–9. DOI: 10.47485/2832-9384.1041

Babatunde, D. E., Denwigwe, I. H., Babatunde, O. M., Gbadamosi, S. L., Babalola, I. P., & Agboola, O. (2019). Environmental and societal impact of nanotechnology. *IEEE Access : Practical Innovations, Open Solutions*, *8*, 4640–4667. DOI: 10.1109/ACCESS.2019.2961513

Babi, M., Fatona, A., Li, X., Cerson, C., Jarvis, V. M., Abitbol, T., & Moran-Mirabal, J. M. (2022). Efficient labeling of nanocellulose for high-resolution fluorescence microscopy applications. *Biomacromolecules*, *23*(5), 1981–1994. DOI: 10.1021/acs.biomac.1c01698 PMID: 35442640

Babu, P. J., Saranya, S., Longchar, B., & Rajasekhar, A. (2022). Nanobiotechnology-mediated sustainable agriculture and post-harvest management. *Current Research in Biotechnology*, *4*, 326–336. DOI: 10.1016/j.crbiot.2022.07.004

Bagheri Hashkavayi, A., Cha, B. S., Lee, E. S., Kim, S., & Park, K. S. (2020). Advances in Exosome Analysis Methods with an Emphasis on Electrochemistry. *Analytical Chemistry*, *92*(19), 12733–12740. DOI: 10.1021/acs.analchem.0c02745 PMID: 32902258

Bahl, E., Jyoti, A., Singh, A., Siddqui, A., Upadhyay, S. K., Jain, D., Shah, M. P., & Saxena, J. (2024). Nanomaterials for intelligent CRISPR-Cas tools: Improving environment sustainability. *Environmental Science and Pollution Research International*, *31*(60), 67479–67495. Advance online publication. DOI: 10.1007/s11356-024-32101-x PMID: 38291210

Bahrulolum, H., Nooraei, S., Javanshir, N., Tarrahimofrad, H., Mirbagheri, V. S., Easton, A. J., & Ahmadian, G. (2021). Green synthesis of metal nanoparticles using microorganisms and their application in the agrifood sector. *Journal of Nanobiotechnology 2021 19:1*, *19*(1), 1–26. DOI: 10.1186/s12951-021-00834-3

Baker, C. J., Mock, N. M., Whitaker, B. D., Roberts, D. P., Rice, C. P., Deahl, K. L., & Aver'yanov, A. A. (2005). Involvement of acetosyringone in plant-pathogen recognition. *Biochemical and Biophysical Research Communications*, *328*(1), 130–136. DOI: 10.1016/j.bbrc.2004.12.153 PMID: 15670760

Baker, C., Shah, S. I., & Hasanain, S. K. (2004). Magnetic behavior of iron and iron-oxide nanoparticle/polymer composites. *Journal of Magnetism and Magnetic Materials*, *280*(2–3), 412–418. DOI: 10.1016/j.jmmm.2004.03.037

Baker, S., Volova, T., Prudnikova, S. V., Satish, S., & Prasad, M. N. N. (2017). Nanoagroparticles emerging trends and future prospect in modern agriculture system. *Environmental Toxicology and Pharmacology*, *53*, 10–17. DOI: 10.1016/j.etap.2017.04.012 PMID: 28499265

Bakki, M., Banane, B., Marhane, O., Esmaeel, Q., Hatimi, A., Barka, E. A., Azim, K., & Bouizgarne, B. (2024). Phosphate solubilizing *Pseudomonas* and *Bacillus* combined with rock phosphates promoting tomato growth and reducing bacterial canker disease. *Frontiers in Microbiology*, *15*, 1289466. DOI: 10.3389/fmicb.2024.1289466 PMID: 38765677

Balan, B. K., & Kurungot, S. (2012). Tuning the functionality of a carbon nanofiber-Pt-RuO2 system from charge storage to electrocatalysis. *Inorganic Chemistry*, *51*(18), 9766–9774. DOI: 10.1021/ic301147s PMID: 22946658

Balan, K., Qing, W., Wang, Y., Liu, X., Palvannan, T., Wang, Y., Ma, F., & Zhang, Y. (2016). Antidiabetic activity of silver nanoparticles from green synthesis using Lonicera japonica leaf extract. *RSC Advances*, *6*(46), 40162–40168. DOI: 10.1039/C5RA24391B

Balasubramaniam, T., Shen, G., Esmaeili, N., & Zhang, H. (2023). Plants' Response Mechanisms to Salinity Stress. *Plants*, *12*(12), 2253. DOI: 10.3390/plants12122253 PMID: 37375879

Balaure, P. C., Gudovan, D., & Gudovan, I. (2017). Nanopesticides: A new paradigm in crop protection. In *New Pesticides and Soil Sensors* (pp. 129–192). Academic Press. DOI: 10.1016/B978-0-12-804299-1.00005-9

Baldassarre, F., Stradis, A. D., Altamura, G., Vergaro, V., Citti, C., Cannazza, G., Capodilupo, A. L., Dini, L., & Ciccarella, G. (2020). Application of calcium carbonate nanocarriers for controlled release of phytodrugs against *Xylella fastidiosa* pathogen. *Pure and Applied Chemistry*, *92*(3), 429–444. DOI: 10.1515/pac-2018-1223

Balla, A., Silini, A., Cherif-Silini, H., Chenari Bouket, A., Alenezi, F. N., & Belbahri, L. (2022). Recent advances in encapsulation techniques of plant growth-promoting microorganisms and their prospects in the sustainable agriculture. *Applied Sciences (Basel, Switzerland)*, *12*(18), 9020. DOI: 10.3390/app12189020

Baloch, N. (2025). Microbial contributions to maize crop production: A comprehensive review of challenges and future perspectives. *Discover Agriculture*, *3*(1), 10. DOI: 10.1007/s44279-025-00164-0

Baltes, N. J., Gil-Humanes, J., & Voytas, D. F. (2017). Genome Engineering and Agriculture: Opportunities and Challenges. *Progress in Molecular Biology and Translational Science*, *149*, 1–26. DOI: 10.1016/bs.pmbts.2017.03.011 PMID: 28712492

Banerjee, S., Mazumder, S., Chatterjee, D., Bose, S., & Majee, S. B. (2022). Nanotechnology for cargo delivery with a special emphasis on pesticide, herbicide, and fertilizer. In *Nano-enabled Agrochemicals in Agriculture* (pp. 105–144). Academic Press. DOI: 10.1016/B978-0-323-91009-5.00002-1

Banerjee, S., Roy, P., Nandi, S., & Roy, S. (2023). Advanced biotechnological strategies towards the development of crops with enhanced micronutrient content. *Plant Growth Regulation*, *100*(2), 355–371. DOI: 10.1007/s10725-023-00968-4 PMID: 36686885

Banik, M., Patra, M., Dutta, D., Mukherjee, R., & Basu, T. (2018). A simple robust method of synthesis of copper-silver core-shell nano-particle: Evaluation of its structural and chemical properties with anticancer potency. *Nanotechnology*, *29*(32), 325102. DOI: 10.1088/1361-6528/aac372 PMID: 29742069

Bano, A., Aziz, M. K., Prasad, B., Ravi, R., Shah, M. P., Lins, P. V. D. S., Meili, L., & Prasad, K. S. (2024). The multifaceted power of biochar: A review on its role in pollution control, sustainable agriculture, and circular economy. *Environmental Chemistry and Ecotoxicology*, *7*, 286–304. DOI: 10.1016/j.enceco.2025.01.004

Bao-Shan, L., shao-qi, D., Chun-hui, L., Li-jun, F., Shu-chun, Q., & Min, Y. (2004). Effect of TMS (nanostructured silicon dioxide) on growth of *Changbai larch* seedlings. *Journal of Forestry Research*, *15*(2), 138–140. DOI: 10.1007/BF02856749

Bao, W., Wang, J., Wang, Q., O'Hare, D., & Wan, Y. (2016). Layered double hydroxide nanotransporter for molecule delivery to intact plant cells. *Scientific Reports*, *6*(1), 26738. DOI: 10.1038/srep26738 PMID: 27221055

Barampuram, S., & Zhang, Z. J. (2011). Recent advances in plant transformation. *Plant chromosome engineering: methods and protocols*, 1-35.

Barbosa Santos, C. H., Frezarin, E. T., Sales, L. R., & Rigobelo, E. C. (2023). Plant Growth-Promoting Rhizobacteria for Sustainable Agricultural Production. *Microorganisms*, *11*(4), 1088. DOI: 10.3390/microorganisms11041088 PMID: 37110511

Bardania, H., Mahmoudi, R., Bagheri, H., Salehpour, Z., Fouani, M. H., Darabian, B., Khoramrooz, S. S., Mousavizadeh, A., Kowsari, M., Moosavifard, S. E., Christiansen, G., Javeshghani, D., Alipour, M., & Akrami, M. (2020). Facile preparation of a novel biogenic silver-loaded Nanofilm with intrinsic anti-bacterial and oxidant scavenging activities for wound healing. *Scientific Reports 2020 10:1, 10*(1), 1–14. DOI: 10.1038/s41598-020-63032-5

Barrena, R., Casals, E., Colón, J., Font, X., Sánchez, A., & Puntes, V. (2009). Evaluation of the ecotoxicity of model nanoparticles. *Chemosphere, 75*(7), 850–857. DOI: 10.1016/j.chemosphere.2009.01.078 PMID: 19264345

Bartczak, D., Nitti, S., Millar, T. M., & Kanaras, A. G. (2012). Exocytosis of Peptide Functionalized Gold Nanoparticles in Endothelial Cells. *Nanoscale, 4*(15), 4470–4472. DOI: 10.1039/c2nr31064c PMID: 22743818

Baruah, S., & Dutta, J. (2009). Nanotechnology applications in pollution sensing and degradation in agriculture: A review. *Environmental Chemistry Letters, 7*(3), 191–204. DOI: 10.1007/s10311-009-0228-8

Basak, S., Bhattacharyya, P., Pudake, R. N., Lokhande, P. E., Rednam, U., & Chakrabarti, S. (2024). Metal-organic framework as nanocarriers for agricultural applications: A review. *Frontiers in Nanotechnology, 6*, 1385981. DOI: 10.3389/fnano.2024.1385981

Bashan, Y. (1986). Alginate beads as synthetic inoculant carriers for slow release of bacteria that affect plant growth. *Applied and Environmental Microbiology, 51*(5), 1089–1098. DOI: 10.1128/aem.51.5.1089-1098.1986 PMID: 16347055

Bashan, Y., de-Bashan, L. E., Prabhu, S. R., & Hernandez, J.-P. (2014). Advances in plant growth-promoting bacterial inoculant technology: Formulations and practical perspectives (1998-2013). *Plant and Soil, 378*(1–2), 1–33. DOI: 10.1007/s11104-013-1956-x

Bashan, Y., & Gonzalez, L. E. (1999). Long-term survival of the plant-growth-promoting bacteria *Azospirillum brasilense* and *Pseudomonas fluorescens* in dry alginate inoculant. *Applied Microbiology and Biotechnology, 51*(2), 262–266. DOI: 10.1007/s002530051391

Bashan, Y., Hernandez, J.-P., Leyva, L., & Bacilio, M. (2002). Alginate microbeads as inoculant carriers for plant growth-promoting bacteria. *Biology and Fertility of Soils, 35*(5), 359–368. DOI: 10.1007/s00374-002-0481-5

Bashir, A. K. H., Razanamahandry, L. C., Nwanya, A. C., Kaviyarasu, K., Saban, W., Mohamed, H. E. A., Ntwampe, S. K. O., Ezema, F. I., & Maaza, M.H. E. A. (2019). Biosynthesis of NiO nanoparticles for photodegradation of free cyanide solutions under ultraviolet light. *Journal of Physics and Chemistry of Solids, 134*, 133–140. DOI: 10.1016/j.jpcs.2019.05.048

Basri, S., Kamarudin, S. K., Daud, W. R. W., & Yaakub, Z. (2010). Nanocatalyst for direct methanol fuel cell (DMFC). In International Journal of Hydrogen Energy (Vol. 35, Issue 15, pp. 7957–7970). Elsevier Ltd. DOI: 10.1016/j.ijhydene.2010.05.111

Bastos, A., Orth, R., Reichstein, M., Ciais, P., Viovy, N., Zaehle, S., Anthoni, P., Arneth, A., Gentine, P., Joetzjer, E., Lienert, S., Loughran, T., McGuire, P. C., O, S., Pongratz, J., & Sitch, S. (2021). Vulnerability of European ecosystems to two compound dry and hot summers in 2018 and 2019. *Earth System Dynamics: ESD, 12*(4), 1015–1035. DOI: 10.5194/esd-12-1015-2021

Bates, K., & Kostarelos, K. (2013). Carbon nanotubes as vectors for gene therapy: Past achievements, present challenges and future goals. *Advanced Drug Delivery Reviews, 65*(15), 2023–2033. DOI: 10.1016/j.addr.2013.10.003 PMID: 24184373

Batiha, G. E. S., Alqahtani, A., Ilesanmi, O. B., Saati, A. A., El-Mleeh, A., & Hetta, H. F.. (2020). Avermectin derivatives, pharmacokinetics, therapeutic and toxic dosages, mechanism of action, and their biological effects. *Pharmaceuticals (Basel, Switzerland)*, *13*(8), 196. DOI: 10.3390/ph13080196 PMID: 32824399

Bavandpour, R., Rajabi, M., Karimi-Maleh, H., & Asghari, A. (2021). Application of deep eutectic solvent and SWCNT-ZrO2 nanocomposite as conductive mediators for the fabrication of simple and rapid electrochemical sensor for determination of trace anti-migration drugs. *Microchemical Journal*, *165*, 106141. DOI: 10.1016/j.microc.2021.106141

Bayda, S., Adeel, M., Tuccinardi, T., Cordani, M., & Rizzolio, F. (2019). The history of nanoscience and nanotechnology: From chemical–physical applications to nanomedicine. *Molecules (Basel, Switzerland)*, *25*(1), 112. DOI: 10.3390/molecules25010112 PMID: 31892180

Bazylinski, D. A., & Frankel, R. B. (2004). Magnetosome formation in prokaryotes. *Nature Reviews Microbiology 2004 2:3*, *2*(3), 217–230. DOI: 10.1038/nrmicro842

Beckers, S. J., Wetherbee, L., Fischer, J., & Wurm, F. R. (2020). Fungicide-loaded and biodegradable xylan-based nanocarriers. *Biopolymers*, *111*(12), e23413. DOI: 10.1002/bip.23413 PMID: 33306838

Beckers, S., Peil, S., & Wurm, F. R. (2020). Pesticide-loaded nanocarriers from lignin sulfonates—A promising tool for sustainable plant protection. *ACS Sustainable Chemistry & Engineering*, *8*(50), 18468–18475. DOI: 10.1021/acssuschemeng.0c05897 PMID: 33381356

Begines, B., Ortiz, T., Pérez-Aranda, M., Martínez, G., Merinero, M., Argüelles-Arias, F., & Alcudia, A. (2020). Polymeric Nanoparticles for Drug Delivery: Recent Developments and Future Prospects. *Nanomaterials (Basel, Switzerland)*, *10*(7), 1403. DOI: 10.3390/nano10071403 PMID: 32707641

Behravan, M., Hossein Panahi, A., Naghizadeh, A., Ziaee, M., Mahdavi, R., & Mirzapour, A. (2019). Facile green synthesis of silver nanoparticles using Berberis vulgaris leaf and root aqueous extract and its antibacterial activity. *International Journal of Biological Macromolecules*, *124*, 148–154. DOI: 10.1016/j.ijbiomac.2018.11.101 PMID: 30447360

Behzadi, S., Serpooshan, V., Tao, W., Hamaly, M. A., Alkawareek, M. Y., Dreaden, E. C., Brown, D., Alkilany, A. M., Farokhzad, O. C., & Mahmoudi, M. (2017). Cellular uptake of nanoparticles: Journey inside the cell. *Chemical Society Reviews*, *46*(14), 4218–4244. DOI: 10.1039/C6CS00636A PMID: 28585944

Beig, B., Niazi, M. B. K., Jahan, Z., Zia, M., Shah, G. A., & Iqbal, Z. (2023). Biodegradable polymer encapsulated nickel nanoparticles for slow release urea promotes Rhode grass yield and nitrogen recovery. *Journal of Polymers and the Environment*, *31*(5), 1866–1883. DOI: 10.1007/s10924-022-02729-2

Bekmezci, M., Altuner, E. E., & Sen, F. (2022). Biological Characterization of Magnetic Hybrid Nanoalloys. *Handbook of Magnetic Hybrid Nanoalloys and Their Nanocomposites*, 1–18. DOI: 10.1007/978-3-030-34007-0_28-1

Bell, I. R., Ives, J. A., & Jonas, W. B. (2014). Nonlinear effects of nanoparticles: Biological variability from hormetic doses, small particle sizes, and dynamic adaptive interactions. *Dose-Response*, *12*(2), 202–232. DOI: 10.2203/dose-response.13-025.Bell PMID: 24910581

Ben-Haim, A. E., Feldbaum, R. A., Belausov, E., Zelinger, E., Maria, R., Nativ-Roth, E., Mani, K. A., Barda, O., Sionov, E., & Mechrez, G. (2024). DNA Delivery to Intact Plant Cells by Casein Nanoparticles with Confirmed Gene Expression (Adv. Funct. Mater. 16/2024). *Advanced Functional Materials*, *34*(16), 2470089. DOI: 10.1002/adfm.202470089

Benjamin, S. R., Lima, F. D., Florean, E. O. P. T., & Guedes, M. I. F. (2019). Current trends in nanotechnology for bioremediation. *International Journal of Environment and Pollution*, *66*(1-3), 19–40. DOI: 10.1504/IJEP.2019.104526

Bennett, E. M., Murray, J. W., & Isalan, M. (2023). Engineering Nitrogenases for Synthetic Nitrogen Fixation: From Pathway Engineering to Directed Evolution. *Biodesign Research, 5*, 0005. https://doi .org/DOI: 10.34133/bdr.0005

Bensadoun, J. C., De Almeida, L. P., Fine, E. G., Tseng, J. L., Déglon, N., & Aebischer, P. (2003). Comparative study of GDNF delivery systems for the CNS: Polymer rods, encapsulated cells, and lentiviral vectors. *Journal of Controlled Release*, *87*(1–3), 107–115. DOI: 10.1016/S0168-3659(02)00353-X PMID: 12618027

Berkner, S., Schwirn, K., & Voelker, D. (2022). Too advanced for assessment? Advanced materials, nanomedicine and the environment. *Environmental Sciences Europe*, *34*(1), 71. Advance online publication. DOI: 10.1186/s12302-022-00647-7 PMID: 35990895

Bernays, E. A. (2009). Phytophagous Insects. In Resh, V. H., & Cardé, R. T. (Eds.), *Encyclopedia of Insects* (2nd ed., pp. 798–800). Academic Press., DOI: 10.1016/B978-0-12-374144-8.00210-1

Bertsch, C., Ramírez-Suero, M., Magnin-Robert, M., Larignon, P., Chong, J., Abou-Mansour, E., Spagnolo, A., Clément, C., & Fontaine, F. (2013). Grapevine trunk diseases: Complex and still poorly understood. *Plant Pathology*, *62*(2), 243–265. DOI: 10.1111/j.1365-3059.2012.02674.x

Bharali, D. J., Klejbor, I., Stachowiak, E. K., Dutta, P., Roy, I., Kaur, N., Bergey, E. J., Prasad, P. N., & Stachowiak, M. K. (2005). Organically modified silica nanoparticles: A nonviral vector for in vivo gene delivery and expression in the brain. *Proceedings of the National Academy of Sciences of the United States of America*, *102*(32), 11539–11544. DOI: 10.1073/pnas.0504926102 PMID: 16051701

Bhardwaj, S., Lata, S., & Garg, R. (2023). Application of nanotechnology for preventing postharvest losses of agriproducts. *The Journal of Horticultural Science & Biotechnology*, *98*(1), 31–44. DOI: 10.1080/14620316.2022.2091488

Bhat, M. A., Mishra, A. K., Shah, S. N., Bhat, M. A., Jan, S., Rahman, S., Baek, K., & Jan, A. T. (2024). Soil and Mineral Nutrients in Plant Health: A Prospective Study of Iron and Phosphorus in the Growth and Development of Plants. *Current Issues in Molecular Biology*, *46*(6), 5194–5222. DOI: 10.3390/cimb46060312 PMID: 38920984

Bhattacharya, T., Soares, G. A. B. E., Chopra, H., Rahman, M. M., Hasan, Z., Swain, S. S., & Cavalu, S. (2022). Applications of Phyto-Nanotechnology for the Treatment of Neurodegenerative Disorders. *Materials 2022, Vol. 15, Page 804, 15*(3), 804. DOI: 10.3390/ma15030804

Bhattacharya, N., Cahill, D. M., Yang, W., & Kochar, M. (2023). Graphene as a nano-delivery vehicle in agriculture–current knowledge and future prospects. *Critical Reviews in Biotechnology*, *43*(6), 851–869. DOI: 10.1080/07388551.2022.2090315 PMID: 35815813

Bieberstein, A., Roosen, J., Marette, S., Blanchemanche, S., & Vandermoere, F. (2013). Consumer choices for nano-food and nano-packaging in France and Germany. In European Review of Agricultural Economics (Vol. 40, Issue 1, pp. 73–94). DOI: 10.1093/erae/jbr069

Bigeard, J., Colcombet, J., & Hirt, H. (2015). Signaling mechanisms in pattern-triggered immunity (PTI). *Molecular Plant*, *8*(4), 521–539. DOI: 10.1016/j.molp.2014.12.022 PMID: 25744358

Bilia, A. R., Piazzini, V., Risaliti, L., Vanti, G., Casamonti, M., Wang, M., & Bergonzi, M. C. (2019). Nanocarriers: A successful tool to increase solubility, stability and optimise bioefficacy of natural constituents. *Current Medicinal Chemistry*, *26*(24), 4631–4656. DOI: 10.2174/09298673256661811011 10050 PMID: 30381065

Bin Hussein, M. Z., Zainal, Z., Yahaya, A. H., & Foo, D. W. V. (2002). Controlled release of a plant growth regulator, alpha-naphthaleneacetate from the lamella of Zn–Al-layered double hydroxide nanocomposite. *Journal of Controlled Release*, *82*(2–3), 417–427. DOI: 10.1016/S0168-3659(02)00172-4 PMID: 12175754

Biondo, F., Baldassarre, F., Vergaro, V., & Ciccarella, G. (2022). Controlled biocide release from smart delivery systems: Materials engineering to tune release rate, biointeractions, and responsiveness. In *Nanotechnology-Based Sustainable Alternatives for the Management of Plant Diseases* (pp. 31–147). Elsevier. DOI: 10.1016/B978-0-12-823394-8.00010-X

Biradar, S., Korrapati, C., Krishna, R. J., & Nanjundeshwara, N. (2025). Spatial Omics and Gene Circuits: The Future in Cancer Biology. *Spatially Variable Genes in Cancer: Development, Progression, and Treatment Response*, 383–408. DOI: 10.4018/979-8-3693-7728-4.ch014

Birnbaum, K. D., & Sánchez Alvarado, A. (2008). Slicing across kingdoms: Regeneration in plants and animals. *Cell*, *132*(4), 697–710. DOI: 10.1016/j.cell.2008.01.040 PMID: 18295584

Boas, U., & Heegaard, H. (2004). Dendrimers in drug research. *Chemical Society Reviews*, *33*(1), 43–63. DOI: 10.1039/b309043b PMID: 14737508

Bobbarala, V., Sahu, S. C., Mishra, B. K., & Jena, B. K. (2012). A Search for Antibacterial Agents. In *A Search for Antibacterial Agents*. IntechOpen., DOI: 10.5772/1085

Bokov, D., Jalil, A. T., Chupradit, S., Suksatan, W., Ansari, M. J., Shewael, I. H., Valiev, G. H., & Kianfar, E. (2020). Nanomaterial by Sol-Gel Method: Synthesis and Application. *Advances in Materials Science and Engineering*, *2021*(1), 5102014. DOI: 10.1155/2021/5102014

Bolsenga, S. J. (1983). Spectral Reflectances of Snow and Fresh-Water Ice from 340 Through 1 100 nm. *Journal of Glaciology*, *29*(102), 296–305. DOI: 10.3189/S0022143000008352

Bonechi, C., Tamasi, G., Donati, A., Bisozzi, F., Baglioni, M., Andreassi, M., Ietta, F., Leone, G., Magnani, A., & Rossi, C. (2025). Cationic liposomes as carriers of natural compounds from plant extract. *Biophysical Chemistry*, *318*, 107381. DOI: 10.1016/j.bpc.2024.107381 PMID: 39733688

Borane, N., Aralekallu, S., Boddula, R., Singh, J., & Kurkuri, M. D. (2024). Recent trends in the "bottom-up" and "top down" techniques in the synthesis and fabrication of myriad carbonaceous nanomaterials. In Biswas, K., Mohanta, Y. K., Mohanta, T. K., & Saravanan, M. (Eds.), *Carbon-Based Nanomaterials in Biosystems* (pp. 91–120). Academic Press., DOI: 10.1016/B978-0-443-15508-6.00002-6

Boretti, A. (2024). Comprehensive risk-benefit assessment of chemicals: A case study on glyphosate. *Toxicology Reports*, *13*, 101803. DOI: 10.1016/j.toxrep.2024.101803 PMID: 39606775

Boros, B. V., & Ostafe, V. (2020). Evaluation of ecotoxicology assessment methods of nanomaterials and their effects. *Nanomaterials (Basel, Switzerland)*, *10*(4), 610. DOI: 10.3390/nano10040610 PMID: 32224954

Bose, B., & Pal, H. (2023). Application of microbial nanotechnology in sustainable agriculture through soil remediation. In *Environmental Applications of Microbial Nanotechnology* (pp. 253–274). Elsevier. DOI: 10.1016/B978-0-323-91744-5.00018-7

Boudh, S., & Singh, J. S. (2019). Pesticide contamination: environmental problems and remediation strategies. *Emerging and eco-friendly approaches for waste management*, 245-269.

Boulaiz, H., Alvarez, P. J., Ramirez, A., Marchal, J. A., Prados, J., Rodríguez-Serrano, F., Perán, M., Melguizo, C., & Aranega, A. (2011). Nanomedicine: Application Areas and Development Prospects. *International Journal of Molecular Sciences*, *12*(5), 3303–3321. DOI: 10.3390/ijms12053303 PMID: 21686186

Bowbrick, P. (2024). Falsehoods and myths in famine research: The Bengal famine and Daoud. *Journal of International Development*, *36*(6), 2513–2536. DOI: 10.1002/jid.3635

Bozbeyoglu, P., & Gündoğdu, A. (2023). Adsorption of hexavalent chromium from aqueous solution onto corn cobs – activated carbon. *Turkish Journal of Analytical Chemistry*, *5*(2), 107–117. DOI: 10.51435/turkjac.1394304

Broda, M. (2020). Natural Compounds for Wood Protection against Fungi—A Review. *Molecules (Basel, Switzerland)*, *25*(15), 3538. DOI: 10.3390/molecules25153538 PMID: 32748877

Brown, P., & Saa, S. (2015). Biostimulants in agriculture. *Frontiers in Plant Science*, *6*, 671. DOI: 10.3389/fpls.2015.00671 PMID: 26379695

Bueno, V., Gao, X., Abdul Rahim, A., Wang, P., Bayen, S., & Ghoshal, S. (2022). Uptake and translocation of a silica nanocarrier and an encapsulated organic pesticide following foliar application in tomato plants. *Environmental Science & Technology*, *56*(10), 6722–6732. DOI: 10.1021/acs.est.1c08185 PMID: 35467849

Buluş, E., Bulus, G. S., & Yakuphanoglu, F. (2020). Production and characterization of novel nature-friendly organic fertilizer covers based on nanotechnology for the agricultural sector. *Journal of Materials and Electronic Devices*, *5*(1), 12–16.

Buonomenna, M. G., Mousavi, S. M., Hashemi, S. A., & Lai, C. W. (2022). Water cleaning adsorptive membranes for efficient removal of heavy metals and metalloids. *Water (Basel)*, *14*(17), 2718. DOI: 10.3390/w14172718

Burlaka, O. M., Pirko, Y. V., Yemets, A. I., & Blume, Y. B. (2015).. . *Tsitologiia i Genetika*, *49*(6), 3–12. PMID: 26841488

Burlaka, O. M., Pirko, Ya. V., Yemets, A. I., & Blume, Ya. B. (2015). Plant genetic transformation using carbon nanotubes for DNA delivery. *Cytology and Genetics*, *49*(6), 349–357. DOI: 10.3103/S009545271506002X PMID: 26841488

Burvey, P., Jain, P., & Singh, A. (2023). Strategic role of casein coated iron oxide nano-particles for delivering nano nutrients for enhance soil quality and their potential. *Materials Today: Proceedings*, *89*, 96–105. DOI: 10.1016/j.matpr.2023.05.739

Bussières, P. (2014). Estimating the number and size of phloem sieve plate pores using longitudinal views and geometric reconstruction. *Scientific Reports*, *4*(1), 4929. DOI: 10.1038/srep04929 PMID: 24816291

Bwambale, E., Abagale, F. K., & Anornu, G. K. (2022). Smart irrigation monitoring and control strategies for improving water use efficiency in precision agriculture: A review. *Agricultural Water Management*, *260*, 107324. DOI: 10.1016/j.agwat.2021.107324

Cabrera-De la Fuente, M., González-Morales, S., Juárez-Maldonado, A., Leija-Martínez, P., & Benavides-Mendoza, A. (2018). Plant nutrition and agronomic management to obtain crops with better nutritional and nutraceutical quality. In Holban, A. M., & Grumezescu, A. M. (Eds.), *Therapeutic Foods* (pp. 99–140). Academic Press. DOI: 10.1016/B978-0-12-811517-6.00004-0

Calabro, R. L., Yang, D. S., & Kim, D. Y. (2019). Controlled Nitrogen Doping of Graphene Quantum Dots through Laser Ablation in Aqueous Solutions for Photoluminescence and Electrocatalytic Applications. *ACS Applied Nano Materials*, *2*(11), 6948–6959. DOI: 10.1021/acsanm.9b01433

Caldeira, K., & Wickett, M. (2003). Oceanography: Anthropogenic carbon and ocean pH. *Nature*, *425*(6956), 365–365. DOI: 10.1038/425365a PMID: 14508477

Camara, M. C., Campos, E. V. R., Monteiro, R. A., do Espirito Santo Pereira, A., de Freitas Proença, P. L., & Fraceto, L. F. (2019). Development of stimuli-responsive nano-based pesticides: Emerging opportunities for agriculture. *Journal of Nanobiotechnology*, *17*(1), 100. DOI: 10.1186/s12951-019-0533-8 PMID: 31542052

Campos, E. V. R., de Oliveira, J. L., & Fraceto, L. F. (2014). Applications of controlled release systems for fungicides, herbicides, acaricides, nutrients, and plant growth hormones: A review. *Advanced Science, Engineering and Medicine*, *6*(4), 373–387. DOI: 10.1166/asem.2014.1538

Campos, E. V. R., do Espirito Santo Pereira, A., de Oliveira, J. L., Villarreal, G. P. U., & Fraceto, L. F. (2022). Nature-based nanocarrier system: An eco-friendly alternative for improving crop resilience to climate changes. *Anthropocene Science*, *1*(3), 396–403. DOI: 10.1007/s44177-022-00029-x

Campos, E. V. R., Oliveira, J. L. D., da Silva, C. M. G., Pascoli, M., Pasquoto, T., Lima, R., Abhilash, P. C., & Fernandes Fraceto, L. (2015). Polymeric and solid lipid nanoparticles for sustained release of carbendazim and tebuconazole in agricultural applications. *Scientific Reports*, *5*(1), 13809. DOI: 10.1038/srep13809 PMID: 26346969

Cañas, J. E., Long, M., Nations, S., Vadan, R., Dai, L., Luo, M., Ambikapathi, R., Lee, E. H., & Olszyk, D. (2008). Effects of functionalized and nonfunctionalized single-walled carbon nanotubes on root elongation of select crop species. *Environmental Toxicology and Chemistry*, *27*(9), 1922–1931. DOI: 10.1897/08-117.1 PMID: 19086209

Canellas, L. P., Olivares, F. L., Okorokova-Facanha, A. L., & Facanha, A. R. (2002). Humic acids isolated from earthworm compost enhance root elongation, lateral root emergence, and plasma membrane H+-ATPase activity in maize roots. *Plant Physiology*, *130*(4), 1951–1957. DOI: 10.1104/pp.007088 PMID: 12481077

Cao, Y., Si, J., Zheng, M., Zhou, Q., & Ge, Z. (2023). X-ray-responsive prodrugs and polymeric nanocarriers for multimodal cancer therapy. Chemical communications.

Cao, Z., Zhou, H., Kong, L., Li, L., Wang, R., & Shen, W. (2020). A novel mechanism underlying multi-walled carbon nanotube-triggered tomato lateral root formation: The involvement of nitric oxide. *Nanoscale Research Letters*, *15*(1), 49. DOI: 10.1186/s11671-020-3276-4 PMID: 32103348

Carmen, B., Roberto, D., Carmen, B., & Roberto, D. (2011). Soil Bacteria Support and Protect Plants Against Abiotic Stresses. *In Abiotic Stress in Plants—Mechanisms and Adaptations*. IntechOpen.

Carmona, F. J., Guagliardi, A., & Masciocchi, N. (2022). Nanosized calcium phosphates as novel macronutrient nano-fertilizers. *Nanomaterials (Basel, Switzerland)*, *12*(15), 2709. DOI: 10.3390/nano12152709 PMID: 35957141

Cartwright, A., Jackson, K., Morgan, C., Anderson, A., & Britt, D. W. (2020). A review of metal and metal-oxide nanoparticle coating technologies to inhibit agglomeration and increase bioactivity for agricultural applications. *Agronomy (Basel)*, *10*(7), 1018. DOI: 10.3390/agronomy10071018

Castro, R. A., Dourado, M. N., de Almeida, J. R., Lacava, P. T., Nave, A., de Melo, I. S., de Azevedo, J. L., & Quecine, M. C. (2018). Mangrove endophyte promotes reforestation tree (*Acacia polyphylla*) growth. *Brazilian Journal of Microbiology : Publication of the Brazilian Society for Microbiology*, *49*(1), 59–66. DOI: 10.1016/j.bjm.2017.04.002 PMID: 28774638

Cerdan, M., Sanchez-Sanchez, A., Oliver, M., Juarez, M., & Sanchez-Andreu, J. J. (Eds.). Effect of foliar and root applications of amino acids on iron uptake by tomato plants. In: 4th Balkan symposium on vegetables and potatoes; 2009 June 17; Plovdiv, BULGARIA2009

Chabbi, J., El Assimi, T., Blazic, R., El Haskouri, J., Kaddami, H., Youcef, H. B., Vidovic, E., El Kadib, A., & Lahcini, M. (2025). Eco-friendly catalysts: Chitosan-silver hybrid nanocarriers for efficient oxidative coupling of anilines to azobenzenes. *International Journal of Biological Macromolecules*, *307*, 141996. DOI: 10.1016/j.ijbiomac.2025.141996 PMID: 40081687

Chadha, S. (2021). Recent advances in nano-encapsulation technologies for controlled release of biostimulants and antimicrobial agents. *Advances in nano-fertilizers and nano-pesticides in agriculture*, 29-55.

Chandrashekar, H. K., Singh, G., Kaniyassery, A., Thorat, S. A., Nayak, R., Murali, T. S., & Muthusamy, A. (2023). Nanoparticle-mediated amelioration of drought stress in plants: A systematic review. *3 Biotech*, *13*(10), 336. https://doi.org/DOI: 10.1007/s13205-023-03751-4

Chang, F. P., Lin-Yun, K., Chia-An, H., Wann-Neng, J., Yann, H., Yueie, C. H., & Chung-Yuan, M. (2013). A simple plant gene delivery system using mesoporous silica nanoparticles as carriers. *Journal of Materials Chemistry. B*, *1*(39), 5279. DOI: 10.1039/c3tb20529k PMID: 32263331

Chang, M., Chou, J. C., & Lee, H. J. (2005). Cellular internalization of fluorescent proteins via arginine-rich intracellular delivery peptide in plant cells. *Plant & Cell Physiology*, *46*(3), 482–488. DOI: 10.1093/pcp/pci046 PMID: 15695452

Chang, P. R., Jian, R., Yu, J., & Ma, X. (2010). Fabrication and characterisation of chitosan nanoparticles/plasticised-starch composites. *Food Chemistry*, *120*(3), 736–740. DOI: 10.1016/j.foodchem.2009.11.002

Chanratana, M., Joe, M. M., Roy Choudhury, A., Anandham, R., Krishnamoorthy, R., Kim, K., Jeon, S., Choi, J., Choi, J., & Sa, T. (2019). Physiological response of tomato plant to chitosan-immobilized aggregated Methylobacterium oryzae CBMB20 inoculation under salinity stress. 3 Biotech, 9(11), 1–13. DOI: 10.1007/s13205-019-1923-1

Chanratana, M., Han, G. H., Melvin Joe, M., Roy Choudhury, A., Sundaram, S., Halim, M. A., & Sa, T. (2018). Evaluation of chitosan and alginate immobilized Methylobacterium oryzae CBMB20 on tomato plant growth. *Archives of Agronomy and Soil Science*, *64*(11), 1489–1502. DOI: 10.1080/03650340.2018.1440390

Charbgoo, F., Ramezani, M., & Darroudi, M. (2017). Bio-sensing applications of cerium oxide nanoparticles: Advantages and disadvantages. *Biosensors & Bioelectronics*, *96*, 33–43. DOI: 10.1016/j.bios.2017.04.037 PMID: 28458132

Chariou, P. L., Ortega-Rivera, O. A., & Steinmetz, N. F. (2020). Nanocarriers for the delivery of medical, veterinary, and agricultural active ingredients. *ACS Nano*, *14*(3), 2678–2701. DOI: 10.1021/acsnano.0c00173 PMID: 32125825

Chattaraj, S., Samantaray, A., Ganguly, A., & Thatoi, H. (2025). Employing plant growth-promoting rhizobacteria for abiotic stress mitigation in plants: With a focus on drought stress. *Discover Applied Sciences*, *7*(1), 68. DOI: 10.1007/s42452-025-06468-6

Chatzichristos, A., & Hassan, J. (2022). Current Understanding of Water Properties inside Carbon Nanotubes. *Nanomaterials 2022, Vol. 12, Page 174, 12*(1), 174. DOI: 10.3390/nano12010174

Chaudhary, A., Shirodkar, S., & Sharma, A. (2017). Characterization of Nickel Tolerant Bacteria Isolated from Heavy Metal Polluted Glass Industry for its Potential Role in Bioremediation. *Soil & Sediment Contamination*, *26*(2), 184–194. DOI: 10.1080/15320383.2017.1267110

Chaudhary, K., Kumar, K., Venkatesu, P., & Masram, D. T. (2021). Protein immobilization on graphene oxide or reduced graphene oxide surface and their applications: Influence over activity, structural and thermal stability of protein. *Advances in Colloid and Interface Science*, *289*, 102367. DOI: 10.1016/j.cis.2021.102367 PMID: 33545443

Chaudhry, N., Dwivedi, S., Chaudhry, V., Singh, A., Saquib, Q., Azam, A., & Musarrat, J. (2018). Bio-inspired nanomaterials in agriculture and food: Current status, foreseen applications and challenges. *Microbial Pathogenesis*, *123*, 196–200. DOI: 10.1016/j.micpath.2018.07.013 PMID: 30009970

Chaud, M., Souto, E. B., Zielinska, A., Severino, P., Batain, F., Oliveira-Junior, J., & Alves, T. (2021). Nanopesticides in Agriculture: Benefits and Challenge in Agricultural Productivity, Toxicological Risks to Human Health and Environment. *Toxics*, *9*(6), 131. DOI: 10.3390/toxics9060131 PMID: 34199739

Chavda, V. P. (2023). *Advanced Phytochemical-Based Nanocarrier Systems for the Treatment of Breast Cancer* (Vol. 15). Cancers.

Chávez-Hernández, J. A., Velarde-Salcedo, A. J., Navarro-Tovar, G., & Gonzalez, C. (2024). Safe nanomaterials: From their use, application and disposal to regulations. Nanoscale advances.Chen, H., Zhang, M., Zhai, G. (2018). Lipid-based nanoparticles in the delivery of amphotericin B for fungal infections. *International Journal of Nanomedicine*, *13*, 867–883.

Chen, C. J. (2021). *Introduction to Scanning Tunneling Microscopy* (3rd ed., Vol. 69). Oxford university press. DOI: 10.1093/oso/9780198856559.001.0001

Cheng, B.. (2015). Clathrin-mediated endocytosis is involved in nanoparticle uptake in plants. *Nature Nanotechnology*, *10*(9), 791–798. PMID: 26237346

Cheng, Z., Zhang, J., Liu, H., Li, Y., Zhao, Y., & Yang, E. (2010). Central nervous system penetration for small molecule therapeutic agents does not increase in multiple sclerosis- and Alzheimer's disease-related animal models despite reported blood-brain barrier disruption. *Drug Metabolism and Disposition: the Biological Fate of Chemicals*, *38*(8), 1355–1361. DOI: 10.1124/dmd.110.033324 PMID: 20427691

Chen, H. (2018). Metal based nanoparticles in agricultural system: Behavior, transport, and interaction with plants. *Chemical Speciation and Bioavailability*, *30*(1), 123–134. DOI: 10.1080/09542299.2018.1520050

Chen, H., & Yada, R. (2011). Nanotechnologies in agriculture: New tools for sustainable development. *Trends in Food Science & Technology*, *22*(11), 585–594. DOI: 10.1016/j.tifs.2011.09.004

Chen, J., Zeng, X., Yang, W., Xie, H., Ashraf, U., Mo, Z., Liu, J., Li, G., & Li, W. (2021). Seed priming with multiwall carbon nanotubes (MWCNTs) modulates seed germination and early growth of maize under cadmium (Cd) toxicity. *Journal of Soil Science and Plant Nutrition*, *21*(3), 1793–1805. DOI: 10.1007/s42729-021-00480-6

Chen, J., Zhang, X., Millican, R., Sherwood, J., Martin, S., Jo, H., & Jun, H. W. (2021). Recent advances in nanomaterials for therapy and diagnosis for atherosclerosis. *Advanced Drug Delivery Reviews*, *170*, 142–199. DOI: 10.1016/j.addr.2021.01.005 PMID: 33428994

Chen, L. X., Zheng, J. N., Wang, A. J., Wu, L. J., Chen, J. R., & Feng, J. J. (2015). Facile synthesis of porous bimetallic alloyed PdAg nanoflowers supported on reduced graphene oxide for simultaneous detection of ascorbic acid, dopamine, and uric acid. *Analyst*, *140*(9), 3183–3192. DOI: 10.1039/C4AN02200A PMID: 25756085

Chen, M., Zhou, Y., Huang, J., Zhu, P., Peng, X., & Wang, Y. (2012). Liposome-Based Delivery Systems in Plant Polysaccharides. *Journal of Nanomaterials*, *2012*(1), 682545. DOI: 10.1155/2012/682545

Chen, P., Ference, C., Sun, X., Lin, Y., Tan, L., & Zhong, T. (2020). Antimicrobial efficacy of liposome-encapsulated citral and its effect on the shelf life of shatangju Mandarin. *Journal of Food Protection*, *83*(8), 1315–1322. DOI: 10.4315/JFP-20-115 PMID: 32294203

Chen, R., & Riviere, J. E. (2017). Biological and environmental surface interactions of nanomaterials: Characterization, modeling, and prediction. *Wiley Interdisciplinary Reviews. Nanomedicine and Nanobiotechnology*, *9*(3), e1440. DOI: 10.1002/wnan.1440 PMID: 27863136

Chen, T., Peng, Y., Qiu, M., Yi, C., & Xu, Z. (2023). Recent advances in mixing-induced nanoprecipitation: From creating complex nanostructures to emerging applications beyond biomedicine. *Nanoscale*, *15*(8), 3594–3609. DOI: 10.1039/D3NR00280B PMID: 36727557

Chen, Y. F., Wang, Y., & Wu, W. H. (2008). Membrane transporters for nitrogen, phosphate and potassium uptake in plants. *Journal of Integrative Plant Biology*, *50*(7), 835–848. DOI: 10.1111/j.1744-7909.2008.00707.x PMID: 18713394

Chhipa, H., & Joshi, P. (2016). Nanofertilisers, nanopesticides and nanosensors in agriculture. *Nanoscience in food and agriculture 1*, 247-282.

Chhipa, H. (2017). Nanofertilizers and nanopesticides for agriculture. *Environmental Chemistry Letters*, *15*(1), 15–22. DOI: 10.1007/s10311-016-0600-4

Chiozzi, V., & Rossi, F. (2020). Inorganic-organic core/shell nanoparticles: Progress and applications. In Nanoscale Advances (Vol. 2, Issue 11, pp. 5090–5105). Royal Society of Chemistry. DOI: 10.1039/D0NA00411A

Choi, W., Abrahamson, J. T., Strano, J. M., & Strano, M. S. (2010). Carbon nanotube-guided thermopower waves. *Materials Today*, *13*(10), 22–33. DOI: 10.1016/S1369-7021(10)70183-8

Chomicki, G., Beinart, R., Prada, C., Ritchie, K. B., & Weber, M. G. (2022). Editorial: Symbiotic relationships as shapers of biodiversity. *Frontiers in Ecology and Evolution*, *10*, 850572. Advance online publication. DOI: 10.3389/fevo.2022.850572

Chou, L. Y. T., Ming, K., & Chan, W. C. W. (2010). Strategies for the intracellular delivery of nanoparticles. *Chemical Society Reviews*, *40*(1), 233–245. DOI: 10.1039/C0CS00003E PMID: 20886124

Christiaens, O., Niu, J., & Nji Tizi Taning, C. (2020). RNAi in insects: A revolution in fundamental research and pest control applications. *Insects*, *11*(7), 415. DOI: 10.3390/insects11070415 PMID: 32635402

Chung, H., Son, Y., Yoon, T. K., Kim, S., & Kim, W. (2011). The effect of multi-walled carbon nanotubes on soil microbial activity. *Ecotoxicology and Environmental Safety*, *74*(4), 569–575. DOI: 10.1016/j.ecoenv.2011.01.004 PMID: 21310485

Chung, S. A., Langford, C. A., Maz, M., Abril, A., Gorelik, M., Guyatt, G., & Mustafa, R. A. (2021). 2021 American College of Rheumatology/Vasculitis Foundation guideline for the management of antineutrophil cytoplasmic antibody–associated vasculitis. *Arthritis & Rheumatology (Hoboken, N.J.)*, *73*(8), 1366–1383. DOI: 10.1002/art.41773 PMID: 34235894

Ciardi, J. A., Tieman, D. M., Lund, S. T., Jones, J. B., Stall, R. E., & Klee, H. J. (2000). Response to Xanthomonas campestris pv. vesicatoria in tomato involves regulation of ethylene receptor gene expression. *Plant Physiology*, *123*(1), 81–92. DOI: 10.1104/pp.123.1.81 PMID: 10806227

Ciftci, N., Sargin, I., Arslan, G., Arslan, U., & Okudan, A. (2020). Ascorbic acid adsorption-release performance and antibacterial activity of chitosan-ter (GMA-MA-NTBA) polymer microcapsules. *Journal of Polymers and the Environment*, *28*(8), 2277–2288. DOI: 10.1007/s10924-020-01773-0

Cinisli, K. T., Uçar, S., & Dikbaş, N. (2019). Use of nanomaterials in agriculture. *Yuzuncu Yil University Journal of Agricultural Sciences*, *29*(4), 817–831. DOI: 10.29133/yyutbd.595658

Clogston, J. D., & Patri, A. K. (2011). Zeta potential measurement. Characterization of nanoparticles intended for drug delivery, 63-70.

Côa, F., Bortolozzo, L. S., Petry, R., Da Silva, G. H., Martins, C. H., de Medeiros, A. M., ... & Martinez, D. S. T. (2020). Environmental toxicity of nanopesticides against non-target organisms: the state of the art. *Nanopesticides: From research and development to mechanisms of action and sustainable use in agriculture*, 227-279.

Coccia, F., Tonucci, L., Bosco, D., Bressan, M., & d'Alessandro, N. (2012). One pot synthesis of lignin-stabilized platinum and palladium nanoparticles and their catalytic behaviours in oxidation and reduction reactions. *Green Chemistry*, *14*(4), 1073–1078. DOI: 10.1039/c2gc16524d

Colin, C., Akpo, E., Perrin, A., Cornu, D., & Cambedouzou, J. (2024). Encapsulation in Alginates Hydrogels and Controlled Release: An Overview. *Molecules (Basel, Switzerland)*, *29*(11), 11–16. DOI: 10.3390/molecules29112515 PMID: 38893391

Colla, G., Hoagland, L., Ruzzi, M., Cardarelli, M., Bonini, P., Canaguier, R., & Rouphael, Y. (2017). Biostimulant action of protein hydrolysates: Unraveling their effects on plant physiology and microbiome. *Frontiers in Plant Science*, *8*, 2202. DOI: 10.3389/fpls.2017.02202 PMID: 29312427

Contento, A. L., & Bassham, D. C. (2012). Structure and function of endosomes in plant cells. *Journal of Cell Science*, *125*(15), 3511–3518. DOI: 10.1242/jcs.093559 PMID: 22935651

Cornelis, G., Hund-Rinke, K., Kuhlbusch, T., Van den Brink, N., & Nickel, C. (2014). Fate and bioavailability of engineered nanoparticles in soils: A review. *Critical Reviews in Environmental Science and Technology*, *44*(24), 2720–2764. DOI: 10.1080/10643389.2013.829767

Correa, S., Dreaden, E. C., Gu, L., & Hammond, P. T. (2016). Engineering nanolayered particles for modular drug delivery. *Journal of Controlled Release*, *240*, 364–386. DOI: 10.1016/j.jconrel.2016.01.040 PMID: 26809005

Corsi, I., Venditti, I., Trotta, F., & Punta, C. (2023). Environmental safety of nanotechnologies: The eco-design of manufactured nanomaterials for environmental remediation. *The Science of the Total Environment*, *864*, 161181. DOI: 10.1016/j.scitotenv.2022.161181 PMID: 36581299

Cota-Ruiz, K., Ye, Y. Q., Valdes, C., Deng, C. Y., Wang, Y., Hernandez-Viezcas, J. A., Duarte-Gardea, M., & Gardea-Torresdey, J. L. (2020). Copper nanowires as nanofertilizers for alfalfa plants: Understanding nano-bio systems interactions from microbial genomics, plant molecular responses and spectroscopic studies. *The Science of the Total Environment*, *742*, 140572. DOI: 10.1016/j.scitotenv.2020.140572 PMID: 32623177

Courtois, P., Rorat, A., Lemiere, S., Guyoneaud, R., Attard, E., Levard, C., & Vandenbulcke, F. (2019). Ecotoxicology of silver nanoparticles and their derivatives introduced in soil with or without sewage sludge: A review of effects on microorganisms, plants and animals. *Environmental Pollution, 253*, 578–598. DOI: 10.1016/j.envpol.2019.07.053 PMID: 31330350

Cruz-Hernández, M. A., Mendoza-Herrera, A., Bocanegra-García, V., & Rivera, G. (2022). Azospirillum spp. From Plant Growth-Promoting Bacteria to Their Use in Bioremediation. *Microorganisms, 10*(5), 1057. DOI: 10.3390/microorganisms10051057 PMID: 35630499

Cruz-Luna, A. R., Cruz-Martínez, H., Vásquez-López, A., & Medina, D. I. (2021). Metal nanoparticles as novel antifungal agents for sustainable agriculture: Current advances and future directions. *Journal of Fungi (Basel, Switzerland), 7*(12), 1033. DOI: 10.3390/jof7121033 PMID: 34947015

Cui, Y., & Lieber, C. M. (2001). Functional Nanoscale Electronic Devices Assembled Using Silicon Nanowire Building Blocks. *Science, 291*(5505), 851–853. DOI: 10.1126/science.291.5505.851 PMID: 11157160

Cunningham, A. J., Gibson, V. P., Banquy, X., Zhu, X. X., & Jeanne, L. C. (2020). Cholic acid-based mixed micelles as siRNA delivery agents for gene therapy. *International Journal of Pharmaceutics, 578*, 119078. DOI: 10.1016/j.ijpharm.2020.119078 PMID: 31988037

Cunningham, F. J., Goh, N. S., Demirer, G. S., Matos, J. L., & Landry, M. P. (2018). Nanoparticle-Mediated Delivery towards Advancing Plant Genetic Engineering. *Trends in Biotechnology, 36*(9), 882–897. DOI: 10.1016/j.tibtech.2018.03.009 PMID: 29703583

Cvjetko, P., Milošić, A., Domijan, A. M., Vrček, I. V., Tolić, S., Štefanić, P. P., & Balen, B. (2017). Toxicity of silver ions and differently coated silver nanoparticles in *Allium cepa* roots. *Ecotoxicology and Environmental Safety, 137*, 18–28. DOI: 10.1016/j.ecoenv.2016.11.009 PMID: 27894021

da Silva Júnior, A. H., Mulinari, J., de Oliveira, P. V., de Oliveira, C. R. S., & Júnior, F. W. R. (2022). Impacts of metallic nanoparticles application on the agricultural soils microbiota. *Journal of Hazardous Materials Advances, 7*, 100103. DOI: 10.1016/j.hazadv.2022.100103

da Silva, R., Viana, V. E., Avila, L. A., Zotti, M. J., Smagghe, G., Junior, A. M., Camargo, E. R., & Fajardo, A. R. (2024). Advances on polymeric nanocarriers for sustainable agriculture: Enhancing dsRNA/siRNA delivery to combat agricultural pests. *International Journal of Biological Macromolecules, 282*, 137000. DOI: 10.1016/j.ijbiomac.2024.137000 PMID: 39476891

Daghan, H. A. T. İ. C. E. (2018). Effects of TiO_2 nanoparticles on maize (*Zea mays* L.) growth, chlorophyll content and nutrient uptake. *Applied Ecology and Environmental Research, 16*, 6873–6883.

Dahl, J. A., Maddux, B. L. S., & Hutchison, J. E. (2007). Toward greener nanosynthesis. In Chemical Reviews (Vol. 107, Issue 6, pp. 2228–2269). DOI: 10.1021/cr050943k

Dahoumane, S. A., Mechouet, M., Wijesekera, K., Filipe, C. D. M., Sicard, C., Bazylinski, D. A., & Jeffryes, C. (2017). Algae-mediated biosynthesis of inorganic nanomaterials as a promising route in nanobiotechnology-a review. In Green Chemistry (Vol. 19, Issue 3, pp. 552–587). Royal Society of Chemistry. DOI: 10.1039/C6GC02346K

Daisy, P., & Saipriya, K. (2012). Biochemical analysis of *Cassia fistula* aqueous extract and phytochemically synthesized gold nanoparticles as hypoglycemic treatment for diabetes mellitus. *International Journal of Nanomedicine*, *7*, 1189–1202. DOI: 10.2147/IJN.S26650 PMID: 22419867

Dalakouras, A., Wassenegger, M., Dadami, E., Ganopoulos, I., Pappas, M. L., & Papadopoulou, K. (2020). Genetically modified organism-free RNA interference: Exogenous application of RNA molecules in plants. *Plant Physiology*, *182*(1), 38–50. DOI: 10.1104/pp.19.00570 PMID: 31285292

Dangi, K., & Verma, A. K. (2021). Efficient & eco-friendly smart nano-pesticides: Emerging prospects for agriculture. *Materials Today: Proceedings*, *45*, 3819–3824. DOI: 10.1016/j.matpr.2021.03.211

Darabi, R., Karimi-Maleh, H., Altuner, E. E., Gulbagca, F., Tiri, R. N. E., Chinnathambi, A., Brindhadevi, K., Halvaci, E., & Sen, F. (2024). Cobalt Nanoparticles Supported Active Carbon from Chitosan Biopolymer Using Thermal Method: Synthesis, Characterization, and Hydrogen Production. *Topics in Catalysis*, •••, 1–10. DOI: 10.1007/s11244-024-01998-6

Darabi, R., Karimi-Maleh, H., Shahidi, S. A., Gulbagca, F., Nour Elhouda Tiri, R., Monadi Sefidan, A., Dehdashtian, N., Alhrasishawi, W., Aygun, A., Esra Altuner, E., & Sen, F. (2023). Hexamethylenetetramine-based nanomaterial synthesis, characterization, investigation of photocatalytic activity against Rhodamine B dye and kinetic study for hydrogen fuel production. *Fuel*, *352*, 128841. DOI: 10.1016/j.fuel.2023.128841

Das, D., Reddy, M., Upadhyaya, K., & Sopory, S. (2002). An efficient leaf-disc culture method for the regeneration via somatic embryogenesis and transformation of grape (Vitis vinifera L.). *Plant Cell Reports*, *20*(11), 999–1005. DOI: 10.1007/s00299-002-0441-4

Das, M., Sarmah, S., & Sarkar, D. (2019). Photo sensing property of nanostructured CdS-porous silicon (PS):p-Si based MSM hetero-structure. *Journal of Materials Science Materials in Electronics*, *30*(12), 11239–11249. DOI: 10.1007/s10854-019-01470-2

Das, P. R., & Sherif, S. M. (2020). Application of exogenous dsRNAs-induced RNAi in agriculture: Challenges and triumphs. *Frontiers in Plant Science*, *11*, 946. DOI: 10.3389/fpls.2020.00946 PMID: 32670336

Das, P., Ghosh, S., & Nayak, B. (2021). Phyto-fabricated Nanoparticles and Their Anti-biofilm Activity: Progress and Current Status. *Frontiers in Nanotechnology*, *3*, 739286. DOI: 10.3389/fnano.2021.739286

Das, S., Debnath, N., Cui, Y., Unrine, J., & Palli, S. R. (2015). Chitosan, carbon quantum dot, and silica nanoparticle mediated dsRNA delivery for gene silencing in *Aedes aegypti*: A Comparative Analysis. *ACS Applied Materials & Interfaces*, *7*(35), 19530–19535. DOI: 10.1021/acsami.5b05232 PMID: 26291176

Das, S., & Pattanayak, S. (2020). Nanotechnological approaches in sustainable agriculture and plant disease management. *Organic Agriculture*, *2*, 13–32. DOI: 10.5772/intechopen.92463

Dauthal, P., & Mukhopadhyay, M. (2015). Biofabrication, characterization, and possible bio-reduction mechanism of platinum nanoparticles mediated by agro-industrial waste and their catalytic activity. *Journal of Industrial and Engineering Chemistry*, *22*, 185–191. DOI: 10.1016/j.jiec.2014.07.009

Davidson, S. J., Saggese, T., & Krajňáková, J. (2024). Deep learning for automated segmentation and counting of hypocotyl and cotyledon regions in mature *Pinus radiata* D. Don. Somatic embryo images. *Frontiers in Plant Science*, *15*, 1322920. Advance online publication. DOI: 10.3389/fpls.2024.1322920 PMID: 38495377

Davies, M. J., Badcock, T. J., Dawson, P., Kappers, M. J., Oliver, R. A., & Humphreys, C. J. (2013). High excitation carrier density recombination dynamics of InGaN/GaN quantum well structures: Possible relevance to efficiency droop. *Applied Physics Letters*, *102*(2), 022106. DOI: 10.1063/1.4781398

de la Rosa, G., López-Moreno, M. L., de Haro, D., Botez, C. E., Peralta-Videa, J. R., & Gardea-Torresdey, J. L. (2013). Effects of ZnO nanoparticles in alfalfa, tomato, and cucumber at the germination stage: root development and X-ray absorption spectroscopy studies. *Pure and Applied Chemistry 4th*, 85(12), 2161-2174.

De Leo, V., Maurelli, A. M., Giotta, L., & Catucci, L. (2022). Liposomes containing nanoparticles: Preparation and applications. *Colloids and Surfaces. B, Biointerfaces*, *218*, 112737. DOI: 10.1016/j.colsurfb.2022.112737 PMID: 35933888

de Oliveira, C., Mulinari, J., Reichert, F., & Júnior, A. (2020, September). Nano-delivery systems of pesticides active agents for agriculture applications-An overview. In *Ciência, Tecnologia e Inovação: Do Campo à Mesa,Proceeding of International Agribusiness Congress,Recife, Brazil* (pp. 25-27). DOI: 10.31692/ICIAGRO.2020.0051

De Oliveira, J. L., Campos, E. V. R., & Fraceto, L. F. (2019). Recent developments on nanopesticides and nanofertilizers: A path to the future of sustainable agriculture. *Nano Today*, *24*, 100190.

De Schutter, K., Taning, C. N. T., Van Daele, L., Van Damme, E. J., Dubruel, P., & Smagghe, G. (2022). RNAi-based biocontrol products: Market status, regulatory aspects, and risk assessment. *Frontiers in Insect Science*, *1*, 818037. DOI: 10.3389/finsc.2021.818037 PMID: 38468882

de Souza, R. B., & Guimaraes, J. R. (2022). Effects of avermectins on the environment based on their toxicity to plants and soil invertebrates—A review. *Water, Air, and Soil Pollution*, *233*(7), 259. DOI: 10.1007/s11270-022-05744-0 PMID: 35789787

Debernardi, J. M., Tricoli, D. M., Ercoli, M. F., Hayta, S., Ronald, P., Palatnik, J. F., & Dubcovsky, J. (2020). A GRF-GIF chimeric protein improves the regeneration efficiency of transgenic plants. *Nature Biotechnology*, *38*(11), 1274–1279. DOI: 10.1038/s41587-020-0703-0 PMID: 33046875

Demir, E. (2020). Drosophila as a model for assessing nanopesticide toxicity. *Nanotoxicology*, *14*(9), 1271–1279. DOI: 10.1080/17435390.2020.1815886 PMID: 32969292

Demirer, G. S., Zhang, H., Goh, N. S., González-Grandío, E., & Landry, M. P. (2019). Carbon nanotube–mediated DNA delivery without transgene integration in intact plants. *Nature Protocols*, *14*(10), 2954–2971. DOI: 10.1038/s41596-019-0208-9 PMID: 31534231

Demirer, G. S., Zhang, H., Goh, N. S., Pinals, R. L., Chang, R., & Landry, M. P. (2020). Carbon nanocarriers deliver siRNA to intact plant cells for efficient gene knockdown. *Science Advances*, *6*(26), eaaz0495. DOI: 10.1126/sciadv.aaz0495 PMID: 32637592

Demirer, G. S., Zhang, H., Matos, J. L., Goh, N. S., Cunningham, F. J., Sung, Y., Chang, R., Aditham, A. J., Chio, L., Cho, M.-J., Staskawicz, B., & Landry, M. P. (2019). High aspect ratio nanomaterials enable delivery of functional genetic material without DNA integration in mature plants. *Nature Nanotechnology*, *14*(5), 456–464. DOI: 10.1038/s41565-019-0382-5 PMID: 30804481

Deng, X., Cao, M., Zhang, J., Hu, K., Yin, Z., Zhou, Z., & Zeng, Y. (2014). Hyaluronic acid-chitosan nanoparticles for co-delivery of MiR-34a and doxorubicin in therapy against triple negative breast cancer. *Biomaterials*, *35*(14), 4333–4344. DOI: 10.1016/j.biomaterials.2014.02.006 PMID: 24565525

Deng, Z., Xu, S., & Li, S. (2008). Understanding a relaxation behavior in a nanoparticle suspension for drug delivery applications. *International Journal of Pharmaceutics*, *351*(1-2), 236–243. DOI: 10.1016/j.ijpharm.2007.10.001 PMID: 18093763

Desai, S., Singh, M., Chavan, A., Wagh, N. S., & Lakkakula, J. (2022). Micro-and nanoencapsulation techniques in agriculture. In *Agricultural nanobiotechnology* (pp. 297–323). Woodhead Publishing. DOI: 10.1016/B978-0-323-91908-1.00009-2

Destito, G., Schneemann, A., & Manchester, M. (2009). Biomedical nanotechnology using virus-based nanoparticles. In *Current Topics in Microbiology and Immunology* (Vol. 327, pp. 95–122). Springer., DOI: 10.1007/978-3-540-69379-6_5

Destoumieux-Garzón, D., Mavingui, P., Boetsch, G., Boissier, J., Darriet, F., Duboz, P., Fritsch, C., Giraudoux, P., Le Roux, F., Morand, S., Paillard, C., Pontier, D., Sueur, C., & Voituron, Y. (2018). The one health concept: 10 years old and a long road ahead. *Frontiers in Veterinary Science*, *5*(FEB), 14. Advance online publication. DOI: 10.3389/fvets.2018.00014 PMID: 29484301

Dheyab, M. A., Aziz, A. A., Moradi Khaniabadi, P., Jameel, M. S., Oladzadabbasabadi, N., Mohammed, S. A., Abdullah, R. S., & Mehrdel, B. (2022). Monodisperse gold nanoparticles: A review on synthesis and their application in modern medicine. *International Journal of Molecular Sciences*, *23*(13), 7400. DOI: 10.3390/ijms23137400 PMID: 35806405

Dhingra, P., Sharma, S., Singh, K. H., Kushwaha, H. S., Barupal, J. K., Haq, S., Kothari, S. L., & Kachhwaha, S. (2022). Seed priming with carbon nanotubes and silicon dioxide nanoparticles influence agronomic traits of Indian mustard (*Brassica juncea*) in field experiments. *Journal of King Saud University. Science*, *34*(4), 102067. DOI: 10.1016/j.jksus.2022.102067

Di Benedetto, N. A., Corbo, M. R., Campaniello, D., Cataldi, M. P., Bevilacqua, A., Sinigaglia, M., & Flagella, Z. (2017). The role of Plant Growth Promoting Bacteria in improving nitrogen use efficiency for sustainable crop production: A focus on wheat. *AIMS Microbiology*, *3*(3), 413–434. DOI: 10.3934/microbiol.2017.3.413 PMID: 31294169

Di Sotto, A., Paolicelli, P., Nardoni, M., Abete, L., Garzoli, S., Di Giacomo, S., Mazzanti, G., Casadei, M. A., & Petralito, S. (2018). SPC liposomes as possible delivery systems for improving bioavailability of the natural sesquiterpene? -caryophyllene: Lamellarity and drug-loading as key features for a rational drug delivery design. *Pharmaceutics*, *10*(4), 4. Advance online publication. DOI: 10.3390/pharmaceutics10040274 PMID: 30551617

Dimkić, I., Janakiev, T., Petrović, M., Degrassi, G., & Fira, D. (2021). Plant-associated *Bacillus* and *Pseudomonas* antimicrobial activities in plant disease suppression via biological control mechanisms - A review. *Physiological and Molecular Plant Pathology*, *117*, 101754. DOI: 10.1016/j.pmpp.2021.101754

Dimkpa, C. O., Singh, U., Bindraban, P. S., Elmer, W. H., Gardea-Torresdey, J. L., & White, J. C. (2019). Zinc oxide nanoparticles alleviate drought-induced alterations in sorghum performance, nutrient acquisition, and grain fortification. *The Science of the Total Environment*, *688*, 926–934. DOI: 10.1016/j.scitotenv.2019.06.392 PMID: 31726574

Dinesh-Kumar, S. P., Anandalakshmi, R., Marathe, R., Schiff, M., & Liu, Y. (2003). Virus-induced gene silencing. *Methods in Molecular Biology (Clifton, N.J.)*, *236*, 287–294. DOI: 10.1385/1-59259-413-1:287 PMID: 14501071

Ditta, A., Arshad, M., & Ibrahim, M. (2015). Nanoparticles in sustainable agricultural crop production: applications and perspectives. Nanotechnology and plant sciences: nanoparticles and their impact on plants, 55-75.

Ditta, A. (2012). How helpful is nanotechnology in agriculture? *Advances in Natural Sciences: Nanoscience and Nanotechnology*, *3*(3), 033002. DOI: 10.1088/2043-6262/3/3/033002

Divya, K., & Jisha, M. S. (2018). Chitosan nanoparticles preparation and applications. *Environmental Chemistry Letters*, *16*(1), 101–112. DOI: 10.1007/s10311-017-0670-y

Djearamane, S., Xiu, L. J., Wong, L. S., Rajamani, R., Bharathi, D., Kayarohanam, S., De Cruz, A. E., Tey, L.-H., Janakiraman, A. K., Aminuzzaman, M., & Selvaraj, S. (2022). Antifungal properties of zinc oxide nanoparticles on Candida albicans. *Coatings*, *12*(12), 1864. DOI: 10.3390/coatings12121864

Dobrzyński, J., Jakubowska, Z., & Dybek, B. (2022). Potential of Bacillus pumilus to directly promote plant growth. *Frontiers in Microbiology*, *13*, 1069053. DOI: 10.3389/fmicb.2022.1069053 PMID: 36620067

Dolai, S., Maiti, P., Ghorai, A., Bhunia, R., Paul, P. K., & Ghosh, D. (2020). Exfoliated molybdenum disulfide-wrapped CdS nanoparticles as a nano-heterojunction for photo-electrochemical water splitting. *ACS Applied Materials & Interfaces*, *13*(1), 438–448. DOI: 10.1021/acsami.0c16972 PMID: 33356109

Dominguez, A. N., Emmert, G. E., Gil, D. M., & Alvarez, R. M. S. (2021). Experimental and theoretical vibrational study of the fungicide pyraclostrobin. *Spectrochimica Acta. Part A: Molecular Spectroscopy*, *259*, 119888. DOI: 10.1016/j.saa.2021.119888 PMID: 34015601

Doner, N., Burhan, H., Bayat, R., Esra Altuner, E., & Sen, F. (2023). Energy Generation through Thermopower Waves using Multi-walled Carbon Nanotube Yarn. *Fuel*, *331*, 125906. DOI: 10.1016/j.fuel.2022.125906

Dong, J. T., Chen, W., Feng, J. G., Liu, X. Q., Xu, Y., Wang, C., Yang, W. C., & Du, X. Z. (2021). Facile, smart, and degradable metal-organic framework nanopesticides gated with FeIII-Tannic acid networks in response to seven biological and environmental stimuli. *ACS Applied Materials & Interfaces*, *13*(16), 19507–19520. DOI: 10.1021/acsami.1c04118 PMID: 33856772

Dong, Z. Z., & Chang, L. Q. (2021). Recent electroporation-based systems for intracellular molecule delivery. *Nanotechnology and Precision Engineering*, *4*(4), 045001. DOI: 10.1063/10.0005649

Doudna, J. A., & Charpentier, E. (2014). The new frontier of genome engineering with CRISPR-Cas9. *Science*, *346*(6213), 1258096. DOI: 10.1126/science.1258096 PMID: 25430774

Doyle, C., Higginbottom, K., Swift, T. A., Winfield, M., Bellas, C., Benito-Alifonso, D., & Whitney, H. M. (2019). A simple method for spray-on gene editing in planta. *BioRxiv*, 805036.

Du Jardin, P. (2015). Plant biostimulants: Definition, concept, main categories and regulation. *Scientia Horticulturae*, *196*, 3–14. DOI: 10.1016/j.scienta.2015.09.021

Duan, H., Wang, D., & Li, Y. (2015). Green chemistry for nanoparticle synthesis. *Chemical Society Reviews*, *44*(16), 5778–5792. DOI: 10.1039/C4CS00363B PMID: 25615873

Duan, L., Ouyang, K., Xu, X., Xu, L., Wen, C., Zhou, X., Qin, Z., Xu, Z., Sun, W., & Liang, Y. (2021). Nanoparticle delivery of CRISPR/Cas9 for genome editing. *Frontiers in Genetics*, *12*, 673286. DOI: 10.3389/fgene.2021.673286 PMID: 34054927

Duarte, S. F. P., & Cerozi, B. S. (2024). Enhancing plant growth in aquaponic systems via potassium manipulation in fish feeds: A pilot study of tailored feeds bridging nutritional gaps in aquaponics. *Agricultural Systems*, *218*, 104001. DOI: 10.1016/j.agsy.2024.104001

Dubey, S. P., Lahtinen, M., & Sillanpää, M. (2010). Tansy fruit mediated greener synthesis of silver and gold nanoparticles. *Process Biochemistry (Barking, London, England)*, *45*(7), 1065–1071. DOI: 10.1016/j.procbio.2010.03.024

Duhan, J. S., Kumar, R., Kumar, N., Kaur, P., Nehra, K., & Duhan, S. (2017). Nanotechnology: The new perspective in precision agriculture. *Biotechnology Reports (Amsterdam, Netherlands)*, *15*, 11–23. DOI: 10.1016/j.btre.2017.03.002 PMID: 28603692

Du, J., Tang, J., Xu, S., Ge, J., Dong, Y., Li, H., & Jin, M. (2020). ZnO nanoparticles: Recent advances in ecotoxicity and risk assessment. *Drug and Chemical Toxicology*, *43*(3), 322–333. DOI: 10.1080/01480545.2018.1508218 PMID: 30264597

Dunbar, T., Tsakirpaloglou, N., Septiningsih, E. M., & Thomson, M. J. (2022). Carbon nanotube-mediated plasmid DNA delivery in rice leaves and seeds. *International Journal of Molecular Sciences*, *23*(8), 8. Advance online publication. DOI: 10.3390/ijms23084081 PMID: 35456898

Durán-Lobato, M., López-Estévez, A. M., Cordeiro, A. S., Dacoba, T. G., Crecente-Campo, J., Torres, D., & Alonso, M. J. (2021). Nanotechnologies for the delivery of biologicals: Historical perspective and current landscape. *Advanced Drug Delivery Reviews*, *176*, 113899. DOI: 10.1016/j.addr.2021.113899 PMID: 34314784

Durmaz, E., Sertkaya, S., Yilmaz, H., Olgun, C., Ozcelik, O., Tozluoglu, A., & Candan, Z. (2023). Lignocellulosic Bionanomaterials for Biosensor Applications. *Micromachines 2023, Vol. 14, Page 1450*, *14*(7), 1450. DOI: 10.3390/mi14071450

Dutta, S., Pal, S., Panwar, P., Sharma, R. K., & Bhutia, P. L. (2022). Biopolymeric nanocarriers for nutrient delivery and crop biofortification. *ACS Omega*, *7*(30), 25909–25920. DOI: 10.1021/acsomega.2c02494 PMID: 35936412

Du, W., Gardea-Torresdey, J. L., Ji, R., Yin, Y., Zhu, J., Peralta-Videa, J. R., & Guo, H. (2015). Physiological and biochemical changes imposed by CeO2 nanoparticles on wheat: A life cycle field study. *Environmental Science & Technology*, *49*(19), 11884–11893. DOI: 10.1021/acs.est.5b03055 PMID: 26368651

Ebralidze, I. I., Laschuk, N. O., Poisson, J., & Zenkina, O. V. (2019). Colorimetric Sensors and Sensor Arrays. *Nanomaterials Design for Sensing Applications*, 1–39. DOI: 10.1016/B978-0-12-814505-0.00001-1

Ederli, L., Varone, L., & De Gara, L. (2021). Nanotechnology applications in plant nutrition and crop protection: Prospects and challenges. *Plant Physiology and Biochemistry*, *167*, 245–256.

Edison, T., & Sethuraman, M. (2012). Instant green synthesis of silver nanoparticles using *Terminalia chebula* fruit extract and evaluation of their catalytic activity on reduction of Methylene Blue. *Process Biochemistry (Barking, London, England)*, *47*(9), 1351–1357. DOI: 10.1016/j.procbio.2012.04.025

Eftekhari, M., Ma, C., & Orlov, Y. L. (2024). Editorial: Applications of artificial intelligence, machine learning, and deep learning in plant breeding. *Frontiers in Plant Science*, *15*, 1420938. Advance online publication. DOI: 10.3389/fpls.2024.1420938 PMID: 38841285

Egamberdieva, D., Ma, H., Alimov, J., Reckling, M., Wirth, S., & Bellingrath-Kimura, S. D. (2020). Response of soybean to hydrochar-based rhizobium inoculation in loamy sandy soil. *Microorganisms*, *8*(11), 1–10. DOI: 10.3390/microorganisms8111674 PMID: 33126699

Egamberdieva, D., Shurigin, V., Alaylar, B., Ma, H., Müller, M. E. H., Wirth, S., Reckling, M., & Bellingrath-Kimura, S. D. (2020). The effect of biochars and endophytic bacteria on growth and root rot disease incidence of fusarium infested narrow-leafed lupin (Lupinus angustifolius L.). *Microorganisms*, *8*(4), 496. Advance online publication. DOI: 10.3390/microorganisms8040496 PMID: 32244470

Egamberdieva, D., Wirth, S. J., & Alqarawi, A. A., F., E., & Hashem, A. (. (2017). Phytohormones and Beneficial Microbes: Essential Components for Plants to Balance Stress and Fitness. *Frontiers in Microbiology*, *8*, 278255. DOI: 10.3389/fmicb.2017.02104 PMID: 29163398

Eichert, T.. (2008). Stomatal uptake of nanoparticle-based agrochemicals. *Plant Physiology*, *148*(1), 31–38.

Eichert, T., & Goldbach, H. E. (2008). Equivalent pore radii of hydrophilic foliar uptake routes in stomatous and astomatous leaf surfaces – further evidence for a stomatal pathway. *Physiologia Plantarum*, *132*(4), 491–502. DOI: 10.1111/j.1399-3054.2007.01023.x PMID: 18334002

Eichert, T., Kurtz, A., Steiner, U., & Goldbach, H. E. (2008). Size exclusion limits and lateral heterogeneity of the stomatal foliar uptake pathway for aqueous solutes and water-suspended nanoparticles. *Physiologia Plantarum*, *134*(1), 151–160. DOI: 10.1111/j.1399-3054.2008.01135.x PMID: 18494856

Eid, N. A., Ibrahim, A. M. H., Elsharawy, A. A., & Salem, K. F. M. (2025). Nanofertilizers for Plant Viral Disease Management. In Kumar, P., & Dubey, R. C. (Eds.), *Nanofertilizers for Sustainable Agriculture: Assessing Impacts on Health, Environment, and Economy* (pp. 149–187). Springer Nature Switzerland. DOI: 10.1007/978-3-031-78649-5_7

Ekambaram, P., Sathali, A. A. H., & Priyanka, K. (2012). Solid lipid nanoparticles: A review. *Sci. Rev. Chem. Commun.*, *2*, 80–102.

El-Belely, E. F., Farag, M. M. S., Said, H. A., Amin, A. S., Azab, E., Gobouri, A. A., & Fouda, A. (2021). Green Synthesis of Zinc Oxide Nanoparticles (ZnO-NPs) Using Arthrospira platensis (Class: Cyanophyceae) and Evaluation of their Biomedical Activities. *Nanomaterials 2021, Vol. 11, Page 95, 11*(1), 95. DOI: 10.3390/nano11010095

Elena, S. L., Daniela, G., Gerard, E., Lorena, B., Ana, L. L. M., Ruth, G., Amanda, C., Marta, E., Miren, E., Antoni, C., Amélia, M. S., Alessandra, D., Antonello, S., Maria, L. G., & Eliana, B. S. (2020). Metal-Based Nanoparticles as Antimicrobial Agents: An Overview. *Nanomaterials (Basel, Switzerland), 10*(2), 292. DOI: 10.3390/nano10020292 PMID: 32050443

Elmer, W., Ma, C., & White, J. (2018). Nanoparticles for plant disease management. *Current Opinion in Environmental Science & Health, 6*, 66–70. DOI: 10.1016/j.coesh.2018.08.002

El-Saadony, M. T., Saad, A. M., Mohammed, D. M., Fahmy, M. A., Elesawi, I. E., Ahmed, A. E., Algopishi, U. B., Elrys, A. S., Desoky, E. M., Mosa, W. F., Abd El-Mageed, T. A., Alhashmi, F. I., Mathew, B. T., AbuQamar, S. F., & El-Tarabily, K. A. (2024). Drought-tolerant plant growth-promoting rhizobacteria alleviate drought stress and enhance soil health for sustainable agriculture: A comprehensive review. *Plant Stress, 14*, 100632. DOI: 10.1016/j.stress.2024.100632

El-Saadony, M. T., Saad, A. M., Soliman, S. M., Salem, H. M., Ahmed, A. I., Mahmood, M., El-Tahan, A. M., Ebrahim, M., Abd, A. A., El-Mageed, T. A., Negm, S. H., Selim, S., Babalghith, A. O., Elrys, A. S., El-Tarabily, K. A., & AbuQamar, S. F. (2022). Plant growth-promoting microorganisms as biocontrol agents of plant diseases: Mechanisms, challenges and future perspectives. *Frontiers in Plant Science, 13*, 923880. DOI: 10.3389/fpls.2022.923880 PMID: 36275556

El-Saadony, M. T., Saad, A. M., Soliman, S. M., Salem, H. M., Desoky, E. S. M., Babalghith, A. O., El-Tahan, A. M., Ibrahim, O. M., Ebrahim, A. A. M., Abd El-Mageed, T. A., Elrys, A. S., Elbadawi, A. A., El-Tarabily, K. A., & AbuQamar, S. F. (2022). Role of nanoparticles in enhancing crop tolerance to abiotic stress: A comprehensive review. *Frontiers in Plant Science, 13*, 946717. DOI: 10.3389/fpls.2022.946717 PMID: 36407622

Elzoghby, A., Freag, M., Mamdouh, H., & Elkhodairy, K. (2017). Zein-based nanocarriers as potential natural alternatives for drug and gene delivery: Focus on cancer therapy. *Current Pharmaceutical Design, 23*(35), 5261–5271. PMID: 28641543

Erduran, V., Altuner, E. E., & Şen, F. (2023). Stability and validation of bionanomaterials. *Synthesis of Bionanomaterials for Biomedical Applications*, 251–263. DOI: 10.1016/B978-0-323-91195-5.00021-0

Esposito, S. (2019). "Traditional" Sol-Gel Chemistry as a Powerful Tool for the Preparation of Supported Metal and Metal Oxide Catalysts. *Materials (Basel), 12*(4), 668. DOI: 10.3390/ma12040668 PMID: 30813441

Estefanía, C. (2023). Plant-Growth-Promoting Rhizobacteria Improve Germination and Bioactive Compounds in Cucumber Seedlings. *Agronomy (Basel), 13*(2), 315. DOI: 10.3390/agronomy13020315

Etsuyankpa, M. B., Augustine, A. U., Musa, S. T., Mathew, J. T., Ismail, H., Salihu, A. M., & Mamman, A. (2024). An Overview of Wastewater Characteristics, Treatment and Disposal: A Review. *Journal of Applied Science & Environmental Management, 28*(5), 1553–1572. DOI: 10.4314/jasem.v28i5.28

Etxeberria, E., Gonzalez, P., & Pozueta, J. (2009). Evidence for two endocytic transport pathways in plant cells. *Plant Science*, *177*(4), 341–348. DOI: 10.1016/j.plantsci.2009.06.014

Evans, R. G., & Sadler, E. J. (2008). Methods and technologies to improve efficiency of water use. *Water Resources Research*, *44*(7), 2007WR006200. Advance online publication. DOI: 10.1029/2007WR006200

Falcão, L. de S., Coelho, D. B., Veggi, P. C., Campelo, P. H., Albuquerque, P. M., & de Moraes, M. A. (2022). Starch as a Matrix for Incorporation and Release of Bioactive Compounds: Fundamentals and Applications. *Polymers*, *14*(12), 1–16. DOI: 10.3390/polym14122361 PMID: 35745937

Falsafi, S. R., Rostamabadi, H., Assadpour, E., & Jafari, S. M. (2020). Morphology and microstructural analysis of bioactive-loaded micro/nanocarriers via microscopy techniques; CLSM/SEM/TEM/AFM. *Advances in Colloid and Interface Science*, *280*, 102166. DOI: 10.1016/j.cis.2020.102166 PMID: 32387755

Fatollahpour Grangah, M., Rashidi, V., Mirshekari, B., Khalilvand Behrouzyar, E., & Farahvash, F. (2020). Effects of nano-fertilizers on physiological and yield characteristics of pinto bean cultivars under water deficit stress. *Journal of Plant Nutrition*, *43*(19), 2898–2910. DOI: 10.1080/01904167.2020.1799000

Fayaz, A. M., Balaji, K., Girilal, M., Yadav, R., Kalaichelvan, P. T., & Venketesan, R. (2010). Biogenic synthesis of silver nanoparticles and their synergistic effect with antibiotics: A study against gram-positive and gram-negative bacteria. *Nanomedicine; Nanotechnology, Biology, and Medicine*, *6*(1), 103–109. DOI: 10.1016/j.nano.2009.04.006 PMID: 19447203

Fendler, J. H. (Ed.). (2008). *Nanoparticles and nanostructured films: preparation, characterization, and applications*. John Wiley & Sons.

Feng, Y., Lai, S., Yang, L., Riedel, R., & Yu, Z. (2018). Polymer-derived porous Bi2WO6/SiC(O) ceramic nanocomposites with high photodegradation efficiency towards Rhodamine B. *Ceramics International*, *44*(7), 8562–8569. DOI: 10.1016/j.ceramint.2018.02.061

Fernandez-Cornejo, J., Jans, S., & Smith, M. (2014). Issues in the economics of pesticide use in agriculture: A review of the empirical evidence. *Agricultural Economics*, *30*(2), 123–132.

Fernandez-Garcia, M., Martinez-Arias, A., Hanson, J. C., & Rodriguez, J. A. (2004). Nanostructured oxides in chemistry: Characterization and properties. *Chemical Reviews*, *104*(9), 4063–4104. DOI: 10.1021/cr030032f PMID: 15352786

Ferreira, D. C. M., de Souza, A. L., da Silveira, J. V. W., Marim, B. M., Giraldo, G. A. G., Mantovan, J., Mali, S., & Pelissari, F. M. (2020). Chapter 17 - Chitosan nanocomposites for food packaging applications. In K. A. B. T.-M. H. N. for S. A.-F. and E. Abd-Elsalam (Ed.), Micro and Nano Technologies (pp. 393–435). Elsevier. https://doi.org/https://doi.org/10.1016/B978-0-12-821354-4.00017-0

Ferrier, D. C., & Honeychurch, K. C. (2021). Carbon Nanotube (CNT)-based biosensors. *Biosensors (Basel)*, *11*(12), 12. Advance online publication. DOI: 10.3390/bios11120486 PMID: 34940243

Fincheira, P., Hoffmann, N., Tortella, G., Ruiz, A., Cornejo, P., Diez, M. C., Seabra, A. B., Benavides-Mendoza, A., & Rubilar, O. (2023). Eco-Efficient Systems Based on Nanocarriers for the Controlled Release of Fertilizers and Pesticides: Toward Smart Agriculture. *Nanomaterials (Basel, Switzerland)*, *13*(13), 1978. DOI: 10.3390/nano13131978 PMID: 37446494

Firdhouse, M. J., & Lalitha, P. (2013). Biosynthesis of silver nanoparticles using the extract of Alternanthera sessilis-antiproliferative effect against prostate cancer cells. *Cancer Nanotechnology*, *4*(6), 137–143. DOI: 10.1007/s12645-013-0045-4 PMID: 26069509

Fire, A., Xu, S., Montgomery, M. K., Kostas, S. A., Driver, S. E., & Mello, C. C. (1998). Potent and specific genetic interference by double stranded RNA in *Caenorhabditis elegans.Nature*, *391*(6669), 806–881. DOI: 10.1038/35888 PMID: 9486653

Fischer, J., Beckers, S. J., Yiamsawas, D., Thines, E., Landfester, K., & Wurm, F. R. (2019). Targeted drug delivery in plants: Enzyme-responsive lignin nanocarriers for the curative treatment of the worldwide grapevine trunk disease. *Advanced Science (Weinheim, Baden-Wurttemberg, Germany)*, *6*(15), 1802315. DOI: 10.1002/advs.201802315 PMID: 31406660

Fisher, E., Mahajan, R. L., & Mitcham, C. (2015). Midstream modulation of technology: Governance from within. *Bulletin of Science, Technology & Society*, *26*(6), 485–496. DOI: 10.1177/0270467606295402

Florencio, C. S., Brandão, F. A. S., de Mello Teixeira, M., Bocca, A. L., Felipe, M. S. S., Vicente, V. A., & Fernandes, L. (2018). Genetic manipulation of Fonsecaea pedrosoi using particles bombardment and Agrobacterium mediated transformation. *Microbiological Research*, *207*, 269–279. DOI: 10.1016/j.micres.2018.01.001 PMID: 29458863

Fraceto, L. F., Grillo, R., Medeiros, G. A., Scognamiglio, V., de Melo, N. F. S., & Rosa, A. H. (2016). Nanotechnology in agriculture: Which innovation potential does it have? *Frontiers in Environmental Science*, *4*, 20. DOI: 10.3389/fenvs.2016.00020

Franci, G., Falanga, A., Galdiero, S., Palomba, L., Rai, M., Morelli, G., & Galdiero, M. (2015). Silver nanoparticles as potential antibacterial agents. *Molecules (Basel, Switzerland)*, *20*(5), 8856–8874. DOI: 10.3390/molecules20058856 PMID: 25993417

Francino, M. P. (2016). Antibiotics and the human gut microbiome: Dysbioses and accumulation of resistances. *Frontiers in Microbiology*, *6*(JAN), 164577. DOI: 10.3389/fmicb.2015.01543 PMID: 26793178

Francis, S. P., Rene Christena, L., Francis, M. P., & Abdul, M. H. (2023). Nanotechnology: A potential approach for nutraceuticals. *Current Nutrition and Food Science*, *19*(7), 673–681. DOI: 10.2174/1573401319666221024162943

Franco, P., & De Marco, I. (2020). The Use of Poly (N-vinyl pyrrolidone) in the Delivery of Drugs: A Review. *Polymers*, *12*(5), 1114. DOI: 10.3390/polym12051114 PMID: 32414187

Fratoddi, I., Venditti, I., Cametti, C., & Russo, M. V. (2014). Gold nanoparticles and gold nanoparticle-conjugates for delivery of therapeutic molecules. Progress and challenges. *Journal of Materials Chemistry. B*, *2*(27), 4204–4220. DOI: 10.1039/C4TB00383G PMID: 32261559

Freed, C. R., Greene, P. E., Breeze, R. E., Tsai, W.-Y., DuMouchel, W., Kao, R., Dillon, S., Winfield, H., Culver, S., Trojanowski, J. Q., Eidelberg, D., & Fahn, S. (2001). Transplantation of Embryonic Dopamine Neurons for Severe Parkinson's Disease. *The New England Journal of Medicine*, *344*(10), 710–719. DOI: 10.1056/NEJM200103083441002 PMID: 11236774

Fu, B., & Yan, Q. (2023). Exopolysaccharide is required for motility, stress tolerance, and plant colonization by the endophytic bacterium *Paraburkholderia phytofirmans* PsJN. *Frontiers in Microbiology*, *14*, 1218653. Advance online publication. DOI: 10.3389/fmicb.2023.1218653 PMID: 37670984

Fu, J., Wang, C., Chen, X., Huang, Z., & Chen, D. (2018). Classification research and types of slow controlled release fertilizers (SRFs) used-a review. *Communications in Soil Science and Plant Analysis*, *49*(17), 2219–2230. DOI: 10.1080/00103624.2018.1499757

Fu, L., Wang, Z., Dhankher, O. P., & Xing, B. (2020). Nanotechnology as a new sustainable approach for controlling crop diseases and increasing agricultural production. *Journal of Experimental Botany*, *71*(2), 507–519. DOI: 10.1093/jxb/erz314 PMID: 31270541

Gandhi, K., Murugesan, S., & Suppaiah, R. (2024). Replicase gene mediated RNA silencing mechanism confers resistance against groundnut bud necrosis virus in tomato (Lycopersicon esculentum Mill.). *European Journal of Plant Pathology*, *170*(2), 405–416. DOI: 10.1007/s10658-024-02908-y

Ganesan, M., Tenchov, R., Iyer, K. A., Ralhan, K., Diaz, L. L., Bird, R. E., ... & Zhou, Q. (2024). Nanoscale materials at work: Mapping emerging applications in energy, medicine, and beyond.

Ganesh, R., & Pillai, A. B. (2017). Perceptions and attitude effects on nanotechnology acceptance: An exploratory framework. *Journal of Nanoparticle Research*, *19*(2), 19. DOI: 10.1007/s11051-016-3733-2

Gao, H., Fan, P., Xu, Q., Li, Y., Wang, J., Weng, J., Zhou, J., & Feng, B. (2020). In Vitro and in Vivo Antitumor Activity of Silver Nanoparticles on B16 Melanoma. *Nano*, *15*(12), 2050163. Advance online publication. DOI: 10.1142/S1793292020501635

Gao, L., Wei, Y., Wang, T., Zhu, Q., Liu, Y., Zhang, Z., Hu, Z., Zhou, B., Zhang, H., Jin, N., Jiao, D., Yin, M., Shen, J., & Yan, S. (2025). Preparation of a Self-Assembled Nanoantiviral Pesticide with Strong Adhesion Performance for Efficiency Control of Destructive Soybean Mosaic Virus. *Journal of Agricultural and Food Chemistry*, *73*(4), 2342–2351. DOI: 10.1021/acs.jafc.4c11405 PMID: 39817843

Gao, W., Thamphiwatana, S., Angsantikul, P., & Zhang, L. (2014). Nanoparticle-based local antimicrobial drug delivery. *Advanced Drug Delivery Reviews*, *78*, 1–14. PMID: 28939377

Gao, X. D., Shi, F. Y., Peng, F., Shi, X. J., Cheng, C. H., Hou, W. L., Xie, H. C., Lin, X. H., & Wang, X. P. (2021). Formulation of nanopesticide with graphene oxide as the nanocarrier of pyrethroid pesticide and its application in spider mite control. *RSC Advances*, *11*(57), 36089–36097. DOI: 10.1039/D1RA06505J PMID: 35492771

Garg, D., Sridhar, K., Stephen Inbaraj, B., Chawla, P., Tripathi, M., & Sharma, M. (2023). Nano-Biofertilizer formulations for agriculture: A systematic review on recent advances and prospective applications. *Bioengineering (Basel, Switzerland)*, *10*(9), 1010. DOI: 10.3390/bioengineering10091010 PMID: 37760112

Gautam, A., & Van Veggel, F. C. J. M. (2013). Synthesis of nanoparticles, their biocompatibility, and toxicity behavior for biomedical applications. *Journal of Materials Chemistry. B*, *1*(39), 5186–5200. DOI: 10.1039/c3tb20738b PMID: 32263325

Gautam, Y. K., Sharma, K., Tyagi, S., Ambedkar, A. K., Chaudhary, M., & Pal Singh, B. (2021). Nanostructured metal oxide semiconductor-based sensors for greenhouse gas detection: Progress and challenges. *Royal Society Open Science*, *8*(3), 201324. Advance online publication. DOI: 10.1098/rsos.201324 PMID: 33959316

Gayda, G. Z., Demkiv, O. M., Stasyuk, N. Y., Serkiz, R. Y., Lootsik, M. D., Errachid, A., Gonchar, M. V., & Nisnevitch, M. (2019). Metallic Nanoparticles Obtained via "Green" Synthesis as a Platform for Biosensor Construction. *Applied Sciences 2019, Vol. 9, Page 720*, *9*(4), 720. DOI: 10.3390/app9040720

Gelvin, S. B. (2017). Integration of Agrobacterium T-DNA into the Plant Genome. *Annual Review of Genetics*, *51*(1), 195–217. DOI: 10.1146/annurev-genet-120215-035320 PMID: 28853920

Gelvin, S. B. (2021). Plant DNA Repair and *Agrobacterium* T-DNA Integration. *International Journal of Molecular Sciences*, *22*(16), 8458. DOI: 10.3390/ijms22168458 PMID: 34445162

Ghareeb, R. Y., Belal, E. B., El-Khateeb, N. M. M., & Shreef, B. A. (2024). Utilizing bio-synthesis of nanomaterials as biological agents for controlling soil-borne diseases in pepper plants: Root-knot nematodes and root rot fungus I. *BMC Plant Biology*, *24*, 110. https://link.springer.com/article/10.1186/s12870-024-04760-y. DOI: 10.1186/s12870-024-04760-y PMID: 38355449

Ghasempour, A., Dehghan, H., Ataee, M., Chen, B., Zhao, Z., Sedighi, M., Guo, X., & Shahbazi, M. A. (2023). Cadmium Sulfide Nanoparticles: Preparation, Characterization, and Biomedical Applications. *Molecules (Basel, Switzerland)*, *28*(9), 3857. DOI: 10.3390/molecules28093857 PMID: 37175267

Ghodake, V. N., Naik, S. V., Bhukhanwala, K. N., Kande, K. V., Bhor, N. J., & Patravale, V. B. (2018). Nanoengineered Systems for Biopesticides. *Handbook of Nanomaterials for Industrial Applications*, 243–259. DOI: 10.1016/B978-0-12-813351-4.00015-8

Ghodake, G., Shinde, S., Saratale, G. D., Kadam, A., Saratale, R. G., & Kim, D. Y. (2020). Water Purification Filter Prepared by Layer-by-layer Assembly of Paper Filter and Polypropylene-polyethylene Woven Fabrics Decorated with Silver Nanoparticles. *Fibers and Polymers*, *21*(4), 751–761. DOI: 10.1007/s12221-020-9624-2

Ghormade, V., Deshpande, M. V., & Paknikar, K. M. (2011). Perspectives for nano-biotechnology enabled protection and nutrition of plants. *Biotechnology Advances*, *29*(6), 792–803. DOI: 10.1016/j.biotechadv.2011.06.007 PMID: 21729746

Ghosh, S., Nandi, S., & Basu, T. (2022). Nano-Antibacterials Using Medicinal Plant Components: An Overview. *Frontiers in Microbiology*, *12*, 768739. DOI: 10.3389/fmicb.2021.768739 PMID: 35273578

Giannelli, G., Potestio, S., & Visioli, G. (2023). The Contribution of PGPR in Salt Stress Tolerance in Crops: Unravelling the Molecular Mechanisms of Cross-Talk between Plant and Bacteria. *Plants*, *12*(11), 2197. DOI: 10.3390/plants12112197 PMID: 37299176

Giannousi, K., Avramidis, I., & Dendrinou-Samara, C. (2013). Synthesis, characterization and evaluation of copper based nanoparticles as agrochemicals against *Phytophthora infestans.RSC Advances*, *3*(44), 21743–21752. DOI: 10.1039/c3ra42118j

Gilbertson, L. M., Pourzahedi, L., Laughton, S., Gao, X., Zimmerman, J. B., Theis, T. L., Westerhoff, P., & Lowry, G. V. (2020). Guiding the design space for nanotechnology to advance sustainable crop production. *Nature Nanotechnology*, *15*(9), 801–810. DOI: 10.1038/s41565-020-0706-5 PMID: 32572231

Gillings, M. R. (2013). Evolutionary consequences of antibiotic use for the resistome, mobilome, and microbial pangenome. *Frontiers in Microbiology*, *4*(JAN). Advance online publication. DOI: 10.3389/fmicb.2013.00004 PMID: 23386843

Gökdemir, F. Ş., Gündoğdu, M., Muftareviç, S., Sunar, A., & Eyidoğan, F. (2023). *Nanopartiküllerin tarımsal bilimlerdeki önemi ve kullanım alanları. 49*(1), 11–17.

Gokdemir, G., Doner, N., Sert, Z., Sen, F., & Ciddi, K. (2022). Effects of CuO, TiO2 and graphite microparticles on the heat transfer properties of greases. *Engineering Science and Technology, an International Journal*, *30*, 101044. DOI: 10.1016/j.jestch.2021.08.001

Göksu, H., Zengin, N., Burhan, H., Cellat, K., & Şen, F. (2020). A Novel Hydrogenation of Nitroarene Compounds with Multi Wall Carbon Nanotube Supported Palladium/Copper Nanoparticles (PdCu@MWCNT NPs) in Aqueous Medium. *Scientific Reports 2020 10:1*, *10*(1), 1–8. DOI: 10.1038/s41598-020-64988-0

Göksu, H., Çelik, B., Yıldız, Y., Şen, F., & Kılbaş, B. (2016). Superior Monodisperse CNT-Supported CoPd (CoPd@CNT) Nanoparticles for Selective Reduction of Nitro Compounds to Primary Amines with NaBH4 in Aqueous Medium. *ChemistrySelect*, *1*(10), 2366–2372. DOI: 10.1002/slct.201600509

Goldshleger, N. F. (2007). Fullerenes and Fullerene-based materials in catalysis. *Fullerene Science and Technology*, *9*(3), 255–280. DOI: 10.1081/FST-100104493

Goldsmith, M., Abramovitz, L., & Peer, D. (2014). Precision nanomedicine in neurodegenerative diseases. *ACS Nano*, *8*(3), 1958–1965. DOI: 10.1021/nn501292z PMID: 24660817

Golestanipour, A., Nikkhah, M., Aalami, A., & Hosseinkhani, S. (2018). Gene delivery to tobacco root cells with single-walled carbon nanotubes and cell-penetrating fusogenic peptides. *Molecular Biotechnology*, *60*(12), 863–878. DOI: 10.1007/s12033-018-0120-5 PMID: 30203379

Gomes, A. F., Almeida, M. C., Sousa, E., & Resende, D. I. (2024). Siderophores and metallophores: Metal complexation weapons to fight environmental pollution. *The Science of the Total Environment*, *932*, 173044. DOI: 10.1016/j.scitotenv.2024.173044 PMID: 38723971

Gomes, A., & Sobral, P. J. D. A. (2021). Plant protein-based delivery systems: An emerging approach for increasing the efficacy of lipophilic bioactive compounds. *Molecules (Basel, Switzerland)*, *27*(1), 60. DOI: 10.3390/molecules27010060 PMID: 35011292

Gong, P., Li, H., He, X., Wang, K., Hu, J., Tan, W., Tan, S., & Zhang, X. Y. (2007). Preparation and antibacterial activity of Fe3O4@Ag nanoparticles. *Nanotechnology*, *18*(28), 604–611. DOI: 10.1088/0957-4484/18/28/285604

González-García, Y., López-Vargas, E. R., Pérez-Álvarez, M., Cadenas-Pliego, G., Benavides-Mendoza, A., Valdés-Reyna, J., Pérez-Labrada, F., & Juárez-Maldonado, A. (2022). Seed priming with carbon nanomaterials improves the bioactive compounds of tomato plants under saline stress. *Plants*, *11*(15), 15. DOI: 10.3390/plants11151984 PMID: 35956461

González-Grandío, E., Demirer, G. S., Jackson, C. T., Yang, D., Ebert, S., Molawi, K., & Landry, M. P. (2021). Carbon nanotube biocompatibility in plants is determined by their surface chemistry. *Journal of Nanobiotechnology*, *19*(1), 1–15. DOI: 10.1186/s12951-021-01178-8 PMID: 34930290

González-Melendi, P., Fernández-Pacheco, R., Coronado, M. J., Corredor, E., Testillano, P. S., Risueño, M. C., Marquina, C., Ibarra, M. R., Rubiales, D., & Pérez-de-Luque, A. (2008). Nanoparticles as smart treatment-delivery systems in plants: Assessment of different techniques of microscopy for their visualization in plant tissues. *Annals of Botany*, *101*(1), 187–195. DOI: 10.1093/aob/mcm283 PMID: 17998213

González-Reza, R. M., Zambrano-Zaragoza, M. L., & Gutiérrez-Martínez, P. (2023). Introduction to Nanotechnological Applications in the Postharvest Processing of Fresh Horticultural Produce. In *Postharvest Nanotechnology for Fresh Horticultural Produce* (pp. 1–25). CRC Press., DOI: 10.1201/9781003142287-1

Gopalakrishnan, V., & Muniraj, S. (2021). Neem flower extract assisted green synthesis of copper nanoparticles – Optimisation, characterisation and anti-bacterial study. *Materials Today: Proceedings*, *36*, 832–836. DOI: 10.1016/j.matpr.2020.07.013

Gopinath, K., Gowri, S., Karthika, V., & Arumugam, A. (2014). Green synthesis of gold nanoparticles from fruit extract of *Terminalia arjuna*, for the enhanced seed germination activity of *Gloriosa superba*. *Journal of Nanostructure in Chemistry*, *4*(3), 1–11. DOI: 10.1007/s40097-014-0115-0

Gordon, J. E., & Christie, P. J. (2014). The Agrobacterium Ti Plasmids. *Microbiology spectrum, 2*(6), . https://doi.org/DOI: 10.1128/microbiolspec.PLAS-0010-2013

Gotteland, M., Riveros, K., Gasaly, N., Carcamo, C., Magne, F., Liabeuf, G., Beattie, A., & Rosenfeld, S. (2020). The pros and cons of using algal polysaccharides as prebiotics. *Frontiers in Nutrition*, *7*, 163. DOI: 10.3389/fnut.2020.00163 PMID: 33072794

Goudarzi, M., & Salavati-Niasari, M. (2016). Controllable synthesis of new Tl2S2O3 nanostructures via hydrothermal process; Characterization and investigation photocatalytic activity for degradation of some anionic dyes. *Journal of Molecular Liquids*, *219*, 851–857. DOI: 10.1016/j.molliq.2016.04.056

Gouda, S., Kerry, R. G., Das, G., Paramithiotis, S., Shin, H., & Patra, J. K. (2017). Revitalization of plant growth promoting rhizobacteria for sustainable development in agriculture. *Microbiological Research*, *206*, 131–140. DOI: 10.1016/j.micres.2017.08.016 PMID: 29146250

Govender, Y., Riddin, T. L., Gericke, M., & Whiteley, C. G. (2010). On the enzymatic formation of platinum nanoparticles. *Journal of Nanoparticle Research*, *12*(1), 261–271. DOI: 10.1007/s11051-009-9604-3

Govindasamy, V., George, P., Kumar, M. *et al.* Multi-trait PGP rhizobacterial endophytes alleviate drought stress in a senescent genotype of sorghum [*Sorghum bicolor* (L.) Moench]. *3 Biotech* **10**, 13 (2020). https://doi.org/DOI: 10.1007/s13205-019-2001-4

Gramaje, D., Úrbez-Torres, J. R., & Sosnowski, M. R. (2018). Managing grapevine trunk diseases with respect to etiology and epidemiology: Current strategies and future prospects. *Plant Disease*, *102*(1), 12–39. DOI: 10.1094/PDIS-04-17-0512-FE PMID: 30673457

Gravely, M., Safaee, M. M., & Roxbury, D. (2019). Biomolecular Functionalization of a Nanomaterial to Control Stability and Retention within Live Cells. *Nano Letters*, *19*(9), 6203–6212. DOI: 10.1021/acs.nanolett.9b02267 PMID: 31424226

Graziano, S., Caldara, M., Gullì, M., Bevivino, A., Maestri, E., & Marmiroli, N. (2022). A Metagenomic and Gene Expression Analysis in Wheat (T. durum) and Maize (Z. mays) Biofertilized with PGPM and Biochar. *International Journal of Molecular Sciences*, *23*(18), 10376. Advance online publication. DOI: 10.3390/ijms231810376 PMID: 36142289

Greff, B., Sáhó, A., Lakatos, E., & Varga, L. (2023a). Biocontrol Activity of Aromatic and Medicinal Plants and Their Bioactive Components against Soil-Borne Pathogens. *Plants 2023, Vol. 12, Page 706*, *12*(4), 706. DOI: 10.3390/plants12040706

Grillo, R., Mattos, B. D., Antunes, D. R., Forini, M. M. L., Monikh, F. A., & Rojas, O. J. (2021). Foliage adhesion and interactions with particulate delivery systems for plant nanobionics and intelligent agriculture. *Nano Today*, *37*, 101078. DOI: 10.1016/j.nantod.2021.101078

Groves, C., Frater, L., Lee, R. G., & Stokes, E. (2021). Is there room at the bottom for CSR? Corporate social responsibility and nanotechnology in the UK. *Journal of Business Ethics*, *101*(4), 525–552. DOI: 10.1007/s10551-010-0731-7

Gruyer, N., Dorais, M., Bastien, C., Dassylva, N., & Triffault-Bouchet, G. (2013, October). Interaction between silver nanoparticles and plant growth. *InInternational Symposium on New Technologies for Environment Control, Energy-Saving and Crop Production in Greenhouse and Plant* 1037 (pp. 795-800).

Guan, Y. F., Pearce, R. C., Melechko, A. V., Hensley, D. K., Simpson, M. L., & Rack, P. D. (2008). Pulsed laser dewetting of nickel catalyst for carbon nanofiber growth. *Nanotechnology*, *19*(23), 235604. DOI: 10.1088/0957-4484/19/23/235604 PMID: 21825799

Gu, C., Guo, C., Dong, X., Hu, Z., Wu, P., Su, Z., Lu, Y., Xu, B., Yu, Z., & Liu, A. (2019). Core-shell structured iron-containing ceramic nanoparticles: Facile fabrication and excellent electromagnetic absorption properties. *Journal of the American Ceramic Society*, *102*(12), 7098–7107. DOI: 10.1111/jace.16619

Guimarães, D., Cavaco-Paulo, A., & Nogueira, E. (2021). Design of liposomes as drug delivery system for therapeutic applications. *International Journal of Pharmaceutics*, *601*, 120571. DOI: 10.1016/j.ijpharm.2021.120571 PMID: 33812967

Guo, H., White, J. C., Wang, Z., & Xing, B. (2018). Nano-enabled fertilizers to control the release and use efficiency of nutrients. *Current Opinion in Environmental Science & Health*, *6*, 77–83. https://doi.org/https://doi.org/10.1016/j.coesh.2018.07.009. DOI: 10.1016/j.coesh.2018.07.009

Guo, K., Cheng, C., Chen, L., Xie, J., Li, S., He, S., & Xiao, F. (2021). Uranium enrichment performence and uranium stress mechanism of Deinococcus radiodurans. *Journal of Radioanalytical and Nuclear Chemistry*, *330*(3), 1191–1206. DOI: 10.1007/s10967-021-08018-x

Guo, K., Cheng, C., Chen, L., Xie, J., Li, S., He, S., & Xiao, F. (2022). Application of Deinococcus radiodurans in the treatment of environmental pollution by heavy metals and radionuclides. *Journal of Radioanalytical and Nuclear Chemistry*, *331*(2), 655–664. DOI: 10.1007/s10967-021-08141-9

Guo, Y., Jiang, N., Zhang, L., & Yin, M. (2020). Green synthesis of gold nanoparticles from Fritillaria cirrhosa and its anti-diabetic activity on Streptozotocin induced rats. *Arabian Journal of Chemistry*, *13*(4), 5096–5106. DOI: 10.1016/j.arabjc.2020.02.009

Gupta, A., Singh, S. K., Singh, M. K., Singh, V. K., Modi, A., Singh, P. K., & Kumar, A. (2019). Plant growth–promoting rhizobacteria and their functional role in salinity stress management. *Abatement of Environmental Pollutants*, 151-160. https://doi.org/DOI: 10.1016/B978-0-12-818095-2.00007-2

Gupta, R., & Xie, H. (2018). Nanoparticles in Daily Life: Applications, Toxicity and Regulations. *Journal of environmental pathology, toxicology and oncology: official organ of the International Society for Environmental Toxicology and Cancer, 37*(3), 209–230. https://doi.org/DOI: 10.1615/JEnvironPathol-ToxicolOncol.2018026009

Gupta, N. (2013). Novel Applications of Functionally Graded Nano, Optoelectronic and Thermoelectric Materials Article in. *International Journal of Materials. Mechanics and Manufacturing*, *1*(3), 221–224. Advance online publication. DOI: 10.7763/IJMMM.2013.V1.47

Gupta, P. K. (2018). Toxicity of fungicides. In *Veterinary toxicology* (pp. 569–580). Academic Press.

Gurunathan, S., Raman, J., Abd Malek, S. N., John, P. A., & Vikineswary, S. (2013). Green synthesis of silver nanoparticles using Ganoderma neo-japonicum Imazeki: A potential cytotoxic agent against breast cancer cells. *International Journal of Nanomedicine*, *8*, 4399–4413. DOI: 10.2147/IJN.S51881 PMID: 24265551

Guterres, S. S., Alves, M. P., & Pohlmann, A. R. (2007). Polymeric Nanoparticles, Nanospheres and Nanocapsules, for Cutaneous Applications. Https://Doi.Org/10.1177/117739280700200002*, 2*, 117739280700200. DOI: 10.1177/117739280700200002

Gu, X., Liu, L., & Zhang, H. (2021). Transgene-free Genome Editing in Plants. *Frontiers in Genome Editing*, *3*, 805317. DOI: 10.3389/fgeed.2021.805317 PMID: 34927134

Gu, Y., Chen, X., Lee, J. H., Monteiro, D. A., Wang, H., & Lee, W. Y. (2012). Inkjet printed antibiotic- and calcium-eluting bioresorbable nanocomposite micropatterns for orthopedic implants. *Acta Biomaterialia*, *8*(1), 424–431. DOI: 10.1016/j.actbio.2011.08.006 PMID: 21864730

Habiba, K., Makarov, V. I., Weiner, B. R., & Morell, G. (2014). Fabrication of nanomaterials by pulsed laser synthesis. *Manufacturing nanostructures, 10*, 263-292.

Haghighi, M., Afifipour, Z., & Mozafarian, M. (2012). The effect of N-Si on tomato seed germination under salinity levels. *Journal of Biological and Environmental Sciences*, *6*(16), 87–90.

Haghighi, M., & Teixeira da Silva, J. A. (2014a). The effect of carbon nanotubes on the seed germination and seedling growth of four vegetable species. *Journal of Crop Science and Biotechnology*, *17*(4), 201–208. DOI: 10.1007/s12892-014-0057-6

Haider, A., Haider, S., Kang, I. K., Kumar, A., Kummara, M. R., Kamal, T., & Han, S. S. (2018). A novel use of cellulose based filter paper containing silver nanoparticles for its potential application as wound dressing agent. *International Journal of Biological Macromolecules*, *108*, 455–461. DOI: 10.1016/j.ijbiomac.2017.12.022 PMID: 29222019

Hajiahmadi, Z., Shirzadian-Khorramabad, R., Kazemzad, M., & Sohani, M. M. (2019). Enhancement of tomato resistance to Tuta absoluta using a new efficient mesoporous silica nanoparticle-mediated plant transient gene expression approach. *Scientia Horticulturae*, *243*, 367–375. DOI: 10.1016/j.scienta.2018.08.040

Hamada, H., Liu, Y., Nagira, Y., Miki, R., Taoka, N., & Imai, R. (2018). Biolistic-delivery-based transient CRISPR/Cas9 expression enables in planta genome editing in wheat. *Scientific Reports*, *8*(1), 14422. DOI: 10.1038/s41598-018-32714-6 PMID: 30258105

Hamidi, M., Azadi, A., & Rafiei, P. (2008). Hydrogel nanoparticles in drug delivery. *Advanced Drug Delivery Reviews*, *60*(15), 1638–1649. DOI: 10.1016/j.addr.2008.08.002 PMID: 18840488

Hammond, J. L., Formisano, N., Estrela, P., Carrara, S., & Tkac, J. (2016). Electrochemical biosensors and nanobiosensors. *Essays in Biochemistry*, *60*(1), 69–80. DOI: 10.1042/EBC20150008 PMID: 27365037

Handford, C. E., Dean, M., Spence, M., Henchion, M., Elliott, C. T., & Campbell, K. (2015). Awareness and attitudes towards the emerging use of nanotechnology in the agri-food sector. *Food Control*, *57*, 24–34. DOI: 10.1016/j.foodcont.2015.03.033

Hanna, E. A., Astete, C. E., Price, T., Tamez, C., Mendez, O. E., Garcia, A., Kewir, F. V., White, J. C., & Sabliov, C. M. (2024). Antifungal Efficacy of Nanodelivered Azoxystrobin against Rhizoctonia solani in Soybean (Glycine max). *ACS Agricultural Science & Technology*, *4*(3), 330–336. DOI: 10.1021/acsagscitech.3c00469

Hano, C., & Abbasi, B. H. (2021). Plant-based green synthesis of nanoparticles: Production, characterization and applications. *Biomolecules*, *12*(1), 31. DOI: 10.3390/biom12010031 PMID: 35053179

Han, S. E., Kim, K. Y., & Htwe Maung, C. E. (2024). *Bacillus subtilis* PE7-Mediated Alleviation of Phosphate Starvation and Growth Promotion of Netted Melon (*Cucumis melo* L. Var. Reticulatus Naud.). *Microorganisms*, *12*(12), 2384. DOI: 10.3390/microorganisms12122384 PMID: 39770587

Hansen, V., Müller-Stöver, D., Imparato, V., Krogh, P. H., Jensen, L. S., Dolmer, A., & Hauggaard-Nielsen, H. (2017). The effects of straw or straw-derived gasification biochar applications on soil quality and crop productivity: A farm case study. *Journal of Environmental Management*, *186*, 88–95. DOI: 10.1016/j.jenvman.2016.10.041 PMID: 27815006

Hao, L., Gong, L., Chen, L., Guan, M., Zhou, H., Qiu, S., Wen, H., Chen, H., Zhou, X., & Akbulut, M. (2020). Composite pesticide nanocarriers involving functionalized boron nitride nanoplatelets for pH-responsive release and enhanced UV stability. *Chemical Engineering Journal*, *396*, 125233. DOI: 10.1016/j.cej.2020.125233

Hao, Y., Ma, C., Zhang, Z., Song, Y., Cao, W., Guo, J., Zhou, G., Rui, Y., Liu, L., & Xing, B. (2018). Carbon nanomaterials alter plant physiology and soil bacterial community composition in a rice-soil-bacterial ecosystem. *Environmental Pollution*, *232*, 123–136. DOI: 10.1016/j.envpol.2017.09.024 PMID: 28947315

Haris, M., Hussain, T., Mohamed, H. I., Khan, A., Ansari, M. S., Tauseef, A., Khan, A. A., & Akhtar, N. (2023). Nanotechnology–A new frontier of nano-farming in agricultural and food production and its development. *The Science of the Total Environment*, *857*, 159639. DOI: 10.1016/j.scitotenv.2022.159639 PMID: 36283520

Hasan, M., Rafique, S., Zafar, A., Loomba, S., Khan, R., Hassan, S. G., Khan, M. W., Zahra, S., Zia, M., Mustafa, G., Shu, X., Ihsan, Z., & Mahmood, N. (2020). Physiological and anti-oxidative response of biologically and chemically synthesized iron oxide: *Zea mays* a case study. *Heliyon*, *6*(8), 1–7. DOI: 10.1016/j.heliyon.2020.e04595 PMID: 32923707

Hassellöv, M., Readman, J. W., Ranville, J. F., & Tiede, K. (2008). Nanoparticle analysis and characterization methodologies in environmental risk assessment of engineered nanoparticles. *Ecotoxicology (London, England)*, *17*(5), 344–361. DOI: 10.1007/s10646-008-0225-x PMID: 18483764

Hassen, A., Saidi, N., Cherif, M., & Boudabous, A. (1998). Effects of heavy metals on Pseudomonas aeruginosa and Bacillus thuringiensis. *Bioresource Technology*, *65*(1–2), 73–82. DOI: 10.1016/S0960-8524(98)00011-X

Hatami, M., & Ghorbanpour, M. (2013). Effect of nanosilver on physiological performance of pelargonium plants exposed to dark storage. *Journal of Horticultural Research*, *21*(1), 15–20. DOI: 10.2478/johr-2013-0003

Hatfaludi, T., Liska, M., Zellinger, D., Ousman, J. P., Szostak, M., Jalava, K., & Lubitz, W. (2004). Bacterial ghost technology for pesticide delivery. *Journal of Agricultural and Food Chemistry*, *52*(18), 5627–5634. DOI: 10.1021/jf049489w PMID: 15373403

Haydar, M. S., Ghosh, D., & Roy, S. (2024). Slow and controlled release nanofertilizers as an efficient tool for sustainable agriculture: Recent understanding and concerns. *Plant Nano Biology*, *100058*, 100058. Advance online publication. DOI: 10.1016/j.plana.2024.100058

Haynes, C. L., & Van Duyne, R. P. (2001). Nanosphere lithography: A versatile nanofabrication tool for studies of size-dependent nanoparticle optics. *The Journal of Physical Chemistry B*, *105*(24), 5599–5611. DOI: 10.1021/jp010657m

Hazra, S., Das, D., Moulick, D., & Hossain, A. (2023). Endophytes: The treasure house of bioactive compounds with potential applications in sustainable agriculture and other sectors. *Biocontrol Agents for Improved Agriculture*, 477-506. https://doi.org/DOI: 10.1016/B978-0-443-15199-6.00001-4

Heikal, Y. M., Sutan, N. A., Heikal, Y. M., & Sutan, N. A. (2021). Mechanisms of genotoxicity and oxidative stress induced by engineered nanoparticles in plants. *Induc. Genotoxicity Oxidative Stress Plants* 151–197, .DOI: 10.1007/978-981-16-2074-4_6

Hejmady, S., Pradhan, R., Alexander, A., Agrawal, M., Singhvi, G., Gorain, B., Tiwari, S., Kesharwani, P., & Dubey, S. K. (2020). Recent advances in targeted nanomedicine as promising antitumor therapeutics. *Drug Discovery Today*, *25*(12), 2227–2244. DOI: 10.1016/j.drudis.2020.09.031 PMID: 33011342

Hench, L. L., & West, J. K. (1990). The sol-gel process. *Chemical Reviews*, *90*(1), 33–72. DOI: 10.1021/cr00099a003

Hendler-Neumark, A., & Bisker, G. (2019). Fluorescent single-walled carbon nanotubes for protein detection. *Sensors (Basel)*, *19*(24), 5403. DOI: 10.3390/s19245403 PMID: 31817932

Hennig, T. B., Bandeira, F. O., Puerari, R. C., Fraceto, L. F., & Matias, W. G. (2023). A systematic review of the toxic effects of a nanopesticide on non-target organisms: Estimation of protective concentrations using a species sensitivity distribution (SSD) approach–The case of atrazine. *The Science of the Total Environment*, *871*, 162094. DOI: 10.1016/j.scitotenv.2023.162094 PMID: 36764548

Heredia-Guerrero, J. A.. (2014). Plant cuticle penetration by nanomaterials. *Advanced Functional Materials*, *24*(18), 2805–2814.

Hernandez, F. J., & Ozalp, V. C. (2012). Graphene and Other Nanomaterial-Based Electrochemical Aptasensors. *Biosensors 2012, Vol. 2, Pages 1-14,* *2*(1), 1–14. DOI: 10.3390/bios2010001

Hernández-Montiel, L. G., Chiquito-Contreras, C. J., Murillo-Amador, B., Vidal-Hernández, L., Quiñones-Aguilar, E. E., & Chiquito-Contreras, R. G. (2017). Efficiency of two inoculation methods of Pseudomonas putida on growth and yield of tomato plants. *Journal of Soil Science and Plant Nutrition*, *17*(4), 1003–1012. DOI: 10.4067/S0718-95162017000400012

Heydari, S., Masoumi, N., Esmaeeli, E., Ayyoubzadeh, S. M., Ghorbani-Bidkorpeh, F., & Ahmadi, M. (2024). Artificial intelligence in nanotechnology for treatment of diseases. *Journal of Drug Targeting*, *32*(10), 1247–1266. DOI: 10.1080/1061186X.2024.2393417 PMID: 39155708

Hillegass, J. M., Shukla, A., Lathrop, S. A., MacPherson, M. B., Fukagawa, N. K., & Mossman, B. T. (2010). Assessing nanotoxicity in cells in vitro. *Wiley Interdisciplinary Reviews. Nanomedicine and Nanobiotechnology*, *2*(3), 219–231. DOI: 10.1002/wnan.54 PMID: 20063369

Hlihor, R. M., Ro ca, M., Hagiu-Zaleschi, L., Simion, I. M., Daraban, G. M., & Stoleru, V. (2022). Medicinal Plant Growth in Heavy Metals Contaminated Soils: Responses to Metal Stress and Induced Risks to Human Health. *Toxics*, *10*(9), 499. DOI: 10.3390/toxics10090499 PMID: 36136464

Hoang, B. T. L., Fletcher, S. J., Brosnan, C. A., Ghodke, A. B., Manzie, N., & Mitter, N. (2022). RNAi as a foliar spray: Efficiency and challenges to field applications. *International Journal of Molecular Sciences*, *23*(12), 6639. DOI: 10.3390/ijms23126639 PMID: 35743077

Hofmann, T., Lowry, G. V., Ghoshal, S., Tufenkji, N., Brambilla, D., Dutcher, J. R., Gilbertson, L. M., Giraldo, J. P., Kinsella, J. M., Landry, M. P., Lovell, W., Naccache, R., Paret, M., Pedersen, J. A., Unrine, J. M., White, J. C., & Wilkinson, K. J. (2020). Technology readiness and overcoming barriers to sustainably implement nanotechnology-enabled plant agriculture. *Nature Food*, *1*(7), 416–425. DOI: 10.1038/s43016-020-0110-1

Hojjati-Najafabadi, A., Davar, F., Enteshari, Z., & Hosseini-Koupaei, M. (2021). Antibacterial and photocatalytic behaviour of green synthesis of Zn0.95Ag0.05O nanoparticles using herbal medicine extract. *Ceramics International*, *47*(22), 31617–31624. DOI: 10.1016/j.ceramint.2021.08.042

Hong, S., & Myung, S. (2007). Nanotube electronics: A flexible approach to mobility. In Nature Nanotechnology (Vol. 2, Issue 4, pp. 207–208). Nature Publishing Group. DOI: 10.1038/nnano.2007.89

Hong, J., Peralta-Videa, J. R., Rico, C., Sahi, S., Viveros, M. N., Bartonjo, J., Zhao, L., & Gardea-Torresdey, J. L. (2014). Evidence of Trans location and Physiological Impacts of Foliar Applied CeO_2 Nanoparticles on Cucumber (*Cucumis sativus*) Plants. *Environmental Science & Technology*, *48*(8), 4376–4385. DOI: 10.1021/es404931g PMID: 24625209

Hong, J., Wang, C., Wagner, D. C., Gardea-Torresdey, J. L., He, F., & Rico, C. M. (2021). Foliar application of nanoparticles: Mechanisms of absorption, transfer, and multiple impacts. *Environmental Science. Nano*, *8*(5), 1196–1210. DOI: 10.1039/D0EN01129K

Hönig, M., Plíhalová, L., Husičková, A., Nisler, J., & Doležal, K. (2018). Role of Cytokinins in Senescence, Antioxidant Defence and Photosynthesis. *International Journal of Molecular Sciences*, *19*(12), 4045. DOI: 10.3390/ijms19124045 PMID: 30558142

Hoo, C. M., Starostin, N., West, P., & Mecartney, M. L. (2008). A comparison of atomic force microscopy (AFM) and dynamic light scattering (DLS) methods to characterize nanoparticle size distributions. *Journal of Nanoparticle Research*, *10*(S1), 89–96. DOI: 10.1007/s11051-008-9435-7

Hossain, Z., Mustafa, G., Sakata, K., & Komatsu, S. (2016). Insights into the proteomic response of soybean towards Al2O3, ZnO, and Ag nanoparticles stress. *Journal of Hazardous Materials*, *304*, 291–305. DOI: 10.1016/j.jhazmat.2015.10.071 PMID: 26561753

Hosseinnejad, M., & Jafari, S. M. (2016). Evaluation of different factors affecting antimicrobial properties of chitosan. *International Journal of Biological Macromolecules*, *85*, 467–475. DOI: 10.1016/j.ijbiomac.2016.01.022 PMID: 26780706

Howell, T. A. (2001). Enhancing water use efficiency in irrigated agriculture. *Agronomy Journal*, *93*(2), 281–289. DOI: 10.2134/agronj2001.932281x

Hristozov, D., Gottardo, S., Semenzin, E., Oomen, A., Bos, P., Peijnenburg, W., van Tongeren, M., Nowack, B., Hunt, N., Brunelli, A., Scott-Fordsmand, J. J., Tran, L., & Marcomini, A. (2016). Frameworks and tools for risk assessment of manufactured nanomaterials. *Environment International*, *95*, 36–53. DOI: 10.1016/j.envint.2016.07.016 PMID: 27523267

Huang, C.-Y., Wang, H., Hu, P., Hamby, R., & Jin, H. (2019). Small RNAs—Big players in plant-microbe interactions. *Cell Host & Microbe*, *26*(2), 173–182. DOI: 10.1016/j.chom.2019.07.021 PMID: 31415750

Huang, H., & Yang, X. (2005). One-step, shape control synthesis of gold nanoparticles stabilized by 3-thiopheneacetic acid. *Colloids and Surfaces. A, Physicochemical and Engineering Aspects*, *255*(1-3), 11–17. DOI: 10.1016/j.colsurfa.2004.12.020

Huang, L., Luo, S., Tong, S., Lv, Z., & Wu, J. (2024). The development of nanocarriers for natural products. *Wiley Interdisciplinary Reviews. Nanomedicine and Nanobiotechnology*, *16*(3), e1967. DOI: 10.1002/wnan.1967 PMID: 38757428

Hui, M., Shengyan, P., Yaqi, H., Rongxin, Z., Anatoly, Z., & Wei, C. (2018). A highly efficient magnetic chitosan "fluid" adsorbent with a high capacity and fast adsorption kinetics for dyeing wastewater purification. *Chemical Engineering Journal*, *345*, 556–565. DOI: 10.1016/j.cej.2018.03.115

Hu, J., & Xianyu, Y. (2021). When nano meets plants: A review on the interplay between nanoparticles and plants. *Nano Today*, *38*, 101143. DOI: 10.1016/j.nantod.2021.101143

Hu, S., Yan, C., Fei, Q., Zhang, B., & Wu, W. (2023). MOF-based stimuli-responsive controlled release nanopesticide: Mini review. *Frontiers in Chemistry*, *11*, 1272725. DOI: 10.3389/fchem.2023.1272725 PMID: 37767340

Husen, A., & Siddiqi, K. S. (2014). Carbon and fullerene nanomaterials in plant system. *Journal of Nanobiotechnology*, *12*(1), 1–10. DOI: 10.1186/1477-3155-12-16 PMID: 24766786

Husna, N., Budianta, D., Munandar, M., & Napoleon, A. (2019). Evaluation of several biochar types as inoculant carrier for indigenous Phosphate Solubilizing Microoorganism from acid sulphate soil. *Journal of Ecological Engineering*, *20*(6), 1–8. DOI: 10.12911/22998993/109078

Hussain, B., Lin, Q., Hamid, Y., Sanaullah, M., Di, L., Khan, M. B., & Yang, X. (2020). Foliage application of selenium and silicon nanoparticles alleviates Cd and Pb toxicity in rice (*Oryza sativa* L.). *The Science of the Total Environment*, *712*, 136497. DOI: 10.1016/j.scitotenv.2020.136497 PMID: 31945526

Hussain, H. I., Yi, Z., Rookes, J. E., Kong, L. X., & Cahill, D. M. (2013). Mesoporous silica nanoparticles as a biomolecule delivery vehicle in plants. *Journal of Nanoparticle Research*, *15*(6), 1676. DOI: 10.1007/s11051-013-1676-4

Hussain, I., Singh, N. B., Singh, A., Singh, H., & Singh, S. C. (2016). Green synthesis of nanoparticles and its potential application. *Biotechnology Letters*, *38*(4), 545–560. DOI: 10.1007/s10529-015-2026-7 PMID: 26721237

Huston, D. C., Khudhir, M., & Hodda, M. (2022). Reliability and Utility of Standard Gene Sequence Barcodes for the Identification and Differentiation of Cyst Nematodes of the Genus Heterodera. *Journal of Nematology*, *54*(1), 20220024. DOI: 10.2478/jofnem-2022-0024 PMID: 35975224

Hu, Y., Tang, J., & Yang, J. (2024). Introducing artificial intelligence technology to plant disease management for sustainable agriculture. *Crop Protection (Guildford, Surrey)*, *184*, 106764. DOI: 10.1016/j.cropro.2024.106764

Hu, Y., Zhang, P., Zhang, X., Liu, Y., Feng, S., Guo, D., Nadezhda, T., Song, Z., & Dang, X. (2021). Multi-wall carbon nanotubes promote the growth of maize (*Zea mays*) by regulating carbon and nitrogen metabolism in leaves. *Journal of Agricultural and Food Chemistry*, *69*(17), 4981–4991. DOI: 10.1021/acs.jafc.1c00733 PMID: 33900073

Hyder, S., Ul-Nisa, M., Shahid, H., Gohar, F., Gondal, A. S., Riaz, N., & Iqbal, R. (2023). Recent trends and perspectives in the application of metal and metal oxide nanomaterials for sustainable agriculture. *Plant Physiology and Biochemistry*, *202*, 107960. DOI: 10.1016/j.plaphy.2023.107960 PMID: 37591032

Hyzinski-García, M. C., Rudkouskaya, A., & Mongin, A. A. (2014). LRRC8A protein is indispensable for swelling-activated and ATP-induced release of excitatory amino acids in rat astrocytes. [n. pag.]. *The Journal of Physiology*, *592*(22), 592. DOI: 10.1113/jphysiol.2014.278887 PMID: 25172945

Iavicoli, I., Leso, V., Beezhold, D. H., & Shvedova, A. A. (2017). Nanotechnology in agriculture: Opportunities, toxicological implications, and occupational risks. *Toxicology and Applied Pharmacology*, *329*, 96–111. DOI: 10.1016/j.taap.2017.05.025 PMID: 28554660

Ijaz, I., Gilani, E., Nazir, A., & Bukhari, A. (2020). Detail review on chemical, physical and green synthesis, classification, characterizations and applications of nanoparticles. *Green Chemistry Letters and Reviews*, *13*(3), 223–245. DOI: 10.1080/17518253.2020.1802517

Ijaz, M., Khan, F., Ahmed, T., Noman, M., Zulfiqar, F., & Rizwan, M.. (2023). Nanobiotechnology to advance stress resilience in plants: Current opportunities and challenges. *Materials Today. Bio*, *•••*, 100759. DOI: 10.1016/j.mtbio.2023.100759 PMID: 37600356

Ilk, S., Saglam, N., & Özgen, M. (2017). Kaempferol loaded lecithin/chitosan nanoparticles: Preparation, characterization, and their potential applications as a sustainable antifungal agent. *Artificial Cells, Nanomedicine, and Biotechnology*, *45*(5), 907–916. DOI: 10.1080/21691401.2016.1192040 PMID: 27265551

Imran, A., Hakim, S., Tariq, M., Nawaz, M. S., Laraib, I., Gulzar, U., Hanif, M. K., Siddique, M. J., Hayat, M., Fraz, A., & Ahmad, M. (2021). Diazotrophs for Lowering Nitrogen Pollution Crises: Looking Deep Into the Roots. *Frontiers in Microbiology*, *12*, 637815. DOI: 10.3389/fmicb.2021.637815 PMID: 34108945

Ingle, P. U., Shende, S. S., Shingote, P. R., Mishra, S. S., Sarda, V., Wasule, D. L., Rajput, V. D., Minkina, T., Rai, M., Sushkova, S., Mandzhieva, S., & Gade, A. (2022). Chitosan nanoparticles (ChNPs): A versatile growth promoter in modern agricultural production. *Heliyon*, *8*(11), e11893. DOI: 10.1016/j.heliyon.2022.e11893 PMID: 36468119

Inobeme, A., Adetunji, C. O., Maliki, M., Onyeachu, B. I., Kelani, T., Eziukwu, C. A., Olori, E., Mathew, J. T., and Bamigboye, M. O. (2023). Strategies to synthesize, advantages, and disadvantages of pharmaceutical nanoparticles. In book: Nanotechnology for Drug Delivery and Pharmaceuticals. Academic Press is an imprint of Elsevier, 371-402, .DOI: 10.1016/B978-0-323-95325-2.00006-7

Inobeme, A., Adetunji, C. O., Mathew, J. T., & Akram, M. (2025). *Extraction, properties, and modification of chitosan-based nanoparticles. Chitosan-Based Nanoparticles for Biomedical Applications*. Woodhead Publishing., DOI: 10.1016/B978-0-443-13997-0.00001-1

Iqbal, M., Umar, S., & Mahmooduzzafar, F. (2019). Nano-fertilization to enhance nutrient use efficiency and productivity of crop plants. *Nanomaterials and plant potential*, 473-505.

Irshad, M. A., Nawaz, R., Ur Rehman, M. Z., Adrees, M., Rizwan, M., Ali, S., & Tasleem, S. (2021). Synthesis, characterization and advanced sustainable applications of titanium dioxide nanoparticles: A review. *Ecotoxicology and Environmental Safety*, *212*, 111978. DOI: 10.1016/j.ecoenv.2021.111978 PMID: 33561774

Islam, M. R., Youngblood, M., Kim, H. I., González-Gamboa, I., Monroy-Borrego, A. G., Caparco, A. A., & Giraldo, J. P. (2024). DNA delivery by virus-like nanocarriers in plant cells. *Nano Letters*, *24*(26), 7833–7842. DOI: 10.1021/acs.nanolett.3c04735 PMID: 38887996

Ismail, L. M., Soliman, M. I., Abd El-Aziz, M. H., & Abdel-Aziz, H. M. (2022). Impact of silica ions and nano silica on growth and productivity of pea plants under salinity stress. *Plants*, *11*(4), 494. DOI: 10.3390/plants11040494 PMID: 35214827

Jaberzadeh, A., Moaveni, P., Tohidi Moghadam, H. R., & Zahedi, H. (2013). Influence of bulk and nanoparticles titanium foliar application on some agronomic traits, seed gluten and starch contents of wheat subjected to water deficit stress. *Notulae Botanicae Horti Agrobotanici Cluj-Napoca*, *41*(1), 201–207. DOI: 10.15835/nbha4119093

Jabr-Milane, L. S., van Vlerken, L. E., Yadav, S., & Amiji, M. M. (2008). Multi-functional nanocarriers to overcome tumor drug resistance. *Cancer Treatment Reviews*, *34*(7), 592–602. DOI: 10.1016/j.ctrv.2008.04.003 PMID: 18538481

Jacob, J., Haponiuk, J. T., Thomas, S., & Gopi, S. (2018). Biopolymer based nanomaterials in drug delivery systems: A review. *Materials Today. Chemistry*, *9*, 43–55. DOI: 10.1016/j.mtchem.2018.05.002

Jacobs, M., Nagaraj, V. J., Mertz, T., Selvam, A. P., Ngo, T., & Prasad, S. (2013). An electrochemical sensor for the detection of antibiotic contaminants in water. *Analytical Methods*, *5*(17), 4325–4329. DOI: 10.1039/c3ay40994e

Jacques, M. T., Oliveira, J. L., Campos, E. V., Fraceto, L. F., & Ávila, D. S. (2017). Safety assessment of nanopesticides using the roundworm Caenorhabditis elegans. *Ecotoxicology and Environmental Safety*, *139*, 245–253. DOI: 10.1016/j.ecoenv.2017.01.045 PMID: 28160702

Jadoun, S., Arif, R., Jangid, N. K., & Meena, R. K. (2020). Green synthesis of nanoparticles using plant extracts: a review. In Environmental Chemistry Letters (Vol. 19, Issue 1, pp. 355–374). Springer. DOI: 10.1007/s10311-020-01074-x

Jain, K. K. (2005). Advances in the delivery of systemic antifungal agents. *Advanced Drug Delivery Reviews*, *57*(8), 1063–1085.

Jain, S. K., Wettberg, E. J. V., Punia, S. S., Parihar, A. K., Lamichaney, A., Kumar, J., Gupta, D. S., Ahmad, S., Pant, N. C., Dixit, G. P., Sari, H., Sari, D., Ma'ruf, A., Toker, P., & Toker, C. (2023). Genomic-mediated breeding strategies for global warming in chickpeas (Cicer arietinum L.). *Agriculture*, *13*(9), 1721. DOI: 10.3390/agriculture13091721

Jaiswal, S. K., Mohammed, M., Ibny, F. Y., & Dakora, F. D. (2021). Rhizobia as a Source of Plant Growth-Promoting Molecules: Potential Applications and Possible Operational Mechanisms. *Frontiers in Sustainable Food Systems*, *4*, 619676. DOI: 10.3389/fsufs.2020.619676

James, M., & Owen, L. (2022). Harnessing AI and machine learning for microfluidic drug screening in plant pathogen management. *Revista Española de Documentación Científica, 16*(2), 2.

Jampílek, J., & Kráľová, K. (2017). Nanomaterials for delivery of nutrients and growth-promoting compounds to plants. *Nanotechnology: An agricultural paradigm*, 177-226. https://doi.org/DOI: 10.1007/978-981-10-4573-8_9

Jampílek, J., & Kráľová, K. (2017). *Nanopesticides: preparation, targeting, and controlled release.* In New pesticides and soil sensors *(pp. 81-127).* Academic Press.

Jampilek, J., & Kralova, K. (2021). Advances in Drug Delivery Nanosystems Using Graphene-Based Materials and Carbon Nanotubes. *Materials (Basel), 14*(5), 1059. DOI: 10.3390/ma14051059 PMID: 33668271

Janatova, A., Bernardos, A., Smid, J., Frankova, A., Lhotka, M., Kourimská, L., Pulkrabek, J., & Kloucek, P. (2015). Long-term antifungal activity of volatile essential oil components released from mesoporous silica materials. *Industrial Crops and Products, 67*, 216–220. DOI: 10.1016/j.indcrop.2015.01.019

Janot, R., & Guérard, D. (2002). Ball-milling: The behavior of graphite as a function of the dispersal media. *Carbon, 40*(15), 2887–2896. DOI: 10.1016/S0008-6223(02)00223-3

Javed, R., Ain, N. U., Gul, A., Arslan Ahmad, M., Guo, W., Ao, Q., & Tian, S. (2022). Diverse biotechnological applications of multifunctional titanium dioxide nanoparticles: An up-to-date review. *IET Nanobiotechnology / IET, 16*(5), 171–189. DOI: 10.1049/nbt2.12085 PMID: 35411585

Jayaraj, J., Megha, P., & Sreedev, P. (2019). Organochlorine pesticides, their toxic effects on living organisms and their fate in the environment. *Interdisciplinary Toxicology, 9*(3-4), 90–100. DOI: 10.1515/intox-2016-0012 PMID: 28652852

Jia, L., Zhang, Q., Li, Q., & Song, H. (2009). The biosynthesis of palladium nanoparticles by antioxidants in garden *Jasminoids ellis*: Long life time catalyst for p-nitrotoluene hydrogenation. *Nanotechnology, 20*(38), 385601. DOI: 10.1088/0957-4484/20/38/385601 PMID: 19713585

Jiang, L., Ding, L., He, B., Shen, J., Xu, Z., Yin, M., & Zhang, X. (2014). Systemic gene silencing in plants triggered by fluorescent nanoparticle-delivered double-stranded RNA. *Nanoscale, 6*(17), 9965–9969. DOI: 10.1039/C4NR03481C PMID: 25068243

Jiang, M., Song, Y., Kanwar, M. K., Ahammed, G. J., Shao, S., & Zhou, J. (2021). Phytonanotechnology applications in modern agriculture. *Journal of Nanobiotechnology, 19*(1), 1–20. DOI: 10.1186/s12951-021-01176-w PMID: 34930275

Jiang, N., Yan, J., Liang, Y., Shi, Y., He, Z., Wu, Y., Zeng, Q., Liu, X., & Peng, J. (2020). Resistance genes and their interactions with bacterial blight/leaf streak pathogens (*Xanthomonas oryzae*) in Rice (*Oryza sativa* L.)—An updated review. *Rice (New York, N.Y.), 13*(1), 3. DOI: 10.1186/s12284-019-0358-y PMID: 31915945

Jiang, Y., Lawrence, M., Ansell, M. P., & Hussain, A. (2018). Cell wall microstructure, pore size distribution and absolute density of hemp shiv. *Royal Society Open Science, 5*(4), 171945. DOI: 10.1098/rsos.171945 PMID: 29765652

Jiang, Y., Zhou, P., Zhang, P., Adeel, M., Shakoor, N., Li, Y., Li, M., Guo, M., Zhao, W., Lou, B., Wang, L., Lynch, I., & Rui, Y. (2022). Green synthesis of metal-based nanoparticles for sustainable agriculture. *Environmental Pollution*, *309*, 119755. DOI: 10.1016/j.envpol.2022.119755 PMID: 35839973

Jiménez, A. F., Cárdenas, P. F., & Jiménez, F. (2022). Intelligent IoT-multiagent precision irrigation approach for improving water use efficiency in irrigation systems at farm and district scales. *Computers and Electronics in Agriculture*, *192*, 106635. DOI: 10.1016/j.compag.2021.106635

Jimenez-Arias, D., Morales-Sierra, S., Borges, A. A., & Diaz, D. D. (2020). Biostimulant nanoencapsulation: The new keystone to fight hunger. *Journal of Agricultural and Food Chemistry*, *68*(27), 7083–7085. DOI: 10.1021/acs.jafc.0c02872 PMID: 32588622

Jindo, K., Canellas, L. P., Albacete, A., dos Santos, L. F., Rocha, R. L. F., & Baia, D. C.. (2020). Interaction between humic substances and plant hormones for phosphorous acquisition. *Agronomy (Basel)*, *10*(5), 640. DOI: 10.3390/agronomy10050640

Jinturkar, K. A., Rathi, M. N., & Misra, A. (2011). 3—Gene Delivery Using Physical Methods. In Misra, A. (Ed.), *Challenges in Delivery of Therapeutic Genomics and Proteomics* (pp. 83–126). Elsevier. DOI: 10.1016/B978-0-12-384964-9.00003-7

Ji, X., Zhang, H., Zhang, Y., Wang, Y., & Gao, C. (2015). Establishing a CRISPR-Cas-like immune system conferring DNA virus resistance in plants. *Nature Plants*, *1*(10), 15144. DOI: 10.1038/nplants.2015.144 PMID: 27251395

Joe, B., & Keith, D. (2022). Leveraging artificial intelligence and machine learning in microfluidics for high-throughput drug screening against plant pathogens. *Revista Española de Documentación Cientifica*, *16*(4), 4.

Jogaiah, S., Singh, H. B., Fraceto, L. F., & De Lima, R. (Eds.). (2020). *Advances in nano-fertilizers and nano-pesticides in agriculture: A smart delivery system for crop improvement.* Woodhead Publishing.

John, R. P., Tyagi, R. D., Brar, S. K., Surampalli, R. Y., & Prévost, D. (2011). Bio-encapsulation of microbial cells for targeted agricultural delivery. *Critical Reviews in Biotechnology*, *31*(3), 211–226. DOI: 10.3109/07388551.2010.513327 PMID: 20879835

Johnson, P. T. J., Dobson, A., Lafferty, K. D., Marcogliese, D. J., Memmott, J., Orlofske, S. A., Poulin, R., & Thieltges, D. W. (2010). When parasites become prey: Ecological and epidemiological significance of eating parasites. *Trends in Ecology & Evolution*, *25*(6), 362–371. DOI: 10.1016/j.tree.2010.01.005 PMID: 20185202

Joseph, T. M., Al-Hazmi, H. E., Śniatała, B., Esmaeili, A., & Habibzadeh, S. (2023). Nanoparticles and nanofiltration for wastewater treatment: From polluted to fresh water. *Environmental Research*, *238*, 117114. DOI: 10.1016/j.envres.2023.117114 PMID: 37716387

Judy, J. D., & Bertsch, P. M. (2014). Bioavailability, toxicity, and fate of manufactured nanomaterials in terrestrial ecosystems. *Advances in Agronomy*, *123*, 1–64. DOI: 10.1016/B978-0-12-420225-2.00001-7

Jung, G., Mugnier, J., Diem, H. G., & Dommergues, Y. R. (1982). Polymer-entrapped rhizobium as an inoculant for legumes. *Plant and Soil*, *65*(2), 219–231. DOI: 10.1007/BF02374652

Kah, M. (2015). Nanopesticides and nanofertilizers: Emerging contaminants or opportunities for risk mitigation? *Frontiers in Chemistry*, *3*, 64. DOI: 10.3389/fchem.2015.00064 PMID: 26636068

Kah, M.. (2018). Foliar uptake of nanopesticides: Current state and future challenges. *Environmental Science. Nano*, *5*(1), 6–20.

Kah, M., Beulke, S., Tiede, K., & Hofmann, T. (2013). Nanopesticides: State of knowledge, environmental fate, and exposure modeling. *Critical Reviews in Environmental Science and Technology*, *43*(16), 1823–1867. DOI: 10.1080/10643389.2012.671750

Kah, M., Hofmann, T., & Nanosciences, A. (2018). Nanopesticides and nanofertilizers: Emerging contaminants or opportunities for risk mitigation? *Frontiers in Environmental Science*, *6*, 6.

Kah, M., Kookana, R. S., Gogos, A., & Bucheli, T. D. (2019). A critical evaluation of nanopesticides and nanofertilizers against their conventional analogues. *Nature Nanotechnology*, *14*(6), 532–540. DOI: 10.1038/s41565-019-0439-5 PMID: 31168071

Kaiser, B., Vogg, G., Fürst, U. B., & Albert, M. (2015). Parasitic plants of the genus Cuscuta and their interaction with susceptible and resistant host plants. *Frontiers in Plant Science*, *6*(FEB), 118928. DOI: 10.3389/fpls.2015.00045 PMID: 25699071

Kale, S. K., Parishwad, G. V., & Patil, A. S. H. A. S. (2021). Emerging agriculture applications of silver nanoparticles. *ES Food & Agroforestry*, *3*, 17–22.

Kalia, A., & Kaur, G. (2018). *Biosynthesis of Nanoparticles Using Mushrooms*. 351–360. DOI: 10.1007/978-3-030-02622-6_17

Kalia, S., Kango, S., Kumar, A., Haldorai, Y., Kumari, B., & Kumar, R. (2014). Magnetic polymer nanocomposites for environmental and biomedical applications. *Colloid & Polymer Science*, *292*, 2025–2052.

Kalteh, M., Alipour, Z. T., Ashraf, S., Aliabadi, M. M., & Nosratabadi, A. F. (2014). Effect of silica nanoparticles on basil (*Ocimum basilicum*) under salinity stress. *Journal of Chemical Health Risks*, *4*, 49–55.

Kanakari, E., & Dendrinou-Samara, C. (2023). Fighting Phytopathogens with Engineered Inorganic-Based Nanoparticles. *Materials 2023, Vol. 16, Page 2388, 16*(6), 2388. DOI: 10.3390/ma16062388

Kandhol, N., Singh, V. P., Pandey, S., Sharma, S., Zhao, L., Corpas, F. J., Chen, Z.-H., White, J. C., & Tripathi, D. K. (2024). Nanoscale materials and NO-ROS homeostasis in plants: Trilateral dynamics. *Trends in Plant Science*, *29*(12), 1310–1318. DOI: 10.1016/j.tplants.2024.06.009 PMID: 39379242

Kang, S., & Zhang, J. (2004). Controlled alternate partial root-zone irrigation: Its physiological consequences and impact on water use efficiency. *Journal of Experimental Botany*, *55*(407), 2437–2446. DOI: 10.1093/jxb/erh249 PMID: 15361526

Kant, R., & Dasgupta, I. (2017). Phenotyping of VIGS-mediated gene silencing in rice using a vector derived from a DNA virus. *Plant Cell Reports*, *36*(7), 1159–1170. DOI: 10.1007/s00299-017-2156-6 PMID: 28540496

Kapeleka, J. A., & Mwema, M. F. (2024). State of nano pesticides application in smallholder agriculture production systems: Human and environmental exposure risk perspectives. *Heliyon*, *10*(20), e39225. DOI: 10.1016/j.heliyon.2024.e39225 PMID: 39492887

Karavidas, I., Ntatsi, G., Vougeleka, V., Karkanis, A., Ntanasi, T., Saitanis, C., Agathokleous, E., Ropokis, A., Sabatino, L., Tran, F., Iannetta, P. P. M., & Savvas, D. (2022). Agronomic practices to increase the yield and quality of common bean (*Phaseolus vulgaris* L.): A systematic review. *Agronomy (Basel)*, *12*(2), 271. DOI: 10.3390/agronomy12020271

Kari, A., Nagymáté, Z., Romsics, C., Vajna, B., Tóth, E., Lazanyi-Kovács, R., Rizó, B., Kutasi, J., Bernhardt, B., Farkas, É., & Márialigeti, K. (2021). Evaluating the combined effect of biochar and PGPR inoculants on the bacterial community in acidic sandy soil. Applied Soil Ecology, 160(March 2020). DOI: 10.1016/j.apsoil.2020.103856

Karimi, F., Altuner, E. E., Aygun, A., Bayat, R., Rajendran, S., & Sen, F. (2023). Synthesis of Silver Nanoparticles by Biogenic Methods: Characterization and Development of a Sensor Sensible to Pharmaceutical Medicine Paracetamol. *Topics in Catalysis*, *1*, 1–9. DOI: 10.1007/S11244-023-01887-4/TABLES/1

Karimi, F., Altuner, E. E., Gulbagca, F., Tiri, R. N. E., Sen, F., Javadi, A., & Dragoi, E. N. (2023). Facile bio-fabrication of ZnO@AC nanoparticles from chitosan: Characterization, hydrogen generation, and photocatalytic properties. *Environmental Research*, *216*, 114668. DOI: 10.1016/j.envres.2022.114668 PMID: 36397611

Karimi, M., Eslami, M., Sahandi-Zangabad, P., Mirab, F., Farajisafiloo, N., Shafaei, Z., Ghosh, D., Bozorgomid, M., Dashkhaneh, F., & Hamblin, M. R. (2016). pH-Sensitive stimulus-responsive nanocarriers for targeted delivery of therapeutic agents. *Wiley Interdisciplinary Reviews. Nanomedicine and Nanobiotechnology*, *8*(5), 696–716. DOI: 10.1002/wnan.1389 PMID: 26762467

Karimi, N., Ghanbarzadeh, B., Hamishehkar, H., Keivani, F., Pezeshki, A., & Gholian, M. M. (2015). Phytosome and Liposome: The Beneficial Encapsulation Systems in Drug Delivery and Food Application. *Applied Food Biotechnology*, *2*(3), 17–27.

Karny, A., Zinger, A., Kajal, A., Shainsky-Roitman, J., & Schroeder, A. (2018). Therapeutic nanoparticles penetrate leaves and deliver nutrients to agricultural crops. *Scientific Reports*, *8*(1), 1–10. DOI: 10.1038/s41598-018-25197-y PMID: 29773873

Kar, P., Banerjee, S., Chhetri, A., & Sen, A. (2021). Synthesis, physicochemical characterization and biological activity of synthesized Silver and Rajat Bhasma nanoparticles using Clerodendrum inerme. *Journal of Phytology*, *13*, 64–71. DOI: 10.25081/jp.2021.v13.7026

Karpuz, Ö., Baltacı, C., Gül, A., Gülen, J., Bozbeyoğlu, P., & Aydoğan, N. (2024). Green synthesis of iron and silver nanoparticles from aqueous extract of buckwheat husk waste: Antibacterial, cytotoxic, and dye decolorization properties. *Biomass Conversion and Biorefinery*, *2024*, 1–33. DOI: 10.1007/s13399-024-06287-6

Karthik, P. S., Himaja, A. L., & Singh, S. P. (2014). Carbon-allotropes: Synthesis methods, applications and future perspectives. *Carbon Letters*, *15*(4), 219–237. DOI: 10.5714/CL.2014.15.4.219

Karunathilake, E. M. B. M., Le, A. T., Heo, S., Chung, Y. S., & Mansoor, S. (2023). The path to smart farming: Innovations and opportunities in precision agriculture. *Agriculture*, *13*(8), 1593.

Kasana, R. C., Panwar, N. R., Kaul, R. K., & Kumar, P. (2017). Biosynthesis and effects of copper nanoparticles on plants. *Environmental Chemistry Letters*, *15*(2), 233–240. DOI: 10.1007/s10311-017-0615-5

Kashapov, R., Ibragimova, A., Pavlov, R., Gabdrakhmanov, D., Kashapova, N., Burilova, E., Zakharova, L., & Sinyashin, O. (2021). Nanocarriers for biomedicine: From lipid formulations to inorganic and hybrid nanoparticles. *International Journal of Molecular Sciences*, *22*(13), 7055. DOI: 10.3390/ijms22137055 PMID: 34209023

Kashyap, P. L., Xiang, X., & Heiden, P. (2015). Chitosan nanoparticle based delivery systems for sustainable agriculture. *International Journal of Biological Macromolecules*, *77*, 36–51. DOI: 10.1016/j.ijbiomac.2015.02.039 PMID: 25748851

Kasote, D. M., Lee, J. H., Jayaprakasha, G. K., & Patil, B. S. (2019). Seed priming with iron oxide nanoparticles modulate antioxidant potential and defense-linked hormones in watermelon seedlings. *ACS Sustainable Chemistry & Engineering*, *7*(5), 5142–5151. DOI: 10.1021/acssuschemeng.8b06013

Kathait, P., Patel, P. K., & Sahu, A. N. (2024). Harnessing exosomes and plant-derived exosomes as nanocarriers for the efficient delivery of plant bioactives. *Nanomedicine*. https://www.tandfonline.com/doi/abs/10.1080/17435889.2024.2354159

Kaur Takkar, M., & Gumber, K. (2024). Agro-nanocarriers: An emerging trend for better plant nutrition. *Journal of Plant Nutrition*, *47*(2), 332–350. DOI: 10.1080/01904167.2023.2275073

Kaur, P., & Choudhury, D. (2021). Functionality of receptor targeted zinc-insulin quantum clusters in skin tissue augmentation and bioimaging. *Journal of Drug Targeting*, *29*(5), 541–550. DOI: 10.1080/1061186X.2020.1864740 PMID: 33307859

Kaushal, M. (2018). Role of microbes in plant protection using intersection of nanotechnology and biology. In Abd-Elsalam, K. A., & Prasad, R. (Eds.), *Nanobiotechnology Applications in Plant Protection* (pp. 111–135). Springer International Publishing.

Kaushal, P., Maity, D., & Awasthi, R. (2024). Nano-green: Harnessing the potential of plant extracts for sustainable antimicrobial metallic nanoparticles. *Journal of Drug Delivery Science and Technology*, *94*, 105488. DOI: 10.1016/j.jddst.2024.105488

Kavita, V. (2017). DNA Biosensors-A Review. *Journal of Bioengineering & Biomedical Science*, *7*(2). Advance online publication. DOI: 10.4172/2155-9538.1000222

Kaviya, S., Santhanalakshmi, J., & Viswanathan, B. (2011). Biosynthesis of silver nano-flakes by *Crossandra infundibuliformis* leaf extract. *Materials Letters*, *67*(1), 64–66. DOI: 10.1016/j.matlet.2011.09.023

Kaya, S. I., Kurbanoglu, S., Yavuz, E., Demiroglu Mustafov, S., Sen, F., & Ozkan, S. A. (2020). Carbon-based ruthenium nanomaterial-based electroanalytical sensors for the detection of anticancer drug Idarubicin. *Scientific Reports*, *10*(1), 11057.

Kazama, T., Okuno, M., Watari, Y., Yanase, S., Koizuka, C., Tsuruta, Y., & Arimura, S. I. (2019). Curing cytoplasmic male sterility via TALEN-mediated mitochondrial genome editing. *Nature Plants*, *5*(7), 722–730. DOI: 10.1038/s41477-019-0459-z PMID: 31285556

Ke, C. L., Deng, F. S., Chuang, C. Y., & Lin, C. H. (2021a). Antimicrobial actions and applications of chitosan. *Polymers*, *13*, 904. DOI: 10.1007/978-3-319-91161-8_5

Ke, C. L., Liao, Y. T., & Lin, C. H. (2021b). MSS2 maintains mitochondrial function and is required for chitosan resistance, invasive growth, biofilm formation and virulence in *Candida albicans.Virulence*, *12*(1), 281–297. DOI: 10.1080/21505594.2020.1870082 PMID: 33427576

Kehayias, P., Jarmola, A., Mosavian, N., Fescenko, I., Benito, F., Laraoui, A., Smits, J., Bougas, L., Budker, D., Neumann, A., Brueck, S., & Acosta, V. (2017). Solution nuclear magnetic resonance spectroscopy on a nanostructured diamond chip. *Nature Communications*, *8*(1), 8. DOI: 10.1038/s41467-017-00266-4 PMID: 28775280

Ke, P. C., Lin, S., Parakhonskiy, B., & Chirth, C. (2018). Nanoparticles for plant imaging and detection. *Current Opinion in Biotechnology*, *49*, 166–172.

Keum, T., Noh, G., Seo, J. E., Bashyal, S., Sohn, D. H., & Lee, S. (2022). Examination of effective buccal absorption of salmon calcitonin using cell-penetrating peptide-conjugated liposomal drug delivery system. *International Journal of Nanomedicine*, *17*, 697–710. DOI: 10.2147/IJN.S335774 PMID: 35210769

Khalil, I. A. H., Arida, I. A., Ahmed, M., Khalil, I. A. H., Arida, I. A., & Ahmed, M. (2020). *Introductory Chapter: Overview on Nanomedicine Market.* Current and Future Aspects of Nanomedicine., DOI: 10.5772/intechopen.91890

Khan, F., Shariq, M., Asif, M., Siddiqui, M. A., Malan, P., & Ahmad, F. (2022). Green Nanotechnology: Plant-Mediated Nanoparticle Synthesis and Application. *Nanomaterials 2022, Vol. 12, Page 673, 12*(4), 673. DOI: 10.3390/nano12040673

Khan, M. N., Mobin, M., Abbas, Z. K., & Alamri, S. A. (2018). Fertilizers and their contaminants in soils, surface and groundwater. *Encyclopedia of the Anthropocene, 5*, 225-240.

Khan, A. R., Salam, A., Li, G., Iqbal, B., Ulhassan, Z., Liu, Q., Azhar, W., Liaquat, F., Shah, I. H., Ul Hassan, S. S., & Du, D. (2024). Nanoparticles and their crosstalk with stress mitigators: A novel approach towards abiotic stress tolerance in agricultural systems. *The Crop Journal*, *12*(5), 1280–1298. DOI: 10.1016/j.cj.2024.09.010

Khan, A., Ahmad, G., Haris, M., & Khan, A. A. (2023). Bio-organics Management: Novel Strategies to Manage Root-knot Nematode, Meloidogyne incognita Pest of Vegetable Crops. *Gesunde Pflanzen*, *75*(1), 193–209. DOI: 10.1007/s10343-022-00679-2

Khan, A., Khan, A., Ali, A., Fatima, S., & Siddiqui, M. A. (2023). Root-Knot Nematodes (Meloidogyne spp.): Biology, Plant-Nematode Interactions and Their Environmentally Benign Management Strategies. *Gesunde Pflanzen*, *75*(6), 2187–2205. DOI: 10.1007/s10343-023-00886-5

Khan, F., Pandey, P., & Upadhyay, T. K. (2022). Applications of nanotechnology-based agrochemicals in food security and sustainable agriculture: An overview. *Agriculture*, *12*(10), 1672. DOI: 10.3390/agriculture12101672

Khan, F., Siddique, A. B., Shabala, S., Zhou, M., & Zhao, C. (2023). Phosphorus Plays Key Roles in Regulating Plants' Physiological Responses to Abiotic Stresses. *Plants*, *12*(15), 2861. DOI: 10.3390/plants12152861 PMID: 37571014

Khan, I., Hasan, M., Kausar, R., Shehzad, J., & Mustafa, G. (2022). Plant molecular responses to nanoparticle stress. In *Plant and nanoparticles* (pp. 239–264). Springer Nature Singapore. DOI: 10.1007/978-981-19-2503-0_9

Khan, M. A., Naeem, M., Akram, W., & Hussain, I. (2019). Impacts of pesticide exposure on human health and ecosystem. *Pesticide Science and Technology*, *17*(3), 312–328.

Khan, M. S., Vishakante, G. D., & Siddaramaiah, H. (2013). Gold nanoparticles: A paradigm shift in biomedical applications. *Advances in Colloid and Interface Science*, *199–200*, 44–58. DOI: 10.1016/j.cis.2013.06.003 PMID: 23871224

Khan, N. A., Khan, S. U., Ahmed, S., Farooqi, I. H., Dhingra, A., Hussain, A., & Changani, F. (2019). Applications of nanotechnology in water and wastewater treatment: A review. *Asian Journal of Water, Environment and Pollution*, *16*(4), 81–86. DOI: 10.3233/AJW190051

Khan, S. T., & Malik, A. (2019). Engineered nanomaterials for water decontamination and purification: From lab to products. *Journal of Hazardous Materials*, *363*, 295–308. DOI: 10.1016/j.jhazmat.2018.09.091 PMID: 30312926

Khan, Z. S., Rizwan, M., Hafeez, M., Ali, S., Adrees, M., Qayyum, M. F., Khalid, S., ur Rehman, M. Z., & Sarwar, M. A. (2020). Effects of silicon nanoparticles on growth and physiology of wheat in cadmium contaminated soil under different soil moisture levels. *Environmental Science and Pollution Research International*, *27*(5), 4958–4968. DOI: 10.1007/s11356-019-06673-y PMID: 31840219

Khatoon, U. T., & Velidandi, A. (2025). An Overview on the Role of Government Initiatives in Nanotechnology Innovation for Sustainable Economic Development and Research Progress. *Sustainability (Basel)*, *17*(3), 3. Advance online publication. DOI: 10.3390/su17031250

Khaydarov, R. A., Khaydarov, R. R., Gapurova, O., Estrin, Y., & Scheper, T. (2009). Electrochemical method for the synthesis of silver nanoparticles. *Journal of Nanoparticle Research*, *11*(5), 1193–1200. DOI: 10.1007/s11051-008-9513-x

Khodadadi, B., Bordbar, M., Yeganeh-Faal, A., & Nasrollahzadeh, M. (2017). Green synthesis of Ag nanoparticles/clinoptilolite using Vaccinium macrocarpon fruit extract and its excellent catalytic activity for reduction of organic dyes. *Journal of Alloys and Compounds*, *719*, 82–88. DOI: 10.1016/j.jallcom.2017.05.135

Khodakovskaya, M., Dervishi, E., Mahmood, M., Xu, Y., Li, Z., Watanabe, F., & Biris, A. S. (2009). Carbon nanotubes are able to penetrate plant seed coat and dramatically affect seed germination and plant growth. *ACS Nano*, *3*(10), 3221–3227. DOI: 10.1021/nn900887m PMID: 19772305

Khoso, M. A., Wagan, S., Alam, I., Hussain, A., Ali, Q., Saha, S., Poudel, T. R., Manghwar, H., & Liu, F. (2024). Impact of plant growth-promoting rhizobacteria (PGPR) on plant nutrition and root characteristics: Current perspective. *Plant Stress*, *11*, 100341. DOI: 10.1016/j.stress.2023.100341

Khot, L. R., Sankaran, S., Maja, J. M., Ehsani, R., & Schuster, E. W. (2012). Applications of nanomaterials in agricultural production and crop protection: A review. *Crop Protection (Guildford, Surrey)*, *35*, 64–70. DOI: 10.1016/j.cropro.2012.01.007

Khoury, W. E., & Makkouk, K. (2010). Integrated Plant Disease Management in Developing Countries. *Journal of Plant Pathology*, *92*(4), S35–S42.

Kieber, J. J., & Schaller, G. E. (2014). Cytokinins. *The Arabidopsis Book / American Society of Plant Biologists*, *12*, e0168. DOI: 10.1199/tab.0168 PMID: 24465173

Kildegaard, K. R., Arnesen, J. A., Adiego-Pérez, B., Rago, D., Kristensen, M., Klitgaard, A. K., Hansen, E. H., Hansen, J., & Borodina, I. (2021). Tailored biosynthesis of gibberellin plant hormones in yeast. *Metabolic Engineering*, *66*, 1–11. DOI: 10.1016/j.ymben.2021.03.010 PMID: 33746070

Kim, C. J., Park, J. E., Hu, X., Albert, S. K., & Park, S. J. (2020). Peptide-driven shape control of low-dimensional DNA nanostructures. *ACS Nano*, *14*(2), 2276–2284. DOI: 10.1021/acsnano.9b09312 PMID: 31962047

Kim, C., Tonga, G. Y., Yan, B., Kim, C. S., Kim, S. T., Park, M. H., Zhu, Z., Duncan, B., Creran, B., & Rotello, V. M. (2015). Regulating Exocytosis of Nanoparticles via Host-Guest Chemistry. *Organic & Biomolecular Chemistry*, *13*(8), 2474–2479. DOI: 10.1039/C4OB02433H PMID: 25569869

Kim, Y. A., Hayashi, T., Endo, M., Kaburagi, Y., Tsukada, T., Shan, J., Osato, K., & Tsuruoka, S. (2005). Synthesis and structural characterization of thin multi-walled carbon nanotubes with a partially facetted cross section by a floating reactant method. *Carbon*, *43*(11), 2243–2250. DOI: 10.1016/j.carbon.2005.03.039

Kishima, H., Poyot, T., Bloch, J., Dauguet, J., Condé, F., Dollé, F., Hinnen, F., Pralong, W., Palfi, S., Déglon, N., Aebischer, P., & Hantraye, P. (2004). Encapsulated GDNF-producing C2C12 cells for Parkinson's disease: A pre-clinical study in chronic MPTP-treated baboons. *Neurobiology of Disease*, *16*(2), 428–439. DOI: 10.1016/j.nbd.2004.03.012 PMID: 15193299

Klikocka, H., Marks, M., Barczak, B., Szostak, B., Podleśna, A., & Podleśny, J. (2018). Response of spring wheat to NPK and S fertilization. The content and uptake of macronutrients and the value of ionic ratios. *Open Chemistry*, *16*(1), 1059–1065. DOI: 10.1515/chem-2018-0116

Kneuer, C., Sameti, M., Bakowsky, U., Schiestel, T., Schirra, H., Schmidt, H., & Lehr, C. M. (2000). A nonviral DNA delivery system based on surface modified silica-nanoparticles can efficiently transfect cells in vitro. *Bioconjugate Chemistry*, *11*(6), 926–932. DOI: 10.1021/bc0000637 PMID: 11087343

Knödler, M., Opdensteinen, P., Sankaranarayanan, R. A., Morgenroth, A., Buhl, E. M., Mottaghy, F. M., & Buyel, J. F. (2023). Simple plant-based production and purification of the assembled human ferritin heavy chain as a nanocarrier for tumor-targeted drug delivery and bioimaging in cancer therapy. *Biotechnology and Bioengineering*, *120*(4), 1038–1054. DOI: 10.1002/bit.28312 PMID: 36539373

Knox, J. P., & Benitez-Alfonso, Y. (2014). Roles and regulation of plant cell walls surrounding plasmodesmata. *Current Opinion in Plant Biology, 22*, 93–100. DOI: 10.1016/j.pbi.2014.09.009 PMID: 25286000

Kocak, Y., Aygun, A., Altuner, E. E., Gonca, S., Berikten, D., Tiri, R. N. E., & Sen, F. (2024). *Eco-friendly production of platinum nanoparticles: physicochemical properties, evaluation of biological and catalytic activities. 21*, 51–62. doi: DOI: 10.1007/S13762-023-05232-W/FIGURES/7

Kocak, N., Sahin, M., Kücükkolbasi, S., & Erdogan, Z. O. (2012). Synthesis and characterization of novel nano-chitosan Schiff base and use of lead (II) sensor. *International Journal of Biological Macromolecules, 51*(5), 1159–1166. DOI: 10.1016/j.ijbiomac.2012.09.003 PMID: 22982811

Kolodziejczak-Radzimska, A., & Jesionowski, T. (2014). Zinc Oxide—From Synthesis to Application: A Review. *Materials 2014, Vol. 7, Pages 2833-2881, 7*(4), 2833–2881. DOI: 10.3390/ma7042833

Komal, J., Desai, H. R., Samal, I., Mastinu, A., Patel, R. D., Kumar, P. V. D., Majhi, P. K., Mahanta, D. K., & Bhoi, T. K. (2023). Unveiling the Genetic Symphony: Harnessing CRISPR-Cas Genome Editing for Effective Insect Pest Management. *Plants, 12*(23), 3961. DOI: 10.3390/plants12233961 PMID: 38068598

Komarova, T., Ilina, I., Taliansky, M., & Ershova, N. (2023). Nanoplatforms for the Delivery of Nucleic Acids into Plant Cells. *International Journal of Molecular Sciences, 24*(23), 23. Advance online publication. DOI: 10.3390/ijms242316665 PMID: 38068987

Kong, J., Martin-Ortigosa, S., Finer, J., Orchard, N., Gunadi, A., Batts, L. A., Thakare, D., Rush, B., Schmitz, O., Stuiver, M., Olhoft, P., & Pacheco-Villalobos, D. (2020). Overexpression of the Transcription Factor *GROWTH-REGULATING FACTOR5* Improves Transformation of Dicot and Monocot Species. *Frontiers in Plant Science, 11*, 572319. DOI: 10.3389/fpls.2020.572319 PMID: 33154762

Kong, X. P., Zhang, B. H., & Wang, J. (2021). Multiple roles of mesoporous silica in safe pesticide application by nanotechnology: A review. *Journal of Agricultural and Food Chemistry, 69*(24), 6735–6754. DOI: 10.1021/acs.jafc.1c01091 PMID: 34110151

Kookana, R. S., Boxall, A. B., Reeves, P. T., Ashauer, R., Beulke, S., Chaudhry, Q., Cornelis, G., Fernandes, T. F., Gan, J., Kah, M., Lynch, I., Ranville, J., Sinclair, C., Spurgeon, D., Tiede, K., & Van den Brink, P. J. (2014). Nanopesticides: Guiding principles for regulatory evaluation of environmental risks. *Journal of Agricultural and Food Chemistry, 62*(19), 4227–4240. DOI: 10.1021/jf500232f PMID: 24754346

Kookana, R. S., Drechsel, P., Jamwal, P., & Vanderzalm, J. (2014). Urban wastewater: A valuable resource turned a challenge. *Environmental Science & Technology, 48*(14), 7467–7475.

Koo, Y., Wang, J., Zhang, Q., Zhu, H., Chehab, E. W., Colvin, V. L., Alvarez, P. J., & Braam, J. (2015). Fluorescence reports intact quantum dot uptake into roots and translocation to leaves of Arabidopsis thaliana and subsequent ingestion by insect herbivores. *Environmental Science & Technology, 49*(1), 626–632. DOI: 10.1021/es5050562 PMID: 25437125

Korah, B. K., Chacko, A. R., Mathew, S., & Mathew, B. (2022). Carbon Nanotubes as Sensors in Food and Agricultural Applications. In *Carbon Nanotubes for Energy and Environmental Applications* (pp. 115–159). Apple Academic Press. DOI: 10.1201/9781003189619-6

Korrapati, C., Yaraladdi, A. B., Vaddi, H., & Raghavan, R. (2024). Applications of the Saffron European Kitchen Spice Crocus sativus. In *Therapeutic and Pharmacological Applications of Ethnobotany* (pp. 98–127). IGI Global., DOI: 10.4018/979-8-3693-1986-4.ch005

Kralova, K., & Jampilek, J. (2021). Responses of medicinal and aromatic plants to engineered nanoparticles. *Applied Sciences (Basel, Switzerland)*, *11*(4), 1813. DOI: 10.3390/app11041813

Kravanja, G., Primozic, M., Knez, Z., & Leitgeb, M. (2019). Chitosan-based (Nano) materials for novel biomedical applications. *Molecules (Basel, Switzerland)*, *24*(10), 1960. DOI: 10.3390/molecules24101960 PMID: 31117310

Kröger, N., Deutzmann, R., & Sumper, M. (1999). Polycationic peptides from diatom biosilica that direct silica nanosphere formation. *Science*, *286*(5442), 1129–1132. DOI: 10.1126/science.286.5442.1129 PMID: 10550045

Kucuk, N., Primožič, M., Knez, Z., & Leitgeb, M. (2023). Sustainable biodegradable biopolymer-based nanoparticles for healthcare applications. *International Journal of Molecular Sciences*, *24*(4), 3188. DOI: 10.3390/ijms24043188 PMID: 36834596

Kühnel, D., & Nickel, C. (2014). The OECD expert meeting on ecotoxicology and environmental fate: Towards the development of improved OECD guidelines for the testing of nanomaterials. *The Science of the Total Environment*, *472*, 347–353. DOI: 10.1016/j.scitotenv.2013.11.055 PMID: 24461369

Kumah, E. A., Fopa, R. D., Harati, S., Boadu, P., Zohoori, F. V., & Pak, T. (2023). Human and environmental impacts of nanoparticles: A scoping review of the current literature. *BMC Public Health*, *23*(1), 1059. DOI: 10.1186/s12889-023-15958-4 PMID: 37268899

Kumar, H., Bhardwaj, K., Kuča, K., Kalia, A., Nepovimova, E., Verma, R., & Kumar, D. (2020). Flower-Based Green Synthesis of Metallic Nanoparticles: Applications beyond Fragrance. *Nanomaterials 2020, Vol. 10, Page 766*, *10*(4), 766. DOI: 10.3390/nano10040766

Kumar, A., Kanwar, R., & Mehta, S. K. (2021a). Eucalyptus oil-based nanoemulsion: A potent green nanowagon for controlled delivery of emamectin benzoate. *ACS Agricultural Science & Technology*, *1*(2), 76–88. DOI: 10.1021/acsagscitech.0c00061

Kumar, B. G., Sadeghi, S., Melikov, R., Aria, M. M., Jalali, H. B., Ow-Yang, C. W., & Nizamoglu, S. (2018). Structural control of InP/ZnS core/shell quantum dots enables high-quality white LEDs. *Nanotechnology*, *29*(34), 345605. DOI: 10.1088/1361-6528/aac8c9 PMID: 29846177

Kumar, D., Pathak, M., & Chaturvedi, A. (2019). Recent advances in nanoformulations for agrochemical delivery: Current status and future perspectives. *Journal of Nanotechnology Research*, *5*(3), 45–58.

Kumar, H., Venkatesh, N., Bhowmik, H., & Kuila, A. (2018). Metallic nanoparticle: A review. *Biomedical Journal of Scientific & Technical Research*, *4*(2), 3765–3775.

Kumar, I., Bhattacharya, J., & Das, B. K. (2022). Efficacy of silver nanoparticles-based foliar spray application to control plant diseases, its effect on productivity, and risk assessment. *Arabian Journal of Geosciences*, *15*(5), 390. DOI: 10.1007/s12517-022-09652-9

Kumari, A., Rana, V., Yadav, S. K., & Kumar, V. (2023). Nanotechnology as a powerful tool in plant sciences: Recent developments, challenges and perspectives. *Plant Nano Biology*, *5*, 100046. DOI: 10.1016/j.plana.2023.100046

Kumari, R., Suman, K., Karmakar, S., Mishra, V., Lakra, S. G., Saurav, G. K., & Mahto, B. K. (2023). Regulation and safety measures for nanotechnology-based agri-products. *Frontiers in Genome Editing*, *5*, 1200987. DOI: 10.3389/fgeed.2023.1200987 PMID: 37415849

Kumari, S., & Sarkar, L. (2021). A review on nanoparticles: Structure, classification, synthesis & applications. *Journal of Scientific Research*, *65*(8), 42–46. DOI: 10.37398/JSR.2021.650809

Kumar, J. A., Krithiga, T., Manigandan, S., Sathish, S., Renita, A. A., Prakash, P., Prasad, B. S. N., Kumar, T. R. P., Rajasimman, M., Hosseini-Bandegharaei, A., Prabu, D., & Crispin, S. (2021). A focus to green synthesis of metal/metal based oxide nanoparticles: Various mechanisms and applications towards ecological approach. *Journal of Cleaner Production*, *324*, 129198. DOI: 10.1016/j.jclepro.2021.129198

Kumar, J., Ramlal, A., Mallick, D., & Mishra, V. (2021b). An overview of some biopesticides and their importance in plant protection for commercial acceptance. *Plants*, *10*(6), 1185. DOI: 10.3390/plants10061185 PMID: 34200860

Kumar, M., Ahmad, S., & Singh, R. (2022). Plant growth promoting microbes: Diverse roles for sustainable and ecofriendly agriculture. *Energy Nexus*, *7*, 100133. DOI: 10.1016/j.nexus.2022.100133

Kumar, P. R., Vivekanandhan, S., Misra, M., Mohanty, A., & Satyanarayana, N. (2012). Soybean (*Glycine max*) leaf extract based green synthesis of palladium nanoparticles. *Journal of Biomaterials and Nanobiotechnology*, *3*(1), 14–19. DOI: 10.4236/jbnb.2012.31003

Kumar, R., Kumar, N., Rajput, V. D., Mandzhieva, S., Minkina, T., Saharan, B. S., Kumar, D., Sadh, P. K., & Duhan, J. S. (2022). Advances in biopolymeric nanopesticides: A new eco-friendly/eco-protective perspective in precision agriculture. *Nanomaterials (Basel, Switzerland)*, *12*(22), 3964. DOI: 10.3390/nano12223964 PMID: 36432250

Kumar, S., Nehra, M., Dilbaghi, N., Marrazza, G., Hassan, A. A., & Kim, K. H. (2019). Nano-based smart pesticide formulations: Emerging opportunities for agriculture. *Journal of Controlled Release*, *294*, 131–153. DOI: 10.1016/j.jconrel.2018.12.012 PMID: 30552953

Kumar, S., & Randhawa, J. K. (2013). High melting lipid based approach for drug delivery: Solid lipid nanoparticles. *Materials Science and Engineering C*, *33*(4), 1842–1852. DOI: 10.1016/j.msec.2013.01.037 PMID: 23498204

Kumar, S., Sindhu, S. S., & Kumar, R. (2021). Biofertilizers: An ecofriendly technology for nutrient recycling and environmental sustainability. *Current Research in Microbial Sciences*, *3*, 100094. DOI: 10.1016/j.crmicr.2021.100094 PMID: 35024641

Kumar, V., Guleria, P., Kumar, V., & Yadav, S. K. (2013). Gold nanoparticle exposure induces growth and yield enhancement in *Arabidopsis thaliana*. *The Science of the Total Environment*, *461*, 462–468. DOI: 10.1016/j.scitotenv.2013.05.018 PMID: 23747561

Kumar, V., Kumar, M., Shrivastava, N., Bisht, S., Sharma, S., & Varma, A. (2016). *Interaction among rhizospheric microbes, soil, and plant roots: Influence on micronutrient uptake and bioavailability. Plant, Soil and Microbes* (Vol. 2). Mechanisms and Molecular Interactions., DOI: 10.1007/978-3-319-29573-2_8

Kumar, V., Sachdev, D., Pasricha, R., Maheshwari, P. H., & Taneja, N. K. (2018). Zinc-supported multiwalled carbon nanotube nanocomposite: A synergism to micronutrient release and a smart distributor to promote the growth of onion seeds in arid conditions. *ACS Applied Materials & Interfaces, 10*(43), 36733–36745. DOI: 10.1021/acsami.8b13464 PMID: 30286290

Kunte, N., McGraw, E., Bell, S., Held, D., & Avila, L. A. (2020). Prospects, challenges and current status of RNAi through insect feeding. *Pest Management Science, 76*(1), 26–41. DOI: 10.1002/ps.5588 PMID: 31419022

Kurepa, J., Paunesku, T., Vogt, S., Arora, H., Rabatic, B. M., Lu, J., Wanzer, M. B., Woloschak, G. E., & Smalle, J. A. (2010). Uptake and distribution of ultrasmall anatase TiO2 Alizarin red S nanoconjugates in Arabidopsis thaliana. *Nano Letters, 10*(7), 2296–2302. DOI: 10.1021/nl903518f PMID: 20218662

Kutawa, A. B., Ahmad, K., Ali, A., Hussein, M. Z., Abdul Wahab, M. A., Adamu, A., Ismaila, A. A., Gunasena, M. T., Rahman, M. Z., & Hossain, M. I. (2021). Trends in nanotechnology and its potentialities to control plant pathogenic fungi: A review. *Biology (Basel), 10*(9), 881. DOI: 10.3390/biology10090881 PMID: 34571758

Kwak, S. Y., Wong, M. H., Lew, T. T. S., Bisker, G., Lee, M. A., Kaplan, A., & Strano, M. S. (2019). Nanosensor technology applied to living plant systems. *Annual Review of Analytical Chemistry (Palo Alto, Calif.), 12*, 113–140. DOI: 10.1146/annurev-anchem-061318-114910 PMID: 28605605

Kwak, S.-Y., Lew, T. T. S., Sweeney, C. J., Koman, V. B., Wong, M. H., Bohmert-Tatarev, K., Snell, K. D., Seo, J. S., Chua, N.-H., & Strano, M. S. (2019). Chloroplast-selective gene delivery and expression in planta using chitosan-complexed single-walled carbon nanotube carriers. *Nature Nanotechnology, 14*(5), 447–455. DOI: 10.1038/s41565-019-0375-4 PMID: 30804482

Kwon, S. G., & Hyeon, T. (2011). Formation mechanisms of uniform nanocrystals via hot-injection and heat-up methods. *Small, 7*(19), 2685–2702. DOI: 10.1002/smll.201002022 PMID: 21805623

Kwon, S., Singh, R. K., Perez, R. A., Abou Neel, E. A., Kim, H. W., & Chrzanowski, W. (2013). Silica-based mesoporous nanoparticles for controlled drug delivery. *Journal of Tissue Engineering, 4*, 2041731413503357. DOI: 10.1177/2041731413503357 PMID: 24020012

Kwon, Y. J. (2012). Before and after endosomal escape: Roles of stimuli-converting siRNA/polymer interactions in determining gene silencing efficiency. *Accounts of Chemical Research, 45*(7), 1077–1088. DOI: 10.1021/ar200241v PMID: 22103667

L'Espérance, E., Bouyoucef, L. S., Dozois, J. A., & Yergeau, E. (2024). Tipping the plant-microbe competition for nitrogen in agricultural soils. *iScience, 27*(10), 110973. DOI: 10.1016/j.isci.2024.110973 PMID: 39391734

Lakshani, N., Wijerathne, H. S., Sandaruwan, C., Kottegoda, N., & Karunarathne, V. (2023). Release Kinetic Models and Release Mechanisms of Controlled-Release and Slow-Release Fertilizers. *ACS Agricultural Science & Technology, 3*(11), 939–956. DOI: 10.1021/acsagscitech.3c00152

Lallana, E., Sousa-Herves, A., Fernandez-Trillo, F., Riguera, R., & Fernandez-Megia, E. (2012). Click chemistry for drug delivery nanosystems. *Pharmaceutical Research, 29*(1), 1–34. DOI: 10.1007/s11095-011-0568-5 PMID: 21913032

Lalpekhlua, K., Tirkey, A., Saranya, S., & Babu, P. J. (2024). Post-harvest Management Strategies for Quality Preservation in Crops. *International Journal of Vegetable Science*, *30*(5), 587–635. DOI: 10.1080/19315260.2024.2392551

Lal, R. (2008). Soils and sustainable agriculture. A review. *Agronomy for Sustainable Development*, *28*(1), 57–64. DOI: 10.1051/agro:2007025

Lambers, H., Mougel, C., Jaillard, B., & Hinsinger, P. (2009). Plant-microbe-soil interactions in the rhizosphere: an evolutionary perspective.

Lam, S. J., O'Brien-Simpson, N. M., Pantarat, N., Sulistio, A., Wong, E. H. H., Chen, Y.-Y., Lenzo, J. C., Holden, J. A., Blencowe, A., Reynolds, E. C., & Qiao, G. G. (2016). Combating multidrug-resistant Gram-negative bacteria with structurally nanoengineered antimicrobial peptide polymers. *Nature Microbiology*, *1*(11), 16162. DOI: 10.1038/nmicrobiol.2016.162 PMID: 27617798

Landry, M. P., & Mitter, N. (2019). How nanocarriers delivering cargos in plants can change the GMO landscape. *Nature Nanotechnology*, *14*(6), 512–514. DOI: 10.1038/s41565-019-0463-5 PMID: 31168070

Lang, C., Mission, E. G., Fuaad, A. A. H. A., & Shaalan, M. (2021). Nanoparticle tools to improve and advance precision practices in the Agrifoods Sector towards sustainability-A review. *Journal of Cleaner Production*, *293*, 126063. DOI: 10.1016/j.jclepro.2021.126063

Langer, K., Balthasar, S., Vogel, V., Dinauer, N., Von Briesen, H., & Schubert, D. (2003). Optimization of the preparation process for human serum albumin (HSA) nanoparticles. *International Journal of Pharmaceutics*, *257*(1–2), 169–180. DOI: 10.1016/S0378-5173(03)00134-0 PMID: 12711172

Langton, C. G. (2019). Nano Technology With Feynman Machines: Scanning Tunneling Engineering and Artificial Life. *Artificial Life*, *14*(6), 443–500. DOI: 10.4324/9780429032769-21

Lari, S., Jonnalagadda, P. R., Yamagani, P., Medithi, S., Vanka, J., Pandiyan, A., Naidu, M., & Jee, B. (2022). Assessment of dermal exposure to pesticides among farmers using dosimeter and hand washing methods. *Frontiers in Public Health*, *10*, 957774. DOI: 10.3389/fpubh.2022.957774 PMID: 36091511

Larue, C., Castillo-Michel, H., Sobanska, S., Cécillon, L., Bureau, S., Barthès, V., Ouerdane, L., Carrière, M., & Sarret, G. (2014). Foliar exposure of the crop Lactuca sativa to silver nanoparticles: Evidence for internalization and changes in Ag speciation. *Journal of Hazardous Materials*, *264*, 98–106. DOI: 10.1016/j.jhazmat.2013.10.053 PMID: 24275476

Larue, C., Laurette, J., Herlin-Boime, N., Khodja, H., Fayard, B., Flank, A. M., Brisset, F., & Carriere, M. (2012). Accumulation, translocation and impact of TiO2 nanoparticles in wheat (Triticum aestivum spp.): Influence of diameter and crystal phase. *The Science of the Total Environment*, *431*, 197–208. DOI: 10.1016/j.scitotenv.2012.04.073 PMID: 22684121

Lattanzio, V. M. T., & Nivarlet, N. (2017). Multiplex dipstick immunoassay for semiquantitative determination of fusarium mycotoxins in oat in *Methods. Molecular Biology*, *1536*, 137–142. DOI: 10.1007/978-1-4939-6682-0_10 PMID: 28132148

Lau, H. Y., Wu, H., Wee, E. J. H., Trau, M., Wang, Y., & Botella, J. R. (2017). Specific and sensitive isothermal electrochemical biosensor for plant pathogen DNA detection with colloidal gold nanoparticles as probes. *Scientific Reports*, *7*(1), 38896. DOI: 10.1038/srep38896 PMID: 28094255

Law, S. S. Y., Liou, G., Nagai, Y., Giménez-Dejoz, J., Tateishi, A., Tsuchiya, K., & Numata, K. (2022). Polymer-coated carbon nanotube hybrids with functional peptides for gene delivery into plant mitochondria. *Nature Communications*, *13*(1), 2417. DOI: 10.1038/s41467-022-30185-y PMID: 35577779

Lebaschi, S., Hekmati, M., & Veisi, H. (2017). Green synthesis of palladium nanoparticles mediated by black tea leaves (Camellia sinensis) extract: Catalytic activity in the reduction of 4-nitrophenol and Suzuki-Miyaura coupling reaction under ligand-free conditions. *Journal of Colloid and Interface Science*, *485*, 223–231. DOI: 10.1016/j.jcis.2016.09.027 PMID: 27665075

Lee, D. Y., Kim, H. W., Yang, H., & Sung, S. H. (2017). Hydrolyzable tannins from the fruits of Terminalia chebula Retz and their α-glucosidase inhibitory activities. *Phytochemistry*, *137*, 109–116. DOI: 10.1016/j.phytochem.2017.02.006 PMID: 28213992

Lee, H. J., Lee, G., Jang, N. R., Yun, J. H., Song, J. Y., & Kim, B. S. (2011). Biological synthesis of copper nanoparticles using plants extract. *Nanotechnology*, *1*, 371–374.

Lee, J., Chatterjee, D. K., Lee, M. H., & Krishnan, S. (2014). Gold nanoparticles in breast cancer treatment: Promise and potential pitfalls. *Cancer Letters*, *347*(1), 46–53. DOI: 10.1016/j.canlet.2014.02.006 PMID: 24556077

Lee, K., Conboy, M., Park, H. M., Jiang, F., Kim, H. J., Dewitt, M. A., Mackley, V. A., Chang, K., Rao, A., Skinner, C., Shobha, T., Mehdipour, M., Liu, H., Huang, W. C., Lan, F., Bray, N. L., Li, S., Corn, J. E., Kataoka, K., & Murthy, N. (2017). Nanoparticle delivery of Cas9 ribonucleoprotein and donor DNA *in vivo* induces homology-directed DNA repair. *Nature Biomedical Engineering*, *1*(11), 889–901. DOI: 10.1038/s41551-017-0137-2 PMID: 29805845

Lee, S., Kim, A., Song, J., Choe, S., Jang, G., & Kim, Y. (2024). Plant growth-promoting rhizobacterium Bacillus megaterium modulates the expression of antioxidant-related and drought-responsive genes to protect rice (*Oryza sativa* L.) from drought. *Frontiers in Microbiology*, *15*, 1430546. DOI: 10.3389/fmicb.2024.1430546 PMID: 39234545

Lei, W. X., An, Z., Zhang, B., Wu, Q., Gong, W., Li, J., & Chen, W. (2020). Construction of gold-SiRNA NPR1 nanoparticles for effective and quick silencing of NPR1 in Arabidopsis thaliana. *RSC Advances*, *10*(33), 19300–19308. DOI: 10.1039/D0RA02156C PMID: 35515443

Letchumanan, D., Sok, S. P. M., Ibrahim, S., Nagoor, N. H., & Arshad, N. M. (2021). Plant-Based Biosynthesis of Copper/Copper Oxide Nanoparticles: An Update on Their Applications in Biomedicine, Mechanisms, and Toxicity. *Biomolecules 2021, Vol. 11, Page 564, 11*(4), 564. DOI: 10.3390/biom11040564

Lew, T. T. S., Koman, V. B., & Strano, M. S. (2020). The emergence of plant nanobionics and living plants as technology. *Nature Plants*, *6*(5), 557–569. DOI: 10.1038/s41477-020-0655-6

Lew, T. T. S., Wong, M. H., Kwak, S. Y., Sinclair, R., Koman, V. B., & Strano, M. S. (2018). Rational design principles for the transport and subcellular distribution of nanomaterials into plant protoplasts. *Small, 14*(44), 1802086. DOI: 10.1002/smll.201802086 PMID: 30191658

Lhamo, D., & Luan, S. (2021). Potential networks of nitrogen-phosphorus-potassium channels and transporters in Arabidopsis roots at a single cell resolution. *Frontiers in Plant Science, 12*, 689545. DOI: 10.3389/fpls.2021.689545 PMID: 34220911

Li, C., Wang, D., Liang, T., Li, G., Wang, X., Cao, M., & Liang, J. (2003). Oxidation behavior of CNTs and the electric double layer capacitor made of the CNT electrodes. *Science in China Series E: Technological Sciences 2003 46:4, 46*(4), 349–354. DOI: 10.1360/03ye0067

Li, T. (2022). An A-Mnse Nanorod as Anode for Superior Potassium-Ion Storage Via Synergistic Effects of Physical Encapsulation and Chemical Bonding. SSRN *Electronic Journal.*

Liang, Z., Chen, K., Li, T., Zhang, Y., Wang, Y., Zhao, Q., Liu, J., Zhang, H., Liu, C., Ran, Y., & Gao, C. (2017). Efficient DNA-free genome editing of bread wheat using CRISPR/Cas9 ribonucleoprotein complexes. *Nature Communications, 8*(1), 14261. DOI: 10.1038/ncomms14261 PMID: 28098143

Liao, Y., Xu, D., Cao, Y., & Zhu, Y. G. (2023). Advancing sustainable agriculture: Enhancing crop nutrition with next-generation nanotech-based fertilizers. *Nano Research, 16*(12), 13205–13225. DOI: 10.1007/s12274-023-6284-8

Li, C., Mei, H., & Zhou, Z. (2021). Surface-modified nanoparticles in fungal disease management. *Nanomedicine (London), 16*(7), 583–598.

Li, F., Ni, B., Zheng, Y., Huang, Y., & Li, G. (2021). A simple and efficient voltammetric sensor for dopamine determination based on ZnO nanorods/electro-reduced graphene oxide composite. *Surfaces and Interfaces, 26*, 101375. DOI: 10.1016/j.surfin.2021.101375

Li, J.. (2021). Role of endocytosis in nanoparticle uptake in plants. *Journal of Nanobiotechnology, 19*(1), 32–45. PMID: 33499885

Li, J., Chang, P. R., Huang, J., Wang, Y., Yuan, H., & Ren, H. (2013). Physiological effects of magnetic iron oxide nanoparticles towards watermelon. *Journal of Nanoscience and Nanotechnology, 13*(8), 5561–5567. DOI: 10.1166/jnn.2013.7533 PMID: 23882795

Li, J., Qian, J., Xu, Y., Yan, S., Shen, J., & Yin, M. (2019). A facile-synthesized star polycation constructed as a highly efficient gene vector in pest management. *ACS Sustainable Chemistry & Engineering, 7*(6), 6316–6322. DOI: 10.1021/acssuschemeng.9b00004

Li, L. H., Chen, Y., Behan, G., Zhang, H., Petravic, M., & Glushenkov, A. M. (2011). Large-scale mechanical peeling of boron nitride nanosheets by low-energy ball milling. *Journal of Materials Chemistry, 21*(32), 11862–11866. DOI: 10.1039/c1jm11192b

Li, L., Luo, Y., Li, R., Zhou, Q., Peijnenburg, W. J. G. M., Yin, N., Yang, J., Tu, C., & Zhang, Y. (2020). Effective uptake of submicrometre plastics by crop plants via a crack-entry mode. *Nature Sustainability, 3*(11), 929–937. DOI: 10.1038/s41893-020-0567-9

Li, M., Huang, Q., & Wu, Y. (2011). A novel chitosan-poly (lactide) copolymer and its submicron particles as imidacloprid carriers. *Pest Management Science*, *67*(7), 831–836. DOI: 10.1002/ps.2120 PMID: 21370387

Li, M., Liu, J., Deng, M., Ge, Z., Afshan, N., Zuo, X., & Li, Q. (2019). Rapid transmembrane transport of DNA nanostructures by chemically anchoring artificial receptors on cell membranes. *ChemPlusChem*, *84*(4), 323–327. DOI: 10.1002/cplu.201900025 PMID: 31939210

Lin, D., & Xing, B. (2007). Phytotoxicity of nanoparticles: Inhibition of seed germination and root growth. *Environmental Pollution*, *150*(2), 243–250. DOI: 10.1016/j.envpol.2007.01.016 PMID: 17374428

Lin, H., Ma, N., He, L., Xu, P., Wang, F., & You, C. (2024). High deposition and precise stimulus-response release performance of lignin-coated dendritic mesoporous organosilica nanoparticles for efficient pesticide utilization. *International Journal of Biological Macromolecules*, *259*, 129163. DOI: 10.1016/j.ijbiomac.2023.129163 PMID: 38181906

Lin, X., Wu, M., Wu, B., Kuga, S., Endo, T., & Huang, Y. (2011). Platinum nanoparticles using wood nanomaterials: Eco-friendly synthesis, shape control and catalytic activity for p-nitrophenol reduction. *Green Chemistry*, *13*(2), 283–287. DOI: 10.1039/C0GC00513D

Li, R., He, Y., Li, Y., Qin, W., Abbas, A., Ji, R., Li, S., Wu, Y., Sun, X., & Yang, J. (2024). Identification of cotton pest and disease based on CFNet- VoV-GCSP -LSKNet-YOLOv8s: A new era of precision agriculture. *Frontiers in Plant Science*, *15*, 1348402. Advance online publication. DOI: 10.3389/fpls.2024.1348402 PMID: 38444536

Liu, C., Zhang, L., Liu, H., & Cheng, K. (2017). Delivery strategies of the CRISPR-Cas9 gene-editing system for therapeutic applications. *Journal of controlled release: official journal of the Controlled Release Society, 266*, 17–26. https://doi.org/DOI: 10.1016/j.jconrel.2017.09.012

Liu, H., Li, Y., Zhu, M., & Chen, T. (2019). A kind of conical cup-stacked carbon nanotube. *Fullerenes, Nanotubes, and Carbon Nanostructures*, *27*(5), 417–422. DOI: 10.1080/1536383X.2019.1578754

Liu, L., Yahaya, B. S., Li, J., & Wu, F. (2024). Enigmatic role of auxin response factors in plant growth and stress tolerance. *Frontiers in Plant Science*, *15*, 1398818. DOI: 10.3389/fpls.2024.1398818 PMID: 38903418

Liu, P., Yang, J., Liu, J., Duan, Y., & Wang, P. (2025). Multipurpose *β*-Cyclodextrin-Involved nano-spherical Aggregates: Self-Assembly Process, foliar Adhesion, and promising applications for kiwifruit and rice Protections. *Chemical Engineering Journal*, *507*, 160385. DOI: 10.1016/j.cej.2025.160385

Liu, Q.. (2009). Mechanisms of nanoparticle uptake and translocation in plants. *Environmental Toxicology and Chemistry*, *28*(5), 994–1002.

Liu, Q., Chen, B., Wang, Q., Shi, X., Xiao, Z., Lin, J., & Fang, X. (2009). Carbon Nanotubes as Molecular Transporters for Walled Plant Cells. *Nano Letters*, *9*(3), 1007–1010. DOI: 10.1021/nl803083u PMID: 19191500

Liu, Q., Fang, L., Yu, G., Wang, D., Xiao, C. L., & Wang, K. (2019). Detection of DNA base modifications by deep recurrent neural network on Oxford Nanopore sequencing data. *Nature Communications*, *10*(1), 2449. DOI: 10.1038/s41467-019-10168-2 PMID: 31164644

Liu, Q., Kim, Y. J., Im, G. B., Zhu, J., Wu, Y., Liu, Y., & Bhang, S. H. (2021). Inorganic nanoparticles applied as functional therapeutics. *Advanced Functional Materials*, *31*(12), 2008171. DOI: 10.1002/adfm.202008171

Liu, Q., Li, Y., Xu, K., Li, D., Hu, H., Zhou, F., & Li, C. (2020). Clay nanosheet-mediated delivery of recombinant plasmids expressing artificial miRNAs via leaf spray to prevent infection by plant DNA viruses. *Horticulture Research*, *7*(1), 7. DOI: 10.1038/s41438-020-00400-2 PMID: 33328436

Liu, R., & Lal, R. (2015). Potentials of engineered nanoparticles as fertilizers for increasing agronomic productions. *The Science of the Total Environment*, *514*, 131–139. DOI: 10.1016/j.scitotenv.2015.01.104 PMID: 25659311

Liu, R., Lal, R., & Potapov, A. (2020). Nanomaterials for sustainable agricultural practices: An overview. *Environmental Chemistry Letters*, *18*, 523–530.

Liu, R., Zhang, H., & Lal, R. (2019). Effects of stabilized nanoparticles on soil microbial communities and plant growth. *The Science of the Total Environment*, *662*, 296–303.

Liu, S., Zhao, Y., Hao, W., Zhang, X. D., & Ming, D. (2020). Micro- and nanotechnology for neural electrode-tissue interfaces. *Biosensors & Bioelectronics*, *170*, 112645. Advance online publication. DOI: 10.1016/j.bios.2020.112645 PMID: 33010703

Liu, T., Xiao, B., Xiang, F., Tan, J., Chen, Z., Zhang, X., & Deng, J. (2020). Ultrasmall copper-based nanoparticles for reactive oxygen species scavenging and alleviation of inflammation related diseases. *Nature Communications*, *11*(1), 2788. DOI: 10.1038/s41467-020-16544-7 PMID: 32493916

Liu, W., Rudis, M. R., Cheplick, M. H., Millwood, R. J., Yang, J. P., Ondzighi-Assoume, C. A., Montgomery, G. A., Burris, K. P., Mazarei, M., Chesnut, J. D., & Stewart, C. N.Jr. (2020). Lipofection-mediated genome editing using DNA-free delivery of the Cas9/gRNA ribonucleoprotein into plant cells. *Plant Cell Reports*, *39*(2), 245–257. DOI: 10.1007/s00299-019-02488-w PMID: 31728703

Liu, X. (2005). A facile route to preparation of sea-urchinlike cadmium sulfide nanorod-based materials. *Materials Chemistry and Physics*, *91*(1), 212–216. DOI: 10.1016/j.matchemphys.2004.11.015

Liu, Y., Laks, P., & Heiden, P. (2002a). Controlled release of biocides in solid wood. I. Efficacy against brown rot wood decay fungus (*Gloeophyllum trabeum*). *Journal of Applied Polymer Science*, *86*(3), 596–607. DOI: 10.1002/app.10896

Liu, Y., Laks, P., & Heiden, P. (2002b). Controlled release of biocides in solid wood. II. Efficacy against *Trametes versicolor* and *Gloeophyllum trabeum* wood decay fungi. *Journal of Applied Polymer Science*, *86*(3), 608–614. DOI: 10.1002/app.10897

Liu, Y., Laks, P., & Heiden, P. (2002c). Controlled release of biocides in solid wood. III. Preparation and characterization of surfactant-free nanoparticles. *Journal of Applied Polymer Science*, *86*(3), 615–621. DOI: 10.1002/app.10898

Liu, Y., & Simon, J. D. (2003). *Isolation and Biophysical Studies of Natural Eumelanins: Applications of Imaging Technologies and Ultrafast Spectroscopy. Pigment Cell Research.* Wiley.

Li, X. Q., & Zhang, W. X. (2006). Iron nanoparticles: The core– shell structure and unique properties for Ni (II) sequestration. *Langmuir, 22*(10), 4638–4642. DOI: 10.1021/la060057k PMID: 16649775

Li, Y., Kim, W., Zhang, Y., Rolandi, M., Wang, D., & Dai, H. (2001). Growth of single-walled carbon nanotubes from discrete catalytic nanoparticles of various sizes. *The Journal of Physical Chemistry B, 105*(46), 11424–11431. DOI: 10.1021/jp012085b

Li, Z., Zhou, X., Cheng, Q., Zhai, W., Mao, B., Li, Y., & Chen, Z. (2023). An integrated feature selection approach to high water stress yield prediction. *Frontiers in Plant Science, 14*, 1289692. Advance online publication. DOI: 10.3389/fpls.2023.1289692 PMID: 38111876

Lombi, E., Donner, E., Taheri, S., Tavakkoli, E., Jämting, Å. K., McClure, S., Naidu, R., Miller, B. W., Scheckel, K. G., & Vasilev, K. (2013). Transformation of four silver/silver chloride nanoparticles during anaerobic treatment of wastewater and post-processing of sewage sludge. *Environmental Pollution, 176*, 193–197. DOI: 10.1016/j.envpol.2013.01.029 PMID: 23434771

Lopes, R. C., Rocha, B. G., Maçôas, E. M., Marques, E. F., & Martinho, J. M. (2022). Combining metal nanoclusters and carbon nanomaterials: Opportunities and challenges in advanced nanohybrids. *Advances in Colloid and Interface Science, 304*, 102667. DOI: 10.1016/j.cis.2022.102667 PMID: 35462268

López-Cabeza, R. (2020). Methods for understanding the fate of nanopesticides in soil and water. *Nanopesticides: From Research and Development to Mechanisms of Action and Sustainable Use in Agriculture*, 111-136.

López-Vargas, E. R., González-García, Y., Pérez-Álvarez, M., Cadenas-Pliego, G., González-Morales, S., Benavides-Mendoza, A., Cabrera, R. I., & Juárez-Maldonado, A. (2020). Seed priming with carbon nanomaterials to modify the germination, growth, and antioxidant status of tomato seedlings. *Agronomy (Basel), 10*(5), 5. Advance online publication. DOI: 10.3390/agronomy10050639

Lowder, S. K., Skoet, J., & Raney, T. (2016). The number, size, and distribution of farms, smallholder farms, and family farms worldwide. *World Development, 87*, 16–29. DOI: 10.1016/j.worlddev.2015.10.041

Lowe, K., Wu, E., Wang, N., Hoerster, G., Hastings, C., Cho, M. J., Scelonge, C., Lenderts, B., Chamberlin, M., Cushatt, J., Wang, L., Ryan, L., Khan, T., Chow-Yiu, J., Hua, W., Yu, M., Banh, J., Bao, Z., Brink, K., & Gordon-Kamm, W. (2016). Morphogenic Regulators *Baby boom* and *Wuschel* Improve Monocot Transformation. *The Plant Cell, 28*(9), 1998–2015. DOI: 10.1105/tpc.16.00124 PMID: 27600536

Lu ChangMei, L. C., Zhang ChaoYing, Z. C., Wen JunQiang, W. J., Wu GuoRong, W. G., & Tao MingXuan, T. M. (2002). Research of the effect of nanometer materials on germination and growth enhancement of Glycine max and its mechanism.

Luangpipat, T., Beattie, I. R., Chisti, Y., & Haverkamp, R. G. (2011). Gold nanoparticles produced in a microalga. *Journal of Nanoparticle Research, 13*(12), 6439–6445. DOI: 10.1007/s11051-011-0397-9

Lucas, J. A., Hawkins, N. J., & Fraaije, B. A. (2015). The evolution of fungicide resistance. *Advances in Applied Microbiology, 90*, 29–92. DOI: 10.1016/bs.aambs.2014.09.001 PMID: 25596029

Lu, H., Zhang, S., Wang, J., & Chen, Q. (2021). A Review on Polymer and Lipid-Based Nanocarriers and Its Application to Nano-Pharmaceutical and Food-Based Systems. *Frontiers in Nutrition*, *8*, 783831. DOI: 10.3389/fnut.2021.783831 PMID: 34926557

Lü, J., Liu, Z., Guo, W., Guo, M., Chen, S., Li, H., Yang, C., Zhang, Y., & Pan, H. (2019). Feeding delivery of dsHvSnf7 is a promising method for management of the pest *Henosepilachna vigintioctopunctata* (Coleoptera: Coccinellidae). *Insects*, *11*(1), 34. DOI: 10.3390/insects11010034 PMID: 31906124

Lu, L., Wang, H., Zhou, Y., Xi, S., Zhang, H., Hu, J., & Zhao, B. (2002). Seed-mediated growth of large, monodisperse core–shell gold–silver nanoparticles with Ag-like optical properties. *Chemical Communications*, (2), 144–145. DOI: 10.1039/b108473a PMID: 12120344

Luo, H. H., Zhang, Y. L., & Zhang, W. F. (2016). Effects of water stress and rewatering on photosynthesis, root activity, and yield of cotton with drip irrigation under mulch. *Photosynthetica*, *54*(1), 65–73. DOI: 10.1007/s11099-015-0165-7

Luo, K., Adra, H. J., & Kim, Y.-R. (2020). Preparation of starch-based drug delivery system through the self-assembly of short chain glucans and control of its release property. *Carbohydrate Polymers*, *243*, 116385. DOI: 10.1016/j.carbpol.2020.116385 PMID: 32532382

Luo, Y., Shen, S., Luo, J., Wang, X., & Sun, R. (2014). Green synthesis of silver nanoparticles in xylan solution via Tollens reaction and their detection for Hg2+. *Nanoscale*, *7*(2), 690–700. DOI: 10.1039/C4NR05999A PMID: 25429650

Lu, X., Li, C., & Huang, Q. (2019). Combining in vitro digestion model with cell culture model: Assessment of encapsulation and delivery of curcumin in milled starch particle stabilized Pickering emulsions. *International Journal of Biological Macromolecules*, *139*, 917–924. DOI: 10.1016/j.ijbiomac.2019.08.078 PMID: 31401275

Lv, J., Christie, P., & Zhang, S. (2019). Uptake, translocation, and transformation of metal-based nanoparticles in plants: Recent advances and methodological challenges. *Environmental Science. Nano*, *6*(1), 41–59. DOI: 10.1039/C8EN00645H

Lv, J., Christie, P., Zhang, S., & Lu, Y. (2019). Functionalized nanoparticles as advanced agrochemicals for sustainable agriculture: Current status and future trends. *Journal of Agricultural and Food Chemistry*, *67*(13), 3281–3291.

Lv, S., Wu, Y., Cai, K., He, H., Li, Y., Lan, M., Chen, X., Cheng, J., & Yin, L. (2018). High Drug Loading and Sub-Quantitative Loading Efficiency of Polymeric Micelles Driven by Donor-Receptor Coordination Interactions. *Journal of the American Chemical Society*, *140*(4), 1235–1238. DOI: 10.1021/jacs.7b12776 PMID: 29332390

Lv, Z., Jiang, R., Chen, J., & Chen, W. (2020). Nanoparticle-mediated gene transformation strategies for plant genetic engineering. *The Plant Journal*, *104*(4), 880–891. DOI: 10.1111/tpj.14973 PMID: 32860436

Ly, H. Q., Chen, Y.-J., Nguyen, V. T., & Tseng, C.-L. (2025). Optimization of the poloxamer 407-conjugated gelatin to synthesize pH-sensitive nanocarriers for controlled paclitaxel delivery. *Journal of Polymer Research*, *32*(2), 38. DOI: 10.1007/s10965-024-04248-1

Lykogianni, M., Bempelou, E., Karamaouna, F., & Aliferis, K. A. (2021). Do pesticides promote or hinder sustainability in agriculture? The challenge of sustainable use of pesticides in modern agriculture. *The Science of the Total Environment, 795*, 148625. DOI: 10.1016/j.scitotenv.2021.148625 PMID: 34247073

Ma, C., Li, G., Xu, W., Qu, H., Zhang, H., Bahojb Noruzi, E., & Li, H. (2024). Recent advances in stimulus-responsive nanocarriers for pesticide delivery. *Journal of Agricultural and Food Chemistry, 72*(16), 8906–8927. DOI: 10.1021/acs.jafc.4c00997 PMID: 38602422

Machado, T. O., Beckers, S. J., Fischer, J., Müller, B., Sayer, C., de Araújo, P. H. H., Landfester, K., & Wurm, F. R. (2020). Bio-based lignin nanocarriers loaded with fungicides as a versatile platform for drug delivery in plants. *Biomacromolecules, 21*(7), 2755–2763. DOI: 10.1021/acs.biomac.0c00487 PMID: 32543851

Machado, T. O., Beckers, S. J., Fischer, J., Sayer, C., de Araújo, P. H. H., Landfester, K., & Wurm, F. R. (2021). Cellulose nanocarriers via miniemulsion allow Pathogen-Specific agrochemical delivery. *Journal of Colloid and Interface Science, 601*, 678–688. DOI: 10.1016/j.jcis.2021.05.030 PMID: 34091315

Machado, T. O., Grabow, J., Sayer, C., de Araújo, P. H., Ehrenhard, M. L., & Wurm, F. R. (2022). Biopolymer-based nanocarriers for sustained release of agrochemicals: A review on materials and social science perspectives for a sustainable future of agri-and horticulture. *Advances in Colloid and Interface Science, 303*, 102645. DOI: 10.1016/j.cis.2022.102645 PMID: 35358807

Mad'Atari, M. F. B., & Folta, K. M. (2019). Transformation improvement with the Standardized Pressure A grobacterium Infiltration Device (SPAID). *BMC Research Notes, 12*, 1–6. PMID: 30876440

Madathil, A. N. P., Vanaja, K. A., & Jayaraj, M. K. (2007). Synthesis of ZnO nanoparticles by hydrothermal method. In *Nanophotonic materials IV* (Vol. 6639, pp. 47–55). SPIE.

Madou, M. J. (Ed.). (2011). *Fundamentals of Microfabrication and Nanotechnology, Three-Volume Set*. CRC press.

Mafuné, F., Kohno, J. Y., Takeda, Y., & Kondow, T. (2003). Formation of Stable Platinum Nanoparticles by Laser Ablation in Water. *The Journal of Physical Chemistry B, 107*(18), 4218–4223. DOI: 10.1021/jp021580k

Magar, H. S., Ghica, M. E., Abbas, M. N., & Brett, C. M. A. (2017). A novel sensitive amperometric choline biosensor based on multiwalled carbon nanotubes and gold nanoparticles. *Talanta, 167*, 462–469. DOI: 10.1016/j.talanta.2017.02.048 PMID: 28340746

Mahalakshmi, S., Hema, N., & Vijaya, P. P. (2020). In Vitro Biocompatibility and Antimicrobial activities of Zinc Oxide Nanoparticles (ZnO NPs) Prepared by Chemical and Green Synthetic Route— A Comparative Study. *BioNanoScience, 10*(1), 112–121. DOI: 10.1007/s12668-019-00698-w

Mahanta, S., Shree, J., Santra, S. C., Moulick, D., & Hossain, A. (2023). Deciphering of mycogenic nanoparticles by spectroscopic methods. In Kuddus, M., Ahmad, I. Z., & Hussain, C. M. (Eds.), *Myconanotechnology and application of nanoparticles in biology*. Academic Press., DOI: 10.1016/B978-0-443-15262-7.00003-6

Mahapatra, M., Mohapatra, S., & Jena, G. K. (2025). Nanostructured Lipid Carriers: A Revolutionary Approach in Antifungal Drug Delivery. *Biomedical Materials & Devices*. Advance online publication. DOI: 10.1007/s44174-025-00277-6

Maharramov, A. M., Hasanova, U. A., Suleymanova, I. A., Osmanova, G. E., & Hajiyeva, N. E. (2019). The engineered nanoparticles in food chain: Potential toxicity and effects. *SN Applied Sciences*, *1*(11), 1362. DOI: 10.1007/s42452-019-1412-5

Maharshi, A., Rachana, D., Pavantheja, V., Geethikalal, V., Harikrishna, K., Pranitha, C., & Dunna, M. (2025). Challenges and Future Directions of Nanotechnology in Plant Sciences. In *Plant-Based Nanoparticle Synthesis for Sustainable Agriculture* (pp. 307–320). CRC Press. DOI: 10.1201/9781003477730-22

Maher, M. F., Nasti, R. A., Vollbrecht, M., Starker, C. G., Clark, M. D., & Voytas, D. F. (2020). Plant gene editing through de novo induction of meristems. *Nature Biotechnology*, *38*(1), 84–89. DOI: 10.1038/s41587-019-0337-2 PMID: 31844292

Mahmoodi Chalbatani, G., Dana, H., Gharagouzloo, E., Grijalvo, S., Eritja, R., Logsdon, C. D., Memari, F., Miri, S. R., Rad, M. R., & Marmari, V. (2019). Small interfering RNAs (siRNAs) in cancer therapy: A nano-based approach. *International Journal of Nanomedicine*, *14*, 3111–3128. DOI: 10.2147/IJN.S200253 PMID: 31118626

Mahmoodzadeh, H., Nabavi, M., & Kashefi, H. (2013). Effect of nanoscale titanium dioxide particles on the germination and growth of canola (*Brassica napus*). *Ornamental Plants*, *3*, 25–32.

Mahmoud, K., Swidan, S., El-Nabarawi, M., & Teaima, M. (2022). Lipid based nanoparticles as a novel treatment modality for hepatocellular carcinoma: A comprehensive review on targeting and recent advances. *Journal of Nanobiotechnology*, *20*(1), 109. DOI: 10.1186/s12951-022-01309-9 PMID: 35248080

Mahmoud, L. M., Kaur, P., Stanton, D., Grosser, J. W., & Dutt, M. (2022). A cationic lipid mediated CRISPR/Cas9 technique for the production of stable genome edited citrus plants. *Plant Methods*, *18*(1), 33. DOI: 10.1186/s13007-022-00870-6 PMID: 35303912

Mahmud, K., Missaoui, A., Lee, K., Ghimire, B., Presley, H. W., & Makaju, S. (2021). Rhizosphere microbiome manipulation for sustainable crop production. *Current Plant Biology*, *27*, 100210. DOI: 10.1016/j.cpb.2021.100210

Mahra, S., Tripathi, S., Tiwari, K., Sharma, S., Mathew, S., Kumar, V., & Sharma, S. (2025). Harnessing nanotechnology for sustainable agriculture: From seed priming to encapsulation. *Plant Nano Biology*, *11*, 100124. DOI: 10.1016/j.plana.2024.100124

Maity, D., Gupta, I., & Saha, S. (2022). Biosynthesized metal oxide nanoparticles for sustainable agriculture: Next-generation nanotechnology for crop production, protection and management. *Nanoscale*, *14*(38), 38. DOI: 10.1039/D2NR03944C PMID: 36124943

Majeed, N., Panigrahi, K. C. S., Sukla, L. B., John, R., & Panigrahy, M. (2020a). Application of carbon nanomaterials in plant biotechnology. *Materials Today: Proceedings*, *30*, 340–345. DOI: 10.1016/j.matpr.2020.01.618

Maksimova, Yu. G. (2019). Microorganisms and carbon nanotubes: Interaction and applications [Review]. *Applied Biochemistry and Microbiology*, *55*(1), 1–12. DOI: 10.1134/S0003683819010101

Malgioglio, G., Rizzo, G. F., Nigro, S., du Prey, V. L., Herforth-Rahmé, J., Catara, V., & Branca, F. (2022). Plant-Microbe Interaction in Sustainable Agriculture: The Factors That May Influence the Efficacy of PGPM Application. *Sustainability (Basel)*, *14*(4), 1–28. DOI: 10.3390/su14042253

Maliga, P. (2022). Engineering the plastid and mitochondrial genomes of flowering plants. *Nature Plants*, *8*(9), 996–1006. DOI: 10.1038/s41477-022-01227-6 PMID: 36038655

Malik, A. (2004). Metal bioremediation through growing cells. *Environment International*, *30*(2), 261–278. DOI: 10.1016/j.envint.2003.08.001 PMID: 14749114

Mali, S., Sakanaka, L. S., Yamashita, F., & Grossmann, M. V. E. (2005). Water sorption and mechanical properties of cassava starch films and their relation to plasticizing effect. *Carbohydrate Polymers*, *60*(3), 283–289. DOI: 10.1016/j.carbpol.2005.01.003

Malozyomov, B., Martyushev, N., Sorokova, S., Efremenkov, E. A., & Qi, M. (2023). Mathematical Modeling of Mechanical Forces and Power Balance in Electromechanical Energy Converter. *Mathematics*, *11*(10), 2394. DOI: 10.3390/math11102394

Maluin, F. N., & Hussein, M. Z. (2020). Chitosan-based agronanochemicals as a sustainable alternative in Crop Protection. *Molecules (Basel, Switzerland)*, *25*(7), 1611. DOI: 10.3390/molecules25071611 PMID: 32244664

Maluin, F. N., Mohd, Z. H., Nor Azah, Y., Sharida, F., Idris, A. S., Nur Hailini, Z. H., & Leona, D. J. D. (2019). Enhanced fungicidal efficacy on *Ganoderma boninense* by simultaneous co-delivery of hexaconazole and dazomet from their chitosan nanoparticles. *RSC Advances*, *9*(46), 27083–27095. DOI: 10.1039/C9RA05417K PMID: 35528577

Malysheva, A., Lombi, E., & Voelcker, N. H. (2015). Bridging the divide between human and environmental nanotoxicology. *Nature Nanotechnology*, *10*(10), 835–844. DOI: 10.1038/nnano.2015.224 PMID: 26440721

Ma, M., Jiang, X., Wang, Q., Guan, D., Li, L., Ongena, M., & Li, J. (2018). Isolation and identification of PGPR strain and its effect on soybean growth and soil bacterial community composition. *International Journal of Agriculture and Biology*, *20*(6), 1289–1297. DOI: 10.17957/IJAB/15.0627

Manawi, Y. M., Ihsanullah, , Samara, A., Al-Ansari, T., & Atieh, M. A. (2018). A Review of Carbon Nanomaterials' Synthesis via the Chemical Vapor Deposition (CVD) Method. *Materials (Basel)*, *11*(5), 822. DOI: 10.3390/ma11050822 PMID: 29772760

Mandal, B. K., & Suzuki, K. T. (2002). Arsenic round the world: A review. *Talanta*, *58*(1), 201–235. DOI: 10.1016/S0039-9140(02)00268-0 PMID: 18968746

Manikandan, V., & Min, S. C. (2023). Roles of polysaccharides-based nanomaterials in food preservation and extension of shelf-life of food products: A review. *International Journal of Biological Macromolecules*, *252*, 126381. DOI: 10.1016/j.ijbiomac.2023.126381 PMID: 37595723

Mansour, R. S. H., Tekade, R. K., & Ali, A. A. (2025). PEGylated nanocarriers for solubilization. In Tekade, R. K., & Jain, N. K. (Eds.), *PEGylated Nanocarriers in Medicine and Pharmacy* (pp. 137–169). Springer Nature., DOI: 10.1007/978-981-97-7316-9_5

Mansuriya, B. D., & Altintas, Z. (2021). Carbon Dots: Classification, Properties, Synthesis, Characterization, and Applications in Health Care-An Updated Review (2018-2021). *Nanomaterials (Basel, Switzerland)*, *11*(10), 2525. DOI: 10.3390/nano11102525 PMID: 34684966

Mapossa, A., da Silva Júnior, A. H., de Oliveira, C. R. S., & Mhike, W. (2023). Thermal, Morphological and Mechanical Properties of Multifunctional Composites Based on Biodegradable Polymers/Bentonite Clay: A Review. *Polymers*, *15*(16), 15. DOI: 10.3390/polym15163443 PMID: 37631500

Marcelino, P. R. F., Milani, K. M. L., Mali, S., dos Santos, O. J. A. P., & de Oliveira, A. L. M. (2016). Formulations of polymeric biodegradable low-cost foam by melt extrusion to deliver plant growth-promoting bacteria in agricultural systems. *Applied Microbiology and Biotechnology*, *100*(16), 7323–7338. DOI: 10.1007/s00253-016-7566-9 PMID: 27147530

Marcon, L., Oliveras, J., & Puntes, V. F. (2021). In situ nanoremediation of soils and groundwaters from the nanoparticle's standpoint: A review. *The Science of the Total Environment*, *791*, 148324. DOI: 10.1016/j.scitotenv.2021.148324 PMID: 34412401

Marquiegui - Alvaro, A., Kottara, A., Chacón, M., Cliffe, L., Brockhurst, M., & Dixon, N.Marquiegui – Alvaro. (2025). Genetic Bioaugmentation-Mediated Bioremediation of Terephthalate in Soil Microcosms Using an Engineered Environmental Plasmid. *Microbial Biotechnology*, *18*(1), e70071. DOI: 10.1111/1751-7915.70071 PMID: 39801293

Martínez-Ballesta, Mc., Gil-Izquierdo, Á., García-Viguera, C., & Domínguez-Perles, R. (2018). Nanoparticles and Controlled Delivery for Bioactive Compounds: Outlining Challenges for New "Smart-Foods" for Health. *Foods 2018, Vol. 7, Page 72, 7*(5), 72. DOI: 10.3390/foods7050072

Martinez-Ballesta, M. C., Chelbi, N., Lopez-Zaplana, A., & Carvajal, M. (2020). Discerning the mechanism of the multiwalled carbon nanotubes effect on root cell water and nutrient transport. *Plant Physiology and Biochemistry*, *146*, 23–30. DOI: 10.1016/j.plaphy.2019.11.008 PMID: 31722266

Martínez-Cano, B., Mendoza-Meneses, C. J., García-Trejo, J. F., Macías-Bobadilla, G., Aguirre-Becerra, H., Soto-Zarazúa, G. M., & Feregrino-Pérez, A. A. (2022). Review and Perspectives of the Use of Alginate as a Polymer Matrix for Microorganisms Applied in Agro-Industry. *Molecules (Basel, Switzerland)*, *27*(13), 1–20. DOI: 10.3390/molecules27134248 PMID: 35807492

Martinez, L. R., Casadevall, A., & Fries, B. C. (2018). Nanotechnology in antifungal therapies. *Fungal Biology Reviews*, *32*(2), 80–90.

Martone, P. T., Estevez, J. M., Lu, F. C., Ruel, K., Denny, M. W., Somerville, C., & Ralph, J. (2009). Discovery of lignin in seaweed reveals convergent evolution of cell wall architecture. *Current Biology*, *19*(2), 169–175. DOI: 10.1016/j.cub.2008.12.031 PMID: 19167225

Mathew, J. T., Inobeme, A., Adetunji, C. O., & Oyewole, O. A. (2025). *Recent applications of chitosan-based nanoformulated metal and metal oxide nanoparticles. Chitosan-Based Nanoparticles for Biomedical Applications*. Elsevier., DOI: 10.1016/B978-0-443-13997-0.00029-1

Mathew, J. T., Inobeme, A., Azeh, Y., Musah, M., Abdulkadir, A., Shaba, E. Y., Etsuyankpa, M. B., Musa, T. S., Muhammad, A. I., Ismail, H., Kanwa, A. M., Mamman, A., & Hussaini, J. (2024). c. Carbon Capture and Storage Via Electrochemical and Bioelectrochemical Techniques: A Review. *Caliphate Journal of Science & Technology*, *2*(2), 141–158. DOI: 10.4314/cajost.v6i2.3

Mathew, J. T., Inobeme, A., Musah, M., Azeh, Y., Abdullahi, A., Shaba, E. Y., Salihu, A. M., Muhammad, E. B., Josiah, J. G., Jibrin, N. A., Ismail, H., Muhammad, A. I., Maurice, J., Mamman, A., & Ndamitso, M. M. (2024). a. A Critical Review of Green Approach on Wastewater Treatment Strategies. *Journal of Applied Science & Environmental Management*, *28*(2), 363–391. DOI: 10.4314/jasem.v28i2.9

Mathew, J. T., Musah, M., Azeh, Y., & Muhammed, M. (2023). a. Adsorptive Removal of Selected Toxic Metals from Pharmaceutical Wastewater using Fe_3O_4/ZnO Nanocomposite. *Dutse Journal of Pure and Applied Sciences*, *9*(4a), 236–248. DOI: 10.4314/dujopas.v9i4a.22

Mathew, J. T., Musah, M., Azeh, Y., & Muhammed, M. (2023). b. Kinetic Study of Heavy Metals Removal from Pharmaceutical Wastewater Using Geopolymer/Fe_3O_4 Nanocomposite. *Bima Journal of Science and Technology*, *7*(4), 152–163. DOI: 10.56892/bima.v7i4.539

Mathew, J. T., Musah, M., Azeh, Y., & Muhammed, M. (2024). b. Development of Fe3O4 Nanoparticles for the Removal of Some Toxic Metals from Pharmaceutical Wastewater. [CaJoST]. *Caliphate Journal of Science & Technology*, *6*(1), 26–34. DOI: 10.4314/cajost.v6i1.4

Mathew, S., Tiwari, D. K., & Tripathi, D. (2021). Interaction of carbon nanotubes with plant system: A review. *Carbon Letters*, *31*(2), 167–176. DOI: 10.1007/s42823-020-00195-1

Ma, W., Tang, S., Dengzeng, Z., Zhang, D., Zhang, T., & Ma, X. (2022). Root exudates contribute to belowground ecosystem hotspots: A review. *Frontiers in Microbiology*, *13*, 937940. DOI: 10.3389/fmicb.2022.937940 PMID: 36274740

Ma, X.. (2010). Uptake and transport of engineered nanoparticles in plants. *Journal of Environmental Quality*, *39*(6), 1893–1903. PMID: 21284305

Ma, X., Geisler-Lee, J., Deng, Y., & Kolmakov, A. (2010). Interactions between engineered nanoparticles (ENPs) and plants: Phytotoxicity, uptake and accumulation. *The Science of the Total Environment*, *408*(16), 3053–3061. DOI: 10.1016/j.scitotenv.2010.03.031 PMID: 20435342

Ma, X., Geisler-Lee, J., Deng, Y., & Kolmakov, A. (2015). Interactions between engineered nanoparticles and plants: Phytotoxicity and plant uptake. *Environmental Science & Technology*, *49*(5), 2757–2771. PMID: 25664567

Ma, X., Zhang, X., Liu, H., & Li, Z. (2020). Highly efficient DNA-free plant genome editing using virally delivered CRISPR-Cas9. *Nature Plants*, *6*(7), 773–779. DOI: 10.1038/s41477-020-0704-5 PMID: 32601419

Ma, Z., Zheng, Y., Chao, Z., Chen, H., Zhang, Y., Yin, M., Shen, J., & Yan, S. (2022). Visualization of the process of a nanocarrier-mediated gene delivery: Stabilization, endocytosis and endosomal escape of genes for intracellular spreading. *Journal of Nanobiotechnology*, *20*(1), 124. DOI: 10.1186/s12951-022-01336-6 PMID: 35264206

Mazaheri-Tirani, M., & Dayani, S. (2020). In vitro effect of zinc oxide nanoparticles on *Nicotiana tabacum* callus compared to ZnO micro particles and zinc sulfate ($ZnSO_4$). [PCTOC]. *Plant Cell, Tissue and Organ Culture, 140*(2), 279–289. DOI: 10.1007/s11240-019-01725-0

Mazhar, T., Shrivastava, V., & Tomar, R. S. (2017). Green synthesis of bimetallic nanoparticles and its applications: A review. *Journal of Pharmaceutical Sciences and Research, 9*(2), 102.

Mazumdar, H., & Ahmed, G. U. (2011). Phytotoxicity effect of silver nanoparticles on *Oryza sativa. International Journal of Chemtech Research, 3*(3), 1494–1500.

Mazzoni-Putman, S. M., Brumos, J., Zhao, C., Alonso, J. M., & Stepanova, A. N. (2021). Auxin Interactions with Other Hormones in Plant Development. *Cold Spring Harbor Perspectives in Biology, 13*(10), a039990. DOI: 10.1101/cshperspect.a039990 PMID: 33903155

McEwen, S. A., & Collignon, P. J. (2018). Antimicrobial resistance: A One Health perspective. *Microbiology Spectrum, 6*(2), 6.2.10. Advance online publication. DOI: 10.1128/microbiolspec.ARBA-0009-2017 PMID: 29600770

ME Trenkel, T. (2021). *Slow-and controlled-release and Stabilized Fertilizers: an option for enhancing nutrient use effiiency in agriculture*. International Fertilizer Industry Association (IFA).

Meena, J., Gupta, A., Ahuja, R., Singh, M., Bhaskar, S., & Panda, A. K. (2020). Inorganic nanoparticles for natural product delivery: A review. *Environmental Chemistry Letters, 18*(6), 2107–2118. DOI: 10.1007/s10311-020-01061-2

Mejía, S. P., Sánchez, A., Vásquez, V., & Orozco, J. (2021). Functional nanocarriers for delivering itraconazole against fungal intracellular infections. *Frontiers in Pharmacology, 12*, 685391. DOI: 10.3389/fphar.2021.685391 PMID: 34262456

Melaiye, A., Sun, Z., Hindi, K., Milsted, A., Ely, D., Reneker, D. H., Tessier, C. A., & Youngs, W. J. (2005). Silver(I)-imidazole cyclophane gem-diol complexes encapsulated by electrospun tecophilic nanofibers: Formation of nanosilver particles and antimicrobial activity. *Journal of the American Chemical Society, 127*(7), 2285–2291. DOI: 10.1021/ja040226s PMID: 15713108

Melanie, M., Miranti, M., Kasmara, H., Malini, D. M., Husodo, T., Panatarani, C., Joni, I. M., & Hermawan, W. (2022). Nanotechnology-Based Bioactive Antifeedant for Plant Protection. *Nanomaterials (Basel, Switzerland), 12*(4), 630. DOI: 10.3390/nano12040630 PMID: 35214959

Men, W., Zhu, P., Dong, S., Liu, W., Zhou, K., Bai, Y., & Zhang, S. (2020). Layer-by-layer pH-sensitive nanoparticles for drug delivery and controlled release with improved therapeutic efficacy in vivo. *Drug Delivery, 27*(1), 180–190. DOI: 10.1080/10717544.2019.1709922 PMID: 31924103

Merino, D., Tomadoni, B., Salcedo, M. F., Mansilla, A. Y., Casalongué, C. A., & Alvarez, V. A. (2020). Nanoclay as carriers of bioactive molecules applied to agriculture. *Handbook of nanomaterials and nanocomposites for energy and environmental applications*, 1-22.

Mgadi, K., Ndaba, B., Roopnarain, A., Rama, H., & Adeleke, R. (2024). Nanoparticle applications in agriculture: Overview and response of plant-associated microorganisms. *Frontiers in Microbiology, 15*, 1354440. DOI: 10.3389/fmicb.2024.1354440 PMID: 38511012

Michielse, C. B., Hooykaas, P. J., van den Hondel, C. A., & Ram, A. F. (2005). Agrobacterium-mediated transformation as a tool for functional genomics in fungi. *Current Genetics*, *48*(1), 1–17. DOI: 10.1007/s00294-005-0578-0 PMID: 15889258

Miethke, M., & Marahiel, M. A. (2007). Siderophore-Based Iron Acquisition and Pathogen Control. *Microbiology and Molecular Biology Reviews : MMBR*, *71*(3), 413–451. DOI: 10.1128/MMBR.00012-07 PMID: 17804665

Mikušová, V., & Mikuš, P. (2021). Advances in Chitosan-Based Nanoparticles for Drug Delivery. *International Journal of Molecular Sciences*, *22*(17), 9652. DOI: 10.3390/ijms22179652 PMID: 34502560

Milani, M. H., Gohari, G., Manganaris, G. A., & Fotopoulos, V. (2024). 10—Impacts of nanocomposites on the postharvest physiology and shelf life of agricultural crops. In Thamaraiselvan, C., Lau, W. J., Maldonado, A. J., & Othman, N. H. (Eds.), *Nanocomposites for Environmental, Energy, and Agricultural Applications* (pp. 315–347). Woodhead Publishing., DOI: 10.1016/B978-0-443-13935-2.00010-3

Miller, E. M., Samec, T. M., & Alexander-Bryant, A. A. (2021). Nanoparticle delivery systems to combat drug resistance in ovarian cancer. *Nanomedicine; Nanotechnology, Biology, and Medicine*, *31*, 102309. DOI: 10.1016/j.nano.2020.102309 PMID: 32992019

Minakawa, A. F. K., Faria-Tischer, P. C. S., & Mali, S. (2019). Simple ultrasound method to obtain starch micro- and nanoparticles from cassava, corn and yam starches. Food Chemistry, 283(May 2018), 11–18. DOI: 10.1016/j.foodchem.2019.01.015

Miralles, P., Church, T. L., & Harris, A. T. (2012). Toxicity, Uptake, and Translocation of Engineered Nanomaterials in Vascular plants. *Environmental Science & Technology*, *46*(17), 9224–9239. DOI: 10.1021/es202995d PMID: 22892035

Miri, A., Shahraki Vahed, H. O., & Sarani, M. (2018). Biosynthesis of silver nanoparticles and their role in photocatalytic degradation of methylene blue dye. *Research on Chemical Intermediates*, *44*(11), 6907–6915. DOI: 10.1007/s11164-018-3529-3

Mirzajani, F., Askari, H., Hamzelou, S., Farzaneh, M., & Ghassempour, A. (2013). Effect of silver nanoparticles on *Oryza sativa* L. and its rhizosphere bacteria. *Ecotoxicology and Environmental Safety*, *88*, 48–54. DOI: 10.1016/j.ecoenv.2012.10.018 PMID: 23174269

Mishra, R., Rana, S., & Gupta, S. (2022). Stimuli-responsive nanoparticles for fungal therapy. *Advanced Drug Delivery Reviews*, *182*, 114120.

Mishra, S., & Sundaram, B. (2023). Fate, transport, and toxicity of nanoparticles: An emerging pollutant on biotic factors. *Process Safety and Environmental Protection*, *174*, 595–607. DOI: 10.1016/j.psep.2023.04.037

Miteu, G. D., Emmanuel, A. A., Addeh, I., Ojeokun, O., Olayinka, T., Godwin, J. S., Adeyemo, O. I., & Benneth, E. O. (2023). Nanoscience and technology as a pivot for sustainable agriculture and its One Health approach awareness. *Science in One Health*, *2*, 100020. DOI: 10.1016/j.soh.2023.100020 PMID: 39077037

Mittal, A. K., Chisti, Y., & Banerjee, U. C. (2013). Synthesis of metallic nanoparticles using plant extracts. *Biotechnology Advances*, *31*(2), 346–356. DOI: 10.1016/j.biotechadv.2013.01.003 PMID: 23318667

Mittal, D., Kaur, G., Singh, P., Yadav, K., & Ali, S. A. (2020). Nanoparticle-based sustainable agriculture and food science: Recent advances and future outlook. *Frontiers in Nanotechnology*, *2*, 579954. DOI: 10.3389/fnano.2020.579954

Mitter, N., Worrall, E. A., Robinson, K. E., Li, P., Jain, R. G., Taochy, C., Fletcher, S. J., Carroll, B. J., Lu, G. Q., & Xu, Z. P. (2017). Clay nanosheets for topical delivery of RNAi for sustained protection against plant viruses. *Nature Plants*, *3*(2), 16207. DOI: 10.1038/nplants.2016.207 PMID: 28067898

Mitter, N., Worrall, E. A., Robinson, K. E., Xu, Z. P., & Carroll, B. J. (2017). Induction of virus resistance by exogenous application of double-stranded RNA. *Current Opinion in Virology*, *26*, 49–55. DOI: 10.1016/j.coviro.2017.07.009 PMID: 28778033

Mittler, R. (2006). Abiotic stress, the field environment and stress combination. *Trends in Plant Science*, *11*(1), 15–19. DOI: 10.1016/j.tplants.2005.11.002 PMID: 16359910

Miyamoto, N., Sakuragi, M., & Kitade, Y. (2024). Advanced Nanotechnology-Based Nucleic Acid Medicines. *Pharmaceutics*, *16*(11), 1367. DOI: 10.3390/pharmaceutics16111367 PMID: 39598491

Miyamoto, T., Tsuchiya, K., & Numata, K. (2020). Dual peptide-based gene delivery system for the efficient transfection of plant callus cells. *Biomacromolecules*, *21*(7), 2735–2744. DOI: 10.1021/acs.biomac.0c00481 PMID: 32432860

Miyamoto, T., Tsuchiya, K., Toyooka, K., Goto, Y., Tateishi, A., & Numata, K. (2022). Relaxation of the Plant Cell Wall Barrier via Zwitterionic Liquid Pretreatment for Micelle-Complex-Mediated DNA Delivery to Specific Plant Organelles. *Angewandte Chemie*, *134*(32), e202204234. DOI: 10.1002/ange.202204234 PMID: 35670289

Miyata, K., Ramaseshadri, P., Zhang, Y., Segers, G., Bolognesi, R., & Tomoyasu, Y. (2014). Establishing an in vivo assay system to identify components involved in environmental RNA interference in the western corn rootworm. *PLoS One*, *9*(7), e101661. DOI: 10.1371/journal.pone.0101661 PMID: 25003334

Mody, V. V., Cox, A., Shah, S., Singh, A., Bevins, W., & Parihar, H. (2014). Magnetic nanoparticle drug delivery systems for targeting tumor. *Applied Nanoscience*, *4*(4), 385–392. DOI: 10.1007/s13204-013-0216-y

Mohamed, G. G., Ibrahim, N. A., & Attia, H. A. (2009). Synthesis and anti-fungicidal activity of some transition metal complexes with benzimidazole dithiocarbamate ligand. *Spectrochimica Acta. Part A: Molecular and Biomolecular Spectroscopy*, *72*(3), 610–615. DOI: 10.1016/j.saa.2008.10.051 PMID: 19119060

Mohamed, Y. M. A., & Elshahawy, I. E. (2022). Antifungal activity of photo-biosynthesized silver nanoparticles (AgNPs) from organic constituents in orange peel extract against phytopathogenic Macrophomina phaseolina. *European Journal of Plant Pathology*, *162*(3), 725–738. DOI: 10.1007/s10658-021-02434-1

Mohammadinejad, R., Moosavi, M. A., Tavakol, S., Vardar, D. Ö., Hosseini, A., Rahmati, M., Dini, L., Hussain, S., Mandegary, A., & Klionsky, D. J. (2019). Necrotic, apoptotic and autophagic cell fates triggered by nanoparticles. *Autophagy*, *15*(1), 4–33. DOI: 10.1080/15548627.2018.1509171 PMID: 30160607

Mohammed, N., Nawar, S. H., Etawy, M. S., Nassar, G. E., & Hassabo, A. G. (2024). Nanotechnology and its applications in industry and product design. *Journal of Textiles. Coloration and Polymer Science*, *21*(2), 273–284.

Mohanraj, V. J., & Chen, Y. (2006). Nanoparticles – A review. *Tropical Journal of Pharmaceutical Research*, *5*(1), 561–573.

Mohanta, D., Patnaik, S., Sood, S., & Das, N. (2019). Carbon nanotubes: Evaluation of toxicity at bio interfaces. *Journal of Pharmaceutical Analysis*, *9*(5), 293–300. DOI: 10.1016/j.jpha.2019.04.003 PMID: 31929938

Mohanta, Y. K., Mishra, A. K., Nayak, D., Patra, B., Bratovcic, A., Avula, S. K., Mohanta, T. K., Murugan, K., & Saravanan, M. (2022). Exploring Dose-Dependent Cytotoxicity Profile of Gracilaria edulis-Mediated Green Synthesized Silver Nanoparticles against MDA-MB-231 Breast Carcinoma. *Oxidative Medicine and Cellular Longevity*, *2022*(1), 3863138. DOI: 10.1155/2022/3863138 PMID: 35251470

Molahalli, V., Sharma, A., Bijapur, K., Soman, G., Shetty, A., Sirichandana, B., & Hegde, G. (2024). Properties, synthesis, and characterization of cu-based nanomaterials. In *Copper-Based Nanomaterials in Organic Transformations* (pp. 1–33). American Chemical Society. DOI: 10.1021/bk-2024-1466.ch001

Molla, K. A., Sretenovic, S., Bansal, K. C., & Qi, Y. (2021). Precise plant genome editing using base editors and prime editors. *Nature Plants*, *7*(9), 1166–1187. DOI: 10.1038/s41477-021-00991-1 PMID: 34518669

Molnár, Z., Bódai, V., Szakacs, G., Erdélyi, B., Fogarassy, Z., Sáfrán, G., Varga, T., Kónya, Z., Tóth-Szeles, E., Szucs, R., & Lagzi, I. (2018). Green synthesis of gold nanoparticles by thermophilic filamentous fungi. *Scientific Reports*, *8*(1), 3943. Advance online publication. DOI: 10.1038/s41598-018-22112-3 PMID: 29500365

Mondéjar-López, M., García-Simarro, M. P., Navarro-Simarro, P., Gómez-Gómez, L., Ahrazem, O., & Niza, E. (2024). A review on the encapsulation of "eco-friendly" compounds in natural polymer-based nanoparticles as next generation nano-agrochemicals for sustainable agriculture and crop management. *International Journal of Biological Macromolecules*, *280*, 136030. DOI: 10.1016/j.ijbiomac.2024.136030 PMID: 39332563

Mondello, V., Songy, A., Battiston, E., Pinto, C., Coppin, C., Trotel-Aziz, P., Clément, C., Mugnai, L., & Fontaine, F. (2018). Grapevine trunk diseases: A review of fifteen years of trials for their control with chemicals and biocontrol agents. *Plant Disease*, *102*(7), 1189–1217. DOI: 10.1094/PDIS-08-17-1181-FE PMID: 30673583

Montazer, M., Shamei, A., & Alimohammadi, F. (2014). Synthesis of nanosilver on polyamide fabric using silver/ammonia complex. *Materials Science and Engineering C*, *38*, 170–176. DOI: 10.1016/j.msec.2014.01.044 PMID: 24656365

Mony, C., Kaur, P., Rookes, J. E., Callahan, D. L., Eswaran, S. V., Yang, W., & Manna, P. K. (2022). Nanomaterials for enhancing photosynthesis: Interaction with plant photosystems and scope of nanobionics in agriculture. *Environmental Science. Nano*, *9*(10), 3659–3683. DOI: 10.1039/D2EN00451H

Mookkan, M. (2018). Particle bombardment–mediated gene transfer and GFP transient expression in Setaria viridis. *Plant Signaling & Behavior*, *13*(4), e1441657. DOI: 10.1080/15592324.2018.1441657 PMID: 29621423

Mora-Godínez, S., Abril-Martínez, F., & Pacheco, A. (2022). Green synthesis of silver nanoparticles using microalgae acclimated to high CO2. *Materials Today: Proceedings*, *48*, 5–9. DOI: 10.1016/j.matpr.2020.04.761

Mora-Huertas, C. E., Fessi, H., & Elaissari, A. (2010). Polymer-based nanocapsules for drug delivery. *International Journal of Pharmaceutics*, *385*(1–2), 113–142. DOI: 10.1016/j.ijpharm.2009.10.018 PMID: 19825408

Morales, F., Ancín, M., Fakhet, D., González-Torralba, J., Gámez, A. L., Seminario, A., Soba, D., Ben Mariem, S., Garriga, M., & Aranjuelo, I. (2020). Photosynthetic metabolism under stressful growth conditions as a bases for crop breeding and yield improvement. *Plants*, *9*(1), 88. DOI: 10.3390/plants9010088 PMID: 31936732

Morcillo, L., & Manzanera, M. (2021). The Effects of Plant-Associated Bacterial Exopolysaccharides on Plant Abiotic Stress Tolerance. *Metabolites*, *11*(6), 337. DOI: 10.3390/metabo11060337 PMID: 34074032

Morgan, K. T., Cushman, K. E., & Sato, S. (2009). Release mechanisms for slow-and controlled-release fertilizers and strategies for their use in vegetable production. *HortTechnology*, *19*(1), 10–12. DOI: 10.21273/HORTTECH.19.1.10

Moriarty, P. (2022). *Nanotechnology: A Very Short Introduction*. Oxford University Press. DOI: 10.1093/actrade/9780198841104.001.0001

Morton, V., & Staub, T. (2008). A short history of fungicides. *APSnet Features, 308*(1).

Mosa, M. A., & Youssef, K. (2021). Topical delivery of host induced RNAi silencing by layered double hydroxide nanosheets: An efficient tool to decipher pathogenicity gene function of Fusarium crown and root rot in tomato. *Physiological and Molecular Plant Pathology*, *115*, 101684. DOI: 10.1016/j.pmpp.2021.101684

Mostafavi, E., & Zare, H. (2022). Carbon-based nanomaterials in gene therapy. *OpenNano*, *7*, 100062. DOI: 10.1016/j.onano.2022.100062

Motsaathebe, P. C., & Fayemi, O. E. (2021). Serotonin electrochemical detection in tomatoes at MWCNT-AONP nanocomposite modified electrode. *Materials Research Express*, *8*(11), 115004. DOI: 10.1088/2053-1591/ac31fd

Moulick, D., Khan, F. R., Siddiqui, N., Pradhan, J., Khamari, P., & Mahanta, S. (2024). *Applications of nanotechnology in abiotic stress management and biofortification. The Nanotechnology Driven Agriculture*. CRC Press.

Moulick, D., Mukherjee, A., Mahanta, S., Chowdhara, B., Pattnaik, B. K., Mukhopadhyay, A. K., Gujre, N., Shree, J., Santra, S. C., & Hazra, S. (2025). Nano-composites and nano-biochar: a novel solution to alleviate heavy metal stress in crops. In *Nanomaterials and nano-biochar in reducing soil stress* (pp. 183–224). Apple Academic Press.

Mourdikoudis, S., Pallares, R. M., & Thanh, N. T. (2018). Characterization techniques for nanoparticles: Comparison and complementarity upon studying nanoparticle properties. *Nanoscale*, *10*(27), 12871–12934. DOI: 10.1039/C8NR02278J PMID: 29926865

Mousavi, S. R., Jahandideh Mahjenabadi, V. A., Khoshru, B., & Rezaei, M. (2024). Spatial prediction of winter wheat yield gap: Agro-climatic model and machine learning approaches. *Frontiers in Plant Science*, *14*, 1309171. Advance online publication. DOI: 10.3389/fpls.2023.1309171 PMID: 38264030

Mouton, J. W., Theuretzbacher, U., & Craig, W. A.. (2016). Tissue concentrations: Their relevance for antimicrobial efficacy. *Clinical Microbiology and Infection*, *14*(2), 109–118.

Mout, R., Ray, M., Yesilbag Tonga, G., Lee, Y. W., Tay, T., Sasaki, K., & Rotello, V. M. (2017). Direct Cytosolic Delivery of CRISPR/Cas9-Ribonucleoprotein for Efficient Gene Editing. *ACS Nano*, *11*(3), 2452–2458. DOI: 10.1021/acsnano.6b07600 PMID: 28129503

Muda, M. S., Kamari, A., Bakar, S. A., Yusoff, S. N. M., Fatimah, I., Phillip, E., & Din, S. M. (2020). Chitosan-graphene oxide nanocomposites as water-solubilising agents for rotenone pesticide. *Journal of Molecular Liquids*, *318*, 114066. DOI: 10.1016/j.molliq.2020.114066

Mujtaba, M., Khawar, K. M., Camara, M. C., Carvalho, L. B., Fraceto, L. F., Morsi, R. E., Elsabee, M. Z., Kaya, M., Labidi, J., Ullah, H., & Wang, D. (2020). Chitosan-based delivery systems for plants: A brief overview of recent advances and future directions. *International Journal of Biological Macromolecules*, *154*, 683–697. DOI: 10.1016/j.ijbiomac.2020.03.128 PMID: 32194112

Mujtaba, M., Wang, D., Carvalho, L. B., Oliveira, J. L., Espirito Santo Pereira, A. D., Sharif, R., Jogaiah, S., Paidi, M. K., Wang, L., Ali, Q., & Fraceto, L. F. (2021). Nanocarrier-mediated delivery of miRNA, RNAi, and CRISPR-Cas for plant protection: Current trends and future directions. *ACS Agricultural Science & Technology*, *1*(5), 417–435. DOI: 10.1021/acsagscitech.1c00146

Mukherjee, S., Chowdhury, D., Kotcherlakota, R., Patra, S., Vinothkumar, B., Bhadra, M. P., Sreedhar, B., & Patra, C. R. (2014). Potential Theranostics Application of Bio-Synthesized Silver Nanoparticles (4-in-1 System). *Theranostics*, *4*(3), 316–335. DOI: 10.7150/thno.7819 PMID: 24505239

Mumtaz, I., Majeed, Z., Ajab, Z., Ahmad, B., Khurshid, K., & Mubashir, M. (2019). Optimized tuning of rosin adduct with maleic anhydride for smart applications in controlled and targeted delivery of urea for higher plant's uptake and growth efficiency. *Industrial Crops and Products*, *133*, 395–408. DOI: 10.1016/j.indcrop.2019.02.036

Munir, A., Salah, M. A., Ali, M., Ali, B., Saleem, M. H., Samarasinghe, K. G. B. A., De Silva, S. I. S., Ercisli, S., Iqbal, N., & Anas, M. (2024). Advancing Agriculture: Harnessing Smart Nanoparticles for Precision Fertilization. *BioNanoScience*, *14*(4), 3846–3863. DOI: 10.1007/s12668-024-01597-5

Munir, N., Gulzar, W., Abideen, Z., Hancock, J. T., El-Keblawy, A., & Radicetti, E. (2023). Nanotechnology improves disease resistance in plants for food security: Applications and challenges. *Biocatalysis and Agricultural Biotechnology*, *51*, 102781. DOI: 10.1016/j.bcab.2023.102781

Munster, M. (2022). Diseases and disorders. NC State Extension Publications | *Browse Popular Publications*.https://content.ces.ncsu.edu/extension-gardener-handbook/5-diseases-and-disorders

Murshed, S. M. S., & Nieto De Castro, C. A. (2014). Superior thermal features of carbon nanotubes-based nanofluids – A review. *Renewable & Sustainable Energy Reviews*, *37*, 155–167. DOI: 10.1016/j.rser.2014.05.017

Musah, M., Mohammed, M., Mathew, J. T., & Umar, M. T. (2023). Evaluation of Seed Extracts of Azadirachta indica. Juss (Meliaceae) as Pesticide for the Treatment of Stored Bean Seeds. [DUJOPAS]. *Dutse Journal of Pure and Applied Sciences*, *9*(2b), 98–108. DOI: 10.4314/dujopas.v9i2b.11

Muscolo, A., Sidari, M., Francioso, O., Tugnoli, V., & Nardi, S. (2007). The auxin-like activity of humic substances is related to membrane interactions in carrot cell cultures. *Journal of Chemical Ecology*, *33*(1), 115–129. DOI: 10.1007/s10886-006-9206-9 PMID: 17111220

Mydin, R. S. M. N., Zahidi, I. N. M., Ishak, N. N., Nik Ghazali, N. S. S., Moshawih, S., & Siddiquee, S. (2018). Potential of Calcium Carbonate Nanoparticles for Therapeutic Applications. *Malays. J. Med. Health Sci.*, *14*(Suppl. 1), 201–206.

Nadaf, S. J., Jadhav, N. R., Naikwadi, H. S., Savekar, P. L., Sapkal, I. D., Kambli, M. M., & Desai, I. A. (2022). Green synthesis of gold and silver nanoparticles: Updates on research, patents, and future prospects. *OpenNano*, *8*, 100076. DOI: 10.1016/j.onano.2022.100076

Nadagouda, M. N., & Varma, R. S. (2008). Green synthesis of silver and palladium nanoparticles at room temperature using coffee and tea extract. *Green Chemistry*, *10*(8), 859–862. DOI: 10.1039/b804703k

Nafees, M., Ali, S., Rizwan, M., Aziz, A., Adrees, M., Hussain, S. M., ... & Junaid, M. (2020). Effect of nanoparticles on plant growth and physiology and on soil microbes. *Nanomaterials and Environmental Biotechnology*, 65-85. https://doi.org/DOI: 10.1007/978-3-030-34544-0_5

Na, G., Zhang, W., Gao, H., Wang, C., Li, R., Zhao, F., Zhang, K., & Hou, C. (2021). Occurrence and antibacterial resistance of culturable antibiotic-resistant bacteria in the Fildes Peninsula, Antarctica. *Marine Pollution Bulletin*, *162*, 111829. DOI: 10.1016/j.marpolbul.2020.111829 PMID: 33243441

Nagashri, N., Archana, L., & Raghavan, R. (2024). Microbial Volatile Compounds (MVOCs) in Food Industries and Food Safety Applications. In *Innovations, Securities, and Case Studies Across Healthcare, Business, and Technology* (pp. 75–96). DOI: 10.4018/979-8-3693-1906-2.ch005

Nair, R., Tomar, M., & Thevuthasan, S. (2021). Nanotechnology in plant biotechnology and agriculture: Prospects and challenges. *Nanotechnology Reviews*, *10*(1), 259–274. DOI: 10.1515/ntrev-2020-0105

Nair, R., Varghese, S. H., Nair, B. G., Maekawa, T., Yoshida, Y., & Kumar, D. S. (2010). Nanoparticulate material delivery to plants. *Plant Science*, *179*(3), 154–163. DOI: 10.1016/j.plantsci.2010.04.012

Nakazato, I., Okuno, M., Yamamoto, H., Tamura, Y., Itoh, T., Shikanai, T., & Arimura, S. I. (2021). Targeted base editing in the plastid genome of Arabidopsis thaliana. *Nature Plants*, *7*(7), 906–913. DOI: 10.1038/s41477-021-00954-6 PMID: 34211131

Nancarrow, N., Aftab, M., Hollaway, G., Rodoni, B., & Trębicki, P. (2021). Yield Losses Caused by Barley Yellow Dwarf Virus-PAV Infection in Wheat and Barley: A Three-Year Field Study in South-Eastern Australia. *Microorganisms*, *9*(3), 645. DOI: 10.3390/microorganisms9030645 PMID: 33808907

Nandety, R. S., Kuo, Y. W., Nouri, S., & Falk, B. W. (2015). Emerging strategies for RNA interference (RNAi) applications in insects. *Bioengineered*, *6*(1), 8–19. DOI: 10.4161/21655979.2014.979701 PMID: 25424593

Naqvi, S., Maitra, A. N., Abdin, M. Z., Akmal, M., Arora, I., & Samim, M. (2012). Calcium phosphate nanoparticle mediated genetic transformation in plants. *Journal of Materials Chemistry*, *22*(8), 3500. DOI: 10.1039/c2jm11739h

Narang, C., Kant, R., & Mandal, S. (2024). Nanosensors for Soil and Nutrients Analytics. In Shahzad, R., Fiaz, S., Qayyum, A., Ul Islam, M., & Lee, I.-J. (Eds.), *Revolutionizing Agriculture: A Comprehensive Exploration of Agri-Nanotechnology* (pp. 177–200). Springer Nature Switzerland., DOI: 10.1007/978-3-031-76000-6_8

Narayanan, K. B., & Sakthivel, N. (2008). Coriander leaf mediated biosynthesis of gold nanoparticles. *Materials Letters*, *62*(30), 4588–4590. DOI: 10.1016/j.matlet.2008.08.044

Narayanan, M., Divya, S., Natarajan, D., Senthil-Nathan, S., Kandasamy, S., Chinnathambi, A., Alahmadi, T. A., & Pugazhendhi, A. (2021). Green synthesis of silver nanoparticles from aqueous extract of Ctenolepis garcini L. and assess their possible biological applications. *Process Biochemistry (Barking, London, England)*, *107*, 91–99. DOI: 10.1016/j.procbio.2021.05.008

Nardi, S., Concheri, G., Pizzeghello, D., Sturaro, A., Rella, R., & Parvoli, G. (2000). Soil organic matter mobilization by root exudates. *Chemosphere*, *41*(5), 653–658. DOI: 10.1016/S0045-6535(99)00488-9 PMID: 10834364

Nardi, S., Pizzeghello, D., Muscolo, A., & Vianello, A. (2002). Physiological effects of humic substances on higher plants. *Soil Biology & Biochemistry*, *34*(11), 1527–1536. DOI: 10.1016/S0038-0717(02)00174-8

Nardi, S., Schiavon, M., & Francioso, O. (2021). Chemical structure and biological activity of humic substances define their role as plant growth promoters. *Molecules (Basel, Switzerland)*, *26*(8), 2256. DOI: 10.3390/molecules26082256 PMID: 33924700

Naseer, A., Ali, A., Ali, S., Mahmood, A., Kusuma, H. S., Nazir, A., Yaseen, M., Khan, M. I., Ghaffar, A., Abbas, M., & Iqbal, M. (2020). Biogenic and eco-benign synthesis of platinum nanoparticles (Pt NPs) using plants aqueous extracts and biological derivatives: Environmental, biological and catalytic applications. *Journal of Materials Research and Technology*, *9*(4), 9093–9107. DOI: 10.1016/j.jmrt.2020.06.013

Nasrollahi, S., & Alizadeh, N. (2024). Synthesis of functionalized superparamagnetic iron oxide nanoparticles for ibuprofen and naproxen hydrophobic drugs delivery. *Materials Chemistry and Physics*, *326*, 129775. DOI: 10.1016/j.matchemphys.2024.129775

Nayak, S., Das, P., & Singh, M. K. (2021). Carbon dot with aggregation induced emission and pH triggered disintegration. *Colloid and Interface Science Communications*, *45*, 100537. DOI: 10.1016/j.colcom.2021.100537

Negi, P., & Anand, S. (2024). Plant disease detection, diagnosis, and management: Recent advances and future perspectives. In Pandey, K., Kushwaha, N. L., Pande, C. B., & Singh, K. G. (Eds.), *Artificial Intelligence and Smart Agriculture: Technology and Applications* (pp. 413–436). Springer Nature., DOI: 10.1007/978-981-97-0341-8_20

Neikov, O. D. (2009). Mechanical crushing and grinding. In *Handbook of non-ferrous metal powders* (pp. 45-62). DOI: 10.1016/B978-1-85617-422-0.00002-1

Nellore, J., Pauline, C., & Amarnath, K. (2013). Bacopa monnieri Phytochemicals Mediated Synthesis of Platinum Nanoparticles and Its Neurorescue Effect on 1-Methyl 4-Phenyl 1,2,3,6 Tetrahydropyridine-Induced Experimental Parkinsonism in Zebrafish. *Journal of Neurodegenerative Diseases*, *2013*, 1–8. DOI: 10.1155/2013/972391 PMID: 26317003

Neme, K., Nafady, A., Uddin, S., & Tola, Y. B. (2021). Application of nanotechnology in agriculture, postharvest loss reduction and food processing: Food security implication and challenges. *Heliyon*, *7*(12), e08539. DOI: 10.1016/j.heliyon.2021.e08539 PMID: 34934845

Nene, S., Shah, S., Rangaraj, N., Mehra, N. K., Singh, P. K., & Srivastava, S. (2021). Lipid based nanocarriers: A novel paradigm for topical antifungal therapy. *Journal of Drug Delivery Science and Technology*, *62*, 102397. DOI: 10.1016/j.jddst.2021.102397

Nett, R. S., Bender, K. S., & Peters, R. J. (2022). Production of the plant hormone gibberellin by rhizobia increases host legume nodule size. *The ISME Journal*, *16*(7), 1809–1817. DOI: 10.1038/s41396-022-01236-5 PMID: 35414717

Ng, K. K., Lovell, J. F., & Zheng, G. (2011). Lipoprotein-inspired nanoparticles for cancer theranostics. *Accounts of Chemical Research*, *44*(10), 1105–1113. DOI: 10.1021/ar200017e PMID: 21557543

Nguyen, M. H., Tran, T. N. M., & Vu, N. B. D. (2022). Antifungal activity of essential oil-encapsulated lipid nanoemulsions formulations against leaf spot disease on tomato caused by Alternaria alternata. *Archiv für Phytopathologie und Pflanzenschutz*, *55*(2), 235–257. DOI: 10.1080/03235408.2021.2015887

Nhani, G. B. B., Di Filippo, L. D., de Paula, G. A., Mantovanelli, V. R., da Fonseca, P. P., Tashiro, F. M., Monteiro, D. C., Fonseca-Santos, B., Duarte, J. L., & Chorilli, M. (2024). High-tech sustainable beauty: Exploring nanotechnology for the development of cosmetics using plant and animal by-products. *Cosmetics*, *11*(4), 112. DOI: 10.3390/cosmetics11040112

Niazian, M., & Niedbała, G. (2020). Machine learning for plant breeding and biotechnology. *Agriculture*, *10*(10), 10. Advance online publication. DOI: 10.3390/agriculture10100436

Nikolova, M. P., & Chavali, M. S. (2020). Metal Oxide Nanoparticles as Biomedical Materials. *Biomimetics 2020, Vol. 5, Page 27*, *5*(2), 27. DOI: 10.3390/biomimetics5020027

Nischwitz, V., Berthele, A., & Michalke, B. (2008). Speciation analysis of selected metals and determination of their total contents in paired serum and cerebrospinal fluid samples: An approach to investigate the permeability of the human blood-cerebrospinal fluid-barrier. *Analytica Chimica Acta, 627*(2), 258–269. DOI: 10.1016/j.aca.2008.08.018 PMID: 18809082

Nizamani, M. M., Hughes, A. C., Zhang, H. L., & Wang, Y. (2024). Revolutionizing agriculture with nanotechnology: Innovative approaches in fungal disease management and plant health monitoring. *The Science of the Total Environment, 172473*, 172473. Advance online publication. DOI: 10.1016/j.scitotenv.2024.172473 PMID: 38615773

Nnadi, E. O., Newman, A. P., Coupe, S. J., & Mbanaso, F. U. (2015). Stormwater harvesting for irrigation purposes: An investigation of chemical quality of water recycled in pervious pavement system. *Journal of Environmental Management, 147*, 246–256. DOI: 10.1016/j.jenvman.2014.08.020 PMID: 25256837

Noman, M., Ahmed, T., Wang, J., Ijaz, M., Shahid, M., Islam, M. S., Azizullah, , Manzoor, I., Li, D., & Song, F. (2023). Nano-enabled crop resilience against pathogens: Potential, mechanisms and strategies. *Crop Health, 1*(1), 15. DOI: 10.1007/s44297-023-00015-8

Nonaka, S., & Ezura, H. (2014). Plant-Agrobacterium interaction mediated by ethylene and super-Agrobacterium conferring efficient gene transfer. *Frontiers in Plant Science, 5*, 681. DOI: 10.3389/fpls.2014.00681 PMID: 25520733

Nonaka, S., Someya, T., Zhou, S., Takayama, M., Nakamura, K., & Ezura, H. (2017). An Agrobacterium tumefaciens Strain with Gamma Aminobutyric Acid Transaminase Activity Shows an Enhanced Genetic Transformation Ability in Plants. *Scientific Reports, 7*(1), 42649. DOI: 10.1038/srep42649 PMID: 28220841

Noori, A., Donnelly, T., Colbert, J., Cai, W., Newman, L. A., & White, J. C. (2020). Exposure of tomato (*Lycopersicon esculentum*) to silver nanoparticles and silver nitrate: Physiological and molecular response. *International Journal of Phytoremediation, 22*(1), 40–51. DOI: 10.1080/15226514.2019.1634000 PMID: 31282192

Novoselov, K. S., Mishchenko, A., Carvalho, A., & Castro Neto, A. H. (2016). 2D materials and van der Waals heterostructures. In Science (Vol. 353, Issue 6298). American Association for the Advancement of Science. DOI: 10.1126/science.aac9439

Nowara, D., Gay, A., Lacomme, C., Shaw, J., Ridout, C., Douchkov, D., Hensel, G., Kumlehn, J., & Schweizer, P. (2010). HIGS: Host-induced gene silencing in the obligate biotrophic fungal pathogen *Blumeria graminis.The Plant Cell, 22*(9), 3130–3141. DOI: 10.1105/tpc.110.077040 PMID: 20884801

Nowicka, B., Ciura, J., Szymańska, R., & Kruk, J. (2018). Improving photosynthesis, plant productivity and abiotic stress tolerance–current trends and future perspectives. *Journal of Plant Physiology, 231*, 415–433. DOI: 10.1016/j.jplph.2018.10.022 PMID: 30412849

Nozik, A. J. (2002). Quantum dot solar cells. *Physica E, Low-Dimensional Systems and Nanostructures, 14*(1–2), 115–120. DOI: 10.1016/S1386-9477(02)00374-0

Nugroho, H., Chew, J. X., Eswaran, S., & Tay, F. S. (2024a). Resource-optimized cnns for real-time rice disease detection with ARM cortex-M microprocessors. *Plant Methods*, *20*(1), 159. DOI: 10.1186/s13007-024-01280-6 PMID: 39415246

Numata, K. (2015). Poly (amino acid) s/polypeptides as potential functional and structural materials. *Polymer Journal*, *47*(8), 537–545. DOI: 10.1038/pj.2015.35

Numata, K., & Kaplan, D. L. (2010). Silk-based delivery systems of bioactive molecules. *Advanced Drug Delivery Reviews*, *62*(15), 1497–1508. DOI: 10.1016/j.addr.2010.03.009 PMID: 20298729

Nussinovitch, A. (2010). Polymer macro- and micro-gel beads: Fundamentals and applications. Polymer Macro- and Micro-Gel Beads: Fundamentals and Applications, 1–303. DOI: 10.1007/978-1-4419-6618-6

O'Shaughnessy, S. A., Evett, S. R., Andrade, A., Workneh, F., Price, J. A., & Rush, C. M. (2015). Site-specific variable rate irrigation as a means to enhance water use efficiency. In *2015 ASABE/IA Irrigation Symposium: Emerging Technologies for Sustainable Irrigation-A Tribute to the Career of Terry Howell, Sr. Conference Proceedings* (pp. 1-21). American Society of Agricultural and Biological Engineers. https://doi.org/DOI: 10.13031/irrig.20152140448

Oberdörster, G., Oberdörster, E., & Oberdörster, J. (2005). Nanotoxicology: An emerging discipline evolving from studies of ultrafine particles. *Environmental Health Perspectives*, *113*(7), 823–839. DOI: 10.1289/ehp.7339 PMID: 16002369

Ogidi, O. I. (2024). Recent advances in anticancer activity and Bioinformatics Approach from Potential Plants. In: Computational Approaches in Biomaterials, Bioinformatics and Biomedical Engineering Applications. *CRC Press, Taylor and Francis Group, Florida, USA*. DOI: 10.1201/9781003354437-3

Ogidi, O. I., & Emaikwu, N. G. (2024). Plant Phenolic compound isolation and its Bioinformatics Approaches of Molecular Mechanisms in Antimicrobial activities and Resistance. In: Computational Approaches in Biomaterials, Bioinformatics and Biomedical Engineering Applications. *CRC Press, Taylor and Francis Group, Florida, USA*. DOI: 10.1201/9781003354437-8

Ogidi, O. I., Izah, S. C., & Akpan, U. M. (2023). Food Quality and Agrochemical Use: Integrated Monitoring, Assessment, and Management Policies. In: One Health implications of Agrochemicals and their sustainable Alternative. *Springer Nature Singapore Pte Ltd.*

Ogidi, O. I., & Akpan, U. M. (2023). Impacts of Chemical Use in Agricultural Practices: Perspectives of Soil Microorganism and Vegetation. In *One Health implications of Agrochemicals and their sustainable Alternative*. Springer Nature Singapore Pte Ltd. DOI: 10.1007/978-981-99-3439-3_27

Ogoun, T. R., & Ogidi, O. I. (2022). Evaluation of natural and Artificial ripened Pineapple (*Ananascosmosus* L. Merr.) fruits on Biochemical parameters of wistar rats. *World Journal of Pharmaceutical and Medical Research,*8(8):41-46. www.wjpmr.com

Oh, N., & Park, J. H. (2014). Surface Chemistry of Gold Nanoparticles Mediates Their Exocytosis in Macrophages. *ACS Nano*, *8*(6), 6232–6241. DOI: 10.1021/nn501668a PMID: 24836308

Okagu, I. U., Okeke, E. S., Ezeorba, W. C. F., Ndefo, J. C., & Ezeorba, T. P. C. (2023). Overhauling the ecotoxicological impact of synthetic pesticides using plants' natural products: A focus on Zanthoxylum metabolites. *Environmental Science and Pollution Research International*, *30*(26), 67997–68021. DOI: 10.1007/s11356-023-27258-w PMID: 37148518

Oladosu, Y., Rafii, M. Y., Arolu, F., Chukwu, S. C., Salisu, M. A., Fagbohun, I. K., Muftaudeen, T. K., Swaray, S., & Haliru, B. S. (2022). Superabsorbent polymer hydrogels for sustainable agriculture: A review. *Horticulturae*, *8*(7), 605. DOI: 10.3390/horticulturae8070605

Olasehinde, T. A., Olaniran, A. O., Okoh, A. I., & Koulen, P. (2017). Therapeutic potentials of microalgae in the treatment of Alzheimer's disease. *Molecules (Basel, Switzerland)*, *22*(3), 480. Advance online publication. DOI: 10.3390/molecules22030480 PMID: 28335462

Oliveira, H. C., Gomes, B. C. R., Pelegrino, M. T., & Seabra, A. B. (2016). Nitric oxide-releasing chitosan nanoparticles alleviate the effects of salt stress in maize plants. *Nitric Oxide*, *61*, 10–19. DOI: 10.1016/j.niox.2016.09.010 PMID: 27693703

Oliveira, L. S. B. L., & Ristroph, D. K. (2024). Critical Review: Uptake and translocation of organic nanodelivery vehicles in plants. *Environmental Science & Technology*, *58*(13), 5646–5669. DOI: 10.1021/acs.est.3c09757 PMID: 38517744

Olorunsola, A., Adetunji, C. O., Olaniyan, O., Adetunji, J. B., Akinbo, K., Adetuyi, B. O., Inobeme, A., Mathew, J. T., & Ajenifujah-Solebo, S. O. (2024). Application of Nanotechnology for the Production of Biopesticides, Bioinsecticides, Bioherbicides, Mosquitoe Repellants and Biofungicides. In book: Handbook of Agricultural Biotechnology, John Wiley & Sons, 293-309. Doi: DOI: 10.1002/9781119836483.ch11

Omran, B. A., & Omran, B. A. (2020). Fundamentals of nanotechnology and nanobiotechnology. *Nanobiotechnology: A Multidisciplinary Field of Science*, 1-36.

Omran, B. A., & Baek, K. H. (2022). Control of phytopathogens using sustainable biogenic nanomaterials: Recent perspectives, ecological safety, and challenging gaps. *Journal of Cleaner Production*, *372*, 133729. DOI: 10.1016/j.jclepro.2022.133729

Osama, K., Mishra, B. N., & Somvanshi, P. (2015). Machine learning techniques in plant biology. In Barh, D., Khan, M. S., & Davies, E. (Eds.), *PlantOmics: The Omics of Plant Science* (pp. 731–754). Springer India., DOI: 10.1007/978-81-322-2172-2_26

Osman, Y. A., Eltanahy, E., Hegazy, A. S., & Razak, A. A. (2023). Nanoformulations of agrochemicals used to apply pesticides and fertilizers for crop improvement. *Nanoformulations for sustainable agriculture and environmental risk mitigation*, 103-117.

Osman, A. I., Zhang, Y., Farghali, M., Rashwan, A. K., Eltaweil, A. S., Abd El-Monaem, E. M., Mohamed, I. M. A., Badr, M. M., Ihara, I., Rooney, D. W., & Yap, P. S. (2024). Synthesis of green nanoparticles for energy, biomedical, environmental, agricultural, and food applications: A review. *Environmental Chemistry Letters*, *22*(2), 841–887. DOI: 10.1007/s10311-023-01682-3

Oturan, M. A., Aaron, J.-J., Oturan, M. A., & Aaron, J.-J. (2014). Advanced Oxidation Processes in Water/Wastewater Treatment: Principles and Applications. *Critical Reviews in Environmental Science and Technology*, *44*(23), 2577–2641. Https://Doi.Org/10.1080/10643389.2013.829765. DOI: 10.1080/10643389.2013.829765

Oviedo, L. R., Oviedo, V. R., Martins, M. O., Fagan, S. B., & da Silva, W. L. (2022). Nanoarchitectonics: The role of artificial intelligence in the design and application of nanoarchitectures. *Journal of Nanoparticle Research*, *24*(8), 157. DOI: 10.1007/s11051-022-05535-y

Oyewole, O. A., Chimbekujwo, K. I., Oniha, M., Omoregie, I. P., Ayanda, O. I., Adetunji, C. O., & Mathew, J. T. (2024). Application of Nanobiofertilization for Bioremediation and Ecorestoration of Polluted Soil/Farmland. In Book: Handbook of Agricultural Biotechnology. Wiley, 5, 1-36. Doi:DOI: 10.1002/9781394211548.ch1

Ozalp, V. C., Karabiyik, G., Bayrac, A. T., & Guvenc Tuna, B. (2019). Graphene-Based Electrochemical Aptasensors. In B. Palys (Ed.), Handbook of Graphene, Biosensors and Advanced Sensor: Vol. vol.6 (University of Warsaw, p. 465). Warsaw: Scrivener Publishing. DOI: 10.1002/9781119468455.ch101

Ozdemir, S., Turkan, Z., Kilinc, E., Altuner, E. E., & Sen, F. (2022). Anoxybacillus flavithermus loaded ɣ-Fe2O3 magnetic nanoparticles as an efficient magnetic sorbent for the preconcentrations of Cu(II) and Mn(II). *Food and Chemical Toxicology*, *168*, 113334. DOI: 10.1016/j.fct.2022.113334 PMID: 35952822

Özdemir, S., Turkan, Z., Kılınc, E., Altuner, E. E., & Sen, F. (2023). Bioaccumulation, resistance, and remediation of Mn(II) and Cu(II) and their impacts on antioxidant enzymes of Anoxybacillus flavithermus. *International Journal of Environmental Science and Technology*, *20*(10), 10823–10834. DOI: 10.1007/s13762-023-05133-y

Ozyigit, I. I. (2020). Gene transfer to plants by electroporation: Methods and applications. *Molecular Biology Reports*, *47*(4), 3195–3210. DOI: 10.1007/s11033-020-05343-4 PMID: 32242300

Ozyigit, I. I., & Yucebilgili Kurtoglu, K. (2020). Particle bombardment technology and its applications in plants. *Molecular Biology Reports*, *47*(12), 9831–9847. DOI: 10.1007/s11033-020-06001-5 PMID: 33222118

Page, M. J., McKenzie, J. E., Bossuyt, P. M., Boutron, I., Hoffmann, T. C., Mulrow, C. D., Shamseer, L., Tetzlaff, J. M., Akl, E. A., Brennan, S. E., Chou, R., Glanville, J., Grimshaw, J. M., Hróbjartsson, A., Lalu, M. M., Li, T., Loder, E. W., Mayo-Wilson, E., McDonald, S., & Moher, D. (2021). The PRISMA 2020 statement: An updated guideline for reporting systematic reviews. *Journal of Clinical Epidemiology*, *134*, 178–189. DOI: 10.1016/j.jclinepi.2021.03.001 PMID: 33789819

Palchik, O., Zhu, J., & Gedanken, A. (2000). Microwave assisted preparation of binary oxide nanoparticles. *Journal of Materials Chemistry*, *10*(5), 1251–1254. DOI: 10.1039/a908795h

Palchoudhury, S., Jungjohann, K. L., Weerasena, L., Arabshahi, A., Gharge, U., Albattah, A., Miller, J., Patel, K., & Holler, R. A. (2018). Enhanced legume root growth with pre-soaking in α-Fe_2O_3 nanoparticle fertilizer. *RSC Advances*, *8*(43), 24075–24083. DOI: 10.1039/C8RA04680H PMID: 35539206

Pal, G. D., Shaikh, M., Forsyth, C. B., Ouyang, B., Keshavarzian, A., & Shannon, K. M. (2015). Abnormal lipopolysaccharide binding protein as marker of gastrointestinal inflammation in Parkinson disease. *Frontiers in Neuroscience*, *9*(SEP), 152896. DOI: 10.3389/fnins.2015.00306 PMID: 26388718

Pal, G., Ingole, K. D., Yavvari, P. S., Verma, P., Kumari, A., Chauhan, C., Chaudhary, D., Srivastava, A., Bajaj, A., & Vemanna, R. S. (2024). Exogenous application of nanocarrier-mediated double-stranded RNA manipulates physiological traits and defence response against bacterial diseases. *Molecular Plant Pathology*, *25*(1), e13417. DOI: 10.1111/mpp.13417 PMID: 38279851

Pal, S., Sharma, M. K., Danielsson, B., Willander, M., Chatterjee, R., & Bhand, S. (2014). A miniaturized nanobiosensor for choline analysis. *Biosensors & Bioelectronics*, *54*, 558–564. DOI: 10.1016/j.bios.2013.11.057 PMID: 24325984

Pandey, K., Lahiani, M. H., Hicks, V. K., Hudson, M. K., Green, M. J., & Khodakovskaya, M. (2018). Effects of carbon-based nanomaterials on seed germination, biomass accumulation and salt stress response of bioenergy crops. *PLoS One*, *13*(8), e0202274. DOI: 10.1371/journal.pone.0202274 PMID: 30153261

Pandey, S., Goswami, G. K., & Nanda, K. K. (2013). Green synthesis of polysaccharide/gold nanoparticle nanocomposite: An efficient ammonia sensor. *Carbohydrate Polymers*, *94*(1), 229–234. DOI: 10.1016/j.carbpol.2013.01.009 PMID: 23544533

Pandey, S., & Kumari, N. (2023). Valorisation of Agricultural and Food Waste Biomass for Production of Bioenergy. In *Recent Technologies for Waste to Clean Energy and its Utilization* (pp. 197–218). Springer Nature Singapore. DOI: 10.1007/978-981-19-3784-2_10

Pandit, M. A., Bhardwaj, K., & Kaur, J. (2023). Microbial nanotechnology: a potential tool for a sustainable environment. In *Environmental applications of microbial nanotechnology* (pp. 217–230). Elsevier.

Panichikkal, J., Prathap, G., Nair, R. A., & Krishnankutty, R. E. (2021). Evaluation of plant probiotic performance of Pseudomonas sp. encapsulated in alginate supplemented with salicylic acid and zinc oxide nanoparticles. *International Journal of Biological Macromolecules*, *166*, 138–143. DOI: 10.1016/j.ijbiomac.2020.10.110 PMID: 33096173

Panichikkal, J., Puthiyattil, N., Raveendran, A., Nair, R. A., & Krishnankutty, R. E. (2021). Application of Encapsulated Bacillus licheniformis Supplemented with Chitosan Nanoparticles and Rice Starch for the Control of Sclerotium rolfsii in Capsicum annuum (L.) Seedlings. *Current Microbiology*, *78*(3), 911–919. DOI: 10.1007/s00284-021-02361-8 PMID: 33580333

Panigrahi, S., Kundu, S., Ghosh, S. K., Nath, S., & Pal, T. (2004). General method of synthesis for metal nanoparticles. *Journal of Nanoparticle Research*, *6*(4), 411–414. DOI: 10.1007/s11051-004-6575-2

Pan, L., & Cai, B. (2023). Phosphate-Solubilizing Bacteria: Advances in Their Physiology, Molecular Mechanisms and Microbial Community Effects. *Microorganisms*, *11*(12), 2904. DOI: 10.3390/microorganisms11122904 PMID: 38138048

Pantidos, N., & Horsfall, L. E. (2014). Biological synthesis of metallic nanoparticles by bacteria, fungi and plants. *Journal of Nanomedicine & Nanotechnology*, *5*(5), 1. DOI: 10.4172/2157-7439.1000233

Pan, Z., Rao, H., Mora-Seró, I., Bisquert, J., & Zhong, X. (2018). Quantum dot-sensitized solar cells. *Chemical Society Reviews*, *47*(20), 7659–7702. DOI: 10.1039/C8CS00431E PMID: 30209490

Paramo, L. A., Feregrino-Pérez, A. A., Guevara, R., Mendoza, S., & Esquivel, K. (2020). Nanoparticles in agroindustry: Applications, toxicity, challenges, and trends. *Nanomaterials (Basel, Switzerland)*, *10*(9), 1–33. DOI: 10.3390/nano10091654 PMID: 32842495

Parashar, S., Sharma, M. K., Garg, C., & Garg, M. (2021). Green Synthesized Silver Nanoparticles as Silver Lining in Antimicrobial Resistance: A Review. *Current Drug Delivery*, *19*(2), 170–181. DOI: 10.2174/1567201818666210331123022 PMID: 33797368

Paret, M. L., Vallad, G. E., Averett, D. R., Jones, J. B., & Olson, S. M. (2013). Photocatalysis: Effect of light-activated nanoscale formulations of TiO_2 on *Xanthomonas perforans* and control of bacterial spot of tomato. *Phytopathology*, *103*(3), 228–236. DOI: 10.1094/PHYTO-08-12-0183-R PMID: 23190116

Parisi, C., Vigani, M., & Rodríguez-Cerezo, E. (2015). Agricultural nanotechnologies: What are the current possibilities? *Nano Today*, *10*(2), 124–127. DOI: 10.1016/j.nantod.2014.09.009

Parkinson, S. J., Tungsirisurp, S., Joshi, C., Richmond, B. L., Gifford, M. L., Sikder, A., Lynch, I., O'Reilly, R. K., & Napier, R. M. (2022). Polymer nanoparticles pass the plant interface. *Nature Communications*, *13*(1), 7385. DOI: 10.1038/s41467-022-35066-y PMID: 36450796

Park, J., Choi, S., Park, S., Yoon, J., Park, A. Y., & Choe, S. (2019). DNA-Free Genome Editing via Ribonucleoprotein (RNP) Delivery of CRISPR/Cas in Lettuce. *Methods in Molecular Biology (Clifton, N.J.)*, *1917*, 337–354. DOI: 10.1007/978-1-4939-8991-1_25 PMID: 30610648

Park, S.-J., Kim, S., Lee, S., Khim, Z. G., Char, K., & Hyeon, T. (2000). Synthesis and magnetic studies of uniform iron nanorods and nanospheres. *Journal of the American Chemical Society*, *122*(35), 8581–8582. DOI: 10.1021/ja001628c

Park, S., Safdar, M., Kim, W., Seol, J., Kim, D., Lee, K.-H., Son, H. I., & Kim, J. (2024). Gelatin Nanoparticles can Improve Pesticide Delivery Performance to Plants. *Small*, *20*(42), 2402899. DOI: 10.1002/smll.202402899 PMID: 38949406

Park, W., Shin, H., Choi, B., Rhim, W. K., Na, K., & Han, D. K. (2020). Advanced hybrid nanomaterials for biomedical applications. *Progress in Materials Science*, *114*, 100686. DOI: 10.1016/j.pmatsci.2020.100686

Parra-Torrejón, B., Cáceres, A., Sánchez, M., Sainz, L., Guzmán, M., Bermúdez-Perez, F. J., & Delgado-López, J. M. (2023). Multifunctional nanomaterials for biofortification and protection of tomato plants. *Environmental Science & Technology*, *57*(40), 14950–14960. DOI: 10.1021/acs.est.3c02559 PMID: 37753594

Parveen, S., Misra, R., & Sahoo, S. K. (2012). Nanoparticles: A boon to drug delivery, therapeutics, diagnostics, and imaging. *Nanomedicine; Nanotechnology, Biology, and Medicine*, *8*(2), 147–166. DOI: 10.1016/j.nano.2011.05.016 PMID: 21703993

Pascoli, M., Jacques, M. T., Agarrayua, D. A., Avila, D. S., Lima, R., & Fraceto, L. F. (2019). Neem oil based nanopesticide as an environmentally-friendly formulation for applications in sustainable agriculture: An ecotoxicological perspective. *The Science of the Total Environment*, *677*, 57–67. DOI: 10.1016/j.scitotenv.2019.04.345 PMID: 31051383

Pascoli, M., Lopes-Oliveira, P. J., Fraceto, L. F., Seabra, A. B., & Oliveira, H. C. (2018). State of the art of polymeric nanoparticles as carrier systems with agricultural applications: A minireview. *Energy, Ecology & Environment*, *3*(3), 137–148. DOI: 10.1007/s40974-018-0090-2

Pashirova, T., Salah-Tazdaït, R., Tazdaït, D., & Masson, P. (2024). Applications of Microbial Organophosphate-Degrading Enzymes to Detoxification of Organophosphorous Compounds for Medical Countermeasures against Poisoning and Environmental Remediation. *International Journal of Molecular Sciences*, *25*(14), 7822. DOI: 10.3390/ijms25147822 PMID: 39063063

Pasupathy, K., Lin, S., Hu, Q., Luo, H., & Ke, P. C. (2008). Direct plant gene delivery with a poly(amidoamine) dendrimer. *Biotechnology Journal*, *3*(8), 1078–1082. DOI: 10.1002/biot.200800021 PMID: 18543240

Patel, A., Tiwari, S., Parihar, P., Singh, R., & Prasad, S. M. (2019). Carbon nanotubes as plant growth regulators: Impacts on growth, reproductive system, and soil microbial community. In Tripathi, D. K., Ahmad, P., Sharma, S., Chauhan, D. K., & Dubey, N. K. (Eds.), *Nanomaterials in Plants, Algae and Microorganisms* (pp. 23–42). Academic Press., DOI: 10.1016/B978-0-12-811488-9.00002-0

Patel, D. K., Kim, H.-B., Dutta, S. D., Ganguly, K., & Lim, K.-T. (2020). Carbon nanotubes-based nanomaterials and their agricultural and biotechnological applications. *Materials (Basel)*, *13*(7), 7. Advance online publication. DOI: 10.3390/ma13071679 PMID: 32260227

Pathak, V. M., Verma, V. K., Rawat, B. S., Kaur, B., Babu, N., Sharma, A., Dewali, S., Yadav, M., Kumari, R., Singh, S., Mohapatra, A., Pandey, V., Rana, N., & Cunill, J. M. (2022). Current status of pesticide effects on environment, human health and it's eco-friendly management as bioremediation: A comprehensive review. *Frontiers in Microbiology*, *13*, 962619. DOI: 10.3389/fmicb.2022.962619 PMID: 36060785

Pathania, D., Gupta, D., Agarwal, S., Asif, M., & Gupta, V. K. (2016). Fabrication of chitosan-g-poly(acrylamide)/CuS nanocomposite for controlled drug delivery and antibacterial activity. *Materials Science and Engineering C*, *64*, 428–435. DOI: 10.1016/j.msec.2016.03.065 PMID: 27127073

Patil, S., Das, S., & Dhavale, D. (2019). Nanoparticles for antifungal drug delivery: An emerging paradigm. *Pharmaceutics*, *11*(6), 305.

Patra, J. K., Das, G., Fraceto, L. F., Campos, E. V. R., Rodriguez-Torres, M. D. P., Acosta-Torres, L. S., Diaz-Torres, L. A., Grillo, R., Swamy, M. K., Sharma, S., Habtemariam, S., & Shin, H.-S. (2018). Nano based drug delivery systems: Recent developments and future prospects. *Journal of Nanobiotechnology*, *16*(1), 1–33. DOI: 10.1186/s12951-018-0392-8 PMID: 30231877

Pattnaik, B. K., Sahu, C., & Moulick, D. (2024). Nanoparticles in soil–water–plant inference. In Roy, S., & Hossain, A. (Eds.), *Nanotechnol driven agric*. CRC Press. DOI: 10.1201/9781003376446-5

Pattoo, T. A. (2023). *Flora to Nano: Sustainable Synthesis of Nanoparticles via Plant-Mediated Green Chemistry*. Plant Science Archives.

Pavlicevic, M., Abdelraheem, W., Zuverza-Mena, N., O'Keefe, T., Mukhtar, S., Ridge, G., Ranciato, J., Haynes, C., Elmer, W., Pignatello, J., Pagano, L., Caldara, M., Marmiroli, M., Maestri, E., Marmiroli, N., & White, J. C. (2022). Engineered Nanoparticles, Natural Nanoclay and Biochar, as Carriers of Plant-Growth Promoting Bacteria. *Nanomaterials (Basel, Switzerland)*, *12*(24), 4474. Advance online publication. DOI: 10.3390/nano12244474 PMID: 36558327

Pawlak, K., & Kołodziejczak, M. (2020). The Role of Agriculture in Ensuring Food Security in Developing Countries: Considerations in the Context of the Problem of Sustainable Food Production. *Sustainability 2020, Vol. 12, Page 5488, 12*(13), 5488. DOI: 10.3390/su12135488

Pechyen, C., Tangnorawich, B., Toommee, S., Marks, R., & Parcharoen, Y. (2024). Green synthesis of metal nanoparticles, characterization, and biosensing applications. *Sensors International*, *5*, 100287. DOI: 10.1016/j.sintl.2024.100287

Peer, D., Karp, J. M., Hong, S., Farokhzad, O. C., Margalit, R., & Langer, R. (2007). Nanocarriers as an emerging platform for cancer therapy. *Nature Nanotechnology*, *2*(12), 751–760. DOI: 10.1038/nnano.2007.387 PMID: 18654426

Pei, Z., Chen, S., Ding, L., Liu, J., Cui, X., Li, F., & Qiu, F. (2022). Current perspectives and trend of nanomedicine in cancer: A review and bibliometric analysis. *Journal of Controlled Release*, *352*, 211–241. DOI: 10.1016/j.jconrel.2022.10.023 PMID: 36270513

Peng, Fei, Xiuping Wang, Wenjing Zhang, Xuejuan Shi, Caihong Cheng, Wenlong Hou, Xiaohu Lin, Xiaolu Xiao, & Jun Li. (2022). Nanopesticide formulation from pyraclostrobin and graphene oxide as a nanocarrier and application in controlling plant fungal pathogens. *Nanomaterials 12*, 7, 1112. .DOI: 10.3390/nano12071112

Peralta-Videa, J. R., Huang, Y., Parsons, J. G., Zhao, L., Lopez-Moreno, L., Hernandez-Viezcas, J. A., & Gardea-Torresdey, J. L. (2016). Plant-based green synthesis of metallic nanoparticles: Scientific curiosity or a realistic alternative to chemical synthesis? *Nanotechnology for Environmental Engineering*, *1*(1), 1–29. DOI: 10.1007/s41204-016-0004-5

Pereira, A. D. E. S., Oliveira, H. C., & Fraceto, L. F. (2019). Polymeric nanoparticles as an alternative for application of gibberellic acid in sustainable agriculture: A field study. *Scientific Reports*, *9*(1), 7135. DOI: 10.1038/s41598-019-43494-y PMID: 31073210

Pereira, J. F., Oliveira, A. L. M., Sartori, D., Yamashita, F., & Mali, S. (2023). Perspectives on the Use of Biopolymeric Matrices as Carriers for Plant-Growth Promoting Bacteria in Agricultural Systems. *Microorganisms*, *11*(2), 1–17. DOI: 10.3390/microorganisms11020467 PMID: 36838432

Pérez-de-Luque, A. (2017). Interaction of nanomaterials with plants: What do we need for real applications in agriculture? *Frontiers in Environmental Science*, *5*(Apr), 12. DOI: 10.3389/fenvs.2017.00012

Perez-de-Luque, A., & Rubiales, D. (2009). Nanotechnology for parasitic plant control. *Pest Management Science*, *65*(5), 540–545. DOI: 10.1002/ps.1732 PMID: 19255973

Perez, J. J., Francois, N. J., Maroniche, G. A., Borrajo, M. P., Pereyra, M. A., & Creus, C. M. (2018). A novel, green, low-cost chitosan-starch hydrogel as potential delivery system for plant growth-promoting bacteria. *Carbohydrate Polymers*, *202*, 409–417. https://doi.org/https://doi.org/10.1016/j.carbpol.2018.07.084. DOI: 10.1016/j.carbpol.2018.07.084 PMID: 30287017

Perez, M. J., Falque, E., & Dominguez, H. (2016). Antimicrobial action of compounds from marine seaweed. *Marine Drugs*, *14*(3), 52. DOI: 10.3390/md14030052 PMID: 27005637

Pérez-Moreno, A., Fabián, F. L., Hermes, P. H., Edgar, V. N., Ileana, V. R., & Ali, A. (2021). Nanoscience and nanotechnology regarding food packaging and nanomaterials to extending the postharvest life and the shelf life of foods. *Food Losses, Sustainable Postharvest and Food Technologies*, 313–384. DOI: 10.1016/B978-0-12-821912-6.00001-8

Perlatti, B., de Souza Bergo, P. L., Fernandes, J. B., & Forim, M. R. (2013). Polymeric nanoparticle-based insecticides: a controlled release purpose for agrochemicals. In *Insecticides-Development of safer and more effective technologies*. IntechOpen. DOI: 10.5772/53355

Peyret, H., Shah, S. N., Meshcheriakova, Y., Saunders, K., & Lomonossoff, G. P. (2025). How do RNA viruses select which RNA to package? The plant virus experience. *Virology*, *604*, 110435. DOI: 10.1016/j.virol.2025.110435 PMID: 39893746

Pezzella, A., D'Ischia, M., Napolitano, A., Misuraca, G., & Prota, G. (1997). Iron-mediated generation of the neurotoxin 6-hydroxydopamine quinone by reaction of fatty acid hydroperoxides with dopamine: A possible contributory mechanism for neuronal degeneration in parkinson's disease. *Journal of Medicinal Chemistry*, *40*(14), 2211–2216. DOI: 10.1021/jm970099t PMID: 9216840

Pindiga, N. Y., Ya'u, K. A., Abubakar, A., & Madugu, Y. D. (2022). Green Synthesis and Characterization of Cobalt, Iron and Copper Nanoparticles Derived from the Stem-Bark Extract of Khaya Senegalensis (Mahogany) and Its Antimirobial Activity. *Online Journal of Chemistry*, *2*(1), 1–14. DOI: 10.31586/ojc.2022.158

Pinton, R., Varanini, Z., & Nannipieri, P. (2000). The rhizosphere as a site of biochemical interactions among soil components, plants, and microorganisms. In *The rhizosphere* (pp. 17–34). CRC press.

Pinto, T. V., Silva, C. A., Siquenique, S., & Learmonth, D. A. (2022). Micro-and nanocarriers for encapsulation of biological plant protection agents: A systematic literature review. *ACS Agricultural Science & Technology*, *2*(5), 838–857. DOI: 10.1021/acsagscitech.2c00113

Pizzeghello, D., Schiavon, M., Francioso, O., Dalla Vecchia, F., Ertani, A., & Nardi, S. (2020). Bioactivity of size-fractionated and unfractionated humic substances from two forest soils and comparative effects on N and S metabolism, nutrition, and root anatomy of *Allium sativum* L. *Frontiers in Plant Science*, *11*, 1203. DOI: 10.3389/fpls.2020.01203 PMID: 32922415

Polat, V. (2013). *Polikarbazol/Kitosan iletken kompozitlerinin elektrokimyasal sentezi ve karakterizasyonu*. Fen Bilimleri Enstitüsü.

Pollak, P., Fraix, V., Krack, P., Moro, E., Mendes, A., Chabardes, S., Koudsie, A., & Benabid, A. L. (2002). Treatment results: Parkinson's disease. *Movement Disorders*, *17*(S3, SUPPL. 3), S75–S83. Advance online publication. DOI: 10.1002/mds.10146 PMID: 11948759

Poreddy, S., Li, J., & Baldwin, I. T. (2017). Plant-mediated RNAi silences midgut-expressed genes in congeneric lepidopteran insects in nature. *BMC Plant Biology*, *17*(1), 199. DOI: 10.1186/s12870-017-1149-5 PMID: 29132300

Pormohammad, A., Monych, N. K., Ghosh, S., Turner, D. L., & Turner, R. J. (2021). Nanomaterials in Wound Healing and Infection Control. *Antibiotics (Basel, Switzerland)*, *10*(5), 473. DOI: 10.3390/antibiotics10050473 PMID: 33919072

Poschenrieder, C., Cabot, C., Martos, S., Gallego, B., & Barceló, J. (2013). Do toxic ions induce hormesis in plants? *Plant Science*, *212*, 15–25. DOI: 10.1016/j.plantsci.2013.07.012 PMID: 24094050

Pourhashem, S., Ghasemy, E., Rashidi, A., & Vaezi, M. R. (2020). A review on application of carbon nanostructures as nanofiller in corrosion-resistant organic coatings. *Journal of Coatings Technology and Research*, *17*(1), 19–55. DOI: 10.1007/s11998-019-00275-6

Pour, M. M., Saberi-Riseh, R., Mohammadinejad, R., & Hosseini, A. (2019). Investigating the formulation of alginate- gelatin encapsulated Pseudomonas fluorescens (VUPF5 and T17-4 strains) for controlling Fusarium solani on potato. *International Journal of Biological Macromolecules*, *133*, 603–613. DOI: 10.1016/j.ijbiomac.2019.04.071 PMID: 31004642

Prabhu, P., & Patravale, V. (2015). Polymeric nanoparticles for targeted treatment in oncology: Current insights. *International Journal of Nanomedicine*, *10*, 1001–1018. PMID: 25678788

Prabhu, S., Vinodhini, S., Elanchezhiyan, C., & Rajeswari, D. (2018). Evaluation of antidiabetic activity of biologically synthesized silver nanoparticles using Pouteria sapota in streptozotocin-induced diabetic rats. *Journal of Diabetes*, *10*(1), 28–42. DOI: 10.1111/1753-0407.12554 PMID: 28323393

Pradhan, S., Durgam, M., & Mailapalli, D. R. (2021). Urea loaded hydroxyapatite nanocarrier for efficient delivery of plant nutrients in rice. *Archives of Agronomy and Soil Science*, *67*(3), 371–382. DOI: 10.1080/03650340.2020.1732940

Prajapati, D., Jabborova, D., Saharan, B. S., Singh, N., Patani, A., Singh, S., & Joshi, C. (2024). Bionanotechnology: A Paradigm for Advancing Environmental Sustainability. *Indian Journal of Microbiology*, *2024*, 1–27. DOI: 10.1007/s12088-024-01389-1 PMID: 40371027

Prajitha, N., Athira, S. S., & Mohanan, P. V. (2019). Bio-interactions and risks of engineered nanoparticles. *Environmental Research*, *172*, 98–108. DOI: 10.1016/j.envres.2019.02.003 PMID: 30782540

Pramanik, P., Krishnan, P., Maity, A., Mridha, N., Mukherjee, A., & Rai, V. (2020). Application of nanotechnology in agriculture. *Environmental Nanotechnology*, *4*, 317–348.

Prasad, R., Bhattacharyya, A., & Nguyen, Q. D. (2017). Nanotechnology in sustainable agriculture: Recent developments, challenges, and perspectives. *Frontiers in Microbiology*, *8*, 1014. DOI: 10.3389/fmicb.2017.01014 PMID: 28676790

Prasad, T. N. V. K. V., Sudhakar, P., Sreenivasulu, Y., Latha, P., Munaswamy, V., Reddy, K. R., Sreeprasad, T. S., Sajanlal, P. R., & Pradeep, T. (2012). Effect of nanoscale zinc oxide particles on the germination, growth and yield of peanut. *Journal of Plant Nutrition*, *35*(6), 905–927. DOI: 10.1080/01904167.2012.663443

Presentato, A., Piacenza, E., Scurria, A., Albanese, L., Zabini, F., Meneguzzo, F., Nuzzo, D., Pagliaro, M., Martino, D. C., Alduina, R., & Ciriminna, R. (2020). A new water-soluble bactericidal agent for the treatment of polymicrobial infections. *Preprints, 9*(9), 2020080106. DOI: 10.20944/preprints202008.0106.v1

Proença, P. L. de F., Campos, E. V. R., Costa, T. G., de Lima, R., Preisler, A. C., de Oliveira, H. C., da Rocha, C. M., de Andrade, D. J., Goncalves, K. C., Polanczyk, R. A., & Fraceto, L. F. (2024). Curcumin and carvacrol co-loaded zein nanoparticles: Comprehensive preparation and assessment of biological activities in pest control. *Plant Nano Biology*, *8*, 100067. DOI: 10.1016/j.plana.2024.100067

Pulingam, T., Foroozandeh, P., Chuah, J. A., & Sudesh, K. (2022). Exploring Various Techniques for the Chemical and Biological Synthesis of Polymeric Nanoparticles. *Nanomaterials (Basel, Switzerland)*, *12*(3), 576. DOI: 10.3390/nano12030576 PMID: 35159921

Punniyakotti, P., Vinayagam, S., Rajamohan, R., Priya, S. D., Moovendhan, M., & Sundaram, T. (2024). Environmental Fate and Ecotoxicological Behaviour of Pesticides and Insecticides in Non-Target Environments: Nanotechnology-Based Mitigation Strategies. *Journal of Environmental Chemical Engineering*, *113349*(5), 113349. Advance online publication. DOI: 10.1016/j.jece.2024.113349

Pu, S. N., Yin, W. Y., Mao, J. F., & Liu, Q. H. (2009). Crosstalk prediction of single- and double-walled carbon-nanotube (SWCNT/ DWCNT) bundle interconnects. *IEEE Transactions on Electron Devices*, *56*(4), 560–568. DOI: 10.1109/TED.2009.2014429

Qian, L., & Yang, X. (2006). Polyamidoamine dendrimers-assisted electrodeposition of gold-platinum bimetallic nanoflowers. *The Journal of Physical Chemistry B*, *110*(33), 16672–16678. DOI: 10.1021/jp063302h PMID: 16913805

Qian, Y. T., Qin, C. D., Chen, M. M., & Lin, S. J. (2020). Nanotechnology in soil remediation— Applications vs. implications. *Ecotoxicology and Environmental Safety*, *201*, 110815. DOI: 10.1016/j.ecoenv.2020.110815 PMID: 32559688

Qiao, H., Chen, J., Dong, M., Shen, J., & Yan, S. (2024). Nanocarrier-Based Eco-Friendly RNA Pesticides for Sustainable Management of Plant Pathogens and Pests. *Nanomaterials (Basel, Switzerland)*, *14*(23), 1874. DOI: 10.3390/nano14231874 PMID: 39683262

Qiao, H., Zhao, J., Wang, X., Xiao, L., Zhu-Salzman, K., Lei, J., Xu, D., Xu, G., Tan, Y., & Hao, D. (2023). An oral dsRNA delivery system based on chitosan induces G protein-coupled receptor kinase 2 gene silencing for *Apolygus lucorum* control. *Pesticide Biochemistry and Physiology*, *194*, 105481. DOI: 10.1016/j.pestbp.2023.105481 PMID: 37532313

Qiao, L., Lan, C., Capriotti, L., Ah-Fong, A., Nino Sanchez, J., Hamby, R., Heller, J., Zhao, H., Glass, N. L., Judelson, H. S., Mezzetti, B., Niu, D., & Jin, H. (2021). Spray-induced gene silencing for disease control is dependent on the efficiency of pathogen RNA uptake. *Plant Biotechnology Journal*, *19*(9), 1756–1768. DOI: 10.1111/pbi.13589 PMID: 33774895

Qin, H., Zhang, H., Li, L. X., Zhou, X. T., Li, J. P., & Kan, C. Y. (2017). Preparation and properties of lambda-cyhalothrin/polyurethane drug-loaded nanoemulsions. *RSC Advances*, *7*(83), 52684–52693. DOI: 10.1039/C7RA10640H

Qin, M., Li, J., & Song, Y. (2022). Toward High Sensitivity: Perspective on Colorimetric Photonic Crystal Sensors. *Analytical Chemistry*, *94*(27), 9497–9507. DOI: 10.1021/acs.analchem.2c01804 PMID: 35759455

Qiu, F., Xing, S., Xue, C., Liu, J., Chen, K., Chai, T., & Gao, C. (2022). Transient expression of a TaGRF4-TaGIF1 complex stimulates wheat regeneration and improves genome editing. *Science China. Life Sciences*, *65*(4), 731–738. DOI: 10.1007/s11427-021-1949-9 PMID: 34406572

Qi, Y., Liu, H., Wang, J., & Wang, Y. (2021). Effects of different straw biochar combined with microbial inoculants on soil environment in pot experiment. *Scientific Reports*, *11*(1), 1–13. DOI: 10.1038/s41598-021-94209-1 PMID: 34282227

Qureshi, A., Singh, D. K., & Dwivedi, S. (2018). Nano-fertilizers: A novel way for enhancing nutrient use efficiency and crop productivity. *International Journal of Current Microbiology and Applied Sciences*, *7*(2), 3325–3335. DOI: 10.20546/ijcmas.2018.702.398

Raaijmakers, J. M., Paulitz, T. C., Steinberg, C., Alabouvette, C., & Moënne-Loccoz, Y. (2009). The rhizosphere: a playground and battlefield for soilborne pathogens and beneficial microorganisms. https://doi.org/DOI: 10.1007/s11104-008-9568-6

Radu, D. R., Lai, C. Y., Jeftinija, K., Rowe, E. W., Jeftinija, S., & Lin, V. S. (2004). A polyamidoamine dendrimer-capped mesoporous silica nanosphere-based gene transfection reagent. *Journal of the American Chemical Society*, *126*(41), 13216–13217. DOI: 10.1021/ja046275m PMID: 15479063

Rafiq, H., Aftab, Z. E. H., Anjum, T., Ali, B., Akram, W., Bashir, U., Mirza, F. S., Aftab, M., Ali, M. D., Li, G., & Li, G. (2022). Bio-fabrication of zinc oxide nanoparticles to rescue mung bean against cercospora leaf spot disease. *Frontiers in Plant Science*, *13*, 1052984. DOI: 10.3389/fpls.2022.1052984 PMID: 36523618

Rahimi, D., Kartoolinejad, D., Nourmohammadi, K., & Naghdi, R. (2016). Increasing drought resistance of *Alnus subcordata* C.A. Mey. Seeds using a nano priming technique with multi-walled carbon nanotubes. *Journal of Forest Science*, *62*(6), 269–278. DOI: 10.17221/15/2016-JFS

Rahman, F. U., Zhu, Q., Wu, Z., Li, X., Chen, W., Xiong, T., & Zhu, X. (2024). Current insights into the biocontrol and biotechnological approaches for postharvest disease management of *Botrytis cinerea*. *Postharvest Biology and Technology*, *216*, 113055. DOI: 10.1016/j.postharvbio.2024.113055

Rahman, M. H., Haque, K. S., & Khan, M. Z. H. (2021). A review on application of controlled released fertilizers influencing the sustainable agricultural production: A Cleaner production process. *Environmental Technology & Innovation*, *23*, 101697. DOI: 10.1016/j.eti.2021.101697

Rahmatizadeh, R., Arvin, S. M. J., Jamei, R., Mozaffari, H., & Reza Nejhad, F. (2019). Response of tomato plants to interaction effects of magnetic (Fe_3O_4) nanoparticles and cadmium stress. *Journal of Plant Interactions*, *14*(1), 474–481. DOI: 10.1080/17429145.2019.1626922

Rai, M., Ingle, A. P., Birla, S., Yadav, A., & Santos, C. A. D. (2016). Strategic role of selected noble metal nanoparticles in medicine. *Critical Reviews in Microbiology*, *42*(5), 696–719. PMID: 26089024

Rai, M., Ingle, A. P., Pandit, R., Paralikar, P., Anasane, N., Santos, C. A., & Duran, N. (2018). Nanotechnology for agriculture: Impact and applications. *The Science of the Total Environment*, *636*, 741–755.

Rai, M., & Kratosova, G. (2015). Management of phytopathogens by application of green nanobiotechnology: Emerging trends and challenges. *Acta Agraria Debreceniensis*, (66), 15–22. DOI: 10.34101/actaagrar/66/1884

Rai, M., Yadav, A., & Gade, A. (2009). Silver nanoparticles as a new generation of antimicrobials. *Biotechnology Advances*, *27*(1), 76–83. DOI: 10.1016/j.biotechadv.2008.09.002 PMID: 18854209

Rai, P. K., Kumar, V., Lee, S., Raza, N., Kim, K. H., Ok, Y. S., & Tsang, D. C. (2018). Nanoparticle-plant interaction: Implications in energy, environment, and agriculture. *Environment International*, *119*, 1–19. DOI: 10.1016/j.envint.2018.06.012 PMID: 29909166

Rai, V., Acharya, S., & Dey, N. (2012). Implications of nanobiosensors in agriculture. *Journal of Biomaterials and Nanobiotechnology*, *03*(2), 315–324. DOI: 10.4236/jbnb.2012.322039

Rajam, B. M., Ramasamy, P., & Mahalingam, U. (2019). Antimicrobial activity of green synthesized plasmonic nanoparticles. *Materials for Biomedical Engineering: Bioactive Materials for Antimicrobial, Anticancer, and Gene Therapy*, 117–151. DOI: 10.1016/B978-0-12-818435-6.00005-0

Rajasekar, S., Akhil, V., & Pattanayak, S. P. (2018). Eugenol-loaded nanoparticles for antifungal activity. *Journal of Biomedical Materials Research*, *106*(12), 3046–3054.

Rajesaheb, K. S., Subramanian, S., Boominathan, P., Thenmozhi, S., & Gnanachitra, M. (2024). Biostimulant in improving crop yield and soil health. *Communications in Soil Science and Plant Analysis*, •••, 1–36.

Raj, M., Moon, J. M., Goyal, R. N., Park, D. S., & Shim, Y. B. (2019). Simultaneous detection of ATP metabolites in human plasma and urine based on palladium nanoparticle and poly(bromocresol green) composite sensor. *Biosensors & Bioelectronics*, *126*, 758–766. DOI: 10.1016/j.bios.2018.11.056 PMID: 30554097

Rajput, P., Singh, A., Agrawal, S., Ghazaryan, K., Rajput, V. D., Movsesyan, H., Mandzhieva, S., Minkina, T., & Alexiou, A. (2024). Effects of environmental metal and metalloid pollutants on plants and human health: Exploring nano-remediation approach. *Stress Biology*, *4*(1), 27. DOI: 10.1007/s44154-024-00156-y PMID: 38777953

Rajput, V. D., Minkina, T., Sushkova, S., & Mandzhieva, S. (2022). Nanotechnology in agriculture: Opportunities, toxicological risks, and ethical concerns. *Environmental Geochemistry and Health*, *44*, 247–262. DOI: 10.1007/s10653-021-01007-5

Rajput, V. D., Minkina, T., Sushkova, S., Tsitsuashvili, V., Mandzhieva, S., Gorovtsov, A., Nevidomskyaya, D., & Gromakova, N. (2018). Effect of nanoparticles on crops and soil microbial communities. *Journal of Soils and Sediments*, *18*(6), 2179–2187. DOI: 10.1007/s11368-017-1793-2

Rajput, V. D., Singh, A., Minkina, T. M., Verma, K. K., & Singh, A. K. (Eds.). (2023). *Nanotechnology for Sustainable Agriculture: An Innovative and Eco-Friendly Approach* (1st ed.). Apple Academic Press., DOI: 10.1201/9781003333128

Rajput, V., Minkina, T., Mazarji, M., Shende, S., Sushkova, S., Mandzhieva, S., Burachevskaya, M., Chaplygin, V., Singh, A., & Jatav, H. (2020). Accumulation of nanoparticles in the soil-plant systems and their effects on human health. *Annals of Agricultural Science*, *65*(2), 137–143. DOI: 10.1016/j.aoas.2020.08.001

Raliya, R., Franke, C., Chavalmane, S., Nair, R., Reed, N., & Biswas, P. (2016). Quantitative Understanding of Nanoparticle Uptake in Watermelon Plants. *Frontiers in Plant Science*, *7*, 1288. DOI: 10.3389/fpls.2016.01288 PMID: 27617020

Raliya, R., Nair, R., Chavalmane, S., Wang, W. N., & Biswas, P. (2016). Mechanistic evaluation of translocation and physiological impact of titanium dioxide and zinc oxide nanoparticles on wheat. *Scientific Reports*, *6*, 22936.

Raliya, R., Nair, R., Chavalmane, S., Wang, W. N., & Biswas, P. (2017). Mechanistic evaluation of translocation and physiological impact of titanium dioxide and zinc oxide nanoparticles on the tomato (*Solanum lycopersicum L.*) plant. *Metallomics*, *9*(7), 964–978. PMID: 26463441

Raliya, R., Saharan, V., Dimkpa, C., & Biswas, P. (2018). Nanofertilizer for Precision and Sustainable Agriculture: Current State and Future Perspectives. *Journal of Agricultural and Food Chemistry*, *66*(26), 6487–6503. DOI: 10.1021/acs.jafc.7b02178 PMID: 28835103

Raliya, R., & Tarafdar, J. C. (2013). ZnO nanoparticle biosynthesis and its effect on phosphorous-mobilizing enzyme secretion and gum contents in Clusterbean (*Cyamopsis tetragonoloba* L.). *Agricultural Research*, *2*(1), 48–57. DOI: 10.1007/s40003-012-0049-z

Rameez, M., Khan, N., Ahmad, S., & Ahmad, M. M. (2024). Bionanocomposites: A new approach for fungal disease management. *Biocatalysis and Agricultural Biotechnology*, *103115*, 103115. Advance online publication. DOI: 10.1016/j.bcab.2024.103115

Ramesh, M., Palanisamy, K., Babu, K., & Sharma, N. K. (2014). Effects of bulk & nano-titanium dioxide and zinc oxide on physio-morphological changes in *Triticum aestivum* Linn. *Jouranl of Global Biosciences*, *3*(2), 415–422.

Rana, R. A., Siddiqui, M. N., Skalicky, M., Brestic, M., Hossain, A., Kayesh, E., Popov, M., Hejnak, V., Gupta, D., Mahmud, N., & Islam, T. (2021). Prospects of nanotechnology in improving the productivity and quality of horticultural crops. *Horticulturae*, *7*(10), 332. DOI: 10.3390/horticulturae7100332

Rani, N., Kumari, K., & Hooda, V. (2025). The role of nanoparticles in transforming plant genetic engineering: Advancements, challenges and future prospects. *Functional & Integrative Genomics*, *25*(1), 1–22. DOI: 10.1007/s10142-025-01528-x PMID: 39841261

Rani, N., Kumari, K., Sangwan, P., Barala, P., Yadav, J., Vijeta, , Rahul, , & Hooda, V. (2022). Nano-iron and nano-zinc induced growth and metabolic changes in Vigna radiata. *Sustainability (Basel)*, *14*(14), 8251. DOI: 10.3390/su14148251

Rani, S., Kumari, N., & Sharma, V. (2022). Uptake, translocation, transformation and physiological effects of nanoparticles in plants. *Archives of Agronomy and Soil Science*, •••, 1–21.

Rao, B. S., Kumar, B. R., Reddy, V. R., Rao, T. S., & Chalapathi, G. J. C. L. (2011). Preparation and characterization of CdS nanoparticles by chemical co-precipitation technique. *Chalcogenide Letters, 8*(3), 177–185.

Rao, N. N. (2019). *Floral Dip Transformation of Camelina Sativa L. With a Myo-Inositol Oxygenase Gene to Potentially Improve Oil Yield*. Arkansas State University.

Rashtbari, S., Dehghan, G., Khataee, S., Amini, M., & Khataee, A. (2021). Dual enzymes-mimic activity of nanolayered manganese-calcium oxide for fluorometric determination of metformin. *Chemosphere, 133063*. Advance online publication. DOI: 10.1016/j.chemosphere.2021.133063 PMID: 34848228

Raskar, S. V., & Laware, S. L. (2014). Effect of zinc oxide nanoparticles on cytology and seed germination in onion. *International Journal of Current Microbiology and Applied Sciences, 3*(2), 467–473.

Rastogi, A., Tripathi, D. K., Yadav, S., Chauhan, D. K., Živčák, M., Ghorbanpour, M., ... & Brestic, M. (2019). Application of silicon nanoparticles in agriculture. *3 Biotech, 9*, 1-11.

Rastogi, A., Tripathi, D. K., Yadav, S., Chauhan, D. K., Zivcak, M., Ghorbanpour, M., et al. (2019). Application of silicon nanoparticles in agriculture. *3 Biotech, 9*, 1-11.

Rastogi, A., Tripathi, D. K., Yadav, S., Chauhan, D. K., Živčák, M., Ghorbanpour, M., et al., (2019). Application of silicon nanoparticles in agriculture. 3 Biotech 9:90. DOI: 10.1007/s13205-019-1626-7

Rastogi, A., Zivcak, M., Sytar, O., Kalaji, H. M., He, X., Mbarki, S., & Brestic, M. (2017). Impact of metal and metal oxide nanoparticles on plant: A critical review. *Frontiers in Chemistry, 5*, 78. DOI: 10.3389/fchem.2017.00078 PMID: 29075626

Ratcliff, F., Martin-Hernandez, A. M., & Baulcombe, D. C. (2001). Technical Advance. Tobacco rattle virus as a vector for analysis of gene function by silencing. *The Plant journal: for cell and molecular biology, 25*(2), 237–245. https://doi.org/DOI: 10.1046/j.0960-7412.2000.00942.x

Raula, J., Kuivanen, A., Lähde, A., Jiang, H., Antopolsky, M., Kansikas, J., & Kauppinen, E. I. (2007). Synthesis of L-leucine nanoparticles via physical vapor deposition at varying saturation conditions. *Journal of Aerosol Science, 38*(12), 1172–1184. DOI: 10.1016/j.jaerosci.2007.08.009

Raveendran, P., Fu, J., & Wallen, S. L. (2003). Completely "Green" Synthesis and Stabilization of Metal Nanoparticles. *Journal of the American Chemical Society, 125*(46), 13940–13941. DOI: 10.1021/ja029267j PMID: 14611213

Ray, M. K., Mishra, A. K., Mohanta, Y. K., Mahanta, S., Chakrabartty, I., Kungwani, N. A., Avula, S. K., Panda, J., & Pudake, R. N. (2023). Nanotechnology as a promising tool against phytopathogens: A futuristic approach to agriculture. *Agriculture, 13*(9), 1856. DOI: 10.3390/agriculture13091856

Ray, P., Sahu, D., Aminedi, R., & Chandran, D. (2022). Concepts and considerations for enhancing RNAi efficiency in phytopathogenic fungi for RNAi-based crop protection using nanocarrier-mediated dsRNA delivery systems. *Frontiers in Fungal Biology, 3*, 977502. DOI: 10.3389/ffunb.2022.977502 PMID: 37746174

Raza, A., Charagh, S., Salehi, H., Abbas, S., Saeed, F., Poinern, G. E., Siddique, K. H. M., & Varshney, R. K. (2023). Nano-enabled stress-smart agriculture: Can nanotechnology deliver drought and salinity-smart crops? *Journal of Sustainable Agriculture and Environment*, *2*(3), 189–214. DOI: 10.1002/sae2.12061

Razzaghi, M., Karimi, A., Aghdasinia, H., & Joghataei, M. T. (2017). Oxidase-Peroxidase sequential polymerization for removal of a dye from contaminated water by horseradish peroxidase (HRP)/glucose oxidase (GOx)/polyurethane hybrid catalyst. *Korean Journal of Chemical Engineering*, *34*(11), 2870–2878. DOI: 10.1007/s11814-017-0183-1

Reddy, K. R. (2017). Green synthesis, morphological and optical studies of CuO nanoparticles. *Journal of Molecular Structure*, *1150*, 553–557. DOI: 10.1016/j.molstruc.2017.09.005

Rehan, M., Al-Turki, A. A., Abdelmageed, A. H., Abdelhameid, N. M., & Omar, A. F. (2023). Performance of Plant-Growth-Promoting Rhizobacteria (PGPR) Isolated from Sandy Soil on Growth of Tomato (*Solanum lycopersicum* L.). *Plants*, *12*(8), 1588. DOI: 10.3390/plants12081588 PMID: 37111812

Rehman, A., Qunyi, T., Usman, M., Manzoor, M. F., Zhao, L., & Feng, J.. (2022). Nano-enabled agrochemicals for sustainable agriculture. In *Nano-enabled agrochemicals in agriculture* (pp. 291–306). Academic Press. DOI: 10.1016/B978-0-323-91009-5.00020-3

Ren, L., Deng, S., Chu, Y., Zhang, Y., Zhao, H., Chen, H., & Zhang, D. (2020). Single-wall carbon nanotubes improve cell survival rate and reduce oxidative injury in cryopreservation of *Agapanthus praecox* embryogenic callus. *Plant Methods*, *16*(1), 130. DOI: 10.1186/s13007-020-00674-6 PMID: 32973916

Rico, C. M.. (2011). Effect of nanoparticles on plant growth and development. *Journal of Nanoparticle Research*, *13*(11), 4001–4014.

Rico, C. M., Majumdar, S., Duarte-Gardea, M., Peralta-Videa, J. R., & Gardea-Torresdey, J. L. (2011). Interaction of nanoparticles with edible plants and their possible implications in the food chain. *Journal of Agricultural and Food Chemistry*, *59*(8), 3485–3498. DOI: 10.1021/jf104517j PMID: 21405020

Rico, C. M., Ofoegbu, P. C., Kirwa, N. J., Wagner, D. C., Abolade, O. M., & Jia, S.. (2022). Changes in metabolite profile of plants exposed to engineered nanomaterials. In *Plant Exposure to Engineered Nanoparticles* (pp. 119–149). Academic Press. DOI: 10.1016/B978-0-323-85032-2.00003-8

Rioux, L. E., Turgeon, S. L., & Beaulieu, M. (2007). Characterization of polysaccharides extracted from brown seaweeds. *Carbohydrate Polymers*, *69*(3), 530–537. DOI: 10.1016/j.carbpol.2007.01.009

Riseh, R. S., Ebrahimi-Zarandi, M., Gholizadeh Vazvani, M., & Skorik, Y. A. (2021). Reducing drought stress in plants by encapsulating plant growth-promoting bacteria with polysaccharides. *International Journal of Molecular Sciences*, *22*(23), 12979. Advance online publication. DOI: 10.3390/ijms222312979 PMID: 34884785

Riseh, R. S., Tamanadar, E., Hajabdollahi, N., Vatankhah, M., Thakur, V. K., & Skorik, Y. A. (2022). Chitosan microencapsulation of rhizobacteria for biological control of plant pests and diseases: Recent advances and applications. *Rhizosphere*, *23*, 100565. DOI: 10.1016/j.rhisph.2022.100565

Ritchie, J. D., & Perdue, E. M. (2008). Analytical constraints on acidic functional groups in humic substances. *Organic Geochemistry*, *39*(6), 783–799. DOI: 10.1016/j.orggeochem.2008.03.003

Rizwan, M., Ali, S., Qayyum, M. F., Ok, Y. S., Adrees, M., Ibrahim, M., & Abbas, F. (2017). Effect of metal and metal oxide nanoparticles on growth and physiology of globally important food crops: A critical review. *Journal of Hazardous Materials, 322*, 2–16. DOI: 10.1016/j.jhazmat.2016.05.061 PMID: 27267650

Rocha, V., Lago, A., Silva, B., Barros, Ó., Neves, I. C., & Tavares, T. (2024). Immobilization of biogenic metal nanoparticles on sustainable materials – green approach applied to wastewater treatment: A systematic review. *Environmental Science. Nano, 11*(1), 36–60. DOI: 10.1039/D3EN00623A

Rodríguez-González, V., Terashima, C., & Fujishima, A. (2019). Applications of photocatalytic titanium dioxide-based nanomaterials in sustainable agriculture. *Journal of Photochemistry and Photobiology C, Photochemistry Reviews, 40*, 49–67. DOI: 10.1016/j.jphotochemrev.2019.06.001

Rodríguez-Rodríguez, R., Espinosa-Andrews, H., Velasquillo-Martínez, C., & García-Carvajal, Z. Y. (2020). Composite hydrogels based on gelatin, chitosan and polyvinyl alcohol to biomedical applications: A review. *International Journal of Polymeric Materials, 69*(1), 1–20. DOI: 10.1080/00914037.2019.1581780

Rogozea, E. A., Petcu, A. R., Olteanu, N. L., Lazar, C. A., Cadar, D., & Mihaly, M. (2017). Tandem adsorption-photodegradation activity induced by light on NiO-ZnO p–n couple modified silica nanomaterials. *Materials Science in Semiconductor Processing, 57*, 1–11. DOI: 10.1016/j.mssp.2016.10.006

Rohman, S., Kaewtatip, K., Kantachote, D., & Tantirungkij, M. (2021). Encapsulation of Rhodopseudomonas palustris KTSSR54 using beads from alginate/starch blends. *Journal of Applied Polymer Science, 138*(12), 1–9. DOI: 10.1002/app.50084

Romero, G., & Moya, S. E. (2012). Synthesis of organic nanoparticles. In *Frontiers of nanoscience* (Vol. 4, pp. 115-141). DOI: 10.1016/B978-0-12-415769-9.00004-2

Roopan, S. M., Bharathi, A., Kumar, R., Khanna, V. G., & Prabhakarn, A. (2011). Acaricidal, insecticidal, and larvicidal efficacy of aqueous extract of *Annona squamosa* L peel as biomaterial for the reduction of palladium salts into nanoparticles. *Colloids and Surfaces. B, Biointerfaces, 92*, 209–212. DOI: 10.1016/j.colsurfb.2011.11.044 PMID: 22205064

Rosegrant, M. W., Ringler, C., Benson, T., Diao, X., Resnick, D., Thurlow, J., Torero, M., & Orden, D. (2005). *Agriculture and achieving the Millennium Development Goals*. World Bank. Retrieved from https://hdl.handle.net/10568/158139

Roshan, A. B., Venkatesh, H. N., Dubery, N. K., & Mohana, D. C. (2022). Chitosan-based nanoencapsulation of *Toddalia asiatica* (L.) Lam essential oil to enhance antifungal and aflatoxin B-1 inhibitory activities for safe storage of maize. *International Journal of Biological Macromolecules, 204*, 476–484. DOI: 10.1016/j.ijbiomac.2022.02.026 PMID: 35151710

Rosova, E., Smirnova, N., Dresvyanina, E., Smirnova, V., Vlasova, E., Ivan'kova, E., Sokolova, M., Maslennikova, T., Malafeev, K., Kolbe, K., Kanerva, M., & Yudin, V. (2021). Biocomposite materials based on chitosan and lignin: Preparation and characterization. *Cosmetics, 8*(1), 24. DOI: 10.3390/cosmetics8010024

Rossi, L., Sharifan, H., Zhang, W., Schwab, A. P., & Ma, X. (2018). Mutual effects and in planta accumulation of co-existing cerium oxide nanoparticles and cadmium in hydroponically grown soybean (Glycine max (L.) Merr.). *Environmental Science. Nano*, *5*(1), 150–157. DOI: 10.1039/C7EN00931C

Roy, A., Sharma, A., Yadav, S., Jule, L. T., & Krishnaraj, R. (2021). Nanomaterials for Remediation of Environmental Pollutants. *Bioinorganic Chemistry and Applications*, *1764647*, 1–16. Advance online publication. DOI: 10.1155/2021/1764647 PMID: 34992641

Ruano, M., Mateos-Maroto, A., Ortega, F., Ritacco, H., Rubio, J. E. F., Guzman, E., & Rubio, R. G. (2021). Fabrication of robust capsules by sequential assembly of polyelectrolytes onto charged liposomes. *Langmuir*, *37*(20), 6189–6200. DOI: 10.1021/acs.langmuir.1c00341 PMID: 33945690

Rudakiya, D., Patel, Y., Chhaya, U., & Gupte, A. (2019). Carbon nanotubes in agriculture: production, potential, and prospects. *Nanotechnology for Agriculture: Advances for Sustainable Agriculture*, 121-130.

Rui, M., Ma, C., Hao, Y., Guo, J., Rui, Y., Tang, X., Zhao, Q., Fan, X., Zhang, Z., Hou, T., & Zhu, S. (2016). Iron oxide nanoparticles as a potential iron fertilizer for peanut (*Arachis hypogaea*). *Frontiers in Plant Science*, *7*, 815. DOI: 10.3389/fpls.2016.00815 PMID: 27375665

Ryan, R. P., Germaine, K., Franks, A., Ryan, D. J., & Dowling, D. N. (2008). Bacterial endophytes: Recent developments and applications. *FEMS Microbiology Letters*, *278*(1), 1–9. DOI: 10.1111/j.1574-6968.2007.00918.x PMID: 18034833

Saberi Riseh, R., Hassanisaadi, M., Vatankhah, M., & Kennedy, J. F. (2023a). Encapsulating biocontrol bacteria with starch as a safe and edible biopolymer to alleviate plant diseases: A review. Carbohydrate Polymers, 302(November 2022), 120384. DOI: 10.1016/j.carbpol.2022.120384

Saberi-Riseh, R., & Moradi-Pour, M. (2021). A novel encapsulation of *Streptomyces fulvissimus* Uts22 by spray drying and its biocontrol efficiency against *Gaeumannomyces graminis*, the causal agent of take-all disease in wheat. *Pest Management Science*, *77*(10), 4357–4364. DOI: 10.1002/ps.6469 PMID: 33960105

Saberi-Rise, R., & Moradi-Pour, M. (2020). The effect of Bacillus subtilis Vru1 encapsulated in alginate – bentonite coating enriched with titanium nanoparticles against Rhizoctonia solani on bean. *International Journal of Biological Macromolecules*, *152*, 1089–1097. DOI: 10.1016/j.ijbiomac.2019.10.197 PMID: 31751702

Saeid, A., Prochownik, E., & Dobrowolska-Iwanek, J. (2018). Phosphorus Solubilization by Bacillus Species. *Molecules (Basel, Switzerland)*, *23*(11), 2897. DOI: 10.3390/molecules23112897 PMID: 30404208

Safaepour, M., Shahverdi, A. R., Shahverdi, H. R., Khorramizadeh, M. R., & Gohari, A. R. (2009). Green synthesis of small silver nanoparticles using geraniol and its cytotoxicity against Fibrosarcoma–Wehi 164. *Avicenna Journal of Medical Biotechnology*, *1*, 111–115. PMID: 23407598

Safdar, M., Kim, W., Park, S., Gwon, Y., Kim, Y.-O., & Kim, J. (2022a). Engineering plants with carbon nanotubes: A sustainable agriculture approach. *Journal of Nanobiotechnology*, *20*(1), 275. DOI: 10.1186/s12951-022-01483-w PMID: 35701848

Sahin, O. G., Gulce, H., & Gulce, A. (2013). Polyvinylferrocenium based platinum electrodeposited amperometric biosensors for lysine detection. *Journal of Electroanalytical Chemistry (Lausanne, Switzerland)*, *690*, 1–7. DOI: 10.1016/j.jelechem.2012.11.021

Sahoo, L., Mohapatra, D., Raghuvanshi, H. R., Kumar, S., Kaur, R., Anshika, , Sapna, , Chawla, R., & Afreen, N. (2024). Transforming agriculture through Artificial Intelligence: Advancements in plant disease detection, Applications, and challenges. *Journal of Advances in Biology & Biotechnology*, *27*(5), 381–388. DOI: 10.9734/jabb/2024/v27i5796

Sahu, S., Gautam, S., Singh, A., Lohani, P., Sharma, C., Pathak, P., Kumar, A., & Singh, H. (2023). Synthesis and characterization of chitosan-zinc-salicylic acid nanoparticles: A plant biostimulant. *International Journal of Biological Macromolecules*, *253*, 127602. DOI: 10.1016/j.ijbiomac.2023.127602 PMID: 37875188

Saifi, R.. (2025). Challenges and Opportunities of Nanomaterials in Sustainable Agriculture. In Rajan, R., Ahmad, F., Khoshariya, A., & Seth, C. S. (Eds.), *Sustainable Era of Nanomaterials. Sustainable Development and Biodiversity* (Vol. 40). Springer., DOI: 10.1007/978-981-96-4471-1_3

Saini, R. K., Patel, S., Bajpai, J., & Bajpai, A. K. (2020). Advanced controlled nanopesticide delivery systems for managing insect pests. *Controlled Release of Pesticides for Sustainable Agriculture*, 155-184.

Sajid, M., Płotka-Wasylka, J., & Andruch, V. (2020). Nanomaterials as a sustainable choice for enhancing agricultural productivity. *Environmental Science. Nano*, *7*(12), 3545–3568.

Salama, H. M. (2012). Effects of silver nanoparticles in some crop plants, common bean (*Phaseolus vulgaris* L.) and corn (*Zea mays* L.). *International Research Journal of Biotechnology*, *3*(10), 190–197.

Salamanca-Buentello, F., Persad, D. L., Court, E. B., Martin, D. K., Daar, A. S., & Singer, P. A. (2005). Nanotechnology and the developing world. *PLoS Medicine*, *2*(5), e97. DOI: 10.1371/journal.pmed.0020097 PMID: 15807631

Salazar-Cerezo, S., Martínez-Montiel, N., García-Sánchez, J., Pérez-y-Terrón, R., & Martínez-Contreras, R. D. (2018). Gibberellin biosynthesis and metabolism: A convergent route for plants, fungi and bacteria. *Microbiological Research*, *208*, 85–98. DOI: 10.1016/j.micres.2018.01.010 PMID: 29551215

Saleh, M. C., van Rij, R. P., Hekele, A., Gillis, A., Foley, E., O'Farrell, P. H., & Andino, R. (2006). The endocytic pathway mediates cell entry of dsRNA to induce RNAi Silencing. *Nature Cell Biology*, *8*(8), 793–802. DOI: 10.1038/ncb1439 PMID: 16862146

Samadi, S., Asgari Lajayer, B., Moghiseh, E., & Rodríguez-Couto, S. (2021). Effect of carbon nanomaterials on cell toxicity, biomass production, nutritional and active compound accumulation in plants. *Environmental Technology & Innovation*, *21*, 101323. DOI: 10.1016/j.eti.2020.101323

Samadi, S., Saharkhiz, M. J., Azizi, M., Samiei, L., Karami, A., & Ghorbanpour, M. (2021). Single-wall carbon nano tubes (SWCNTs) penetrate *Thymus daenensis* Celak. Plant cells and increase secondary metabolite accumulation in vitro. *Industrial Crops and Products*, *165*, 113424. DOI: 10.1016/j.indcrop.2021.113424

Sampat, B. N. (2010). When do applicants search for prior art? *The Journal of Law & Economics*, *53*(2), 399–416. DOI: 10.1086/651959

Sampathkumar, K., Tan, K. X., & Loo, S. C. J. (2020). Developing Nano-Delivery Systems for Agriculture and Food Applications with Nature-Derived Polymers. *iScience*, *23*(5), 101055. DOI: 10.1016/j.isci.2020.101055 PMID: 32339991

Sánchez-Hernández, E., Santiago-Aliste, A., Correa-Guimarães, A., Martín-Gil, J., Gavara-Clemente, R. J., & Martín-Ramos, P. (2024). Carvacrol encapsulation in chitosan–carboxymethylcellulose–alginate nanocarriers for postharvest tomato protection. *International Journal of Molecular Sciences*, *25*(2), 2. Advance online publication. DOI: 10.3390/ijms25021104 PMID: 38256176

Sanghi, A. K., Kumar, R., Chaubey, K. K., Pal, A. K., Sharma, S., Dayal, D., Kumar, A., & Verma, D. K. (2025). Application of Chitin/Chitosan Based Agro Nano Chemicals in Crop Protection. In Bachheti, R. K., Bachheti, A., & Husen, A. (Eds.), *Chitin-Based Nanoparticles for the Agriculture Sectors* (pp. 155–177). Springer Nature., DOI: 10.1007/978-981-96-0920-8_9

Santana, I., Jeon, S. J., Kim, H. I., Islam, M. R., Castillo, C., Garcia, G. F., & Giraldo, J. P. (2022). Targeted carbon nanostructures for chemical and gene delivery to plant chloroplasts. *ACS Nano*, *16*(8), 12156–12173. DOI: 10.1021/acsnano.2c02714 PMID: 35943045

Santana, I., Wu, H., Hu, P., & Giraldo, J. P. (2020). Targeted delivery of nanomaterials with chemical cargoes in plants enabled by a biorecognition motif. *Nature Communications*, *11*(1), 2045. DOI: 10.1038/s41467-020-15731-w PMID: 32341352

Santiago-Aliste, A., Sánchez-Hernández, E., Buzón-Durán, L., Marcos-Robles, J. L., Martín-Gil, J., & Martín-Ramos, P. (2023). Uncaria tomentosa-Loaded Chitosan Oligomers–Hydroxyapatite–Carbon Nitride Nanocarriers for Postharvest Fruit Protection. *Agronomy (Basel)*, *13*(9), 9. Advance online publication. DOI: 10.3390/agronomy13092189

Santos, A., Aguilar, M., Torres, F., & Ruiz, L. (2020). Advances in nanotechnology for agricultural sustainability: Challenges and future prospects. *Journal of Agricultural Science and Technology*, *22*(2), 89–103.

Saravanan, A., Kumar, P. S., Hemavathy, R. V., Jeevanantham, S., Jawahar, M. J., Neshaanthini, J. P., & Saravanan, R. (2022). A review on synthesis methods and recent applications of nanomaterial in wastewater treatment: Challenges and future perspectives. *Chemosphere*, *307*, 135713. DOI: 10.1016/j.chemosphere.2022.135713 PMID: 35843436

Sargazi, S., Laraib, U., Er, S., Rahdar, A., Hassanisaadi, M., Zafar, M. N., Díez-Pascual, A. M., & Bilal, M. (2022). Application of Green Gold Nanoparticles in Cancer Therapy and Diagnosis. *Nanomaterials 2022, Vol. 12, Page 1102, 12*(7), 1102. DOI: 10.3390/nano12071102

Saruchi, K., Kumar, V., Ghfar, A. A., & Pandey, S. (2023). Controlled release of harmful pesticide dichlorvos through synthesized biodegradable aloe vera–acrylic acid-based hydrogel and its utilization in soil water management. *Journal of Analytical Science and Technology*, *14*(1), 12. DOI: 10.1186/s40543-023-00374-5

Sarvarian, P., Samadi, P., Gholipour, E., Shams Asenjan, K., Hojjat-Farsangi, M., Motavalli, R., Motavalli Khiavi, F., & Yousefi, M. (2022). Application of Emerging Plant-Derived Nanoparticles as a Novel Approach for Nano-Drug Delivery Systems. *Immunological Investigations*, *51*(4), 1039–1059. DOI: 10.1080/08820139.2021.1891094 PMID: 33627016

Sasson, Y., Levy-Ruso, G., Toledano, O., & Ishaaya, I. (2007). Nanosuspensions: emerging novel agrochemical formulations. In Insecticides design using advanced technologies (pp. 1-39). Springer Berlin Heidelberg.

Satchanska, G., Davidova, S., & Petrov, P. D. (2024). Natural and Synthetic Polymers for Biomedical and Environmental Applications. *Polymers*, *16*(8), 1159. DOI: 10.3390/polym16081159 PMID: 38675078

Sathishkumar, M., Sneha, K., Kwak, I. S., Mao, J., Tripathy, S. J., & Yun, Y. S. (2009). Phyto-crystallization of palladium through reduction process using *Cinnamon zeylanicum* bark extract. *Journal of Hazardous Materials*, *171*(1-3), 400–404. DOI: 10.1016/j.jhazmat.2009.06.014 PMID: 19576689

Sathishkumar, P., Preethi, J., Vijayan, R., Mohd Yusoff, A. R., Ameen, F., Suresh, S., Balagurunathan, R., & Palvannan, T. (2016). Anti-acne, anti-dandruff and anti-breast cancer efficacy of green synthesised silver nanoparticles using Coriandrum sativum leaf extract. *Journal of Photochemistry and Photobiology. B, Biology*, *163*, 69–76. DOI: 10.1016/j.jphotobiol.2016.08.005 PMID: 27541567

Savithramma, N., Ankanna, S., & Bhumi, G. (2012). Effect of nanoparticles on seed germination and seedling growth of *Boswellia ovalifoliolata* an endemic and endangered medicinal tree taxon. *Nano Vision*, *2*, 61–68.

Scaglia, B., Nunes, R. R., Rezende, M. O. O., Tambone, F., & Adani, F. (2016). Investigating organic molecules responsible of auxin-like activity of humic acid fraction extracted from vermicompost. *The Science of the Total Environment*, *562*, 289–295. DOI: 10.1016/j.scitotenv.2016.03.212 PMID: 27100009

Schaafsma, G. (2009). Safety of protein hydrolysates, fractions thereof and bioactive peptides in human nutrition. *European Journal of Clinical Nutrition*, *63*(10), 1161–1168. DOI: 10.1038/ejcn.2009.56 PMID: 19623200

Schnell, M. J., Johnson, J. E., Buonocore, L., & Rose, J. K. (1997). Construction of a novel virus that targets HIV-1-infected cells and controls HIV-1 infection. *Cell*, *90*(5), 849–857. DOI: 10.1016/S0092-8674(00)80350-5 PMID: 9298897

Schnoor, B., Elhendawy, A., Joseph, S., Putman, M., Chacón-Cerdas, R., Flores-Mora, D., Bravo-Moraga, F., Gonzalez-Nilo, F., & Salvador-Morales, C. (2018). Engineering atrazine loaded poly (lactic-co-glycolic acid) nanoparticles to ameliorate environmental challenges. *Journal of Agricultural and Food Chemistry*, *66*(30), 7889–7898. DOI: 10.1021/acs.jafc.8b01911 PMID: 30039704

Schoebitz, M., López, M. D., & Roldán, A. (2013). Bioencapsulation of microbial inoculants for better soil-plant fertilization. A review. *Agronomy for Sustainable Development*, *33*(4), 751–765. DOI: 10.1007/s13593-013-0142-0

Schoebitz, M., Simonin, H., & Poncelet, D. (2012). Starch filler and osmoprotectants improve the survival of rhizobacteria in dried alginate beads. *Journal of Microencapsulation*, *29*(6), 532–538. DOI: 10.3109/02652048.2012.665090 PMID: 22372947

Schwab, F.. (2015). Nanoparticle interactions with plants: Uptake and mechanisms. *Journal of Agricultural and Food Chemistry*, *63*(14), 3839–3850.

Schwab, F., Zhai, G., Kern, M., Turner, A., Schnoor, J. L., & Wiesner, M. R. (2016). Barriers, pathways and processes for uptake, translocation and accumulation of nanomaterials in plants—Critical review. *Nanotoxicology*, *10*(3), 257–278. DOI: 10.3109/17435390.2015.1048326 PMID: 26067571

Schwartz-Duval, A. S., Konopka, C. J., Moitra, P., Daza, E. A., Srivastava, I., Johnson, E. V., & Pan, D. (2020). Intratumoral generation of photothermal gold nanoparticles through a vectorized biomineralization of ionic gold. *Nature Communications*, *11*(1), 4530. DOI: 10.1038/s41467-020-17595-6 PMID: 32913195

Schwartz, S. H., Hendrix, B., Hofer, P., Sanders, R. A., & Zheng, W. (2020). Carbon dots for eficient small interfering RNA delivery and gene silencing in plants. *Plant Physiology*, *184*(2), 647–657. DOI: 10.1104/pp.20.00733 PMID: 32764133

Seankamsorn, A., Cherdthong, A., So, S., & Wanapat, M. (2021). Influence of chitosan sources on intake, digestibility, rumen fermentation, and milk production in tropical lactating dairy cows. *Tropical Animal Health and Production*, *53*(2), 241. DOI: 10.1007/s11250-021-02697-0 PMID: 33811551

Seankamsorn, A., Cherdthong, A., & Wanapat, M. (2020). Combining crude glycerin with chitosan can manipulate in vitro ruminal efficiency and inhibit methane synthesis. *Animals (Basel)*, *10*(1), 37. DOI: 10.3390/ani10010037 PMID: 31878076

Sebastian, A., Nangia, A., & Prasad, M. N. V. (2019). Cadmium and sodium adsorption properties of magnetite nanoparticles synthesized from *Hevea brasiliensis* Muell. Arg. bark: Relevance in amelioration of metal stress in rice. *Journal of Hazardous Materials*, *371*, 261–272. DOI: 10.1016/j.jhazmat.2019.03.021 PMID: 30856436

Seddighinia, F. S., Iranbakhsh, A., Oraghi Ardebili, Z., Nejad Satari, T., & Soleimanpour, S. (2020). Seed priming with cold plasma and multi-walled carbon nanotubes modified growth, tissue differentiation, anatomy, and yield in bitter melon (*Momordica charantia*). *Journal of Plant Growth Regulation*, *39*(1), 87–98. DOI: 10.1007/s00344-019-09965-2

Sedghi, M., Hadi, M., & Toluie, S. G. (2013). Effect of nano zinc oxide on the germination parameters of soybean seeds under drought stress. *Annals of West University of Timisoara. Series of Biology*, *16*(2), 73.

Sehaole, T. P. (2012). *Evaluation of different media formulations on spore production and toxicity of Bacillus thuringiensis subsp. aizawai* (Doctoral dissertation, North-West University).

Selby, L. I., Cortez-Jugo, C. M., Such, G. K., & Johnston, A. P. R. (2017). Nanoescapology: progress toward understanding the endosomal escape of polymeric nanoparticles. *Wiley interdisciplinary reviews. Nanomedicine and nanobiotechnology, 9*(5), . https://doi.org/DOI: 10.1002/wnan.1452

Semaltianos, N. G. (2010). Nanoparticles by laser ablation. *Critical Reviews in Solid State and Material Sciences*, *35*(2), 105–124. DOI: 10.1080/10408431003788233

Septiani, P., Pramesti, Y., Ghildan, M., Aprilia, K. Z., Awaludin, R., Medina, S., Subandiyah, S., & Meitha, K. (2025). RNAi-based biocontrol for crops: A revised expectation for a non-recent technology. *Planta*, *261*(2), 44. DOI: 10.1007/s00425-025-04625-0 PMID: 39862243

Serag, M. F., Kaji, N., Gaillard, C., Okamoto, Y., Terasaka, K., Jabasini, M., Tokeshi, M., Mizukami, H., Bianco, A., & Baba, Y. (2011). Trafficking and subcellular localization of multiwalled carbon nanotubes in plant cells. *ACS Nano*, *5*(1), 493–499. DOI: 10.1021/nn102344t PMID: 21141871

Serag, M. F., Kaji, N., Habuchi, S., Bianco, A., & Baba, Y. (2013). Nanobiotechnology meets plant cell biology: Carbon nanotubes as organelle targeting nanocarriers. *RSC Advances*, *3*(15), 4856. DOI: 10.1039/c2ra22766e

Serov, D. A., Gritsaeva, A. V., Yanbaev, F. M., Simakin, A. V., & Gudkov, S. V. (2024). Review of antimicrobial properties of titanium dioxide nanoparticles. *International Journal of Molecular Sciences*, *25*(19), 10519. DOI: 10.3390/ijms251910519 PMID: 39408848

Shahid, M. A., Sarkhosh, A., Khan, N., Balal, R. M., Ali, S., Rossi, L., Gómez, C., Mattson, N., Nasim, W., & Garcia-Sanchez, F. (2020). Insights into the physiological and biochemical impacts of salt stress on plant growth and development. *Agronomy (Basel)*, *10*(7), 938. DOI: 10.3390/agronomy10070938

Shah, M., Fawcett, D., Sharma, S., Tripathy, S. K., & Poinern, G. E. J. (2015). Green synthesis of metallic nanoparticles via biological entities. *Materials (Basel)*, *8*(11), 7278–7308. DOI: 10.3390/ma8115377 PMID: 28793638

Shah, V., & Belozerova, I. (2009). Influence of metal nanoparticles on the soil microbial community and germination of lettuce seeds. *Water, Air, and Soil Pollution*, *197*(1-4), 143–148. DOI: 10.1007/s11270-008-9797-6

Shahwar, D., Mushtaq, Z., Mushtaq, H., Alqarawi, A. A., Park, Y., Alshahrani, T. S., & Faizan, S. (2023). Role of microbial inoculants as bio fertilizers for improving crop productivity: A review. *Heliyon*, *9*(6), e16134. DOI: 10.1016/j.heliyon.2023.e16134 PMID: 37255980

Shakiba, S., Astete, C. E., Paudel, S., Sabliov, C. M., Rodrigues, D. F., & Louie, S. M. (2020). Emerging investigator series: polymeric nanocarriers for agricultural applications: synthesis, characterization, and environmental and biological interactions. *Environmental Science. Nano*, *7*(1), 37–67. DOI: 10.1039/C9EN01127G

Shang, Y., Hasan, A., & Ahammed, G. (2022). Nanocarrier-mediated plant biostimulation and defense enhancement: New frontiers for agricultural sustainability. *Advances in Plant Biology*, *16*, 75–92.

Shang, Y., Hasan, M. K., & Ahammed, G. J. (2020). Applications of nanotechnology in plant growth and crop protection: A mini-review. *Journal of Plant Growth Regulation*, *39*(4), 1381–1393.

Shang, Y., Hasan, M. K., Ahammed, G. J., Li, M., Yin, H., & Zhou, J. (2019). Applications of Nanotechnology in Plant Growth and Crop Protection: A Review. *Molecules (Basel, Switzerland)*, *24*(14), 2558. DOI: 10.3390/molecules24142558 PMID: 31337070

Shankar, S. S., Ahmad, A., & Sastry, M. (2003). Geranium leaf assisted biosynthesis of silver nanoparticles. *Biotechnology Progress*, *19*(6), 1627–1631. DOI: 10.1021/bp034070w PMID: 14656132

Shankar, S., & Rhim, J. W. (2015). Amino acid mediated synthesis of silver nanoparticles and preparation of antimicrobial agar/silver nanoparticles composite films. *Carbohydrate Polymers*, *130*, 353–363. DOI: 10.1016/j.carbpol.2015.05.018 PMID: 26076636

Shankramma, K., Yallappa, S., Shivanna, M. B., & Manjanna, J. (2016). Fe_2O_3 magnetic nanoparticles to enhance *S. lycopersicum* (tomato) plant growth and their biomineralization. *Applied Nanoscience*, *6*(7), 983–990. DOI: 10.1007/s13204-015-0510-y

Shao, C., Zhao, H., & Wang, P. (2022). Recent development in functional nanomaterials for sustainable and smart agricultural chemical technologies. *Nano Convergence*, *9*(1), 11. DOI: 10.1186/s40580-022-00302-0 PMID: 35235069

Sha, R., Zhu, S., Wu, L., Li, X., Zhang, H., Yao, D., Lv, Q., Wang, F., Zhao, F., Li, P., & Yu, K. (2022). Pre-Harvest application of multi-walled carbon nanotubes improves the antioxidant capacity of 'Flame Seedless' grapes during storage. *Sustainability (Basel)*, *14*(15), 15. Advance online publication. DOI: 10.3390/su14159568

Sharkawy, A., Barreiro, M. F., & Rodrigues, A. E. (2020). Chitosan-based Pickering emulsions and their applications: A review. *Carbohydrate Polymers*, *250*(August), 116885. DOI: 10.1016/j.carbpol.2020.116885 PMID: 33049878

Sharma, A., & Das, J. (2019). Small molecules derived carbon dots: Synthesis and applications in sensing, catalysis, imaging, and biomedicine. In Journal of Nanobiotechnology (Vol. 17, Issue 1, p. 92). BioMed Central Ltd. DOI: 10.1186/s12951-019-0525-8

Sharma, D., Dhuriya, Y. K., Sharma, J., & Gupta, M. (2019). Nanoelements: An agricultural paradigm for targeted plant nutrition therapeutic approach. *Nanotechnology for Agriculture: Crop Production & Protection*, 73-83.

Sharma, L., & Tagad, C. (2022). Biosynthesis of polysaccharides-stabilized metal nanoparticles for chemical and biosensing applications. *Polysaccharide Nanoparticles: Preparation and Biomedical Applications*, 553–583. DOI: 10.1016/B978-0-12-822351-2.00021-8

Sharma, A., Jain, N., & Reddy, A. D.. (2020). Advances in the development of nanomedicines for antifungal therapeutics. *Expert Opinion on Drug Delivery*, *17*(7), 913–928.

Sharma, M. M. M., Kapoor, D., Loyal, A., Kumar, R., & Sharma, P. (2025). Role of carbon-based nanomaterials in biotic stress managements. In *Smart Nanomaterials Technology* (pp. 255–269). Springer Nature Singapore., DOI: 10.1007/978-981-97-5104-4_13

Sharma, M. M. M., Kapoor, D., Loyal, A., Kumar, R., Sharma, P., & Husen, A. (2025). Plant response to silver nanoparticles in terms of growth, development, production, and protection: An overview. In Husen, A. (Ed.), *Plant Response to Silver Nanoparticles: Plant Growth, Development, Production, and Protection* (pp. 1–22). Springer Nature., DOI: 10.1007/978-981-97-7352-7_1

Sharma, P., Kumar, S., & Mittal, A. (2021). In situ polymerization of nanoparticles for antifungal applications. *Materials Today. Communications*, *26*, 102182.

Sharma, S., Singh, N., & Rastogi, M. (2022). Nanotechnology to Address Micronutrient and Macronutrient Deficiency. In *Nanotechnology* (pp. 35–59). CRC Press. DOI: 10.1201/9781003220350-3

Sharma, V. K., Ria, A. Y., & Lin, Y. (2009). Silver nanoparticles: Green synthesis and their antimicrobial activities. *Advances in Colloid and Interface Science*, *145*(1-2), 83–96. DOI: 10.1016/j.cis.2008.09.002 PMID: 18945421

Shastri, D. H., & Gandhi, S. (2024). Nanorevolution Unleashing the Power of Nanotechnology. *Current Nanomedicine*, *14*(3), 227–246. DOI: 10.2174/0124681873279660231226070118

Shaw, D. S., & Honeychurch, K. C. (2022). Nanosensor applications in plant science. *Biosensors (Basel)*, *12*(9), 675. DOI: 10.3390/bios12090675 PMID: 36140060

Shehzad, J., & Mustafa, G. (2023). Mechanism of reactive oxygen species regulation in plants. In *Reactive Oxygen Species: Prospects in Plant Metabolism* (pp. 17–41). Springer Nature Singapore. DOI: 10.1007/978-981-19-9794-5_2

Shekhar, S., Sharma, S., Kumar, A., Taneja, A., & Sharma, B. (2021). The framework of nanopesticides: A paradigm in biodiversity. *Materials Advances*, *2*(20), 6569–6588. DOI: 10.1039/D1MA00329A

Shelar, A., Nile, S. H., Singh, A. V., Rothenstein, D., Bill, J., Xiao, J., Chaskar, M., Kai, G., & Patil, R. (2023). Recent advances in nano-enabled seed treatment strategies for sustainable agriculture: Challenges, risk assessment, and future perspectives. *Nano-Micro Letters*, *15*(1), 54. DOI: 10.1007/s40820-023-01025-5 PMID: 36795339

Shelar, G. B., & Chavan, A. M. (2015). Myco-synthesis of silver nanoparticles from *Trichoderma harzianum* and its impact on germination status of oil seed. *Biolife : An International Quarterly Journal of Biology and Life Sciences*, *3*(1), 109–113.

Shen, D., Zhou, F., Xu, Z., He, B., Li, M., Shen, J., Yin, M., & An, C. (2014). Systemically interfering with immune response by a fluorescent cationic dendrimer delivered gene suppression. *Journal of Materials Chemistry. B*, *2*(29), 4653–4659. DOI: 10.1039/C4TB00411F PMID: 32262277

Shende, S. S., Rajput, V. D., Gorovtsov, A. V., Harish, Saxena, P., Minkina, T. M., ... & Kizilkaya, R. (2021). Interaction of nanoparticles with microbes. *Plant-Microbes-Engineered Nano-particles (PM-ENPs) Nexus in Agro-Ecosystems: Understanding the Interaction of Plant, Microbes and Engineered Nano-particles (ENPS)*, 175-188.

Sheng, W., Liu, J., Liu, S., Lu, Q., Kaplan, D. L., & Zhu, H. (2014). One-step synthesis of biocompatible magnetite/silk fibroin core-shell nanoparticles. *Journal of Materials Chemistry. B*, *2*(42), 7394–7402. DOI: 10.1039/C4TB01125B PMID: 32261964

Shen, M., Liu, S., Jiang, C., Zhang, T., & Chen, W. (2023). Recent advances in stimuli-response mechanisms of nano-enabled controlled-release fertilizers and pesticides. *Eco-Environment & Health*, *2*(3), 161–175. DOI: 10.1016/j.eehl.2023.07.005 PMID: 38074996

Shew, A. M., Nalley, L. L., Snell, H. A., Nayga, R. M.Jr, & Dixon, B. L. (2018). CRISPR versus GMOs: Public acceptance and valuation. *Global Food Security*, *19*, 71–80. DOI: 10.1016/j.gfs.2018.10.005

Shibata, T., Ishii, T., Nozawa, H., & Tamamura, T. (1997). High-aspect-ratio nanometer-pattern fabrication using fullerene-incorporated nanocomposite resists for dry-etching application. *Japanese Journal of Applied Physics*, *36*(12, SUPPL. B), 7642–7645. DOI: 10.1143/JJAP.36.7642

Shi, J., & Avramidis, S. (2019). Evolution of wood cell wall nanopore size distribution in the hygroscopic range. *Holzforschung*, *73*(10), 899–910. DOI: 10.1515/hf-2018-0198

Shin, W. K., Cho, J., Kannan, A. G., Lee, Y. S., & Kim, W. (2016). Cross- linked composite gel polymer electrolyte using mesoporous methacrylate functionalized SiO2 nanoparticles for lithium- ion polymer batteries. *Scientific Reports*, *6*(1), 26332. DOI: 10.1038/srep26332 PMID: 27189842

Shishir, M. R. I., Gowd, V., Suo, H., Wang, M., Wang, Q., Chen, F., & Cheng, K.-W. (2021). Advances in smart delivery of food bioactive compounds using stimuli-responsive carriers: Responsive mechanism, contemporary challenges, and prospects. *Comprehensive Reviews in Food Science and Food Safety*, *20*(6), 5449–5488. DOI: 10.1111/1541-4337.12851 PMID: 34668321

Shivaji, S., Madhu, S., & Singh, S. (2011). Extracellular synthesis of antibacterial silver nanoparticles using psychrophilic bacteria. *Process Biochemistry (Barking, London, England)*, *46*(9), 1800–1807. DOI: 10.1016/j.procbio.2011.06.008

Shojaei, T. R., Salleh, M. A. M., Tabatabaei, M., Mobli, H., Aghbashlo, M., Rashid, S. A., & Tan, T. (2019). Applications of nanotechnology and carbon nanoparticles in agriculture. In Rashid, S. A., Raja Othman, R. N. I., & Hussein, M. Z. (Eds.), *Synthesis, Technology and Applications of Carbon Nanomaterials* (pp. 247–277). Elsevier., DOI: 10.1016/B978-0-12-815757-2.00011-5

Shou, H., Frame, B. R., Whitham, S. A., & Wang, K. (2004). Assessment of transgenic maize events produced by particle bombardment or Agrobacterium-mediated transformation. *Molecular Breeding*, *13*(2), 201–208. DOI: 10.1023/B:MOLB.0000018767.64586.53

Shukla, P., Chaurasia, P., Younis, K., Qadri, O. S., Faridi, S. A., & Srivastava, G. (2019). Nanotechnology in sustainable agriculture: Studies from seed priming to post-harvest management. *Nanotechnology for Environmental Engineering*, *4*(1), 1–15. DOI: 10.1007/s41204-019-0058-2

Siddiqui, M. H., & Al-Whaibi, M. H. (2014). Role of nano-SiO_2 in germination of tomato (*Lycopersicum esculentum* seeds Mill.). *Saudi Journal of Biological Sciences*, *21*(1), 13–17. DOI: 10.1016/j.sjbs.2013.04.005 PMID: 24596495

Sidhu, R. K., Kumar, R., Rana, P. S., & Jat, M. L. (2021). Automation in drip irrigation for enhancing water use efficiency in cereal systems of South Asia: Status and prospects. *Advances in Agronomy*, *167*, 247–300. DOI: 10.1016/bs.agron.2021.01.002

Siegel, R. A., & Rathbone, M. J. (2011). Overview of controlled release mechanisms. In *Fundamentals and applications of controlled release drug delivery* (pp. 19–43). Springer US., DOI: 10.1007/978-1-4614-0881-9_2

Sigmon, L. R., Adisa, I. O., Liu, B., Elmer, W. H., White, J. C., Dimkpa, C. O., & Fairbrother, D. H. (2021). Biodegradable polymer nanocomposites provide effective delivery and reduce phosphorus loss during plant growth. *ACS Agricultural Science & Technology*, *1*(5), 529–539. DOI: 10.1021/acsagscitech.1c00149

Silva, A. T., Nguyen, A., Ye, C., Verchot, J., & Moon, J. H. (2010). Conjugated polymer nanoparticles for effective siRNA delivery to tobacco BY-2 protoplasts. *BMC Plant Biology*, *10*(1), 291. DOI: 10.1186/1471-2229-10-291 PMID: 21192827

Silva, G. A. (2006). Neuroscience nanotechnology: Progress, opportunities and challenges. *Nature Reviews. Neuroscience*, *7*(1), 65–74. DOI: 10.1038/nrn1827 PMID: 16371951

Silva, L. I., Pereira, M. C., Carvalho, A. M., Buttrós, V. H., Pasqual, M., & Dória, J. (2023). Phosphorus-Solubilizing Microorganisms: A Key to Sustainable Agriculture. *Agriculture*, *13*(2), 462. DOI: 10.3390/agriculture13020462

Singewar, K., & Fladung, M. (2023). Double-stranded RNA (dsRNA) technology to control forest insect pests and fungal pathogens: Challenges and opportunities. *Functional & Integrative Genomics*, *23*(2), 185. DOI: 10.1007/s10142-023-01107-y PMID: 37243792

Singh, Rajender. (2024). Introduction to Transmission Electron Microscopy (TEM). *Transmission Electron Microscopy Sample Preparation*, 1–20. DOI: 10.1007/978-3-031-82967-3_1

Singh, A., Dhiman, N., Kar, A. K., Singh, D., Purohit, M. P., Ghosh, D., & Patnaik, S. (2020). Advances in controlled release pesticide formulations: Prospects to safer integrated pest management and sustainable agriculture. *Journal of Hazardous Materials*, *385*, 121525. DOI: 10.1016/j.jhazmat.2019.121525 PMID: 31740313

Singh, A., Hua Hsu, M., Gupta, N., Khanra, P., Kumar, P., Prakash Verma, V., & Kapoor, M. (2020). Derivatized carbon nanotubes for gene therapy in mammalian and plant cells. *ChemPlusChem*, *85*(3), 466–475. DOI: 10.1002/cplu.201900678 PMID: 32159284

Singh, A., Singh, N. B., Hussain, I., Singh, H., & Singh, S. C. (2015). Plant-nanoparticle interaction: An approach to improve agricultural practices and plant productivity. *International Journal of Pharmaceutical Science Invention*, *4*(8), 25–40.

Singh, D., & Gurjar, B. R. (2022). Nanotechnology for agricultural applications: Facts, issues, knowledge gaps, and challenges in environmental risk assessment. *Journal of Environmental Management*, *322*, 116033. DOI: 10.1016/j.jenvman.2022.116033

Singh, G., Ramadass, K., Sooriyakumar, P., Hettithanthri, O., Vithange, M., Bolan, N., Tavakkoli, E., Van Zwieten, L., & Vinu, A. (2022). Nanoporous materials for pesticide formulation and delivery in the agricultural sector. *Journal of Controlled Release*, *343*, 187–206. DOI: 10.1016/j.jconrel.2022.01.036 PMID: 35090962

Singh, H., Sharma, A., Bhardwaj, S. K., Arya, S. K., Bhardwaj, N., & Khatri, M. (2021). Recent advances in the applications of nano-agrochemicals for sustainable agricultural development. *Environmental Science. Processes & Impacts*, *23*(2), 213–239. DOI: 10.1039/D0EM00404A PMID: 33447834

Singh, J., & Lee, B. K. (2016). Influence of nano-TiO_2 particles on the bioaccumulation of Cd in soybean plants (*Glycine max*): A possible mechanism for the removal of Cd from the contaminated soil. *Journal of Environmental Management*, *170*, 88–96. DOI: 10.1016/j.jenvman.2016.01.015 PMID: 26803259

Singh, M., Malhotra, N., & Sharma, K. (2020). Buckwheat (Fagopyrum sp.) genetic resources: What can they contribute towards nutritional security of changing world? *Genetic Resources and Crop Evolution*, *67*(7), 1639–1658. DOI: 10.1007/s10722-020-00961-0

Singh, P., Chauhan, P. K., Upadhyay, S. K., Singh, R. K., Dwivedi, P., Wang, J., Jain, D., & Jiang, M. (2022). Mechanistic Insights and Potential Use of Siderophores Producing Microbes in Rhizosphere for Mitigation of Stress in Plants Grown in Degraded Land. *Frontiers in Microbiology*, *13*, 898979. DOI: 10.3389/fmicb.2022.898979 PMID: 35898908

Singh, R. P., Handa, R., & Manchanda, G. (2021). Nanoparticles in sustainable agriculture: An emerging opportunity. *Journal of Controlled Release*, *329*, 1234–1248. DOI: 10.1016/j.jconrel.2020.10.051 PMID: 33122001

Singh, R., Choudhary, P., Kumar, S., & Daima, H. K. (2024). Mechanistic approaches for crosstalk between nanomaterials and plants: Plant immunomodulation, defense mechanisms, stress resilience, toxicity, and perspectives. *Environmental Science. Nano*, *11*(6), 2324–2351. DOI: 10.1039/D4EN00053F

Singh, R., Mehra, R., Walia, A., Gupta, S., Chawla, P., kumar, H., Thakur, A., Kaushik, R., & Kumar, N. (2023). Colorimetric sensing approaches based on silver nanoparticles aggregation for determination of toxic metal ions in water sample: A review. *International Journal of Environmental Analytical Chemistry*, *103*(6), 1361–1376. DOI: 10.1080/03067319.2021.1873315

Singh, R., Singh, S., Parihar, P., Singh, V. P., & Prasad, S. M. (2015). Arsenic contamination, consequences and remediation techniques: A review. *Ecotoxicology and Environmental Safety*, *112*, 247–270. DOI: 10.1016/j.ecoenv.2014.10.009 PMID: 25463877

Singh, S. (2019). Zinc oxide nanoparticles impacts: Cytotoxicity, genotoxicity, developmental toxicity, and neurotoxicity. *Toxicology Mechanisms and Methods*, *29*(4), 300–311. DOI: 10.1080/15376516.2018.1553221 PMID: 30489211

Singh, S. (2025). Optimizing Nutrient Uptake: Nano-Fertilizers and Soil Amendments. In *Food and Industry 5.0: Transforming the Food System for a Sustainable Future* (pp. 229–248). Springer Nature Switzerland. DOI: 10.1007/978-3-031-76758-6_15

Singh, S. C., Mishra, S. K., Srivastava, R. K., & Gopal, R. (2010). Optical properties of selenium quantum dots produced with laser irradiation of water suspended Se nanoparticles. *The Journal of Physical Chemistry. C, Nanomaterials and Interfaces*, *114*(41), 17374–17384. DOI: 10.1021/jp105037w

Sinha, D., & Tandon, P. K. (2020). An overview of nitrogen, phosphorus and potassium: Key players of nutrition process in plants. *Sustainable solutions for elemental deficiency and excess in crop plants*, 85-117. https://doi.org/DOI: 10.1007/978-981-15-8636-1_5

Sinha, T., Bhagwatwar, P., Krishnamoorthy, C., & Chidambaram, R. (2019). Polymer based micro-and nanoencapsulation of agrochemicals. *Polymers for agri-food applications*, 5-28.

Slomberg, D. L., & Schoenfisch, M. H. (2012). Silica nanoparticle phytotoxicity to Arabidopsis thaliana. *Environmental Science & Technology*, *46*(18), 10247–10254. DOI: 10.1021/es300949f PMID: 22889047

Slowing, I. I., Vivero-Escoto, J. L., Wu, C. W., & Lin, V. S. (2008). Mesoporous silica nanoparticles as controlled release drug delivery and gene transfection carriers. *Advanced Drug Delivery Reviews*, *60*(11), 1278–1288. DOI: 10.1016/j.addr.2008.03.012 PMID: 18514969

Smith, A. M., & Gilbertson, L. M. (2018). Rational ligand design to improve agrochemical delivery efficiency and advance agriculture sustainability. *ACS Sustainable Chemistry & Engineering*, *6*(11), 13599–13610. DOI: 10.1021/acssuschemeng.8b03457

Smith, D. K., Hirst, A. R., Love, C. S., Hardy, J. G., Brignell, S. V., & Huang, B. (2005). Self-assembly using dendritic building blocks—Towards controllable nanomaterials. *Progress in Polymer Science*, *30*(3–4), 220–293. DOI: 10.1016/j.progpolymsci.2005.01.006

Soenen, S. J. H., Illyes, E., Vercauteren, D., Braeckmans, K., Majer, Z., De Smedt, S. C., & De Cuyper, M. (2009). The role of nanoparticle concentration-dependent induction of cellular stress in the internalization of non-toxic cationic magnetoliposomes. *Biomaterials*, *30*(36), 6803–6813. DOI: 10.1016/j.biomaterials.2009.08.050 PMID: 19765821

Sohlot, M., Das, S., & Debnath, N. (2022). Recent developments in silica nanoparticle-based drug delivery system. In Hameed, S., & Rehman, S. (Eds.), *Nanotechnology for Infectious Diseases* (pp. 237–261). Springer., DOI: 10.1007/978-981-16-9190-4_11

Solanki, P., Bhargava, A., Chhipa, H., Jain, N., & Panwar, J. (2015). Nano-fertilizers and their smart delivery system. *Nanotechnologies in food and agriculture*, 81-101.

Solomon, N. O., Kanchan, S., & Kesheri, M. (2024). Nanoparticles as detoxifiers for industrial wastewater. *Water, Air, and Soil Pollution*, *235*(3), 214. DOI: 10.1007/s11270-024-07016-5

Sonawane, H., Shelke, D., Chambhare, M., Dixit, N., Math, S., Sen, S., Borah, S. N., Islam, N. F., Joshi, S. J., Yousaf, B., Rinklebe, J., & Sarma, H. (2022). Fungi-derived agriculturally important nanoparticles and their application in crop stress management–Prospects and environmental risks. *Environmental Research*, *212*, 113543. DOI: 10.1016/j.envres.2022.113543 PMID: 35613631

Songa, E. A., & Okonkwo, J. O. (2016). Recent approaches to improving selectivity and sensitivity of enzyme-based biosensors for organophosphorus pesticides: A review. *Talanta*, *155*, 289–304. DOI: 10.1016/j.talanta.2016.04.046 PMID: 27216686

Song, B., & Cho, C.-W. (2024). Applying polyvinyl alcohol to the preparation of various nanoparticles. *Journal of Pharmaceutical Investigation*, *54*(3), 249–266. DOI: 10.1007/s40005-023-00649-4

Song, H., Fan, Y., Zhang, J., Cooper, A. M., Silver, K., Li, D., Li, T., Ma, E., Zhu, K. Y., & Zhang, J. (2019a). Contributions of dsRNases to differential RNAi efficiencies between the injection and oral delivery of dsRNA in *Locusta migratoria.Pest Management Science*, *75*(6), 1707–1717. DOI: 10.1002/ps.5291 PMID: 30525311

Song, J. Y., Kwon, E. Y., & Kim, B. S. (2010). Biological synthesis of platinum nanoparticles using *Diopyros kaki* leaf extract. *Bioprocess and Biosystems Engineering*, *33*(1), 159–164. DOI: 10.1007/s00449-009-0373-2 PMID: 19701776

Song, S. J., Jiang, X. F., Shen, H., Wu, W. N., Shi, Q. Q., Wan, M. H., Zhang, J., Mo, H., & Shen, J. (2021). MXene (Ti_3C_2) Based pesticide delivery system for sustained release and enhanced pest control. *ACS Applied Bio Materials*, *4*(9), 6912–6923. DOI: 10.1021/acsabm.1c00607 PMID: 35006991

Song, S. J., Wang, Y. L., Xie, J., Sun, B. H., Zhou, N. L., Shen, H., & Shen, J. (2019b). Carboxymethyl chitosan modified carbon nanoparticle for controlled emamectin benzoate delivery: Improved solubility, pH-responsive release, and sustainable pest control. *ACS Applied Materials & Interfaces*, *11*(37), 34258–34267. DOI: 10.1021/acsami.9b12564 PMID: 31461267

Soni, R. A., Rizwan, M. A., & Singh, S. (2022). Opportunities and potential of green chemistry in nanotechnology. *Nanotechnology for Environmental Engineering*, *7*(3), 661–673. DOI: 10.1007/s41204-022-00233-5

Sosnowski, J., Truba, M., & Vasileva, V. (2023). The Impact of Auxin and Cytokinin on the Growth and Development of Selected Crops. *Agriculture*, *13*(3), 724. DOI: 10.3390/agriculture13030724

Soumare, A., Diedhiou, A. G., Thuita, M., Hafidi, M., Ouhdouch, Y., Gopalakrishnan, S., & Kouisni, L. (2020). Exploiting Biological Nitrogen Fixation: A Route Towards a Sustainable Agriculture. *Plants*, *9*(8), 1011. DOI: 10.3390/plants9081011 PMID: 32796519

Soundararajan, D., Ramana, L. N., Shankaran, P., & Krishnan, U. M. (2022). Nanoparticle-based strategies to target HIV-infected cells. *Colloids and Surfaces. B, Biointerfaces*, *213*, 112405. DOI: 10.1016/j.colsurfb.2022.112405 PMID: 35255375

Soundarrajan, C., Sankari, A., Dhandapani, P., Maruthamuthu, S., Ravichandran, S., Sozhan, G., & Palaniswamy, N. (2012). Rapid biological synthesis of platinum nanoparticles using *Ocimum sanctum* for water electrolysis applications. *Bioprocess and Biosystems Engineering*, *35*(5), 827–833. DOI: 10.1007/s00449-011-0666-0 PMID: 22167464

Sousa de Almeida, M., Susnik, E., Drasler, B., Taladriz-Blanco, P., Petri-Fink, A., & Rothen-Rutishauser, B. (2021). Understanding nanoparticle endocytosis to improve targeting strategies in nanomedicine. *Chemical Society Reviews*, *50*(9), 5397–5434. DOI: 10.1039/D0CS01127D PMID: 33666625

Sozer, N., & Kokini, J. L. (2009). Nanotechnology and its applications in the food sector. *Trends in Biotechnology*, *27*(2), 82–89. DOI: 10.1016/j.tibtech.2008.10.010 PMID: 19135747

Sprenger, S. (2013). Epoxy resin composites with surface-modified silicon dioxide nanoparticles: A review. *Journal of Applied Polymer Science*, *130*(3), 1421–1428. DOI: 10.1002/app.39208

Squire, H. J., Tomatz, S., Voke, E., González-Grandío, E., & Landry, M. (2023). The emerging role of nanotechnology in plant genetic engineering. *Nature Reviews Bioengineering*, *1*(5), 314–328. DOI: 10.1038/s44222-023-00037-5

Sridhar, A., Ponnuchamy, M., Kumar, P. S., & Kapoor, A. (2021). Food preservation techniques and nanotechnology for increased shelf life of fruits, vegetables, beverages and spices: A review. *Environmental Chemistry Letters*, *19*(2), 1715–1735. DOI: 10.1007/s10311-020-01126-2 PMID: 33192209

Stampoulis, D., Sinha, S. K., & White, J. C. (2009). Assay-dependent phytotoxicity of nanoparticles to plants. *Environmental Science & Technology*, *43*(24), 9473–9479. DOI: 10.1021/es901695c PMID: 19924897

Steinwand, M. A., & Ronald, P. C. (2020). Crop biotechnology and the future of food. *Nature Food*, *1*(5), 273–283. DOI: 10.1038/s43016-020-0072-3

Stylios, G. K. (2013). There is plenty of room at the bottom, R.P. Feynman. *International Journal of Clothing Science and Technology, 25*(5). Advance online publication. DOI: 10.1108/IJCST-06-2013-0067

Subhankari, I., & Nayak, P. L. (2013). Synthesis of copper nanoparticles using *Syzygium aromaticum* (Cloves) aqueous extract by using green chemistry. *World Journal of Nano Science and Technology, 2*, 14–17.

Subramanian, K. S., Manikandan, A., Thirunavukkarasu, M., & Rahale, C. S. (2015). Nano-fertilizers for balanced crop nutrition. *Nanotechnologies in food and agriculture*, 69-80.

Subramani, T., & Ganapathyswamy, H. (2020). An overview of liposomal nano-encapsulation techniques and its applications in food and nutraceutical. *Journal of Food Science and Technology, 57*(10), 3545–3555. DOI: 10.1007/s13197-020-04360-2 PMID: 32903987

Su, C., Li, Y., Liu, S., Feng, H., Wang, J., & Yan, S. (2023). Star polymer soil delivery nanoplatform for applying biological agents in the field to control plant rhizosphere diseases. *Journal of Controlled Release, 364*, 406–419. DOI: 10.1016/j.jconrel.2023.10.053 PMID: 37924956

Sujatha, P., Karunakaran, G., & Rajendran, S. (2022). Smart nanocarrier-based targeted delivery of pesticides: A review. *Environmental Nanotechnology, Monitoring & Management, 17*, 100628.

Sulak, M., & Yardımcı, B. K. (2025). *Green synthesis of tin and titanium nanoparticles using edible plant extracts: exploring their anticancer and antifungal activities*. Biomass Conv. Bioref., DOI: 10.1007/s13399-025-06745-9

Sun, C., & Gallaghar, R. G., E. caspersun@comcast. ne. (2008). *Nanomaterials: new challenges in environmental health and safety.*

Sun, Z., Martinez, A., & Wang, F. (2016). Optical modulators with 2D layered materials. In Nature Photonics (Vol. 10, Issue 4, pp. 227–238). Nature Publishing Group. DOI: 10.1038/nphoton.2016.15

Sun, D.. (2020). Foliar uptake and distribution of silica nanoparticles in plants. *Scientific Reports, 10*(1), 12345. PMID: 32704070

Sun, H., Cao, Y., Kim, D., & Marelli, B. (2022). Biomaterials technology for agrofood resilience. *Advanced Functional Materials, 32*(30), 2201930. DOI: 10.1002/adfm.202201930

Sun, H., Lei, C., Xu, J., & Li, R. (2021). Foliar uptake and leaf-to-root translocation of nanoplastics with different coating charge in maize plants. *Journal of Hazardous Materials, 416*, 125854. DOI: 10.1016/j.jhazmat.2021.125854 PMID: 33892383

Sun, Q., Li, J., & Le, T. (2018). Zinc oxide nanoparticle as a novel class of antifungal agents: Current advances and future perspectives. *Journal of Agricultural and Food Chemistry, 66*(43), 11209–11220. DOI: 10.1021/acs.jafc.8b03210 PMID: 30299956

Sun, Q., Liu, Y., Liu, H., & Dumroese, R. K. (2020). Interaction of biochar type and rhizobia inoculation increases the growth and biological nitrogen fixation of robinia pseudoacacia seedlings. *Forests, 11*(6), 1–12. DOI: 10.3390/f11060711

Sun, Y., Ma, L., Ma, J., Li, B., Zhu, Y., & Chen, F. (2022). Combined application of plant growth-promoting bacteria and iron oxide nanoparticles ameliorates the toxic effects of arsenic in Ajwain (*Trachyspermum ammi* L.). *Frontiers in Plant Science*, *13*, 1098755. Advance online publication. DOI: 10.3389/fpls.2022.1098755 PMID: 36643291

Suriyaprabha, R., Karunakaran, G., Yuvakkumar, R., Rajendran, V., & Kannan, N. (2012). Silica nanoparticles for increased silica availability in maize (*Zea mays*. L) seeds under hydroponic conditions. *Current Nanoscience*, *8*(6), 902–908. DOI: 10.2174/157341312803989033

Su, Y., Ashworth, V., Kim, C., Adeleye, A. S., Rolshausen, P., Roper, C., White, J., & Jassby, D. (2019). Delivery, uptake, fate, and transport of engineered nanoparticles in plants: A critical review and data analysis. *Environmental Science. Nano*, *6*(8), 2311–2331. DOI: 10.1039/C9EN00461K

Su, Y., Zhou, X., Meng, H., Xia, T., Liu, H., Rolshausen, P., Roper, C., McLean, J. E., Zhang, Y., Keller, A. A., & Jassby, D. (2022). Cost–benefit analysis of nanofertilizers and nanopesticides emphasizes the need to improve the efficiency of nanoformulations for widescale adoption. *Nature Food*, *3*(12), 1020–1030. DOI: 10.1038/s43016-022-00647-z PMID: 37118298

Su, Z., Yao, C., Tipper, J., Yang, L., Xu, X., Chen, X., & Zheng, Y. (2023). Nanostrategy of targeting at embryonic trophoblast cells using CuO nanoparticles for female contraception. *ACS Nano*, *17*(24), 25185–25204. DOI: 10.1021/acsnano.3c08267 PMID: 38088330

Szopa, D., Mielczarek, M., Skrzypczak, D., Izydorczyk, G., Mikula, K., Chojnacka, K., & Witek-Krowiak, A. (2022). Encapsulation efficiency and survival of plant growth-promoting microorganisms in an alginate-based matrix – A systematic review and protocol for a practical approach. *Industrial Crops and Products*, *181*(March), 114846. Advance online publication. DOI: 10.1016/j.indcrop.2022.114846

Tagad, C. K., Dugasani, S. R., Aiyer, R., Park, S., Kulkarni, A., & Sabharwal, S. (2013). Green synthesis of silver nanoparticles and their application for the development of optical fiber based hydrogen peroxide sensor. *Sensors and Actuators. B, Chemical*, *183*, 144–149. DOI: 10.1016/j.snb.2013.03.106

Takemura, K. (2021). Surface plasmon resonance (SPR)-and localized SPR (LSPR)-based virus sensing systems: Optical vibration of nano-and micro-metallic materials for the development of next-generation virus detection technology. *Biosensors (Basel)*, *11*(8), 250. DOI: 10.3390/bios11080250 PMID: 34436053

Takeshita, V., Campos, E. V., Rodrigues, J. S., & Fraceto, L. F. (2023). Opinion: Hybrid nanoparticle systems–Two-way delivery approach for agriculture. *Plant Nano Biology*, *6*, 100053. DOI: 10.1016/j.plana.2023.100053

Takkar, M. K., & Gumber, K. (2024). Engineered Nano-Carriers for Stimuli Controlled Release of Micronutrient in Plants. https://doi.org/DOI: 10.21203/rs.3.rs-4677814/v1

Tamayo, L. A., Zapata, P. A., & Vejar, N. D. (2014). Nanoparticle-loaded hydrogels for antifungal treatments. *Journal of Biomedical Nanotechnology*, *10*(7), 1419–1430.

Tamjidi, F., Shahedi, M., Varshosaz, J., & Nasirpour, A. (2013). Nanostructured lipid carriers (NLC): A potential delivery system for bioactive food molecules. *Innovative Food Science & Emerging Technologies*, *19*, 29–43. DOI: 10.1016/j.ifset.2013.03.002

Tang, S., Liang, J., Li, O., Shao, N., Jin, Y., Ni, J., Fei, X., & Li, Z. (2023). Morphology-Tailored Hydroxyapatite Nanocarrier for Rhizosphere-Targeted Phosphorus Delivery. *Small, 19*(14), 2206954. DOI: 10.1002/smll.202206954 PMID: 36599675

Taning, C. N., Arpaia, S., Christiaens, O., Dietz-Pfeilstetter, A., Jones, H., Mezzetti, B., Sabbadini, S., Sorteberg, H. G., Sweet, J., Ventura, V., & Smagghe, G. (2020). RNA-based biocontrol compounds: Current status and perspectives to reach the market. *Pest Management Science*, *76*(3), 841–845. DOI: 10.1002/ps.5686 PMID: 31743573

Tao, R., You, C., Qu, Q., Zhang, X., Deng, Y., Ma, W., & Huang, C. (2023). Recent advances in the design of controlled-and sustained-release micro/nanocarriers of pesticide. *Environmental Science. Nano*, *10*(2), 351–371. DOI: 10.1039/D2EN00446A

Tarannum, N., Pooja, K., Jakhar, S., & Mavi, A. (2024). Nanoparticles assisted intra and transdermic delivery of antifungal ointment: An updated review. *Discover Nano*, *19*(1), 11. DOI: 10.1186/s11671-023-03932-3 PMID: 38195832

Tavares, T. D., Antunes, J. C., Padrão, J., Ribeiro, A. I., Zille, A., Amorim, M. T. P., Ferreira, F., & Felgueiras, H. P. (2020). Activity of Specialized Biomolecules against Gram-Positive and Gram-Negative Bacteria. *Antibiotics 2020, Vol. 9, Page 314, 9*(6), 314. DOI: 10.3390/antibiotics9060314

Tawfik, M. M., Mohamed, M. H., Sadak, M. S., & Thalooth, A. T. (2021). Iron oxide nanoparticles effect on growth, physiological traits and nutritional contents of *Moringa oleifera* grown in saline environment. *Bulletin of the National Research Center*, *45*(1), 1–9. DOI: 10.1186/s42269-021-00624-9

Tawiah, B., Ofori, E. A., & George, S. C. (2024). Nanotechnology in societal development. In *Nanotechnology in Societal Development* (pp. 1–64). Springer Nature Singapore. DOI: 10.1007/978-981-97-6184-5_1

Tester, R. F., Karkalas, J., & Qi, X. (2004). Starch - Composition, fine structure and architecture. *Journal of Cereal Science*, *39*(2), 151–165. DOI: 10.1016/j.jcs.2003.12.001

Thakur, A., Bharti, R., & Sharma, R. (2022). Carbon nanotubes: Types, synthesis, cytotoxicity and applications in biomedical. *Materials Today: Proceedings*, *50*, 2256–2268. DOI: 10.1016/j.matpr.2021.10.002

Thakur, A., & Kumar, A. (2022). Recent advances on rapid detection and remediation of environmental pollutants utilizing nanomaterials-based (bio) sensors. *The Science of the Total Environment*, *834*, 155219. DOI: 10.1016/j.scitotenv.2022.155219 PMID: 35421493

Thanh, V. M., Bui, L. M., Bach, L. G., Nguyen, N. T., Thi, H. L., & Hoang Thi, T. T. (2019). Origanum majorana L. Essential Oil-Associated Polymeric Nano Dendrimer for Antifungal Activity against *Phytophthora infestans.Materials (Basel)*, *12*(9), 1446. DOI: 10.3390/ma12091446 PMID: 31060208

Thapa, D., Kulkarni, N., Mishra, S. N., Paulose, P. L., & Ayyub, P. (2010). Enhanced magnetization in cubic ferrimagnetic CuFe2O4 nanoparticles synthesized from a citrate precursor: The role of Fe2+. *Journal of Physics. D, Applied Physics*, *43*(19), 195004. DOI: 10.1088/0022-3727/43/19/195004

Thavaseelan, D., & Priyadarshana, G. (2021). Nanofertilizer use for sustainable agriculture. *J Res Technol Eng*, *2*(1), 41–59.

Thomas, P., Rumjit, N. P., George, P. J., Lai, C. W., Tyagi, P., Johan, M. R. B., & Saravanakumar, M. P. (2020). Remediation of heavy metal Ions using nanomaterials sourced from wastewaters. *Nanotechnology for Food. Agriculture, and Environment*, •••, 255–296.

Thombare, N.. (2021). *Guar gum based hydrogel as controlled micronutrient delivery system: Mechanism and kinetics of boron release for agricultural applications* (Vol. 112). Biopolymers.

Tian, J., Wong, K. K. Y., Ho, C. M., Lok, C. N., Yu, W. Y., Che, C. M., Chiu, J. F., & Tam, P. K. H. (2007). Topical delivery of silver nanoparticles promotes wound healing. *ChemMedChem*, *2*(1), 129–136. DOI: 10.1002/cmdc.200600171 PMID: 17075952

Tian, L. (2002). Development of a sensor-based precision herbicide application system. *Computers and Electronics in Agriculture*, *36*(2-3), 133–149. DOI: 10.1016/S0168-1699(02)00097-2

Tichapondwa, S. M., Newman, J. P., & Kubheka, O. (2020). Effect of TiO2 phase on the photocatalytic degradation of methylene blue dye. *Physics and Chemistry of the Earth Parts A/B/C*, *118–119*, 102900. DOI: 10.1016/j.pce.2020.102900

Tijing, L. D., Dizon, J. R. C., Ibrahim, I., Nisay, A. R. N., Shon, H. K., & Advincula, R. C. (2020). 3D printing for membrane separation, desalination and water treatment. In *Applied Materials Today* (Vol. 18). Elsevier Ltd., DOI: 10.1016/j.apmt.2019.100486

Tilman, D., Cassman, K. G., Matson, P. A., Naylor, R., & Polasky, S. (2002). Agricultural sustainability and intensive production practices. *Nature*, *418*(6898), 671–677. DOI: 10.1038/nature01014 PMID: 12167873

Timofeeva, A. M., Galyamova, M. R., & Sedykh, S. E. (2022). Bacterial Siderophores: Classification, Biosynthesis, Perspectives of Use in Agriculture. *Plants*, *11*(22), 3065. DOI: 10.3390/plants11223065 PMID: 36432794

Timuralp, C., Ercan, E., Ercetin, U., Aygun, A., Esra Altuner, E., & Sen, F. (2024). Nanoencapsulation of phase change material with CuO nanoparticles: Development of thermal properties for energy storage. *Inorganic Chemistry Communications*, *170*, 113258. DOI: 10.1016/j.inoche.2024.113258

Tiwari, S., Dubey, N. K., & Kumar, C. (2023). Phytochemicals as an Eco-Friendly Source for Sustainable Management of Soil-Borne Plant Pathogens in Soil Ecosystem—Agroecological Approaches for Sustainable Soil Management—*Wiley Online Library*. https://onlinelibrary.wiley.com/doi/abs/10.1002/9781119911999.ch13

Tiwari, P., Nayak, P., Prusty, S. K., & Sahu, P. K. (2018). Phytochemistry and pharmacology of tinospora cordifolia: A review. *Systematic Reviews in Pharmacy*, *9*(1), 70–78. DOI: 10.5530/srp.2018.1.14

Tleuova, A. B., Wielogorska, E., Talluri, V. P., Štěpánek, F., Elliott, C. T., & Grigoriev, D. O. (2020). Recent advances and remaining barriers to producing novel formulations of fungicides for safe and sustainable agriculture. *Journal of Controlled Release*, *326*, 468–481. DOI: 10.1016/j.jconrel.2020.07.035 PMID: 32721524

Tomar, B., Patle, T., Parihar, S. S., Singh, P., & Tomar, S. S. (2024). Enhancing Nutrient Uptake With Nano Fertilizers and Soil Amendments. In *Harnessing NanoOmics and Nanozymes for Sustainable Agriculture* (pp. 43–59). IGI Global., DOI: 10.4018/979-8-3693-1890-4.ch003

Tombuloglu, H., Ercan, I., Alshammari, T., Tombuloglu, G., Slimani, Y., Almessiere, M., & Baykal, A. (2020). Incorporation of micro-nutrients (nickel, copper, zinc, and iron) into plant body through nanoparticles. *Journal of Soil Science and Plant Nutrition*, *20*(4), 1872–1881. DOI: 10.1007/s42729-020-00258-2

Tong, Y. J., Shao, L. H., Li, X. L., Lu, J. Q., Sun, H. L., Xiang, S., Zhang, Z. H., Wu, Y., & Wu, X. M. (2018). Adhesive and stimulus sesponsive polydopamine-coated graphene oxide system for pesticide-loss control. *Journal of Agricultural and Food Chemistry*, *66*(11), 2616–2622. DOI: 10.1021/acs.jafc.7b05500 PMID: 29485869

Torney, F., Trewyn, B. G., Lin, V. S.-Y., & Wang, K. (2007). Mesoporous silica nanoparticles deliver DNA and chemicals into plants. *Nature Nanotechnology*, *2*(5), 295–300. DOI: 10.1038/nnano.2007.108 PMID: 18654287

Torres, J. F., Rubilar, O., Tortella, G., Diez, M. C., & Carrión, O. (2020). Nanotechnology and plant biotechnology: Advances and challenges in agroindustry. *Plant Science*, *297*, 110503.

Torti, S., Schlesier, R., Thümmler, A., Bartels, D., Römer, P., Koch, B., Werner, S., Panwar, V., Kanyuka, K., Wirén, N. V., Jones, J. D. G., Hause, G., Giritch, A., & Gleba, Y. (2021). Transient reprogramming of crop plants for agronomic performance. *Nature Plants*, *7*(2), 159–171. DOI: 10.1038/s41477-021-00851-y PMID: 33594264

Tortorici, S., Cimino, C., Ricupero, M., Musumeci, T., Biondi, A., Siscaro, G., Carbone, C., & Zappalà, L. (2022). Nanostructured lipid carriers of essential oils as potential tools for the sustainable control of insect pests. *Industrial Crops and Products*, *181*, 114766. DOI: 10.1016/j.indcrop.2022.114766

Tripathi, D. K., Shweta, Singh, S., Singh, S., Pandey, R., Singh, V. P., Sharma, N. C., Prasad, S. M., Dubey, N. K., & Chauhan, D. K. (2017). An overview on manufactured nanoparticles in plants: Uptake, translocation, accumulation and phytotoxicity. *Plant physiology and biochemistry: PPB*, *110*, 2–12. https://doi.org/DOI: 10.1016/j.plaphy.2016.07.030

Tripathi, D., Singh, M., & Pandey-Rai, S. (2022). Crosstalk of nanoparticles and phytohormones regulate plant growth and metabolism under abiotic and biotic stress. *Plant Stress*, *6*, 100107. DOI: 10.1016/j.stress.2022.100107

Trivedi, P., Leach, J. E., Tringe, S. G., Sa, T., & Singh, B. K. (2020). Plant–microbiome interactions: From community assembly to plant health. *Nature Reviews. Microbiology*, *18*(11), 1–15. DOI: 10.1038/s41579-020-0412-1 PMID: 32788714

Trivedi, P., & Pandey, A. (2008). Recovery of plant growth-promoting rhizobacteria from sodium alginate beads after 3 years following storage at 4°C. *Journal of Industrial Microbiology & Biotechnology*, *35*(3), 205–209. DOI: 10.1007/s10295-007-0284-7 PMID: 18087737

Truong, H., Fournier, C., Hichami, S., Dole, L., Leroy, F., Wendehenne, D., Nicolas-Francès, V., & Besson-Bard, A. (2024). Apo-siderophores promote growth of iron-deficient Arabidopsis plants by mobilizing iron from roots to shoots and reducing oxidative stress in roots. *Plant Stress*, *12*, 100488. DOI: 10.1016/j.stress.2024.100488

Tudi, M., Daniel Ruan, H., Wang, L., Lyu, J., Sadler, R., Connell, D., Chu, C., & Phung, D. T. (2021). Agriculture Development, Pesticide Application and Its Impact on the Environment. *International Journal of Environmental Research and Public Health, 18*(3), 1112. DOI: 10.3390/ijerph18031112 PMID: 33513796

Tu, L., He, Y., Yang, H., Wu, Z., & Yi, L. (2015). Preparation and characterization of alginate-gelatin microencapsulated *Bacillus subtilis* SL-13 by emulsification/internal gelation. *Journal of Biomaterials Science. Polymer Edition, 26*(12), 735–749. DOI: 10.1080/09205063.2015.1056075 PMID: 26159659

Tunca, E. Ü. (2016). Nanoteknolojinin Temeli Nanopartiküller Ve Nanopartiküllerin Fitoremediasyon U. *Ordu University Journal of Science and Tecnology, 5*(2), 23–34. Retrieved from https://dergipark.org.tr/en/pub/ordubtd/issue/11070/132200

Tungare, K., Gupta, J., Bhori, M., Garse, S., Kadam, A., Jha, P., Jobby, R., Amanullah, M., & Vijayakumar, S. (2024). Nanomaterial in controlling biofilms and virulence of microbial pathogens. *Microbial Pathogenesis, 192*, 106722. DOI: 10.1016/j.micpath.2024.106722 PMID: 38815775

Türkaslan, B., Aktaş, M., & Akar, S.TÜRKASLAN. (2020). B., AKTAŞ, M., & AKAR, S. (2020). Fonksiyonelleştirilmiş Titanyum Dioksit Nanotüp Sentezi. *Süleyman Demirel Üniversitesi Fen Edebiyat Fakültesi Fen Dergisi, 15*(1), 167–176. DOI: 10.29233/sdufeffd.666020

Tysnes, O. B., & Storstein, A. (2017). Epidemiology of Parkinson's disease. *Journal of Neural Transmission (Vienna, Austria), 124*(8), 901–905. DOI: 10.1007/s00702-017-1686-y PMID: 28150045

Tzeng, S. Y., & Green, J. J. (2013). Subtle changes to polymer structure and degradation mechanism enable highly effective nanoparticles for siRNA and DNA delivery to human brain cancer. *Advanced Healthcare Materials, 2*(3), 468–480. DOI: 10.1002/adhm.201200257 PMID: 23184674

Ullah, H., Khan, I., Mustafa, G., Shehzad, J., Hasan, M., Saeed, U., & Chaudhari, S. K. (2023). Molecular characterization and mitigative role of silicon dioxide nanoparticles in Ocimum basilicum under lead (Pb) stress. *Silicon, 15*(6), 2551–2565. DOI: 10.1007/s12633-022-02178-5

Ullah, J., Gul, A., Khan, I., Shehzad, J., Kausar, R., Ahmed, M. S., & Mustafa, G. (2024). Green synthesized iron oxide nanoparticles as a potential regulator of callus growth, plant physiology, antioxidative and microbial contamination in Oryza sativa L. *BMC Plant Biology, 24*(1), 939. DOI: 10.1186/s12870-024-05627-y PMID: 39385076

Ulvila, J., Parikka, M., Kleino, A., Sormunen, R., Ezekowitz, R. A., Kocks, C., & Ramet, M. (2006). Double-stranded RNA is internalized by scavenger receptor-mediated endocytosis in Drosophila S2 Cells. *The Journal of Biological Chemistry, 281*(20), 14370–14375. DOI: 10.1074/jbc.M513868200 PMID: 16531407

Umavathi, S., Mahboob, S., Govindarajan, M., Al-Ghanim, K. A., Ahmed, Z., Virik, P., Al-Mulhm, N., Subash, M., Gopinath, K., & Kavitha, C. (2021). Green synthesis of ZnO nanoparticles for antimicrobial and vegetative growth applications: A novel approach for advancing efficient high quality health care to human wellbeing. *Saudi Journal of Biological Sciences*, *28*(3), 1808–1815. DOI: 10.1016/j.sjbs.2020.12.025 PMID: 33732066

Umbarkar, M., Thakare, S., Surushe, T., Giri, A., & Chopade, V. (2021). Formulation and Evaluation of Liposome by Thin Film Hydration Method: Formulation of Liposome. *Journal of Drug Delivery and Therapeutics*, *11*(1), 72–76. DOI: 10.22270/jddt.v11i1.4677

Umegaki, T., Yan, J. M., Zhang, X. B., Shioyama, H., Kuriyama, N., & Xu, Q. (2009). Hollow Ni-SiO2 nanosphere-catalyzed hydrolytic dehydrogenation of ammonia borane for chemical hydrogen storage. *Journal of Power Sources*, *191*(2), 209–216. DOI: 10.1016/j.jpowsour.2009.02.054

Upadhyay, S. K., Srivastava, A. K., Rajput, V. D., Chauhan, P. K., Bhojiya, A. A., Jain, D., Chaubey, G., Dwivedi, P., Sharma, B., & Minkina, T. (2022). Root Exudates: Mechanistic Insight of Plant Growth Promoting Rhizobacteria for Sustainable Crop Production. *Frontiers in Microbiology*, *13*, 916488. DOI: 10.3389/fmicb.2022.916488 PMID: 35910633

Usmani, A., Mishra, A., Jafri, A., Arshad, M., & Siddiqui, M. A. (2019). Green Synthesis of Silver Nanocomposites of Nigella sativa Seeds Extract for Hepatocellular Carcinoma. *Current Nanomaterials*, *4*(3), 191–200. DOI: 10.2174/2468187309666190906130115

Usman, M., Farooq, M., Wakeel, A., Nawaz, A., Cheema, S. A., Rehman, H., Ashraf, I., & Sanaullah, M. (2020). Nanotechnology in agriculture: Current status, challenges and future opportunities. *The Science of the Total Environment*, *721*, 137778. DOI: 10.1016/j.scitotenv.2020.137778 PMID: 32179352

Utazi, E. B., Oyewole, O. A., Yakubu, J. G., Tsado, P. Y., Omoregie, I. P., Adetunji, C. O., Mathew, J. T., Igiku, V., Eniola, K. I. T., & Yerima, M. B. (2024). Toxicological Effects of Nanobiofertilizer on Water Body, Water Quality, Lower Plants, Zooplanktons, and Beneficial Microorganisms. In Book: Handbook of Agricultural Biotechnology. Wiley, 5, 201-232. Doi: DOI: 10.1002/9781394211548.ch8

Valenstein, J. S., Lin, V. S.-Y., Lyznik, L. A., Martin-Ortigosa, S., Wang, K., & Peterson, D. J.. (2013). Mesoporous silica nanoparticle-mediated intracellular cre protein delivery for maize genome editing via loxP site excision. *Plant Physiology*, *164*(2), 537–547. DOI: 10.1104/pp.113.233650 PMID: 24376280

Van Dijk, A. D. J., Kootstra, G., Kruijer, W., & De Ridder, D. (2021). Machine learning in plant science and plant breeding. *iScience*, *24*(1), 101890. DOI: 10.1016/j.isci.2020.101890 PMID: 33364579

van Droffelaar, B., & Jacobs, M. (2018). Nature-Based Training Program Fosters Authentic Leadership. *Journal of Leadership Studies*, *12*(3), 7–18. DOI: 10.1002/jls.21569

Vance, C. P. (2001). Symbiotic Nitrogen Fixation and Phosphorus Acquisition. Plant Nutrition in a World of Declining Renewable Resources. *Plant Physiology*, *127*(2), 390–397. DOI: 10.1104/pp.010331 PMID: 11598215

Vanti, G. L., Nargund, V. B., Basavesha, K. N., Vanarchi, R., Kurjogi, M., & Mulla, S. I.. (2019). Synthesis of *Gossypium hirsutum*-derived silver nanoparticles and their antibacterial efficacy against plant pathogens. *Applied Organometallic Chemistry*, *33*(1), e4630. DOI: 10.1002/aoc.4630

Varadachari, C., & Goertz, H. M. (2010). *Slow-release and controlled-release nitrogen fertilizers*. Indian Nitrogen Group, Society.

Varkouhi, A. K., Scholte, M., Storm, G., & Haisma, H. J. (2011). Endosomal escape pathways for delivery of biologicals. *Journal of Controlled Release*, *151*(3), 220–228. DOI: 10.1016/j.jconrel.2010.11.004 PMID: 21078351

Varlamov, V. P., & Mysyakina, I. S. (2018). Chitosan in biology, microbiology, medicine, and agriculture. *Microbiology*, *87*(5), 712–715. DOI: 10.1134/S0026261718050168

Vasanthkumar, R., Baskar, V., Vinoth, S., Roshna, K., Mary, T. N., Alagupandi, R., Saravanan, K., Radhakrishnan, R., Arun, M., & Gurusaravanan, P. (2024). Biogenic carbon quantum dots from marine endophytic fungi (*Aspergillus flavus*) to enhance the curcumin production and growth in *Curcuma longa* L. *Plant Physiology and Biochemistry*, *211*, 108644. DOI: 10.1016/j.plaphy.2024.108644 PMID: 38710114

Vauthier, C., & Bouchemal, K. (2009). Methods for the preparation and manufacture of polymeric nanoparticles. *Pharmaceutical Research*, *26*(5), 1025–1058. DOI: 10.1007/s11095-008-9800-3 PMID: 19107579

Vega-Vásquez, P., Mosier, N. S., & Irudayaraj, J. (2020). Nanoscale drug delivery systems: From medicine to agriculture. *Frontiers in Bioengineering and Biotechnology*, *8*, 79. DOI: 10.3389/fbioe.2020.00079 PMID: 32133353

Veglió, F., Beolchini, F., & Gasbarro, A. (1997). Biosorption of toxic metals: An equilibrium study using free cells of Arthrobacter sp. *Process Biochemistry (Barking, London, England)*, *32*(2), 99–105. DOI: 10.1016/S0032-9592(96)00047-7

Vejlupkova, Z., Warman, C., Sharma, R., Scheller, H. V., Mortimer, J. C., & Fowler, J. E. (2020). No evidence for transient transformation via pollen magnetofection in several monocot species. *Nature Plants*, *6*(11), 1323–1324. DOI: 10.1038/s41477-020-00798-6 PMID: 33139861

Venditti, I. (2017). Morphologies and functionalities of polymeric nanocarriers as chemical tools for drug delivery: A review. *Journal of King Saud University. Science*.

Venkatachalam, M., Singaravelu, G., Govindaraju, K., & Ahn, J. S. (2013). *PTP 1B inhibitory action of a phytochemical propanoic acid, 2-(3-acetoxy-4,4,14-trimethylandrost-8-en-17-yl)*. Current Science, JSTOR. Retrieved from https://www.jstor.org/stable/24097522

Venugopal, P. D., & Dively, G. P. (2017). Climate change, transgenic corn adoption and field-evolved resistance in corn earworm. *Royal Society Open Science*, *4*(6), 170210. DOI: 10.1098/rsos.170210 PMID: 28680673

Vercelheze, A. E. S., Marim, B. M., Oliveira, A. L. M., & Mali, S. (2019). Development of biodegradable coatings for maize seeds and their application for Azospirillum brasilense immobilization. *Applied Microbiology and Biotechnology*, *103*(5), 2193–2203. DOI: 10.1007/s00253-019-09646-w PMID: 30685816

Verma, K. K., Song, X. P., Joshi, A., Tian, D. D., Rajput, V. D., Singh, M., Arora, J., Minkina, T., & Li, Y. R. (2022). Recent Trends in Nano-Fertilizers for Sustainable Agriculture under Climate Change for Global Food Security. *Nanomaterials 2022, Vol. 12, Page 173*, *12*(1), 173. DOI: 10.3390/nano12010173

Verma, P., & Ratan, J. K. (2020). Assessment of the negative effects of various inorganic water pollutants on the biosphere—an overview. *Inorganic Pollutants in Water*, 73–96. DOI: 10.1016/B978-0-12-818965-8.00005-6

Vessey, J. K. (2003). Plant growth promoting rhizobacteria as biofertilizers. *Plant and Soil, 255*(2), 571–586. DOI: 10.1023/A:1026037216893

Victoria, J., Tripathi, S., Prakash, V., Tiwari, K., Mahra, S., Sharma, A., Rana, S., Kandhol, N., Sahi, S., Tripathi, D. K., & Sharma, S. (2024). Encapsulated nanopesticides application in plant protection: Quo vadis? *Plant Physiology and Biochemistry, 206*, 108225. DOI: 10.1016/j.plaphy.2023.108225 PMID: 38147708

Vijayakumar, P. S., Abhilash, O. U., Khan, B. M., & Prasad, B. L. V. (2010). Nanogold-loaded sharp edged carbon bullets as plant-gene carriers. *Advanced Functional Materials, 20*(15), 2416–2423. DOI: 10.1002/adfm.200901883

Vijayaraghavan, K., & Ashokkumar, T. (2017). Plant-mediated biosynthesis of metallic nanoparticles: A review of literature, factors affecting synthesis, characterization techniques and applications. *Journal of Environmental Chemical Engineering, 5*(5), 4866–4883. DOI: 10.1016/j.jece.2017.09.026

Vinzant, K., Rashid, M., Clouse, D. E., Ghosh, P., Quadir, M., Davis, V. A., & Khodakovskaya, M. V. (2025). From Plants to Plants: Plant-Derived Biological Polymers as Sustainable and Safe Nanocarriers for Direct Delivery of DNA to Plant Cells. *Nano Letters, 25*(14), 5572–5581. DOI: 10.1021/acs.nanolett.4c05489 PMID: 40139733

Vinzant, K., Rashid, M., & Khodakovskaya, M. V. (2023). Advanced applications of sustainable and biological nano-polymers in agricultural production. *Frontiers in Plant Science, 13*, 1081165. DOI: 10.3389/fpls.2022.1081165 PMID: 36684740

Vithanage, M., Seneviratne, M., Ahmad, M., Sarkar, B., & Ok, Y. S. (2017). Contrasting effects of engineered carbon nanotubes on plants: A review. *Environmental Geochemistry and Health, 39*(6), 1421–1439. DOI: 10.1007/s10653-017-9957-y PMID: 28444473

Vogel, E., Santos, D., Mingels, L., Verdonckt, T. W., & Broeck, J. V. (2019). RNA interference in insects: Protecting beneficials and controlling pests. *Frontiers in Physiology, 9*, 1912. DOI: 10.3389/fphys.2018.01912 PMID: 30687124

Voko, M. P., Aremu, A. O., Makunga, N. P., Nisler, J., Doležal, K., & Masondo, N. A. (2024). The potential applications of cytokinins and cytokinin oxidase/dehydrogenase inhibitors for mitigating abiotic stresses in model and non-model plant species. *Current Plant Biology, 40*, 100398. DOI: 10.1016/j.cpb.2024.100398

Voukkali, I., Papamichael, I., Loizia, P., Economou, F., Stylianou, M., Naddeo, V., & Zorpas, A. A. (2024). Fashioning the Future: Green chemistry and engineering innovations in biofashion. *Chemical Engineering Journal, 497*, 155039. DOI: 10.1016/j.cej.2024.155039

Vurro, M., Miguel-Rojas, C., & Pérez-de-Luque, A. (2019). Safe nanotechnologies for increasing the effectiveness of environmentally friendly natural agrochemicals. *Pest Management Science, 75*(9), 2403–2412. DOI: 10.1002/ps.5348 PMID: 30672106

Vygovsky, Y. N., & Malov, A. N. (2003). *The Physics of Laser Biostimulation*. JSC MILTA-PKP GIT.

Wahab, M., & Janaswamy, S. (2023). A review on biogenic silver nanoparticles as efficient and effective antidiabetic agents. *Functional Food Science - Online ISSN: 2767-3146, 3*(7), 93–107. DOI: 10.31989/ffs.v3i7.1119

Wahab, A., Muhammad, M., Ullah, S., Abdi, G., Shah, G. M., Zaman, W., & Ayaz, A. (2024). Agriculture and environmental management through nanotechnology: Eco-friendly nanomaterial synthesis for soil-plant systems, food safety, and sustainability. *The Science of the Total Environment, 926*, 171862. DOI: 10.1016/j.scitotenv.2024.171862 PMID: 38527538

Waller, F., Achatz, B., Baltruschat, H., Fodor, J., Becker, K., Fischer, M., Heier, T., Hückelhoven, R., Neumann, C., Von Wettstein, D., Franken, P., & Kogel, K. H. (2005). The endophytic fungus *Piriformospora indica* reprograms barley to salt-stress tolerance, disease resistance, and higher yield. *Proceedings of the National Academy of Sciences of the United States of America, 102*(38), 13386–13391. DOI: 10.1073/pnas.0504423102 PMID: 16174735

Waltz, E. (2018). With a free pass, CRISPR-edited plants reach market in record time. *Nature Biotechnology, 36*(1), 6–7. DOI: 10.1038/nbt0118-6b PMID: 29319694

Wang, S., Yu, M., Jiang, J., Zhang, W., Guo, X., Chang, S., & Campbell, M. (2017). Evidence aggregation for answer re-ranking in open-domain question answering. *arXiv preprint arXiv:1711.05116.*

Wang, Zhantong, Gao, H., Zhang, Y., Liu, G., Niu, G., & Chen, X. (2017). Functional ferritin nanoparticles for biomedical applications. *Frontiers of Chemical Science and Engineering 2017 11:4, 11*(4), 633–646. DOI: 10.1007/s11705-017-1620-8

Wang, C., Li, C., Han, Q., Wu, F., & Zou, X. (2023). A performance analysis of a litchi picking robot system for actively removing obstructions, Using an artificial intelligence algorithm. *Agronomy (Basel), 13*(11), 2795. DOI: 10.3390/agronomy13112795

Wang, D., Saleh, N. B., Byro, A., Zepp, R., Sahle-Demessie, E., Luxton, T. P., Ho, K. T., Burgess, R. M., Flury, M., White, J. C., & Su, C. (2022). Nano-enabled pesticides for sustainable agriculture and global food security. *Nature Nanotechnology, 17*(4), 347–360. DOI: 10.1038/s41565-022-01082-8 PMID: 35332293

Wang, H., Qiao, X., Chen, J., & Ding, S. (2005). Preparation of silver nanoparticles by chemical reduction method. *Colloids and Surfaces. A, Physicochemical and Engineering Aspects, 256*(2-3), 111–115. DOI: 10.1016/j.colsurfa.2004.12.058

Wang, J. W., Grandio, E. G., Newkirk, G. M., Demirer, G. S., Butrus, S., Giraldo, J. P., & Landry, M. P. (2019). Nanoparticle-mediated genetic engineering of plants. *Molecular Plant, 12*(8), 1037–1040. DOI: 10.1016/j.molp.2019.06.010 PMID: 31283997

Wang, J., Hao, K., Yu, F., Shen, L., Wang, F., Yang, J., & Su, C. (2022). Field application of nanoliposomes delivered quercetin by inhibiting specific hsp70 gene expression against plant virus disease. *Journal of Nanobiotechnology, 20*(1), 16. DOI: 10.1186/s12951-021-01223-6 PMID: 34983536

Wang, J., Zhang, S., & Zhang, Y. (2010). Fabrication of chronocoulometric DNA sensor based on gold nanoparticles/poly(l-lysine) modified glassy carbon electrode. *Analytical Biochemistry*, *396*(2), 304–309. DOI: 10.1016/j.ab.2009.10.004 PMID: 19818728

Wang, K., Peng, Y., Pu, J., Fu, W., Wang, J., & Han, Z. (2016). Variation in RNAi efficacy among insect species is attributable to dsRNA degradation in vivo. *Insect Biochemistry and Molecular Biology*, *77*, 1–9. DOI: 10.1016/j.ibmb.2016.07.007 PMID: 27449967

Wang, L., Hu, C., & Shao, L. (2018). The antimicrobial activity of nanoparticles: Present situation and prospects for the future. *International Journal of Nanomedicine*, *13*, 1227–1249. PMID: 28243086

Wang, L., Ning, C., Pan, T., & Cai, K. (2022). Role of silica nanoparticles in abiotic and biotic stress tolerance in plants: A review. *International Journal of Molecular Sciences*, *23*(4), 1947. DOI: 10.3390/ijms23041947 PMID: 35216062

Wang, M., Hu, Z., Yang, T., Pei, H., & Zhang, F. (2023). A dual pesticide–fertilizer silicon-base nanocomposite to synergistically control fungal disease and provide nutrition. *Environmental Science. Nano*, *10*(12), 3462–3475. DOI: 10.1039/D3EN00500C

Wang, N., Cheng, X. J., Li, N., Wang, H., & Chen, H. Y. (2019). Nanocarriers and their loading strategies. *Advanced Healthcare Materials*, *8*(6), 1801002. DOI: 10.1002/adhm.201801002 PMID: 30450761

Wang, P.. (2016). Transport and fate of nanoparticles in plants: Current knowledge and future perspectives. *Environmental Science & Technology*, *50*(10), 6001–6014.

Wang, P., Lombi, E., Zhao, F. J., & Kopittke, P. M. (2016). Nanotechnology: A New Opportunity in Plant Sciences. *Trends in Plant Science*, *21*(8), 699–712. DOI: 10.1016/j.tplants.2016.04.005 PMID: 27130471

Wang, P., Lombi, E., Zhao, F. J., & Kopittke, P. M. (2021). Nanotechnology for plant nutrition: A review. *Nature Nanotechnology*, *16*(3), 247–260. DOI: 10.1038/s41565-021-00893-3

Wang, P., Lombi, E., Zhao, F. J., & Kopittke, P. M. (2021). Nanotechnology for plant nutrition: Opportunities and challenges. *Trends in Plant Science*, *26*(6), 590–600.

Wang, P., Lombi, E., Zhao, F. J., & Kopittke, P. M. (2022). Nanotechnology: A new opportunity in plant sciences. *Trends in Plant Science*, *27*(1), 33–43. DOI: 10.1016/j.tplants.2021.08.008 PMID: 27130471

Wang, P., Zhang, L., Zheng, W., & Guo, Z. (2020). Lipid nanoparticles for nucleic acid delivery in plants. *Journal of Controlled Release*, *323*, 628–649. DOI: 10.1016/j.jconrel.2020.05.007

Wang, Q., Ebbs, S. D., Chen, Y., & Ma, X. (2013). Trans-generational impact of cerium oxide nanoparticles on tomato plants. *Metallomics*, *5*(6), 753–759. DOI: 10.1039/c3mt00033h PMID: 23689668

Wang, W. N., Tarafdar, J. C., & Biswas, P. (2016). Nanoparticle synthesis and their applications in agriculture. *Journal of Nanobiotechnology*, *14*(1), 40. PMID: 27278090

Wang, W., Raliya, R., Sahle-Demessie, E., & Biswas, P. (2016). Nanoparticle transport in plants: Current knowledge and future perspectives. *Environmental Science. Nano*, *3*(6), 1044–1060.

Wang, X., Han, H., Liu, X., Gu, X., Chen, K., & Lu, D. (2012). Multi-walled carbon nanotubes can enhance root elongation of wheat (*Triticum aestivum*) plants. *Journal of Nanoparticle Research*, *14*(6), 841. DOI: 10.1007/s11051-012-0841-5

Wang, X., Xie, H., Wang, P., & Yin, H. (2023). Nanoparticles in plants: Uptake, transport and physiological activity in leaf and root. *Materials (Basel)*, *16*(8), 3097. DOI: 10.3390/ma16083097 PMID: 37109933

Wang, X., & Zhang, W. (2022). The Janus of Protein Corona on nanoparticles for tumor targeting, immunotherapy and diagnosis. *Journal of Controlled Release*, *345*, 832–850. DOI: 10.1016/j.jconrel.2022.03.056 PMID: 35367478

Wang, X., Zheng, K., Cheng, W., Li, J., Liang, X., Shen, J., Dou, D., Yin, M., & Yan, S. (2021). Field application of star polymer-delivered chitosan to amplify plant defense against potato late blight. *Chemical Engineering Journal*, *417*, 129327. DOI: 10.1016/j.cej.2021.129327

Wang, Y., Li, C. N., Wang, T., Li, X. G., & Li, X. (2020). Polylactic Acid–Graphene Oxide-based Materials for loading and sustained release of poorly soluble pesticides. *Langmuir*, *36*(41), 12336–12345. DOI: 10.1021/acs.langmuir.0c02320 PMID: 32988210

Wang, Y., Li, M., Ying, J., Shen, J., Dou, D., Yin, M., Whisson, S. C., Birch, P. R., Yan, S., & Wang, X. (2023a). High-efficiency green management of potato late blight by a self-assembled multicomponent nano-bioprotectant. *Nature Communications*, *14*(1), 5622. DOI: 10.1038/s41467-023-41447-8 PMID: 37699893

Wang, Y., Li, X., Liu, L., Liu, B., Wang, F., & Chen, C. (2020). Tissue targeting and ultrasound-targeted microbubble destruction delivery of plasmid DNA and transfection in vitro. *Cellular and Molecular Bioengineering*, *13*(1), 99–112. DOI: 10.1007/s12195-019-00597-w PMID: 32030111

Wang, Z. Y., Yu, X. L., Gao, D. M., Feng, W. Q., Xing, B. S., & Li, F. M. (2010). Effect of nano-rutile TiO_2 and multiwalled carbon nanotubes on the growth of maize (*Zea mays* L.) seedlings and the relevant antioxidant response. *Huan Jing ke Xue=. Huan Jing Ke Xue*, *31*(2), 480–487. PMID: 20391721

Wang, Z., Liao, F., Guo, T., Yang, S., & Zeng, C. (2012). Synthesis of crystalline silver nanoplates and their application for detection of nitrite in foods. *Journal of Electroanalytical Chemistry (Lausanne, Switzerland)*, *664*, 135–138. DOI: 10.1016/j.jelechem.2011.11.006

Wang, Z., Liu, G., Zheng, H., & Chen, X. (2014). Rigid nanoparticle-based delivery of anti-cancer siRNA: Challenges and opportunities. *Biotechnology Advances*, *32*(4), 831–843. DOI: 10.1016/j.biotechadv.2013.08.020 PMID: 24013011

Wang, Z., Li, Y., Zhang, B., Gao, X., Shi, M., Zhang, S., Zhong, S., Zheng, Y., & Liu, X. (2023b). Functionalized carbon dot-delivered RNA nano fungicides as superior tools to control Phytophthora pathogens through plant RdRP1 mediated spray-induced gene silencing. *Advanced Functional Materials*, *33*(22), 2213143. DOI: 10.1002/adfm.202213143

Wang, Z., Thavarajah, N., & Fernando, X. (2024). Exploring Silica Nanoparticles: A Sustainable Solution for Pest Control in Sri Lankan Rice Farming. *Technologies*, *12*(11), 210. DOI: 10.3390/technologies12110210

Wang, Z., Zhu, W., Qiu, Y., Yi, X., von dem Bussche, A., Kane, A., Gao, H., Koski, K., & Hurt, R. (2016). Biological and environmental interactions of emerging two-dimensional nanomaterials. *Chemical Society Reviews*, *45*(6), 1750–1780. DOI: 10.1039/C5CS00914F PMID: 26923057

Wani, T. A., Masoodi, F. A., Baba, W. N., Ahmad, M., Rahmanian, N., & Jafari, S. M. (2019). Nano-encapsulation of agrochemicals, fertilizers, and pesticides for improved plant production. In *Advances in Phytonanotechnology* (pp. 279–298). Academic Press. DOI: 10.1016/B978-0-12-815322-2.00012-2

Wanyika, H. (2013). Sustained release of fungicide metalaxyl by mesoporous silica nanospheres. *Journal of Nanoparticle Research*, *15*(8), 1831. DOI: 10.1007/s11051-013-1831-y

Warghane, A., Saini, R., Shri, M., Andankar, I., Ghosh, D. K., & Chopade, B. A. (2024). Application of nanoparticles for management of plant viral pathogen: Current status and future prospects. *Virology*, *592*, 109998. DOI: 10.1016/j.virol.2024.109998 PMID: 38301447

Wasule, D. L., Shingote, P. R., & Saxena, S. (2024). Exploitation of functionalized green nanomaterials for plant disease management. *Discover Nano*, *19*(1), 118. DOI: 10.1186/s11671-024-04063-z PMID: 39023655

Weinand, T., El-Hasan, A., & Asch, F. (2023). Role of Bacillus spp. Plant Growth Promoting Properties in Mitigating Biotic and Abiotic Stresses in Lowland Rice (*Oryza sativa* L.). *Microorganisms*, *11*(9), 2327. DOI: 10.3390/microorganisms11092327 PMID: 37764171

Wei, X., Fu, T., He, G., Zhong, Z., Yang, M., Lou, F., & He, T. (2023). Characteristics of rhizosphere and bulk soil microbial community of Chinese cabbage (*Brassica campestris*) grown in Karst area. *Frontiers in Microbiology*, *14*, 1241436. DOI: 10.3389/fmicb.2023.1241436 PMID: 37789857

Welch, R. M., & Shuman, L. (1995). Micronutrient nutrition of plants. *Critical Reviews in Plant Sciences*, *14*(1), 49–82. DOI: 10.1080/07352689509701922

Wen, X. (2019). One-pot route to graft long-chain polymer onto silica nanoparticles and its application for high-performance poly (L -lactide) nanocomposites. *RSC Advances*, *9*(24), 13908–13915. DOI: 10.1039/C9RA01360A PMID: 35519579

Westermayr, J., & Marquetand, P. (2021). Machine Learning for Electronically Excited States of Molecules. *Chemical Reviews*, *121*(16), 9873–9926. DOI: 10.1021/acs.chemrev.0c00749 PMID: 33211478

Whyard, S., Singh, A. D., & Wong, S. (2009). Ingested double-stranded RNAs can act as species-specific insecticides. *Insect Biochemistry and Molecular Biology*, *39*(11), 824–832. DOI: 10.1016/j.ibmb.2009.09.007 PMID: 19815067

Wilson, M. P., & Schwarzman, M. R. (2009). Toward a new US chemicals policy: Rebuilding the foundation to advance new science, green chemistry, and environmental health. *Environmental Health Perspectives*, *117*(8), 1202–1209. DOI: 10.1289/ehp.0800404 PMID: 19672398

Winston, W. M., Molodowitch, C., & Hunter, C. P. (2002). Systemic RNAi in *C. Elegans* requires the putative transmembrane protein SID-1. *Science*, *295*(5564), 2456–2459. DOI: 10.1126/science.1068836 PMID: 11834782

Wong, A. C., & Wright, D. W. (2016). Size-dependent cellular uptake of DNA functionalized gold nanoparticles. *Small, 12*(40), 5592–5600. DOI: 10.1002/smll.201601697 PMID: 27562251

Wong, I. Y., Bhatia, S. N., & Toner, M. (2013). Nanotechnology: Emerging tools for biology and medicine. *Genes & Development, 27*(22), 2397–2408. DOI: 10.1101/gad.226837.113 PMID: 24240230

Wong, M. H., Misra, R. P., Giraldo, J. P., Kwak, S. Y., Son, Y., Landry, M. P., Swan, J. W., Blankschtein, D., & Strano, M. S. (2016). Lipid Exchange Envelope Penetration (LEEP) of Nanoparticles for Plant Engineering: A Universal Localization Mechanism. *Nano Letters, 16*(2), 1161–1172. DOI: 10.1021/acs.nanolett.5b04467 PMID: 26760228

Woo, J. W., Kim, J., Kwon, S. I., Corvalán, C., Cho, S. W., Kim, H., Kim, S. G., Kim, S. T., Choe, S., & Kim, J. S. (2015). DNA-free genome editing in plants with preassembled CRISPR-Cas9 ribonucleoproteins. *Nature Biotechnology, 33*(11), 1162–1164. DOI: 10.1038/nbt.3389 PMID: 26479191

Worrall, E. A., Bravo-Cazar, A., Nilon, A. T., Fletcher, S. J., Robinson, K. E., Carr, J. P., & Mitter, N. (2019). Exogenous application of RNAi-inducing double-stranded RNA inhibits aphid-mediated transmission of a plant virus. *Frontiers in Plant Science, 10*, 265. DOI: 10.3389/fpls.2019.00265 PMID: 30930914

Worrall, E. A., Hamid, A., Mody, K. T., Mitter, N., & Pappu, H. R. (2018). Nanotechnology for plant disease management. *Agronomy (Basel), 8*(12), 285. DOI: 10.3390/agronomy8120285

Wu, Y., Feng, J., Hu, G., Zhang, E., & Yu, H. H. (2023). Colorimetric Sensors for Chemical and Biological Sensing Applications. *Sensors 2023, Vol. 23, Page 2749, 23*(5), 2749. DOI: 10.3390/s23052749

Wu, C. F., Smith, D. A., Lai, E. M., & Chang, J. H. (2018). The Agrobacterium Type VI Secretion System: A Contractile Nanomachine for Interbacterial Competition. *Current Topics in Microbiology and Immunology, 418*, 215–231. DOI: 10.1007/82_2018_99 PMID: 29992360

Wu, H., & Li, Z. (2022). Recent advances in nano-enabled agriculture for improving plant performance. *The Crop Journal, 10*(1), 1–12. DOI: 10.1016/j.cj.2021.06.002

Wu, H., Nißler, R., Morris, V., Herrmann, N., Hu, P., Jeon, S.-J., Kruss, S., & Giraldo, J. P. (2020). Monitoring plant health with near-infrared fluorescent H_2O_2 nanosensors. *Nano Letters, 20*(4), 2432–2442. DOI: 10.1021/acs.nanolett.9b05159 PMID: 32097014

Wurm, F. R., & Weiss, C. K. (2014). Nanoparticles from renewable polymers. *Frontiers in Chemistry, 2*, 49. DOI: 10.3389/fchem.2014.00049 PMID: 25101259

Wu, S., Zhu, H., Liu, J., Yang, Q., Shao, X., Bi, F., Hu, C., Huo, H., Chen, K., & Yi, G. (2020). Establishment of a PEG-mediated protoplast transformation system based on DNA and CRISPR/Cas9 ribonucleoprotein complexes for banana. *BMC Plant Biology, 20*(1), 425. DOI: 10.1186/s12870-020-02609-8 PMID: 32933485

Wu, T., Bi, Y., Yu, Y., Zhou, Z., Yuan, B., Ding, X., Zhang, Q., Chen, X., Yang, H., Liu, H., & Chu, Z. (2023). Activated Expression of Rice DMR6-like Gene OsS3H Partially Explores the Susceptibility to Bacterial Leaf Streak Mediated by Knock-Out OsF3H04g. *International Journal of Molecular Sciences, 24*(17), 13263. DOI: 10.3390/ijms241713263 PMID: 37686066

Wu, W., Du, K., Kang, X., & Wei, H. (2021). The diverse roles of cytokinins in regulating leaf development. *Horticulture Research*, *8*(1), 118. DOI: 10.1038/s41438-021-00558-3 PMID: 34059666

Wu, W., Xiao, X., Zhang, S., Ren, F., & Jiang, C. (2011). Facile method to synthesize magnetic iron oxides/TiO_2 hybrid nanoparticles and their photodegradation application of methylene blue. *Nanoscale Research Letters*, *6*(1), 1–15. DOI: 10.1186/1556-276X-6-533 PMID: 21961891

Wynant, N., Santos, D., Van Wielendaele, P., & Vanden Broeck, J. (2014). Scavenger receptor-mediated endocytosis facilitates RNA interference in the desert locust, *Schistocerca gregaria.Insect Molecular Biology*, *23*(3), 320–329. DOI: 10.1111/imb.12083 PMID: 24528536

Xiang, H., Meng, J., Shao, W., Zeng, D., Ji, J., Wang, P., Zhou, X., Qi, P., Liu, L., & Yang, S. (2023). Plant protein-based self-assembling core–shell nanocarrier for effectively controlling plant viruses: Evidence for nanoparticle delivery behavior, plant growth promotion, and plant resistance induction. *Chemical Engineering Journal*, *464*, 142432. DOI: 10.1016/j.cej.2023.142432

Xiao, D., Gao, X., Xu, J., Liang, X., Li, Q., Yao, J., & Zhu, K. Y. (2015). Clathrin-dependent endocytosis plays a predominant role in cellular uptake of double-stranded RNA in the red flour beetle. *Insect Biochemistry and Molecular Biology*, *60*, 68–77. DOI: 10.1016/j.ibmb.2015.03.009 PMID: 25863352

Xie, B., Wei, X., Wan, C., Zhao, W., Song, R., Xin, S., & Song, K. (2023). Exploring the Biological Pathways of Siderophores and Their Multidisciplinary Applications: A Comprehensive Review. *Molecules (Basel, Switzerland)*, *29*(10), 2318. DOI: 10.3390/molecules29102318 PMID: 38792179

Xie, Y., Li, B., Zhang, Q., Zhang, C., Lu, K., & Tao, G. (2011). Effects of nano-TiO_2 on photosynthetic characteristics of *Indocalamus barbatus.Dongbei Linye Daxue Xuebao*, *39*, 22–25.

Xing, M., Liu, Y., Guo, Z., & Tang, Y. (2018). Engineered nanomaterials in sustainable agriculture: From design to applications. *Environmental Science & Technology*, *52*(12), 7361–7375.

Xing, R., Liu, S., Guo, Z., Yu, H., Wang, P., Li, C., & Li, P. (2018). Biofunctionalization of nanoparticles for plant imaging and targeted pesticide delivery: A review. *Journal of Agricultural and Food Chemistry*, *66*(28), 7066–7073.

Xin, X. P., Judy, J. D., Sumerlin, B. B., & He, Z. L. (2020). Nano-enabled agriculture: From nanoparticles to smart nanodelivery systems. *Environmental Chemistry*, *17*(6), 413–425. DOI: 10.1071/EN19254

Xu, C., & Qu, X. (2014). Cerium oxide nanoparticle: a remarkably versatile rare earth nanomaterial for biological applications. *NPG Asia Materials 2014 6:3*, *6*(3), e90–e90. DOI: 10.1038/am.2013.88

Xuan, L., Ju, Z., Skonieczna, M., Zhou, K., & Huang, R. (2023). Nanoparticles-induced potential toxicity on human health: Applications, toxicity mechanisms, and evaluation models. *MedComm*, *4*(4), e327. DOI: 10.1002/mco2.327 PMID: 37457660

Xu, C. L., Cao, L. D., Bilal, M., Cao, C., Zhao, P. Y., Zhang, H. J., & Huang, Q. L. (2021). Multifunctional manganese-based carboxymethyl chitosan hydrogels for pH-Triggered pesticide release and enhanced fungicidal activity. *Carbohydrate Polymers*, *262*, 117933. DOI: 10.1016/j.carbpol.2021.117933 PMID: 33838810

Xu, C. L., Shan, Y. P., Bilal, M., Xu, B., Cao, L. D., & Huang, Q. L. (2020). Copper Ions Chelated mesoporous silica nanoparticles via dopamine chemistry for controlled pesticide release regulated by coordination bonding. *Chemical Engineering Journal*, *395*, 125093. DOI: 10.1016/j.cej.2020.125093

Xu, F. X., Zhou, X. B., Zhang, L., & Guo, X. Y., M., L., Zhu, Y. C., & Jiang. (2024). Effects of nitrogen application rate on grain yield of hybrid rice and soil nutrient under rice-fish co-culture system in Southeast of Sichuan Province. *Zhiwu Yingyang Yu Feiliao Xuebao*, *30*(5), 909–921.

Xu, L., Cao, L. D., Li, F. M., Wang, X. J., & Huang, Q. L. (2014). Utilization of chitosan-lactide copolymer nanoparticles as controlled release pesticide carrier for pyraclostrobin against *Colletotrichum gossypii* Southw. *Journal of Dispersion Science and Technology*, *35*(4), 544–550. DOI: 10.1080/01932691.2013.800455

Xu, T., Wang, Y., Aytac, Z., Zuverza-Mena, N., Zhao, Z., Hu, X., Ng, K. W., White, J. C., & Demokritou, P. (2022). Enhancing agrichemical delivery and plant development with biopolymer-based stimuli responsive core–shell nanostructures. *ACS Nano*, *16*(4), 6034–6048. DOI: 10.1021/acsnano.1c11490 PMID: 35404588

Xu, Z. L., Tang, T., Lin, Q., Yu, J. Z., Zhang, C. P., Zhao, X. P., Kah, M., & Li, L. (2022). Environmental risks and the potential benefits of nanopesticides: A review. *Environmental Chemistry Letters*, *20*(3), 2097–2108. DOI: 10.1007/s10311-021-01338-0

Yadav, A., & Yadav, K. (2018). Nanoparticle-based plant disease management: tools for sustainable agriculture. Nanobiotechnology applications in plant protection, 29-61.

Yadav, A., Alhan, S., & Dhania, G. (2024). Nanobiostimulants for Enhancing Plant Growth and Yield. *Nanobiostimulants: Emerging Strategies for Agricultural Sustainability*, 111-134.

Yadav, A., Yadav, K., Ahmad, R., & Abd-Elsalam, K. A. (2023). Emerging frontiers in nanotechnology for precision agriculture: Advancements, hurdles and prospects. *Agrochemicals*, *2*(2), 220–256. DOI: 10.3390/agrochemicals2020016

Yadav, N., Garg, V. K., Chhillar, A. K., & Rana, J. S. (2023). Recent advances in nanotechnology for the improvement of conventional agricultural systems: A review. *Plant Nano Biology*, *4*, 100032. DOI: 10.1016/j.plana.2023.100032

Yan, A., & Chen, Z. (2019). Impacts of silver nanoparticles on plants: A focus on the phytotoxicity and underlying mechanism. *International Journal of Molecular Sciences*, *20*(5), 1003. DOI: 10.3390/ijms20051003 PMID: 30813508

Yang, B., Wang, L., Zhang, M., Luo, J., Lu, Z., & Ding, X. (2020). Fabrication, Applications, and Prospects of Aramid Nanofiber. *Advanced Functional Materials*, *30*(22), 2000186. DOI: 10.1002/adfm.202000186

Yang, C., Powell, C. A., Duan, Y., Shatters, R., & Zhang, M. (2015). Antimicrobial Nanoemulsion Formulation with Improved Penetration of Foliar Spray through Citrus Leaf Cuticles to Control Citrus Huanglongbing. *PLoS One*, *10*(7), e0133826. DOI: 10.1371/journal.pone.0133826 PMID: 26207823

Yang, F., Hong, F., You, W., Liu, C., Gao, F., Wu, C., & Yang, P. (2006). Influence of nano-anatase TiO_2 on the nitrogen metabolism of growing spinach. *Biological Trace Element Research, 110*(2), 179–190. DOI: 10.1385/BTER:110:2:179 PMID: 16757845

Yang, T., & Duncan, T. V. (2021). Challenges and Potential Solutions for Nanosensors Intended for Use with Foods. *Nature Nanotechnology, 163*(16), 251–265. DOI: 10.1038/s41565-021-00867-7 PMID: 33712739

Yang, W., Gan, L., & Han, S. (2019). Advances in polymeric nanoparticles for antifungal treatments. *Current Drug Targets, 20*(5), 451–460.

Yang, Y., Hou, X., Ma, S., Huang, S., Chen, J., Fang, Z., Nie, G., Xu, B., Serra, C., & Ding, S. (2022). Tetraphenylethylene-Based Nanogels by Physical Encapsulation Technology: An AIEgen Transparent Film Thermometers. *ACS Applied Polymer Materials, 4*(3), 1974–1982. DOI: 10.1021/acsapm.1c01826

Yan, H., Zhao, G., Lu, W., Hu, C., Wang, X., Liu, G., Shi, N., Dieu Thuy, N. T., & Zhang, L. (2023). A flexible and wearable paper-based chemiresistive sensor modified with SWCNTs-PdNPs-polystyrene microspheres composite for the sensitive detection of ethylene gas: A new method for the determination of fruit ripeness and corruption. *Analytica Chimica Acta, 1239*, 340724. DOI: 10.1016/j.aca.2022.340724 PMID: 36628724

Yan, S., Ren, B. Y., & Shen, J. (2021). Nanoparticle-mediated double-stranded RNA delivery system: A promising approach for sustainable pest management. *Insect Science, 28*(1), 21–34. DOI: 10.1111/1744-7917.12822 PMID: 32478473

Yan, S., Yin, M.-Z., & Shen, J. (2022a). Nanoparticle-based nontransformative RNA insecticides for sustainable pest control: Mechanisms, current status and challenges. *Entomologia Generalis, 43*(1), 21–30. DOI: 10.1127/entomologia/2022/1618

Yan, Y., Zhu, X., Yu, Y., Li, C., Zhang, Z., & Wang, F. (2022). Nanotechnology Strategies for Plant Genetic Engineering. *Advanced Materials, 34*(7), e2106945. DOI: 10.1002/adma.202106945 PMID: 34699644

Yao, K. S., Li, S. J., Tzeng, K. C., Cheng, T. C., Chang, C. Y., Chiu, C. Y., Liao, C. Y., Hsu, J. J., & Lin, Z. P. (2009). Fluorescence silica nanoprobe as a biomarker for rapid detection of plant pathogens. *Advanced Materials Research, 79–82*, 513–516. . DOI: 10.4028/www.scientific.net/AMR.79-82.513

Yaqoob, S. B., Adnan, R., Rameez Khan, R. M., & Rashid, M. (2020). Gold, Silver, and Palladium Nanoparticles: A Chemical Tool for Biomedical Applications. *Frontiers in Chemistry, 8*, 520643. DOI: 10.3389/fchem.2020.00376 PMID: 32582621

Yaron, P. N., Holt, B. D., Short, P. A., Lösche, M., Islam, M. F., & Dahl, K. N. (2011). Single wall carbon nanotubes enter cells by endocytosis and not membrane penetration. *Journal of Nanobiotechnology, 9*(1), 1–15. DOI: 10.1186/1477-3155-9-45 PMID: 21961562

Ye, C., Hu, X. S., Wang, Z. W., Wei, D., Smagghe, G., Christiaens, O., Niu, J., & Wang, J. J. (2021). Involvement of clathrin-dependent endocytosis in cellular dsRNA uptake in aphids. *I. Insect Biochemistry and Molecular Biology, 132*, 103557. DOI: 10.1016/j.ibmb.2021.103557 PMID: 33639241

Yiamsawas, D., Beckers, S. J., Lu, H., Landfester, K., & Wurm, F. R. (2017). Morphology-controlled synthesis of lignin nanocarriers for drug delivery and carbon materials. *ACS Biomaterials Science & Engineering*, *3*(10), 2375–2383. DOI: 10.1021/acsbiomaterials.7b00278 PMID: 33445295

Yildiz, T., Gu, R., Zauscher, S., & Betancourt, T. (2018). Doxorubicin-loaded protease-activated near-infrared fluorescent polymeric nanoparticles for imaging and therapy of cancer. *International Journal of Nanomedicine*, *13*, 6961–6986. DOI: 10.2147/IJN.S174068 PMID: 30464453

Ying, S., Guan, Z., Ofoegbu, P. C., Clubb, P., Rico, C., He, F., & Hong, J. (2022a). Green synthesis of nanoparticles: Current developments and limitations. *Environmental Technology & Innovation*, *26*, 102336. DOI: 10.1016/j.eti.2022.102336

Yin, J., Su, X., Yan, S., & Shen, J. (2023). Multifunctional Nanoparticles and Nanopesticides in Agricultural Application. *Nanomaterials (Basel, Switzerland)*, *13*(7), 1255. DOI: 10.3390/nano13071255 PMID: 37049348

Yin, L., & Zhong, Z. (2020). Nanoparticles. In *Biomaterials science* (pp. 453–483). Academic Press. DOI: 10.1016/B978-0-12-816137-1.00031-3

Yonathan, K., Mann, R., Mahbub, K. R., & Gunawan, C. (2022). The impact of silver nanoparticles on microbial communities and antibiotic resistance determinants in the environment. *Environmental Pollution*, *293*, 118506. DOI: 10.1016/j.envpol.2021.118506 PMID: 34793904

Yong, J., Wu, M., Zhang, R., Bi, S., Mann, C. W., Mitter, N., & Xu, Z. P. (2022). Clay nanoparticles efficiently deliver small interfering RNA to intact plant leaf cells. *Plant Physiology*, *190*(4), 2187–2202. DOI: 10.1093/plphys/kiac430 PMID: 36135825

Yong, J., Xu, W., Wu, M., Zhang, R., Mann, C. W. G., Liu, G., Brosnan, C. A., Mitter, N., Carroll, B. J., & Xu, Z. P. (2025). Lysozyme-coated nanoparticles for active uptake and delivery of synthetic RNA and plasmid-encoded genes in plants. *Nature Plants*, *11*(1), 131–144. DOI: 10.1038/s41477-024-01882-x PMID: 39747606

Yong, J., Zhang, R., Bi, S., Li, P., Sun, L., Mitter, N., & Xu, Z. P. (2021). Sheet-like clay nanoparticles deliver RNA into developing pollen to efficiently silence a target gene. *Plant Physiology*, *187*(2), 886–899. DOI: 10.1093/plphys/kiab303 PMID: 34608968

Yoo, S. D., Cho, Y. H., & Sheen, J. (2007). Arabidopsis mesophyll protoplasts: A versatile cell system for transient gene expression analysis. *Nature Protocols*, *2*(7), 1565–1572. DOI: 10.1038/nprot.2007.199 PMID: 17585298

Younas, A.. (2024). Applicability of Nanomaterials in Agriculture. In Shahzad, R., Fiaz, S., Qayyum, A., Ul Islam, M., & Lee, I. J. (Eds.), *Revolutionizing Agriculture: A Comprehensive Exploration of Agri-Nanotechnology. Nanotechnology in the Life Sciences*. Springer., DOI: 10.1007/978-3-031-76000-6_2

Younis, S. A., Kim, K. H., Shaheen, S. M., Antoniadis, V., Tsang, Y. F., Rinklebe, J., Deep, A., & Brown, R. J. (2021). Advancements of nanotechnologies in crop promotion and soil fertility: Benefits, life cycle assessment, and legislation policies. *Renewable & Sustainable Energy Reviews*, *152*, 111686. DOI: 10.1016/j.rser.2021.111686

Youssef, A. M., Abdel-Aziz, M. S., & El-Sayed, S. M. (2014). Chitosan nanocomposite films based on Ag-NP and Au-NP biosynthesis by Bacillus Subtilis as packaging materials. *International Journal of Biological Macromolecules, 69*, 185–191. DOI: 10.1016/j.ijbiomac.2014.05.047 PMID: 24875320

Yuvaraj, M., Subramanian, K. S., & Cyriac, J. (2023). Efficiency of zinc oxide nanoparticles as controlled release nanofertilizer for rice (*Oryza sativa* L). *Journal of Plant Nutrition, 46*(18), 4477–4493. DOI: 10.1080/01904167.2023.2233561

Zago, S. L., dos Santos, M. F., Konrad, D., Fiorini, A., Rosado, F. R., Missio, R. F., & Vendruscolo, E. C. G. (2019). Shelf Life of Azospirillum brasilense in Alginate Beads Enriched With Trehalose and Humic Acid. *Journal of Agricultural Science (Toronto), 11*(6), 269. DOI: 10.5539/jas.v11n6p269

Zaheer, S., Mustafa, G., Shehzad, J., Tariq, T., Chaudhari, S. K., & Hasan, M. (2022). Green Synthesis of SiO 2 NPs Promote Vigna radiata L Germination, Membrane Damage, Antioxidant enzymes and Physiological activities.

Zahoor, M., Nazir, N., Iftikhar, M., Naz, S., Zekker, I., Burlakovs, J., Uddin, F., Kamran, A. W., Kallistova, A., Pimenov, N., & Khan, F. A. (2021). A Review on Silver Nanoparticles: Classification, Various Methods of Synthesis, and Their Potential Roles in Biomedical Applications and Water Treatment. *Water 2021, Vol. 13, Page 2216, 13*(16), 2216. DOI: 10.3390/w13162216

Zaidi, S. S., Vanderschuren, H., Qaim, M., Mahfouz, M. M., Kohli, A., Mansoor, S., & Tester, M. (2019). New plant breeding technologies for food security. *Science, 363*(6434), 1390–1391. DOI: 10.1126/science.aav6316 PMID: 30923209

Zaim, N. S. H. B. H., Tan, H. L., Rahman, S. M. A., Abu Bakar, N. F., Osman, M. S., Thakur, V. K., & Radacsi, N. (2023). Recent advances in seed coating treatment using nanoparticles and nanofibers for enhanced seed germination and protection. *Journal of Plant Growth Regulation, 42*(12), 7374–7402. DOI: 10.1007/s00344-023-11038-4

Zamore, P. D., Tuschl, T., Sharp, P. A., & Bartel, D. P. (2000). RNAi: Double-stranded RNA directs the ATP-dependent cleavage of mRNA at 21 to 23 nucleotide intervals. *Cell, 101*(1), 25–33. DOI: 10.1016/S0092-8674(00)80620-0 PMID: 10778853

Zare, N., Karimi, F., Altuner, E. E., Mashkoor, H., Tiri, R. N. E., Gulbagca, F., Bijad, M., Cherif, A., & Sen, F. (2024). Synthesis of Bimetallic Nickel-Cobalt Nanoparticles Supported by Functional Multi-Walled Carbon Nanotubes by Chemical Methods: Characterization and Hydrogen Energy Generation. *Topics in Catalysis, 67*(9–12), 606–614. DOI: 10.1007/s11244-024-01907-x

Zayed, O., Hewedy, O. A., Abdelmoteleb, A., Ali, M., Youssef, M. S., Roumia, A. F., Seymour, D., & Yuan, C. (2023). Nitrogen Journey in Plants: From Uptake to Metabolism, Stress Response, and Microbe Interaction. *Biomolecules, 13*(10), 1443. DOI: 10.3390/biom13101443 PMID: 37892125

Zhang, C., & Hu, H. (2014). High-efficiency nuclear transformation of the diatom Phaeodactylum tricornutum by electroporation. *Marine Genomics, 16*, 63–66. DOI: 10.1016/j.margen.2013.10.003 PMID: 24269346

Zhang, D., Yao, X., Wang, J., & Wu, Y. (2021). Advances in nanotechnology for plant genetic engineering. *Biotechnology Advances, 48*, 107712. DOI: 10.1016/j.biotechadv.2021.107712

Zhang, H., Cao, Y., Xu, D., Goh, N. S., Demirer, G. S., Cestellos-Blanco, S., Chen, Y., Landry, M. P., & Yang, P. (2021a). Gold-Nanocluster-Mediated Delivery of siRNA to Intact Plant Cells for Efficient Gene Knockdown. *Nano Letters*, *21*(13), 5859–5866. DOI: 10.1021/acs.nanolett.1c01792 PMID: 34152779

Zhang, H., Demirer, G. S., Zhang, D., Ye, T., Goh, N. S., & Landry, M. P. (2019). DNA nanostructures coordinate gene silencing. *Proceedings of the National Academy of Sciences of the United States of America*, *116*(15), 7543–7548. DOI: 10.1073/pnas.1818290116 PMID: 30910954

Zhang, H., Demirer, G. S., Zhang, D., Ye, T., Goh, N. S., & Landry, M. P. (2021c). Engineering DNA nanostructures for delivery of genetic material into plants. *Nature Protocols*, *16*(3), 1704–1722. DOI: 10.1038/s41596-020-00506-1

Zhang, H., Goh, N. S., Wang, J. W., Pinals, R. L., González-Grandío, E., Demirer, G. S., Butrus, S., Fakra, S. C., Del Rio Flores, A., Zhai, R., Zhao, B., Park, S. J., & Landry, M. P. (2022a). Nanoparticle cellular internalization is not required for RNA delivery to mature plant leaves. *Nature Nanotechnology*, *17*(2), 197–205. DOI: 10.1038/s41565-021-01018-8 PMID: 34811553

Zhang, H., Zhang, H., Demirer, G. S., González-Grandío, E., Fan, C., & Landry, M. P. (2020b). Engineering DNA nanostructures for siRNA delivery in plants. *Nature Protocols*, *15*(9), 3064–3087. DOI: 10.1038/s41596-020-0370-0 PMID: 32807907

Zhang, J., Khan, S. A., Heckel, D. G., & Bock, R. (2017). Next-generation insect-resistant plants: RNAi-mediated crop protection. *Trends in Biotechnology*, *35*(9), 871–882. DOI: 10.1016/j.tibtech.2017.04.009 PMID: 28822479

Zhang, M., Li, G., Huang, L., Ran, P., Huang, J., Yu, M., Yuqian, H., Guo, J., Liu, Z., & Ma, X. (2021). Versatile fabrication of liquid metal nano-ink based flexible electronic devices. *Applied Materials Today*, *22*, 100903. DOI: 10.1016/j.apmt.2020.100903

Zhang, M., Ma, W., Tao, R., Fan, Q., Zhang, M., Qin, D., Cao, X., Li, J., Xiong, R., & Huang, C. (2024). Nanomaterials: Recent advances in plant disease diagnosis and treatment. *Nano Today*, *57*, 102326. DOI: 10.1016/j.nantod.2024.102326

Zhang, P., Guo, Z., Zhang, Z., Fu, H., White, J. C., & Lynch, I. (2020). Nanomaterial transformation in the soil–plant system: Implications for food safety and application in agriculture. *Small*, *16*(21), 2000705. DOI: 10.1002/smll.202000705 PMID: 32462786

Zhang, Q., Gong, M., Xu, X., Li, H., & Deng, W. (2022). Roles of Auxin in the Growth, Development, and Stress Tolerance of Horticultural Plants. *Cells*, *11*(17), 2761. DOI: 10.3390/cells11172761 PMID: 36078168

Zhang, Q., Jia, A., Wan, Y., Liu, H., Wang, K., Peng, H., Dong, Z., & Hu, J. (2014). Occurrences of three classes of antibiotics in a natural river basin: Association with antibiotic-resistant Escherichia coli. *Environmental Science & Technology*, *48*(24), 14317–14325. DOI: 10.1021/es503700j PMID: 25427067

Zhang, Q., Zhao, L., Shen, M., Liu, J., Li, Y., Xu, S., Chen, L., Shi, G., & Ding, Z. (2022b). Establishment of an Efficient Polyethylene Glycol (PEG)-Mediated Transformation System in *Pleurotus eryngii* var. *ferulae* Using Comprehensive Optimization and Multiple Endogenous Promoters. *Journal of Fungi (Basel, Switzerland)*, *8*(2), 186. DOI: 10.3390/jof8020186 PMID: 35205941

Zhang, R., Han, Y., Xie, W., Liu, F., & Chen, S. (2022). Advances in protein-based nanocarriers of bioactive compounds: From microscopic molecular principles to macroscopical structural and functional attributes. *Journal of Agricultural and Food Chemistry*, *70*(21), 6354–6367. DOI: 10.1021/acs.jafc.2c01936 PMID: 35603429

Zhang, W., Zeng, Z., Liu, Z., Huang, J., Xiao, R., Shao, B., Liu, Y., Liu, Y., Tang, W., Zeng, G., Gong, J., & He, Q. (2020). Effects of carbon nanotubes on biodegradation of pollutants: Positive or negative? *Ecotoxicology and Environmental Safety*, *189*, 109914. DOI: 10.1016/j.ecoenv.2019.109914 PMID: 31761551

Zhang, W., Zhao, J., & Fang, J.. (2020). Nanotechnology-based strategies for treatment of fungal infections. *Journal of Materials Chemistry. B*, *8*(5), 994–1007.

Zhang, X. (2019). Environmental DNA Shaping a New Era of Ecotoxicological Research. *Environmental Science & Technology*, *53*(10), 5605–5612. DOI: 10.1021/acs.est.8b06631 PMID: 31009204

Zhang, X. H., Tee, L. Y., Wang, X. G., Huang, Q. S., & Yang, S. H. (2015). Off-target effects in CRISPR/Cas9-mediated genome engineering. *Molecular Therapy. Nucleic Acids*, *4*, 4. DOI: 10.1038/mtna.2015.37 PMID: 26575098

Zhang, X. Z., Ervin, E. H., & Schmidt, R. E. (2003). Physiological effects of liquid applications of a seaweed extract and a humic acid on creeping bentgrass. *Journal of the American Society for Horticultural Science*, *128*(4), 492–496. DOI: 10.21273/JASHS.128.4.0492

Zhang, Y., Cao, Y., Jiang, W., Ma, Q., Shin, J., Sun, H., & Marelli, B. (2024). Polymeric Nanocarriers Autonomously Cross the Plant Cell Wall and Enable Protein Delivery for Stress Sensing. *Advanced Materials*, *36*(41), 2409356. DOI: 10.1002/adma.202409356 PMID: 39149770

Zhang, Y., Huang, C., & Xiong, R. (2024). Advanced materials for intracellular delivery of plant cells: Strategies, mechanisms and applications. *Materials Science and Engineering R Reports*, *160*, 100821. DOI: 10.1016/j.mser.2024.100821

Zhang, Y., Liang, Z., Zong, Y., Wang, Y., Liu, J., Chen, K., Qiu, J. L., & Gao, C. (2016). Efficient and transgene-free genome editing in wheat through transient expression of CRISPR/Cas9 DNA or RNA. *Nature Communications*, *7*(1), 12617. DOI: 10.1038/ncomms12617 PMID: 27558837

Zhang, Y., Long, H., Wang, M. Y., Li, Y., Ma, L., Chen, K., Zheng, Y., & Jiang, T. (2020). The hidden mechanism of chemical fertiliser overuse in rural China. *Habitat International*, *102*, 102210. DOI: 10.1016/j.habitatint.2020.102210

Zhang, Y., Shin, J., Sun, H., Chang, H. F., Martinez, M. R., Perkins, L. A., & Lowry, G. V. (2024). High Aspect Ratio Polymer Nanocarriers for Gene Delivery and Expression in Plants. *Nano Letters*, *25*(2), 681–690. DOI: 10.1021/acs.nanolett.4c04704 PMID: 39810730

Zhang, Y., & Zhang, S. (2017). Development of nanoparticles for antifungal delivery. *Nanomedicine; Nanotechnology, Biology, and Medicine*, *13*(7), 2303–2312. PMID: 28712917

Zhang, Y., Zhang, Y., & Jin, S. (2020). Public acceptance of genetically modified food: Socio-demographic variables, general attitudes, and information framing. *Science Communication*, *42*(2), 248–273. DOI: 10.1177/1075547019898143

Zhang, Z., Zhou, Z., Feng, S., Guo, P., Wang, Y., Hao, B., Guo, W., & Li, F. Y. (2024). Synergistic effects of AMF and PGPR on improving saline-alkaline tolerance of *Leymus chinensis* by strengthening the link between rhizosphere metabolites and microbiomes. *Environmental Technology & Innovation*, *36*, 103900. DOI: 10.1016/j.eti.2024.103900

Zhan, Y., Ma, W., Zhang, Y., Mao, C., Shao, X., Xie, X., & Lin, Y. (2020). Diversity of DNA nanostructures and applications in oncotherapy. *Biotechnology Journal*, *15*(1), 1900094. DOI: 10.1002/biot.201900094 PMID: 31464361

Zhao, F., Xin, X., Cao, Y., Su, D., Ji, P., Zhu, Z., & He, Z. (2021). Use of carbon nanoparticles to improve soil fertility, crop growth and nutrient uptake by corn (*Zea mays* L.). *Nanomaterials (Basel, Switzerland)*, *11*(10), 2717. DOI: 10.3390/nano11102717 PMID: 34685156

Zhao, G., Zhao, Y., Lou, W., Abdalmegeed, D., Guan, R., & Shen, W. (2020). Multi-walled carbon nanotubes can promote Brassica napus L. and Arabidopsis thaliana L. root hair development through nitric oxide and ethylene pathways. *International Journal of Molecular Sciences*, *21*(23), 9109. DOI: 10.3390/ijms21239109 PMID: 33266061

Zhao, H., Zhi, Y. L., Lara, Y., Arvind, D., Xin, Z., & Jun, W. (2016). Polymer-based nanoparticles for protein delivery: Design, strategies and applications. *Journal of Materials Chemistry. B*, *4060*(23), 4060–4071. DOI: 10.1039/C6TB00308G PMID: 32264607

Zhao, J., Song, R., Li, H., Zheng, Q., Li, S., Liu, L., Li, X., Bai, L., & Liu, K. (2022). New formulation to accelerate the degradation of pesticide residues: Composite nanoparticles of imidacloprid and 24-epibrassinolide. *ACS Omega*, *7*(33), 29027–29037. DOI: 10.1021/acsomega.2c02820 PMID: 36033692

Zhao, J., Yan, S., Li, M., Sun, L., Dong, M., Yin, M., Shen, J., & Zhao, Z. (2023). NPFR regulates the synthesis and metabolism of lipids and glycogen via AMPK: Novel targets for efficient corn borer management. *International Journal of Biological Macromolecules*, *247*, 125816. DOI: 10.1016/j.ijbiomac.2023.125816 PMID: 37451386

Zhao, L., Lu, L., Wang, A., Zhang, H., Huang, M., Wu, H., Xing, B., Wang, Z., & Ji, R. (2020). Nano-biotechnology in agriculture: Use of nanomaterials to promote plant growth and stress tolerance. *Journal of Agricultural and Food Chemistry*, *68*(7), 1935–1947. DOI: 10.1021/acs.jafc.9b06615 PMID: 32003987

Zhao, M., Li, P., Zhou, H., Hao, L., Chen, H., & Zhou, X. (2022). pH/redox dual responsive from natural polymer-based nanoparticles for on-demand delivery of pesticides. *Chemical Engineering Journal*, *435*, 134861. DOI: 10.1016/j.cej.2022.134861

Zhao, Q., Lin, Y., Han, N., Li, X., Geng, H., Wang, X., & Wang, S. (2017). Mesoporous carbon nanomaterials in drug delivery and biomedical application. *Drug Delivery*, *24*(2), 94–107. DOI: 10.1080/10717544.2017.1399300 PMID: 29124979

Zhao, Q., Ma, C., White, J. C., Dhankher, O. P., Zhang, X., Zhang, S., & Xing, B. (2017). Quantitative evaluation of multi-wall carbon nanotube uptake by terrestrial plants. *Carbon, 114*, 661–670. DOI: 10.1016/j.carbon.2016.12.036

Zhao, X., Meng, Z., Wang, Y., Chen, W., Sun, C., Cui, B., Cui, J., Yu, M., Zeng, Z., Guo, S., Luo, D., Cheng, J. Q., Zhang, R., & Cui, H. (2017). Pollen magnetofection for genetic modification with magnetic nanoparticles as gene carriers. *Nature Plants, 3*(12), 956–964. DOI: 10.1038/s41477-017-0063-z PMID: 29180813

Zhao, Y. (2012). Auxin Biosynthesis: A Simple Two-Step Pathway Converts Tryptophan to Indole-3-Acetic Acid in Plants. *Molecular Plant, 5*(2), 334–338. DOI: 10.1093/mp/ssr104 PMID: 22155950

Zhao, Z., Lacombe, J., Simon, L., Sanchez-Ballester, N. M., Khanishayan, A., Shaik, N., Case, K., Dugas, P. Y., Repellin, M., Lollo, G., Soulairol, I., Harris, A. F., Gordon, M., Begu, S., & Zenhausern, F. (2024). Physical, biochemical, and biological characterization of olive-derived lipid nanovesicles for drug delivery applications. *Journal of Nanobiotechnology, 22*(1), 453–550. DOI: 10.1186/s12951-024-02964-w PMID: 39558361

Zheng, Y., Chen, X., Joseph, N.D., Zhang, Y., Chen, H., & Gao, B. (2022). Occurrences and impacts of engineered nanoparticles in soils and groundwater. *Emerg. Contam. Soil Groundw. Syst.* 165–204, .DOI: 10.1016/B978-0-12-824088-5.00005-7

Zheng, L., Hong, F., Lu, S., & Liu, C. (2005). Effect of nano-TiO_2 on strength of naturally aged seeds and growth of spinach. *Biological Trace Element Research, 104*(1), 83–91. DOI: 10.1385/BTER:104:1:083 PMID: 15851835

Zheng, Y., Hu, Y., Yan, S., Zhou, H., Song, D., Yin, M., & Shen, J. (2019). A polymer/detergent formulation improves dsRNA penetration through the body wall and RNAi-induced mortality in the soybean aphid *Aphis glycines.Pest Management Science, 75*(7), 1993–1999. DOI: 10.1002/ps.5313 PMID: 30610748

Zhou, H., Wan, F., Jian, Y., Guo, F., Zhang, M., Shi, S., Yang, L., Li, S., Liu, Y., & Ding, W. (2023). Chitosan/dsRNA polyplex nanoparticles advance environmental RNA interference efficiency through activating clathrin-dependent endocytosis. *International Journal of Biological Macromolecules, 253*, 127021. DOI: 10.1016/j.ijbiomac.2023.127021 PMID: 37741481

Zhou, J., Shum, K. T., Burnett, J. C., & Rossi, J. J. (2013). Nanoparticle-based delivery of RNAi therapeutics: Progress and challenges. *Pharmaceuticals (Basel, Switzerland), 6*(1), 85–107. DOI: 10.3390/ph6010085 PMID: 23667320

Zhou, Q., & Tang, D. (2020). Recent advances in photoelectrochemical biosensors for analysis of mycotoxins in food. *Trends in Analytical Chemistry, 124*, 115814. DOI: 10.1016/j.trac.2020.115814

Zhou, X. Q., Hayat, Z., Zhang, D. D., Li, M. Y., Hu, S., Wu, Q., Cao, Y.-F., & Yuan, Y. (2023). Zinc oxide nanoparticles: Synthesis, characterization, modification, and applications in food and agriculture. *Processes (Basel, Switzerland), 11*(4), 1193. DOI: 10.3390/pr11041193

Zhu, S., Song, Y., Zhao, X., Shao, J., Zhang, J., & Yang, B. (2015). The photoluminescence mechanism in carbon dots (graphene quantum dots, carbon nanodots, and polymer dots): current state and future perspective. *Nano Research 2014 8:2, 8*(2), 355–381. DOI: 10.1007/s12274-014-0644-3

Zhu, K. Y., & Palli, S. R. (2020). Mechanisms, applications, and challenges of insect RNA interference. *Annual Review of Entomology*, *65*(1), 293–311. DOI: 10.1146/annurev-ento-011019-025224 PMID: 31610134

Zhu, T., Wang, L., Sun, W., Wang, M., Yang, Z., Ji, T., Wang, S., Wang, Y., Xia, L., & Liu, G. (2018). Origin of Aragonite Scale Deposition on Carbon Steel at Ambient Circumstances. *Industrial & Engineering Chemistry Research*, *57*(1), 401–413. DOI: 10.1021/acs.iecr.7b03399

Zhu, X. Q., Zhang, J. Y., Xie, Q., & Hou, Z. L. (2020). High Sensitivity and Ultrafast Response Ethanol Sensors Based on Graphene Oxide. *ACS Applied Materials & Interfaces*, *12*(34), 38708–38713. DOI: 10.1021/acsami.0c12196 PMID: 32805966

Zhu, Y., Liu, X., Hu, Y., Wang, R., Chen, M., Wu, J., Wang, Y., Kang, S., Sun, Y., & Zhu, M. (2019). Behavior, remediation effect and toxicity of nanomaterials in water environments. *Environmental Research*, *174*, 54–60. DOI: 10.1016/j.envres.2019.04.014 PMID: 31029942

Zhu, Y., Zhao, Y., Li, B., Huang, C.-L., Zhang, S.-Y., Yu, S., Chen, Y.-S., Zhang, T., Gillings, M. R., & Su, J.-Q. (2017). Continental-scale pollution of estuaries with antibiotic resistance genes. *Nature Microbiology*, *2*(4), 16270. DOI: 10.1038/nmicrobiol.2016.270 PMID: 28134918

Zhu, Z. J., Wang, H., Yan, B., Zheng, H., Jiang, Y., Miranda, O. R., Rotello, V. M., Xing, B., & Vachet, R. W. (2012). Effect of surface charge on the uptake and distribution of gold nanoparticles in four plant species. *Environmental Science & Technology*, *46*(22), 12391–12398. DOI: 10.1021/es301977w PMID: 23102049

Zia-ur-Rehman, M., Mfarrej, M. F. B., Usman, M., Anayatullah, S., Rizwan, M., Alharby, H. F., Abu Zeid, I. M., Alabdallah, N. M., & Ali, S. (2023). Effect of iron nanoparticles and conventional sources of Fe on growth, physiology and nutrient accumulation in wheat plants grown on normal and salt-affected soils. *Journal of Hazardous Materials*, *458*, 131861. DOI: 10.1016/j.jhazmat.2023.131861 PMID: 37336110

Zia-ur-Rehman, M., Naeem, A., Khalid, H., Rizwan, M., Ali, S., & Azhar, M. (2018). Responses of plants to iron oxide nanoparticles. In *Nanomaterials in plants, algae, and microorganisms* (pp. 221–238). Academic Press.

Ziemienowicz, A. (2014). Agrobacterium-mediated plant transformation: Factors, applications and recent advances. *Biocatalysis and Agricultural Biotechnology*, *3*(4), 95–102. DOI: 10.1016/j.bcab.2013.10.004

Zotti, M., Dos Santos, E. A., Cagliari, D., Christiaens, O., Taning, C. N. T., & Smagghe, G. (2018). RNA interference technology in crop protection against arthropod pests, pathogens and nematodes. *Pest Management Science*, *74*(6), 1239–1250. DOI: 10.1002/ps.4813 PMID: 29194942

Zou, L., Cheng, M., Hu, K., Feng, J., & Tu, L. (2024). Vesicular drug delivery systems for oral absorption enhancement. *Chinese Chemical Letters*, *35*(7), 109129. DOI: 10.1016/j.cclet.2023.109129

Zou, T., Sipponen, M. H., & Österberg, M. (2019). Natural shape-retaining microcapsules with shells made of chitosan-coated colloidal lignin particles. *Frontiers in Chemistry*, *7*, 370. DOI: 10.3389/fchem.2019.00370 PMID: 31192192

Zuverza-Mena, N., Martínez-Fernández, D., Du, W., Hernandez-Viezcas, J. A., Bonilla-Bird, N., López-Moreno, M. L., Komárek, M., Peralta-Videa, J. R., & Gardea-Torresdey, J. L. (2017). Exposure of engineered nanomaterials to plants: Insights into the physiological and biochemical responses-A review. *Plant Physiology and Biochemistry*, *110*, 236–264. DOI: 10.1016/j.plaphy.2016.05.037 PMID: 27289187

Zuverza-Mena, N., Martínez-Fernández, D., Du, W., Hernandez-Viezcas, J. A., Bonilla-Bird, N., López-Moreno, M. L., Komárek, M., Peralta-Videa, J. R., & Gardea-Torresdey, J. L. (2017). Exposure of engineered nanomaterials to plants: Insights into the physiological and biochemical responses-A review. *Plant physiology and biochemistry. PPB*, *110*, 236–264. PMID: 27289187

About the Contributors

Jen-Tsung Chen is a professor of cell biology at the National University of Kaohsiung in Taiwan. He teaches cell biology, genomics, proteomics, plant physiology, and plant biotechnology. Dr. Chen's research interests include bioactive compounds, chromatography techniques, plant molecular biology, plant biotechnology, bioinformatics, and systems pharmacology. He is an active editor of academic books and journals to advance the exploration of multidisciplinary knowledge involving plant physiology, plant biotechnology, nanotechnology, ethnopharmacology, and systems biology. He serves as an editorial board member in reputed journals, including Plant Methods, GM Crops & Food, Plant Nano Biology, Biomolecules, International Journal of Molecular Sciences, and as a guest editor in Frontiers in Plant Science, Frontiers in Pharmacology, Journal of Fungi, and Current Pharmaceutical Design. Dr. Chen published books in collaboration with international publishers on computational biology, drug discovery, nanotechnology, bioengineering, plant functional genomics, plant speed breeding, and CRISPR-based plant genome editing. In 2023 and 2024, Dr. Chen was recognized as one of the "World's Top 2% Scientists" by Elsevier and Stanford University.

***

Maryam Abdul Ajees is an emerging biotechnology researcher and undergraduate student at the Manipal School of Life Sciences, Manipal Academy of Higher Education (MAHE), India. With a deep-seated interest in molecular biology, nanotechnology, stem cell therapy, and molecular docking, she has demonstrated academic excellence and early research initiative throughout her academic journey. Maryam has gained hands-on research experience through her role as a research assistant in the Biophysics Laboratory at the Department of Atomic and Molecular Physics, MAHE, and is currently contributing to ongoing projects under. Her technical skill set spans key laboratory techniques such as DNA/RNA extraction, gel electrophoresis, UV and fluorescence spectroscopy, and the use of bioinformatics tools including the Schrödinger Suite and AutoDock Vina. She is also proficient in programming basics in Python, PERL, and HTML/CSS. Her scholarly achievements include the publication of a review article titled *"Diatoms: Harnessing Nature's Microscopic Marvels for Biosensing and Multifaceted Applications,"* and participation in the 2024 Biotechnology Hackathon, where she secured third prize. Maryam also earned recognition as a top 10 finalist in the Hult Prize On-Campus competition for her entrepreneurial pitch. Known for her leadership and collaborative mindset, Maryam has served as a class representative, a school parliament speaker, and a coordinator during her school years. Her blend of academic rigor, technical skills, and soft skills, such as communication, teamwork, and critical thinking, positions her as a promising contributor to the future of interdisciplinary research in biotechnology and sustainable innovation.

Shafique Ahmad is working as an Associate Professor in the Department of Botany, University of Poonch Rawalakot, Rawalakot, Jammu & Kashmir, Pakistan. His research group is composed of an interdisciplinary team of scientists who are passionate about environmental science research with a particular focus on investigating the complex interactions between plants and their environment. His research group study plant physiological and biogeochemical responses under abiotic stresses (i.e., cold, heavy metals, salt and temperature stress) to natural environmental variation and to global and regional environmental changes in natural ecosystems, with emphasis on fundamental aspects of plant growth, structural and physiological responses to environmental stress and climate change, and production of secondary metabolites. The group has also been interested in synthesis of metals and their oxide nanoparticles and their application in agriculture, medicine and environment. In addition to this, the is also interested in ethnopharmacological research which involves plant toxicity in livestock and traditional societies that incorporate wild plants into their daily life and takes the ethnobotanical approach to drug discovery.

Shruti Awasthi, after two years of corporate experience as Biochemist, has chosen to grow in the field of teaching, research & development. Her career in research & training started in the year 2003. She completed her Ph.D in Medical Biochemistry from Indian Holistic Medical Academy, in the year 2012; MBA in HRM; M.Phil in Biochemistry and M.Sc in Biochemistry and B.Sc in Chemistry and got merit in Biology in S.Y. B.Sc in South Gujarat University. She has 10 years of experience in overseeing research projects; published research papers in National and International Journals of great repute. She has presented several research papers which were acknowledged in the form of awards and titles. She was conferred the "Young Scientist Award" by IHMA in the year 2012 for her research work. She has won 12 gold medals for presentation in National & International Symposiums. Further she has handled various training session for students (SSEP i.e. Scientific skill enhancement programme). Her key strengths are good communication, presentation and analytical skills. She is registered life member of Indian Science Congress and IHMA. The areas of her research interests are – Molecular Characterization of medicinal plants and synthesis and characterization of nanomaterials.

Deepak is pursuing an M.Tech in Industrial Biotechnology (2023-2025) with research area in the field of Nanobiotechnology from Guru Gobind Singh Indraprastha University, Dwarka, Delhi, India. In 2013, Deepak successfully qualified for the UGC National Eligibility Test (UGC NET). His academic credentials include a B.Sc. from Himachal Pradesh University, Shimla, Himachal Pradesh; an M.Sc. in Biotechnology from IK Gujral Punjab Technical University; an M.Sc. in Information Technology from the same university; a B.Ed. from Guru Gobind Singh Indraprastha University, Dwarka; and an M.Ed. from Maharshi Dayanand University, Rohtak, Haryana. With a professional background that includes serving as an Assistant Professor at Risemax College of Education from 2017 to 2022 in Faridabad, Haryana, Deepak has made significant contributions to the academic community. He is the author of several books, including "Pedagogy of Physical Sciences" (ISBN: 978-93-82073-53-6), "Understanding Discipline and Subject" (ISBN: 978-93-82073-75-8), and "Creating An Inclusive School" (Balaji Publication of Education Pvt Ltd., Faridabad, Haryana, English Medium, ISBN: 978-93-82073-72-3). Additionally, he has contributed a chapter titled "ICT in Service of Humanity" to the book "Information Communication Technology in Education" (ISBN: 978-93-82073-36-9). Deepak's research interests encompass both the pure and applied aspects of educational psychology as well as scientific research. Through his scholarly work, he continues to advance knowledge and understanding in these fields, making a lasting impact on education, science and technology.

Abel Inobeme is a Senior Lecturer and Head of Department of Chemistry at Edo State University Uzairue, Nigeria. He holds a PhD and Masters in Analytical Chemistry, both from the Federal University of Technology Minna Nigeria; a B.Sc (Ed) Chemistry from Ahmadu Bello University Zaria, Nigeria and a Nigerian Certificate in Education from the Federal College of Education Zaria. He also holds a Certificate in PhET from the University of Colorado Boulder. His research areas include environmental toxicological assessments, food analysis and toxicology, wastewater treatment, material chemistry, and nanotechnology. He has contributed a total of 120 chapters in more than 90 different books in several top scientific publishers such as Elsevier, Springer Nature, Taylor and Francis, American Chemical Society, Frontiers and Wiley. He has over 200 publications. He has 3 published books to his credit. He is currently the National Publication Secretary of the Material Science and Engineering Association of Nigeria (MSN). He is also a member of the Chemical Society Nigeria (CSN) and the Nigerian Institute of Management (NIM). He was recently ranked among the top 500 scientists in Nigeria based on SCOPUS/SCIVAL output for the year 2022.

Ravi Kant is currently an Assistant Professor at the University School of Biotechnology at Guru Gobind Singh Indraprastha University, Dwarka, New Delhi, a position he has held since September 2023. In this role, he instructs and mentors B.Tech. and M.Tech. students in both teaching and research in Biotechnology. He has a keen interest in the area of Nanobiotechnology and is also focusing on Green and Chemical Synthesis of Nanoparticles. Overall, he has published over 15 papers in reputed journals such as JACS, mAbs, JBC, eLife, Journal of Virology, Nature Scientific Reports, ACS Biochemistry, and several international book chapters. Dr. Kant completed his Ph.D. in Biochemistry at Montana State University, Bozeman, USA, under Prof. Brian Bothner, focusing on structural biophysical virology. Previously, Dr. Kant was an Assistant Professor at Guru Gobind Singh Indraprastha University in the University School of Chemical Technology for two years, teaching courses related to Biochemical Engineering. He guided B.Tech. and M.Tech. biochemical engineering students in their final year research projects and contributed to the GGSIPU University Compendium and seminar series. Before his Ph.D., he taught Biotechnology courses at Meerut Institute of Engineering and Technology and Biochemistry to Medical Lab Technology students at Integrated Institute of Technology, Govt. NCT Delhi. During his Ph.D., he served as a Teaching Assistant for nearly four years. As a postdoctoral researcher under Prof. Michael L. Gross at Washington University in St. Louis, Dr. Kant's research spanned from studying simple protein systems to complex supramolecular protein complexes. His notable projects included characterizing the SARS-CoV-2 nsp1 protein, epitope mapping, and identifying the binding sites and effects of drugs on Transthyretin proteins. He has extensive experience in Hydrogen Deuterium Exchange Mass Spectrometry (HDX-MS), Fast Photochemical Oxidation of Proteins (FPOP), Native Mass Spectrometry (NMS), and Covalent Labeling (CL). He also contributed a book chapter for Elsevier titled "Protein Footprinting by Mass Spectrometry: Hydrogen/Deuterium Exchange, Specific Amino Acid Labeling, and Fast Photochemical Oxidation of Proteins." In addition to his academic work, Dr. Kant is keen on community outreach and served as a Community Assistant in family and graduate housing at MSU, Bozeman for over four years. His dedication to teaching and research makes him a valuable asset to the academic community.

Ilham Khan holds an M.Sc. and M.Phil. in Plant Sciences from Quaid-i-Azam University, Islamabad. He is currently pursuing a Ph.D. in Plant Proteomics at the same institution, focusing on the therapeutic potential of medicinal plants. His research integrates metabolomics, transcriptomics, and proteomics, utilizing various bioinformatics tools to explore the antidiabetic properties of medicinal plants through in vitro, in silico, and in vivo studies. In addition, Mr. Khan has expertise in nanoparticle synthesis, investigating their biological activities and their role in mitigating biotic and abiotic stress in crops. He has contributed to the scientific community through several research articles and book chapters published in international journals.

Archana Lakshmaiah is a Research Scholar in the Department of Life Sciences at Sri Sathya Sai University for Human Excellence, Kalaburagi, Karnataka, India. Her research focuses on the behavior on multi omics approaches to congenital heart diseases, mitochondrial diseases.

Gicy Mathew, currently pursuing a Master of Science in Biotechnology at Christ University, Bangalore, with a Bachelor's degree in Biotechnology, Botany, and Chemistry. Research interests include marine biotechnology, plant-derived bioactive compounds, and nanobiotechnology. Ongoing work focuses on the extraction and microencapsulation of bioactive metabolites from plant and seaweed sources, with a particular emphasis on their application in nanoparticle synthesis aiming to contribute to sustainable innovations in life sciences.

John Tsado Mathew serves as a lecturer within the Department of Chemistry at Ibrahim Badamasi Babangida University Lapai, situated in Nigeria. He boasts a distinguished academic background, holding a PhD and Masters in Analytical Chemistry, as well as a B.Tech. in Industrial Chemistry, all earned from the Federal University of Technology Minna Nigeria. His scholarly pursuits are focused on various domains including food analysis and toxicology, environmental toxicological assessments, environmental remediation, wastewater treatment, and nanotechnology. Dr. Mathew has made significant contributions to the field, with an impressive tally of 49 chapters contributed to over 20 different books across esteemed scientific publishers such as Elsevier, Springer Nature, Taylor and Francis, American Chemical Society, Frontiers, and Wiley. Additionally, he has authored over 70 publications in peer-reviewed journals, spanning both local and international platforms. Notably, he has been acknowledged for his scholarly impact, being ranked among the top 500 scientists in Nigeria for the year 2022 based on SCOPUS/SCIVAL output.

Annamalai Muthusamy obtained MSc and MPhil degree in Botany from University of Madras, India (1991) and Bharathidasan University, India (1993) respectively. He experienced as PhD Scholar at Dept. of Plant Science and received PhD from Bharathidasan University, India (2002) in the area of in Plant Breeding/Biotechnology. Dr. A Muthusamy joined Manipal School of Life Sciences, MAHE, Manipal as Assistant Professor (2006-2009), later promoted as Associate Professor (2009-2015), Associate Professor and Head (2015-2018), Professor and Head (2018-2024) and currently working as Professor in Dept. of Plant Sciences, MSLS, MAHE.. Dr. A Muthusamy teaches Botany/Plant Sciences, Environmental Sciences, Developmental Biology, Plant Biotechnology and Recombinant DNA technology for both UG and PG programs and also coordinator for Alumni Relations at MSLS. His research interests gradually expanded from plant cell, tissue and organ culture, soma/gametoclonal variations, laser-assisted morphogenesis and breeding, physio-biochemical and molecular aspects of stress responses, DNA methylation & stress responses, molecular identification of plant diseases and diagnosis, in vitro breeding for novel secondary metabolites and natural products.

Odangowei Inetiminebi Ogidi is a Lecturer, Acting Head, Department of Biochemistry and Sub-Dean, Faculty of Basic Medical Sciences, Bayelsa Medical University, Yenagoa, Bayelsa State, Nigeria. He holds a Bachelor of Science degree in Biochemistry, Master of Science and Doctor of Philosophy degrees in Biotechnology. He has also served as a Lecturer, Head of Departments of Science Laboratory Technology and Biochemistry, and Deputy Director, Centre for Continuing Education in Federal Polytechnic Ekowe, Bayelsa State. He has authored over 80 peer-reviewed academic articles in reputable local and international journals, published over 40 book chapters and presented over 35 conference papers. He is a member of various Journal Editorial Boards. He is indexed in World Scientist and University Ranking (AD Scientific Index ID 4306173) amongst other international bodies. He have supervised over 30 ND and HND students and currently supervising over 10 undergraduate students. His research interest has been Environmental Biotechnology, Phytomedicine, Phytochemistry, Human and Environmental health risk assessments and ecotoxicology amongst others. He is a member of West Africa Society of Toxicology, Nigerian Society of Biochemistry and Molecular Biology, Foundation for African Development through International Biotechnology and Nigerian Environmental Society.

Ramya Raghavan is a researcher working on molecular physiology and oncogenesis, where the research team delved into epigenetic factors leading to intellectual disability with severe speech impairment and behavioral problems. Following doctoral studies, Dr Raghavan employed a transcriptomics to study the expression of neural markers in the developing brain. This research provided valuable insights into how the brain builds its complex circuitry. Her interest then shifted towards the practical application of scientific knowledge in a clinical setting. Here, she focused on developing and optimizing novel diagnostic and research methods ultimately contributing to novel diagnoses and better patient care.

Supreeth H. S. is currently pursuing masters degree in Biotechnology at Christ University, Bengaluru with a strong passion for scientific research and literature writing. Their field of interest involves microbiology, plant science and agricultural biology and is currently working on a research project that aims sustainable degradation of agricultural waste through microbial intervention and PGRs production. Apart from this, he is working on enzyme biology and many literature works.

Ansar Mehmood currently works at the Botany, University of Poonch Rawalakot. Ansar does research in Nanobiotechnology, Biotechnology, ethnopharmacology and wild edibles. Investigation of nutritional profile, phytochemical analysis and antioxidant activities of wild plant food resources of Azad Jammu and Kashmir.

Junaid Shehzad completed his M.Phil. in the Department of Plant Sciences at Quaid-i-Azam University, Islamabad, where he is currently pursuing a Ph.D. His research focuses on the proteomic responses of crops to abiotic stress, with a particular interest in understanding the mitigative role of nanoparticles under stresses such as heavy metal toxicity. Mr. Shehzad has authored research articles and book chapters published in international journals, contributing to the fields of plant proteomics and stress tolerance mechanisms.

Divya Shukla is currently pursuing her Master's in Industrial Biotechnology at the University School of Biotechnology, GGSIPU. Her academic journey is driven by a profound fascination with molecular biology and the transformative potential of nanotechnology. She began her hands-on research experience at Vasantrao Naik College of Agricultural Biotechnology, where she investigated the synthesis of nanoparticles using _Trichoderma_ spp. in a microbiology laboratory, effectively bridging the fields of mycology and nanomaterials. Her master's thesis focused on a detailed comparative study of chemically synthesized zinc oxide nanoparticles and their chitosan composites, meticulously examining the impact of pH on their properties. This project solidified her interest in the application of nanobiotechnology in medicine and industry. Divya is dedicated to contributing to the advancement of biotechnological solutions that address critical global issues, particularly in healthcare and sustainable technologies. Her commitment to research and innovation in these fields makes her a valuable asset in the scientific community.

Index

D

E

F

S

T

www.ingramcontent.com/pod-product-compliance
Lightning Source LLC
LaVergne TN
LVHW080403011026
849161LV00009B/2337
* 9 7 9 8 3 6 9 3 9 8 6 9 2 *